Handbook of
Imaging Materials

Handbook of Imaging Materials

Second Edition, Revised and Expanded

edited by

Arthur S. Diamond
Diamond Research Corporation
Ojai, California

David S. Weiss
Heidelberg Digital L.L.C.
Rochester, New York

MARCEL DEKKER, INC.   NEW YORK · BASEL

The first edition of this book was edited by Arthur S. Diamond.

ISBN: 0-8247-8903-2

This book is printed on acid-free paper.

Headquarters
Marcel Dekker, Inc.
270 Madison Avenue, New York, NY 10016
tel: 212-696-9000; fax: 212-685-4540

Eastern Hemisphere Distribution
Marcel Dekker AG
Hutgasse 4, Postfach 812, CH-4001 Basel, Switzerland
tel: 41-61-261-8482; fax: 41-61-261-8896

World Wide Web
http://www.dekker.com

The publisher offers discounts on this book when ordered in bulk quantities. For more information, write to Special Sales/Professional Marketing at the headquarters address above.

Current printing (last digit):
10 9 8 7 6 5 4 3 2 1

PRINTED IN THE UNITED STATES OF AMERICA

Preface to the Second Edition

Since its publication in 1991, the *Handbook of Imaging Materials* has taken its place on the reference shelf of many leading scientists, engineers, and executives in the imaging industry. It is also well known in the academic world as both a sourcebook and textbook. There is no comparable reference available that is as highly focused or as comprehensive in defining the field of digital imaging technology.

To compile this second edition, we have worked with the authors to expand and update their original chapters while bringing in new contributors to cover the many advances that have occurred over the past decade. This second edition was enlarged and enriched by new ink jet, thermal, and electrophotographic technologies. These technologies, the digital color press, and new thermal dye diffusion methods all promise to challenge silver halide photography in amateur, commercial, and professional applications.

Although the use of diazotype materials is also fading into obsolescence, this process retains a foothold in the marketplace, primarily for wide format engineering and architectural blueprints and drawings. For these reasons and for the excellence of its content, Chapter 3 on diazo papers, films, and chemicals was again included in this volume.

Among the new chapters are: color photographic materials (Chapter 2), electrophotography (Chapter 4), photothermographic and thermographic imaging materials (Chapter 13), and papers and films for ink jet printing (Chapter 15).

In order to incorporate these topics effectively, the chapter on cylithography was excised. Despite major technical advances that have been made in cylith technology since 1991, that process has not had a significant impact in the imaging materials marketplace.

Finally, we are deeply indebted to those who contributed their knowledge, their writing skills, and their patience to this revised edition. We are grateful for their efforts in enabling us to expand the scope of this text and to bring it up-to-date in a field of science that is so rapidly advancing.

Arthur S. Diamond
David S. Weiss

Preface to the First Edition

The field of imaging technology has undergone massive changes over the past 50 years, from the 1940s, when photography served the entire gamut of reprographic applications, to its narrower focus in the imaging industry today. A half-century ago, silver halide–based papers and films performed every task from image capture to document copying.

Diazo processes took on special importance during the war years in the 1940s, growing rapidly in the 1950s to become the predominant method for engineering-drawing reproduction. Electrophotography grabbed the spotlight in the late 1950s and in the 1960s, when it revolutionized office procedures by increasing worker productivity through automation. Exploitation of this technology continues apace in facsimile, in desktop publishing, and in other areas of business communication.

Today, silver halide emulsion is still the primary image-capture medium for visible light, infrared, and x-ray photography. Amateur photographers, carrying some 250 million 35mm cameras worldwide, depend on this technology. In the United States alone, the retail photofinishing market reported revenues of $5.34 billion in 1989, based on almost 15 billion exposures on silver film.

Although it is now a commercial reality, the digital, electronic camera is not expected to equal the number of photographs taken on silver film until the turn of the century, when an estimated 23 billion exposures will be made in this country.

While silver photography continues to hold onto the most demanding imaging applications—those that depend on its high-speed, high-resolution, and continuous-tone capabilities—it has been essentially abandoned in document copying and largely replaced in many other areas of black-and-white and color reproduction, in microfilm duplicating, and in the graphic arts.

To serve the many markets that comprise the imaging industry, a host of new ink jet, thermal, electrophotographic, electrographic, ion deposition, and microencapsulation processes have emerged. This handbook was compiled for the purpose of sorting out these imaging methods and applications, while describing the materials used in each process, and their properties and performance characteristics. Each chapter has been written by a leading authority in a particular field of imaging technology and contains a discussion of the imaging materials that are being used with an eye toward future technological advances.

Many changes still lie ahead, particularly in the area of color printing and color copying, and in a possible revolution of the printing industry. As the turn of the century draws near, we can expect to be surrounded by color images, at home, at the office, in our shops, airports, and other public places. Digital scanning, image processing, and economical, nonphotographic processes will make large-format color affordable and accessible to everyone. It is likely that our walls will be hung with family portraits, beautiful landscapes, and the world's art treasures, scanned, digitized, and reproduced with breathtaking clarity, colossal in size, and vibrant with color.

Liquid toners, by virtue of their extremely fine particle size, highly saturated colors, and adaptability to digital electrostatic imaging systems, are making large-format color a reality. But the largest potential application for liquid toner is as a replacement for printer's ink in the computer-driven press of tomorrow.

These are some of the forces that are shaping the future for imaging technology, a field that is concerned with the visualization of our knowledge and culture, our literature and art. This text was prepared with the objective of bringing together the latest information in this dynamic field, recognizing that advances in imaging science and technology will inevitably improve the efficiency of our industry and the quality of our lives.

It was a suggestion by Dr. Maurits Dekker that prompted this work, an effort that spanned three years and drew on the expertise of 19 authors. Dr. Dekker recognized the need for a comprehensive text to serve as a reference for scientists and engineers in the imaging industry who are responsible for the design and development of imaging hardware and for the manufacture of the papers and films, toners and developers, inks and coatings, and other materials needed to produce hard copy. He also saw the evolution of imaging science curricula as an increasing number of colleges and universities. This reference handbook was thus planned to serve the needs of both the academic and the industrial world.

I am grateful to Dr. Dekker for his inspiration, enthusiasm, and patience in commissioning the effort that resulted in this unprecedented text. I also would like to acknowledge the efforts of the authors who contributed to the successful completion of this formidable task and to the unflagging optimism of my dear wife, Becky, whose winning smile has seen me through the most ambitious undertakings of my career.

Arthur S. Diamond

Contents

Contributors

Michael A. Andreottola *American Ink Jet Corporation, Billerica, Massachusetts*

Paul M. Borsenberger† *Eastman Kodak Company, Rochester, New York*

Douglas E. Bugner *Eastman Kodak Company, Rochester, New York*

P. J. Cowdery-Corvan *Eastman Kodak Company, Rochester, New York*

Paul F. Doll *American Ink Jet Corporation, Billerica, Massachusetts*

L. E. Friedrich *Eastman Kodak Company, Rochester, New York*

George A. Gibson *Xerox Corporation, Webster, New York*

Robert J. Gruber *Xerox Corporation, Webster, New York*

J. F. Hamilton* *Eastman Kodak Company, Rochester, New York*

Lewis O. Jones* *Consultant, Ontario, New York*

Robert Joslyn *Kyocera Industrial Ceramics Corp., Vancouver, Washington*

Paul C. Julien *Xerox Corporation, Webster, New York*

J. A. Kapecki *Eastman Kodak Company, Rochester, New York*

† *Deceased*
* *Retired*

S. O. Kasap   *University of Saskatchewan, Saskatoon, Saskatchewan, Canada*

Klaus B. Kasper   *Boulder Consultants, Boulder, Colorado*

Sean M. Kelly   *American Ink Jet Corporation, Billerica, Massachusetts*

James R. Larson   *Xerox Corporation, Webster, New York*

Lubo Michaylov   *Worldwide Images, Carmel Valley, California*

J. Mort   *Xerox Corporation, Webster, New York*

Henry Mustacchi   *Consultant, Port Washington, New York*

Steven P. Schmidt   *Dade Behring, Glasgow, Delaware*

B. E. Springett*   *Xerox Corporation, Webster, New York*

Dene H. Taylor   *Specialty Papers & Films, New Hope, Pennsylvania*

David S. Weiss   *Heidelberg Digital L.L.C., Rochester, New York*

D. R. Whitcomb   *Eastman Kodak Company, Rochester, New York*

Walter J. Wnek   *DuPont, Inc., Wilmington, Delaware*

---

* *Current affiliation*: Fingerpost Advisers, Rochester, New York

Handbook of
Imaging Materials

1

Conventional Photographic Materials

J. F. HAMILTON*

Eastman Kodak Company, Rochester, New York

1.1 INTRODUCTION

1.1.1 Advantages

It is an interesting exercise to imagine that conventional silver halide photographic film had only recently been invented, after a century or so in which the only imaging systems known to the world were electrostatic and electronic imaging such as charge-coupled devices. What would the creative advertising agencies of today tout as its superior improvements over previously existing technologies?

Surely they would welcome the liberation from external power sources, the small size and light weight of a piece of film, its flexibility, allowing it to be wrapped on a compact spool, and the fact that it can sit for months, even years, on a shelf and be ready for use instantaneously. They would point out that the sensitive material of this very inexpensive product is itself converted to the recorded image by a relatively simple chemical treatment requiring, in its basic form, no complex auxiliary equipment. They would marvel that it can be made sensitive to select spectral ranges from the ultraviolet well into the infrared and to high energy particles or radiation, and over the remarkable extent to which it can be made to yield a true (or, if desired, false) color rendition of an original, or a black-and-white image with archival stability. They would extoll the superior image quality of this new product, emphasizing its extremely high information packing density, its sharpness, its uniformity, its low level background noise, and its freedom from defects.

They would hail its ability to integrate (albeit not with constant efficiencies) over exposure times from hours to the picosecond range, and they would emphasize that over

---

* Retired.

the most commonly used range of times it has virtually no temperature dependence or sensitivity, at least within the range of normal ambients.

In addition, they would emphasize the versatile ability of photographic manufacturers to tailor this material for special uses, producing negative or positive images, instant or thermal processing, etc. And no doubt this represents only a scratching of the surface of the virtues that could be attributed to this remarkable new accomplishment of modern chemical ingenuity. Even discounting the obvious excesses of this tongue-in-cheek scenario, it would have to be admitted that the conventional silver halide imaging process has set the standards to be aimed at by designers of rival technologies and that it still remains superior in a number of respects.

1.1.2 Uses

Alternative processes have certain attributes, however, that make them better adapted for some applications, in which they are replacing silver halide materials. Even in these fields, however, which we now think of as the exclusive domain of electrostatic or electronic imaging (e.g., document copying, news and amateur cinematography), there was often a silver halide special-purpose material that at one time served the need. The current major classes of conventional silver halide based products manufactured can be found in any of a number of financial reports on the industry, but the total listing of special-purpose materials, either current or superseded, is astoundingly large.

1.2 FILM COMPOSITION

Photographic films, papers, and other media consist of one or more sensitive layers along with certain ancillary layers (filters, chemical barriers, etc.) coated on a suitable flexible or (occasionally) rigid support. Typical color films contain at least six sensitive layers (high and low sensitivity of each of the three color sensitivities) and several other nonsensitive ones.

The principal component of the sensitive layers is individual crystals (often called *grains*, in the photographic literature) of silver bromide, silver chloride, or mixed crystals of the two, frequently containing also up to several percent silver iodide in solid solution. Gelatin is normally used as a binder. Linear crystal dimensions vary greatly among special-purpose materials, covering the range from about 0.03 μm to several micrometers. A single sensitive layer normally has a thickness averaging a number of monolayers of grains.

Depending on the type of material, the sensitive layers will also contain other components as well. Chemical sensitizers are almost always applied to the grains, as are sensitizing dyes in all materials having spectral sensitivity to longer-than-blue wavelengths (see later sections). The most common color negative materials also contain color couplers, organic precursors of the dyes that form the color image. Spreading or wetting agents, chemical hardeners for the gelatin, fungicides, and other active chemical components are also sometimes incorporated, and, with the great variety of special-purpose materials offered, the combinations are extensive.

1.3 SILVER HALIDE STRUCTURE

1.3.1 Crystal Structure

Silver bromide and silver chloride are the simple monovalent ionic salts of the two components. They both exhibit sixfold symmetry and have the sodium chloride crystal structure under ambient conditions with lattice constants of 0.5547 nm for AgCl and 0.5775 nm for AgBr (Wyckoff, 1964). These values are slightly lower than twice the sum of the standard ionic radii, indicating some inadequacy of the hard-sphere ionic model. This feature is one of several manifestations of the presence of some covalent component of bonding in these materials, based on ionicity rankings (Phillips, 1970).

Silver chloride and silver bromide form coherent mixed crystals in all proportions (Chateau, 1959), by simple substitution on the halide sublattice. The iodide ion is reported (Chateau et al., 1958) to be soluble in silver bromide to a maximum concentration of about 30 mol% and in silver chloride to only a few mol%. The limited solubility reflects the fact that pure silver iodide has fourfold symmetry and crystallizes in either the sphalerite or the wurtzite crystal structure (Wyckoff, 1964), but not the sodium chloride form. The standard ionic radius of the iodide ion is 19% greater than that of chloride and 11% greater than that of bromide.

1.3.2 Electronic Characterization

These crystals are classified as electronic insulators. The filled valence bands and the empty conduction bands are separated by forbidden gaps of 3.25 eV in the chloride and 2.68 eV in the bromide. From simple Fermi statistics, the expected free carrier concentrations in intrinsic materials at room temperatures are 10^{-28} and $10^{-23}\,\mathrm{cm}^{-3}$, respectively.

1.3.3 Electronic Structure of the Elements

The electronic structures of the four principal atoms and ions along with those of several related species are given in Table 1.1. The halogen ions Cl⁻, Br⁻, and I⁻ form by acquiring a single electron to fill the 3p, 4p, and 5p shells, respectively, which are fivefold occupied in the atomic state.

The silver atom consists of a rare gas Kr core, a filled 4d electron shell, and a single 5s electron. The positive ion forms by loss of the 5s electron.

The alkali atom Rb has the same Kr core and 5s electron but is missing the 4d shell. This difference, as subsequent discussions will show, is responsible for many of the unique properties of the silver halides, particularly in contrast with those of the superficially similar alkali halides.

The closely related noble metal, Au, differs from silver mainly in the order of the rare gas core. Gold is in many ways similar to silver and has become an important ingredient in the photographic process. As indicated in Table 1.1, the first ionization potential of gold

Table 1.1 Electronic Structure of Selected Elements

Element	Z	Electronic structure (atom)	Ionization potentials (eV) First	Ionization potentials (eV) Second	Electron affinity (eV)
Cl	17	[Ne]$3s^2 3p^5$			3.63
Br	35	[A]$3d^{10} 4s^2 4p^5$			3.38
I	53	[Kr]$4d^{10} 5s^2 5p^5$			3.08
Ag	47	[Kr]$4d^{10} 5s^1$	7.57	21.48	
Rb	37	[Kr]$5s^1$	4.18	27.5	
Au	79	[Xe]$4f^{14} 5d^{10} 6s^1$	9.22	20.5	

is larger than that of silver, and this feature is largely responsible for its photographic consequence.

1.3.4 Emulsion Precipitation

For commercial photographic purposes, the silver halide crystals, being very sparingly soluble in water, are formed by precipitation from aqueous solutions of more soluble salts of these elements (Duffin, 1966; Wey, 1985). Typically, two solutions of reactants such as KBr and $AgNO_3$ are introduced with vigorous mixing in separate streams to a reaction vessel containing some of the halide solution and dilute gelatin or another peptizing agent. The finished suspension of crystals in the gelatin solution is generally referred to as an *emulsion*, a terminology clearly not consistent with general chemical use.

In modern equipment the electrochemical potentials of the reagent ions are monitored continuously at all stages of the precipitation, and these data are fed to automated flow equipment that controls the concentrations of the two reacting ions in the solution over the growing crystals.

Extensive exploration of the variables involved in precipitation has produced a wealth of technological data that allows rigid control of such features as size, size distribution, crystal habit, halide compositional structure, and defect content of the grain population. For example, silver bromide crystals grown with only slight excess bromide ion concentration have stable [200] faces and thus are cubic, whereas a higher bromide ion excess results in stable [111] faces and octahedral crystal shape. Still greater halide excess forms complex ions such as $AgBr_3^{2-}$ in solution. These ions lead to the formation of growth twins in the precipitate and more complex crystal shapes. One such form is that of a triangular or hexagonal tabular platelet, which results when the crystal contains two or more parallel twins. Precipitates containing almost exclusively grains of this form can be made, and these are now used in some commercial materials. When the thickness-to-breadth ratio is made quite small, such grains have a large specific surface and therefore certain practical advantages (Berg, 1983).

Many of the basic physical properties of the silver halides have been determined from experiments using macroscopic melt-grown crystals or from vacuum-deposited thin films (Berry et al., 1963).

1.3.5 Chemical Sensitization

Virtually all commercial emulsions are subjected to chemical sensitization treatments between precipitation and coating to improve their sensitivity to light (Duffin, 1966; Harbison and Spencer, 1977).

The most commonly used procedure involves the addition of compounds containing labile sulfur in combination with gold salts and usually certain other modifying chemicals, followed by digestion at elevated temperature. This practice results in the formation of adsorbed molecules of Ag_2S, AgAuS, and/or Au_2S. Following the chemical conversion, there is evidence (Roth and Simpson, 1980; Corbin et al., 1980) for a second essential stage of the process, thought to involve surface migration and reorganization of the sulfide molecules into aggregates. Some features of this process have been interpreted (Keevert and Gokhale, 1987) to be consistent with the formation of dimers as the critical step.

Even in the relatively small amounts normally used, the conversion of labile sulfur to insoluble sulfide can be followed quantitatively by extraction (Sturmer and Blackburn, 1979) using the radioactive species ^{35}S. The dependence of sensitivity (log reciprocal exposure for some fixed photographic response) on the amount of converted sulfide is typified by the results given in Fig. 1.1. The sensitivity first increases with added sulfide but eventually reaches a plateau and usually declines if the sulfur is further increased. When expressed as a surface coverage, the sulfur content at the plateau position is reported to vary little among preparations, and an optimum coverage of 2×10^4 sulfide molecules per square micrometer of silver halide surface has been reported (Sturmer and Blackburn, 1979). In a typical high-speed film, this value translates to the order of a few milligrams of sodium thiosulfate per mole of silver halide. The gold salt (e.g., $KAuCl_4$) is typically present in a similar weight ratio.

Neither the position of the subsequent decreasing portion of the curve (see Fig. 1.1) nor the extent of the decline is as reproducible among preparations. Nevertheless, it is encountered with enough regularity that it can definitely be attributed to an excess of sensitizer and not to some other spurious cause.

The treatment also invariably causes an increase in fog—a fraction of the crystallites that develop without exposure. The fogged fraction increases with increasing sensitizer content.

Historically, the sensitizing effect of the labile sulfur compounds was discovered some years before the important effect of adding gold salts was revealed. During that period, therefore, many materials were made and many mechanistic studies undertaken using only *sulfur sensitization*. Scientific studies frequently still involve such preparations, to separate the complex effects of the two components in what is commonly called *sulfur-plus-gold sensitization*. However, even without absolute knowledge of the practices of all manufacturers, it is reasonably safe to conjecture that all high-speed camera films today

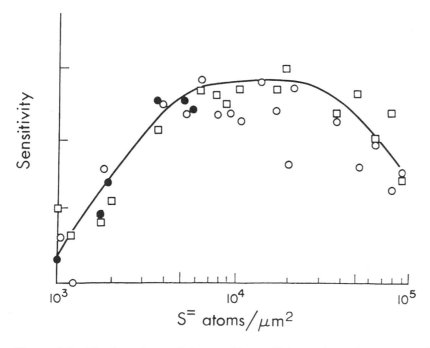

Figure 1.1 The dependence of photographic sensitivity on the surface concentration of sulfide converted in chemical sensitization.

are sensitized with sulfur and gold treatments. This generalization cannot be made for other materials for other applications.

1.4 THE LATENT IMAGE AND DEVELOPMENT

1.4.1 The Development Reaction

After a photographic material has been exposed to a sufficiently strong pattern of light or ionizing radiation, it is said to bear a *latent image*. This latent image is capable of being transformed by a *developer* into a visible image corresponding to the light pattern. The essential active ingredient of a photographic developer is a moderately strong chemical reducing agent. The reducing agent transfers electrons to grains that have received some minimum exposure, causing them to begin to be reduced to metallic silver, but not to those that have received less than the minimum requirement. It is now broadly agreed that the latent image consists of centers at which the absorption of radiation has caused the formation of small clusters of mixed Ag/Au metal atoms—the latent image centers. When gold has been used in the sensitization, the latent image centers are composed at least partly of gold atoms. These centers in turn control the development reaction so that there is *discrimination* between exposed and unexposed grains.

1.4.2 Fog and Discrimination

The energy level pattern of the silver halide and silver phases, shown in Fig. 1.2, leads directly to an electronic description of discrimination in development. Injection of electrons into the silver halide conduction band would lead to indiscriminate reduction even of unexposed grains and thus what is known as fog. A developing agent must therefore be chosen whose chemical potential places its frontier electrons well below the silver halide conduction band in energy. The rate of the thermally activated injection will then be given by the product of a preexponential factor including a proportionality to the number of potential surface injection sites and a Boltzmann term involving the energy discrepancy. The developer must be chosen so that the fog rate is tolerable.

In energy, the silver Fermi level lies far enough below the silver halide conduction band that nonfogging developers can be found whose chemical potential allows thermodynamically favored electron transfer to bulk silver. Thus silver acts as an electrode (Pontius et al., 1972; Pontius and Willis, 1973), accepting electrons from developer molecules and transferring them to defect silver ions within the exposed crystal, thus producing more silver, in an autocatalytic reaction. For every electron transferred to the developing grain, a halide ion passes into solution thus maintaining charge balance and assuring the continued supply of defect silver ions.

1.4.3 Size Dependence of Latent Image Centers

There is extensive evidence for a size requirement for a latent image center capable of initiating development. In fact, this evidence leads to the conclusion that there is a smallest range of sizes that is unable to initiate development, a range of larger sizes for which this occurs very slowly (i.e., with activation), and finally a size limit beyond which the reaction proceeds very rapidly. These observations are rationalized (Trautweiler 1968; Jaenicke, 1972) by a proposed regime of size-dependent energy levels for silver clusters, allowing for the lowest unfilled level of the smallest clusters to lie well above the bulk silver Fermi

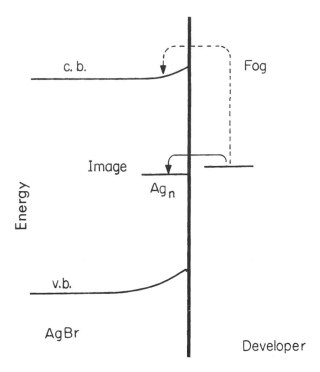

Figure 1.2 The energetics of electron transfer in photographic development. Transfer to silver is thermodynamically favored, but that to the conduction band (c.b.) of an unexposed crystal, to produce fog, requires considerable thermal activation; v.b. = valence band.

level and also above the level of the developer electrons, those of the intermediate size class to lie marginally above the developer level, and finally a size limit beyond which all clusters can favorably accept electrons from the developer. In the intermediate range, the initiation of both image and fog development is thermally activated and, even though the activation barrier is much lower for the image, discrimination is nonetheless problematic, owing to the much higher density of sites in the preexponential of the fog reaction.

Basic molecular orbital concepts (Hamilton and Baetzold, 1981; Tani, 1983) predict that there is an odd—even oscillation of the lowest unfilled level with size in simple geometric forms of clusters of atoms such as silver with a single s-electron, and experimental studies have confirmed this principle for vapor phase copper and silver clusters (Powers et al., 1983). If this is true also for photographic clusters, all odd-sized clusters should easily accept an electron and grow to the next larger size; all limiting reactions should occur at the even sizes (Hailstone and Hamilton, 1987). It is possible that geometric constraints and/or electronic interaction with the host silver halide would modify or even totally eliminate this pattern, and neither experimental nor theoretical analyses have been able to resolve the uncertainty.

However, a number of photographic (Hailstone and Hamilton, 1985, 1987; Hailstone et al., 1987; Hada et al., 1980; Hada and Kawasaki, 1985) and model studies (Hamilton and Logel, 1974; Fayet et al., 1985) agree on some general aspects of the size dependence of developability. The smallest size class of clusters consists of the metal dimer, which appears not to be able to exert discrimination. This feature can be rationalized with no difficulty in terms of the expected energy level of the lowest unfilled level of the dimer. The excitation energy of the silver dimer in vacuum—that is, the energy to raise an electron from the highest occupied to the lowest unfilled level—is 2.85 eV (Brown and Ginter, 1978), larger than the AgBr forbidden gap. The lowest unfilled level must lie near the bottom of the conduction band. Thus the activation energy for electron transfer is not greatly smaller and the Boltzmann probability of transfer is not greatly larger than that of the fogging reaction. The differences in the density-of-sites factor has the effect of making fog more likely and eliminating any possibility of discrimination by the dimer.

When gold is used in the sensitization and therefore becomes incorporated in the latent image center, the next size class, the three-atom cluster (or perhaps three and four atoms, if the oscillating energy level pattern does apply) is developable with only a very small thermal activation, so that under most practical development conditions grains containing these centers are all developable. All larger centers are developed with no detectable activation delay.

Without gold in the sensitization, the latent image centers are exclusively silver and the characteristics are different. Clusters of three (or three and four) atoms have higher activation barriers to electron transfer and have only a very marginal chance of initiating discriminating development before fog becomes excessive. The next size class, four atoms (or five and six), have lower but still significant activation barriers and develop, though slowly. Only by the next size class, five atoms (or seven and eight), does the activation delay become undetectable.

The effect of gold on decreasing the developable size of the latent image is the result of the greater electron affinity of gold-containing clusters. Replacement of one or more silver atoms in a marginally developable cluster by gold causes the lowest unfilled level to be nearer to the developer level and thus more accessible to the electrons of the developer molecules.

To summarize, for practical development conditions the minimum size of the developable latent image is three atoms in emulsions sensitized with sulfur plus gold. In materials without gold, the development probability of sizes between perhaps seven and four atoms increases with development time, all contributing significantly to discrimination.

1.5 THE DEVELOPED IMAGE

1.5.1 Composition and Structure

Once initiated, the development of a given center is autocatalytic and proceeds with an increasing rate unless it becomes inhibited by exhaustion of either the silver halide or the reducing agent or unless some other species interferes with the reaction.

In most black-and-white materials, the developed silver is the light-absorbing material of the final image. On a microscopic scale, the silver is a tangled mass of silver filaments, for reasons that have never been adequately explained. In color materials, on the other hand, the image consists of dyes formed by reaction of the oxidized developer species with precursor molecules called couplers, which are either incorporated into the coating or dissolved into the developer solution. The technology employs couplers of varying molecular structure such that the three necessary image dyes are selectively formed from a common oxidized developer species. The developed silver is removed from the color image by dissolution along with the undeveloped silver halide.

1.5.2 Sensitometry

For a developed silver image, the optical density in any region is reasonably well given by the Nutting equation (Nutting, 1913):

$$D = 0.434n\bar{a} \tag{1.1}$$

where n is the number of developed grains per unit area, $\bar{a}$ is the mean projected area per grain, and 0.434 is $\log_{10} e$. This equation holds rigidly for the so-called random–opaque-dot model (Picinbono, 1955). In real materials the value of $\bar{a}$ must actually be the optical cross section of a developed grain, including the effect of light scattering in the coating (Farnell and Solman, 1963). The optical cross section of a tangle of silver filaments is not easily accessible except by Eq. (1.1), but it is obviously different from the measurable area of the undeveloped grain from which it is derived. When the undeveloped grain area is used in Eq. (1.1), as is often done, the expression is only approximately valid.

Photographic response is customarily depicted by plotting the developed optical density versus the logarithm of the incident energy of the exposing radiation. Such a plot produces a *characteristic curve* having the general form shown schematically in Fig. 1.3. Typically there is a background D_{min} value or *fog*, followed by a gradually rising *toe*, frequently an approximately linear portion, and finally a *shoulder* at the D_{max} value. The slope of the linear portion is the *gamma* γ or *contrast* of the material, and the log E difference between the toe and shoulder regions gives the exposure latitude. The reciprocal of the log exposure required, usually to produce some fixed low density, gives a measure

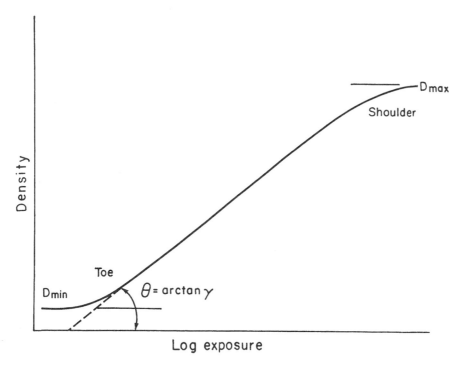

Figure 1.3 Sketch of a typical characteristic curve (D versus log E) for a photographic material.

of the *sensitivity S* or speed of the materials. A variety of conventions have been proposed for measuring sensitivity, and that applied commercially is specified in detail by the International Standardization Organization (ISO).

1.5.3 Reciprocity Law Failure

In general, photographic sensitivity is not independent of the exposure time/light intensity combination. This dependence is termed reciprocity failure. The most common form is a loss of sensitivity for very weak images that require long exposure times, an inefficiency called low-intensity reciprocity failure (LIRF) (Webb, 1950). Most common photographic materials show quite significant failure for exposure times greater than perhaps 0.1 to 1 second, and for applications such as astronomy, which require much longer exposure, very specialized techniques have been developed to minimize the inefficiency (Babcock et al., 1974). Some types of material also exhibit high-intensity reciprocity failure (HIRF), a loss of sensitivity for very short exposure times. The common practice of using gold salts in the chemical sensitization has all but eliminated this defect in modern commercial camera films (Hailstone and Hamilton, 1985).

1.5.4 Quantum Sensitivity

Under certain limited conditions it is possible to measure the quantities necessary to convert a conventional characteristic curve of density versus log incident energy into an absolute one whose variables are the fraction of grains made developable and the number of absorbed photons per grain. This allows for specification of the *quantum sensitivity* of the material, that is, the number of absorbed photons to make some particular fraction of the grains developable. Such measurements have been made on a number of materials by now (see Hamilton, 1988, Refs. 1–11), and the results indicate that the best conventional chemical sensitization methods are able to approach but not exceed a quantum sensitivity value of about 8 absorbed photons per grain at the 50% developable level. To be sure, some commercial and experimental materials fall far short of this position, but the clustering of results at about this value is strongly indicative of some kind of limit there. The limit is not strongly dependent on grain size up to a linear dimension of 1 or 2 μm, but above that value a deterioration of quantum sensitivity appears to be normally observed (Farnell, 1969; Tani, 1985).

1.5.5 Grain Size and Sensitivity

Below this limit, however, where the quantum sensitivity is nearly independent of grain size, the size directly affects conventional sensitivity, owing to the larger collecting power of larger grains. The range of ISO speeds available among commercial camera films and other materials is in fact achieved largely by changes in grain size. In spectral regions where the light is absorbed by the silver halide, it is approximately true that

$$S \propto l^3 \tag{1.2}$$

where l is a mean grain linear dimension; but when the absorption is by dyes at the grain surface,

$$S \propto l^2 \tag{1.3}$$

It is obvious that a narrow spread of crystal sizes in general produces high contrast and short latitude, whereas a broad size distribution gives extended latitude and lower contrast.

In spite of the reservations expressed earlier, there is a general correlation between the undeveloped grain size l^2 and the optical cross section $\bar{a}$ of the developed grains in Eq. (1.1). Thus it also follows that for a given mass or volume of silver halide per unit area ($nl^3 = $ Ag) the maximum density, or in fact the density corresponding to any given fraction of grains developed, is

$$D = 0.434 \frac{\text{Ag}}{l^3} \bar{a} \qquad (1.4)$$

or

$$D/\text{Ag} \approx \text{const} \left(\frac{1}{l} \right) \qquad (1.5)$$

The density per unit mass of silver, termed the covering power, is greater for small grain size and decreases in roughly inverse proportion to grain size. To achieve a given $D_{\max}$, therefore, more silver coverage is required for larger grain size.

1.5.6 Graininess

It is also true that as the crystal size and therefore the film speed increases, so does the familiar *graininess* pattern of the image. This property is only indirectly related to the size of the crystals, however, for even the largest of those used are far below the resolution limit of the eye or any other readout system, under normal conditions. The subjective perception of graininess has been shown to correlate with the measured *granularity* G (Selwyn, 1935, 1942), which is defined as the root-mean-square fluctuation of optical density in a nominally uniform image area, when measured with an aperture of specified area. The density fluctuation results principally from the variance of the number of image elements within the aperture area. With random statistics, the variance is given by $n^{1/2}$. Thus in an ideal black-and-white material, granularity is expected to increase with the $1/2$ power of the image density. Among materials of different crystal size, the foregoing relationships may be combined to predict that speed and granularity are related by

$$G \propto S^{1/3} \qquad (1.6)$$

in the intrinsic absorption region and by

$$G \propto S^{1/2} \qquad (1.7)$$

in the region of dye absorption.

For color development, other considerations make the situation more complex. Nevertheless, it is still true that the perception of graininess results from the condition that within the image areas (pixels) resolved by the visual system the number of contributing image centers is small enough that the variance of that number among nominally identical pixels is perceptible.

1.6 ABSORPTION OF IMAGING LIGHT

1.6.1 Electronic Band Structure

The conduction bands of the silver halides in question are simple isotropic bands with minima at the center of the Brillouin zone. However, the valence bands of both AgCl and AgBr are significantly influenced by the strong admixture of the silver 4d atomic levels with either the chlorine 3p or bromine 4p levels in that energy range. Owing to the inversion symmetry of the crystal, these levels do not mix at the Brillouin zone center, but at the zone boundaries they hybridize and spread to produce an inverted valence band (Kunz, 1982) as shown in Fig. 1.4.

The minimum energy transition corresponding to the long wavelength limit of light absorption is an indirect or nonvertical transition from the zone boundary of the valence band to the center of the zone of the conduction band. To conserve momentum, one or more phonons is either absorbed or emitted in the transition.

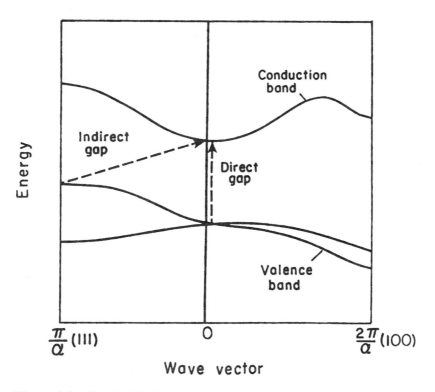

Figure 1.4   Sketch of the forms of the uppermost valence band level and the lowest conduction band level of silver bromide in reciprocal space. The valence band is inverted, having its highest energy point at the Brillouin zone boundary. Thus the indirect gap is smaller in energy than the direct gap.

At room temperature the indirect absorption edges extend through the near-ultraviolet spectral region for AgCl and well into the blue for AgBr (Moser and Urbach, 1956), but with absorption coefficients orders of magnitude lower than for the higher energy direct or vertical transitions, as shown in Fig. 1.5.

Studies of the quantum yield of free carriers indicate that absorption of a photon produces an exciton, which is autoionized to give a free electron–hole pair with near unit efficiency.

1.6.2 Spectral Sensitization with Dyes

In all except the ultraviolet and blue spectral region, the photographic process depends on absorption of the incident light by organic sensitizing dyes adsorbed to the surfaces of the grains. This process has been known and studied in the silver halides (West, 1974; West and Gilman, 1977) for more than a hundred years.

Relative Quantum Efficiency

A procedure to measure the relative efficiencies of photons in the spectral region of dye absorption and those absorbed by the silver halide itself has become standard. The expo-

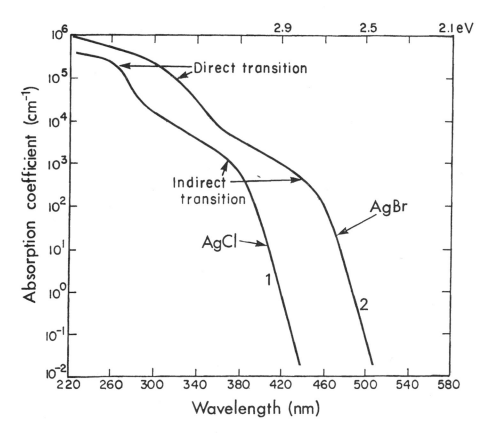

Figure 1.5 Absorption coefficients of silver chloride and silver bromide. The exponential form of the edges is apparent below the indirect transition energy.

sure (in energy density) to produce some fixed developed density is measured for monochromatic light of 400 nm wavelength (E_{400}) and also for light at the wavelength λ of the dye absorption maximum (E_λ). The absorption of the film at the two wavelengths (A_{400}, A_λ) is also measured. When the exposure values are converted to quanta, the ratio of the required number of photons at the two wavelengths is given by

$$\phi_r = \frac{400 E_{400} A_{400}}{\lambda E_\lambda A_\lambda} \tag{1.8}$$

The quantity ϕ_r is termed the *relative quantum efficiency* (RQE) and has the significance that when $\phi_r = 1$, a photon absorbed by the dye is as effective in producing latent image as one absorbed by the silver halide. Dyes with ϕ_r between 0.8 and 1 are not uncommon.

Electron Transfer

By far the majority of practical sensitizing dyes are of the cyanine or merocyanine classes, consisting of a conjugated carbon chain linking cyclic end groups. In the ground state of such a molecule, the bonding highest occupied (HO) orbital is filled, and absorption of light promotes an electron to the lowest vacant (LV) antibonding orbital, usually in a $\pi \rightarrow \pi^*$ transition. The energy levels involved depend on molecular structure, and most of the relationships are now well understood.

A feature of major practical importance is the spectral location of the maximum of the dye absorption band, which is a measure of the energy difference between the HO and LV energy levels. The spectral characteristics, however, give no direct information on the absolute positions of either of the two levels involved. Such information is obtained either by theoretical calculations or experimentally, by photoemission measurements, or, more commonly, by electrochemical studies of dyes in solution.

A great many tests have been made in search of a correlation between sensitizing efficiency (i.e., RQE) and oxidation and/or reduction potentials. The result is that almost all effective sensitizing dyes for silver bromide have reduction potentials more negative than about -1.1 V relative to a silver–silver chloride electrode. This correlation is taken to indicate that the lowest unfilled electronic level (LV) of this group of dyes lies above the bottom of the silver bromide conduction band. The mechanism of the effective spectral sensitization process is therefore concluded to be the direct transfer of an electron from the excited state of an adsorbed sensitizing dye molecule into the silver halide conduction band, where it becomes indistinguishable from one formed by intrinsic absorption.

1.7   ELECTRONIC TRANSPORT

1.7.1   Electrons

Classic time-of-flight measurements have been employed with single crystal materials to measure electron drift mobility μ_D, for comparison with values of Hall mobility μ_H obtained from the photo-Hall effect (Brown, 1976; Evrard, 1984).

As related to the normal applications of the photographic process, it suffices to say that in high-quality single-crystal materials the room temperature electron drift mobility values agree with those for Hall mobility and are given along with effective mass values determined from cyclotron resonance in Table 1.2. The mobilities increase with decreasing temperature as expected from simple phonon-scattering theory.

Table 1.2 Transport Properties of Electronic Carriers in the Silver Halides

	Mobility $(cm^2 V^{-1} s^{-1})$	Polaron effective mass
Electrons		
AgCl	50	0.32
AgBr	60	0.288
Holes		
AgCl	10^{-2}	[a]
AgBr	1	1.71, 0.79[b]

[a] Unmeasured owing to self-trapping.
[b] Double valued because of band anisotropy.

1.7.2 Holes

Silver Chloride

The low temperature transport properties of holes in the silver halides are markedly different from those of electrons. The situation is more clearly understood for the hole in AgCl, which at liquid helium temperature is self-trapped (Höhne and Stasiw, 1968; Toyozawa, 1961).

Electron spin resonance (ESR) studies (Höhne and Stasiw, 1968) have revealed that the carrier is localized on a silver ion and that the nearest-neighbor chloride ions are displaced in the tetragonal Jahn–Teller mode, as illustrated in Fig. 1.6.

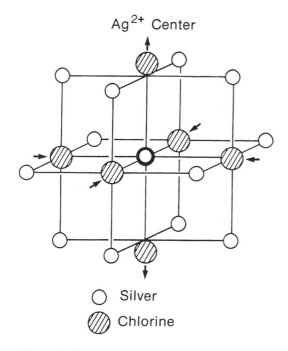

Figure 1.6 The Jahn–Teller distortion of the six chloride ions around a self-trapped hole (Ag^{2+}) in silver chloride.

A broad optical absorption centered at 1.2 eV has been attributed to the optical excitation of the AgCl self-trapped hole center (Ulrichi, 1970). The thermal ionization energy has been estimated (Laredo et al., 1983) as 0.12 eV, indicating that at room temperature the self-trapped state is transient only. Nevertheless, it is important enough to limit severely the drift mobility of the hole in silver chloride. Direct electrical measurements have not been successful, but indirect measurements of diffusion effects (Müller et al., 1970) give a room temperature mobility value of about 10^{-2} cm^2 V^{-1} s^{-1}, thus nearly four powers of 10 lower than that of the electron.

Silver Bromide

Studies show that in silver bromide the hole is not self-trapped at liquid helium temperature (Hodby, 1969). Nevertheless, the room temperature drift and Hall mobilities of the hole in this material are about 1 cm^2 V^{-1} s^{-1}, about 50 times lower than that of the electron, and exhibit a relatively steep negative temperature dependence, as shown in Fig. 1.7. Interpretations generally (Toyozawa and Sumi, 1974) attribute these mobility characteristics also to the self-trapped state, which is regarded as metastable in the bromide salt but stable in the chloride. A weak photoinduced transient absorption centered at 0.88 eV has

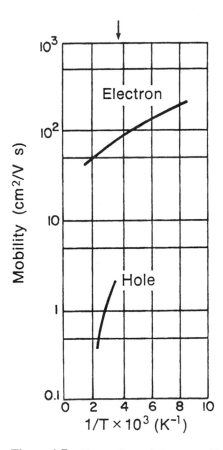

Figure 1.7   Comparison of electron and hole mobilities in silver bromide in the temperature range near room temperature.

been observed (Ulrichi, 1970) at 20 K in AgBr and attributed to the metastable self-trapped hole, by analogy to the stable 1.2 eV absorption in AgCl.

1.8 CHARGED POINT DEFECTS

1.8.1 The Dielectric Constant

Electrostatically charged defect sites and ionic charge carriers play crucial roles in the photographic process, and the presence of these species is strongly influenced by the dielectric properties of the silver halide materials. Table 1.3 lists the room temperature values of the static ε_0 and high frequency ε_∞ dielectric constants of silver chloride and bromide (Lowndes, 1966; Lowndes and Martin, 1969) along with those of select alkali halides. It is clear that the values for the silver salts are considerably higher than those of the related alkali salts. Since electrostatic energies are related to the inverse square of the dielectric constant, and the charged-defect concentration depends on the formation energy in a Boltzmann-type factor, it follows that defects are present in far higher concentration in the silver halides than in many other ionic crystals.

The high ionic component (i.e., $\varepsilon_0 - \varepsilon_\infty$) of the silver halide dielectric constant is attributable to the particular electronic structure of the silver ion. Resonant coupling between the d- and s-electronic levels of the silver ion results in a prominent quadrupolar deformation mode (Bilz and Weber, 1984). Because the silver ion is not rigidly restrained to a spherical shape, the lattice of the silver halide crystals is unusually ''soft'' and particularly compliant to certain stress modes. Among those are the ionic displacements induced by electrostatic forces. Thus the high dielectric constants.

The quadrupolar deformability also strongly influences the static elastic constants and the lattice dynamics of the silver halides and the ionic transport properties as well.

1.8.2 Static Defects

The classic stationary charged defect is the surface kink site on the [200] face (Seitz, 1951), as illustrated in Fig. 1.8. The formal charge of such a site is $\pm e/2$ depending upon its occupancy by a cation or an anion (Amelinckx, 1979). A dislocation jog is the volume analogue of the surface kink, and the electrostatic formalism is precisely the same, although the energetics are modified because of the very different considerations of lattice strain.

Strictly analogous sites cannot be envisaged for the photographically important [111] surface of the silver halides. However, a [111] surface bounded by a perfect ionic plane has an effective charge of either plus or minus e/2 for each ion in that plane. Only for a boundary layer of exactly half the number of ions in a normal plane is charge neutral-

Table 1.3 Static (ε_0) and High-Frequency (ε_∞) Dielectric Constants of the Silver Halides and Selected Alkali Halides at Room Temperature

| Constant | Halides | | | | |
	AgCl	AgBr	RbCl	RbBr	KBr
ε_0	11.15	12.5	4.92	4.86	4.90
ε_∞	3.92	4.62	2.18	2.34	2.36

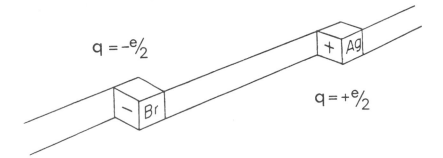

Figure 1.8 Negative and positive kink site structures on the [200] face of a cubic ionic crystal.

ity achieved. Rearrangement of half of the ions in a plane in register with the threefold symmetry of the underlying lattice is not straightforward, and some type of rather severe reconstruction is necessary (Hamilton and Brady, 1970). At best some statistical discontinuities are to be expected. Such sites will have formal partial electronic charges, as do the [200] kink sites, and can therefore be referred to as *kinklike* sites.

The Coulomb energy of these partially charged structural defects is particularly low in the silver halides owing to the high dielectric constants. They are therefore present in relatively high dynamic concentrations, and arguments can be and have been made that they play a major role in the photochemical processes of practical significance in these materials. They interact with mobile charge carriers, both electronic and ionic, and because of their nonintegral electrostatic charge impart unique properties to the centers so formed.

1.8.3 Intrinsic Frenkel Disorder

The dominant mobile ionic defects in the silver halides (Friauf, 1984) are of the Frenkel type on the cation sublattice (i.e., interstitial silver ions Ag_i^+ and silver ion vacancies V_{Ag}^-). These are illustrated in Fig. 1.9. Owing to the larger ionic radii, defects on the halide sublattice are many powers of 10 lower in concentration. The thermodynamic equilibrium is expressed by

$$n_F^2 = n_i n_v = 2N^2 \exp\left(-\frac{\Delta G_F}{kT}\right) \tag{1.9}$$

where n_i, n_v, and N are the volume concentrations of interstitials, vacancies, and lattice sites, and ΔG_F is the Gibbs free energy of formation of a Frenkel pair. The equilibrium intrinsic value of n_i and n_v is designated as n_F. Extensive studies of the thermodynamic quantities have been made by measurements of the temperature dependence of conductivity and radiotracer diffusion.

The conductivity σ_i is given by the expression

$$\sigma_i = (n_i \mu_i + n_v \mu_v)e \tag{1.10}$$

where μ_i and μ_v are the mobilities of the interstitials and vacancies.

Currently accepted values (Friauf, 1984) for the formation enthalpies and entropies in AgCl and AgBr are given in Table 1.4 along with the room temperature concentrations

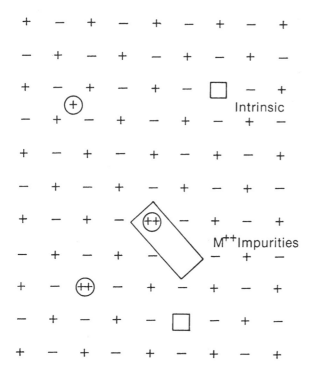

Figure 1.9 Schematic diagram of intrinsic and impurity-induced Frenkel disorder in a silver halide crystal. The diagram shows a [200] plane, but the interstitial silver ion is in fact displaced by ¼ [200] out of the plane.

of defects in each. Comparison with numbers given in Section 1.3.2 emphasizes that the concentration of ionic carriers is some 15 powers of 10 greater than that of electronic carriers in AgBr and 18 powers of 10 greater in AgCl.

1.8.4 Surface-Induced Defects

In addition to the classical Frenkel mechanism for pairwise formation of interstitials and vacancies, each of these point defects can be produced individually at a crystal surface of an extended internal defect (Hoyen, 1984). Kinklike sites on the surface or jogs on

Table 1.4 Thermodynamic Constants for Frenkel Disorder Formation in Silver Halides

Constant	Halides	
	AgCl	AgBr
Formation enthalpy, h_F (eV)	1.49 ± 0.021	1.163 ± 0.023
Formation entropy, S/k (e.u.)	11.13 ± 0.20	7.28 ± 0.58
Mol fraction defects at room temperature	4.92×10^{-11}	4.71×10^{-9}
Defect concentration, n_F at room temperature (cm^{-3})	1.15×10^{12}	9.80×10^{13}

Source: R. J. Friauf (1984).

dislocations are usually envisioned as the sources and sinks for silver ions in these extrinsic processes. From a simple thermodynamic cycle it follows that

$$\Delta G_F = \Delta G_i + \Delta G_v \qquad (1.11)$$

where the three terms represent the Gibbs free energies for formation of a Frenkel pair, for formation of a silver ion interstitial at some defect site, and for formation of a silver ion vacancy at the complementary site. In general, ΔG_i and ΔG_v are unequal, and the more readily formed defect is produced in excess. This leaves the surface (or dislocation) with a nonzero charge and a diffuse space charge region of the opposite sign near the surface or around the dislocation.

The spatial distribution of defects within the space charge region may be derived by solution of the Poisson equation with appropriate boundary conditions, and thorough theoretical analyses of this situation have been made.

Because of the relatively large specific surface of the microcrystalline dispersions of photographic materials, the surface generation process has a dominant effect on the ionic defect concentrations, which may differ by two or more powers of 10 from those in the volume of macroscopic samples. Among the techniques used to explore these properties experimentally, the most common is a contactless measurement of conductivity by means of the frequency dependence of dielectric loss, applied directly to photographic coatings (Van Biesen, 1970). These measurements are supported by more conventional conductivity measurements on thin films of the pure or doped silver halide materials, and a very few direct investigations of the surface potential or the profile of the potential distribution near the surface.

Results consistently indicate that on silver bromide microcrystals or thin films, ΔG_i is less than ΔG_v, leaving the surface with a net negative charge, compensated by a corresponding excess concentration of subsurface interstitial silver ions, as much as two or three powers of 10 greater than the intrinsic value. From the concentrations given in Table 1.4, it is easily seen that in a photographic microcrystal of silver bromide with a typical volume of 0.1 μm^3, there may be of the order 10^3 or more silver interstitials and on average far fewer than one silver ion vacancy.

Schematic diagrams of the defect concentration and electrostatic potential profiles are shown in Fig. 1.10. The most recent determinations (Hudson et al., 1987) of surface potential give a room temperature value for AgBr of about -0.1 V.

1.8.5 Impurity Effects

In impure materials or at reduced temperatures, the volume defect content can be determined by impurities. The most common controlling impurities are polyvalent metal ions, by which are introduced a corresponding concentration of silver vacancies (Fig. 1.9). Thus at low temperatures

$$[M^{2+}] \approx n_v \gg n_i \qquad (1.12)$$

Less frequently, divalent anionic impurities can be present in excess concentrations such that the defect balance is dominated by the corresponding excess silver interstitials.

Divalent cations bind the corresponding silver vacancies in nearest-neighbor lattice sites with energies of 0.2 or 0.3 eV. Thus at room temperature, depending on the impurity level, there may be finite concentrations of both species: the electrically neutral bound impurity–vacancy pairs and the isolated charged impurities. Higher valent cations have

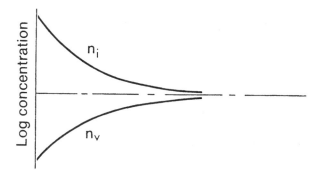

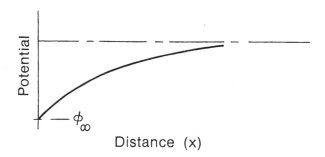

Distance (x)

Figure 1.10 Schematic plot of interstitial and vacancy concentrations and electrostatic potential versus distance from a silver halide surface.

correspondingly higher numbers of charge-compensating silver vacancies, and it is generally true that the dissociation energy increases for removal of successive vacancies. The corresponding association considerations apply for the polyvalent anions and the compensating silver interstitials.

1.8.6 Transport

The mobility μ of an ionic carrier is quite simply derived as

$$\mu = \frac{el^2 v_0}{6kT} \exp\left(-\frac{\Delta G_\mu}{kT}\right)$$ (1.13)

where l is the distance of an individual jump, v_0 is the vibrational frequency of the defect, ΔG_μ is the Gibbs free energy of the jump, and the other symbols have their usual significance. The enthalpy and entropy contributions to ΔG_μ are also obtained from computer fitting of conductivity data. Values for both carriers determined in this way (Friauf, 1984) are given in Table 1.5, along with the corresponding room temperature mobilities, as obtained from Eq. (13). The room temperature mobility of the interstitial silver ion is many times higher than that of the vacancy in both materials. For this carrier the dominant mechanism for motion is the collinear interstitialcy jump. In this process a silver interstitial

Table 1.5 Thermodynamic Constants for Frenkel Defect Motion in Silver Halides

Constant	Halides	
	AgCl	AgBr
Vacancy/jump		
enthalpy (eV)	0.306 ± 0.008	0.325 ± 0.011
entropy (e.u.)	-0.65 ± 0.12	1.01 ± 0.28
Room temperature mobility		
$(cm^2 V^{-1} s^{-1})$	6.1×10^{-7}	1.7×10^{-6}
Collinear interstitialcy jump		
enthalpy (eV)	0.018 ± 0.008	0.042 ± 0.011
entropy (e.u.)	-3.83 ± 0.12	-3.34 ± 0.28
Room temperature mobility		
$(cm^2 V^{-1} s^{-1})$	3.3×10^{-3}	7.7×10^{-4}

Source: R. J. Friauf (1984).

moves by $\frac{1}{4} \langle 111 \rangle$ into a lattice position, the lattice silver ion being similarly displaced by $\frac{1}{4} \langle 111 \rangle$ into the adjacent interstitial position.

Both the silver ions involved in the interstitialcy jump (i.e., the initial interstitial and the lattice ion that is displaced) encounter identical energetic barriers as they pass through the center of the triad of halide ions between the interstitial and lattice sites. There is a sizable spatial overlap, and the low jump enthalpy is possible only because of the quadrupolar deformability of the silver ion, as discussed above.

The enthalpies for the collinear interstitialcy jumps, 0.02 eV in silver chloride and 0.04 eV in silver bromide, are unusually low, making these among the most highly conducting ionic solids known.

1.8.7 Shallow Electron States

Charged point defects provide electrostatic potential wells at which electronic carriers of the opposite charge are bound. The binding at such a center is reasonably approximated by simple effective mass theory (Stoneham, 1975). If q is the charge of the defect, m^* the effective mass of the carrier, and ε_0 the static dielectric constant of the medium, the total binding energy W is given by

$$W = -\frac{q^2 e^2 m^*}{2\hbar^2 \varepsilon_0^2} \tag{1.14}$$

The carrier is bound in an orbit of radius r given by

$$r = \frac{\varepsilon_0 \hbar^2}{qem^*} \tag{1.15}$$

Calculated values of binding energy and orbital radius for electrons and holes at Coulomb centers in AgCl and AgBr are given in Table 1.6. The most significant feature to be seen is that, owing to the high dielectric constants, the binding energies are very small, comparable to room temperature thermal energy, and the orbital radii several lattice constants.

Table 1.6 Coulombic Levels in the Silver
Halides from Effective Mass Theory

	Binding energy W (meV)	Orbital radius r (nm)
AgCl		
electron	39	1.53
hole[a]		
AgBr		
electron	23	2.39
hole	140	0.41

[a] Input data not available.

Because of the difference in effective mass, the binding energies of shallow hole states are expected to be larger and the radii smaller. Calculated effective mass values for AgBr give a binding energy of 140 meV and an orbital radius of 4.1 Å. However, the effective mass model may be inapplicable in this range of radii. Because of the dominant self-trapping, input data for the hole in AgCl are not available.

These shallow electron states have been observed and identified at low temperature in both materials, as exposure-induced changes in absorption, luminescence, or conductivity in the infrared spectral region (Brandt and Brown, 1969; Sakuragi and Kanzaki, 1977). Experimental values of binding energies are about 24 meV in AgBr and 36 to 40 meV in AgCl, in remarkably good agreement with the predictions of effective mass theory.

Coulombic trapping centers in the silver halides are expected to have simple Langevin (1903) cross sections σ_c, with values given by the expression

$$\sigma_c = \frac{4\pi q \mu_m}{V_t \varepsilon_0} \tag{1.16}$$

in which q is the charge of the trapping center, μ_m is the microscopic mobility of the carrier, V_t is its thermal velocity, and ε_0 is the static dielectric constant. Substitution of appropriate values for these quantities gives cross sections as indicated in Table 1.7 for electrons and holes in two halides.

Thus the shallow electron levels in both AgCl and AgBr appear to be simply explained. Any closed-shell, positively charged center provides a shallow potential well that will bind an electron in a diffuse orbit with a radius of several lattice spacings and a binding energy of some 20 to 40 meV. In spite of the weak binding, however, they have capture cross sections many times larger than lattice dimensions.

Table 1.7 Calculated Langevin Cross
Sections σ_c for Charged Trapping Centers in the
Silver Halides (nm)

	AgCl	AgBr
Electron	6.4×10^2	5.9×10^2
Hole		2×10^1

1.9 DEEP ELECTRON LEVELS

Certain localized centers, when incorporated within or on the surface of silver halide crystals, introduce specific electronic energy levels within the forbidden gap owing to the chemical energy levels of the particular centers. These levels can act as deep trapping levels for electrons and holes. Many chemical impurity effects have been studied in single-crystal samples (Eachus and Spoonhower, 1987). However, only a few types of foreign center play important roles in the photographic process, and others will not be treated here.

1.9.1 Two-Step Capture

Carriers are usually trapped at deep levels (i.e., with binding energy greater than a few times the phonon energy) by a two-step mechanism, as illustrated in Fig. 1.11 (Gibb et al., 1977). A direct one-step capture is unlikely because the moving carrier does not remain in the vicinity of a deep trapping center long enough for any of the various mechanisms for losing the required energy to occur with any significant probability. A shallow level at the same center, accessible by the emission of no more than a few phonons, is required if the carrier is to be localized long enough for this to occur. In charged centers, the Coulombic binding provides the initial shallow level. Under these conditions, the overall cross section of the deep center is given by

$$\sigma = \sigma_c \eta \tag{1.17}$$

where

$$\eta = \frac{\nu}{\nu + \nu_i} \tag{1.18}$$

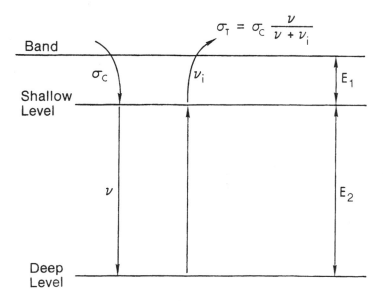

Figure 1.11 Schematic diagram of the transitions in the two-step trapping of a carrier at a deep level in silver halide. The overall cross section is reduced from the shallow level value σ_c by the factor $\eta = \nu/(\nu + \nu_i)$, the deexcitation efficiency, owing to the possibility of reionization to the band.

σ_c is the cross section for the shallow level (Section 1.8.7), ν is the rate of deexcitation from the shallow level to the deep, and ν_i is the ionization rate of the carrier from the shallow level to the band.

For the cross section to be large, not only must there be a shallow level but also η must approach unity. This can happen only if there is an efficient mechanism for the carrier in the shallow level to lose its excess energy in the deexcitation to the deep level. The deexcitation comes about by coupling to the host lattice, which relaxes locally and absorbs the energy as emitted phonons. The ''soft'' crystal lattice of the silver halides (Section 1.8.1) serves well in this regard.

The required lattice distortion is far smaller, the smaller the energy loss in a transition. Thus the deexcitation rate is increased significantly if there are one or more intermediate allowed levels between the shallow and deep ones shown in Fig. 1.11. This principle becomes important for the photographic process.

1.9.2 Halide Impurities

The impurity halide ions are one notable case of an impurity with practical importance. Owing to the differences in the electron affinities of the halogen atoms (Table 1.1), the higher atomic number halide ions, when substitutionally inserted in a silver halide of lower atomic number, introduce a filled level just above the valence band of the crystal. Thus I^- in AgBr and both I^- and Br^- in AgCl act as shallow hole traps. Evidence is that the binding energy of holes at these centers is in the range from 40 to 50 meV, thus similar to that expected from Coulombic defects. They do not constitute deep levels, and probably do not require the two-step trapping mechanism. Since the halide impurities simply replace the host halide ion with the same charge, they have no long-range attraction for holes and are expected to have capture cross sections of the order of lattice dimensions. Their location at kinklike defects, however, would confer the charge of the defect and increase the cross section. There is reason to expect that at the high iodide content used in many commercial materials there would be hopping of holes between impurity centers or even impurity banding.

1.9.3 Atomic Silver

A silver atom in or on the surface of a silver halide crystal has a singly occupied 5s electronic level near the middle of the silver halide forbidden gap. It is capable, therefore, of acting as a deep trap for either an electron or a hole. A neutral atom would be expected to have only a geometric cross section for either carrier, but Ag located at either a positive or a negative kinklike defect would have the larger Langevin cross section for the oppositely charged carrier and a much reduced cross section for the like-charged one. Once again, the charge also provides the shallow level for the two-step trapping process. Calculations show (Hamilton and Baetzold, 1981) that location at a charged site affects also the energy of the 5s level.

1.9.4 Silver/Gold Clusters

As the preceding discussions of photographic development have intimated, similar considerations apply for larger silver clusters as well. In the silver dimer, Ag_2, the atomic 5s levels of the two atoms combine to give a filled bonding molecular orbital and an empty antibonding one. An electron from the conduction band could in principle be captured into the antibonding orbital or a hole from the valence band to the bonding one. Vapor phase studies (Brown and Ginter, 1978) and model calculations (Hamilton and Baetzold, 1981) indicate that the energy difference between these two levels, the excitation energy of the dimer, is comparable with or perhaps even greater than the silver halide band gaps, so that it can be imagined that one or the other of these levels, but not both, would lie within the band gap of the host crystal. The calculations, in fact, indicate that this is the case and that the energy manifold is shifted by the electrostatic charge of the defect site at which the dimer resides so as to make it a favorable trap for an electron at a positive site or for a hole at a neutral or negative site but to be incapable of trapping the opposite carrier. Note that this is an energetic argument, over and above the difference in cross section, which affects the propensity in the same sense.

The silver trimer, Ag_3, again is expected to have a singly occupied orbital near midgap, in addition to the filled bonding and empty antibonding orbitals. Larger clusters have an increasing number of intermediate filled and unfilled levels and should be able to interact in similar manner with electrons and holes. The Fermi energy of bulk silver lies some 1 to 1.2 eV below the conduction band edge. The same effects of charge on cross section are expected to apply. The increasing multiplicity of levels causes the deexcitation step to be more facile in larger clusters.

When the clusters contain one or more atoms of gold, the 6s electron has a larger ionization energy than does the 5s electron of silver (Table 1.1), and this has the effect of depressing the manifold of energy levels, both filled and unfilled. This is in keeping with interpretations of the effect of gold in the sensitization on the photochemical and development processes.

1.9.5 Silver/Gold Sulfide Centers

It has been indicated earlier that aggregates of silver/gold sulfide molecules play a principal role in achieving current sensitivity levels of photographic films. Bulk silver or gold sulfide is a semiconducting solid whose band gap is evidently some 1 to 1.5 eV, for it absorbs throughout the visible spectrum and luminesces in the red and near-infrared regions. The electronic structure of small clusters is unknown, but some indications from luminescence are that they have similar but somewhat smaller energy gaps.

Several different types of measurement (Hamilton et al., 1988; Sturmer et al., 1974; Gilman et al., 1987; Kellogg and Hodes, 1987) place the lowest vacant electronic level of these sensitizer centers some 0.2 to 0.3 eV below the silver bromide conduction band edge and therefore the filled level near midgap. Assuming that the charge affects the cross section for interaction with electronic carriers, as suggested for silver clusters, those aggregates located at positive kinklike defects should provide effective but reversible traps for electrons. This expectation is confirmed by measurements of photoconductivity (Kellogg, 1974).

1.9.6 Sulfide/Metal Centers

When a silver atom forms at charged sensitizer centers, the manifold of allowed energy levels includes the shallow Coulomb level, the 0.2 to 0.3 eV level arising from the sensitizing aggregate, and the deep level associated with the silver atom. A current proposal (Hamilton et al., 1988; Hamilton, 1988) for the mechanism of sensitization (Fig. 1.12) is that the presence of the sulfide level facilitates the deexcitation step of the electron capture at the silver atom, thus increasing the overall cross section at this center. This step in the

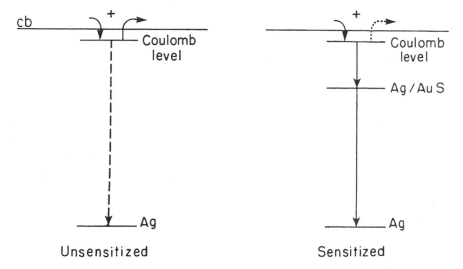

Figure 1.12 A proposed mechanism for the chemical sensitization effect. The intermediate level of the sensitizer material on the right-hand side improves the efficiency of deexcitation to the deep level, compared with the unsensitized case on the left.

process has been shown (Hamilton, 1983a, 1984) to be the limiting inefficiency in unsensitized emulsions. The deexcitation at larger silver clusters is promoted by the multiplicity of silver levels and therefore depends less on the presence of the sulfide level.

1.10 PHOTODECOMPOSITION

1.10.1 Silver

A key feature of the silver halides, which leads directly to their photochemical decomposition as opposed to the simple reversion of other photoconductors to the original state, is the combination of the photoconductive properties and the ionic defect structure. An electron in a localized state, even a shallow trapping level, has a likely possibility of capturing and combining with a mobile interstitial silver ion or a gold ion from the sensitizer. This extra ion, since it is not subject to the normal Madelung potential experienced by a lattice silver ion, has its vacant level well below the conduction band and normally also below the level in which the electron is trapped. In a successful combination, the electron loses the excess energy to the lattice and is deexcited into the metal valence s-level, forming what may be loosely termed a metal atom. Such a process was first clearly stated by Gurney and Mott (1938) and has come to be known (Mitchell, 1957) as the *Gurney–Mott principle*.

When the site at which the electron is trapped is adjacent to a positive kink or jog, the partial residual charge aids in the addition of the metal ion. Since the initial site has a partial positive charge, trapping of the electron does not render it electrically neutral but rather leaves it, for long-range effects, with a partial negative charge. Thus there is an electrostatic attraction for the mobile ion. Under this driving force, it can be shown (Berg, 1940; Hamilton and Brady, 1962) that the mean time for the ionic step corresponds roughly to the dielectric relaxation time of the crystal. For microcrystalline samples with their high concentration of surface-generated interstitial silver ions, this time can be determined by dielectric loss measurements and is found to be of the order 10^{-7} s, depending on crystal size. In this case the ion just at the defect site is also not a normal lattice ion, and the electron must presumably be considered to be shared between it and the added metal ion. It may be that this is the manner in which the gold of the sensitization material becomes part of the metallic center formed.

After electron trapping at any defect-associated silver atom or cluster, the net charge again becomes negative by a partial charge, and the electrostatic driving force for capture of another silver ion is restored. The addition of this silver ion modifies the energy level scheme accordingly, once more reversing the odd-even character of the center.

Thus the repetition of the electronic and ionic steps in a strictly alternating sequence makes possible the continued growth of silver centers at partially charged defects, eventually producing clusters large enough to act as latent image centers. The importance of the unique effective fractionally charged character of kink-type and jog-type structural defects at every stage during this sequence is apparent.

1.10.2 Halogen

A complementary process applies to the formation of halogen. If a silver ion vacancy combines with a trapped hole center, the net effect is the loss of a lattice silver ion and conversion of the corresponding halide ion to a halogen atom. In emulsion grains, where the vacancy concentration is suppressed by the surface effect, the ionic process involved at a trapped hole center has been shown to be the ejection of an adjacent lattice silver ion into an interstitial position and its motion away from the center (Platikanova and Malinowski, 1978).

1.10.3 Reversibility of Atomic Species

Both the silver and the halogen atomic species are thermally reversible in the silver halides, decomposing by the reverse of the ionic steps by which they are formed and ionization of the then weakly bound electronic carriers into their respective electronic bands. This instability is a very significant feature distinguishing the silver halides from other ionic crystals such as alkali halides, in which trapped electron and hole centers of atomic dimensions are thermally stable at room temperature. It is this property, generally attributed to the high dielectric constants of these materials, that permits the extremely efficient concentration of the effect of the spatially dispersed absorption events into a localized photochemical change at one or a few centers per grain.

1.10.4 Nucleation and Growth

When the instability of the single silver atomic center is included, silver formation becomes a classic nucleation-and-growth process (Burton and Berg, 1946). It may be represented (Hamilton, 1984) by the diagram of Fig. 1.13, which emphasizes the reversibility of the initial interactions, the separate nucleation and growth stages, the regularity of the alternating electronic and ionic stages, and the reversal of charge, which provides an electrostatic driving force for each step. The nucleation-and-growth feature of the process is particularly effective in controlling photographic response.

Specifically, if photons arrive at any one crystallite slowly enough, each electron will be consumed individually by growth on the first nucleus formed. The simultaneous

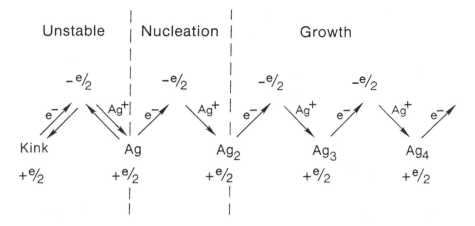

Figure 1.13 The steps in the growth of a silver cluster in silver halide, showing the nucleation-and-growth process, the alternating electronic and ionic steps, and the reversal of sign of the partial charges.

presence of two electrons, the condition necessary for a second nucleation event, never again occurs, and all silver is concentrated into a single cluster. Thus fully efficient concentration requires nothing more than the nucleation-and-growth process resulting from the instability of the single silver atom.

1.11 EFFICIENCY CONSIDERATIONS

The sequence depicted in Fig. 1.13 is a process that could lead to a cluster consisting of one silver or gold atom for each absorbed photon. The numerical analyses of quantum sensitivity and the minimum latent image size cited earlier, however, indicate otherwise. The most efficient photographic emulsions appear to require about 8 absorbed photons to produce a three-atom developable center. Clearly, the process is only about one-third to one-half as efficient as is theoretically possible.

1.11.1 Recombination

The principal residual inefficiency, and also the inefficiency addressed by chemical sensitization, has been shown (Hailstone et al., 1988) to be electron-hole recombination, which proceeds in competition with the reactions of Fig. 1.13.

In these materials direct band-to-band recombination is of negligible importance, being momentum-forbidden as a consequence of the indirect gap. Recombination of any consequence therefore occurs exclusively at lattice singularities, where first one carrier becomes localized, forming what may be termed a recombination center, and then the opposite carrier is also captured. The sequence forms some type of localized exciton, which then is deexcited by loss of its excess energy.

The lower diffusion velocity of the hole has the result that the principal recombination process is the capture of a free electron at a localized-hole center. The defect centers expected to provide recombination sites are the transient self-trapped hole centers, the isoelectronic iodide impurities (and bromide in silver chloride), negative kinklike physical defects, and iodide ions at such defects. Holes localized in the self-trapped states and at isolated halide impurities have a full positive charge and the corresponding Langevin cross sections for subsequent electron capture, whereas holes at either bromide or iodide kinks would have only a partial positive charge. Their cross section for electron capture might be expected to be smaller, but only by perhaps a factor of 2 or so, as would those of partially charged silver centers. However, because the initial charge of species associated with these defect sites is negative, their cross section for the prior hole capture would be far greater than those of isoelectronic traps, and they might well dominate the recombination process.

The opposite process, the capture of a free hole at a trapped-electron center, is also charge favored but in practice less likely because the hole is so much more localized than the electron.

At cryogenic temperatures, a significant mechanism for deexcitation of the exciton so formed is luminescent emission. As the temperature is raised, however, the luminescence is totally quenched well below room temperature, owing to the dominance of a much more rapid thermally activated nonradiative decay.

1.11.2 The Symmetry Principle

If this type of interaction is the principal recombination channel, there is a basis for quantitatively rationalizing the observed approximately half-efficient formation of silver (Hamilton, 1982, 1983b). When two competing processes have equal 50% efficiency, it follows that they have equal rates. In the present case the two rate-determining steps are shown in Fig. 1.14. Both consist of the capture of an electron from the conduction band by a deep center with a positive charge. The initial cross section of each center is the Langevin value as determined by charge, and the deexcitation of both has near unit efficiency, at least in chemically sensitized emulsions (Hamilton, 1990). This provides a logical explanation for the near-equivalent quantum sensitivity limit of the best emulsions, even those

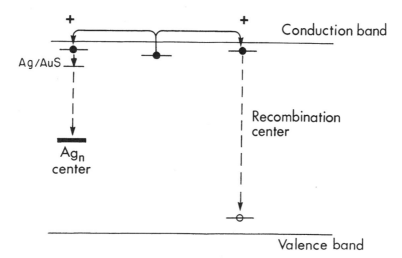

Figure 1.14 The symmetry of the competing steps that determine photographic sensitivity. The electron, in the course of being repeatedly trapped and released, encounters one or the other of the indicated positively charged centers.

from different sources. Their efficiencies fall short of the ultimate limits, but they lie at a secondary statistical limit set by the inherent symmetry of the competing processes.

1.11.3 Dye Holes

All of the discussions above have been cast in terms of recombination at localized hole centers in the silver halide crystal. Yet, as has been emphasized earlier, much of photography depends on electrons injected from sensitizing dyes with ground state energy levels well above the valence band of the crystal. Thus recombination must often occur by capture of a free electron at a dye-localized hole. The argument can be made that this does not significantly alter the points previously stated. If the dye molecule is originally electrostatically neutral or, by virtue of a preferred adsorption site, charged by $\pm e/2$, it will become a positively charged center after injection of the electron, and the same arguments about cross section for electron capture apply. The recombination then is completed when the captured electron is deexcited to the ground state level of the dye.

The similarity of photographic sensitivity in the intrinsic and dye-induced spectral region (i.e., RQE values near unity: see Section 1.6.2), argues strongly that the recombination probabilities are much the same regardless of whether the hole is localized at a crystal defect or a dye molecule. The quantitative aspects of the sensitivity argument presented above depend on the retention of holes in a state subject to recombination by electron capture with a cross section comparable to that of silver centers for a time long compared with the silver-forming steps.

1.12 SUMMARY

Several unique physical properties of the photographically important silver halides (i.e., silver bromide and silver chloride) are intimately involved in the workings of conventional photography. To a surprising degree, these properties can be ascribed to the presence of the 4d outer electron shell of the silver ion. This feature of the electronic structure distinguishes the halide salts of the Group 1B elements of the periodic table from those of the otherwise similar alkali metals.

Alkali halides and silver halides are both electronic insulators with relatively wide forbidden gaps separating the valence and conduction bands. In both types of material the conduction bands are derived from the first vacant s-level of the cation. In the alkali halides, the valence band is also simply related to the appropriate filled halogen p-level. In the silver halides, however, this halogen level is strongly hybridized with the nearly energetically coincident silver d-levels. Thus the valence band is broadened at the Brillouin zone boundary, giving the two silver salts much narrower indirect forbidden gaps of 2.68 eV in the bromide and 3.25 eV in the chloride. Although the population of thermal carriers is still negligible, this shifts the indirect absorption edge of the silver salts strikingly toward the visible part of the spectrum.

This aspect of the electronic structure also introduces a coupling between the d- and s-levels of the silver ion, making it much more easily deformable in a quadrupolar or football-shaped mode. This in turn has profound effects on many ionic and crystalline properties. The dielectric constants are high, as are the elastic compliances of the lattices; the formation energy of charged point defects is low, as is the binding energy of shallow electron states; and the enthalpy for motion of a charged interstitial cation is lower than in any other conventional ionic crystal.

All these features play a role in the photographic process. The sensitive salts are photodecomposed as photoconduction electrons, and mobile interstitial silver ions alternately collect at charged-defect sites to produce clusters of silver atoms. The unoccupied or singly occupied energy levels of the silver clusters provide deep traps for capture of electrons from the conduction band, provided it is possible to dispose of the excess energy to be deexcited to the deep level. The soft crystal lattice provides for strong electron–lattice coupling, which plays a strong part in the energy disposal.

Before the silver atoms begin to accumulate, however, only shallow electronic levels are available, and the earliest stages of the process are totally reversible. This produces in effect a conventional nucleation-and-growth process, which is capable of very efficient concentration of the silver into one or a few clusters.

The current sensitivity position of the best available photographic materials is close to but significantly below the theoretical limit. This position can be understood in terms of a statistically determined limit set by the symmetry of the forward, silver-producing step, on the one hand, and the competing recombination reaction, on the other.

The details of the positioning of the unfilled electronic levels of silver clusters of various size within the silver halide forbidden gap provide a rational explanation for discrimination in photographic development and for the size dependence of activation delays.

REFERENCES

Amelinckx, S. (1979). Dislocations in solids, in *Dislocations in Crystals*, Vol. 2 (F. R. N. Nabarrow, ed.). North Holland, Amsterdam, Chapter 6, p. 67.

Babcock, T. A., Sewell, M. H., Lewis, W. C., and James, T. H. (1974). *Astron. J.*, *79*: 1497.

Berg, W. F. (1940). *Proc. R. Soc. London*, *A174*: 559.

Berg, W. F. (1983). *J. Photogr. Sci.*, *31*: 62.

Berry, C., West, W., and Moser, F. (1963). Chapter 12, in *The Art and Science of Growing Crystals* (J. J. Gilman, ed.). John Wiley, New York.

Bilz, H., and Weber, W. (1984). In *The Physics of Latent Image Formation in Silver Halides* (A. Baldereschi, W. Czaja, E. Tosatti, and M. Tosi, eds.). World Scientific, Singapore, p. 25.

Brandt, R. C., and Brown, F. C. (1969). *Phys. Rev.*, *181*: 1241.

Brown, C. M., and Ginter, M. L. (1978). *J. Mol. Spectrosc.*, *69*: 25.

Brown, F. C. (1976). Reactivity of solids, in *Treatise on Solid State Chemistry*, Vol. 4 (B. Hannay, ed.). Plenum, New York, p. 333.

Burton, P. C., and Berg, W. F. (1946). *Photogr. J.*, *86B*: 2.

Chateau, H. (1959). *C. R. Acad. Sci. Paris*, *248*: 1950.

Chateau, H., Moncet, M. C., and Pouradier, J. (1958). Ergebnisse der International Konferenz für Wissenschaftliche Photographie, Kölin, in *Wissenschaftliche Photographie* (W. Zichler, H. Frieser, and O. Helwich, eds.). Verlag Dr. O. Helwich, Darmstadt, p. 16.

Corbin, D., Gingello, A., MacIntyre, G., and Carroll, B. H. (1980). *Photogr. Sci. Eng.*, *24*: 45.

Duffin, G. F. (1966). *Photographic Emulsion Chemistry*. Focal Press, London.

Eachus, R. S., and Spoonhower, J. P. (1987). In *Progress in Basic Principles of Imaging Systems* (F. Granzer and E. Moisar, eds.). Vieweg, Braunschweig, p. 175.

Evrard, R. (1984). In *The Physics and Chemistry of Latent Image Formation in Silver Halides* (A. Baldereschi, W. Czaja, E. Tosatti, and M. Tosi, eds.). World Scientific, Singapore, p. 57.

Farnell, G. C. (1969). *J. Photogr. Sci.*, *17*: 116.

Farnell, G. C., and Solman, L. R. (1963). *J. Photogr. Sci.*, *11*: 347.

Fayet, P., Granzer, F., Hegenbart, G., Moisar, E., Pischel, B., and Wöste, L. (1985). *Phys. Rev. Lett.*, *55*: 3002.

Friauf, R. J. (1984). In *The Physics of Latent Image Formation in Silver Halides* (A. Baldereschi, W. Czaja, E. Tosatti, and M. Tosi, eds.). World Scientific, Singapore, p. 79.

Gibb, R. M., Rees, G. J., Thomas, B. W., Wilson, B. L. H., Hamilton, B., Wight, D. R., and Mott, N. F. (1977). *Philos. Mag.*, *36*: 1021.

Gilman, P. B., Jr., Penner, T. L., Koszelak, T. D., and Mroczek, S. K. (1987). In *Progress in Basic Principles of Imaging Systems* (F. Granzer and E. Moisar, eds.). Vieweg, Braunschweig, p. 228.

Gurney, R. W., and Mott, N. F. (1938). *Proc. R. Soc. London, Ser. A*, *164*: 151. See also Mott, N. F., and Gurney, R. W. (1948). *Electronic Processes in Ionic Crystals*. Clarendon, Oxford, Chapter VII.

Hada, H., Kawasaki, M., and Fujimoto, H. (1980). *Photogr. Sci. Eng.*, *24*: 232.

Hada, H., and Kawasaki, M. (1985). *J. Imaging Sci.*, *29*: 51.

Hailstone, R. K., and Hamilton, J. F. (1985). *J. Imaging Sci.*, *29*: 125.

Hailstone, R. K., and Hamilton, J. F. (1987). *J. Imaging Sci.*, *31*: 229.

Hailstone, R. K., Liebert, N. B., Levy, M., and Hamilton, J. F. (1987). *J. Imaging Sci.*, *31*: 185, 225.

Hailstone, R. K., Liebert, N. B., Levy, M., McCleary, R. T., Girolmo, S. R., Jeanmaire, D. L., and Boda, C. R. (1988). *J. Imaging Sci.*, *32*: 113.

Hamilton, J. F. (1982). *Photogr. Sci. Eng.*, *26*: 263.

Hamilton, J. F. (1983a). *Photogr. Sci. Eng.*, *27*: 225.

Hamilton, J. F. (1983b). *Radiat. Eff.*, *72*: 103.

Hamilton, J. F. (1984). In *The Physics of Latent Image Formation in Silver Halides* (A. Baldereschi, W. Czaja, E. Tosatti, and M. Tosi, eds.). World Scientific, Singapore, p. 203.

Hamilton, J. F. (1988). *Adv. Phys.*, *37*: 359.

Hamilton, J. F. (1990). *J. Imaging Sci.*, 34. 1.

Hamilton, J. F., and Baetzold, R. C. (1981). *Photogr. Sci. Eng.*, *25*: 189.

Hamilton, J. F., and Brady, L. E. (1962). *J. Phys. Chem.*, *66*: 2384.

Hamilton, J. F., and Brady, L. E. (1970). *Surf. Sci.*, *23*: 389.

Hamilton, J. F., Harbison, J. M., and Jeanmaire, D. L. (1988). *J. Imaging Sci.*, *32*: 17.

Hamilton, J. F., and Logel, P. C. (1974). *Photogr. Sci. Eng.*, *18*: 507.

Harbison, J. M., and Spencer, H. R. (1977). Chapter 5, in *The Theory of the Photographic Process*, 4th ed. (T. H. James, ed.). Macmillan, New York.

Hodby, J. W. (1969). *Solid State Commun.*, *7*: 811.

Höhne, M., and Stasiw, M. (1968). *Phys. Status Sol.*, *28*: 247.

Hoyen, H. (1984). In *The Physics of Latent Image Formation in Silver Halides* (A. Baldereschi, W. Czaja, E. Tosatti, and M. Tosi, eds.). World Scientific, Singapore, p. 151.

Hudson, R. A., Farlow, G. C., and Slifkin, L. M. (1987). *Phys. Rev. B*, *36*: 4651.

Jaenicke, W. (1972). *J. Photogr. Sci.*, *20*: 2.

Keevert, J., and Gokhale, V. (1987). *J. Imaging Sci.*, *31*: 243.

Kellogg, L. M. (1974). *Photogr. Sci. Eng.*, *18*: 378.

Kellogg, L. M., and Hodes, J. (1987). *SPSE Conference*, Rochester, NY, p. 179.

Kunz, A. B. (1982). *Phys. Rev.*, *B26*: 2070.

Langevin, P. (1903). *Ann. Chem. Phys.*, *28*: 433.

Laredo, E., Paul, W. B., Rowan, L. G., and Slifkin, L. (1983). *Phys. Rev. B*, *27*: 2470.

Lowndes, R. P. (1966). *Phys. Lett.*, *21*: 26.

Lowndes, R. P., and Martin, D. H. (1969). *Proc. R. Soc. London. Ser. A.* *308*: 473.

Mitchell, J. W. (1957). *J. Photogr. Sci.*, *5*: 49.

Moser, F., and Urbach, F. (1956). *Phys. Rev.*, *102*: 1519.

Müller, P., Spenke, S., and Teltow, J. (1970). *Phys. Status Sol.*, *41*: 81.

Nutting, P. G. (1913). *Philos. Mag.*, *(6)26*: 423.

Phillips, J. C. (1970). *Rev. Mod. Phys.*, *42*: 317.

Picinbono, B. (1955). *C. R. Acad. Sci. Paris*, *240*: 2296.

Platikanova, V., and Malinowski, J. (1978). *Phys. Status Sol.*, *47*: 683.

Pontius, R. B., Willis, R. G., and Newmiller, R. J. (1972). *Photogr. Sci. Eng.*, *16*: 406.

Pontius, R. B., and Willis, R. G. (1973). *Photogr. Sci. Eng.*, *17*: 21.

Powers, D. E., Hansen, S. G., Geusic, M. E., Michalopoulos, D. L., and Smalley, R. E. (1983). *J. Chem. Phys.*, *78*: 2866.

Roth, P. H., and Simpson, W. H. (1980). *Photogr. Sci. Eng.*, *24*: 133.

Sakuragi, S., and Kanzaki, H. (1977). *Phys. Rev. Lett.*, *38*: 1302.

Seitz, F. (1951). *Rev. Mod. Phys.*, *23*: 328.

Selwyn. E. W. H. (1935). *Photogr. J.*, *75*: 571.

Selwyn, E. W. H. (1942). *Photogr. J.*, *82*: 209.

Stoneham, A. W. (1975). *Theory of Defects in Solids*. Clarendon, Oxford.

Sturmer, D. M., and Blackburn, L. N. (1979). Sulfur chemical sensitizations of monodisperse AgBr emulsion grains using radioactive thiosulfate. Paper F-4, presented at SPSE 32nd Annual Conference, May 13–17, 1979, Boston.

Sturmer, D. M., Gaugh, W. S., and Brushci, B. J. (1974). *Photogr. Sci. Eng.*, *18*: 56.

Tani, T. (1983). *Photogr. Sci. Eng.*, *27*: 75.

Tani, T. (1985). *J. Imaging Sci.*, *29*: 93.

Toyozawa, Y. (1961). *Progr. Theor. Phys.* (*Kyoto*), *26*: 29.

Toyozawa, Y., and Sumi, A. (1974). In *Twelfth International Conference on Physics of Semiconductors* (B. G. Teubner, ed.). Stuttgart, p. 179.

Trautweiler, F. (1968). *Photogr. Sci. Eng.*, *12*: 138.

Ulrichi, W. (1970). *Phys. Status Sol.*, *40*: 557.

Van Biesen. J. (1970). *J. Appl. Phys.*, *41*: 1910.

Webb, J. H. (1950). *J. Opt. Soc. Am.*, *40*: 3, 197.

West, W. (1974). *Photogr. Sci. Eng.*, *18*: 35.

West, W., and Gilman, P. B. (1977). Chapter 10, in *The Theory of the Photographic Process*, 4th ed. (T. H. James, ed.). Macmillan, New York.

Wey, J. S. (1985). *Chem. Eng. Commun.*, *35*: 231.

Wyckoff, R. W. G. (1964). *Crystal Structures*, Vols. 1, 2. Wiley, New York.

2

Color-Forming Photographic Materials

L. E. FRIEDRICH and J. A. KAPECKI

Eastman Kodak Company, Rochester, New York

2.1 INTRODUCTION

2.1.1 The Three-Color System

This chapter reviews the major dye-forming materials that are used in color negative films and papers. For perspective, a brief overview is given of conventional photographic phenomena. More comprehensive reviews of color photographic systems are available (1–3) and can be examined for documentation of concepts that are not referenced in this chapter.

The photographic system is used to capture the spatial and spectral distributions of light so that they can be stored, transmitted, and viewed at a future time. The spatial information is captured through the use of lenses, which focus light from points in the scene to points in the film plane. The spectral information is a continuous distribution of visible wavelengths from the scene, which is the difference between the illuminant spectrum and light that is adsorbed or diffracted by the object within a scene.

Fortunately, a high-quality photographic system does not have to capture the continuous distribution of visible wavelengths from a scene because human eyes are not analog receptors of continuous visible radiation. Retinas have cones with three receptors that absorb light in the violet-blue (ca. 400–500 nm), blue-green (ca. 450–610 nm), and green-red (500–700 nm) regions (4). The three optic signals provide human brains with data that are perceived as color (5)

Because human brains convert analog information in colored light into signals in three optic channels, a high-quality photographic system can be constructed by conversion of continuous wavelengths of light into images in three records of different color. This knowledge led to the invention of the integrated tripack for conventional films and papers, in which the three-color recording and reproducing records are stacked on top of one

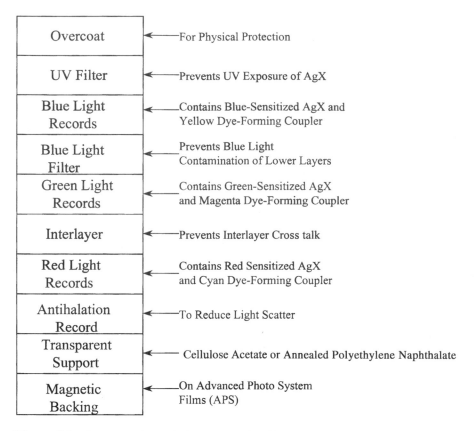

Figure 2.1   Structure of a typical color negative film.

another (see Fig. 2.1) (6). In such systems, the three subtractive primary dyes are generated in processing (cyan, magenta, and yellow, each of which modulates one of the additive primaries, red, green, and blue). Before processing, each color record of the tripack stores information from the scene in the form of a "latent" silver image that is capable of catalyzing the formation of the visible dyes, which are the major subject of this chapter. Reddish light of 600–700 nm gives a cyan dye, greenish light of 500–600 nm gives a magenta dye, and bluish light of 400–500 nm gives a yellow dye. This negative film image is used as a mask between a printer illuminant and a photographic paper that has similar negative-working dye chemistry. The end result is a positive colored image that reflects from the photographic color print paper.

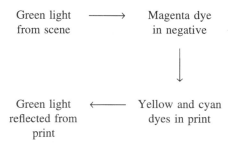

This two-stage negative–negative system has enjoyed widespread use even though it is complex, because color films can be constructed with very wide latitudes in sensitivity to different light intensities. Therefore color negative films can be used in a wide range of situations, from low to high light intensities, and from still images to snapshots of kinetic images with fast shutter speeds. When low light levels expose the negative, the exposure times of the printer are lessened to deliver moderate light exposures to the paper prints. When high light levels generate large amounts of dyes in the negatives, greater printer exposures times are used, also to deliver moderate light exposures to the print materials.

So-called reversal or single-stage systems such as Ektachrome film faithfully map specific exposures from a scene to specific light transmission through the slide medium, yielding a positive image. If such single-stage films are under- or overexposed, these materials yield correspondingly dark or light projected images.

Before describing the classes of color-forming materials used in color negative films, it is useful to review the major sequential chain of events that converts scene light to recorded dye in the negative. Knowledge of these events is basic to appreciation of photographic chemistry and underscores the importance of the couplers or dye-forming materials, which form the detail of this chapter.

2.1.2 The Photographic Chain of Information

Most simply, wherever light from the scene illuminates a silver halide crystal, that light generates a silver atom cluster called the latent image. The overall elementary chemistry of this process is

$$AgX + Light \rightarrow Ag_0 + 1/2 \; X_2$$

The result is a photochemical reaction to effect a disproportionation: Ag^+ goes to Ag_0 and X^- goes to X_0 (as bromine gas in the case of X = Br). Silver atom clusters of only several atoms may be thermally unstable or subject to oxidation by ambient oxidants. Therefore a minimum threshold of light must impinge on the film to store scene information.

The next two stages of the imaging chain occur when the film is processed during the development step.

In the first stage, only those silver halide crystals with a stable silver atom cluster are reduced by the color developer, a substituted paraphenylenediamine (PPD), because the silver atoms are a miniature metallic electrode on which the oxidation–reduction preferentially takes place. As the redox reaction proceeds, further elemental silver is formed, and the reaction becomes autocatalytic. The purpose of this reaction is to transform the original miniscule latent image into an amplified signal of oxidized developer (D_{ox}). In the second stage, the highly electrophilic D_{ox} cation reacts with a dye-forming material or "coupler," to form first an unstable leuco "colorless" dye and then the stable image dye.

As the scheme shows, there are two types of couplers, depending on whether Y is hydrogen or an electronegative leaving group. Examples of leaving groups are halides, or fragments with a nitrogen, oxygen, or sulfur atom attachment. When Y is hydrogen, a cross-oxidation of the leuco dye with a second molecule of D_{ox} occurs to generate the image dye. When Y is a good leaving group, a simple elimination of H-Y takes place to convert the leuco dye into the image dye. The terms "four-equivalent" and "two-equivalent" in the scheme refer to the change in oxidation state as the two classes of couplers are transformed to dyes. These are the same as the number of silver ions that are stoichiometrically reduced to silver metal when one mole of dye is formed.

Following the development or dye-forming step, the silver metal formed in the process is reconverted to a silver salt by an oxidant or "bleach," so called because it removes the black silver metal. Then, usually in another process step, the silver salt is solubilized and removed by a silver complexing agent known as a "fix," leaving only the stable or fixed dye image.

2.1.3   The Integrated Tripack

In a typical color negative film, the uppermost layers contain silver halide sensitized to blue light and a yellow dye-forming coupler. The middle layers contain green-sensitized silver halide and a magenta dye-forming coupler, and the bottom image-forming layers contain red-sensitized silver halide with a cyan dye-forming coupler. The blue-sensitized layer is typically overcoated with a layer to remove ultraviolet light. A blue-light filter is placed between the blue and green records because the silver halide in the lower layers retains its native sensitivity to blue light. The couplers themselves are often dissolved in high boiling organic media called coupler solvents. When the mixture is dispersed in the aqueous gelatin matrix, the dissolving property of the coupler solvent confers chemical reactivity to the couplers. Figure 2.1 (shown previously) gives a schematic of a typical color negative film.

The materials in film are related to their function. The main materials are the so-called "emulsions" of silver halides, their adsorbed sensitizing dyes, and the couplers. Necessary infrastructure includes the polymeric gelatin phase in which the silver halide and coupler are suspended and the flexible support (usually cellulose acetate, though polyethylene 2,6-naphthalene dicarboxylate is used for the Advanced Photo System films). Additional classes of materials include coating aids (that give coatings without bubbles, streaks, or other defects), scavengers of D_{ox} (that are placed between imaging layers to prevent cross talk among sublayers), and antihalation dyes (that prevent internally reflected light within the film from exposing silver halide grains). Also, the film contains abrasion resistant materials.

2.2   IMAGE COUPLERS

2.2.1   Characteristics of Image Couplers

Image couplers are the premier organic materials in color films. The function of these couplers is to convert faithfully the fleeting color information carried by D_{ox} into the permanent storage in dyes. Couplers, therefore, are ionizable and potentially nucleophilic materials that have the all-important ability to form a colored azomethine-containing product. The azomethine group is the link between the developer and coupler halves.

The coupling reaction links an electron-rich moiety with an electron-poor moiety. The link hybridizes the highest occupied molecular orbital (HOMO) of the oxidized developer with the lowest unoccupied molecular orbital (LUMO) of the ionized coupler, which generates a product with a relatively low-energy transition in the visible region (a dye). The different hues of the visible dyes are determined in part by the LUMO levels of the coupler moieties and the conformations of the dyes.

This description suggests that molecular orbital calculations might be useful in the search for couplers with desired dye hues. However, computations of λ-max have not always been accurate, in part because of the nature of the pi–pi* transition (7). The transition involves charge transfer from the paraphenylenediamine HOMO part of the dye to the coupler LUMO part. Such a transition has a large transition dipole moment and a variable solvent effect. Computations of solvent effects, still in their infancy, account for only the bulk dielectric effect of the medium.

Couplers have three primary chemical properties of interest (along with other important factors such as manufacturing cost, coupler and dye stability, and environmental benignancy). The primary properties are

> The pK_a or acidity of the coupler
> The kc or nucleophilicity of the coupler at the processing pH, e.g., pH 10 for many
> > color negative and paper processes
> The visible absorption band of the resulting dye

pK_a is important because most neutral couplers are relatively unreactive to D_{ox}. To be reactive, couplers need to be nucleophilic, and this is achieved by ionization. For ionization, their pK_as in dispersion are optimally near, or less than, the pH of the developer processing solution.

For water-insoluble couplers, there are several methods for determining the pK_as. One is to measure the pK_a in micelle media, such as aqueous Triton X-100. These pK_as are approximations to all-aqueous pK_as. Another is to measure the $pH_{1/2}$ (or pH for half ionization) in liquid dispersion media, such as a 1:2 ratio of coupler:tri-tolylphosphate coupler solvent with gelatin and a surfactant. Such values are approximations to in-film half-ionization points and are typically several units higher in value than pK_as measured in aqueous micelle media. Also, when extraneous absorptions and light diffraction can be tolerated, spectral measurement of in-film $pH_{1/2}$s are sometimes possible.

Partial (or complete) coupler ionization at a processing pH of 10 is not sufficient for reactivity. The resulting anion needs to have reactivity toward D_{ox} (8). This feature is tracked by the rate constant, kc^-, of the fully ionized coupler anion. For some series of couplers, there is a relationship between the pK_as of the couplers and their log kc^-s. As might be expected, as the pK_a increases, the coupler anion becomes more reactive, and kc^- increases (9). Therefore there is often an optimum pK_a near the processing pH (e.g., 10). If substituent changes were to increase the pK_a much above 10, the coupler would not be ionized sufficiently for reaction. If the pK_a were much lower than the processing pH, the coupler anion would have a reduced nucleophilicity.

The last major requirement is a useful visible absorption band for the dye. It is easily appreciated that the hue of the dye in the film must appropriately control light transmission. Hue encompasses not only the λ-max of the absorption but also the shape of the absorption band. For some dyes, the absorption band will be too broad in shape, and the challenge is to narrow or "sharpen up" the band. Theoretical considerations are of marginal use in this research effort, because band broadness relates to the number of dye conformations that are present, dye aggregate formation, and flexibility of the dye in its ground and excited states. Besides hue, there is also covering power, which is defined as the amount of density delivered by a unit amount of dye per unit area in the film. Covering power has many subtleties, but a major factor is the extinction coefficient, or oscillator strength, of the absorption.

2.2.2   A Quantitative Description of the Imaging Chain

A simple model that connects observed density (D) in film negatives to mechanistic quantities is

$$D = Ag\left(\frac{1}{2}\right)\left(\frac{1}{S}\right)(F_{Dye})(C_P) \tag{2.1}$$

Here, Ag is the mole per meter squared of the silver that is formed, which can be independently determined in a process that excludes the silver bleaching step. As described previously, D_{ox} is stoichiometrically one-half the amount of silver and is the reason for the fraction (1/2). If all the D_{ox} were converted to dye, the amount of dye would be Ag/2/S, where S is the stoichiometry of D_{ox} that is theoretically needed to produce one mole of dye. The product of 2 times S is the equivalency of the coupler, which is the moles of silver that stoichiometrically accompany the formation of one mole of dye. The quantity F_{Dye} is the fractional amount of theoretically possible dye that could be formed from the D_{ox}. Not all D_{ox} goes to dye, because of side reactions, such as the reaction of D_{ox} with sulfite in the developer. Lastly, C_P is the covering power of dye that relates the density of a dye deposit to its amount. The units of C_P are square meters per mole.

   The equation is both qualitatively and quantitatively useful because it illustrates the chain of quantities that give dye density: silver and D_{ox}, stoichiometric dye, efficiency of dye formation, and covering power. Different imaging chemistries usually do not have a large impact on the silver and D_{ox} terms. Different coupler chemistries impact the density mainly through the stoichiometry factor S, through the influence of pK_a and coupler reactivity on the term F_{Dye}, and through the molar extinction coefficient of the dye in C_P.

2.2.3   Yellow Image Couplers

Conventional Yellow Image Couplers

One generic structure dominates the yellow couplers that are in use, that of the beta-ketocarboxamides, where R_2 is a coupling-off group (COG) that is eliminated after reaction with D_{ox}. Of the three coupler classes, the yellow dye formers are the only ones that are acyclic. The resulting dyes are not totally planar, which contributes to their short absorption of blue light and leads to a lower extinction coefficient.

Typically, R_1 is either t-butyl or aromatic. The quaternary carbon of the t-butyl group is desired for light stability of the dye (10), and these couplers are therefore widely used in color paper materials. Even though these dyes have greater light stability (11), the couplers are not generally as reactive as when R_1 is aromatic (12). A way to improve the reactivity of the couplers with a t-butyl group is to ''tie'' two of the methyl groups into a cyclopropane ring (13). This preserves the quaternary carbon that is good for light stability, and it reduces the steric hindrance.

A variety of COGs have been used at position R_2. The need is to provide a substituent that is electronegative enough to be a good leaving group, so that the intermediate leuco dye readily forms the yellow dye. However, if the electron withdrawal is too great, then the coupler anion is less nucleophilic toward oxidized developer.

One good compromise for R_2 is the N-1, 5-membered heterocyclic 2,5-diones.

Various combinations of X and Y include the oxazolidine diones (X = O, Y = C) (14), triazolidine diones (X = Y = N) (15), and imidazolidine diones (X = N, Y = C) (16).

The role of R_3 is usually as a ballast group to hold the coupler in its coated layer, although ballasting can also be put into an R_1 aryl group. Various substituents can be used for R_3 including carbamoyl, ester, and sulfamoyl.

The most often used group in the ortho anilino position is a chloro group as shown in the general structure. One property is that the chloro group increases the extinction coefficient of the dye by almost 40% (17).

The Challenge of High Extinction Coefficients for Yellow Dyes

An interesting challenge in yellow coupler chemistry is to improve the low extinction coefficients of the acyclic beta-keto amides. As mentioned earlier, the nonplanarity of yellow dyes results in a reduced extinction coefficient (typically 15,000 to 21,000 L/mol cm) relative to cyan and magenta dyes (>30,000). This means that more silver halide and coupler dispersions need to be coated in yellow layers than in other layers, in order to match their densities with neutral scene exposures.

Two approaches are evolving to achieve higher dye extinctions. The more modest approach is to investigate novel R_1 groups. When R_1 is t-butyl, the extinctions are between 15,000 and 17,000, whereas when R_1 is aryl, the extinctions are between 19,000 and 22,000. Indolinyl substituents for R_1 have been found to give extinctions of between 24,000 and 25,000, and these higher extinctions probably relate in part to the dihedral angle between the beta carbonyl and coupling site carbon (18).

A second, ingenious approach utilizes a preformed but shifted yellow dye as the COG (19,20).

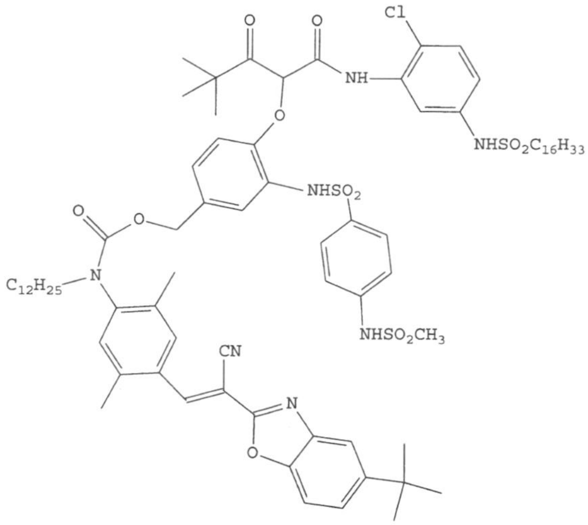

A conventional yellow coupler releases a COG that forms a quinone methide (see discussion of quinone methids in Section 2.4.6). Sequentially, the quinone methide releases a carbamic acid that decarboxylates to give a second yellow dye. In its carbamate form, the second yellow dye hue is shifted hypsochromically mostly out of the visible region. Coupling produces two molecules of dye. The conventional yellow dye has an extinction of ca. 15,000. The dye that is released has an extinction of ca. 50,000. The sum is 65,000, which is truly a breakthrough. With use of these materials, the coupler and silver halide laydowns in the film can be drastically reduced.

2.2.4   Magenta Image Couplers

Two Major Classes of Magenta Image Couplers

There are two major classes of magenta image couplers, the pyrazolones and the pyrazolotriazoles.

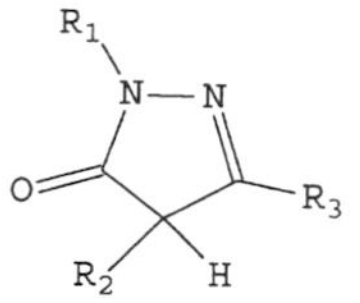

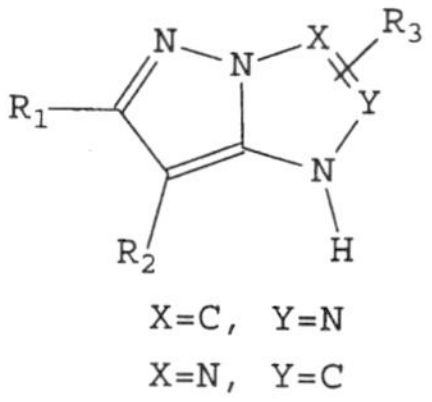

These classes are structurally quite different as well as being different photographically. The pyrazolones have been known since the 1930s (21), while the pyrazolotriazoles were identified in the 1970s (22). As a class, the pyrazolones were more difficult to convert into successful two-equivalent couplers and existed for years as only four-equivalent couplers, whereas the 7-chloropyrazolotriazoles (R_2 = Cl) were readily available once routes to these heterocycles became available.

Pyrazolone Magenta Image Couplers

There are two subclasses of pyrazolones, which are the C-4 arylcarbamoyl and anilino materials, respectively, and are shown below.

Arylcarbamoylpyrazolone   Anilinopyrazolone

By far the most prevalent R_1 group is the 2,4,6,-trichlorophenyl group. Higher chlorinated phenyls (23) have also been used, but the use of highly chlorinated materials can pose environmental concerns.

Three common R_2 groups have been used. The four-equivalent couplers with R_2 = H are active materials toward oxidized developer, but a reaction between the coupler and its dye can lead to dye loss during storage of the negative. This phenomenon was studied in solution by Vittum and Duennebier (24) with the finding of some amazing chemistry. The end products formed from the dye are the starting coupler and paraphenylenediamine (Path A). This is formally a four-electron reduction that splits the azomethine double bond. The source of the electrons (and protons) are four unreacted pyrazolone couplers that oxidize and combine to form two moles of a bis-pyrazolone. An implied intermediate in the reaction is the leuco dye.

These findings explain why fade occurs mainly in the lower density "toe" regions of low green exposure where excess coupler is present as the reductant. If excess coupler is destroyed by postprocess treatment of the film with a reagent like formaldehyde (which makes a bis-methylene pyrazolone by an aldol-elimination-addition process, Path B), then the fade process is eliminated.

Another way to avoid the dye loss is to use two-equivalent pyrazolones with a COG. Much research has been done to produce two-equivalent materials with a leaving group at C-5. There are four potential classes of useful leaving groups: halides, oxy-, nitrogen-, and sulfur-linked materials. All have their limitations, but the one that has been most successful is the arylthio class of coupling-off groups (25). In order for the released thiols not to interfere with silver development, the COG is typically ballasted so that the arylthiol stays in the oil droplet until it is oxidized to its final disulfide product by oxidized developer. Because of the consumption of an extra mole of oxidized developer to form the disulfide, pyrazolones with arylthio COGs do not effectively function as true two-equivalent couplers.

Another class of COG on pyrazolone image couplers is the pyrazolo group (26). Among the various nitrogen heterocycles, the pyrazolo COG provides the right amount of electron withdrawal to be a good leaving group, while not lowering the nucleophilicity of the coupler anion toward oxidized developer.

The choice of R_3 depends on the subclass of pyrazolone. With the C-4 arylcarbamoyl materials, an o-chloro is seldom used in the aryl group (27). A standard ballast is typically substituted in the meta position. With the C-4 anilino materials, an o-chloro is usually used to get the proper hue along with meta ballasting (28).

While the C-4 arylcarbamoyl materials are active couplers, the anilino materials are even more reactive with oxidized developer, because the more electron donating anilino raises both the pK_a and the coupling rate constant (29). If the pK_a (or $pH_{1/2}$) of the coupler were much above 10, there might not be an advantage to this senario because there would be less of a more nucleophilic coupler anion with the possibility of no net benefit. But the anilino pK_as are near 10, so these couplers enjoy high ionization at pH 10 along with high nucleophilicity.

Another property of the anilino pyrazolones is their greater resistance to formaldehyde. Formaldehyde is a ubiquitous pollutant that can destroy pyrazolones in film by forming an unreactive methylene-bis-coupler at the coupling site (the same reaction that can be used post-process to destroy excess coupler; see Path B in the above scheme). Perhaps the lower pK_as of the carbamoyl pyrazolones produce greater concentrations of coupler anion that react with formaldehyde.

A major exception to the use of monomeric coupling materials is the polymeric magenta couplers, which are intrinsically ballasted by attaching a pyrazolone nucleus to a copolymer of propene, butylacrylate, and styrene (30). These novel couplers are also used in tandem with a pyrazolo COG for efficient use of oxidized developer. To achieve the correct hue of the dye, a carbamoyl link is used between the pyrazolone ring at C-4 and the polymer.

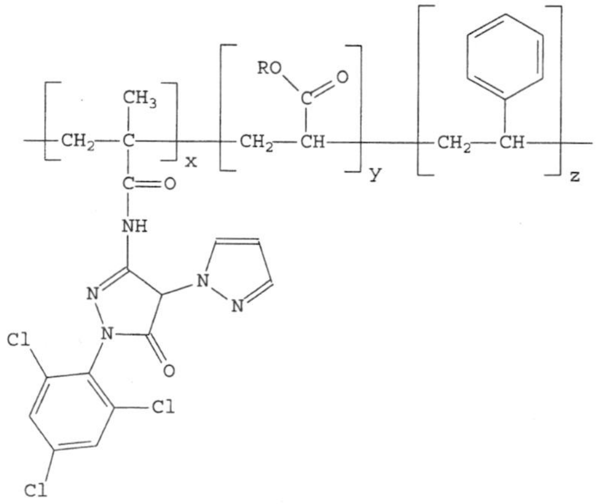

Pyrazolotriazole Image Couplers

Pyrazolotriazoles also have two important subclasses shown below.

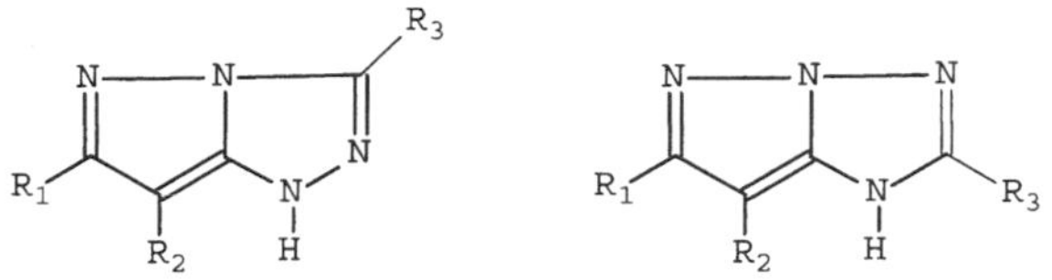

There are two big appeals that these couplers enjoy. First, their magenta dyes have less unwanted blue absorptions (ideal magenta dyes should absorb in the green region only). Unwanted absorption of the dyes can be compensated by the use of colored masking couplers (see below). Such technology cannot be used in color paper where white backgrounds are needed, so pyrazolotriazole couplers are very valuable for imaging the green information in prints.

Second, these couplers are almost always used as efficient two-equivalent materials with a simple chloro COG at R_2, which also addresses the high $pH_{1/2}$s that are usually well above 10 in liquid dispersions. Using a chloro COG lowers the $pH_{1/2}$ closer to 10, and this gives more activity to the coupling reaction.

Another way to improve ionization of these high $pH_{1/2}$ pyrazolotriazoles is to include, usually in the ballast, a carboxylic acid, phenol, sulfonamide, or other group that is capable of ionizing below or near pH 10. This ionization is believed to raise the dielectric constant of the organic phase, perhaps by pulling water into the dispersion droplet. This effect serves to lower the $pH_{1/2}$ of pyrazolotriazole couplers.

The nature of the R_1 substituent at C-6 depends on the use. In products that do not have to endure the stress of dye light stability, such as color negative films, a simple methyl group is usually used, providing reactive couplers. However, since the pyrazolotriazole dyes, particularly those represented by the first structure above, are less light stable than pyrazolone dyes, their use in output products such as color print paper requires a sterically large group in the C-6 position (31).

The reason a group like t-butyl enhances the light stability of a pyrazolotriazole dye is complex. Pyrazolotriazole dyes typically aggregate in the oil droplet dispersions, and it is these aggregates that are particularly photosensitive toward bleaching (32). The t-butyl group is thought to disrupt partially the aggregation of the dyes.

Use of the t-butyl group does not come without a price, however. Its steric bulk dramatically lowers the coupling rate, so that such materials cannot be used with high-competition sulfite-containing developer formulations that are typically used in color negative processing. Fortunately, developers for color papers need not be formulated as high in competition, so these t-butyl substituted materials can be used there for enhanced dye light stability.

For R_3, an electrically neutral alkyl-like group provides the proper hue of the dye for most photographic uses (33). A large variety of linking groups further down the R_3 substituent chain and different ballast groups have been used. In the last decade, 2-arylpyrazolotriazoles have been found to have higher dye extinction and narrower bandwidth dyes (34).

Pursuit of Further Dye Light Stability for Pyrazolotriazoles

An area of active research has been to enhance the steric effect at C-3, just as has been done at C-6 with the t-butyl group. For example, the coupler below has a 6-t-butyl group, a 3-quaternary carbon atom, and on the end of the ballast chain, a group said to impart further light stabilization (35). Often, light stabilizers are coated as separate molecules in a codispersion with the coupler.

2.2.5   Cyan Image Couplers

Naphtholic Cyan Couplers

There are two major classes that have found widespread use in films and papers. Again, each of these comes in two subgroups.

The first type are the naphthols, many of which have ideal in-film $pH_{1/2}$s that are near the developer pH of 10.

This feature makes them more highly ionized than the second type of couplers (phenols, see below), and the naphthols generally enjoy high reactivities with D_{ox}. We speculate that this reactivity may be due to secondary orbital overlap of the D_{ox} cation and naphtholate anion pi orbitals, which may lower the energy of the transition state for coupling.

By far the most used substitutent at R_1 is the carbonamido group. This group creates the opportunity for H-bonding both in the coupler and in the dye. The former affects the pK_a of the coupler and the latter the hue of the dye.

Both two- and four-equivalent (36) naphthol couplers are known. The most useful COGs on two-equivalent couplers are linked through oxygen and include aryloxy and alkyloxy with auxillary functional groups in the alkyl chain (37).

A fundamental problem with naphthol dyes is a potential to reform the leuco dye in the bleach step, commonly an iron complex formulation.

$$Ag + Fe(III) \rightarrow Ag(I) + Fe(II)$$

During the bleaching process, the ratio of Fe(II) to Fe(III) increases in the film, particularly in regions of high silver (high exposure). The increased ratio of Fe(II) to Fe(III) locally lowers the redox potential in the film and allows the possibility of cyan dye reduction. The naphthol dyes are kinetically prone to this reduction, and once formed they are sluggish to reoxidize back to the dye as diffusion reestablishes the bulk ratio of Fe(II) to Fe(III).

One solution is to use a carbamate group ($-NHCO_2R$) at R_3 (38). The carbamate group provides stabilization of the dye by internal hydrogen bonding.

Phenolic Cyan Couplers

A second solution to the problem of leuco dye formation relies on the phenolic class (39) of cyan couplers that come in two major subgroups.

For film uses, the first subgroup usually has R_4 as H with carbamoyl-like groups at R_1 [or ureidoaryl groups with electron withdrawing substituents (40)] and R_3 (41). Both two- and four-equivalent materials are available. Their dyes have good dark stability (42) but have less robust light stability.

In paper, where light stability is more critical, the predominant phenolic coupler comes from the second subgroup, with R_4 as chloro and R_3 as alkyl (and R_1 still carbamoyl) (43). This subclass has greater light stability and a more hypsochromic hue that is better for color rendition in prints. Chloro COGs are common at R_2.

Novel Cyan Couplers

There has been a rush of activity in recent years to invent new heterocyclic cyan couplers that use traditional magenta-like couplers or totally new heterocyclic systems with electron withdrawing groups and extended conjugation. Such substituents lower the LUMO energy of the dye, which produces the batho shift into the cyan. Two examples include conventional-looking pyrazolotriazole couplers. The hue shift is accomplished by the use of further aryl conjugation and a carbamoyl substituent (44).

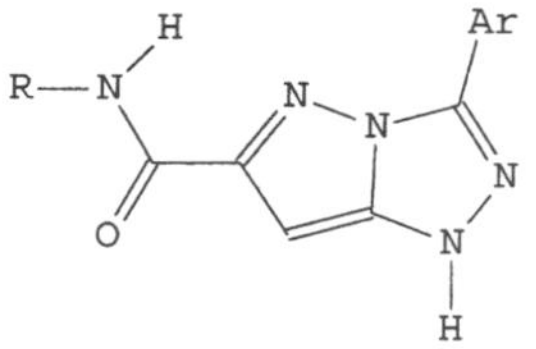

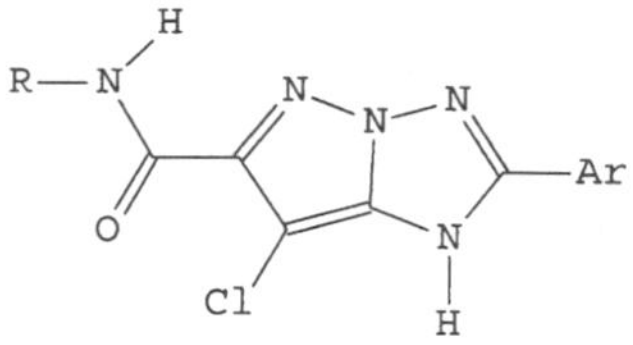

The dyes from these materials are reported to have molar extinction coefficients over 60,000 (45).

Another class includes the pyrrolotriazoles that typically have a cyano group and further conjugative substitutents to move the hue into the cyan (46). The system shown below has a carbamate coupling-off group and is also reported to have a high extinction coefficient for the dye (47).

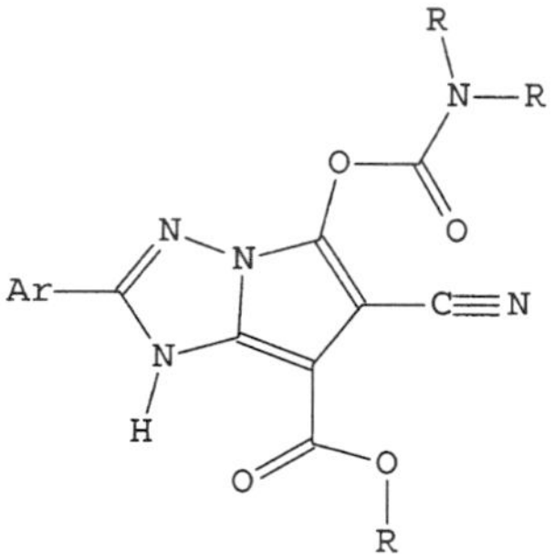

It is challenging to introduce a new class of coupler into photographic products. Not only must the new material be reactive with D_{ox} under development conditions and produce a dye with a correct hue, but more than a dozen other attributes must not be compromised. These include the raw stock keeping of the coupler, the dark and light stabilities of the dye, the granularity and acutance of the image, and a host of other practical aspects. Furthermore, these complex materials must fit within the cost structure of a competitively marketed product.

2.3 MASKING COUPLERS

2.3.1 How Masking Couplers Correct for Unwanted Dye Absorptions

Masking couplers are members of a class called image-modifying couplers. Such couplers are not used primarily to form the color image but rather to modify it, improving attributes like color, grain, or sharpness. They usually are two-equivalent couplers that have a chemically blocked photographically useful group (PUG) in the coupling position. After D_{ox} reacts with the image-modifying coupler, the PUG is eliminated from the leuco dye. In the case of masking couplers, the chemistry modifies the effective hue of the dye image in film to compensate for unwanted absorptions of the image dyes. For example, magenta dyes have a major pi–pi* absorption in the green (500–600 nm) but also have a tail or minor absorption in the blue (400–500 nm). Similarly, cyan dyes tail into or absorb in the green and/or blue regions. This is a fundamental problem. (Masking couplers can also be used to generate interlayer interimage effects (IIEs), which are discussed in the next section.)

Unwanted absorptions in the negative absorb light of a different color than that which generated the image dye, leading to desaturated or muddy colors. For example, suppose a picture is taken of midsummer grass that has a pure green color. Light from the scene exposes the green record, such that during development a magenta dye is formed in the negative. This magenta dye has an absorption that tails into the blue (or yellow) region and it may also have absorption of lower extinction in the blue region. Even though the light from the scene was pure green, the record in the negative acts as though both green and blue light came from the scene. When a print is made from this negative, the grass will appear bluish-green, not green.

Because the film-print system has two stages, adding masking couplers in the sublayers that contain the image couplers bearing unwanted dye absorptions can achieve a very clever masking of these unwanted absorptions (48). Carefully follow the logic. Masking couplers are designed to possess the unwanted color before development. Typically azo chromophores (which can tautomerize to hydrazo forms) are used as coupling-off groups (or PUGs) on a parent coupler.

During development, when D_{ox} reacts with the masking coupler (in competition with image couplers), the azo group is coupled off, and the starting color is destroyed. Ideally, the amount of color destruction is exactly matched by the amount of unwanted color formation caused by formation of the image dye in the coating.

In the example of a green scene exposure given above, the film would be built with both a magenta image coupler and a yellow-colored masking coupler in the same sublayers that are sensitive to green light. Before exposure from the scene, the film has a uniform yellow tint caused by the masking coupler. Green light from the scene produces on development the magenta dye with its unwanted yellow color, as well as a corresponding loss of yellow color from reaction of the masking coupler with D_{ox}. The net result is that the yellow masking coupler and its chemistry with D_{ox} has masked (removed from view) the unwanted yellow *image* of the green grass, replacing it with a uniform yellow color in

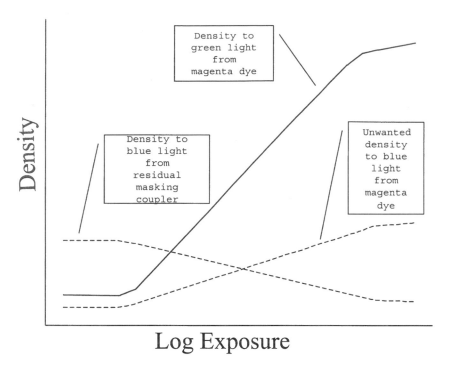

Figure 2.2   Use of yellow-colored coupler to mask the unwanted blue density from a magenta image dye.

both green and nongreen areas of the scene. Figure 2.2 shows how this masking is accomplished.

Though the uniform yellow cast in the resulting negative will filter blue light from the printing lamp, this can be compensated by a longer blue exposure during printing.

The final consideration is the choice of parent coupler that bears the colored masking chromophore as a coupling-off group. There is an obvious choice. If the goal is to mask unwanted absorptions that are originating in the green record, then a magenta coupler is ideal, as further magenta dye is then made as the masking color is destroyed. An obvious need is for the loss of masking color to be matched to the gain of unwanted color from the image dye. This is controlled in part by the choice of the reactivity of the coupler to which the masking dye is attached. The relative rates of D_{ox} coupling with image and masking couplers must be properly matched.

This masking technique is extensively used in color negative films but is inappropriate for directly viewed materials like slides and prints where the color of the mask would be visible. Cyan dyes also have unwanted absorptions in both the green and the blue, so there are both magenta-colored cyan masking couplers and yellow-colored cyan masking couplers. There are also magenta-colored yellow masking couplers (49) in the literature, although there is limited need for them because yellow dyes do not tail badly into the green region.

2.3.2   Yellow-Colored Magenta Couplers

The most common of the masking couplers that is used in the magenta record is based on the anilino pyrazolone magenta couplers (50). They are drawn below in their hydrazone form where there is hydrogen bonding to the C-5 carbonyl group:

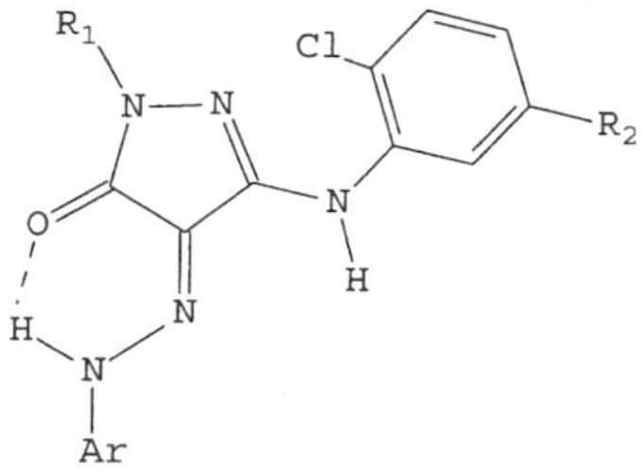

The types of substituents that are used at R_1 and R_2 are typical of the magenta image couplers described earlier. Trichlorophenyl is most common at R_1 along with lesser use of the pentachlorophenyl group. At R_2, the carbonamido group and its variants are common. Often, simple electron donating substituents are used in the Ar group, such as ethers.

The detailed mechanism for coupling of these materials with oxidized developer is uncertain. Certainly, the first step is ionization of the hydrazone proton to form an anion. Collapse of D_{ox} at the C-4 carbon is likely to form an azo leuco intermediate. At this point, one pathway could be elimination of the azo group to form the magenta image dye. D_{ox} would likely further oxidize the released azo group, which is an imide, into an aryl diazonium radical or cation, both of which would give colorless products.

2.3.3 Masking Couplers in the Cyan Layers

Masking couplers in the cyan layers are largely based on the naphthol cyan image couplers.

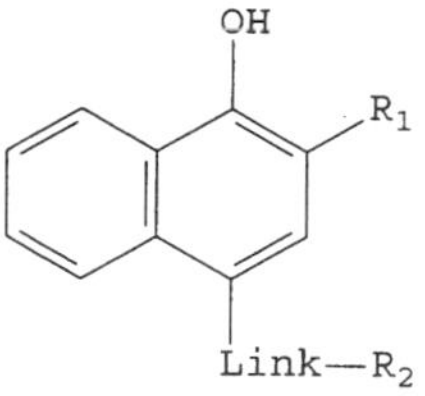

The usual carbamoyl group is used as R_1, and the link is often an ether coupling-off group. Various R_2 groups are used, depending on the color of the mask that is desired. All of them are formally azo chromophores.

The following structure represents the most common type of magenta-colored cyan masking coupler (51).

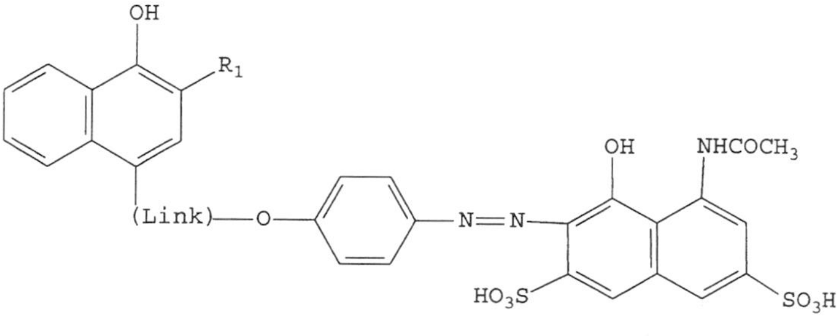

The structure is drawn in its azo form, but it could be a hydrazo tautomer. Coupling of the naphthol anion with D_{ox} releases the link, and the ionizable sulfonic acid groups solubilize the magenta azo chromophore so that it washes out of the film. In this way, red light from the scene removes magenta color.

A yellow-colored naphthol masking coupler (52) that can be used in the cyan layer to account for unwanted blue absorptions in the cyan dye is shown below.

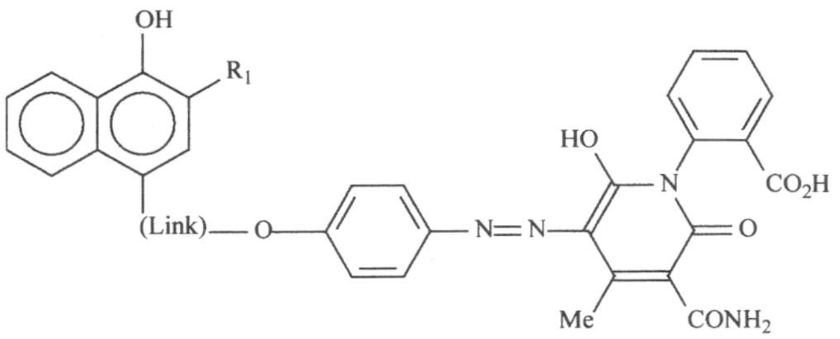

This material can also be drawn in its hydrazo tautomeric form, and it has solubilizing groups that can also cause the released chromophore to wash out of the film.

The use of masking couplers has its drawbacks, and image dyes without unwanted absorptions would be preferred. There are speed penalties to pay, because the preformed dye absorbs light that could expose silver halide. The resulting higher minimum densities in the negative also mean that a lesser exposure range is available before encountering density maximum limitations. Last, when a masking coupler (e.g., yellow colored) is used for IIE (mentioned at the beginning of this section, and see below under ''How Masking Couplers Deliver IIE''), extra yellow image coupler is needed. The higher use of yellow image couplers costs more and may give greater graininess in the image.

2.4 DEVELOPMENT INHIBITOR RELEASING COUPLERS (DIRs)

2.4.1 Interlayer Interimage Effects

What Are DIRs?

DIRs are both development-modifying (formation of D_{ox} and silver) and image-modifying (formation of color) materials. The mechanistic use of these couplers in film is to retard silver development and D_{ox} formation.

It is fair to say that DIRs do not simply fix problems from image coupler chemistry; they deliver more pleasing photographic images over and above what even ideal image couplers can deliver. Two uses will be described, and they are both as creative as the use of masking couplers for unwanted absorptions.

The first use is for enhanced interlayer interimage effects (IIEs), which was mentioned in passing in the masking coupler subsection. IIE refers to chemistry that produces greener greens (grass), bluer blues (sky), and redder reds (Santa Claus) than are true in the scene. This has been found to be desirable by customers. IIEs convert ''muddy'' colors of many real scenes (e.g., bluish-greens) into purer colors (greens).

How does this happen? DIRs are couplers with a silver development inhibitor in the COG position. When chemically blocked in this manner, the inhibitor is inert. When the DIR couples with D_{ox} (in competition with the image and masking couplers), a dye from the parent coupler is formed along with a free inhibitor molecule. A development inhibitor is a molecule that can diffuse to a silver halide grain and can react with surface silver ions of high free energy to form insoluble silver-inhibitor salts on the surface. These high-free energy silver ions are likely those that replenish interstitial silver ions that are reduced at the minielectrode of the latent image. Therefore lowering the free energy of surface silver ions by reaction with inhibitors serves to reduce the overall development rate.

Inhibitors can be designed with appropriate diffusion coefficients so that they diffuse during development to adjacent color records and inhibit development chemistry there. This means that red light from the scene generates unblocked inhibitors that slow development not only in the red sublayers but also in the green sublayers (and to a lesser extent, in the blue sublayers that are farther away). Green light leads to inhibition in the green sublayers as well as in both the blue and red sublayers (which straddle the green sublayers), and so on for blue light.

With inhibitors diffusing within and among all sublayers, the rates of these processes (inhibitor generation, diffusion, and inhibition) need to be balanced. This is necessary so that ''neutral'' or shades of gray light from the scene produces the equal amounts of densities from the cyan, magenta, and yellow image dyes.

Why is it desirable to do this? Consider a segment from the scene that is green (or red or blue). Only the green chrome is exposed. Inhibitors are not released from the red and blue sublayers. Therefore, there will be more silver halide development occurring in the green layer from green light than when the green layer is exposed by neutral light. More magenta dye is produced, which translates to a higher density of green color in the print. Green grass appears greener (more saturated green) than real. Blue skies appear bluer than real. Reds are redder. This potentially desirable effect is the result of IIEs.

It is now easy to understand why the inhibitors need to be blocked as COGs on couplers so that they are released ''imagewise'' with development chemistry. If they were not blocked, inhibitors would be present to operate on the development of red, green, or blue images as well as neutral images. If inhibitors were not blocked, neutral images would have the same saturation as red, green, and blue images.

The features needed for good DIR couplers include a rapid ''imagewise'' release of the inhibitor, which means that free inhibitors are generated by D_{ox} on a time scale that is rapid compared to overall development. This is needed so that the inhibitors have a chance to diffuse to adjacent color records before development is complete. Also needed are high diffusion coefficients for the inhibitors.

How Masking Couplers Deliver IIE

In the prior section on masking couplers, it was mentioned that they also serve to increase IIEs. The use of masking couplers in film would be desirable even if there were no unwanted absorptions from the image dyes. To explain this, focus for example on the blue and green layers, where there are yellow and magenta image couplers and a yellow-colored masking coupler in the green layer. Assume that the magenta image coupler gives a dye with no unwanted yellow color. When a scene contains blue, green, and red light (neutral scene color), the blue light will make a yellow dye, and the green light will initiate development chemistry that partially destroys the yellow mask. The final blue density is the sum that remains from the mask, and that density formed from the yellow image coupler. This blue density combines with the green and red densities in the other layers, and when the film is designed right, the combined colored densities produce a gray patch of some saturation in the print.

When pure blue light of the same amount (as in the neutral exposure above) comes from a different segment of the scene, the same amount of yellow dye will be made, but now the yellow mask in the green sublayers is not removed. Therefore there is more total blue density, which translates to a bluer (more saturated) image in the print. This is why masking couplers are useful, even if image dyes were to have no unwanted absorptions. In the presence of image couplers with unwanted absorptions, the film needs to be built so that the masking density that is lost is greater than the formation of unwanted image dye density.

2.4.2 DIRs for Sharpness

A second use of DIR couplers is to increase the sharpness of scene edges. An edge consists of two different scene colors or densities that can be degraded by light scatter during the exposure step or by the horizontal component of D_{ox} diffusion across the edge during its reaction with couplers. However, if inhibitor is generated imagewise, its horizontal component of diffusion at the edge also generates a changing concentration of inhibitor across the edge. Figure 2.3a illustrates these edge profiles for light exposure and released inhibitor. This horizontal concentration profile for inhibitor across a scene edge also occurs when inhibitors diffuse (vertically) into adjacent color records, because diffusion is isotropic. Therefore adjacent to the higher exposure side of the edge (scene), there is a lesser amount of inhibitor than in the bulk of the scene. This gives greater silver halide development at the edge with the formation of greater amounts of D_{ox} and dye. On the nonscene side of the edge, inhibitor that has diffused from the scene side reduces development and density. The overall result is an edge with a greater gradient of density, which appears as a sharper edge. It is as if a child had taken a colored crayon to enhance the density on the scene side of the edge and an eraser to reduce the density on the nonscene side. Figure 2.3b shows the result. The ''ears'' of the density profile are real.

The same features in a DIR coupler that increase IIE also increase sharpness. DIRs for both IIE and sharpness profit from being more reactive with D_{ox} than image couplers, so that the action of inhibitors on silver halide development can compete with formation of the color image. Inhibitors also enhance IIE and sharpness when they have a long diffusion length. This diffusion path is shortened by high adsorption to silver halide grains, which coincidentally is part of the inhibition mechanism.

2.4.3 Delayed-Release DIRs

A way to prevent temporarily the reversible adsorption of inhibitors to silver halide emulsions so that the inhibitors diffuse farther is to block the inhibitor with an organic linking group, sometimes called a switch. Attached to a coupler, these yield what are called delayed-release DIR couplers. When the $k_{release}$ produces a half-life of the switch–inhibitor combination (that is, seconds to minutes), then the profile of the released inhibitor concentration is broader, and both sharpness and IIE are enhanced.

$$\text{Coupler–switch–inhibitor} \xrightarrow{\ D_{ox}\ } \text{Dye + Switch–inhibitor} \tag{2.2}$$
$$\text{Inhibitor + Switch} \quad \xleftarrow{\quad k_{release}\quad \ \rvert}$$
$$\text{fragment}$$

If the switch releases the inhibitor in a few seconds or less, there are no consequences on IIE and sharpness. Nevertheless, a fast switch may provide a synthetically more accessible coupler. Or the switch may confer more reactivity of the coupler toward D_{ox} than the directly attached inhibitor (DIR) would provide.

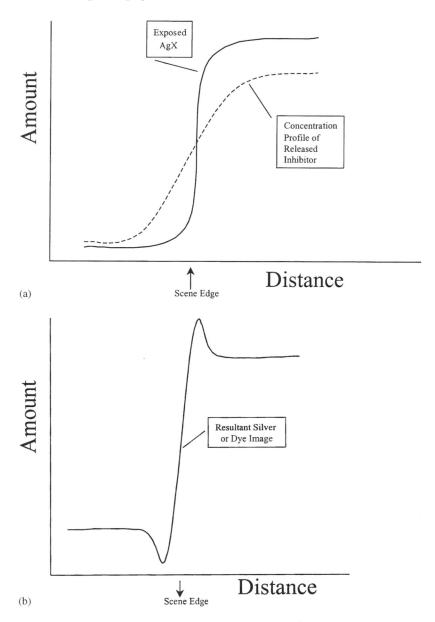

Figure 2.3 (a) The concentration profile of a released inhibitor at a scene edge. (b) The final density profile at a scene edge.

A description of the DIR materials is primarily a description of the switches and inhibitor fragments. All the common coupler parents can be used, although as mentioned, more reactive parents (toward D_{ox}) are needed for DIRs than for image couplers. Furthermore, only ca. 25 to 50 mg/m² of laydown for a DIR coupler is required, compared to tens or scores of milligrams per square foot for an image coupler. This means that changes in the DIR coupler structure can be made to enhance reactivity with D_{ox} or to enhance release of the COG, even if the structural changes generate dyes with less than optimal hue characteristics.

2.4.4 The Inhibitor Classes

There are two fundamental types of development inhibitors, both of which share the common feature of strong binding constants with silver ions. These two types are the mercaptans and nitrogen heterocycles (where one tautomer contains a NH group). The central class of the mercaptans is the mercaptotetrazoles, whereas the nitrogen heterocycles are most represented by the benzotriazoles.

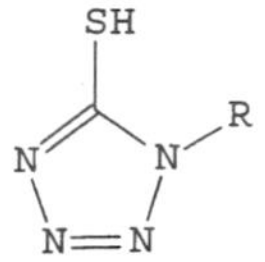

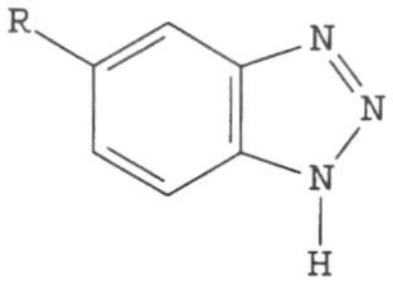

The mercaptotetrazoles have pK_as lower than pH 7, so they are fully ionized in pH 10 color developers. The sulfur anions form extremely strong bonds to silver ions, and their solubility products (pK_{sp}s) are typically ca. 15 (53), well above that of silver bromide at 12.3. However, the simple ability to complex the silver ion does not make a development inhibitor. The low molecular weight mercaptotetrazoles (e.g., R = Me) are not inhibitors, nor are higher homologues that contain an ionizable group such as carboxy. Such inhibitors do form complexes with silver ions, but such complexes are soluble in the aqueous gelatin of the film and dissolve the grain. To be an inhibitor, the silver complex needs to be hydrophobic so that it will stay on the grain surface and lower the free energy of surface silver ions. Typically, the most common R group is phenyl, although alkyl groups are also prevalent. Besides the mercaptotetrazoles (54), systems such as substituted mercapto-oxadiazoles (55) and mercaptothiadiazoles (56) also function as inhibitors.

2.4.5 Self-Destruct Inhibitors

Just as mercapto systems need to be hydrophobic to function as inhibitors, so do benzotriazoles. Unsubstituted benzotriazole is a weak inhibitor. A common linking group on the benzotriazole is a carboxylic acid ester (57) such as shown below.

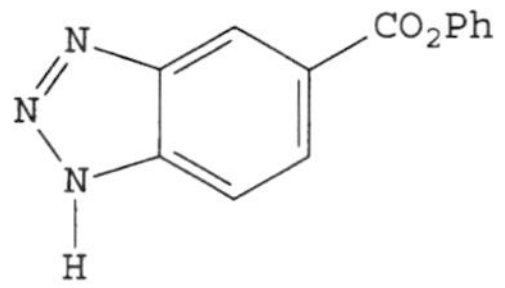

Besides increasing the hydrophobicity, the ester group addresses a potential problem of inhibitor materials referred to as "seasoning" of the developer. Inhibitors that diffuse appropriately in the film also diffuse into the developer bath. Their concentration in the bath can build, and unless there are protective layers built in the top of the film pack, they can enter subsequently processed film and alter development in an uncontrolled fashion. However, the phenyl ester in the benzotriazole shown above can hydrolyze to the carboxylic acid in the developer bath. With a free ionized carboxylic acid group after hydrolysis, the benzotriazole is no longer a moderate inhibitor because it is no longer hydrophobic enough to adsorb and stay on the grain. This feature is called "self-destruct" chemistry and is also used with the mercaptotetrazoles (58). Typically, the self-destruct link is an activated ester group.

As a class, the benzotriazoles do not adsorb to silver halides as strongly as most of the mercapto systems. They have higher pK_as, between 6 and 10, and their pK_{sp}s with silver ion are several units lower than mercaptotetrazoles (59). This enables the benzotriazoles to excel at creating longer range IIEs among sublayers without use of a switch. There are a host of other potential heterocyclic inhibitors, but so far the most useful have been variants on the benzotriazoles. Both 1,2,3- (60) and 1,2,4-triazoles (61) show promise.

2.4.6 Delayed-Release Switches

For delayed-release DIR couplers, there are three general classes of switches that are commonly used, and they come in different flavors. The first invented were the so-called ''carbamate'' switches (62) of the general structure below.

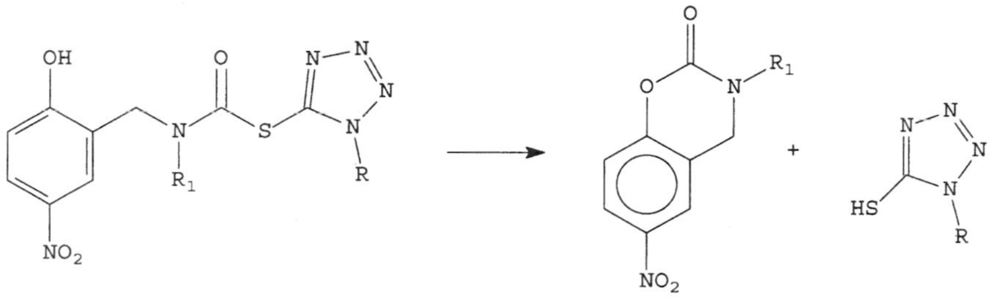

The switch on the left, shown with a blocked mercaptotetrazole inhibitor, is attached to the coupler as a COG through the phenolic OH group. R_1 is typically a small alkyl group. At pH 10 the half-life for release of the mercaptotetrazole inhibitor needs to be seconds for the inhibitor to function properly.

The second major type of switch with many variants is based on vinylagous elimination chemistry. Two examples are

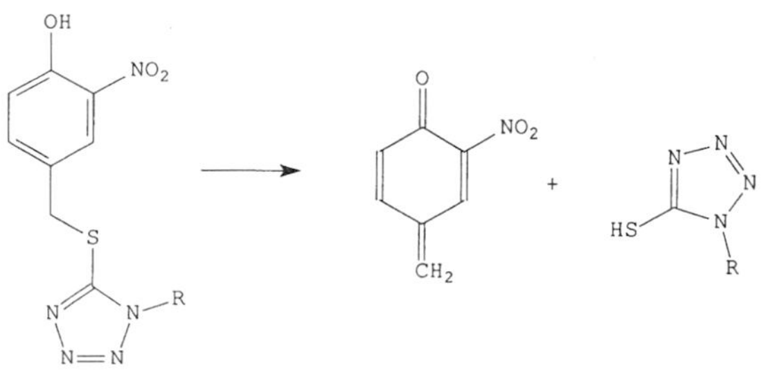

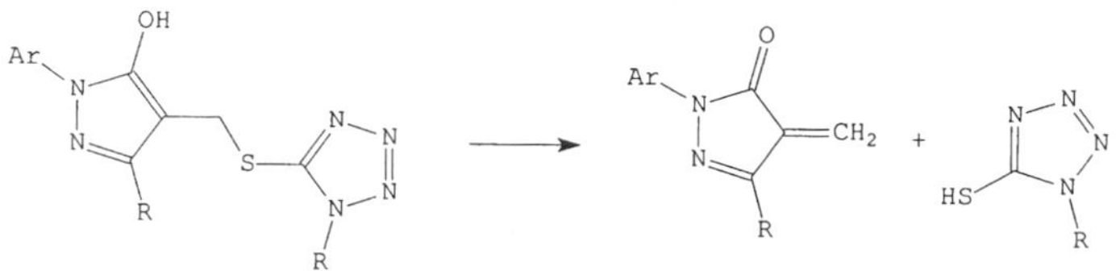

The first of these is called a "quinone methide" switch (63), based on the structure of the fugitive product. This product reacts with a nucleophile to regenerate an innocuous aromatic material. This switch is attached to a coupler through the phenolic group. The second switch based on a pyrazolone nucleus works similarly (64).

A third type of switch, sometimes called a link, is attached to the coupler

$$\text{HOCH}_2 \left[\text{NCH}_2 \right]_n \text{Inhibitor}$$
$$\underset{R}{|}$$

through the oxygen atom (OH group) (65). After release of the COG, formaldehyde is released to give directly the inhibitor (n = 0), or to give a second intermediate that can further fragment to release the free inhibitor (n greater than or equal to 1).

A novel idea for a switch is based on redox and addition–elimination chemistry (66) and is exemplified by

This material is attached to the coupler through the OH group that is para to the inhibitor. This released switch is redox active and can be oxidized by film oxidants (D_{ox}) to an o-quinone. The inhibitor can then be released by nucleophilic addition of sulfite from the developer in a Michael addition–elimination reaction.

2.4.7 Examples of DIR Couplers

The above sections show the pieces of DIRs. Below are some complete examples of DIR couplers (67–69).

DIRs are among the most complex and expensive materials in film. Working from the outside of each structure inward, each contains a ballast ($C_{14}H_{29}$, OAr, or $C_{16}H_{23}$) that holds the coupler in its desired color record. Each contains a coupler so that the COG can be released imagewise with D_{ox} (naphthol, pyrazolone, or pivaloylacetamide). Two contain a switch (quinone-methide or carbamate). Each contains an inhibitor (mercapto-oxadiazole, mercaptotetrazole, or self-destruct mercaptotetrazole). When the function of DIR couplers is understood, so are their structures.

2.4.8 Granularity Advantage from DIR Couplers, Tradeoffs, and DQE

As discussed, image-modifying chemistries (DIRs, colored masking couplers, and other unmentioned materials) can be used to improve sharpness, to correct for unwanted dye absorptions, and to increase color saturation. DIRs and similar materials can also reduce granularity, which is the nonuniformity that is created in a uniform image field by the spatial distribution of silver halide grains and the subsequent dye clouds. The inhibitors from DIRs perform this function by shutting off, or slowing down, D_{ox} production from silver halide grains and thereby forcing more individual grains to contribute to a given density level. The forced use of more information-bearing centers by use of inhibitors contributes to the improved signal-to-noise ratio.

These beneficial effects of image-modifying chemistries do not come without a price. The information recorded during exposure of the silver halide material is fixed. During keeping and processing, this total information can be lost or degraded, but not enhanced or maintained. The roles of image-modifying chemistries are to permit the film designer to trade off this information in various display options to meet the needs of the particular film product.

Photographic scientists have long known of the various trade-offs between, say, sharpness and speed (related to color contrast) or speed and granularity that can be achieved using image-modifying chemistry. More recently, these compromises have been quantified in a metric called Detective Quantum Efficiency (DQE) (70), a measure of the information-carrying capacity of the film, and cast in terms of parameters of interest to the film designer. DQE is a ratio of the measured output signal-to-noise produced by the film (amplification, acutance, and granularity all enter into the output). This permits photographic scientists to assess whether they have optimally used the information recorded by the film and to measure the effectiveness of the chemical trade-offs they have made.

Increasing the information-carrying capacity of the film by designing more efficient silver halide emulsions is one of the major goals of emulsion scientists. This gives the color imaging chemists a greater opportunity to use image-modifying chemistries to create optimally pleasing pictures.

2.5 PUTTING IT ALL TOGETHER

There are over 100 chemicals in a typical modern color negative film. With few exceptions, these chemicals do not act independently but rather in hundreds of two-way and higher-order interactions. A piece of color negative film may be among the most complex man-made chemical devices.

Such complexity creates a challenge for photographic scientists. However, we need to simplify the product, not glory in its complexity. This goal may be achievable, as films can now be digitally scanned and manipulated thereby requiring less chemical intervention.

Even if half the chemistry can be removed, there are many of degrees of freedom left for building a film. There is the freedom to modify the structure of each organic compound in almost an infinite number of ways. There is the freedom to alter the laydown in milligrams per square foot. Each material can be incorporated in different sublayers. Sublayers can be arranged in different orders.

The end use of putting the materials together is to control more than a dozen photographic responses ranging from reciprocity of different exposure times and light levels to the stability of the image dyes. Though digital imaging systems continue to make inroads into traditional photographic applications, silver halide systems with their massive information-carrying capacity continue to set the standard for quality/cost trade-offs. As exemplified by the fusion of chemical and electronic information capture in the Advanced Photo System (APS), the future of imaging may lie with those hybrid systems that make best use of the unique qualities of both technologies.

REFERENCES

1. J Kapecki, J Rodgers. Color photography. In: *Kirk-Othmer Encyclopedia of Chemical Technology*. 4th ed. Vol. 6. John Wiley, 1993.

2. RD Theys, G Sosnovsky. Chemistry and processes of color photography. Chem Rev, 97: 83–132, 1997.

3. TH James, ed. *The Theory of the Photographic Process*. 4th ed. Macmillan, 1977.

4. RWG Hunt. *The Reproduction of Colour in Photography, Printing, and Television*. Fountain Press, 1987.

5. DL MacAdam. *Sources of Color Science*. MIT Press, Cambridge, MA, 1970.

6. DL MacAdam. *Sources of Color Science*. MIT Press, Cambridge, MA, 1970.

7. LE Friedrich, JE Eilers. Progress towards Calculation of the hues of azomethine dyes. International East-West Symposium III, Society for Imaging Science and Technology, and Society of Photographic Science and Technology of Japan, November 8–13, 1992, Proceedings, C-23.

8. LKJ Tong, MC Glesmann. The mechanism of dye formation in color photography. III. Oxidative condensation with p-phenylenediamines in aqueous alkaline solutions. J Amer Chem Soc 79: 583–592, 1957.

9. LKJ Tong, MC Glesmann. Kinetics and mechanism of oxidative coupling of p-phenylenediamines. J Amer Chem Soc 90: 5164–5173, 1968.

10. A Weissberger, CJ Kibler. US Patent 3,265,506, 1966.

11. See Ref. 3, p 354.

12. See Ref. 3, p 355.

13. H Kobayashi, Y Shimura, Y Yoshioka. US Patent 5,359,080, 1994.

14. A Aria, K Nakazyo, Y Oishi, A Okumura. US Patent 4,404,274, 1983.

15. T Endo, W Fujimatsu, M Fujiwhara, H Imamura, T Kojima. US Patent 4,314,023, 1982.

16. S Ichijima, K Nakazyo, A Okumura, K Shiba, A Sugizaki. US Patent 4,022,620, 1977.

17. See Ref. 3, p 355.

18. N Daiba. Coupler aided molecular design for yellow couplers of high extinction coefficient. In: Proceedings of the Society of Photographic Science and Technology of Japan, Presentation 1-A-6, Kyoto, 1998.

19. JB Mooberry, JJ Siefert, D Hoke, ZZ Wu, DT Southby, FD Coms. US Patent 5,457,004, 1995.

20. D Hoke, JB Mooberry, JJ Seifert, DT Southby, ZZ Wu. High-extinction dyes from yellow imaging couplers: the release of a preformed dye from the coupling-off group. J Imaging Sci Technol 42: 528–533, 1998.

21. M Seymour. US Patent 1,969,479, 1934.

22. J Bailey. US Patent 3,705,896, 1972.

23. S Tanaka, M Nagato. European Patent Application EP 877,288, 1998.

24. PW Vittum, FC Duennebier. The reaction between pyrazolones and their azomethine dyes. J Amer Chem Soc 72: 1536–1538, 1950.

25. D Bailey, V Flow, D Giacherio, S Krishnamurthy, J Pawlak, T Rosiek, S Singer. US Patent 5,262,292, 1993.

26. N Furutachi, S Ichijima. US Patent 4,308,343, 1981.

27. A Loria, P Vittum, A Weissberger. US Patent 2,600,788, 1952.

28. S Tanaka, M Magato. European Patent Application EP 877,288, 1998.

29. A Bowne, D Bailey. Eastman Kodak Company, unpublished data.

30. T Hirano, T Ozawa. European Patent Application EP 133,262, 1985.

31. K Hotta, T Iijima, H Kashiwagi, K Katoh, S Kawakatsu, K Kumashiro, N Nakayama, H Ohya, K Shinozaki, T Uchida. European Patent Application EP 197,153, 1986.

32. K Furuya, N Furutachi, S Oda, K Maruyama. Photochemical reactions of 1H-pyrazolo[1,5-b][1,2,4]triazole azomethine dyes. J Chem Soc Perkin Trans 2: 531–536, 1994.

33. K Hotta, T Iijima, H Kashiwagi, K Katoh, S Kawakatsu, K Kumashiro, N Nakayama, H Ohya, K Shinozaki, T Uchida. European Patent Application 197,153, 1986.

34. Y Mizukawa, M Motoki, T Sato, O Takahashi. European Patent Application 571,959, 1993.

35. H Kita, Y Kaneko, N Mizukura, T Kubota. US Patent 5,254,451, 1993.

36. I Salminen, P Vittum, A Weissberger. US Patent 2,474,293, 1949.

37. H Deguchi, T Endo, W Ishikawa, T Kamita, S Kikuchi, H Wada. US Patent 4,134,766, 1979.

38. H Kobayashi, K Mihayashi. European Patent Application EP 307,927, 1989.

39. G Guisto. European Patent Application EP 389,817, 1990.

40. P Lau. US Patent 4,333,999, 1982.

41. H Osborn. US Patent 4,124,396, 1978.

42. P Lau. US Patent 4,333,999, 1982.

43. P Ramello. US Patent 3,772,002, 1973.

44. S Ikesu, VB Rudchenko, M Fukuda, Y Kaneko. European Patent Applications 744,655 and 717,315, 1996.

45. K Miyazawa, S Tanaka. European Patent Application 844,525, 1998.

46. T Ito, Y Shimada, K Matsuoka, Y Yoshioka. European Patent Application 710,881, 1996.

47. K Miyazawa, S Tanaka. European Patent Application 844,525, 1998.

48. JR Thirtle. Inside color photography. Chemtech 9: 25–35, 1979.

49. A Loria. US Patent 3,408,194, 1968.

50. G Lestina. US Patent 3,519,429, 1970.

51. A Loria. US Patent 3,476,563, 1969.

52. K Mihayashi, A Ohkawa. European Patent EP 517,214, 1992.

53. See Ref. 3, p 8.

54. J Abbott, W Coffey. US Patent 3,615,506, 1971.

55. S Kida, K Kishi, S Nakagawa, H Sugita, S Uemura. US Patent 4,421,845, 1983.

56. M Ihama, Y Kume, K Mihayashi, K Tamoto. US Patent 4,933,989, 1990.

57. K Adachi, S Ichijima, H Kobayashi, K Sakanoue. US Patent 4,477,563, 1984.

58. R DeSelms, J Kapecki. US Patent 4,782,012, 1988.

59. TH James, ed. *The Theory of the Photographic Process.* 4th ed., p. 8. Macmillan, 1977.

60. P Bergthaller. US Patent 5,455,149, 1995.

61. U Griesel, H Odenwalder, H Ohlschlager. US Patent 4,579,816, 1986.

62. P Lau. US Patent 4,248,962, 1981.

63. J Poslusny, W Slusarek, R Szajewski. US Patent 4,962,018, 1990.

64. S Kida, K Kishi, S Nakagawa, H Sugita, S Uemura. US Patent 4,421,845, 1983.

65. K Mihayashi, T Obayashi, A Ohkawa. US Patent 5,326,680, 1994.

66. M Ihama, Y Kume, K Mihayashi, K Tamoto. US Patent 4,933,989, 1990.

67. O Ishige, Y Kaneko, T Nakamura, N Sato. US Patent 5,571,661, 1996.

68. J Abbott, W Coffey. US Patent 3,615,506, 1971.

69. K Adachi, S Ichijima, H Kobayashi, K Sakanoue. US Patent 4,477,563, 1984.

70. See Ref. 3, p 636.

3

Diazo Papers, Films, and Chemicals

Henry Mustacchi

Consultant, Port Washington, New York

3.1 DIAZOTYPE PROCESS

The diazotype process, because of its simplicity, versatility, and low cost, has remained the most widely used method for the production of copies of engineering and architectural drawings. The manufacture of diazotype materials is still a strong and well-established industry in the majority of the countries in the world. Although no longer growing in the United States and Europe, the use of diazotype papers and films is continuing to expand in other parts of the globe and in particular in developing countries.

The diazotype process owes its existence to the unique characteristics of aromatic diazonium compounds and more specifically to the following properties:

1. In the presence of certain classes of compounds called couplers, and in a range of pH values, diazonium compounds react to form colored azo dyes.
2. Diazonium compounds are sensitive to light, and when subjected to irradiations of light of a specific wavelength (normally in the UV range), they decompose to give colorless substances that can no longer form azo dyes.

These two properties can be translated by the following chemical reactions:

$$[ArN_2]^+X^- + AR-OH \xrightarrow[\text{MOH}]{\text{pH} > 7} Ar-N=N-AR-OH + X^-M^+ + H_2O$$

$$[ArN_2]^+X^- + H_2O \xrightarrow{h\nu} Ar-OH + N_2 + HCl$$

The photolytic decomposition of the diazonium compound is best accomplished in the presence of traces of water, whereas the reaction between the diazonium compound and the coupler takes place in general in an alkaline environment. Schematically, the diazotype process can then be written as follows:

$$\text{Diazonium compound} \xrightarrow[\text{UV light}]{\text{exposed to}} \text{unreactive colorless product} \qquad (1)$$

$$\text{Diazonium compound} + \text{coupler} \xrightarrow[\text{medium}]{\text{alkaline}} \text{colored azo dye} \qquad (2)$$

It can be seen from the reactions above that the diazotype process, in its conventional form, is positive working.

It should be mentioned at this point that the abbreviation of diazonium as well as diazotype to diazo has come into common language and is generally accepted.

Graphically, the production of a positive image by the diazotype process is illustrated in Fig. 3.1. A graphic original composed of a light-transmitting substrate, such as a transparent or translucent paper, on which are lines opaque to light, is placed in close contact on top of a light-sensitive diazotype material, such as a diazotype paper, and exposed to ultraviolet light. The ultraviolet radiations that pass through the nonimage areas will reach the light-sensitive layer containing the diazonium compound and decompose it proportion-

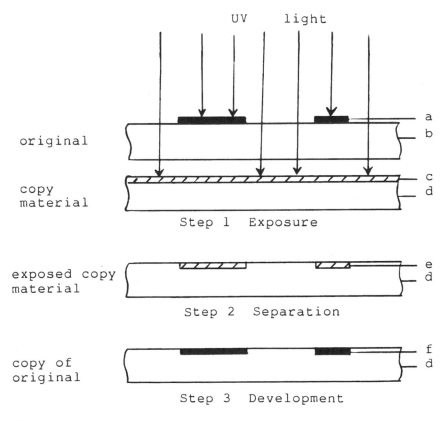

Figure 3.1   Production of a positive image by the diazotype process: a, image area; b, light-transmitting substrate; c, diazotype layer; d, base support; e, latent image area; and f, azo dye image area.

ally to the amount of energy it receives; under the opaque lines, where the light is blocked, the diazonium compound remains undecomposed. The diazotype paper, after exposure, carries a barely visible image formed of the remaining pale yellow diazonium compound on a white background.

To render the image permanently visible, the diazonium compound and the coupling component are allowed to react together to form a highly colored azo dye in the image area. The amount of azo dye formed is proportional to the amount of diazonium compound left undecomposed by the ultraviolet light; it follows that weak actinic opacity originals will give low-density dye images and that high actinic opacity originals will give high-density dye images.

The reaction rate between the diazonium compound and the coupling component is pH dependent. At low pH values, the reaction is either completely inhibited or extremely slow. As the pH value increases, the reaction rate also increases to reach a maximum at pH values above 10 or 11. It can be seen that to prevent the diazonium compound and the coupler from reacting together before exposure, it will be necessary either to separate them physically or to create a strong acidic environment for the diazonium compound and the coupler. Based on these alternatives, two processes emerged.

In one process, the diazonium compound alone, from a solution, is applied to the base support, and only after print exposure is the coupler brought in contact with the diazonium compound; for the foregoing reason, the term one-component is commonly used when referring to this process and its materials.

In another process, the diazonium compound and the coupler, dissolved together in the presence of an acid, are applied to the base support, and after exposure the acid is neutralized with alkali. For this process and its materials, the term two-component is often used.

The choice of the diazonium compound and of the coupling component will determine which process is applicable, which azo dye color will be obtained, and how much energy will be necessary to expose the material. All these elements are discussed later.

The photosensitivity of diazotype materials is limited to a very narrow range of the spectral region, between 350 and 420 nm, which is the region in which most diazonium compounds used in the production of diazotype materials absorb light. Any light source that emits radiations within this spectral range will be suitable for exposing diazotype materials. The most readily available source of light that would decompose the diazo is the sun. Even though solar power is still used practically to expose diazo papers in certain parts of the world, however, the sun is not a controllable source of actinic light and cannot be the base of a commercial system.

Arc lamps and mercury vapor lamps were in the past conventional sources of ultraviolet light and were extensively used in diazo photoprinting equipment. Arc lamps have ceased to be used; but mercury vapor lamps continue to find applications where high-energy outputs are required.

Fluorescent tubes have become very popular as ultraviolet light sources for diazotype printing, since it was found that by doping the tubes (i.e., adding certain metals in the tube) the emission spectrum of the tube can be shifted toward the high wavelengths in the ultraviolet range.

Quantitative studies of the action of light on diazo compounds have shown that the quantum yield of the photolysis of diazos is below unity; the quantum yield is defined as the quotient of the number of molecules activated by light and the number of absorbed photons. If one molecule of diazo is decomposed by one photon, the quantum yield would be equal to unity. However, part of the energy used to excite the diazo molecule is converted to thermal energy or fluorescence, while the remaining part of the energy produces the photochemical decomposition of the diazo.

In contrast, the quantum yield of silver halide after development is many orders of magnitude higher because of development amplification; it is considered that the energy required to form a diazo image is 700,000 to 1,000,000 times greater than that needed to form a silver halide image.

Because of the nonsensitivity of diazonium compounds to green, yellow, and red portions of the visible spectrum, it is possible to handle diazotype materials in tungsten light for long periods or in low-energy blue light for short periods of time.

3.2 DIAZO PAPERS

The most common diazotype material in the world is what is usually called diazo paper, also called heliographic paper, dyeline paper, and, in the United States, blueprint or whiteprint paper. The term heliographic paper seems best to describe this type of photosensitive material because it is derived from the Greek words for "sun" and for "writing"; thus "heliographic" stands for writing with light—a most poetic way of describing diazo paper. In the United States, the term "blueprint" was first used for a type of engineering reproduction paper based on ferrocyanide systems, which gave deep blue negative prints showing white lines on a blue background; the first diazo papers were intended to simulate the negative blueprint paper color and were mistakenly called by the same name.

Different types of diazo paper are available depending on the character, appearance, and properties of the base paper itself. Two main classes of diazo papers dominate the market: the first, with the larger volume and for general purposes, uses an opaque or semiopaque base paper; the second, representing a smaller volume and for specialized applications, uses a more or less translucent or transparent base paper. The opaque diazo papers cost substantially less than the translucent ones.

3.2.1 Base Papers

Opaque Base

To be suitable as a diazo base, the substrate and the coating formulation must be adapted to each other.

The base paper should be receptive to the coating while keeping the coating as close to the surface as possible. Penetration would cause a loss of printing speed due to the filter action of the paper fibers to UV light, and simultaneously a loss of print dye density. The stability of the light-sensitive diazonium compound could also be impaired by a slow reaction with the chemicals contained in the paper and with the cellulosic fibers themselves.

The base paper should be as uniform as possible, and its surface uniformly receptive to the coating solution, to prevent uneven print dye formation.

The base paper should be of a high brightness, to ensure the maximum print contrast. This is achieved frequently with the use of optical brighteners, which are fluorescent dyes. The base paper should be physically strong to withstand processing through the various operations of the diazo material manufacture as well as to resist rough handling of the finished print: it should therefore have good wet and dry strength, folding endurance, tear resistance, and burst strength. The base paper should produce flat prints and therefore should be as inert as possible to changes in relative humidity of the surrounding atmosphere.

The base paper should be free of chemicals that may affect the diazo compound, the coupler, any of the other additives, or the print dye. Oxidizing agents and reducing agents decompose the diazo and the azo dye; various metal salts react with the diazo and, in particular, ferric salts give undesirable color reactions with many of the blue couplers.

The base paper must not be alkaline, lest premature coupling or diazo decomposition occur.

The base paper should be acidic enough to provide an environment that contributes to the shelf life of the diazo material, without being acid enough to affect the physical stability of the cellulosic fibers or the color of the print dye.

Because of the numerous and critical requirements for a diazo base paper, only a limited number of paper mills in the world have undertaken the task of producing this material, and as a consequence the price of a diazo base reflects its stringent quality requirements.

The choices of pulp and sizes as well as other additives determine the character and properties of the finished base paper. The quality of the available water is of major concern to the base paper mill, and the most important factor is its iron content. While 20 ppm is considered the upper limit of iron content in the paper, there is no limit set at the lower end. Currently standard diazotype materials use a base paper of between 70 and 80 g/m^2 basis weight, equivalent to between 18^1/$_2$ and 20 lb reams in the U.S. system. Heavier and lighter base papers are also used for special applications. The conversion between the metric system of base paper weight and the U.S. system is given in Fig. 3.2.

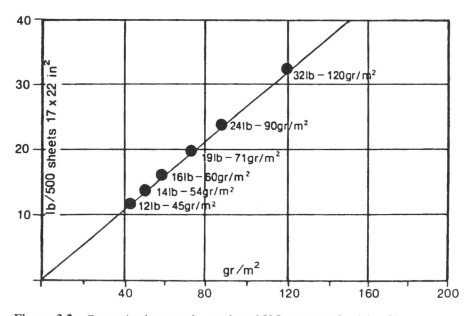

Figure 3.2 Conversion between the metric and U.S. systems of weight of base paper.

A typical specification for a diazo base paper of 20 lb of substance weight is

Basis weight	19–20$\frac{1}{2}$ lb/ream
Caliper	0.0036–0.0040 in.
Moisture	3–4%
Smoothness (Sheffield)	90–140
Brightness (TAPPI-T-452m-48)	90 minimum
Opacity	82 minimum
pH (TAPPI-T-435m-52)	4.2–4.5
Porosity (W. L. Gurley)	30–60 seconds
Iron	20 ppm maximum
Fold (Schopper)	100 minimum
Tear (Elmendorf)	60 minimum
Mullen	35 minimum

The majority of diazotype paper prints are made on a white stock base. However, in certain special circumstances, base papers of various colors are used for rapid identification. Tinted in blue, green, pink, yellow, or salmon, these diazo base papers were in the past produced by the paper mills. As demand decreased, tinted bases have become rarer, and diazotype material manufacturers have turned to tinting the white base stock themselves by the addition of water-soluble dyes to their coating solutions. This approach results in the simultaneous tinting and sensitizing of the base paper.

Transparent and Translucent Base

An important group of diazotype materials available in the market has transparent or translucent substrates sensitized with a diazo–coupler combination, giving azo dyes of substantial actinic absorption. These materials have an application in the production of intermediate prints used to make further diazotype material prints. The need to reproduce an original drawing that is the result of many hours of conceptual thought, design, and drafting, whether manual or computer aided, is self-evident; but the risk of damaging this valuable drawing increases rapidly with the number of times it is manipulated. An intermediate diazotype print of the drawing is generally made and used for further copying instead of the original drawing, which is then stored safely. Such a copy is often called the "second original," "second master," or "intermediate" because it has all the information of the original and can perform exactly like the initial drawing from which it was reproduced. Further copies of the intermediate can be made either on opaque diazo materials or on other intermediate diazo materials. Before making these copies, it would be possible to modify the intermediate by deletions, additions, or corrections to the drawing, creating in this manner a new drawing without having redrawn the original.

One of the major requirements for a diazotype intermediate material is that the substrate be transparent or translucent to actinic radiations.

Paper in its most common forms does not have a good transparency without further treatment because its fibrous nature causes light to be scattered within it. The degree of translucency of a paper will be related to the proportion of parallel light rays the paper is able to transmit in relation to the total amount of light the paper receives. The more light passes through the paper without being absorbed or scattered, the more distinctly we are able to see objects through the paper.

Paper is a fibrous material composed of a multitude of discrete fibers disposed in a sheet configuration with many voids or interstices between the fibers. The disposed fibers scatter the light; however, if the voids between fibers are reduced or if they have an index of refraction very close to that of the fibers, less light scattering takes place and consequently the paper becomes more translucent. The process of causing the paper to allow more light to be transmitted through it is called transparentizing.

Paper can be transparentized during the course of its manufacture, as in the case of glassine or natural tracing paper. This involves beating the fibers or treating them chemically until they become smaller and more plastic and can be compressed to form a dense sheet, which after supercalendering turns transparent. However, such a sheet, unless plasticized, is brittle and has poor tear resistance or folding endurance. Despite its physical limitations, natural tracing paper has found a practical application as a drafting medium and as a substrate for diazotype intermediates, particularly in Europe.

Another method of rendering paper transparent is to fill the voids between the fibers with a material that has an index of refraction identical or very close to that of the paper fibers. This index of refraction of paper fibers varies with the origin of the cellulose fiber and with other factors such as moisture content. In addition, cellulose fibers are birefringent: they have different values in the axial direction of the fiber and in the transverse direction, and these values vary between a minimum of 1.50 and a maximum of 1.58. It is considered that materials with an index of refraction close to the range of 1.50–1.58 are potentially suitable for transparentizing, with the best ones having an index of refraction around 1.55.

F. V. E. Vaurio (1) gives a list of commercial products that might be useful in controlling the transparency of paper by chemical treatment; the list indicates the index of refraction values in ascending order, giving the commercial product name and the manufacturer when appropriate.

Among the most common materials used for transparentizing paper are mineral oils or waxes, polymeric thermoplastic resins (e.g., polystyrenes, polybutenes, polypropenes), various resin derivatives, and various polymers or copolymers of acrylic monomers and styrene, with or without plasticizers.

Transparentizing materials are applied to the paper either in their original state or in the form of hot melt or organic solvent solution or aqueous emulsion. The application is performed by tub dipping, roller coating, or any other conventional method. The excess material is removed by a doctoring process using solid or air scrapers, squeeze rollers, or a size press. To improve the distribution of the transparentizer within the fibers of the paper and to displace the air in the voids, the paper is sometimes ''wet packed,'' that is, allowed to stand for a given period of time (from a few hours to a few days) and then passed through a drying oven either to remove the solvent vehicle or condense, cross-link, or polymerize the active ingredients in the transparentizer.

In situ polymerization of transparentized materials is done by heat, photo, or electron beam action immediately after impregnation in a continuing web pass.

The paper base provided for transparentizing is of great importance for achieving the best results, and although a wide variety of papers may be employed, those used in the diazo industry are limited to papers prepared from rags, cotton fibers, and chemical pulp. In the United States, the requirements for high tear and fold strength, as well as permanency, have led to the exclusive use of 100% rag base or mixtures of rag and sulfite pulp with high rag fiber content. A typical specification for a 100% rag content base paper of 14 lb substance is:

Basis weight	13.3–14.7 lb/ream
Caliper	0.024–0.0026 in
Brightness	78–82
Bursting strength	27–30
Smoothness (Sheffield)	110–140
Fold endurance	
Before aging	
MD	500–800
CD	350–500
After aging	
MD	300–500
CD	225–300
Opacity	
Before aging	72–74
After aging	72–74
Tear strength	
MD	40–45
CD	43–50
Moisture	4.0–6%
Porosity (Geimer test)	3.5 cm^3, min
pH	4.0–4.5

It should be noted that after transparentizing many of the properties change, and it is expected in general that tear and fold strength will be increased rather than decreased.

Natural tracing paper and transparentized rag paper are the two major types of translucent papers used as drafting media and as diazotype intermediates. Each has some advantages over the other and some disadvantages, which are worth mentioning because the world is passionately divided as to which is preferred.

Natural tracing paper is more transparent to light both by reflection and by transmission; this leads to better look-through and better reprint properties.

Natural tracing paper is also heavier in general (between 80 and 110 g/m^2) and therefore is stiffer and handles better. However, because of its relatively short fiber consistency, it is more brittle and cracks easily; in addition, creases form readily upon handling and show prominently in reprints.

Natural tracing paper is more sensitive to atmospheric moisture and absorbs it, causing the paper to curl excessively because of expansion of the fibers and surface distortions.

Natural tracing paper accepts pencil and ink well, and because of its hard surface, the pencil and ink lines can be obliterated readily by physical means (e.g., with an eraser or a razor blade).

Transparentized rag base paper is more opaque to light; to increase translucency, the paper is kept at a low basis weight (between 50 and 70 g/m^2).

Transparentized rag base paper has substantially better physical characteristics, because of its long fibers; it shows relatively high fold and tear strength values, is much less sensitive to humidity, and remains more dimensionally stable under a wide range of atmospheric conditions. Pencil and ink lines are also readily accepted, but eradication is more difficult.

Both types of base paper can be coated to overcome some of their limitations and are widely used throughout the world as drafting media and as substrates for diazo intermediates.

3.2.2 Two-Component Diazo Papers

In two-component diazo papers, both the aromatic diazo compound and the coupling component necessary to produce the azo dye image are present in the light-sensitive layer. In addition, to prevent a reaction between the diazo and the coupler prior to exposure and development, acids and/or acidic salts must be present; they create an environment in which the coupling reaction is inhibited.

After exposure, the diazo paper is subjected to an alkaline medium that neutralizes the acids and changes the environment to one in which the coupling reaction proceeds rapidly.

It can be seen that the chemical reactivities of the diazo compound and of the coupling component are of fundamental importance in determining whether the two can coexist in an acidic medium without reacting prematurely with each other. If the reactivities are too high, it will not be possible to stabilize the system against premature coupling, even at low pH values. If the reactivities are too low, it will take too long for the system to produce the azo dye even at high pH values.

Among the very large number of diazos and couplers that could potentially give azo dyes, only a few have been found to have the necessary chemical reactivities to be of practical use in two-component papers. We will review the specific diazos and couplers commercially used in Section 3.3.

The development of a two-component paper is carried out by raising the pH value of the diazo–coupler system until the azo dye formation is complete. This can be achieved by different methods.

Ammonia Development

The method most frequently used for developing two-component diazo papers consists in subjecting the exposed material to ammonia vapor; since the presence of water favors the coupling reaction, the ammonia gas is always mixed with some water vapor when in contact with the paper.

In practice, the development takes place in the developing chamber of a photoprinting machine, where the concentrations of ammonia gas and water vapor are controlled as well as the temperature and the contact time with the paper. During the initial stage in the development, the acids or acid salts are neutralized and subsequently, as the pH value of the paper rises rapidly, the diazo reacts with the coupler to give the azo dye in the image area; at the end of the development, when the paper is saturated with ammonia gas, the pH of the layer may reach a value over 13. Once the paper has been removed from its alkaline environment, the excessive ammonia tends to dissipate from the paper and the pH regresses toward lower values until an equilibrium is reached.

The residual ammonium salts in the layer tend to be from slightly acidic to slightly alkaline according to the type of acid used in the stabilization of the system.

Because the diazo paper remains dry during the entire ammonia development, it is often referred to as "dry developing paper."

Ammonia fumes, above 25 ppm concentrations, are objectionable to breathe and in large concentrations can be harmful. For these reasons, considerable efforts have been made to find other ways to develop two-component diazo papers. These efforts, however, have led to only modest commercial successes in two other types of development, which are reviewed below. However, thanks to great improvements in the concept, design, and construction of ammonia developing photoprinting equipment, and because of the high quality of two-component diazotype materials, the ammonia process has become the most practical and most widely used diazo process in the world.

Amine Development

A different way to develop two-component diazo papers consists in the application of an alkaline liquid to the surface of the paper. Theoretically, any solution of an alkaline chemical product, inorganic or organic, would be functional in neutralizing the acids in the diazo layer and in raising the pH value sufficiently to produce the azo coupling reaction. However, it was established at an early stage that aqueous solutions cause the azo dyes of two-component systems to bleed profusely, giving unacceptable blurred images. Focus was then turned to nonaqueous alkaline solutions, and it was found that when an amine solution is applied in a thin layer to the surface of a two-component diazo paper, development is achieved at an acceptable rate and the images are sharp and dense.

The C. Bruning division of Addressograph Multigraph Corporation (2) in the United States first proposed and commercialized a nonaqueous amine developing system. They called it pressure development because the application of the liquid to the paper was under pressure between rollers to reduce the amount of liquid applied to just enough to achieve full development.

Various amines have been proposed, but the ones that are mostly used are monoethanolamine, diethanolamine, and triethanolamine. The developing liquid has to be formulated to meet different requirements such as spreading, wetting, having low volatility, and having an extended shelf life. The following example of an amine developer is taken from U.S. Patent 3,446,620, published in 1969.

Diethanolamine	40%
Diethylene glycol	30%
4-Methoxy-4-methyl-pentanol-2	20%
Water	10%
pH	11.7

The amine development system, which in principle overcomes the drawbacks of the ammonia system, has itself so many limitations that it has not been widely accepted, so only a small proportion of the total amount of diazo paper is currently developed by this method. A comparison between ammonia and amine development shows that the overall quality of the prints is superior with ammonia development. This is because after development, the surface of the amine-developed paper remains alkaline, whereas the surface of the ammonia-developed paper loses its high alkalinity after the ammonia has evaporated. Most of the couplers and the decomposition products of the diazos tend to discolor more rapidly in an alkaline environment than in a neutral one. Moreover, their oxidation products are colored, which means that the background of an amine-developed print will, upon aging, yellow more rapidly than that of a similar ammonia-developed print. Because of the difference in pH value of ammonia- and amine-developed prints, the azo dye colors obtained are slightly different and in general are less bright and less dense with amine development. Although blue azo dyes of acceptable standard are easily obtained with amine development, black and brown azo dyes are much more difficult to achieve and lead often to unacceptable results. Finally, after development with the amine solution, the liquid applied to the surface diffuses slowly in the substrate. This is possible when the substrate is porous as in the case of paper; with film products, the applied liquid cannot be absorbed satisfactorily by the impervious coating, and the surface consequently remains tacky for a long period of time. Even under the best conditions, amine-developed materials are more delicate to handle.

Thermal Development

Many attempts have been made over the years to create a diazo material whose development would be the result of the application of heat alone.

Most methods of diazo thermal development are based on the fact that upon heating, the pH value of a light-sensitive layer changes progressively to reach levels at which the coupling reaction takes place. The change in pH value is rendered possible by the ability of certain compounds to decompose at high temperatures with the liberation of ammonia or other substances of basic character.

Among the chemicals proposed for pH changes in thermal development are salts of strong bases and weak volatile acids, salts of weak bases and weak volatile acids, and ureas, thioureas, and their derivatives. In conjunction with these alkali-generating substances, it has also been proposed to use heat-decomposable acids and acid salts as stabilizers (3).

Another approach to thermal development has been to use a chemical that is not a coupler at room temperature but when heated to a given temperature decomposes into another chemical, which is a coupler (4).

The thermal developing diazo system is normally applied to the substrate from an aqueous solution containing all the functional chemicals (i.e., diazos, couplers, stabilizing acids, and alkali generators).

At the developing temperature, which is often above 130°C, alkali is generated, causing neutralization of the acidic stabilizers and formation of the azo dye image; simultaneously, the diazonium compounds start decomposing, since most diazonium salts are thermally unstable above a certain temperature. It follows that for such systems, it will be necessary to use a diazo compound of high thermal stability, a coupler of high coupling energy, and an alkali-generating agent that is stable at normal temperatures but decomposes rapidly and completely at the developing temperature.

While workable compositions could be formulated with the above-mentioned chemicals and within the constraints of the system, none have in practice led to commercially suitable materials because of many observed disadvantages, notably very poor shelf life, poor image density and quality, slow printing speed, and background discoloration.

To overcome the stability problems caused mostly by the intimate proximity of the chemicals, various suggestions were made to physically separate the alkali generator from the rest of the chemicals and bring them together by the action of heat. In one type of physical separation requiring a multiple coating technique, a wax layer isolates the diazo layer from the alkali-containing layer. Under heat, the wax melts, allowing the alkali to come in contact with the diazo layer and promote the coupling reaction.

In another type of physical separation, the diazo chemical, the coupler, and/or the alkali are encapsulated in microcapsules whose outer skin protects them from reacting at room temperature. Under the action of heat, however, the capsule skin breaks, to permit the coupling reaction to occur.

In another type of separation, the diazo, the coupler, and the alkali are chosen to be totally water insoluble; they are separately ground with a fusable material, dispersed in an aqueous matrix, and coated as separate layers. Under heat, the fusable material melts, bringing all the reacting chemicals together. Such a method has the advantage of giving materials with good shelf life but, unfortunately, with poor densities and poor line definition.

A further type of physical separation consists in using the base paper as the barrier between the diazo chemicals and the alkaline developing agent. A solution of ammonia-generating salt is coated on the reverse side of the two-component diazo paper and subsequently overcoated with a gas barrier layer. Application of heat liberates ammonia gas, which cannot go through the gas-impermeable barrier layer and, therefore, must pass through the base paper to cause development of the material.

Thermal development of papers has stirred the imagination of a vast number of researchers, and hundreds of patents were taken on this subject. Although in principle, all the patents are valid in their teaching, in practice, none has led to a material that is entirely suitable commercially, either because of unacceptable shelf life or because of lower quality.

3.2.3 One-Component Diazo Papers

In one-component diazo papers, only the aromatic diazo compound, and various stabilizers, are present in the light-sensitive layer. The coupling component is contained in a separate aqueous developing solution, which is applied to the exposed diazo paper. On contact, the diazo and the coupler react to form the azo dye. Because the diazo paper is wet during the application of the developer, this system is often referred to by one of the terms "semiwet," "semidry," or "moist."

The first one-component diazo paper, which was invented by the van der Grintens (5) in 1932, used a slightly alkaline developer solution to promote the azo dye formation.

Later, W. P. Leuch (6) established that with the use of higher coupling energy diazos, it was possible to obtain adequate development with a neutral or even slightly acidic developer solution.

In both acid- and alkali-type development, upon application of the developer, the azo dye formation must take place very rapidly and must be complete before the thin film of liquid dries up. Also of great importance is the need for the azo dyes formed to be insoluble in the developer, lest they tend to diffuse in the liquid film before it has completely dried, causing blurred images. It is said in such cases that the line "bleeds." To be satisfactory for this process, the coupling activity of the diazo compounds and the couplers must be very high, and their reaction products must be water fast.

For the acid-type development, the diazo compounds must be of higher coupling energy than for the alkali-type development. There is an optimum pH value at which each diazo–coupler pair performs best; therefore, depending on the system used, the developer pH could vary between 4.5 and 11.5.

With increasing pH values, the shelf life of the developer decreases because oxidation of the coupling component is produced by alkaline conditions.

With decreasing pH values, the rate of reaction between the diazo and the coupler decreases, causing incomplete development. After alkaline development, the surface of the paper remains alkaline; a rapid discoloration of the print background takes place as a result of oxidation of the coupler and the light-decomposed diazo compound. Efforts to improve the background by the use of antioxidants have not been entirely successful; and of the two moist developing systems, only the neutral/acid one is now widely used.

From a practical point of view, a serious limitation of the one-component process is that to obtain different color papers, it is necessary to use different developers, whereas with the two-component process, different color papers are obtained with the same developer.

Under normal conditions, one-component papers require drying after the application of the aqueous liquid developers. To overcome this limitation, it has been suggested (7) that a very small amount of a very concentrated developer be applied; in such an instance, the paper emerges from the developer practically touch dry, without the need of heat to remove the excess water.

Of the four major development methods for diazo papers, thermal is the least used commercially. Each of the other three (ammonia, amine, and moist) has its advantages and drawbacks, and each has its place in industry.

The manufacturers of ammonia-developed papers have reduced the inconvenience and objections associated with the handling of ammonia by offering equipment designed to dispense ammonia only when needed or to use cylinders of compressed anhydrous ammonia requiring infrequent changes. In many cases, ammonia extraction has been eliminated by the use of ammonia-absorbing systems, and residual ammonia smell on the prints has been minimized by print vacuuming inside the machine before delivery.

The manufacturers of amine-developed papers offer convenience and low maintenance with extended shelf life developers.

The manufacturers of moist-developed papers have reduced the inconvenience of their system by offering equipment designed to dispense dry prints and to eliminate in some cases the need to clean the machine at the end of each working day.

Because of the greater versatility of the ammonia system over the moist one, which is translated into a greater choice of available materials and in many cases better quality, ammonia diazo papers are more widely used. This is particularly true in the United States, where more than 95% of all diazo materials are for ammonia development. In Europe and in some other parts of the world, the dominance of the ammonia process is not as great, and moist diazo materials enjoy some degree of popularity.

3.2.4 Intermediate Diazo Papers

Original drawings often need to be reproduced on a diazo paper many times, and to prevent their damage through constant handling, a diazo paper intermediate copy of the original drawing is made to be used exactly in the same way while the original drawing is stored safely away. Such an intermediate copy is also called the second original, submaster, or simply diazo reproduction intermediate.

The requirements of an intermediate diazo paper are many and start with the substrate, which needs to be translucent or transparent to UV light. We have seen in the preceding section on base papers that both natural tracing paper and transparentized rag-based papers, with or without some chemical pulp, meet the requirement of light transmission. These bases are all used in the manufacture of diazo intermediate papers.

The azo dye formed on diazo intermediates must be resistant to the passage of UV light to ensure the satisfactory reproduction of the original drawing. The azo dye must have a high absorption rate (actinic opacity) in the wavelength range emitted by the UV light sources. Yellow azo dye has the highest absorption rate, whereas blue azo dye has the lowest. Orange, brown, and red azo dyes are in between.

Additionally, the azo dye must have a high visual density; in this case, the blue azo dyes have the maximum visual density and the yellow ones the lowest. Therefore, since both actinic and visual densities of the azo dyes are necessary, a compromise is often required, and intermediate diazo papers have azo dyes varying in shade between sepia and chocolate brown. Occasionally for some special application a black diazo intermediate is produced that has the maximum visual density but not the maximum actinic opacity. The primary function of a diazo intermediate is to act as a second original for the production of diazo copies; a secondary function, not less important, is to allow modifications of the original drawing by additions or deletions. To achieve this, it is necessary to erase or eradicate a part of the azo dye image and replace it by new pencil or ink lines. Since not all drawings require further corrections, it was found necessary to offer two different classes of diazo intermediate papers: those that are nonerasable and those that are easily erasable.

Nonerasable intermediates are available on natural tracing paper and on transparentized paper. On such materials the azo dye tends to penetrate within the fibrous structure of the substrate, and it cannot be removed mechanically (with an eraser or with a razor blade) without destroying part of the substrate. The azo dye can, however, be chemically destroyed (rendered invisible) by the action of chemical eradicators that contain strong reducing agents.

In erasable intermediates, a barrier layer is applied, between the light-sensitive layer and the substrate, to isolate the azo dye image from the base and allow its removal by mechanical means without affecting the substrate.

With natural tracing paper, a resin lacquer, applied from an aqueous or a solvent medium, prevents penetration of the light-sensitive chemicals in the paper. Such intermediates are often called lacquered tracing papers.

With transparentized papers, an impervious aqueous barrier layer is applied prior to the sensitized layer to prevent migration of the azo dye into the substrate.

Diazo lacquered tracing papers are popular in Europe, but not in the United States, whereas the opposite is true with diazo transparentized papers and in particular all-rag erasable transparentized papers.

3.3   CHEMICALS, AQUEOUS SYSTEMS

The preparation of a diazo paper consists primarily in the application to the paper of an aqueous chemical solution that contains the following compounds:

> One or more light-sensitive diazonium salts
> Zero or more coupling components
> Stabilizing agents
> Development accelerators
> Solution flow modifiers
> Auxiliary chemicals

Each of these classes of chemicals will be reviewed in detail.

3.3.1   Diazonium Salts

Many thousands of diazo compounds (8) have been synthesized and have at one time or another been mentioned in connection with the diazotype process. To be suitable for use in the preparation of diazo papers, a diazonium compound must meet a number of requirements. It must have adequate light sensitivity; it must be thermally stable enough to be handled and stored safely; it must be soluble in water to a degree allowing its use. Moreover, its photolytic decomposition products must be colorless and must have a reasonable resistance to oxidation; it must have adequate coupling energy (not too high when used in two-component papers and sufficiently high when used in one-component papers); and it must give with specific couplers azo dyes of desirable shades and brightness and, preferably, of good light and water fastness. In addition, the diazonium compound must be easy to manufacture, ecologically acceptable, and relatively low in cost. With so many requirements, it is no wonder that out of the thousands of diazos listed in the literature, only very few (between one and two dozen) are currently in use for the manufacture of all diazo papers.

The useful water-soluble diazo compounds can be classified according to their structure, their light sensitivity, or their coupling reactivity. However, since we have already distinguished between two-component and one-component diazo papers, it would seem natural to consider the diazo compounds suitable for each type of material, remembering that those with low-to-medium reactivity are used in two-component systems and those with high reactivity are used in one-component systems.

Diazos Used for Two-Component Papers

The commercial diazonium salts used for two-component diazo papers all have the following structure:

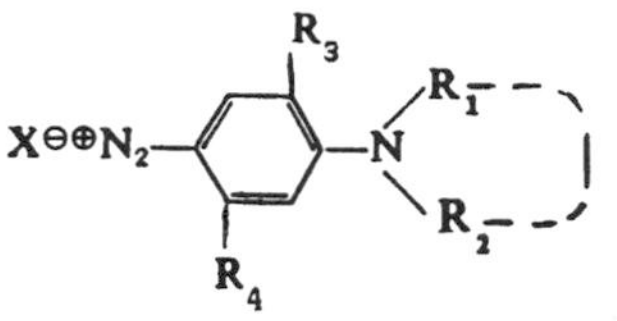

where R_1, and R_2 could be chosen from the groups methyl, ethyl, propyl, butyl, isopropyl, benzyl, cyclohexyl, methoxy, ethoxy or be part of a ring, which could be morpholine, pyrrolidine, or piperidine; R_3 and R_4 could be chosen from hydrogen, chlorine, methyl, ethyl, methoxy, ethoxy, isopropoxy, and butoxy. X, an anion, could be chosen from chloride (zinc chloride double salt) bisulfate and sulfoisophthalate. The diazonium ion forms with the anion a stable salt that can be isolated as a crystalline yellow-to-orange solid.

The light sensitivity and the reactivity of the diazonium compound will be greatly affected by the choice of the different radicals attached to the aromatic ring.

For a specific diazonium ion configuration, the flammability and the thermal stability will be greatly affected by the choice of the anion that forms the diazonium salt. For the above-mentioned anions, the flammability decreases and the thermal stability increases when passing from the zinc chloride double salt to the bisulfate and to the sulfoisophthalate. In fact, it has recently been shown that some diazonium zinc chloride salts are too flammable or thermally unstable, and therefore too hazardous to be used in their pure form. The addition of a substantial amount (20–30%) of acid or inorganic salt diluent to these unstable diazonium zinc chloride salts renders them less hazardous and acceptable for practical use.

The following diazo compounds have a low-to-medium light sensitivity and are generally used in standard speed two-component blue, black, red, and sepia diazo papers:

1. 1-Diazo-4-*N*-*N*-dimethylaminobenzene chloride, zinc chloride; also used as the half-zinc-chloride salt (stabilized by mixing with 30% tartaric or citric acid), or as the 5-sulfoisophthalate salt

2. 1-Diazo-4-*N,N*-diethylaminobenzene chloride, zinc chloride; also used as the half-zinc-chloride salt (stabilized by mixing with 30% tartaric or citric acid), or as the 5-sulfoisophthalate salt
3. 1-Diazo-4-(*N*-methyl, *N*-hydroxyethyl)aminobenzene chloride, half-zinc-chloride; also used as 5-sulfoisophthalate salt
4. 1-Diazo, *N,N*-ethyl, *N*-hydroxyethylaminobenzene chloride, half-zinc-chloride
5. 2-Diazo-1-naphthol-5-sulfonic acid sodium salt

The following diazo compounds have a high light sensitivity and are generally used in fast or superfast two-component blue, black, red, and sepia diazo papers.

6. 1-Diazo-2,5-diethoxy-4-morpholinobenzene chloride, half-zinc-chloride (because of its flammability and relatively poor shelf life, this diazo is currently used only at 70% or lower strength; it is also used as the bisulfate salt at 70% strength or as the 5-sulfoisophthalate salt)
7. 1-Diazo-2,5-dibutoxy-4-morpholinobenzene, bisulfate (80% strength); also available as the half-zinc-chloride salt at 70% strength and as the 5-sulfoisophthalate
8. 1-Diazo-3-methyl-4-pyrrolidinobenzene chloride, zinc chloride
9. 1-Diazo-2,5-dimethoxy-4-morpholinobenzene chloride, half-zinc-chloride

Diazos Used for One-Component Papers

The following diazo compounds have high reactivity and are generally used in one-component diazo papers.

10. 1-Diazo-4-(*N*-ethyl-*N*-benzyl)aminobenzene chloride, half-zinc-chloride
11. 1-Diazo-2,5-dimethoxy-4-*p*-tolylmercaptobenzene chloride, half-zinc-chloride
12. 1-Diazo-2,5-diethoxy-4-*p*-tolylmercaptobenzene chloride, half-zinc-chloride (70% strength)
13. 1-Diazo-2-chloro-5-(*p*-chlorophenoxy)-4-*N,N*-dimethylaminobenzene chloride, half-zinc-chloride (70% strength)
14. 1-Diazo-2-chloro-5-(*p*-chlorophenoxy)-4-*N,N*-diethylaminobenzene chloride, half-zinc-chloride (70% strength)
15. 1-Diazo-3-chloro-4-*N,N*-diethylaminobenzene chloride, half-zinc-chloride
16. 1-Diazo-3-chloro-4-(*N*-methyl-*N*-cyclohexyl)aminobenzene chloride, half-zinc-chloride (70% strength)

It should be pointed out that not all the diazo compounds in one group have equal reactivity: some are considerably more active than others.

In a few cases, it is possible to use a one-component diazo in a two-component system by selecting a low-to-very-low reactivity coupler to form the azo dye, and a two-component diazo in a one-component system by using a strong alkaline developer.

Diazonium salts are relatively complicated chemicals to synthesize, and their cost is somewhat high. In addition, for safety as well as for environmental control reasons, more restrictions are placed on the use and disposal of certain classes of chemicals (e.g., zinc chloride), with the results that a smaller range of diazos are being manufactured today and, whenever possible, the zinc chloride salts are being replaced by bisulfate salts or sulfoisophthalate salts.

3.3.2 Couplers

Couplers are the main color-determining components of diazotype materials and, as in the case of diazonium salts, many thousands (9) have been synthesized and could theoretically be considered to form colored azo dyes.

To be suitable for use in the preparation of diazo papers, a coupler must meet a number of requirements: it must have sufficient reactivity to couple rapidly with the diazo compound to give an intensely colored azo dye, but not too rapidly (for this would cause premature coupling in two-component systems); it should be water soluble to a degree allowing its use, and compatible in solution with the diazo; it should be stable to light and oxidation; it should not absorb ultraviolet radiations in the region where the diazos are light sensitive. The azo dyes formed should be bright, intense with desirable shades, stable to light, and preferably waterfast. In addition, the coupler must be easy to manufacture, ecologically acceptable, and relatively low in cost.

As for the diazos, the existence of so many requirements has limited the choice of couplers currently used to just a few (between two and three dozen).

Couplers belong to the chemical classes of aromatic amines, phenols, phenol ethers, or aliphatic compounds containing active methylene groups. Of these classes, only a few hydroxy and polyhydroxy compounds of the benzene and naphthalene series, in addition to some compounds with active methylene groups, are of practical importance.

Rather than classifying these couplers according to their structure, we list them according to the dye color they generally give with most diazos.

Blue Dyes

The following couplers are used to give blue dyes in two-component diazo papers.

1. 2,3-Dihydroxynaphthalene-6-sulfonic acid sodium salt
2. 2,7-Dihydroxynaphthalene-3,6-disulfonic acid disodium salt
3. 2-Hydroxynaphthalene-3-carboxylic acid-3′-*N*-morpholinopropylamide
4. 2-Hydroxynaphthalene-3-carboxylic acid ethanolamide
5. 2-Hydroxynaphthalene-3-carboxylic acid diethanolamide
6. 2,3-Dihydroxynaphthalene
7. 1-Hydroxynaphthalene-4-sulfonic acid sodium salt
8. 2-Hydroxynaphthalene-3-carboxylic acid-*N*-diethylenetriamine, hydrochloride salt
9. 2-Hydroxynaphthalene-3,6-disulfonic acid sodium salt

Not all the couplers above have the same coupling energy or give the same shade of blue, and some are considerably more popular than others in practical applications.

Yellow-to-Brown Dyes

The following couplers are used to give yellow-to-brown dyes in two-component diazo papers.

10. Resorcinol
11. Diresorcinol sulfide
12. Resorcinol monohydroxyethyl ether
13. 4-Chlororesorcinol
14. 2,4,3′-Trihydroxydiphenyl
15. 2,4-Dihydroxybenzylamine, methane sulfonic acid salt, 60% strength

16. β-Resorcylic acid ethanolamide
17. 2,5-Dimethyl-4-morpholinomethylphenol hydrochloride
18. Catechol monohydroxyethyl ether
19. 3-Hydroxyphenyl urea
20. Cyanoacet-morpholide
21. 1,10-Dicyanoacet-triethylenetetramine hydrochloride
22. Cyanoacetamide
23. Acetoacetanilide
24. Acetoacet-*o*-toluidide
25. Acetoacet-*o*-anisidide
26. Acetoacet-benzylamide
27. 1,4-Bis(acetoacet-ethylenediamine)
28. α-Resorcylic acid

Not all the couplers above have the same coupling energy, and they can give a range of colors varying between pale yellow and dark maroon; only a few are used in large amounts, whereas most of the others are used as minor constituents of a more complex coupler system in diazo papers.

Red Dyes

The following couplers are used to give red dyes in two-component diazo papers.

29. α-Resorcylic ethanolamide
30. 4-Bromo-α resorcylic acid
31. 4-Bromo-α resorcylic acid amide
32. 4-Bromo-α resorcylic acid methylamide
33. 1-Phenyl-3-methyl-pyrazolone

Couplers Used in Developers of One-Component Diazotype Papers

It has been shown that the coupling energy of couplers suitable for the one-component diazo system must be extremely high, particularly in the case of the acid or neutral development. A single coupler meets all the requirements for this application and is therefore universally used; the azo dye colors obtained with it vary from dark magenta to greenish-black; quasi-black colors have also been achieved. This coupler is

34. Phloroglucinol

For one-component alkaline development, in addition to phloroglucinol, various couplers of high coupling energy, chosen among the couplers listed for two-component papers, are used; for instance, resorcinol is used for brown, 2,3-dihydroxynaphthalene for blue, and 1-phenyl-3-methyl-pyrazolone for red.

3.3.3 Auxiliary Additives

In addition to diazos, couplers, and acid stabilizers, diazotype preparations require special chemicals to facilitate the coating application and to optimize the different properties of the finished diazo paper.

To improve the quality of the azo dye image given by a particular diazo sensitizing solution, the diazo paper is often precoated before the application of the sensitizer, and the precoat itself is a preparation that requires a number of special chemicals. These chemicals, which play a fundamental role in achieving the best results, are an integral part of the diazo paper technology.

Acid Stabilizers

Necessary to stabilize the solution against deterioration and to extend the shelf life of the diazo paper, acid stabilizers in the sensitizing solution serve to lower the pH to a level at which coupling is inhibited. Most frequently used are the following organic and inorganic acids: citric, tartaric, oxalic, boric, sulfuric, phosphoric, 5-sulfosalicylic, methanesulfonic, and *p*-toluenesulfonic; acetic and formic acids are used to stabilize the sensitizing solution only, since they are volatile and would be eliminated during the drying process.

An excess of stabilizing acid would tend to slow down the developing speed of the paper, whereas an insufficient amount would shorten the shelf life. Strong inorganic acids have a better stabilizing effect but can be used only in very small amounts; weaker organic acids allow a better compromise to be reached between shelf life and development rate.

Each particular acid stabilizer has also an effect on the shade of the azo dye and on the stability of the azo dye to atmospheric pH changes. Boric acid, which is a good stabilizer in one-component systems, must be used with caution in two-component systems, because it can substantially reduce the coupling activity of certain blue couplers.

Zinc Chloride

Zinc chloride has a special place in two-component sensitizing systems, because of its multiple advantageous effects. It has a stabilizing effect on the diazonium salts, and it also extends considerably the shelf life of the diazo paper; in addition, it often promotes faster coupling, increases the azo dye brightness, and minimizes the printing speed loss of the paper on aging. For these reasons, zinc chloride is included in practically all ammonia developing papers. Zinc chloride is not usually used in one-component sensitizing systems because it shows no specific benefit. It affects developer receptivity, and sometimes it even decreases the stability of the one-component diazo.

Stabilizing Salts

Some salts have a positive effect on improving the stability of the diazo paper, while other salts increase the resistance of the diazonium compound to thermal decomposition. One particular salt might be beneficial to one specific diazo and not to others, and this is what makes formulating diazo sensitizers so difficult. Among the salts most frequently used are aluminum sulfate, sodium sulfosalicylate, zinc sulfate, zinc *p*-toluene sulfonate, and the sodium salts of naphthalene mono-, di-, and trisulfonic acids. Some of these salts have also a solubilizing effect, which manifests itself in an improved compatibility of the diazo in the sensitizer.

Thiourea

It was discovered in the early days of diazo technology (10) that the incorporation of derivatives of thiocarbonic acid prevent rapid deterioration of the print background due to the action of air and light on the decomposition products of the diazo. Thiourea proved to be the simplest and most efficient of this class of compounds, and it has become a common ingredient of diazo layers. As a mild reducing agent effective in acidic conditions, thiourea counteracts print background yellowing; in addition, it facilitates dye formation during development and improves the brightness of the azo dye.

Thiourea has also some negative aspects: it reduces the stability of the diazo compounds in general, and it promotes precoupling with some blue couplers.

Solubilizers

Solubilizers are necessary to increase the compatibility of different chemicals in the sensitizing solution. Many diazonium salts have limited compatibility with certain couplers, particularly in the presence of zinc chloride, and give rise to insoluble dark tarry products that cause coating defects such as black or white spots. To prevent this situation, chemicals with solubilizing properties are incorporated in the sensitizer preparation; these compounds have either a general effect on all systems or a selective effect on a particular combination of a diazo and a coupler. The following are the main solubilizers used:

Caffeine is an effective solubilizer in both blueline and blackline diazo preparations, in two-component as well as in one-component systems.

Theophilline is a similarly effective solubilizer when couplers of the resorcinol family are present.

Caprolactam is a general solubilizer with strong effect.

Acetic acid, in addition to lowering the solution pH, has a solubilizing effect.

Alcohols such as ethyl alcohol and isopropyl alcohol can be used in moderate amounts as solubilizers.

Acrylamide was at one time a popular solubilizer, but because of the severe health hazard it presents, its use in diazo papers has been completely discontinued.

1,3,6–1,3,7-Naphthalene trisulfonic acid, a sodium salt, is also frequently used as a solubilizer in one-component and in two-component systems.

Development Promoters

To reduce to a minimum the dwell time of a two-component diazo paper in an ammonia-developing chamber, development must be as rapid as possible. Without the presence of traces of water, the coupling reaction between the diazo and the coupler would not take place; for this reason, the sensitized paper should have, after drying, a moisture content of between 3 and 5%. If the moisture content is below this range, the paper develops slowly; if it is above this range, the paper will spoil in a short time.

To retain moisture, humectants are frequently used. These are compounds such as ethylene glycol, diethylene glycol, triethylene glycol, glycerine, dipropylene glycol, and polyols; the amount and type of humectant affect the shade of the azo dye as well as the shelf life of the paper.

Other chemicals were also found to accelerate greatly the rate of development, but it is not clear why they have this effect. Among these chemicals, urea and its derivatives play in important role; widely used are urea, dimethyl urea, and 1-allyl-3-β-hydroxyethyl-2-thiourea, also know as AETH. In addition to promoting development, some of these compounds improve the compatibility of diazos and couplers; they also have a substantial effect on the azo dye shade.

Tetrahydrofurfural alcohol has occasionally been used as a development accelerator.

Flow Modifiers

Sensitizers and precoat solutions applied to paper must not only be chemically sound. They also must have the right physical properties to give the best results. Temperature, surface tension, viscosity, foam tendency, homogeneity, and other physical parameters are all important, and they must be adjusted to ensure that the solutions can be coated, without problems, to give defect-free materials. As a consequence, many chemicals that act as flow modifiers are added either to the sensitizer or to the precoat. The following are among those frequently used.

Saponin

This natural product with spreading properties is an important ingredient of most diazo coating preparations. While used in only very small amounts, saponin increases greatly the uniformity of the liquid coating on the paper web before drying. Care should be exercised to avoid excessive solution agitation, because the chemical has a strong foaming tendency.

Wetting and Spreading Agents

These are used in conjunction with or as a replacement for saponin; they can be selected to have good compatibility and spreading properties and low foaming action. Particularly useful are the acetylenic glycols and the nonionic nonfoaming surfactants such as 3,6-di-methyl-4-octyne-3,6-diol. These chemicals must be used in very small amounts, other wise they could cause penetration of the solutions into the fibrous structure of the base paper.

Antifoam Agents

Diazo solutions and precoat preparations containing dispersions of pigments and resins tend to foam when subjected to mechanical movement. During the coating operation, stirrers and pumps often introduce enough air to create foam problems. If foam is carried onto the paper, it causes coating streaks and other imperfections. One should minimize foam through the selection of nonfoaming systems. When foam does occur, antifoam agents should be used to eliminate it. Such chemicals must be selected carefully and used sparingly, since they themselves could be the source of white spot defects.

Wax Dispersions

These are used in precoats and in diazo solutions to lubricate the paper surface, hence to facilitate the handling of the diazo paper through the photoprocessing machines. The elevated temperature encountered in the ammonia-developing chamber of large machines softens the chemicals on the paper and could cause them to stick. High-melting-point waxes correct or reduce this problem.

Dispersing and antisettling agents are required to aid the dispersion of pigments in the solution and to prevent rapid sedimentation of the heavier particles of pigment.

Dyes

The print background of diazo papers contains slightly yellowish components resulting from the light decomposition products of the diazonium salts, their oxidation products, and the oxidation products of the couplers; this background discoloration increases from extended exposure to daylight.

To minimize any yellow cast and sometimes to tint the print background distinctively, dyes are added in minute amounts not only to the base paper but also to the coating solutions. The best compensating effects are obtained from mixtures of blue-violet and blue-green dyes or from single dyes generating both these hues. To be suitable, the dyes must also be water soluble, compatible with all the ingredients in the coating solutions, and lightfast.

Two of the most commonly used dyes are methyl violet and methylene blue; the first is blue-violet, and the second is blue-green. These dyes, however, do not have a good stability to light or to pH changes, and other proprietary commercial dyes are often preferred.

Sometimes, to give the diazo paper a distinctive strong tint, larger concentrations of colored dyes are added to the sensitizer and to the backcoat solution; in such instances, the diazo paper becomes intentionally tinted in red, blue, green, or yellow.

Pigments

To maximize the quality of the azo dye image, insoluble inert materials of mineral or organic polymeric nature are added either to the sensitizing solution or, more frequently, to the precoat layer. Each particle of pigment acts as a receptor for the azo dye, which absorbs the incident light, and as a reflector for the nonabsorbed part of the incident light, to improve the total brilliancy of the print image.

The role of the pigment is also to create a more even surface for a finer image grain.

Among the many pigments that have been suggested are silicas, blanc fixe, uncooked rice starch and dextrines, aluminum oxides, silicates of calcium, magnesium, or aluminum, clays, and diatomaceous earths.

Particle size and surface area play an important part in the performance of the pigment, and those with particle sizes from 0.1 to 5 µm have given the best results.

Silicas and rice starch are among the most popular additives for azo dye enhancement. Colloidal silicas of low particle size and noncolloidal silicas of slightly larger particle size have a pronounced effect on image dye density, but care must be exercised when using them, because they increase solution viscosity and can cause coating difficulties. Rice starch produces less dye enhancement, but it can be used in large amounts with very little effect on coating preparation viscosity; it is preferred over silicas for one-component systems.

Binders

The role of binders is first and foremost to fix all the chemicals firmly to the paper and prevent them from coming off during processing and handling of the diazo paper. By forming a continuous layer, they also contribute to the azo dye enhancement and to the general quality of the coating.

Binders are used in the precoat layer, in the sensitizing layer, or in both. To be suitable, binders must be compatible with the chemicals used in each layer, have an affinity for the azo dye, be permeable to ammonia gas when used in two-component papers and to water in the case of one-component papers, and have a high softening point to prevent sticking problems during coating and processing through hot ammonia developing machines. In addition, they must be colorless and inert, have good binding properties for the pigments and other chemicals, and have good adhesion to the paper surface.

Appropriate binders that meet many of the requirements above are chosen from the class of natural proteins and the group of synthetic materials.

Among the most frequently used protein binders are casein, an animal protein soluble in the form of sodium or ammonium caseinate, and a vegetal protein extracted from soy that is also soluble in alkaline media.

Among the synthetic materials, every possible polymer and copolymer has been tested in diazotype preparations, and a vast number of synthetic resin dispersions and solutions have been suggested with various degrees of success. The most commonly used are stabilized emulsions of polymers and copolymers of vinyl acetate, acrylic acid, vinyl chloride, styrene, and vinylidene chloride. Also frequently used are aqueous solutions of polyvinyl alcohol, fully or partially hydrolized, polyvinyl pyrrolidone, and methyl-, hydroxypropyl-, and other cellulose ethers.

Developer Chemicals

In addition to the essential coupling components, developers for one-component diazo papers contain a variety of chemicals whose functions are to create the best environmental condition for the azo coupling reaction, to wet the surface of the paper evenly, to reduce the spreading or bleeding of the azo dye, and to extend the useful life of the developing liquid.

Alkaline developers contain alkalies or preferably alkaline salts to set and maintain the pH value of the solution between 9 and 12. The following are commonly used in varying combinations: sodium or potassium hydroxide, sodium carbonate, potassium tetraborate, trisodium phosphate, borax, potassium metaborate, and similar salts.

Neutral or slightly acidic developers contain acids and acid salts to set and maintain the pH value of the solution between 5 and 7. The following are commonly used in varying combinations: citric acid, sodium formate, sodium benzoate, sodium acetate, sodium tartrate, sodium citrate, and similar salts.

Sodium or potassium hydroquinone monosulfonate is generally used to reduce oxidation of the couplers, and so are sodium thiosulfate, sodium hydrosulfite, and thiourea dioxide.

Sodium lauryl sulfate, sodium isopropylnaphthalene sulfonate, sodium dibutyl sulfosuccinate, and similar surface active agents are used to lower the surface tension of the developer for improved wetting of the paper.

Sodium sulfate is sometimes used to reduce spreading of the azo dye.

Miscellaneous Chemicals

Some chemicals that do not fall into any of the preceding categories are used in diazotype preparations for specific purposes; they have a unique effect on a single property and are effective only in a particular set of conditions. Many of these chemicals are directed at controlling the azo dye shade. For example, magnesium chloride is frequently used to increase the brightness of some blue azo dyes, whereas sodium chloride, sodium monophosphate, and calcium chloride have a beneficial effect on the shade of blackline papers by eliminating the reddish hue of some blue azo dyes.

3.4 FORMULATIONS, AQUEOUS SYSTEMS

Diazo paper manufacturers offer a wide range of materials that differ according to color, developing method, speed, substrate surface appearance, and application.

For instance, diazo papers are available in black, blue, red, and brown; in standard, fast, and superfast speeds; on opaque papers of basis weight from 45 to 220 g/m^2 or on translucent paper; for ammonia, amine, moist, or thermal development; with matte or glossy surface; for simple drawing reproduction or for continuous tone photographic duplication; and for proofing purposes or for the permanent record, with an almost endless list of applications.

Each type of paper necessitates a special formulation, and since many of the features and properties of diazo papers are judged subjectively, no single formulation for a given type of diazo paper can meet everyone's requirements. There are as many formulation possibilities as there are permutations of the major constituents. The role of the diazo paper formulator is to develop a recipe that is as simple and as economical as possible that will fulfill as many of the basic requirements as possible and have the minimum number of drawbacks.

A formulator, when designing a diazo paper, will aim at achieving optimum results with regard to the following properties:

Coating evenness
Shelf life
Color shade in full tones and halftones
Printing speed
Optical density and contrast
Development rate
Actinic opacity (for sepia intermediate papers)
Reprint speed translucency (for sepia intermediate papers)
Background appearance and resistance to daylight exposure
Coating smoothness
Resistance to rub-off
Freedom from curl
Azo dye water and light fastness
Azo dye resistance to color shift at different pH values
Ease of azo dye eradication (mostly for sepia intermediate papers)
Acceptance of pencil and ink lines
Ease of paper processing through photoprinting equipment
Handling characteristics

In addition, the formulator will have to take into consideration features relating to the coating solutions, such as compatibility with chemicals, solubility, viscosity, pot life, ease of coating, and drying.

Not least, the formulator will select chemicals that are readily available, not hazardous to health, ecologically safe, and economically acceptable.

Considering the complexity of the problems facing a diazo paper formulator, it is no surprise that few comprehensive recipes have been published outside the patent literature. Most diazo formulations are the result of extensive trials by companies trying to obtain a technical advantage over their competitors, and this explains why they are kept confidential.

The different formulations given in this section have been developed and tested in the laboratory of Andrews Paper and Chemical Corporation, a New York–based company not involved in the manufacture of diazo papers but in the supply of raw materials, paper, and chemicals to the diazo industry worldwide. Some of the chemicals in the formulations are mentioned by their Andrews code references; a complete explanation of the codes appears in the appendix to this chapter.

3.4.1 Precoat Formulations

Precoat D258

This precoat is suitable for all ammonia-developed diazotype papers on opaque base paper; it is an alkaline precoat.

Mix with a high-speed stirrer or a homogenizer for 30 minutes

Water	5	liters
Coating aid 200	3.5	g
Ammonia	35	cm^3
Pigment 2820	493	g

Add under moderate stirring

8% Antifoam A emulsion	13	cm^3
5% Antifoam T dispersion	130	cm^3
Resin VP	500	cm^3
Dispersion F	152	cm^3
Water to make a total of	10	liters

The pH value of this precoat is adjusted and maintained between 8.5 and 9.5.

Precoat D265

This precoat is suitable for all ammonia-developed diazotype papers on opaque base paper; it is an acid/neutral precoat.

Mix with a high-speed stirrer or a homogenizer for 30 minutes

Water	5	liters
Citric acid	11	g
Pigment 2820	515	g

Add under moderate stirring

5% Antifoam T dispersion	160	cm^3
Resin VN	680	cm^3
Dispersion F	160	cm^3
Water to make a total of	10	liters

The pH value of this precoat is adjusted and maintained between 5 and 7

Precoat D278

This precoat is suitable for all amine-developed diazotype papers on opaque base paper.

Mix with a high-speed stirrer for 30 minutes

Water	5	liters
Pigment 65	1500	g
Pigment R	750	g

Add under moderate stirring

Resin VN	1	liter
Dispersion F	150	cm^3
Water to make total of	10	liters

Precoat D261

This precoat is suitable for all ammonia developed diazotype papers on opaque base paper; it uses a protein binder.

Mix with a high-speed stirrer or a homogenizer for 30 minutes

Water	5	liters
Pigment 2820	493	g

Add under moderate stirring

5% Antifoam T dispersion	130	cm^3
Binder IQ	175	g
Ammonia	35	cm^3

Stir for 30 minutes; then add

Resin VP	175	cm^3
Dispersion F	150	cm^3
Salicylic acid	4	g
Water to make a total of	10	liters

The pH value of this precoat is adjusted and maintained between 8 and 9.

In this precoat, the vegetal protein Binder IQ can be replaced by the animal protein Binder C.

Precoat D316

This precoat is suitable for all moist-developed diazotype papers on opaque base paper; it is an acid precoat.

Mix with a high-speed stirrer or a homogenizer for 30 minutes

Water	5	liters
Citric acid	15	g
Dye AC-1	1.5	g
Pigment 2820	250	g
Pigment R	500	g

Add under moderate stirring

5% Antifoam T dispersion	150	cm^3
Resin VN	600	cm^3
Dispersion F	250	cm^3
Water to make a total of	10	liters

The pH value of this precoat is to be adjusted and maintained between 5 and 7.

Precoat D274

This precoat is suitable to produce a glossy finish on all ammonia developed diazo papers.

Mix under moderate stirring to avoid foaming

Water	2.5	liters
Resin VG	6.7	liters
Resin VP	500	cm^3
20% Antifoam L dispersion	75	cm^3
Dispersion F	100	cm^3
Water to make a total of	10	liters

Precoat D260

This precoat is suitable as a barrier layer for erasable ammonia developed diazo intermediate papers.

Mix under moderate stirring to avoid foaming

Water	2.5	liters
Ammonia	10	cm^3
Dye AC-1	2	g
Resin PS-75N	3.5	liters
Antifoam L	10	cm^3
Antifoam M	6	cm^3

Resin VK-2	1.7	liters
Water to make a total of	10	liters

The specific gravity and viscosity of all precoats must be kept constant during the entire period of coating by the continuous addition of a compensating solution containing water, ammonia, and resins in the case of alkaline precoats, and water with resins in the case of acid precoats.

3.4.2 Sensitizer

Blueline Formulations

Blueline Ammonia, Standard Speed, D7128

This sensitizer is to be used in conjunction with precoats D258, D265, D261, or D274.
Mix with a high-speed stirrer, in sequence

Water at 60 to 65°C	7.5	liters
Pigment 2820	25	g
Citric acid	100	g
Thiourea	400	g
Caffeine	100	g
Accelerator ST	600	cm^3
Coupler 111	200	g
Isopropyl alcohol	125	cm^3
Diazo 48NF	225	g
Zinc chloride	500	g
Wetter 27	5	g
Water to make a total of	10	liters

A faster printing version of this system is obtained by reducing the amount of Diazo 48NF to 150 g.

Blueline Ammonia, Standard Speed, D7120

This sensitizer is to be used in conjunction with precoats D258, D265, D261, or D274. The shade of blue obtained is less violet than with Sensitizer D7128.
Mix with a medium-speed stirrer, in sequence

Water at 50 to 55°C	5	liters
Citric acid	50	g
Stabilizer TT	150	g
Accelerator LM	500	g
Thiourea	400	g
Coupler 111	175	g
Isopropyl alcohol	100	cm^3
Diazo 49NF	180	g
Developaid	100	g
Zinc chloride	600	g
Saponin	2.5	g
Water to make a total of	10	liters

A faster printing version of this system is obtained by reducing the amount of Diazo 49NF to 120 g.

Blueline Ammonia, Fast Speed, D7137

This sensitizer is to be used in conjunction with precoats D258, D265, D261, or D274. The shade of blue obtained is neutral blue with little or no red hue; this color is often referred to as "blueprint blue" in the United States.

Mix with a medium-speed stirrer, in sequence

Water at 30 to 35°C	5	liters
Sulfuric acid (concentrated)	15	cm^3
Stabilizer AB	150	g
Solubilizer HI	300	g
Caffeine	100	g
Coupler 144	125	g
Dipropylene glycol	300	cm^3
Diazo 59S	175	g
Zinc chloride	75	g
Stabilizer CD	150	g
Wetter 27	5	g
Water to make a total of	10	liters

A superfast printing version of this system is obtained by reducing the amount of Diazo 59S to 125 g.

Blueline, Fast Speed, for Amine Development, D7122

This sensitizer is to be used in conjunction with Precoat D278. Mix with a medium-speed stirrer, in sequence

Water at 30 to 35°C	5	liters
Citric acid	100	g
Stabilizer TT	50	g
Acetic acid	100	cm^3
Thiourea	100	g
Caffeine	200	g
Coupler 144	100	g
Isopropyl alcohol	100	cm^3
Accelerator ST	400	cm^3
Diazo 54S	150	g
Stabilizer CD	400	g
Wetter 27	5	g
Water to make a total of	10	liters

A superfast printing version of this system is obtained by reducing the amount of Diazo 54S to 100 g.

Blueline Ammonia, Standard Speed, for Use without Precoat, D7131

This sensitizer does not require a separate precoat. Density enhancement is achieved by adding the pigment and the binder to the diazo solution; such a system is often called a pseudo-precoat sensitizer or a one-pot sensitizer.

Mix with a high-speed stirrer for 20 minutes

Water at 30 to 35°C	7.5	liters
Pigment 2820	400	g
Citric acid	100	g

Reduce stirring speed and then add in sequence

Caffeine	50	g
Thiourea	200	g
Accelerator ST	300	cm^3
Coupler 111	200	g
Diazo 48L	180	g
Zinc chloride	250	g
Wetter 27	5	g
Resin VC-1	1	liter
Water to make a total of	10	liters

A faster printing version of this system is obtained by reducing the amount of Diazo 48L to 120 g.

Blueline Ammonia, Standard Speed, for Tropical Climate for Use Without Precoat, D132

This sensitizer, which does not require a precoat, is advocated when the paper is to be used in an environment of high temperature and humidity and when it is established that the water for the sensitizer, the base paper, or any of the other chemicals, contains iron at a level higher than 25 ppm.

Mix with a high-speed stirrer for 20 minutes

Water at 60 to 65°C	5	liters
Pigment 2820	400	g
Citric acid	200	g

Reduce stirring speed and then add in sequence

Water at 60 to 65°C	2.5	liters
Thiourea	300	g
Accelerator LM	200	g
Coupler O	125	g
Diazo 48NF	140	g
Zinc chloride	250	g
Wetter 27	5	g
Resin VC-1	1	liter
Water to make a total of	10	liters

Blackline Formulations

Diazotype blackline papers have gained enormous popularity in the past decade, and although blueline paper is still used more than blackline paper in the United States and in Japan, this is not the case in most other parts of the world.

Many single couplers give with diazonium salts blue azo dyes, but from a practical point of view, not a single coupler gives a black azo dye. To obtain a black image, it is necessary to use a minimum of two couplers, one blue and one brown, and more often three or more couplers chosen from those giving blue, brown, or yellow azo dyes, preferably at the same coupling rate. Each auxiliary chemical in the formulation could have an influence on the shade of the azo dye or the coupling reactivity of the couplers, thus affecting the final blackline color.

The main objective is to achieve a neutral black in the full tones as well as in the halftones of a gray scale image.

Blackline Ammonia, Standard Speed, D9156

This sensitizer is to be used in conjunction with precoats D258, D265, D261, or D274.
  Mix with a high-speed stirrer in sequence

Water at 50 to 55°C	5	liters
Pigment 2820	25	g
Stabilizer AB	400	g
Thiourea	500	g
Sodium monophosphate	200	g
Accelerator LM	250	g
Coupler O	95	g
Coupler 950	100	g
Isopropyl alcohol	100	cm^3
Developaid	100	cm^3
Diazo 48NF	250	g
Zinc chloride	425	g
Wetter 27	5	g
Water to make a total of	10	liters

A faster printing version of this system is obtained by reducing the amount of Diazo 48NF to 180 g.

Blackline Ammonia, Fast speed, D9102

This sensitizer is to be used in conjunction with precoats D258, D265, or D261.
  Mix with a medium-speed stirrer, in sequence

Water at 30 to 35°C	7.5	liters
Citric acid	200	g
Sulfosalicylic acid	50	g
Thiourea	200	g
Accelerator LM	400	g
Caffeine	100	g
Coupler 166	100	g
Coupler 670	18	g
Coupler 690	60	g
Isopropyl alcohol	100	cm^3
Dipropylene glycol	150	cm^3
Diazo 59S	250	g
Zinc chloride	300	g
Saponin	2.5	g
Water to make a total of	10	liters

A faster printing version of this system is obtained by reducing the amount of Diazo 59S to 200 g.

Blackline Ammonia, Superfast Speed, D9155

This sensitizer is to be used in conjunction with precoats D258, D265, or D261.
  Mix with a medium-speed stirrer, in sequence

Water at 55 to 60°C	3	liters
Stabilizer AB	200	g

Citric Acid	100	g
Theophylline	150	g
Coupler 195	50	g
Coupler O	100	g
Coupler 620	25	g
Coupler 690	165	g
Coupler 950	30	g
Thiourea	400	g
Aluminum sulfate	125	g
Isopropyl alcohol	100	cm^3
Dipropylene glycol	200	cm^3
Diazo 59NF	200	g
Zinc chloride	250	g
Stabilizer CD	250	g
Wetter 27	5	g
Water to make a total of	10	liters

Blackline, Fast Speed, for Amine Development, D9152

This sensitizer is to be used in conjunction with Precoat D278.

Mix with a medium-speed stirrer, in sequence

Water at 30 to 35°C	7.5	liters
Sulfuric acid (concentrated)	15	cm^3
Stabilizer AB	150	g
Solubilizer HI	300	g
Stabilizer CD	150	g
Caffeine	100	g
Coupler 144	100	g
Coupler 195	5	g
Coupler 690	250	g
Coupler 620	5	g
Coupler 950	15	g
Diazo 59S	200	g
Developaid	100	cm^3
Zinc chloride	75	g
Wetter 27	5	g
Water to make a total of	10	liters

Blackline Ammonia, Standard Speed, for Use Without Precoat, D9-144

This sensitizer is a one-pot sensitizer that does not require a separate precoat.

Mix with a high-speed stirrer, in sequence

Water at 55 to 60°C	7.5	liters
Pigment 2820	400	g
Citric acid	100	g
Thiourea	200	g
Theophylline	100	g
Allyl hydroxyethyl thiourea	100	g
Coupler O	70	g

Coupler 111	20	g
Coupler 950	110	g
Isopropyl alcohol	100	cm^3
Zinc chloride	250	g
Diazo 48NF	250	g
Wetter 27	5	g
Resin VC-1	1	liter
Water to make a total of	10	liters

A faster printing version of this system is obtained by reducing the amount of Diazo 48NF to 150 g.

Redline Formulations

Redline Ammonia, Standard Speed, D1220

This sensitizer is to be used in conjunction with precoats D258, D265, D261, or D274.
  Mix with a medium-speed stirrer, in sequence

Water at 50 to 55°C	7.5	liters
Sulfosalicylic acid	150	g
Thiourea	500	g
Coupler RG	150	g
Isopropyl alcohol	100	cm^3
Developaid	100	cm^3
Diazo 49L	130	g
Zinc chloride	500	g
Saponin	2.5	g
Water to make a total of	10	liters

Redline Ammonia, Fast Speed, D1228

This sensitizer is to be used in conjunction with precoats D258, D265, D261, or D274.
  Mix with a medium-speed stirrer, in sequence

Water 70 to 75°C	2	liters
Stabilizer TT	175	g
Caffeine	25	g
Coupler 480	35	g
Coupler 166	40	g

Add

Water at room temperature	6	liters
Thiourea	200	g
Allyl hydroxyethyl thiourea	200	g
Isopropyl alcohol	100	cm^3
Developaid	100	cm^3
Diazo 88	100	g
Zinc chloride	500	g
Saponin	2.5	g
Water to make a total of	10	liters

Bicolor Formulation

Blue/Red Ammonia, D374

This formulation is to be used in conjunction with precoats D258, D265, D261, or D274.

With this system, the fully opaque markings of the original are reproduced in blue by the coupling reaction of Diazo 59S and Coupler 144, whereas semiopaque markings, such as pencil or gray ink, are reproduced in red by the coupling reaction of Diazo 67 with its own decomposition product.

Mix with a medium-speed stirrer, in sequence

Water at 30 to 35°C	7.5	liters
Citric acid	200	g
Sulfosalicylic acid	75	g
Caffeine	150	g
Coupler 144	100	g
Accelerator LM	100	g
Thiourea	200	g
Isopropyl alcohol	100	cm^3
Dipropylene glycol	300	cm^3
Diazo 59S	175	g
Diazo 67	125	g
Zinc chloride	400	g
Wetter 27	5	g
Water to make a total of	10	liters

Brownline Formulations

Brownline Ammonia, Standard Speed, for Opaque Base Paper, D1080

This sensitizer is to be used in conjunction with precoats D258, D265, D261, or D274.

Mix with a medium-speed stirrer, in sequence

Water at 30 to 35°C	7.5	liters
Citric acid	300	g
Sulfosalicylic acid	50	g
Thiourea	300	g
Accelerator LM	200	g
Theophylline	50	g
Coupler RX	150	g
Coupler 950	50	g
Isopropyl alcohol	100	cm^3
Dipropylene glycol	100	cm^3
Diazo 49L	225	g
Zinc chloride	300	g
Saponin	2.5	g
Water to make a total of	10	liters

Brownline Ammonia, Standard Speed, for Natural Transparent Base Paper, D1081

This sensitizer requires no precoat.

Mix with a high-speed stirrer

Water at 30 to 35°C	3	liters
Pigment 2820	250	g

Reduce stirring speed and add, in sequence

Ethylene glycol monobutyl ether	1.6	liters
Isopropyl alcohol	1.6	liters
Stabilizer TT	125	g
Sulfosalicylic acid	40	g
Boric acid	150	g
Diethylene glycol	200	cm^3
Thiourea	90	g
Coupler RX	160	g
Chlororesorcinol	120	g
Coupler 320	22	g
Diazo 88	500	g
Zinc chloride	450	g
20% Polyvinyl alcohol solution	2	liters
Dispersion F	50	cm^3
Water to make a total of	10	liters

Brownline Ammonia, Fast Speed, for Transparentized Base Paper, D1089

This sensitizer is to be used in conjunction with precoats D258, D265, or D261.
Mix with a high-speed stirrer, in sequence

Water at 30 to 35°C	7.5	liters
Pigment 2820	25	g
Citric acid	125	g
Thiourea	400	g
Theophylline	200	g
Coupler 603	300	g
Isopropyl alcohol	100	cm^3
Dipropylene glycol	200	cm^3
Diazo 59S	350	g
Diazo 88	200	g
Zinc sulfate	200	g
Wetter 27	20	g
Water to make a total of	10	liters

Brownline Ammonia, Fast Speed, Glossy Finish, for Transparentized Base Paper, D1088

This sensitizer is to be used in conjunction with Precoat D274.
Mix with a high-speed stirrer, in sequence

Water at 30 to 35°C	7.5	liters
Pigment 2820	25	g
Citric acid	200	g
Acetic acid	200	cm^3
Thiourea	100	g
Coupler 950	250	g
Coupler 603	120	g
Coupler 111	20	g
Isopropyl alcohol	100	cm^3
Ethylene glycol monobutyl ether	250	cm^3

Theophylline	100	g
Diazo 59S	175	g
Diazo 10	100	g
Zinc chloride	100	g
20% Polyvinyl alcohol solution	500	cm^3
Wetter 27	10	g
Water to make a total of	10	liters

Brownline Ammonia, Fast Speed, Erasable, for Transparentized Base Paper D1098

This sensitizer is to be used in conjunction with Precoat D260.

    Mix with a high-speed stirrer

Water at room temperature	3	liters
Stabilizer TT	150	g
Pigment 2820	300	g
Pigment 65	1500	g

Reduce stirring speed and add in sequence

Isopropyl alcohol	2.5	liters
Ethylene glycol monobutyl ether	1	liter
Thiourca	75	g
Theophylline	150	g
Coupler 950	250	g
Chlororesorcinol	75	g
Coupler 320	10	g
Diazo 59S	250	g
Diazo 88	88	g
Zinc chloride	150	g
20% Polyvinyl alcohol solution	2	liters
Saponin	5	g
Wetter 27	25	g
Water to make a total of	10	liters

Filter through 100 μm polypropylene filter bag before use.

Moist Formulations

Sensitizer, Fast Speed, for Alkaline Development, D1431

This sensitizer is to be used in conjunction with Precoat D316. The line color depends on the developer used; black and blue are the most popular.

    Mix with a medium-speed stirrer, in sequence

Water at 30 to 35°C	7.5	liters
Sulfamic acid	15	g
Caffeine	60	g
Diazo 54S	180	g
Ammonium oxalate	35	g
Aluminum sulfate	35	g
Solubilizer 1,3,6–1,3,7	50	g
Wetter 27	5	g
Water to make a total of	10	liters

A superfast printing version of this system is obtained by reducing the amount of Diazo 54S to 125 g.

Alkaline Developers

To be used with Sensitizer D1431.

Dissolve in water to make one liter of developer:

Potassium carbonate	35	g
Boric acid	17	g
Sodium hyposulfite	4	g
Sodium benzoate	12	g
Tetrasodium pyrophosphate	0.25	g
Potassium hydroquinone monosulfonate	0.5	g
Thiourea dioxide	0.2	g
Wetter 27	0.5	g

Add to solution above different coupler systems for different colors.

For black:	Phloroglucinol	4.8	g
	Resorcinol	4.2	g
For blue:	Coupler 122	4.75	g
	Coupler dinol	2	g
For red:	1-Phenyl-3-methylpyrazolone	4	g
For yellow:	Coupler EBA	10	g
For brown:	Coupler Dinol	4	g
	Resorcinol	6	g
	1-Phenyl-3-methylpyrazolone	1.5	g

Sensitizer, Standard Speed, for Neutral Development, D2060

This sensitizer is to be used in conjunction with Precoat D316. The line color is black with a standard acid/neutral moist developer.

Mix with a medium-speed stirrer, in sequence

Water at room temperature	7.5	liters
Tartaric acid	100	g
Stabilizer TT	200	g
Caffeine	100	g
Aluminum sulfate	100	g
Diazo 72	150	g
0.5% Methyl violet solution	50	g
Water to make a total of	10	liters

Sensitizer, Fast Speed, for Neutral Development, D2058

This sensitizer is to be used in conjunction with Precoat D316. The line color is black with a standard acid/neutral moist developer.

Mix with a medium-speed stirrer, in sequence

Water, at room temperature	7.5	liters
Tartaric acid	100	g
Stabilizer TT	100	g
Diazo 72	20	g

Diazo 78	90	g
0.5% Methyl violet solution	50	cm^3
Water to make a total of	10	liters

A superfast printing version of this system is obtained by omitting the Diazo 72.

Sensitizer, Fast Speed, for Neutral Development, D2065

This sensitizer is a one-pot sensitizer that does not require a separate precoat. The line color is black with a standard acid/neutral moist developer.

Mix with a high-speed stirrer, in sequence

Water, at room temperature	6.5	liters
Pigment R	850	g
Tartaric acid	100	g
Boric acid	45	g
Sulfuric acid (concentrated)	5	cm^3
Caffeine	50	g
Diazo 72	40	g
Diazo 78	75	g
0.5% Methyl violet solution	50	cm^3

Reduce stirring speed and add

20% Polyvinyl alcohol solution	500	cm^3
Resin VW-2	500	cm^3
Dispersion F	50	cm^3
Water to make a total of	10	liters

Sensitizer, Standard Speed, Neutral Development for Transparentized Base Paper, D2059

This sensitizer is to be used in conjunction with Precoat D316. The line color is black with a standard acid/neutral moist developer.

Mix with a medium-speed stirrer, in sequence

Water, at room temperature	7.5	liters
Tartaric acid	50	g
Stabilizer CD	400	g
Stabilizer TT	25	g
Diazo 72	160	g
Diazo 87	160	g
Wetter 27	25	g
0.5% Methyl violet solution	75	cm^3
Water to make a total of	10	liters

A violet-brown color is obtained with this system when Diazo 72 is deleted and the amount of Diazo 87 doubled.

Acid/Neutral Developer for Blackline

Dissolve in water to make one liter of developer

| Sodium formate | 60 | g |
| Sodium benzoate | 15 | g |

Sodium tartrate	5	g
Potassium bitartrate	1	g
Phloroglucinol	4	g
Potassium hydroquinone monosulfonate	0.5	g
Wetter 27	0.5	g

A blueline version of this developer is obtained by replacing phloroglucinol by a mixture of:

| Coupler 122 | 4 | g |
| Coupler dinol | 2 | g |

3.4.3 Backcoats

The application of a precoat and/or a sensitizer on one side of the paper causes the paper to curl strongly toward the coated side after drying. The cellulosic fibers on the surface of the paper first swell and expand upon contact with the aqueous solution, then shrink and contract upon removal of the water by drying; this creates an imbalance between the coated and uncoated sides of the paper which makes a curl. To counterbalance this effect and to produce a flat sheet, it is necessary to apply a backcoat to the back side of the paper.

The backcoat can have functions other than controlling the curl; it can be designed to increase the shelf life, to decrease absorption of a moist developer, to increase slippage through processing machines, to improve drawing properties of the back surface, or to color the back side of the paper for identification purposes. In one rare instance, a backcoat was designed to supply ammonia for the thermal development of two-component papers.

Whatever other function is to be served, the backcoat must, first and foremost, allow the production of a diazo paper that remains flat before, during, and after exposure and development.

The composition of the backcoat will depend on the degree of control required. Some base papers, such as natural tracing paper, tend to give very pronounced curl on coating, whereas other base papers, such as all-rag transparentized paper, show very little curl; opaque base paper for diazo coating gives in general a medium degree of curl.

The simplest possible backcoat is plain water. Although the liquid form is commonly used, steam has the same effect and is occasionally applied as a backcoat.

In general, the degree of back curl is proportional to the amount of water applied to the paper surface; it follows that if the minimum amount of plain water that can be applied still causes some back curl, it will be necessary to take further corrective action by the addition of various chemicals to the backcoat. It was established that glycols, humectants, zinc chloride, urea, and similar compounds relax the tension in the cellulosic fibers by moisture retention.

Backcoat formulations are arrived at by trial and error, varying first the amount of liquid applied and then the chemical composition.

A typical curl-control backcoat might contain one or more chemicals for the following effects:

To relax curl:
Citric acid	0–3%
Zinc chloride	0–5%
Diethylene glycol	0–5%
Urea	0–5%

To increase curl:

Polyvinyl alcohol	0–1%
Stabilizer TT	0–0.1%

A backcoat to reduce the absorption of moist developer might contain

Polyvinyl acetate emulsion	0–5%
Polystyrene emulsion	0–5%
Polyvinylidene chloride emulsion	0–3%

A backcoat to facilitate print passage in processing machines might contain

Dispersion JB-1	0–3%

A coloring backcoat might contain one or more of the following chemicals:

Diazotint red	0–2%
Diazotint blue	0–2%
Diazotint yellow	0–2%

3.5 DIAZO FILMS

The making of intermediate prints on translucent or transparent materials plays an extremely important role in the safeguarding of original drawings and in the creation of new ones.

We have seen in Section 3.2.4 how diazo intermediate papers allow the production of "second originals," which are used instead of the initial drawing to make copies and can be modified by the addition or deletion of further graphic information.

Paper, however, good as it may be, because of its fibrous structure, absorbs and diffracts light appreciably; it also has a limited physical strength, is often sensitive to atmospheric conditions, and is not dimensionally stable.

An early search to overcome many of the limitations of paper has led to the use of films as substrates for diazotype intermediates.

Even before World War II, cellulose acetate film was suggested for diazo coating. This film had excellent clarity and could be sensitized with solvent systems to give bright and dense images. Later, the film was made suitable for aqueous sensitizing and moist development by saponification of its surface, which rendered it hydrophilic. Athough a great improvement in light transmission and in strength over paper resulted, this film had serious drawbacks: if unplasticized, the film was brittle and could crack or tear easily; if plasticized, the film was soft and the plasticizer exuded. Cellulose triacetate film, because of its improved physical characteristics, replaced for a while the earlier cellulose acetate film; but only in the middle of the 1950s, with the commercial availability of polyester film, was a major step forward made in the production of substantially improved diazotype films.

Polyester films are a range of biaxially drawn films made from polyethylene terephthalate polymer. The process for manufacturing polyester film involves first extrusion of the polymer through a slot die as a continuous sheet and cooling the sheet rapidly to prevent it from crystallizing. The next step consists of heating the film and drawing it equally in two directions at right angles; this has the effect of orienting the molecules of the polymer so that they lie in the plane of the film, but with no particular order in this plane. Finally the plane-oriented film is held under tension and heat set. These treatments confer to the film's outstanding strength and a high degree of dimensional stability at room temperature. In addition, the film has good resistance to most common chemicals. Clarity, strength, and dimensional stability make polyester films an ideal candidate as a substrate for a diazo film intermediate. The following are typical physical and thermal properties of a general-purpose polyester film 25 μm thick:

Density	1.39 g/m
Tensile strength	>1600 kg/m^2
Tear strength (Elmendorf)	20 g · cm/m
Fold endurance (MIT test)	>200,000 cycles
Elongation at break	60–110%
Refractive index	1.50–1.60 D
Water absorption after prolonged immersion	<0.6%
Softening point	165°C
Melting point	250–265°C
Coefficient of expansion	27 × 10°C
Coefficient of thermal conductivity	4 × 10 cal cm/m^2 sec °C
Operational temperature range	−60 to 150°C
Change in dimensions with humidity per 10% change in RH at 20°C	0.007%

Changes in polymer formulation, surface treatment, and manufacturing conditions are made to produce variations in optical, physical, and surface properties for the different grades of polyester film available. In addition, polyester films for diazotype coatings are available in thickness gauges varying between 36 and 175 μm, with the most popular gauges being 50 and 75 μm for engineering films, and 75 and 125 μm for microfilms and microfiches. Although the thinner gauge general-purpose polyester films are quite clear, for the thicker gauges the films can present a certain degree of haze that is detrimental for certain applications (e.g., microfilm); special optically clear films are specifically made for these applications.

Polyester film is particularly inert to chemicals, with practically no absorption for water or solvents.

Whereas the fibrous structure of paper allows penetration of the coating solution within the fibers, in the case of polyester film, the coating solution remains on the surface of the film upon evaporation of the solvents, and the chemicals are deposited in the form of a coated layer; to ensure retention during handling, this coated layer must have cohesion and a good adhesion to the surface of the film. The inertness of polyester film frequently causes adhesion problems, and it is generally necessary to treat the surface of the film to promote the adhesion of coated layers. This problem has received much attention early, at the introduction of polyester film for diazo coating, and a number of imaginative solutions were offered. In particular, three different types of surface treatment have produced acceptable results.

1. *Chemical etching*. Polyester film is attacked by strong acids and strong alkalies, and when these are applied at the correct concentration and under the right conditions, the surface of the film is sufficiently modified to cause improved adhesion of subsequently applied resin layers.

Among the chemicals suggested for chemical etching are some phenols and chlorinated phenols, trichloracetic acid (TCA) and chromosulfuric acid, sodium and potassium hydroxide, hydrogen peroxide, and other oxidizing agents; of these, only TCA has found a wide application in this field. The TCA is dissolved at a concentration of 5 to 10% in a suitable solvent such as methyl ethyl ketone or xylene, applied to the film in a very thin layer, and dried to leave 1 to 2 g/m^2 of chemical.

It should be mentioned that this process presents handling hazards and causes severe corrosion of equipment, but it is nevertheless used when other alternatives are not available.

2. *Resin coating*. This method consists of replacing the polyester surface by a new chemically different surface with good adhesion properties. Various resins or mixtures of resins have been cited in the patent literature to improve adhesion to polyester film; they range from vinylidene chloride to acrylates, from epoxy to isocyanates, and from acrylonitrile to polyesters, with each system being most effective in a particular coating case.

Such a priming method was popular until polyester film manufacturers found a way to incorporate in their manufacturing procedure a step that results in a film with built-in adhesion promoting properties.

A resin composition is applied very thinly to the polyester film after it is cast, but before it is stretched to become biaxially oriented; the coated resin, which is now an integral part of the film surface, improves adhesion without affecting any other properties of the polyester film.

Special grades of surface-treated films are now commercially available and have found a large use in diazo coating.

3. *Corona discharge*. When polyester film is subjected to a corona discharge from high-potential electrodes, the polarity, frictional, and adhesion properties of the film surface change; possibly some polymer chain breakage takes place as well, with the formation of hydroxyl, carboxylic, and carbonyl groups, which would be responsible for adhesion promotion.

The treatment is performed either in line on the coating machine, prior to further coatings, or as a separate step, in which case the next coating must follow rapidly (within a few hours), so that the effect of the corona discharge is not lost.

The range of resins that adhere well to corona-pretreated film is more limited than the range of resins that adhere to polyester films pretreated at source by the manufacturer. For instance, although acrylic resins adhere well to both types of treated film, cellulosic ester resins adhere well only to the second type. In addition to the resins, the solvents in which they are dissolved play an important role in the level of adhesion reached.

3.5.1 Engineering Films

Diazo engineering films, also called diazo film intermediates, reproduction films, or simply diazo films, find their major application in the reproduction of engineering graphics, in the same way as diazo paper intermediates. They were born as a direct result of the shortcomings of the translucent intermediates, particularly the poor strength and lack of dimensional stability of the substrate and the poor definition of the image. Polyester film afforded all the superior properties of the base plus those conveyed by high-quality coatings. Diazo intermediate films have excellent tear strength and dimensional stability under varying temperature and humidity conditions; good handling characteristics; water, stain, and shelf resistance; and good printing and reprinting speeds. The prints have good line definition, visual density, and actinic opacity, with a clear background that resists oxidation and light discoloration; the front and back surfaces of the film offer good drafting qualities. Finally, the diazo films possess prolonged shelf life before printing and, after printing, retain their performance for many decades.

With such features, diazo film intermediates frequently perform better than the originals they are meant to replace.

An extremely large range of diazo films are available worldwide. The vast majority of them are made for ammonia development, with a few rare instances of diazo films made for moist development and even fewer made for amine development.

The technical problems associated with the design of ammonia films are considerably simpler than those relating to making moist or amine development films. For an ammonia diazo film, the layer containing the diazo simply needs to be permeable to ammonia gas; appropriate resins give layers that meet this requirement in addition to all the other requirements expected from the layer.

In the case of moist and amine development films, a liquid is applied to the surface of the films. Whereas with paper this developing liquid can penetrate into the fibers of the paper and the fibrous structure can accommodate the different chemicals of the developer, with film, which is a totally nonabsorbing medium, the liquid must be absorbed by a thin resin layer containing the diazo chemicals, but without softening it prior to drying to the point at which damage by physical contact might occur during processing of the film. Very few resins come close to fulfilling the requirement for a suitable moist or amine development layer.

Diazo films are commercially available in different gauges. For a long time the gauge relating to the particular diazo film was that of the polyester film substrate, not that of the coated product. Today, manufacturers are less specific, and the gauge may be either that of the uncoated film or that of the coated film, which includes the thickness of the different layers applied. Because polyester film is relatively costly, the price of the finished product will greatly depend on the thickness of the film.

The thinnest diazo film guage is 36 µm or 1.5 mils (thousandths of an inch); below this value, films are flimsy and difficult to handle, at both the manufacturing and the processing stages, and as finished prints.

Diazo films of 50 µm or (2 mils) are extremely popular; they have sufficient body and weight to be handled easily, and they offer economic advantage over thicker gauges.

For extra strength and dimensional stability, diazo films of 75 µm (3.1 mils) are chosen. Although they are more costly than the thinner gauges, their superior handling properties make them the choice of quality-conscious users.

Two thicker diazo films are also commercially available for special applications requiring extra rigidity, strength, or dimensional stability. They are 100 µm (4 mils) and 125 µm (5 mils). These thick films are used when true-to-scale reproduction of engineering drawings is called for; in these cases, printing is done on flatbed printing equipment to avoid the slight distortion caused by printing around the cylinder of conventional print machines.

Diazo films are also made with different surface finishes; the side of the film with the sensitized layer, which is generally considered to be the front side, can be matte, semimatte, deglossed or glossy; the first three types of finish contain pigments in decreasing levels, while the glossy finish contains no pigment at all. The choice of the surface finish depends on taste as well as on the intended function of the film. A matte surface is more suited for pencil and ink additions than a glossy surface, but it will also absorb and diffract light more, reducing look-through and reprint speed.

A deglossed surface has just enough pigment to make it appear hazy and has a very low degree of roughness; the infinitesimal surface irregularity allows better contact during printing between a smooth surface original and the diazo film. Upon exposure to ultraviolet light, the diazonium salt decomposes with liberation of nitrogen; the nitrogen gas is trapped between the original and the reproduction film, and if both films are perfectly smooth, they will be lifted apart by the gas, causing poor contact and therefore image blur. This, of course, takes place at a microscopic scale, which explains why a slight surface roughness accommodates the nitrogen gas formed and increases image sharpness.

The coating layer on the side of the diazo film opposite the sensitized side is called the backcoat. This side is, in general, matte, and is designed to have good drafting qualities. For a completely clear diazo film with maximum transparency, both sides of the film are glossy.

A drafting film is a non-light-sensitive material in which one or both sides of the film have a matte layer, which is conceived for its drafting properties.

For the same reasons mentioned in the case of diazo intermediate papers, diazo films are available in sepia or blackline colors; sepia covers a range of shades varying from yellow to chocolate brown, and blackline can appear through transmitted light very dark green, brown, or bluish. Sepia film is indicated for maximum UV absorption with reasonable visual contrast, whereas blackline is indicated for maximum visual contrast with reasonable actinic opacity. Blackline is also chosen when the prints have to be photographed for the generation of microfilms.

As second originals, diazo films are intended to be used for the production of multiple copies of the original. In this process, they are frequently exposed to ultraviolet light, which as we know degrades the azo dye and catalyzes the oxidation of the phenolic couplers and the decomposition products of the diazonium salts, thus causing increased background discoloration. The more the film is exposed to light, the greater the possibility for the image to fade and the background to discolor; this would result in lower visual contrast and poorer reproduction performance.

To overcome the problems above, diazo films with specific stability to ultraviolet light are made available; these are called UV-stable diazo films and are generally sepia.

Diazo films, irrespective of their color, are offered in two printing speed ranges qualified as standard and fast. Standard speed films are expected to have a higher actinic opacity than fast speed films because they use either slow speed diazonium salts, giving dense azo dyes, or a high concentration of fast speed diazonium salts. In both cases, the result is a slower printing speed.

However, the relationship between printing speed and actinic opacity is not always as indicated above: by judiciously selecting the diazonium salts and the couplers, fast speed films can show as good an opacity as standard speed films.

A number of diazo films are also made with an emphasis on a particular property or feature; among these films are erasable diazo films, nonreproducible blueline diazo films, and bluebase diazo films.

Usually when a correction is needed on a diazo film, part of the layer containing the image is removed by physical or chemical action, leaving the clear polyester film visible underneath. Pencil or ink additions are applied to the reverse side of the film; for this purpose, it is common to print in reverse so that the image becomes right reading when looked through the back of the film.

In erasable diazo films, when the layer containing the image is removed by mechanical erasure or chemical eradication, instead of the clear polyester film, a matte layer, which readily accepts pencil and ink lines, is bared underneath the removed image area. In this instance it is not necessary to print in reverse because the corrections and additions are made on the same side as the image. Erasable diazo films require that the diazo layer be coated on top of a drafting layer without intermingling with it.

Nonreproducible blueline diazo films give faint but distinctly visible blueline images that have such a low actinic opacity that they do not reproduce when reprinted on diazo papers. The blue image is used as a guide for outlining or modifying with pencil or ink lines some particular part of the image. Upon reprinting, only the drawn pencil or ink lines reproduce and are visible on the print.

The majority of diazo films show the image on a quasi-white background. Although dyes are used to tint the background, their aim is to mask any yellow cast caused by the chemicals or to give a more pleasing shade to the white background by making it appear bluish, greenish, or grayish.

In some instances, the background is deliberately given a pronounced blue color that does not affect any of the functional properties of the material. These films, described as bluebase diazo films, were popular in Europe because they simulated the color of traditional blue natural tracing paper and blue diazo lacquered paper, which existed before diazo films. They are still used in some parts of the world but are almost nonexistent in the United States.

3.5.2 Diazo Microfilms

The production of microfilms has long been an established practice in the world, stimulated by the need to preserve more and more records and the recognition that size reduction can lead not only to economies in storage space but also to the establishment of an organized record system having good accessibility, retrieval, and reproduction capabilities.

The originals that have to be microfilmed vary between small documents such as bank checks or printed pages of a book with dense black type on a clean reflecting background, to large engineering drawings sometimes having weak pencil lines on low reflectance paper.

The initial transference of graphic information to microfilm is done exclusively on silver halide films because of their high light sensitivity, which allows the formation of images through an optical reduction system at very short exposure times. To record the wide range of line densities, and thickness, the silver microfilm is designed to have a high contrast (a gamma value γ of more than 4) and a high resolution (fine grain); this last requirement follows from the fact that if the original is capable of separating 10 lines/mm, after an optical reduction of 20, the film should be able to have a resolution of 200 lines/mm for the enlarged image to be readable.

Diazo films, which because of their lack of sensitivity are not suitable for the initial recording of microimages from originals, have been found to be eminently suitable for the duplication of silver and other microfilms at lower costs per frame. The resolution of diazo microfilm is vastly in excess of that of silver halide films because the image is molecular instead of granular, and it can reach values in excess of 1000 lines/mm.

Positive working ammonia-developed diazotype films specifically intended for use in microreproduction systems are commercially produced by a relatively small number of companies in the world. Although in principle the manufacture of diazo microfilm is not much different from that of engineering film, in practice it involves more sophisticated coating and converting equipment and a degree of cleanliness of a different order of magnitude. Since the smallest dust particle or impurity speck in the coating would be magnified to appear like a boulder on the reproduced image, such contaminants must be absent at every stage of the operation. The entire manufacturing and converting process must be conducted in a clean room air environment, which requires costly equipment and operational procedures.

Diazo microfilms are available in the same standard sizes as the silver halide microfilms they are duplicating; they are made in rolls, fiches, and aperture card microformats.

Rolls of diazo microfilm are supplied in widths of 16, 35, 70, and 105 mm and in varying lengths of 100 to 600 m. Diazo microfiches are supplied in standard size sheets of 105 mm $\times$ 148 mm or 180 mm $\times$ 240 mm and in packets of between 100 and 500 fiches.

The aperture card format is used mainly in engineering applications; in an aperture card, a single frame of 35 mm microfilm is mounted in a cutout window in the card.

Diazo microfilms are manufactured on different gauges of optically clear polyester film; these are 65 to 75, 100, 125, and 175 µm. The thin and medium gauges are generally used for rolls or aperture cards, while the thick gauges are used for fiches.

The image color in diazo microfilms is predominantly blue or black. It is desirable to have high-density azo dyes both in the visual and in the actinic regions of the spectrum so that the film can be reviewed directly or reprinted onto other ultraviolet-sensitive materials. However, some diazo microfilms are manufactured that are not intended to be duplicated, and such films require only good visual contrast.

The properties of two commercial diazo microfilms are as follows:

Blueline Microfilm Suitable for Use in Readers Only

Maximum dye density	1.9 –2.0
Minimum background density	0.02–0.03
$\bar{\gamma}$	2.2 –2.3
$\bar{\gamma}$ ratio	0.7 –0.8

*Blackline Microfilm Suitable for Use in Readers and for the Production
of Further Generations of Diazo Duplicates*

Maximum dye density	1.8 –1.9
Minimum background density	0.02–0.03
$\bar{\gamma}$	1.8 –1.9
$\bar{\gamma}$ ratio	1.4 –1.6

The γ ratio is the ratio of the $\bar{\gamma}$ of the second diazotype generation and the $\bar{\gamma}$ of the first diazotype generation.

Diazo microfilms fall in one of two groups: the first group is intended to be used as ''work'' or ''use'' copies and is generally found in libraries, record centers, or working environments. The value of these films lies in their being available for ready reference, and because of their frequent handling, they have a relatively limited life expectancy.

The second group is intended for use as medium- and long-term storage copies. Medium-term diazo microfilm is indicated for the preservation of records for a minimum of 10 years, when stored under proper conditions; long-term diazo microfilm is indicated for the preservation of records for a minimum of 100 years, when stored under proper conditions. Of vital importance for satisfactory extended preservation is the control of light exposure, temperature, and humidity, and the protection of the film from dirt and impurities.

Diazo images, as is well known, show density changes after exposure to light; for equal exposures, the changes could be slight or severe depending on the chemicals used. In the case of diazo microfilms, density changes are critical, especially when copies are viewed in commercial readers for extended periods of time; this is because the microfilm is exposed not only to light but also to heat from the light source. The degree to which the image quality is retained under such usage conditions will determine the acceptability of the diazo microfilm. Light fading tests using a fadeometer are necessary during the formulating phase of diazo microfilms; low- and high-density areas of the image are placed in the fadeometer for a specific length of time (8 hours), and the changes in density are measured with a densitometer; the film is considered to be unacceptable if the optical density difference between maximum and minimum densities is less than 0.8.

3.5.3 Vesicular Films

The vesicular process is based on the use of diazonium salts in the same manner as the diazotype process, but the image generation relies on a light-scattering phenomenon, not on an azo dye formation.

The first commercial vesicular films were produced by the Kalvar Corporation (11) in 1959 on the basis of its U.S. Patent 2,911,299, but the principle of light-scattering photography was disclosed much earlier, by Kalle (12) in 1932 and by GAF (13) in 1955. In vesicular films, a thermoplastic resin layer containing a light-sensitive diazonium salt is applied very thinly to a transparent polyester film base. The thermoplastic layer consists of a random mixture of amorphous and crystalline polymer areas, which give the system a uniform index of refraction and make it appear transparent.

Upon exposure to ultraviolet light, the diazo compound in the thermoplastic layer photolyzes with release of nitrogen gas; when the exposed film is immediately thereafter subjected to heat, the gaseous nitrogen expends to form microscopic vesicles, usually spherical, with a high concentration of crystalline areas on the surface. These vesicles have an index of refraction different from the nonorganized surrounding media, hence they scatter the light incident upon them, causing an image pattern in the exposed areas. Figure 3.3 illustrates the vesicle formation. Since the light-sensitive diazonium compound is still present in the unexposed areas, a fixing step is required to obtain a permanent image. This is accomplished by exposing the entire film once more to ultraviolet light to decompose the residual diazo and allowing the nitrogen gas that is formed to diffuse slowly from the layer at room temperature.

In the exposed and developed vesicular film, image contrast is the result of incident light reflection, refraction, and transmission: when viewed in reflected light, the exposed areas in which there are light-scattering vesicles appear white, and the nonexposed or clear areas appear dark. However, when viewed by transmitted light, the exposed areas, in which there are scattering vesicles, appear dark and the unexposed areas appear clear. It follows that the exposed and developed vesicular film gives a negative-mode image in all light transmitted applications—for example, when the image is projected on the screen of a microfilm reader.

A positive-mode image is also possible if a black layer is applied to the film underneath the diazo–resin layer; in this case, in reflected light, the black of the underneath

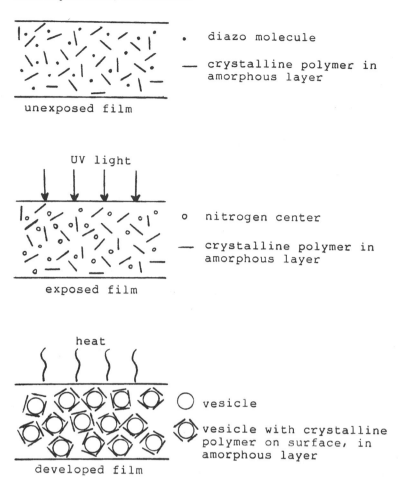

unexposed film

- • diazo molecule
- —— crystalline polymer in amorphous layer

UV light

exposed film

- o nitrogen center
- —— crystalline polymer in amorphous layer

heat

developed film

- ◯ vesicle
- ◈ vesicle with crystalline polymer on surface, in amorphous layer

Figure 3.3 Principle of vesicle formation.

layer may clearly be seen through the unexposed transparent areas, whereas the exposed areas appear opaque white.

The sensitivity of vesicular films depends on the choice of the diazonium salt; it can vary with commercial diazo compounds between 350 and 420 nm. Exposure times are similar to those of engineering diazo films, but since the vesicular process is negative mode, images can already be obtained at very low exposure levels; as exposure time increases, more light-scattering centers are formed and a greater image density results. Above a certain point in the exposure scale, density is no longer directly proportional to the exposure time because the nitrogen gas generated diffuses during the prolonged exposure; the reciprocity law is valid over a range of exposure times from 0.05 to 60 seconds.

The temperature of the film during exposure should not be allowed to exceed 45°C, as this would result in premature development and a much higher rate of diffusion of the image-forming gas centers.

Development of the vesicular film is obtained by applying sufficient heat, for a short period of time, preferably by conduction or convection, to soften the resin layer and allow the nitrogen gas nucleus to expand into a sizable vesicle, which can reach up to micrometers in diameter. Maximum development is a function of temperature and time, with temperatures ranging from 85 to 150°C and times ranging from milliseconds to a few seconds. Because the latent image decays rapidly through gas diffusion, development must follow immediately after exposure. After development and image formation, the residual unexposed diazonium compound is destroyed by the action of ultraviolet light, and the film is allowed to stand for several hours away from any source of heat to permit the gas diffusion process to take place.

To obtain reproducible results with so many parameters affecting the vesicular image quality, initial exposure and development temperature must be critically controlled in vesicular processing equipment, to a fraction of a second and to less than 1°C.

The resolution of vesicular films depends on the developing conditions but is in general higher than 150 lines/mm and can reach 500 lines/mm with special coating compositions. These high resolutions make vesicular film highly suitable for microfilm duplication.

Because they contain only the diazonium salt and no azo coupler, in a hydrophobic environment, vesicular films, when packed adequately, have a shelf life of many years, well in excess of diazo microfilms. In addition, a properly exposed, developed, and cleared vesicular image is stable under normal usage conditions and will withstand temperatures up to 90°C for extended periods of time; such a vesicular image will also show less density change after exposure to light than an equivalent diazo microfilm image.

The manufacture of vesicular films requires the same type of sophisticated equipment and clean environment conditions as diazo microfilms. Both types of film are made in the same formats and for the same applications; both are indicated for the preservation of records and for use as "work" copies. It appears, however, that diazo microfilm and in particular blueline diazo microfilm are gaining popularity for use as working copies, whereas vesicular film is still considered to be the standard for medium- and long-term storage duplicates.

3.5.4   Other Diazo Films

The simplicity and versatility of the diazo process has led to many other applications for diazo films; some are extensions of engineering film technology and others are extensions of diazo microfilm technology; all use polyester film base, require solvent coating, and are ammonia developed. Together, all other diazo films represent a commercially small volume of materials, either because of their limited industrial use or because, in some instances, better materials directed at the same application exist.

The following are the major specialty diazo films that can be found in the market.

Overhead Projection Films

The ability of diazonium compounds and coupling components to give different color azo dye images coupled with the possibility of producing these color images on optically clear film has given rise to the range of color films used in overhead projection. These films are designed to produce high-contrast color images on a completely clear background. By superimposing on the overhead projector two or more films, each with a different pattern in a different color, a composite multicolor image is progressively built up on the screen. Such a tool is extremely valuable in educational lectures, where most of the overhead projection films find their applications.

A careful selection of diazos and couplers allows the production of red, blue, yellow, brown, green, and black films. Each color film must be printed and developed separately, and the printing must be done in register to obtain perfect superimposition of the different images.

The production of overhead projection film is very similar to that of diazo microfilm, and although the magnification of the image on the projection screen is not as great as that of a microfilm on a reader screen, the same degree of cleanliness is required during all the manufacturing operations.

Color Proofing Films

Color proofing is a very well established technique in the printing industry.

Without entering into the details of color printing, it should be briefly explained that color photographs that are to be printed by a printing process are first converted into four screened color separation negatives, which are subsequently used to produce four printing plates: one for each of the primary colors (represented by cyan, magenta, and yellow), and one for black. Each plate is used to apply one of the four color inks to reconstitute the color photograph. The appearance of the final print will depend on the density and amount of each of the four inks applied during the printing process.

To assess the overall look of the final print, a color proof is made. There are different ways of producing this color proof and one of them is to make first a positive of each of the four color separation negatives and print the four positive films on a diazo film of the correct color, that is cyan, magenta, yellow, and black. The azo dye color of each diazo film must be of the exact density and hue to ensure that when the films are superimposed in register, the composite image will be a perfect color reproduction of the initial color image. It must be recognized that it is extremely difficult to find a combination of diazo and coupler that will give a good cyan or a good magenta color; it is easier to obtain a good yellow color. This limitation has prevented the wide use of diazo color proofing films; moreover, other, better materials are available.

Diazo Masking Films

One of the major requirements of an engineering diazo film is to produce an azo dye image of good actinic opacity, generally sepia, to give good reproduction prints on other diazo materials. In certain applications, an even greater opacity than that of an engineering film is necessary. In these applications, longer than usual exposures to ultraviolet light are needed to complete the photochemical reaction producing the image, as for instance, in the cases of photopolymerization or diazo–resin photomolecular rearrangement. The fabrication of printed circuit boards and the preparation of lithographic plates fall into such categories.

Silver halide films have proved most satisfactory for producing images having an almost total blocking power to ultraviolet light. Whenever possible, however, special diazo films, also called masking films, are made to substitute for silver halide films in these applications. The azo dye, which generally is yellow to orange, absorbs light almost completely between 320 and 450 nm but allows light of higher wavelengths (in the visible range) to pass through. This property is particularly useful when the masking film print is placed over an original drawing to view and check the image.

Diazo masking films are usually clear films of relatively thick gauges (125–175 nm) because of the need for high dimensional stability and accuracy in the printing step. The printing speed of such films is, most of the time, lower than that of diazo engineering films because of the use of high concentrations of slow-to-medium-speed diazonium compounds.

The production of diazo masking films is similar to that of a clear diazo engineering film, but with a higher quality standard.

Adhesive-Backed Diazo Films

It is sometimes necessary in engineering graphics to transfer part of a drawing or image to another area and to make copies of this new composite drawing or image. It is also taken for granted that all the information to be reproduced is on translucent or transparent material. Adhesive-backed diazo films were specifically designed to meet the requirement of this particular kind of application.

A thin polyester film (25–35 μm) carries on one side a sepia or blackline reproduction diazo layer and on the other side an adhesive layer; this film is laminated to a silicone release paper to protect the adhesive layer during processing and handling. In usage, a print is first made on the adhesive-backed diazo film and developed; it is then cut out for transfer. The silicone release paper is peeled off the film, and the print is finally stuck down in its new place on the transfer film. The composite image is subsequently reprinted or photographed.

Manufacturers of diazotype materials do not in general involve themselves with laminating operations, and therefore they secure a film that carries the adhesive layer and is laminated to the silicone release paper. Their involvement consists in applying to the polyester film surface, from solvents, the diazo layer, which for all practical purposes is identical to the one used for diazo engineering films.

Opaque Diazo Films

All the diazo films we have seen so far were intended for use as reproduction materials. The base, of necessity, was either left clear or made slightly matte but still transparent to ultraviolet light. Visual contrast was always secondary to actinic opacity. For some applications, visual contrast is the most important criterion for the film, which is not meant to be reprinted but only viewed, or, in rare cases, photographed. Such films must have the densest possible image on the whitest possible background. They are usually intended for the production of display prints of engineering or architectural drawings.

The diazo light-sensitive layer is generally applied on one side of the polyester film base, while the opaque white layer is applied on the other side of the film. Diazos and couplers are selected to give optimal visual contrast and, in the case of a blackline, the most neutral tones. The whiteness and opaqueness of the back side of the film are achieved by the use of white pigments, such as titanium dioxide, bound by a resin matrix. Manufacture of opaque diazo film, printing, and developing are the same as for all other engineering diazo films.

3.6 CHEMICALS, SOLVENT SYSTEMS

The hydrophobicity of polyester film makes it very difficult to coat aqueous preparations evenly. In addition, water-soluble and water-dispersible resins do not possess all the properties required for a satisfactory coating layer, whereas many solvent-soluble resins do.

With almost no exceptions, all quality diazo films are produced from systems containing only organic solvents or a predominant amount of organic solvents. The diazos, couplers, and auxiliary chemicals used in the sensitizing layer must be soluble in solvents, such as alcohols, ketones, and glycol ethers; the resins and other ingredients must be soluble in the same polar solvents or in a mixture of polar and nonpolar solvents such as hydrocarbons. Pigments do not dissolve but should be easily dispersed in these systems.

3.6.1 Solvent-Soluble Diazos

Not all diazonium salts are soluble in the above-cited polar solvents. Some diazonium zinc chloride salts have sufficient solvent solubility to be used for film coatings, but most of the other have too low a solubility to be of practical use. Two zinc chloride diazos with acceptable solvent solubility are

1. 1-Diazo-4-*N*-*N*-diethylaminobenzene chloride; zinc chloride
2. 1-Diazo-2,5-dibutoxy-4-morpholinobenzene chloride, half-zinc-chloride

Among the diazo salts with increased solvent solubility are the borofluoride salts and the hexafluorophosphate salts. Both types of diazo salt are soluble in many polaric solvents at relatively high concentrations, the borofluoride salts being more soluble in alcohols and the hexafluorophosphate salts in ketones.

The commercially available solvent-soluble diazonium salts can be used for film coatings.

3. 1-Diazo-4-(*N*-ethyl-*N*-benzyl)aminobenzene borofluoride
4. 1-Diazo-2-ethoxy-4-*N*,*N*-diethylaminobenzene borofluoride
5. 1-Diazo-2,5-dibutoxy-4-morpholinobenzene borofluoride
6. 1-Diazo-2,5-diethoxy-4-morpholinobenzene borofluoride
7. 1-Diazo-4-*N*,*N*-diethylaminobenzene borofluoride
8. 1-Diazo-3-chloro-4-*N*,*N*-dibutylaminobenzene borofluoride
9. 1-Diazo-2,5-diethoxy-4-*p*-tolymercaptobenzene borofluoride
10. 1-Diazo-3-chloro-4-(*N*-methyl-*N*-cyclohexyl)aminobenzene borofluoride
11. 1-Diazo-3-methyl-4-pyrrolidinobenzene borofluoride
12. 1-Diazo-2,5-dibutoxy-4-morpholinobenzene hexafluorophosphate
13. 1-Diazo-2,5-diethoxy-4-morpholinobenzene hexafluorophosphate
14. 1-Diazo-4-*N*,*N*-diethylaminobenzene hexafluorophosphate
15. 1-Diazo-3-chloro-4-*N*,*N*-diethylaminobenzene hexafluorophosphate
16. 1-Diazo-3-chloro-4-pyrrolidinobenzene hexafluorophosphate
17. 1-Diazo-3-methyl-4-pyrrolidinobenzene hexafluorophosphate

3.6.2 Solvent-Soluble Couplers

Many of the couplers that are soluble in water and are extensively used in aqueous diazo coatings are not soluble in solvents; such couplers include the sodium salts of dihydroxy-naphthalene sulfonic acids, which are used for blueline and blackline papers.

Other couplers that have little solubility or are totally insoluble in water are highly soluble in solvents and are most suitable for diazo film coatings.

A few couplers are soluble in both water and solvents.

The following commercially available solvent soluble couplers can be used for film coatings.

Blue Couplers

1. 2,3-Dihydroxynaphthalene-6-sulfonic acid
2. 2-Hydroxynaphthalene-3-carboxylic acid ethanolamide
3. 2-Hydroxynaphthalene-3-carboxylic acid diethanolamide
4. 2,3-Dihydroxynaphthalene
5. 2-Hydroxynaphthalene-3-carboxylic acid-2'-methylanilide
6. 2-Hydroxynaphthalene-3-carboxylic acid-2'-methoxyanilide
7. 2-Hydroxynaphthalene-3-carboxylic acid α-naphthylamide
8. 2-Hydroxynaphthalene-3-carboxylic acid-3'-nitroanilide
9. 2-Hydroxynaphthalene-3-carboxylic acid-2'-ethoxyanilide

Yellow Couplers

10. Catechol monohydroxyethyl ether
11. *m*-Hydroxyphenylurea
12. β-Resorcylic acid
13. Bis(2,1-dimethylethyl)5-methyl-4-hydroxyphenyl thioether
14. Phenyl phenol
15. *p*-Dihydroxyphenyl thioether
16. Cyanoacetamide
17. Cyanoacet morpholide
18. Acetoacet benzylamide
19. Acetoacetanilide
20. Acetoacet-*o*-toluidide
21. Acetoacet-*o*-anisidide
22. Acetoacetoxyethyl methacrylate
23. 3,3'-Methylene bisacetoacetanilide

Brown Couplers

24. Resorcinol
25. 4-Chlororesorcinol
26. 4,6-Dichlororesorcinol
27. Methyl resorcinol
28. Diresorcinol sulfide
29. Diresorcinol sulfoxide
30. β-Resorcylic acid ethanolamide
31. Resorcinol monohydroxyethyl ether

Red Couplers

32. α-Resorcylic acid
33. α-Resorcylic acid ethanolamide
34. 1-Phenyl-3-methyl-5-pyrazolone
35. 4,4'-Methylene bis(3-methyl-1-phenyl-5-pyrazolone)
36. 4-Bromo-α-resorcylic acid
37. 4-Bromo-α-resorcylic acid amide
38. 4-Bromo-α-resorcylic acid methylamide

3.6.3 Auxiliary Solvent-Soluble Additives

Solvent diazo preparations for film coatings, like aqueous ones for paper coatings, require in addition to the diazos and couplers special chemicals to promote shelf life, development, background stability, and other properties. Many of these chemicals are the same as those used in aqueous systems because they are sufficiently soluble in solvents; some are suitable for use in solvent systems only (e.g., the solvent-soluble resins).

Acid stabilizers are necessary to ensure the stability of the diazo solution as well as to extend the shelf life of the film. The same acids used in aqueous diazo solutions can be used for solvent ones. In addition, acids are used as catalysts for cross-linking of thermosetting resins incorporated in the sensitizing coating or applied as separate coatings on the front or back of the film. Strong acids are preferred over weak acids because they can be effective in much smaller amounts. Among the most commonly used acids are 5-sulfosalicylic acid and *p*-toluene sulfonic acid.

Zinc chloride and thiourea are used for the same reasons as in diazo papers except that in general their levels are much lower in diazo films.

Stannic chloride, which cannot be used in aqueous diazo systems because it causes precipitation of the diazo, can be used in solvent because the diazo stannic salt is soluble in this medium. Solubilizers are very seldom necessary because few compatibility problems arise in solvents.

Development promoters are useful, and those mentioned for diazo papers, which are soluble in solvents, such as allyl hydroxyethyl thiourea and caprolactam, are frequently used. Humectants, like polyglycols, improve development but must be used only in moderation because their action tends to soften the resin layer adversely.

Coating aids such as wetting agents and defoamers, which are so important in aqueous coatings, are very rarely used in solvent coatings, since their effects on the coating performance of the solution are secondary to those of the solvents themselves.

Dyes play similar roles in diazo films and diazo papers. They are chosen for their solvent solubility and for their ability to mask the yellow cast in the print background. Methylene blue and methyl violet are two common dyes that are sometimes used in film coatings, but their sensitivity to pH changes and to ultraviolet light reduces their effectiveness; commercially available proprietary dyes are often required. Occasionally, blue dyes are used to give a distinct blue tint to the diazo film background, and they can be added either to the front coat or to the backcoat with similar results.

In addition to diazos, couplers, and other additives used in the sensitizer, three categories of chemicals are of critical importance for formulating film coatings: solvents, resins, and pigments.

Solvents

The choice of solvents affects the degree of solubility of each of the chemicals, as well as the viscosity, flow, and wetting properties of the solution, the evaporation rate of the system, the level of adhesion of the coated layer to the polyester film surface, and the overall physical quality of the finished product.

The solvents in film coatings are selected from the groups of alcohols, ketones, glycol ethers, and hydrocarbons. Table 3.1 lists these solvents with their main properties.

Low-boiling solvents have in general better solubility properties than high-boiling solvents, and they give lower resin solution viscosities. However, they evaporate too

Table 3.1 Properties of Solvents Used in Film Coatings

Solvent	Molecular weight	Specific gravity at 20/20°C	Boiling point range °C at 760 mm Hg	Evaporation rate (vs. but- Acet = (1)	Flash point tag, closed cup (°F)	Explosive limits (vol % in air) Lower	Upper
Methyl alcohol	32.04	0.793	64–65	3.5	54	6.7	36.0
Ethyl alcohol, anhydrous	46.07	0.792	74–80	1.9	54	3.3	19.0
Ethyl alcohol 95%	46.07	0.812	74–80	1.7	58	3.3	19.0
Isopropyl alcohol anhydrous	60.09	0.787	82–83	1.7	53	2.0	12.0
Isopropyl alcohol 91%	60.09	0.818	79.7–80.7	1.6	61	2.0	12.0
n-Butyl alcohol	74.12	0.811	117–118	0.46	97	1.2	10.9
						(100°C)	(100°C)
Acetone	58.08	0.792	55.5–56.5	7.7	−4	2.6	12.8
Methyl ethyl ketone	72.10	0.806	79–80	4.6	16	1.8	10.0
Methyl isobutyl ketone	100.16	0.802	114–117	1.6	60	1.2	
Ethylene glycol methyl ether	76.1	0.965	123.5–125	0.5	102		
Ethylene glycol ethyl ether	90.1	0.931	134–136	0.2	110		
Ethylene glycol n-butyl ether	118.2	0.902	169–172	0.1	143		
Ethylene glycol methyl ether acetate	118.2	1.005	140–147	0.2	120		
Toluene	93.13	0.870	110–111	1.5	45	1.2	7.0
Xylene	106.16	0.869	138–140	0.75	83	1.0	7.0

quickly before coating and during the drying process, causing viscosity changes in the solution and possible surface defects or adhesion failure. High-boiling solvents give high resin solution viscosities, which may make coating difficult, and they require higher drying temperatures; possible solvent retention after drying may cause softening of the layer.

Alcohols are good solvents for the diazo chemicals but are often poor solvents for the resins; the opposite is frequently the case with ketones and glycol ethers. Hydrocarbons are generally poor solvents for both the diazo chemicals and the resins, but in small amounts they have beneficial effects on the flow properties of the solution.

Because of the conflicting behaviors of the different solvents in relation to the various needs of the diazo system, it is very unusual to use a single solvent in a film coating solution. Mixtures of solvents are more common, with each solvent contributing one aspect of the total package requirement.

Drying conditions and application methods also determine the choice of solvents. Coating machines with high drying capacity allow the use of a greater amount of higher boiling solvents than machines with limited drying capacity or low oven temperatures. Multiple reverse roll coating applications require lower viscosity solutions than applications with a single roller followed by a wire wound rod metering device.

From a practical point of view, a mixture of 50% methyl alcohol and 50% acetone, which is frequently used in film coatings, can be considered to be a low-boiling solvent blend with excellent solvent power, giving low-viscosity solutions and having a very fast evaporation rate.

A mixture of 40% ethyl alcohol, 40% methyl ethyl ketone, 10% ethylene glycol monoethyl ether, and 10% xylene can be considered to be a medium-boiling solvent blend; such a mixture or similar ones are commonly used as general-purpose resin solution blends for wire wound rod metering coating systems.

Mixtures with predominantly high-boiling solvent blends are used only exceptionally.

The use of solvents in film coatings carries a severe fire and explosion risk, which should not be ignored or minimized. The flash point of a solvent is a measure of its flammability, and the lower the flash point, the greater the flammability risk. Table 3.2 relates the flammability risk to the flash point of liquids, or mixture of liquids or liquids containing solids in solution or in suspension.

The use of flammable volatile liquids also involves the risk of explosion when the concentration of solvent vapor in air is between a lower and an upper flammability limit; in this range, the application of a flame or spark of sufficient thermal intensity can ignite the vapor–air mixture and cause an explosion.

Table 3.2 Flammability Risk in Relation to Flash Point and Initial Boiling Point of Liquids

Flammability risk	Flash point, closed cup	Initial boiling point
Very high	<23°C (73°F)	≤37.8°C (100°F)
High	≥23°C (73°F)	≤37.8°C (100°F)
Medium	≥37.8°C	<61°C (141°F)
Low	≥61°C	<93.4°C (200°F)

It can be seen from Tables 3.1 and 3.2 that low-flash-point solvents, such as acetone and methyl alcohol, present the highest fire and explosion hazard, whereas high-flash-point solvents, such as the glycol ethers, have reduced risks.

A solvent-coating film operation requires extraordinary safety measures: all electrical equipment must be explosion proof, and all personnel must be thoroughly trained in the handling and use of flammable liquids. It is mainly for these reasons, and because costly special equipment is required, that the manufacture of solvent-coated products is limited to a much smaller number of companies than the manufacture of diazotype papers from aqueous solutions.

With the enhanced consciousness of health hazards and environmental pollution by volatile solvents, the use of many solvents has been restricted or even banned in some parts of the world; reformulating efforts aimed at the use of less hazardous solvents are constantly being made.

Resins

Resins are indispensable in binding the diazo chemicals or the pigments into a coherent, continuous film layer with good flexibility, hardness, and adhesion to the polyester base. The ideal resin layer must also be colorless, non-UV-absorbing, and permeable to the developing media (ammonia gas or liquid developer); it must have a high softening point, and it must be water insoluble, light stable, and resistant to oxidation. Resins must be easily soluble in the solvent blends chosen and must give reasonable viscosities at the required concentrations. Some resins must give crystal clear coatings (e.g., when intended for use in microfilm), some must be extremely hard (e.g., for use in drafting film), while others must be able to soften under heat (e.g., for use in vesicular film).

Few single resins, if any, meet all the different requirements of the various film applications, and a particular resin coating often consists of two or more resins, each contributing some specific property.

Thermoplastic as well as thermosetting resins have been used in film coatings. Thermoplastic resins are preferred for the diazo sensitizing layer because they are more permeable to ammonia gas; thermosetting resins are often selected for drafting films because of their final hardness.

Among the various thermoplastic resins available, the following classes of resins are the most frequently encountered in film coatings.

Polyvinyl Acetates. These resins give layers of good flexibility and adhesion but of relatively low softening point; the layers have also a good gas permeability. Free hydroxyl groups in the resins allow cross-linking and hardening of the layer. Many grades with good solvent solubilities and a wide range of viscosities are available.

Polyvinyl Butyrals. These resins give layers of good flexibility and adhesion with higher softening points than the polyvinyl acetates, but with lower gas permeability; free hydroxyl groups allow some degree of cross-linking. These resins are rarely used in diazo layers but are frequently used, with other resins, in drafting layers.

Polyvinyl Acetate and Polyvinyl Chloride Copolymers. These resins combine the properties of the two classes of polymers; they are sometimes used to improve adhesion to polyester film.

Cellulose Acetates. These resins have a high softening point, excellent clarity, good flexibility when properly plasticized, and good gas permeability, but unfortunately poor adhesion to the polyester film surface. They have an acceptable adhesion level to natural tracing paper. Cellulose acetate resins are easily saponified by strong alkalies; this property has been utilized to render the surface of cellulose acetate layers water receptive for sensitizing or for the absorption of moist developers. These resins are mostly used in the preparation of moist developing diazo films and natural tracing intermediates.

Cellulose Acetate Propionates and Butyrates. These two classes of cellulosic ester resins are prominent in film coatings because they singly meet the greatest number of requirements for good results. They are clear and flexible, have adequate hardness and softening points, offer acceptable solution viscosity at the right concentration for diazo sensitizers, and have good adhesion, good binding properties, and good gas permeability. Their single main drawback is their tendency slowly to hydrolyze under the influence of strong acids and liberate unpleasant-smelling propionic or butyric acid; the butyric acid odor is found to be more objectionable than the propionic acid one, and as butyrates are more easily hydrolyzed than propionates, their use is also less recommended.

Cellulose mixed acetate esters are particularly popular for diazo film layers; they are manufactured by Eastman Chemical Corporation in the United States and are available in a number of grades, which vary according to the molecular weights, solubility characteristics, and viscosities. Three cellulose ester resins are commonly used in diazo film coatings:

Eastman Cellulose Acetate Propionate, grade 482-0.5
Eastman Cellulose Acetate Propionate, alcohol-soluble grade 504-0.2
Eastman Cellulose Acetate Butyrate, grade 381-0.1

A special type of thermoplastic resin is required for diazo vesicular films; such a resin must be sufficiently rigid and sparingly permeable to gas, and must exhibit some gas diffusivity. Four groups of resins are often cited in vesicular films:

Polyvinylidene chloride resins
Polyacrylonitrile and polymethacrylonitrile resins
Polystyrene resins
Epoxide resins

Among the thermosetting resins available, the following classes of resin are frequently used in film coatings.

Amino–Formaldehyde Resins. These urea–formaldehyde and malamine–formaldehyde resins, modified or unmodified, often are used as self-cross-linking resins or as cross-linking agents either for alkyd-type resins or for resins containing reactive groups such as free hydroxyl groups in polyvinyl acetates or polyvinyl alcohols. Their use in small amounts increases the hardness and softening points of thermoplastic resins, allowing these materials to find applications in drafting films.

Pigments

When added to resins in sufficient concentration, pigments produce matte layers; at low concentration, they simply reduce light reflection and glare by the film surface. Pigments also provide antiblocking properties to the coated layer without impairing its quality.

Pigments used in reproduction films or in drafting films must not absorb much ultraviolet light; they must be sufficiently hard to give pencil abrasive properties to the layer, but not hard enough to cause rapid wear of the pencil lead. They must have a surface area sufficient to give adequate ink absorption in the case of a drafting layer, but not so great as to allow ink line spreading and feathering. Particle size distribution must be as even as possible, with the ultimate particle size not too small (which would adversely affect viscosity) and not too large (which would adversely affect coating appearance).

Pigments must easily be dispersible in the coating solution medium without giving thixotropic conditions; they must have a low iron content for diazo layers.

No single pigment meets all the requirements for all film applications, and it is common to use mixtures of pigments, each contributing some properties to the final product.

Among the pigments most frequently used in film coatings are the natural crystalline and amorphous silicas, the synthetic amorphous silicas, and calcined aluminum silicates. In special applications, such as the production of white opaque film, titanium dioxide is used as an opacifying pigment.

The silicas, either natural or synthetic, are the major pigments used in engineering film coatings. The acceptable range of particle size of silica pigments for diazo or drafting films is between 1 and 10 μm; the best results are obtained with particle sizes of 3 to 5 μm.

Commercial microcrystalline silicas, which are special grades of ground quartz, crystobalite, or tridymite, are available in 5 to 10 μm sizes with a relatively large particle size distribution; the larger size grades need to be further ground (e.g., in a ball mill) to reduce their abrasiveness. The popularity of crystalline silicas, which resulted from their low cost, has decreased considerably as a consequence of growing health consciousness; it is known today that crystalline silicas may cause lung damage (silicosis), and they recently have been classified as probable carcinogens.

Amorphous silicas are considered to be only mild irritants because of their drying properties; they are manufactured in a great variety of grades with different properties. Many commercial grades are available in the right particle size range (3–5 μm), the right surface area range (300–500 m^2/g), and with particular ease of dispersion. It is often found that two or more amorphous silicas with high and low property values give better matte results than a single silica with medium property values.

Diatomaceous earth, or diatomite, is primarily a form of natural amorphous silica that typically contains minor amounts of quartz. It has been suggested for use, with other pigments, in film coatings.

Calcined aluminum silicates, which are calcined clays, have shown interesting properties when used in resin systems for matte layers. They are easily dispersible and give adequate matte surfaces, but they also have a low hardness; they are frequently used in combination with harder silicas.

Titanium dioxide is rarely found in diazo films because of its opaqueness to ultraviolet light; occasionally minute amounts are used to increase the whiteness of the layer. Only in opaque films is titanium dioxide the pigment of choice.

3.7 FORMULATIONS, SOLVENT SYSTEMS

Many types of diazo film and drafting film are offered in the market for different applications, and manufacturers tend to specialize in particular fields, such as engineering graphics, diazo microfilms, or vesicular films. Basically, most diazo films for reproduction purposes aspire to excellence in printing speed, actinic opacity, visual density, development rate, reprinting speed, shelf life, background and dye stability, drafting characteristics, or other less important features.

Diazo films are available in different colors, different printing speeds, different surface finishes, and different gauges.

Since perfect adhesion of any coated layer to the film is expected, the first task of any manufacturer is to ensure that the film of his choice is surface treated for adhesion promotion, before applying a coating. The second task will be to decide whether the diazo layer is to be achieved in a single step or in two separate steps.

In the two-step method, the resins, which will act as binders for the chemicals, are first applied from a solvent solution in a well-defined layer, often called a lacquer or precoat lacquer, and dried. In a second step, the diazos, couplers, and auxiliary chemicals are applied from a solvent or mainly solvent solution on top of the resin layer, with the aim of combining with it after drying to form a single, homogeneous, light-sensitive layer. In the one-step method, the resins, diazos, couplers, and auxiliary chemicals are all mixed together, applied from a solvent solution in one operation to the film, and dried to give the final light-sensitive layer. Such a system is often referred to as one-pot film sensitizing, in the same way as the one-pot coating or pseudo-precoat in diazotype papers.

With the two-coating-layers approach, the formulator has more latitude in the choice of resins and solvents, and need have less concern for solubility problems or compatibility with the diazo chemicals. In addition, variations in the thickness of the precoat lacquer do not affect the functional properties of the sensitizing layer, such as printing speed and density, and greater manufacturing control of each layer is possible. The main disadvantage in this case is the higher cost involved in two applications instead of one.

With the one-coating-layer approach, the formulator has to ensure that common solvents for resins and diazo chemicals are chosen, and that the concentration and thickness of the resin layer remain constant from beginning to end of the coating operation, to avoid variations in print speed and other fundamental properties of the diazo film.

The coating method selected will determine the type of coating equipment to be used; the more sophisticated the coating machine, the easier it is to control the one-step application.

Whatever the coating method, the concentration of resin must be such as to bind perfectly all the other chemicals; in addition, some chemicals in the sensitizer, by a side effect, act as plasticizers for the resin and tend to lower its softening point. Care should be exercised to ensure, by a judicious choice of resins, a nonsticking condition of the surface of the diazo film, at room temperature as well as at the developing temperature. Sticking is promoted by solvent retention, and it is therefore necessary that all the solvents, particularly the high boiling solvents, be eliminated during drying.

To avoid surface tackiness or chalkiness caused by insufficient binding of the chemicals, it is generally accepted that the ratio of resin solids to all other chemical solids be at least 1.4 or 1.5.

For practical reasons, the solvent-coated layer applied must not be too thick or too thin; suitable results have been achieved with wet coating weights of 40 to 60 g/m^2 and dry weights of 8 to 12 g/m^2. Care should be exercised to avoid solution viscosity changes during coating, which would inevitably lead to variations in the amounts applied.

It is common in film coatings to distinguish between glossy layers in which no pigment is used and matte layers in which sufficient pigment is used to render the film visibly matte. Between these two extremes, depending on the ratio of pigment to resin, the surface of the film can have a deglossed, pearly, or semimatte appearance. Diazo films with all these degrees of surface finish are being manufactured. Most of the engineering diazo films have a matte-drafting layer on their backside.

A film with a glossy diazo layer on one side and a matte layer on the other is generally called a single-matte diazo film. A film with a matte diazo layer on one side and a matte layer on the other is known as a double-matte diazo film. A film with a glossy diazo layer on one side and no layer or a glossy layer on the other is generally called a clear diazo film.

When a film has a drafting layer on one side, but no diazo layer on the other side, it is called a drafting film. Frequently, drafting films have drafting layers on both sides.

Drafting films are used by draftsmen, architects, and engineers to produce the original drawing, either by hand or by computer-aided drafting (CAD) equipment. They play an extremely important role in the reprographic industry, and their development has been parallel to that of engineering diazotype films.

Drafting films have to meet stringent and varied requirements because of the many techniques used by those who draw. Ideally, a drafting layer must have all the following properties:

> It must accept lead or plastic pencils of different hardness, from the softest to the hardest, without difficulty and without being damaged.
>
> Pencil lines must be easy to erase without leaving a ghost line.
>
> Pencil line corrections applied on an erased area must be accepted without loss of density.
>
> Pencil lines must appear to be continuous and must retain the same thickness from beginning to end; the surface must not be too abrasive to wear out the pencil point too rapidly, but abrasive enough to cause a dense line.
>
> A drafting layer must accept ink lines from a variety of inks without ink spread or feathering and with rapid ink drying.
>
> Ink lines must have good adhesion, to ensure against removal by frequent handling or lifting by adhesive tape; they must also be easy to erase and to reapply on an erased area without loss of line quality.
>
> A drafting layer must have a high ultraviolet light translucency.
>
> It must have a resistance to water and to certain solvents used for cleaning or degreasing the surface.
>
> A drafting film must be antistatic, to prevent the attraction and retention of dust and to facilitate separation from the copy film during printing.

Frequently, a drafting film that is eminently suitable for pencil work is less suitable for ink work, and it is common to apply a separate ink-receptive layer to a pencil drafting layer for optimum performance.

Most film formulators require an extensive knowledge and expertise outside the normal field of diazotype paper formulations. In addition, special equipment is required for solvent coating, and this explains why only a few of the many diazo paper manufacturers have ventured into the diazo film field.

The different examples of formulations given in this section have been developed and tested in the laboratory of Andrews Paper and Chemical Corporation in the same manner as the aqueous formulations. Some of the chemicals are mentioned by their Andrews code references; a complete explanation of the codes appears in the appendix.

All equipment for solvent coating must be explosion proof and grounded.

3.7.1 Glossy Lacquers

Glossy Film Precoat Lacquer, D4410

Mix with a high-speed stirrer or a homogenizer until completely dissolved

Methyl ethyl ketone	7.5	liters
Ethylene glycol ethyl ether	2.5	liters
Xylene	2.5	liters
Resin SB-10	1	kg
Resin SA-60	1	kg

Recommended dry coating weight is 9 to 11 g/m^2.

Glossy Film Precoat Lacquer, D4414

Mix with a high-speed stirrer or a homogenizer until completely dissolved

Methyl ethyl ketone	3	liters
Methanol	3	liters
Ethylene glycol ethyl ether	3	liters
n-Butyl alcohol	400	cm^3
Resin CP-50	1.35	kg

Dry coating weight is 8 to 12 g/m^2.

Glossy Natural Tracing Paper Precoat Lacquer, D447

Mix with a high-speed stirrer or a homogenizer until completely dissolved

Methyl ethyl ketone	6	liters
Ethylene glycol ethyl ether	3	liters
Isopropyl alcohol	3	liters
Resin CP-50	2	kg
Resin AY	250	g

Dry coating weight is 9 to 11 g/m^2.

3.7.2 Matte Lacquers

Matte Film Precoat Lacquer, D442

Mix with a high-speed turbine mixer or in a ball mill

(A)	Methyl ethyl ketone	1	liter
	Ethylene glycol ethyl ether	1.5	liters
	Resin AY	100	g
	Amorphous silica, 5 μm	400	g
	Pigment 65	200	g

Dissolve with a high-speed stirrer

(B)	Methyl ethyl ketone	2	liters
	Ethylene glycol ethyl ether	2	liters
	Ethyl alcohol	2.5	liters
	Resin CP-50	800	g
	Resin AY	300	g

Add part A to part B and maintain under gentle stirring during usage. Dry coating weight is 9 to 11 g/m^2.

Matte Film Backcoat Lacquer, D2175 (Drafting Matte)

Mix with a homogenizer or any size-reducing equipment

Ethylene glycol ethyl ether	1	liter
Isopropyl alchohol	1	liter
Methyl ethyl ketone	0.5	liter
Resin CP-50	100	g
Amorphous silica, 9 μm	200	g
Amorphous silica, 4 μm	400	g

Dissolve with a high-speed stirrer

Methyl ethyl ketone	2	liters
Ethylene glycol ethyl ether	3	liters
Ethylene glycol butyl ether	0.5	liter
Isopropyl alcohol	2	liters
Resin SA-60	1	g
Proresin C8000	300	g
p-Toluenesulfonic acid	30	g

Maintain under gentle stirring during usage. Dry coating weight is 9 to 12 g/m^2

3.7.3 Sensitizers

Sepialine Sensitizer for Glossy and Matte Lacquers, D4011
Mix with a high-speed stirrer for 30 minutes

Isopropyl alcohol	6	liters
Ethylene glycol ethyl ether	2.5	liters
Methanol	1.5	liters
Citric acid	100	g
Thiourea	50	g
50% Stannic chloride solution	100	cm^3
Chlororesorcinol	150	g
Coupler 950	100	g
Coupler RX	100	g
Diazo 55	280	g

Blackline Sensitizer for Glossy and Matte Lacquers, D408
Mix with a high-speed stirrer for 30 minutes

Ethyl alcohol	3	liters
Methanol	3	liters
Ethylene glycol ethyl ether	3	liters
Methyl ethyl ketone	1	liter
Acetic acid (glacial)	150	cm^3
Tartaric acid	300	g
Sulfosalicylic acid	50	g
Thiourea	100	g
Diresorcyl sulfide	20	g
Coupler 122	60	g
Coupler 603	100	g
Coupler 670	200	g
Diazo 88	250	g
Zinc chloride	100	g

One-Pot Glossy Sepia Film Sensitzer, D401
Mix with a high-speed stirrer or homogenizer until completely dissolved

Methyl ethyl ketone	4	liters
Methanol	4	liters
Ethylene glycol ethyl ether	2	liters

Formic acid	100	cm³
p-Toluenesulfonic acid	100	g
Thiourea	50	g
Resorcinol	180	g
Diazo 60	160	g
Zinc chloride	70	g
0.5% Methyl violet solution	10	cm³
Resin CP-50	900	g

Dry coating weight is 9 to 11 g/m².

One-Pot Matte Sepia Film Sensitizer, D4031

Mix with a high-speed turbine mixer

(A)	Methyl ethyl ketone	2.5	liters
	Ethylene glycol ethyl ether	0.5	liter
	Methanol	1	liter
	Pigment 65	500	g
	Resin CP-50	80	g

Mix separately with a high-speed turbine stirrer

(B)	Methyl ethyl ketone	2.5	liters
	Ethylene glycol ethyl ether	1.2	liters
	Methanol	1.5	liters
	Ethylene glycol butyl ether	300	cm³
	Formic acid	100	cm³
	Sulfosalicylic acid	20	g
	Solubilizer HI	40	g
	Coupler 615	70	g
	Chlororesorcinol	70	g
	Coupler 950	40	g
	Diazo 55	160	g
	Zinc chloride	75	g
	0.5% Methyl violet solution	10	cm³
	Resin CP-50	720	g

Add part A to part B and maintain under gentle stirring during usage. Dry coating weight is 9 to 11 g/m².

One-Pot Glossy Blackline Film Sensitizer, D4025

Mix with a high-speed stirrer or homogenizer

Methyl ethyl ketone	4	liters
Methanol	4	liters
Ethylene glycol ethyl ether	2	liters
Citric acid	80	g
p-Toluenesulfonic acid	25	g
Thiourea	70	g
Solubilizer HI	50	g
Coupler 950	150	g
Coupler 1134	35	g

Coupler 122	30	g
Coupler 615	80	g
Diazo 69	160	g
Zinc chloride	50	g
0.5% Methyl violet solution	10	cm^3
Resin SA-60	250	g
Resin CP-60	750	g

Dry coating weight is 9 to 11 g/m^2.

One-Pot Matte Blackline Film Sensitizer, D4032
Mix with a high-speed turbine stirrer

(A) Methyl ethyl ketone	2.5	liters
Ethylene glycol ethyl ether	0.5	liter
Methanol	1	liter
Pigment 65	500	g
Resin CP-50	80	g

Mix separately with a high-speed turbine stirrer

(B) Methyl ethyl ketone	2.5	liters
Ethylene glycol ethyl ether	1.2	liters
Methanol	1.5	liters
Ethylene glycol butyl ether	300	cm^3
Sulfosalicylic acid	120	g
Citric acid	40	g
Thiourea	80	g
Coupler 660	80	g
Coupler 640	20	g
Coupler 1134	36	g
Diazo 54 (ZnCl$_2$ salt)	120	g
Zinc chloride	30	g
0.5% Methyl violet solution	20	cm^3
Resin CP-50	720	g

Add part A to part B and maintain under gentle stirring during usage. Dry coating weight is 9 to 11 g/m^2.

One-Pot Glossy Sensitizers for Color Films, D2698
Mix with a high-speed turbine stirrer

(A) Methyl ethyl ketone	3.5	liters
Ethylene glycol ethyl ether	1.7	liters
Methanol	0.5	liter
Resin CP-50	700	g

Mix separately with a high-speed stirrer

(B.1) For magenta color		
Methyl ethyl ketone	1.5	liters

	Methanol	3	liters
	p-Toluenesulfonic acid	50	g
	Coupler 120	25	g
	Diazo 78	100	g
(B.2)	For cyan color		
	Methyl ethyl ketone	1.5	liters
	Methanol	3	liters
	p-Toluenesulfonic acid	20	g
	Citric acid	20	g
	Solubilizer HI	40	g
	Coupler 1167	40	g
	Diazo 39	40	g
(B.3)	Zinc chloride	20	g
	For yellow color		
	Methyl ethyl ketone	1.5	liters
	Methanol	3	liters
	p-Toluenesulfonic acid	50	g
	Coupler 601	100	g
	Diazo 76	100	g
	50% Stannic chloride solution	20	cm^3

Add part A to parts B.1, B.2, or B.3. Dry coating weight is 7 to 9 g/m^2.

3.8 DIAZO PAPER AND FILM MANUFACTURE

In the manufacture of diazotype materials, the initial step consists in selecting the appropriate formulation and the corresponding necessary raw materials. The next step involves mixing the chemicals according to the formulation, thus preparing the coating solutions, which are applied, with the use of a coating machine, to the chosen substrate. After coating, the resulting diazo material is converted from its mill-size roll into consumer-size rolls or sheets, which are stored for shipping to users.

Although the same sequence of activities applies to the manufacture of both diazo paper and diazo film, there are major differences in the equipment used and the handling conditions.

Diazo paper is a lower cost product with wider quality tolerances than diazo film. In the case of paper, the fibrous nature of the substrate minimizes coating irregularities, whereas with clear film the slightest coating fault becomes obvious and is generally unacceptable. For these reasons, more sophisticated equipment is needed for mixing and for applying solutions for diazo films and, since most of them are produced with flammable solvent systems, all equipment used must meet rigorous explosion hazard standards. In addition, the cleanliness requirements in film coating make it imperative to have a low-dust or dust-free environment.

Whatever the type of material produced, a diazo coating operation must have a plant layout that allows the raw materials to move conveniently to the unwind end of the coating machine and the coated products to be taken from the windup end of the machine to the finishing room, without crisscrossing or backtracking. A layout example for a two-machine coating plant for diazotype paper is shown in Fig. 3.4.

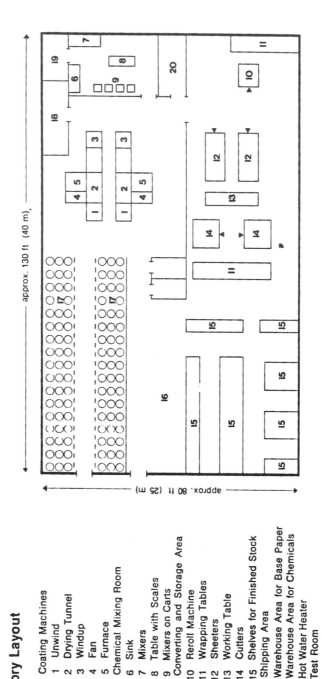

Factory Layout

1-5 Coating Machines
  1 Unwind
  2 Drying Tunnel
  3 Windup
  4 Fan
  5 Furnace
6-9 Chemical Mixing Room
  6 Sink
  7 Mixers
  8 Table with Scales
  9 Mixers on Carts
10-15 Converting and Storage Area
  10 Reroll Machine
  11 Wrapping Tables
  12 Sheeters
  13 Working Table
  14 Cutters
  15 Shelves for Finished Stock
  16 Shipping Area
  17 Warehouse Area for Base Paper
  18 Warehouse Area for Chemicals
  19 Hot Water Heater
  20 Test Room

Figure 3.4 Layout for a two-machine diazotype coating plant.

3.8.1 Coating Equipment

At the heart of the manufacture of diazotype materials is the coating machine.

Aqueous Coating Machine

A coating machine is composed of different sections, each designed to perform a specific task. These sections are

> Unwind stand
> Precoat station
> Sensitizing station
> Backcoat station
> Drying tunnels after each coating application
> Windup stand

The diazo coating operation consists in placing a mill roll of diazo base paper on the unwind stand, threading the web through the machine, then applying to one side of the moving web, with a rotating applicator roller that is partly immersed in the coating preparation, an excess of precoat solution; after a short period of imbibition, the excess coating is removed with an air knife or other metering device. The operation proceeds with the elimination of the remaining water in the paper by passage through a drying tunnel in which the wet paper is subjected to a hot airstream. The web of precoated paper emerges from the dryer and meets the second coating station, in which an applicator roller applies, on top of the dry precoat layer, the sensitizing solution, followed by metering and drying, in the same manner as described for the precoat. Upon leaving the drying tunnel, the dry coated paper meets, face up, a third coating station; here the backcoat is applied to the backside of the material and dried. As the coated web leaves the drying tunnel for the last time, it may be cooled by contact with a water-cooled roller and wound up on the rewind stand.

Variations on the procedure above are frequent. On some machines, for example, the coating sequence is altered; the backcoat is applied after the precoat and before the sensitizing. Experience has shown, however, that better curl control is achieved when backcoating is done last. Occasionally, only two coatings are applied, the first being a combination of the precoat and the sensitizer (pseudo-precoat), the second the backcoat. This is often necessary when the coating machine has only two coating stations.

In designing and building coating machines, a balance must be struck between efficiency, automation, and cost. Aqueous coaters are made in different widths to handle mill rolls of paper from 90 cm up to 2 meters wide; the machines also run at coating speeds varying between 1800 and 9000 m/h, with the majority running at 4000 to 5000 m/h.

For a schematic description of the coating sequence of a commercial diazo aqueous coater, see Fig. 3.5. The roll of paper as received from the mill is secured on a metal shaft and mounted in the unwind stand; the shaft may be a simple bar with metal chucks or collars holding the roll in position or a more elaborate air shaft. The unwind stand may be of a single or double fixed position style; with such unwinds, the machine must be brought to a standstill or slowed down considerably to change rolls; manual splicing is frequently done at slow coating speed on a double-shaft unwind stand. For automatic flying splice at full machine speed, a two-position turret unwind is recommended. Tension of the web during unwinding of the roll is maintained by use of a manual, hydraulic, or electromagnetic brake on the unwind shaft. Too little tension reveals itself by a slack in

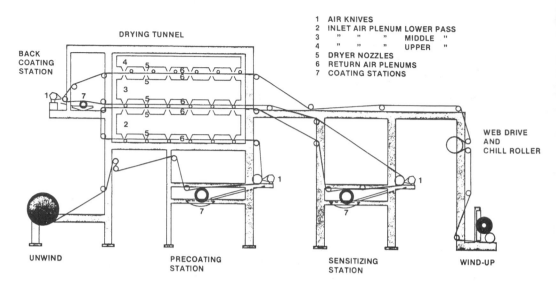

Figure 3.5 Coating sequence of a commercial diazo aqueous coater.

the web, while too much tension causes a taut web with eventual creases and possible break.

Each of the coating stations in an aqueous coating machine is composed of the following parts:

Applicator roller with web contact control
Coating pan, drip pan, and recirculation system
Air knife with backing roller

The applicator roller picks up the coating preparation from the coating pan and transfers it, by contact, to the traveling web. To achieve an even contact along the entire width of the applicator roller, the web is held down by adjustable drop rollers kept absolutely parallel with it.

Large-diameter rubber-covered coating rollers of 8 to 10 in are preferred because they minimize the risk of solution penetration into the base paper and reduce the need for high revolution speeds, which could cause foam generation and coating defects. The variable speed applicator roller should be able to rotate with or against the web direction, depending on which way gives the best result with a particular type of coating.

The coating pan receives the solution to be applied, and its design has to meet a number of requirements such as

Prevention of dead corners.
Provision for solution movement at the bottom of the pan to minimize settling in pigmented coatings.
Separation of the runback solution from the air knife to minimize foam generation.
Feeding the fresh solution from the bottom of one end of the pan and overflowing both in front and back of the coating roller at the other end of the pan.

The coating solution is continuously circulated between a reservoir tank and the coating pan, with the aid of a recirculation pump. A larger holding tank feeds the reservoir. The coating solution needs to be constantly adjusted for concentration and viscosity. The reason behind this is simple. In the coating pan, as the coating proceeds, the solution is enriched because of evaporation of water and mixing with the more concentrated runback from the air knife. If no countermeasures were taken, the solution in the entire recirculation system would become more and more concentrated. Therefore a diluted solution is fed from the holding tank into the reservoir tank. The section between the applicator roller and the air knife is called the runback zone, and as the web travels through this section, the coating solution diffuses partially into the base paper, while the runback solution from the air knife mixes with fresh solution. The imbibition period, which is the time the web remains in this section, is important because many of the properties of the finished material are influenced by this parameter. Imbibition periods of 1 to 2 seconds are in general satisfactory.

The excess solution applied by the applicator roller is metered by the air knife, which controls the wet coating weight left on the paper after the excess has been metered off. This wet coating weight is a function of the air knife pressure, the viscosity of the coating preparation, the web speed, and the absorptivity of the paper. Other factors, such as air knife angle, distance between air knife and web, and lip opening, affect the metering action. The aerodynamic design of the air knife lips provides for a nonturbulent airflow evenly across the total length of the air knife. A backing roller, generally Teflon coated to facilitate cleaning, keeps the web at a controlled angle and distance from the air knife lips.

After each coating station, the wet web enters the drying tunnel at one end and emerges dry at the other end.

In the process, dry hot air is passed through the tunnel, transferring part of its energy to the web, causing temperature rise and evaporation of water and/or other solvent, and removing the vapors from the drying web.

In modern coating machines, the drying efficiency is maximized by recycling the dryer air four or five times, raising its dynamic energy through powerful blowers to linear velocities of 5000 ft/min, and regenerating its calorific capacity by passing the air after each cycle through a direct-fired modulating gas furnace.

To accomplish this objective, the drying tunnel is equipped with plenum chambers to achieve even dryer air distribution across the web and to force the air through a series of spaced nozzles. The hot air jets impinge at short distances on the wet web surface, transfer caloric energy, and remove evaporated moisture from the web. Part of this moisture-laden air is then drawn into an exhaust stack by an extraction fan; the remaining part passes through the heater and is then recirculated into the plenum chamber.

The drying performance of the oven is governed by the velocity, volume, and temperature of the heated air. The temperature setting itself depends on a number of factors, such as ambient temperature and humidity in the coating room, width and basis weight of the web, linear speed of the web, and wet coating weight of the paper. The diazo paper is in general dried to a predetermined moisture content for optimum shelf life; such a moisture content averages 3 to 4% for two-component diazo papers, and 2.5 to 3% for one-component papers. Underdrying can cause poor shelf life and ''pick-off'' of the coating, while overdrying can cause diazo decomposition, loss of print contrast, and paper brittleness.

When the finished coated paper leaves the drying tunnel, it may be cooled by contact with a large-diameter water-cooled roller, which often doubles as a drive roller.

The paper is finally rewound in the windup stand on a cardboard core mounted on a metal shaft, of the same design as the unwind shaft. The windup can include single or double fixed position stands, single-drum or double-drum surface units, or two-position center wind turrets, with or without automatic splicing. Many aqueous coating machines have supplementary features designed to improve quality, efficiency, and automation. Such features are in general very costly and are justified only when very large production volumes are possible. Among these features are automatic moisture control and web guide control systems.

Solvent Coating Machine

A diazo solvent coating machine is in many ways similar to a diazo aqueous coating machine, with unwind and rewind stands, coating stations, and a drying tunnel. However, each of these sections is specially designed for the handling of polyester film and of solvent solutions.

The unwind and rewind can be of the single or dual type, but in every case, center-winding is used, with fine tension control. Unlike paper, polyester film under excessive tension does not break and can cause serious equipment damage.

Probably the most critical part of a solvent coater is the coating station and in particular the coating head. Figure 3.6 shows a selection of coating heads for diazotype layers: 1, a kiss coating roller with scraper bar metering; 2, a kiss coating roller with air knife metering; and 3, a web feeder table with lateral flow and air knife metering. All three methods have been used for coating diazo paper, but the method of Figure 3.6(2) dominates the industry.

Solvent resin solutions are generally viscous and, polyester film being impervious, any amount of solution applied remains on the surface. Under these conditions, air knife metering is not suitable, and other ways of controlling the layer thickness become neces-

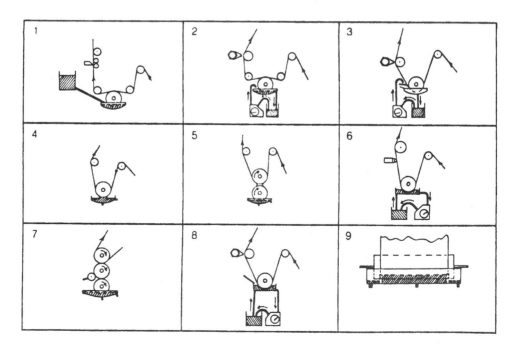

Figure 3.6   Coating heads for diazotype layers.

sary. One way, applicable for slow coating speeds, consists in creating a meniscus between the web and the solvent solution: as the web travels, it carries some of the solution with it. On the web, an equilibrium is soon reached between the fresh solution applied and the gravity flowback; no metering takes place after application, and the coating weight applied is controlled by the web speed as well as the solid content and viscosity of the coating solution. Figures 3.6(4) and 3.6(5) illustrate this type of application.

Increased production speed requirements have led to the introduction of wire wound bar and reverse roller coating techniques, respectively, Figs. 3.6(6) and 3.6(7).

In the wire wound bar method, the solvent solution is applied to the film web with an applicator roller dipping in it; the excess solution is subsequently metered with a rotating wire wound rod in close contact with the film. The amount of solution left on the film after metering is the amount that is allowed to pass the rod through the interstices or grooves between the tightly wound wire around the rod. A section of wire wound rod is illustrated in Fig. 3.7.

The coating weight of applied solution is controlled by the diameter of the wire used. The wire wound rods, often called Mayer rods or Mayer bars, are generally made of stainless steel and vary in diameter between $1/4$ and 1 in; the wire, also made of stainless steel, varies in gauge between 0.01 and 0.05 in. To ensure perfect contact with the full width of the web, the Mayer rod is supported by a rigid, low friction cradle, generally Teflon covered; the rod is rotated very slowly, with or against the web, to avoid wear in one spot and to allow any obstruction accidentally lodged in a groove of the rod to be freed.

The wire wound rod metering technique is frequently used in the production of glossy and matte diazo engineering films with viscous resin solutions. However, with low-viscosity solutions, as in the case of a diazo sensitizer containing no resin, the air knife metering method, shown in Fig. 3.6(8), is preferred.

For the production of microfilm, the use of a Mayer rod is not recommended because the film could be scratched by the wire pressing against the moving web. To avoid any physical damage of the film, a different coating technique is utilized. The solvent solution is metered onto the application roller before it is applied to the film; the metering is done by a multiplicity of rollers, accurately distanced from each other, transferring a controlled amount of liquid to one another and ultimately to the web. These coating techniques are called reverse roller coating, and a reverse three-roll system is illustrated in Figure 3.6(7).

Most solvent coating machines have two coating stations, one for each side of the

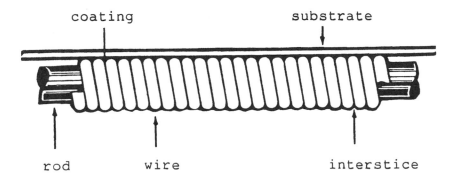

Figure 3.7   Section of a wire wound rod.

film or paper. Machines with only one station need two separate passes to produce a material that is coated front and back.

To achieve good coating results, it is important to maintain solution homogeneity by constant circulation, and concentration constancy by viscosity adjustments, during the course of the operation. Filtering the coating solution through cartridge filters is beneficial, especially in the production of microfilm. Drying conditions for solvent coating are also different from those for aqueous coating, since many solvents have a lower boiling point and a higher evaporation rate than water. The solvents must be eliminated during drying without disturbing the resin layer; impinging the wet web with high velocity hot air is not recommended because it is likely to evaporate the low-boiling solvents too quickly, causing surface defects. Larger volumes of air at lower velocity are preferred, in particular in the first drying phase. If resin curing or cross-linking is required after the solvent has evaporated, sufficient energy, in the form of convection or radiation heat, is supplied to the layer, taking care however that the transition temperature of the polyester film is not reached.

Substantial fire hazards and explosion risks are associated with the use of flammable solvents. It is imperative to prevent the buildup of static electricity and to eliminate it before it can cause an electric discharge near a flammable liquid or solvent vapor environment. The grounding of equipment and the fitting of antistatic devices in strategic positions is of critical importance in solvent coating operations.

Converting Equipment

Diazo coated materials are usually wound on the rewind stand in jumbo rolls as large as the original rolls; very occasionally they are wound on the coating machine into smaller rolls. For sale to consumers, the jumbo rolls must be converted into customer-size rolls or sheets.

Converting is an important part of the manufacturing process, and the various machines needed (rewinders, sheeters, slitters, cutters, prefolders, punchers, wrappers, etc.) account for a substantial part of the investment for a coating plant.

During converting, the diazo materials are exposed to the atmospheric environment for a longer period of time than during coating. For this reason, the areas of converting must be dry and cool, and not exposed to sun or other actinic light. Preferably, the areas should be air conditioned, dust free, and temperature and humidity controlled.

Rewinding or rerolling machines produce, from jumbo rolls, small rolls varying in size between a few meters and a few hundred meters. Rolls of 20 to 50 meters or yards are the most popular for use on manual operating photoprinting machines; larger rolls of 300 to 500 meters or yards are produced for automatic feeding photoprinting equipment.

Manually operated rewinders are used for a small volume output of rolls, such as a few hundred rolls per day, but for volume production, automatic rewinding machines can produce, nonstop, a few thousand rolls of diazo paper per day.

Diazo film is generally rewound on a cardboard core at a slower speed than diazo paper, to permit one more careful inspection for coating defects.

Sometimes rewinding into customer rolls is accompanied by an edge trimming operation, which produces rolls to an exact predetermined width. More often, the trimming is conducted separately by simultaneously rewinding and edge slitting the entire jumbo roll; this is particularly necessary when diazo film is not coated to the full width of the film, but is left with a few millimeters uncoated at both edges of the film.

Diazo microfilm is converted into narrow width rolls in one operation on slitter/rewinder machines, which can be semiautomatic or fully automatic.

A very large proportion of diazo material is supplied to users in sheet form. The material is first cut from jumbo rolls, by means of a sheeter, into large sheets of the same width as the coated roll, and stacked to piles of 100, 250, or 500 sheets. Then, with the aid of a guillotine cutter, the cut sheets are squared, trimmed, and cut to the required size in stacks.

In view of the infinite number of possible cut sheet sizes, standardizing efforts have led in Europe to the standard of the Deutsche Institut für Normung (DIN) for cut sheets; this German standard, which is followed in most countries of the world, excluding the United States, is based on a sheet size of 1 m^2, being the area of a sheet 841 mm $\times$ 1189 mm; all succeeding sizes are half the larger sheet size, as shown in Table 3.3.

The American standard sheets for diazotype papers given in Federal Standard 00131E lists the following sizes, in inches: 8 $\times$ 10½, 8½ $\times$ 11, 8 $\times$ 13, 8½ $\times$ 14, 11 $\times$ 17, 17 $\times$ 22, 18 $\times$ 24, 22 $\times$ 24, 22 $\times$ 34, 23 $\times$ 25, 24 $\times$ 30, 24 $\times$ 36, 24 $\times$ 40, 28 $\times$ 40, and 30 $\times$ 42. Other sizes are also supplied on request.

For some special applications, diazo paper is used prefolded in continuous lengths; such a format is produced with specialized prefolding machines.

After being converted, the sensitized diazo materials must be protected against ultraviolet light and outside moisture. Wrapping materials opaque to actinic light and impermeable to moisture must be used. While formerly, multiple wrappers consisting of an inner liner and outer light and moisture barriers were used, today, single wrappers, either laminated or all plastic, are preferred. Black, red, or green thick pigmented polyethylene wrappers are popular and are used with manual or automatic wrapping machines.

3.8.2 Quality Control and Test Methods

Maintaining the quality standard of diazotype materials during manufacture is of primary importance, and systematic quality control must be performed in the course of each coating. Frequently, tests are made at the beginning and at the end of each mill roll to permit, if necessary, corrective measures before a new roll is coated. Some properties can be checked very rapidly, and the results allow operators to make immediate adjustments in the coating, while other properties require lengthy testing procedures, and the results cannot be used to make changes during the coating operation. Testing the first group of properties

Table 3.3 International Standards for Paper Sheet Sizes

DIN Number	Size (mm^2)
A0	841 $\times$ 1189
A1	594 $\times$ 841
A2	420 $\times$ 594
A3	297 $\times$ 420
A4	210 $\times$ 297
A5	148 $\times$ 210
A6	105 $\times$ 148

forms the basis of a comprehensive quality control procedure. The following immediate tests are routinely performed:

General Coating Quality

Using two large strips over the full width of the web, visual inspection of the coated material indicates obvious defects such as splashes, scratches, or holes. One strip is developed after being completely or partially burnt out by light exposure, and the other is developed fully without exposure; both are then checked for imperfections such as white or black spots, dark or light streaks, mottle, and uneven density. The cause for each imperfection must be found and eliminated.

Black or dark spots are often caused by pinhole penetration of the sensitizer into the base paper, by insoluble particles in the solution, or by tar formed through solution incompatibility. Better paper sizing, solution filtration, and the addition of solubilizers must be considered to tackle this type of defect.

White spots are generally caused by lack of wetting of the paper surface and local repellency; spreading or wetting agents, finely dispersed silica, and polyvinyl acetate resin additions often correct this type of defect.

Streaks are frequently the result of a dirty air knife or of foam in the solution; frequent cleaning of the air knife lips and use of antifoam agents usually solve these problems.

Producing visually defect-free coatings is every diazo coating operator's goal, and this requires a great deal of vigilance and experience.

Printing Speed

Using a reliable photoprinting machine and a control chart, a test print is made on a tear-out of coated material. Control charts, such as the Andrews Reproduction Control Chart, shown in Fig. 3.8, or the Kodak Projection Print Scale, are transparent film positives having a gray scale and other test designs.

A reference sample material is printed at the same exposure setting on the photoprinting machine and with the same control chart as the tear-out material. A comparison of the gray scale or step wedge on the two prints allow a judgment whether the newly coated material is faster, equal, or slower printing than the reference material. If faster or slower, corrective measures can be taken immediately by increasing or decreasing the concentration of diazo in the sensitizing solution. This is frequently done by adding a more or less concentrated diazo solution to the coating solution until a suitable match of the step wedge is achieved. A stock of reference material, made previously from a coating with correct printing speed, is usually kept in a refrigerator, when not in use, to reduce any potential speed loss on aging.

Primary printing speed standards are set by sensitometric measurements on calibrated and drift-free exposure instruments.

Color and Shade

The determination of the correct full tone and halftone color and shade is mostly done visually, on a print, against a control material; divergence from expected azo dye color is frequently an indication of an error in the making of the sensitizing solution, and corrective action can quickly be taken. A shift in the shade of a blackline material is often the consequence of an imbalance in the coupler ratios, taking place during continuous coating;

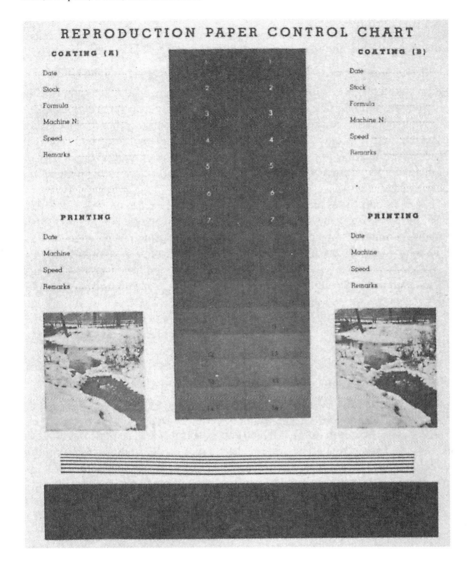

Figure 3.8 Reproduction control chart.

this is remedied by a readjustment of the coupler concentrations through addition of a certain amount of the coupler that has been most depleted.

Moisture Content

The final moisture content of a diazo paper has a great influence on its keeping qualities and its development rate. The moisture content must be kept within a very narrow range. Moisture reading or recording instruments allow a reliable measure of the moisture content either continuously during coating or by taking a measurement on a tear-out. If the moisture value found is too high, the drying temperature of the coating machine must be increased or its running speed must be decreased; the opposite action is taken if the moisture value is too low.

Density and Background

Maximum dye density and minimum background density of a print are both measured with a reflection densitometer; the values found must fall within the upper and lower values set for this type of material. Frequently, a visual assessment against a control material is sufficient to make a correct judgment. The density evaluation must be made under consideration of the printing speed, as a low density is often the result of a high printing speed; correcting one will immediately affect the other. Sometimes, a low dye density at the correct printing speed is the result of insufficient precoat application, which can easily be corrected by adjusting the precoat concentration. Minimum background density can be affected by overdrying, as the result of diazo decomposition, or occasionally by solution precoupling. A high background density can also be the result of iron contamination in the paper in water or in one of the chemicals, and all must be checked.

Rub-Off

Coatings that are poorly bound to the base paper are subject to powdering or rub-off. A simple way of testing the degree of rub-off, under either cold or hot conditions, is to rub a white paper towel or similar material, with a constant pressure, over the surface length of 20 to 30 cm of a fully developed sheet of diazo paper and observe on the towel the amount of dye transferred. Well-bound coatings give practically no removal of dye. To reduce rub-off, an increased amount of binding resin should be added to the precoat or to the sensitizer.

Coating Adhesion

The coated layer adhesion to the substrate must be excellent, especially in the case of film coatings. To test adhesion, the material is first fully developed (this is not required when testing a drafting film) and the layer is crisscrossed with a razor blade; this is done carefully, to prevent any cuts in the base itself. Then an adhesive tape is placed on the cut area and pulled away in a fast upward movement—no part of the layer must be removed and transferred to the adhesive tape. Poor adhesion is a serious defect, which might be caused by inadequate adhesion promotion treatment of the film base surface, and corrective measures are necessary.

Coating Weight

Coating weight is routinely tested in the case of film coatings, but not in the case of diazo papers. The amount of solids applied affects the final properties of the material as well as the cost, and it should be kept within a narrow range.

An uncoated and a coated sheet of film of specific size are weighed, and the weight of dry coating per unit area determined. If below standard, a thicker layer is applied, and vice versa; attention should be paid in this instance to printing speed and density, because these properties are affected by layer thickness.

Solvent Resistance of Cured Layers

To ensure full cross-linking in a film drafting layer, the resistance of the surface to attack by solvent is tested. A cotton swab moistened with methyl ethyl ketone is rubbed back and forth a number of times on the surface of the matte drafting layer; if cross-linking of the resins is complete, no removal of the layer is observed; if not, the drying temperature or the catalyst content might have to be increased, or the coating speed reduced.

Curl

An undesirable curl can lead to difficulties in print processing or in the use of the material. Curl, or curvature of the sheet, can be temporary—for instance, if disappearing within a brief time after emerging from the developing section of a machine—or permanent, if present at all times after equilibrium with the surroundings has been achieved. Curl is assessed by placing a sheet on flat impervious surface and, after equilibrium, measuring the greatest lift from the flat surface of any part of the sheet. Curl in paper coating is corrected by varying the amount or nature of the backcoat.

Miscellaneous Tests

The following tests are performed to evaluate other properties of the material. These properties can rarely be altered while the material is being made, but the test results can be most useful in changing the design of the product by reformulation of the coating solution.

Development Rate

Photoprinting machines have synchronized printing and developing sections; therefore, the diazo material must have a good development rate to develop at the same setting as its printing. Development rate is assessed in a comparative way by a test against a control material.

A whole sheet of sensitized material is developed at a speed many times its normal printing speed; then, while most of the sheet is protected by a light-absorbing cover sheet, a section is exposed to light to decompose any unreacted diazo. Thereupon, the sheet is developed again at the same fast speed as before, and a second section is similarly exposed to light, and so on. The result is a gradated print showing the effects of being developed once, twice, and more times, until no further gradation is noticeable at full development; the material with the densest gradation has the fastest development rate. This test can be performed on a machine set with hot or cold developing conditions, to give information on the hot or cold development rate.

Actinic Opacity

This property applies only to diazo paper or film intermediates. The actinic opacity determines the effectiveness of the reproduction material to produce dense images in reprints. To measure actinic opacity, a sheet of diazo intermediate material is fully developed; the developed print is half-covered with a light-absorbing cover sheet, printed on a diazotype paper, preferably slow speed blackline, at the printing speed of this paper, and fully developed. The covered part of the intermediate gives the maximum density obtainable on the print paper, while the uncovered part gives the density obtainable with the intermediate. Optical densities of the two areas are measured with a reflection densitometer, and the actinic opacity is determined by the percent ratio of reprint density and maximum density values.

Actinic Transmission

This property applies only to translucent or transparent materials. The actinic transmission measures the efficiency with which actinic light passes through a material. The reprint speed of a translucent material depends on its actinic transmission. A photoprinting machine, with a dial calibrated in linear speed, is used to measure actinic transmission. In a first step, a sheet of diazo paper, preferably slow speed blackline, is exposed and developed on the photoprinting machine, at a speed setting such as to leave a background density of 0.05 above minimum density; this is represented by a slight dye haze in the background. The linear print speed used is recorded. In a second step, a sheet of translucent material (e.g., a drafting material or a completely exposed and developed intermediate material) is printed on a sheet of the same slow speed blackline diazo paper as before, with the photoprinting machine at various speeds, until a speed setting is established that leaves after development a background density of 0.05 above minimum density. The linear print speed used is also recorded. The actinic transmission is the percent ratio of the linear reprint speed of the translucent material and the linear print speed of the diazo paper.

Accelerated Aging Tests

These tests are designed to give an estimate of the period of time during which a diazotype material can produce satisfactory results. This period of time, also referred to as the shelf life, can be determined by forced aging of the sensitized material, to simulate within a relatively short time the effect of a much longer period of natural aging. The shelf life is evaluated by subjecting samples of sensitized material to an atmosphere of high humidity and/or an elevated temperature for a specific time.

Natural aging over several months (5–6 months at least) can be approximated by suspending a sheet of sensitized material in a closed container maintained at a relative humidity of 43% and a constant temperature of 50°C for a period of 24 hours; a saturated solution of potassium carbonate salt in contact with excess salt, in a closed container, will produce above it, at a temperature of 50°C, a relative humidity of 43%. Similar results would be obtained if a packet of sensitized sheets were wrapped in aluminum foil, sealed, and kept in an oven at 50°C for 7 days. A longer natural aging period (6–9 months at least) can be approximated by changing the saturated salt solution to one consisting of sodium chloride, which would give an atmosphere of 75% relative humidity at a temperature of 50°C.

A comparative assessment of the stability of diazo materials to general deterioration with age can be made simply and quickly by sealing sheets of material in aluminum foil and heating them at 80°C for 1 hour; this test, however, shows less correlation with normal aging at room temperature and should only be used when a fast evaporation is needed.

The effects of natural or accelerated aging on sensitized materials are partial decomposition of the diazo and/or premature formation of azo dye, which would give prints with lower maximum dye densities and higher minimum background densities; values for these properties can be measured before and after aging, and recorded for comparison purpose.

Light Aging

When diazo prints are exposed to light and air, the azo dye image tends to fade and the background tends to discolor. To determine the extent of loss of image density and increase in background color, light aging tests are performed.

For reproducible results, a fadeometer instrument, which emits a controlled amount of ultraviolet light, must be used.

Two sheets of diazo material are half-covered by an opaque material, exposed, and developed. One of them is then placed in the fadeometer and, after a given period of time, the maximum and minimum densities are taken and compared to those of the other sheet that has not been exposed to light.

Since the sun is a rich source of ultraviolet light, it can be used instead of a fadeometer, to make comparative light aging tests on different diazo prints.

In addition to the above-mentioned tests performed routinely or otherwise on diazotype materials, many other properties can also be tested. Among these are water fastness for diazo papers, ink and pencil acceptance and erasure for drafting materials, chemical and physical eradication for intermediate materials, sensitometric characteristics for microfilms and vesicular films, static retention for all film products, and all physical properties of the base materials themselves.

3.9 CONCLUDING REMARKS

Diazotype materials have been in existence for more than half a century, and current Western production must be in excess of 1 billion square meters per year.

What is the future of the diazotype process?

In microfilm duplication, the diazo process is being challenged by electrophotographic imaging systems.

Until not so long ago, plain paper copiers existed only for the duplication of office documents; in the past few years, larger plain paper copiers, capable of reproducing large-size originals, have appeared and are replacing some diazo photoprinting machines. However, the higher costs of plain paper copiers and a certain lack of versatility has so far limited their acceptance in the marketplace.

The future of the diazo process is also being threatened by operational limitations imposed by environmental regulations. Restrictions on usage and disposal of chemicals considered to be hazardous (e.g., heavy metals, solvents, many organic and inorganic compounds) are making the manufacture of diazo materials more difficult and more costly. Sacrifices in quality and in the range of materials available will follow the eventual replacement of hazardous chemicals by nonhazardous ones.

Nevertheless, and as long as it remains economically competitive with other processes, the diazo process, because of its simplicity and versatility, is likely to remain popular and to be widely used for many years to come.

APPENDIX

Andrews code	Chemical description
Accelerator LM	Dimethylurea
Accelerator ST	Zinc methane sulfonate composite solution
Antifoam A	Proprietary mineral oil antifoam agent
Antifoam L	Proprietary acetylenic glycol antifoam agent
Antifoam M	Proprietary alkyl phosphate antifoam agent
Antifoam T	Proprietary fatty acids antifoam agent
Binder C	Casein
Binder IQ	Vegetal protein
Coating Aid 200	Hydrous magnesium silicate
Coupler 111	2,3-Dihydroxynaphthalene-6-sulfonic acid sodium salt
Coupler 120	2-Hydroxynaphthalene-3-carboxylic acid methyl ester
Coupler 122	2-Hydroxynaphthalene-3-carboxylic acid ethanolamide
Coupler 144	2-Hydroxynaphthalene-3-carboxylic acid-3'-*N*-morpholino propylamide
Coupler 166	2-Hydroxynaphthalene-3-carboxylicacid diethanolamide
Coupler 195	Proprietary composite blue coupler
Coupler 375	4-Bromo-α-resorcylic acid
Coupler 480	4-Bromo-α-resorcylic acid amide

APPENDIX (Continued)

Andrews code	Chemical description
Coupler 603	2,5-Dimethyl-4-morpholinomethyl phenol
Coupler 615	Bis(2-1,dimethylethyl)5-methyl 4-hydroxyphenyl thioether
Coupler 620	Proprietary composite yellow coupler
Coupler 640	3,3′-Methylene bis(acetoacetanilide)
Coupler 660	1-Hydroxynaphthalene-2-carboxylic acid-3′-N-morpholino propylamide
Coupler 670	Cyanoacet-morpholide
Coupler 690	1,10-Dicyanoacet-triethylene-tetramine, hydrochloride
Coupler 950	2,4-Dihydroxybenzylamine, alkyl sulfonate salt solution
Coupler 1134	2-Hydroxynaphthalene-3-carboxylic acid-2′-methoxyanilide
Coupler 1167	2-Hydroxynaphthalene-3-carboxylic acid-3′-nitroanilide
Coupler Dinol	2,3-Dihydroxynaphthalene
Coupler EBA	1,4-Bis(acetoacet-ethylenediamine)
Coupler O	2,7-Dihydroxynaphthalene-3,6-disulfonic acid disodium salt
Coupler RG	α-Resorcylic acid ethanolamide
Coupler RX	β-Resorcylic acid ethanolamide
Developaid	Polyglycol
Diazo 10	1-Diazo-4-(N-ethyl-N-benzyl)amino benzene chloride, half-zinc-chloride
Diazo 39	1-Diazo-2-ethoxy-4-N,N-diethylaminobenzene chloride, half-zinc-chloride
Diazo 48L	1-Diazo-4-N,N-dimethylaminobenzene chloride, half-zinc-chloride, 70% strength
Diazo 48NF	1-Diazo-4-N,N-dimethylaminobenzene sulfoisophthalate salt
Diazo 49L	1-Diazo-4-N,N-diethylaminobenzene chloride, half-zinc-chloride, 70% strength
Diazo 49NF	1-Diazo-4-N,N-diethylaminobenzene, sulfoisophthalate salt
Diazo 54	1-Diazo-2,5-dibutoxy-4-morpholinobenzene chloride, half-zinc-chloride
Diazo 54S	1-Diazo-2,5-dibutoxy-4-morpholinobenzene bisulfate, 80% strength
Diazo 55	1-Diazo-2,5-dibutoxy-4-morpholinobenzene borofluoride
Diazo 59S	1-Diazo-2,5-diethoxy-4-morpholinobenzene bisulfate, 77% strength
Diazo 60	1-Diazo-2,5-diethoxy-4-morpholinobenzene borofluoride
Diazo 67	2-Diazo-1-naphthol-5-sulfonic acid sodium salt
Diazo 69	1-Diazo-4-N,N-diethylaminobenzene borofluoride
Diazo 72	1-Diazo-2,5-diethoxy-4-p-tolylmercaptobenzene chloride, half-zinc-chloride
Diazo 76	1-Diazo-2,5-diethoxy-4-p-tolylmercaptobenzene borofluoride
Diazo 78	1-Diazo-2-chloro-5-(4′-chlorophenoxy)-4-N,N-diethylaminobenzene chloride, half-zinc-chloride
Diazo 87	1-Diazo-3-chloro-4-N-methyl-N-cyclohexylaminobenzene chloride, zinc chloride
Diazo 88	1-Diazo-3-methyl-4-pyrrolidinobenzene chloride, zinc chloride
Diazotint blue	Proprietary water-soluble blue dye
Diazotint red	Proprietary water-soluble red dye
Diazotint yellow	Proprietary water-soluble yellow dye
Dispersion F	Wax dispersion
Dispersion JB-1	Rosin aqueous dispersion
Dye AC-1	Proprietary blue dye
Pigment 65	Calcined silicate
Pigment 2820	Amorphous silica
Pigment R	Rice starch
Proresin C8000	Melamine-formaldehyde resin

APPENDIX (Continued)

Andrews code	Chemical description
Resin AY	Polyvinyl acetate resin
Resin CP-50	Cellulose acetate propionate
Resin CP-60	Cellulose acetate propionate, alcohol soluble
Resin PS-75N	Vinyl chloride copolymer dispersion
Resin SA-60	Acrylic resin
Resin SB-10	Vinyl chloride–vinyl acetate copolymer resin
Resin VC-1	Vinyl acetate homopolymer dispersion
Resin VK-2	Vinyl acetate multipolymer dispersion
Resin VG	Styrene polymer dispersion
Resin VN	Vinyl acetate copolymer dispersion
Resin VP	Vinyl acetate homopolymer dispersion
Resin VW-2	Vinyl acetate homopolymer dispersion
Solubilizer 136, 137	Naphthalene trisulfonic acid, sodium salt
Solubilizer HI	Aliphatic lactam
Stabilizer AB	Sodium salicylsulfonate
Stabilizer CD	Zinc toluenesulfonic acid
Stabilizer TT	*p*-Toluenesulfonic acid
Wetter 27	Dihydroxydialkyl hexyne

REFERENCES

1. F. V. E. Vaurio. *Tappi*, *43*(1), 18–24 (1960).
2. K. Parker. U.S. Patent 3,446,620 (1969).
3. J. Kosar. *Photogr. Sci. Eng.* *5*, 239–243 (1961).
4. R. F. Coles and R. A. Miller. U.S. Patent 3,076,721 (1963).
5. L. P. F. van der Grinten and K. J. J. van der Grinten. U.S. Patent 1,841,653 (1932).
6. W. P. Leuch. U.S. Patent 2,113,944 (1938).
7. J. P. Bomers and G. J. Vosbeck. U.S. Patent 4,043,816 (1977).
8. R. Landau. *Les Diazos*. Paris, France, 1960. Distributed by Andrews Paper & Chemical Company.
9. R. Landau. *Les Copulants*. Paris, France, 1962. Distributed by Andrews Paper & Chemical Company.
10. W. Krieger and R. Zahn. U.S. Patent 1,803,906 (1931).
11. A. Baril, Jr., I. H. De Barbieris, R. T. Niesert, and T. Stearns. U.S. Patent 2,911,299 (1959).
12. Kalle Company. British Patent 402,737 (1932).
13. C. E. Herrick, Jr., and A. K. Balk. U.S. Patent 2,699,392 (1955).

4

A Brief Introduction to Electrophotography

B. E. SPRINGETT*

Xerox Corporation, Webster, New York

Much has been written about electrophotography since it burst upon the office-place scene in about 1960, and much development has occurred in all sorts of applications beyond simply copying a document onto one side of a piece of 8.5 $\times$ 11″ paper. The demonstration of the basic invention dates from 1938, and the first manual product, the Xerox Model A, came onto the market in 1949; the first automatic copier, the Xerox 914, appeared in 1959. The subject, then, is barely more than 50 years old. The sheer number of companies and people who have become involved in electrophotography due to its twin attractions of being a profitable business as well as an outstanding example of the need for a multidisciplined approach to unravel the science behind the early empirical methods, has produced a strong competition for both markets and inventions. The number of identifiably different ways of practicing electrophotography is very large and is driven by both a search for improvement and a search for a path not covered by someone else's patents. The general approach is best summed up by a quote from the inimitable Mae West, ''When faced by a choice between two evils, I always pick the one I haven't tried before!''

This said, it is apparent that any brief introduction to the subject must gloss over the many nuances in the differing ways of executing any particular function. It is also true that the majority of the readers of this book will have some passing familiarity with the subject. Accordingly, the subject matter will be reduced to its simplest terms and only passing mention given to alternative methods to the one described. It will be left to the reader to seek out much of the details by consulting the references. There will be very little discussion of image quality in terms of image digital image processing; only the

* *Current affiliation*: Fingerpost Advisers, Rochester, New York

major influences of some of the subsystems will be touched upon. This chapter will follow the order given below.

1. Overview of the process and main applications
2. Discussion of the subsystem elements
3. Applications to color
4. Major alternatives
5. Summary

4.1 OVERVIEW OF THE PROCESS AND MAIN APPLICATIONS

The original motivation of Chester Carlson was to find a simpler, cheaper way to reproduce documents in his work as a patent attorney. The mostly nontechnical history of the subsequent development can be found in Refs. 1–3. Today, the applications of electrophotography can be viewed as capturing some of the aspects of silver halide photography and magnetic recording in order to produce hard-copy output on mostly paper, since a piece of paper remains the single best method of displaying and viewing text and graphics under all circumstances. The resemblances to photography are that the image is captured and written to an intermediate medium optically, the output is on paper, and the image quality is quite high; the resemblances to magnetic recording are that the intermediate media can be constantly refreshed and overwritten, the image spends part of its time in digital form, and the process is fast. Unfortunately the electrophotographic process is not as compact as cameras or tape cassettes, but the economics of creating or displaying an 8.5 × 11″ image are highly favorable compared to photography or CRT monitors. The process can be used to copy an existing image or to print an image created or modified with a computer. The essence of this is shown in Fig. 4.1; on the left is a copier, and on the right a printer (which can be turned into a copier by the addition of a scanner). Externally they are indistinguishable. In the copier, the original image is coupled optically directly into the electrophotographic processor; in a printer, the ''original'' is indirectly coupled, which involves some digital transforms transparent to the user. Thus the process is used for copying and printing and to create multifunction machines that do at least both these functions and perhaps others, such as facsimile or internet communication. In the past, the process would turn colored originals into black-and-white hard copy: today, the process is quite capable of producing colored output from colored input, or coloring black-and-white input. The speed range available today for 8.5 × 11″ pages is from 6 pages per minute to in excess of 400 pages per minute (ppm). Reduction and enlargement from 50 to 200% is also readily available. The prices, at the time of writing, for machines using the electrophotographic process range from less than $300 to more than $500,000. The size (i.e., the floor space required) of the machines tends to become rapidly larger as the number of paper sizes handled by the machine increases, as the amount of paper stored in the machine increases, and as the variety of input and output finishing options increases. It is this, plus the greater amounts of internal computing power necessary at higher speeds, which drives the highly nonlinear rise in price as speed increases; there is, however, as with automobiles, a rough correlation between price and machine physical volume.

Turning to the internal workings of the machines, Fig. 2 shows, on the left, a schematic of all the process elements arranged for cyclic or repetitive hard-copy production, and, on the right, a simplified version of what each element is intended to achieve. The whole process operates in the dark; the only light allowed is strictly for imaging or re-

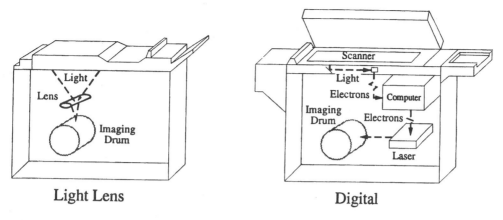

Figure 4.1 To the user these two copiers will look alike. The one on the left is a representation of a light lens copier; i.e., the original document is imaged through a system of moving lenses in a one-to-one or a reduced or an enlarged form directly on to the photoreceptor. The one on the right is a digital copier: the same original is imaged on to an electronic image capture device such as a CCD array, which subsequently processes the data for control of a digital imaging source such as a laser diode or an LED, which then directly writes the image on the photoreceptor. This latter scheme is capable of a much greater range of enlargement or reduction, and the image is accessible to the user for editing in various ways. Such a copier can also be readily connected to computers or modems.

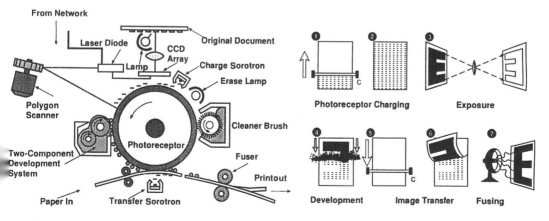

Figure 4.2 This figure shows the complete electrophotographic process. On the right, the steps in the process are displayed so as to illustrate the function of each. On the left is an example of how these various steps are incorporated into a machine; it also gives some idea of the physical cross section of the various subsystems. The steps in order are: charging the photoreceptor, imaging the photoreceptor, developing the image, transferring the image to paper, fusing the image to the paper, and cleaning and refreshing the photoreceptor.

freshing purposes. The system is built to accommodate the width of the output media and there is no restriction on the orientation of the page. The paper can be in the form of cut sheets or a web fed from a roll. Each of the process elements shown in the figure will be addressed in more detail later, but first the way in which they interact will be discussed. As can readily be appreciated from Fig. 4.2, the overall process is a serial one. In effect, each subsystem sets the stage for the next one to perform its specific task. Consequently, understanding the limitations and sources of variance for each subsystem constitutes a large part of the R&D effort in developing new machines.

The first step is to achieve a uniform charge on the photoreceptor; for this to occur properly, the photoreceptor must act like a very uniform, ideal capacitor in the dark. Once charged, the photoreceptor can be discharged by shining light on it. This is done in an image-wise fashion either by reflecting light from an original document or by switching on and off a digital light source such as a laser to create a pattern of light and dark that simulates an original document. This, in turn, creates a pattern of charged and discharged areas on the photoreceptor that constitute a latent (i.e., undeveloped) *electrostatic* image. The electric fields above the surface of the photoreceptor are what interact with the development subsystem to dress this image with toner, which is a dry powder ink consisting of colored particles each of which is less than half the diameter of a human hair. Because the ink is dry we use the word *xerography* (*xero* from the Greek for dry and *graphy* from the Greek for writing) as an alternative term for electrophotography. The toner particles acquire the electric charge necessary for an interaction with the latent electrostatic image within the development unit. Upon exiting the development unit, a fully formed and visible image is on the surface of the photoreceptor. The remaining steps are to transfer this toner image to the media, fix it permanently to the media, and refresh the photoreceptor in preparation for the next cycle of imaging. The transfer is done by bringing a sheet of paper, say, in contact with the image and creating an electric field across the paper–toner–photoreceptor sandwich that is in such a direction as to attract the charged toner to the paper. The paper carrying the toner image, which can still be smudged or disturbed at this point, is then moved to the fusing unit. This unit applies sufficient heat and pressure to cause the toner to melt and coalesce, as well as to flow partially into the paper fibers. The paper is then passed to an output tray for collection by the user. The photoreceptor, now minus the toner image but still carrying the electrostatic image, rotates further to arrive at the erase station where the electrostatic image is removed, usually by shining light over the whole surface of the photoreceptor, which causes the remaining charged areas to discharge. This returns the photoreceptor to its uniform, ideal capacitor state.

The basic process that has just been described has, over the course of time, been applied first to copiers with fixed platens and speeds in the tens of pages per minute, then to duplicators intended for central reproduction departments operating at outputs well in excess of one hundred pages per minute, and moving on to printers both large and small, fast and slow beginning in about 1978. Duplex, or two-sided copying and printing was also introduced in this period. All these machines create predominantly black-and-white hard-copy output. In about 1974 the first color electrophotographic analogue copier showed up; by 2000, over 150 different digital color electrophotographic copiers and printers brands had been introduced representing about 12 different manufacturers and covering speeds ranging from 2 to 130 $8.5'' \times 11''$ impressions per minute. The applications range from simple copying of documents, microfilm, and books, to printing from personal computers, to facsimile machines, to small office multifunction products, to printing from mainframe computers and large databases, to machines one meter wide that print on long rolls of paper, to emulations of four-color offset press output.

The foregoing implies printing on paper: the varieties of paper that can now be used is very large, covering the majority of basis weights and coatings encountered in the commercial printing world—roughly speaking from about 60 gsm (grams per square meter) to about 250 gsm. Besides paper, hard copy can be produced on overhead transparency materials, opaque white film, adhesive labels, card stock, tag and ticket stock, and transfer sheets, which can be used subsequently to transfer the image to textiles, ceramics, or even unwieldy three-dimensional objects. The imaging media that are used and the characteristics of the electrophotographic machine are determined by the customers' applications. Each special application requires some variant of the basic electrophotographic process. To appreciate the breadth of the machines available, consider that DataQuest publishes collections of specification sheets for copiers, printers, and facsimiles: each of these books is about 2 cm thick with one listing per page! (Refs. 4,5).

Another measure of the growth of the usage of the electrophotographic process is to examine the market size of the consumables as a function of time. This can only be done in a very approximate way by estimating the number of 8.5 × 11″ pages produced worldwide and applying algorithms based on toner page yields and photoreceptor lives to arrive at quantities per year (Note that toner page yields and photoreceptor lives are not constant over time!) This process yields compound annual growth rates in the range of 7 to 9% from 1985 to the present for millions of pounds of toner and for millions of photoreceptor units. This growth slowed to 3–5% beginning in 1999. In 2002, the worldwide production of toner is projected to be in the range of 330–380 Mlbs, of which some 25–35 Mlbs is projected to be color toner. Organic photoreceptor production is projected to reach 100 M units by this time.

4.2 DISCUSSION OF THE SUBSYSTEM ELEMENTS

The overall description of the process has been very brief. In order to gain a deeper understanding of black on white printing and copying, the reader is advised to refer to the items listed in Refs. 6–20, and the references to detailed papers contained within them. The references have been chosen with two criteria in mind: (1) that they be reasonably comprehensive from a system point of view, and (2) that they be relatively straightforward and uncomplicated but lead the reader to more detailed or complete specific references. Not all variations on the subsystems will be discussed; again, much more information is available in the references and the extant patent literature. To fix ideas, the discussion will be limited to printers or the digital form of the electrophotographic process. This is because another market trend for the copying function at the time of writing is that few if any clean-sheet, purely analogue or light lens copiers are in the R&D pipeline. The underlying electronic cost trends are such that the cost premium of digital over analogue is becoming vanishingly small. It should be added that another market trend is that the ubiquitous personal computer is causing a significant migration of what was once copied centrally to local printers. Hence the phenomenal growth of the sub-20 ppm printers over the last decade.

4.2.1 Charging

There are two basic methods, which both require the electrical breakdown of air by strong electric fields. Physically, these are very different schemes, however. Figure 4.3 shows the geometry of the *scorotron*, which is a noncontacting device for causing a well-defined electrical breakdown of air to occur immediately surrounding the central fine wire (about

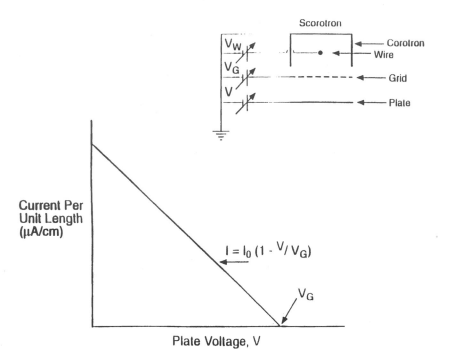

Figure 4.3 This figure shows a schematic cross section of a scorotron and a graph of its performance characteristic. V_W is typically in the range of 3.5–4.0 kilovolts, which causes an air breakdown or corona to be established around the central wire, which is typically 75 μm in diameter; V_G is typically in the range of 500–800 volts. The graph shows how the current decreases in a linear fashion as the voltage beneath the scorotron varies. Since this current ceases to flow at V_G for the ideal case, the photoreceptor will charge up to V_G. The grid is a thin-metal noncorrosible mesh with a typical transmission of 80%.

75 μm in diameter) from which the desired sign of charge is extracted and directed toward the photoreceptor by the application of suitable voltages to the wire and screen grid. For organic photoreceptors, which are the dominant type of photoreceptor in today's printers, negative charge is deposited on the photoreceptor surface producing a negative voltage in the range of −500 to −900 volts. When the electric field in the gap between the scorotron screen and the photoreceptor is zero, no more charge is delivered. Thus the screen voltage controls the photoreceptor surface potential. This is also illustrated in the figure.

The *primary charge roller,* or *biased charging roll* (Fig. 4.4) is a contact device that requires the application of both DC and AC voltages to achieve air breakdown in the nip-forming regions, and the separation and delivery of charge to the photoreceptor. In this case, the photoreceptor voltage is usually in the range from −400 to −600 volts; this voltage is controlled by the DC component of the applied voltages as is also shown in the figure. The roll itself is a multilayered composite of elastomers and other polymers tailored to provide the correct nip shape, electrical impedance, and wear properties.

Both devices produce oxides of nitrogen and ozone; the scorotron more than the biased charging roll. Some form of venting or filtering is required for the former. Both can have deleterious effects on the photoreceptor. The scorotron can deposit a surface conducting layer which in subsequent steps in the process can lead to image blur. The

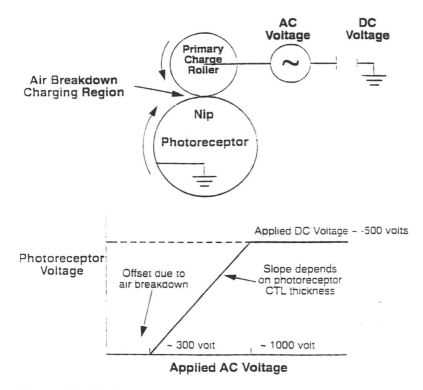

Figure 4.4 This figure shows an alternative charging scheme using a partially conducting elastomeric roller (primary charge roller) that has a typical diameter of 1 cm. In this case, the DC voltage plays the same role as V_G in Fig. 3: it governs the charged photoreceptor voltage. The AC voltage plays the role of V_W in Fig. 3: it creates a steady air breakdown in the nip regions. Typical values of the voltages are shown on the graph, which also shows how charging performance depends on these voltages.

primary charge roller (PCR) creates its discharge right at the surface of the photoreceptor; the reactive species in this discharge cause bond-breaking damage to the photoreceptors' topmost polymer layer, which results in excessive wear. The scorotron is capable of delivering more current but is less compact and so is most often found in the higher speed, larger volume machines. The biased charging roll is a common component in the print cartridges for desktop printers, where it is often called the "primary charge roller" or PCR.

4.2.2 Imaging

There are several types of image sources. If an original document is being copied, a scanning fluorescent or tungsten–halogen lamp is used to illuminate the document, and lenses are used to focus the image on the photoreceptor. This scheme has a speed limitation of some 80 ppm; beyond that speed flash lamps and lenses are used. This change means that a shift from drum photoreceptor architectures to belt photoreceptor architectures is practically mandatory.

In the case of printers, the two main digital image sources are *laser diodes*, whose light output is reflected onto the photoreceptor from a spinning mirror via focusing lenses, and *LED image bars*. In both cases, the solid-state light source is switched on and off at about 35MHz. The LED image bar is a page-wide device consisting of a linear array of individual LEDs plus a focusing lens that is quite compact and is positioned close to the photoreceptor. Current capability is an addressability of 600 dpi (dots per inch). The laser diode systems are not compact, and the source is remote from the photoreceptor, as is shown in Fig. 4.5. The main drawbacks to the LED image bar are that it is expensive to achieve a pixel-to-pixel uniformity of light output greater than ±2%, that careful alignment is required, and that one dead LED pixel will leave a black line on the image. The main drawbacks to the laser diode system are that it is prone to motion quality error due to polygon mirror wobble (and not all mirrors are the same) and that there is an inevitable nonuniformity of pixel size and positioning as the beam is swept across the photoreceptor.

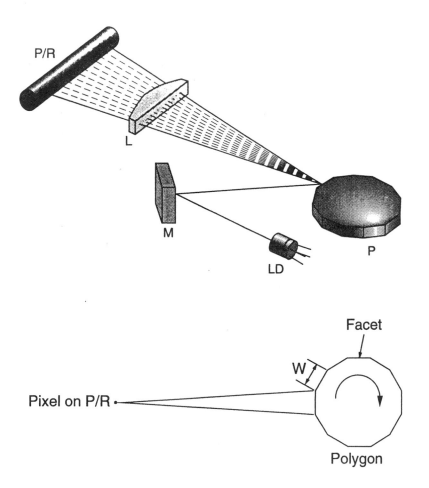

Figure 4.5 This figure is intended to illustrate only one of the possible digital imaging schemes. The main components are a laser diode emitting at a typical wavelength of 780 nm (in the near-infrared part of the spectrum), which can be switched on and off at very high frequencies, a system of lenses that focuses the beam to a spot (pixel) about 42 μm in diameter at the half-power point for 600 dpi, and a rapidly rotating mirror (in the range of 30,000 rpm), which directs the beam across the photoreceptor parallel to its axis and perpendicular to its direction of rotation.

For low-end printers, only a single laser diode is required; for high speed engines, two or four beams that are interlaced are required. Laser diode systems have already reached the 1200 dpi capability plateau. For very wide printing systems, LED image bars are preferred for architectural reasons.

Being solid-state devices, both LEDs and laser diodes are most efficient in the wavelength band 635–820 nm, the most common wavelength in use being 780 nm. The shape of the spot of light from a single pixel, i.e., roughly 40 μm in extent, is such that the pixel is somewhat elliptical; the long axis is in the paper path (or process) direction, and the light intensity falls off from the center in a Gaussian (or bell) shape. This means that pixels have to overlap to achieve reasonable uniformity of exposure on the photoreceptor. This spatial overlap occurs at roughly the point where the light intensity has fallen by half from the central value. It is variations in this overlap that give rise to motion quality artifacts in prints, especially in halftones (or gray levels achieved by systematic patterns of pixels being on or off).

4.2.3 Photoreceptor

The action of the charging system and imaging system combined is to produce a charge pattern on the surface of the photoreceptor that constitutes the *latent electrostatic image*. In copiers, this pattern consists of charge where the original document is dark, and practically none where the original document is light, with variations in between these two states depending on the optical density of the original. This is known as *charged area development* (CAD) or ''write-white.'' In printers, for mostly historical reasons, the opposite convention is used—this is known as *discharged area development* (DAD) or ''write-black.''

The photoreceptor must be designed to respond correctly to the wavelength of the exposure source; it must also be capable of repeated cycles of charge and discharge with no memory of any previous cycle; and it must be capable of supporting the required charge level and achieving the required discharge level over many hundreds of thousands of cycles without these levels varying significantly. The most important characteristic of the photoreceptor is its *photoinduced discharge curve* (PIDC). It is this response that determines the amount of light required to discharge it, and it determines the spatial shape of the charge pattern produced by a single pixel exposure. An example of such a curve is shown in Fig. 4.6. This curve can be approximately described algebraically: this enables computations to be performed of the resulting charge pattern shape when the photoreceptor is exposed to a single pixel, for example. Basically, as the exposure intensity increases, the single pixel charge pattern shape changes from being Gaussian to square. Coincidentally, the width of a line of single pixels will increase on the printed page. Basically, the intensity is set relative to the sharp bend in the PIDC, so that the effects of variations in the pixel overlap due to motion quality problems are minimized while the line resolution on the print is maximized.

The organic photoreceptor is composed of four basic layers as shown in Fig. 4.7. First comes the substrate, which is electrically grounded in operation; next is a *blocking layer,* or *an undercoat layer* (UCL), which prevents charge leakage from the substrate to the top surface; the *charge generating layer* (CGL), which is the layer in which charges are created by the action of light, comes next; and lastly comes the *charge transport layer* (CTL), which allows the movement of charge across the photoreceptor under the action of the electric fields created by the surface charge in order to effect discharge, and which must also support the surface charge in the dark. The top three layers involve various

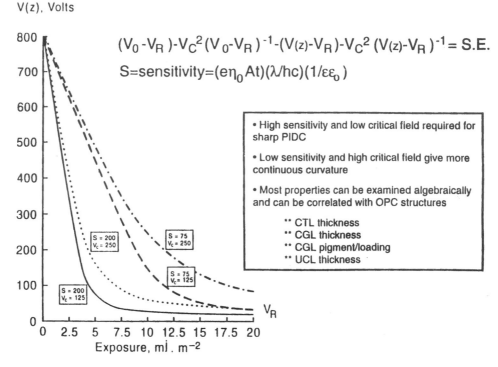

V(z), Volts

$$(V_0 - V_R) - V_C^2 (V_0 - V_R)^{-1} - (V_{(z)} - V_R) - V_C^2 (V_{(z)} - V_R)^{-1} = S.E.$$

$$S = \text{sensitivity} = (e\eta_0 At)(\lambda/hc)(1/\varepsilon\varepsilon_0)$$

- High sensitivity and low critical field required for sharp PIDC

- Low sensitivity and high critical field give more continuous curvature

- Most properties can be examined algebraically and can be correlated with OPC structures

  ** CTL thickness
  ** CGL thickness
  ** CGL pigment/loading
  ** UCL thickness

S = 200
V_c = 250

S = 75
V_c = 250

S = 75
V_c = 125

S = 200
V_c = 125

V_R

Exposure, mJ . m⁻²

Figure 4.6 This figure shows a set of photodischarge curves typical of photoreceptors in the marketplace. As the exposure is increased, the photoreceptor voltage decreases due to the flow of charges created in the CGL by the action of light on the surface. The equation and the comments in the box illustrate how it is possible to set about designing a photoreceptor for a particular performance. The parameter S depends upon the photoreceptor thickness, t, the light wavelength λ, the dielectric constant ε, and the efficiency with which the light is turned into electric charge in the CGL, η; the remaining factors are various physical constants. *Source*: A. Melnyk. In: Proceedings of the Third International Congress on Advances in Non-Impact Printing (1986).

materials embedded or dissolved in polymers. It is these materials that enable each layer to perform its special function. Typically, the photoreceptor will respond to a spectrum of light, not just a single wavelength, which is dictated by the charge-generating pigment used. For digital printers, this pigment is most often a member of the metallophthalocyanine family, since these pigments respond well to 780 nm light. The absorption of light in the CGL creates pairs of oppositely charged entities (electron–hole pairs), which the pre-existing electric field separates. Since an organic photoreceptor is usually charged negatively, the positive charge (the hole) flows to the surface, while its oppositely charged partner (the electron) leaks away to ground through the UCL. This scheme is shown conceptually in Fig. 4.8 for both CAD and DAD schemes.

These photoreceptors come in both drum and belt formats. The drums are dip coated, and the belts are web coated. Typically, the three top layers are coated in these ways in multiple pass operations. The dimensions of each layer are governed in part by the constraints of the coating process and in part by the electrophotographic system needs. For example, if higher voltages or very long life is required, the CTL tends be about 30 μm thick. In order to avoid laser diode coherent light interference effects, the CGL is usually

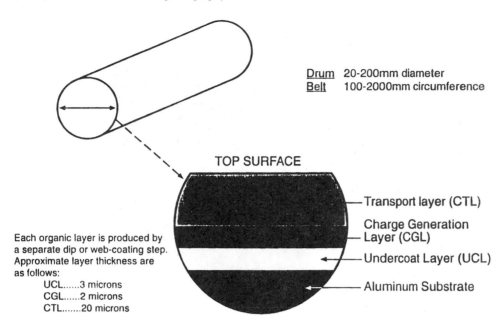

Drum 20-200mm diameter
Belt 100-2000mm circumference

TOP SURFACE

— Transport layer (CTL)

Charge Generation
— Layer (CGL)

— Undercoat Layer (UCL)

— Aluminum Substrate

Each organic layer is produced by
a separate dip or web-coating step.
Approximate layer thickness are
as follows:
    UCL......3 microns
    CGL......2 microns
    CTL.......20 microns

Figure 4.7　This figure illustrate a typical drum organic photoreceptor (OPC) structure; belt photoreceptor structures are quite similar. The UCL serves to block injection of charges from the metallic substrate; the CGL is where light is absorbed and creates electric charges; and the CTL is the layer that gives the photoreceptor its dielectric properties and simultaneously allows transport of one sign of the charge created in the CGL to flow to the surface to discharge the photoreceptor.

coated thick enough to absorb most of the incoming light. However, all photoreceptors generate some small amount of charge in the dark (dark decay), which often depends on the thickness of the CGL. The UCL must not be so insulating as to prevent the escape of charge from the CGL to ground.

The trends in organic photoreceptor development are to achieve longer life, greater electro-optical stability, and better overall coating uniformity. The drivers for these activities are the move toward faster, more robust machines and color printing.

4.2.4　Development

Referring to Fig. 4.8, we see that this step involves attracting the toner to the latent electrostatic image. In the case of typical printers (i.e., the right-hand side of Fig. 4.8), this means depositing toner in the discharged areas by making use of the fringe fields associated with the image. For this to happen, the toner must be negatively charged also. The development subsystem has two essential functions. One is to cause the toner to acquire the desired sign of charge. The other is to deliver the toner into close proximity to the image. There are two main schemes that are in use to achieve these twin goals. There are many variations of these schemes in existence; the reader is referred to Refs. 6, 10, 12, and 15 for additional details on some of these variations.

Figure 4.9 depicts an outline of a *two-component development* (TCD) method. The two components are a carrier and a toner. The carrier is a magnetic material such as steel grit or a ferrite, which is often partially coated with a polymer. The toner is a polymer

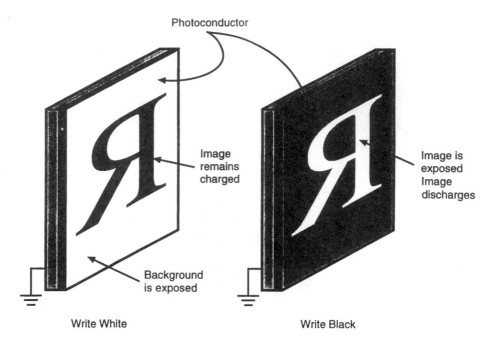

Figure 4.8   This figure shows the basic difference between charged (CAD) and discharged (DAD) area development. In both cases the electrostatic latent image, here shown in the shape of an R, has nonuniform fringing fields associated with it that will attract charged toner. In the write-black case on the right, which is typical of digital printers, the toner is the same sign as the electrostatically charged background areas, from which it is repelled, depositing in the uncharged image areas. If opposite sign toner is used, a reversed image will be developed; i.e. a black-on-white image will become a white-on-black image.

resin containing a colorant and other functional agents. The carrier-to-toner diameter size ratio is roughly 10:1. Feedback systems are used to keep the mass (or volume) ratio of toner to carrier, the *toner concentration*, at a reasonably constant ratio of roughly 3–5 wt%. The surface chemical properties of both the toner and the carrier are such that when they are tumbled together in the agitating area of the development housing, *triboelectric* interactions cause the toner to acquire a negative charge and the carrier to acquire positive charges. The figure shows the opposite situation, which would be true for a light lens copier CAD system. The magnetic forces created by the internal magnets plus the rotation of the shell around them cause the magnetic carrier to be transported into the region of the photoreceptor bearing the electrostatic image. The carrier is, of course, also transporting the toner, which is electrostatically bound to it by Coulomb forces. The magnetic field strengths are arranged so as to make the carrier beads form into hairlike structures as they approach the electrostatic image creating a *magnetic brush*. In order to complete the process, an additional voltage is applied to the rotating shell that is less negative than the image background areas, but much more negative than the image areas. This assures that the electric field in the background areas opposes the deposition of toner and simultaneously creates a stripping force toward the photoreceptor in the image areas. The strength of this electric field determines the quantity of toner deposited, which will in turn dictate the darkness of the final image on the page.

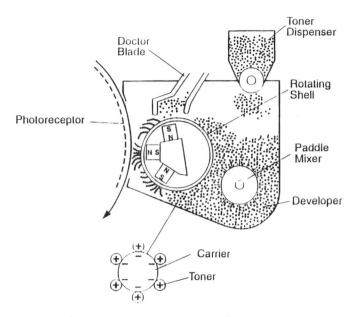

Figure 4.9   Cross section of a typical two-component development subsystem (TCD) showing the toner supply arrangement at the top right, the mixing area where the toner and carrier are blended together and triboelectric charging occurs, and the magnetic transporting scheme on the left. The toner dispenser is typically what the customer buys and replaces; the other component part of the system containing the carrier remains as part of the machine.

The other main scheme is called a single-component development (SCD) system. A schematic of it is shown in Fig. 4.10. SCD is much more compact than TCD. For this reason it is most often used in small printers. There is only toner in this scheme, which in order to be transported from a sump to the electrostatic image is itself magnetic. This is achieved by having small magnetic particles in the toner resin, which often serve as colorants as well. In order to charge the toner particles, some combination of rubbing contact with a triboelectrically active polymer and bias voltages is used. A separate development bias is used on the rotating shell, as for TCD, to ensure deposition of toner on the electrostatic image.

In both cases, the rotating roll that transports the toner from the sump to the development nip must be coated with materials that ensure proper electrical performance. That is, excessive current flows must be prevented, as must excessive accumulation of charge as toner is continually deposited on the electrostatic image.

As was mentioned above, the toner itself contains other functional agents. The role of these agents is severalfold: they help to ensure the acquisition of the correct triboelectric charge (*charge control agents* [CCA]); they help to assure good flow properties that are essential for preventing the toner from agglomerating or clumping (*flow agents*); they help to assure good release properties from the fuser rolls (*release agents*); and they can impart particular properties to the toner on the printed page such as enabling *magnetic ink character recognition* (MICR) for check-writing purposes. A typical SCD toner particle is shown in Fig. 4.11. Often, charge control, flow, and release agents are externally blended into the developer and are collectively known as *external additives*. The toner resin itself

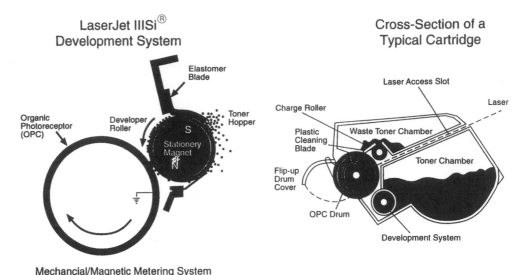

Figure 4.10 Cross sections of a typical single-component development subsystem (SCD) showing the toner supply arrangement, the area where the toner triboelectric charging occurs in conjunction with the magnetic transporting scheme on the lower left in the right-hand sketch. The whole cartridge containing an OPC, a charging system, and a cleaning blade as well as the toner is typically what the customer buys and replaces; the other components such as exposure remain as part of the machine.

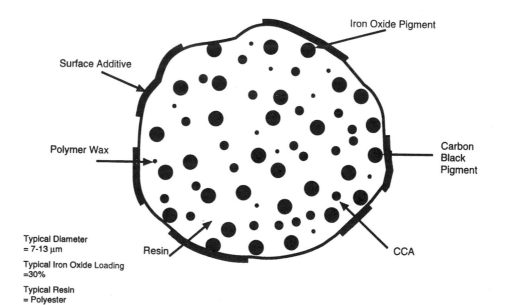

Figure 4.11 Shown in this figure is a typical SCD toner particle. Several of these components are also present in TCD toners. No iron oxide is present in color toners. The typical particle size is in the range of 7 to 9 μm, with a range about the average diameter of about 4 μm. The wax and the resin largely control the basic fusing properties, but the other additives modify the rheological behavior as a function of temperature. The surface additive, the resin, and the particle morphology largely control the flow properties. The CCA and the surface additives largely control the triboelectric charging properties, but the pigment and the resin modify this.

also must satisfy certain melt rheology requirements having to do with the fusing process. The toner particle size also influences the image quality—basically smaller is better—but there are limitations on what is possible that are governed by the manufacturing process, health considerations, and cost considerations. The typical average particle size for conventionally produced toner at the time of writing is between 6 and 7 μm. In terms of colorants, black presents no problem for either TCD or SCD systems. But since magnetic materials tend to be black or dark brown, SCD is particularly unsuitable for process color (or four-color) systems; the magnetic material makes all colors appear muted or even muddy. In order to have single-component development systems without magnetic transport, significant changes to the charging and delivery schemes have to be made (Ref. 15).

The trends in toner development are toward yet smaller particle sizes, toward lower cost manufacturing processes, and toward lower melting materials with good flow capabilities and improved environmental stability. The drivers for these trends are the need to improve color print quality and to decrease color per page costs with much improved reliability.

At this point it is useful to pause and see where in the electrophotographic process we are. Referring to Fig. 4.2, we have completed discussing steps 1–4 in terms of the right-hand side of the figure; and we have traveled about 200° counterclockwise from charging in terms of the left-hand side. Figure 4.12 summarizes the situation pictorially for achieving a single line as the image in both CAD and DAD processes. From this point on, the paper or other media enter the process.

4.2.5 Transfer

As with charging, there are two main approaches, which both involve the creation of an electric field across the photoreceptor–toner image–paper (or medium) sandwich. At the transfer step, the paper or medium must come into intimate contact with the toned image while moving at the same velocity as the photoreceptor; this is to minimize any possibility of image disturbance. It is a mechanical engineering matter beyond the scope of this introduction. Suffice it to say that this is readily accomplished so that upon contact no image disturbance occurs. The goal of the transfer step is to achieve 100% transfer of all the toner in the image areas from the photoreceptor to the medium without any disturbance to the image, while simultaneously preventing any stray toner in the background areas from transferring over. This is not as simple as it sounds, for two reasons. First, small charged particles, each of which has the same sign of charge, have to be physically moved a very small distance. Coulomb repulsion forces act so as to move these particles apart. These forces are significantly reduced when the particles are in contact with the smooth insulating photoreceptor. Second, at small distances, a molecular force, the van der Waals force, is stronger than the Coulomb force holding the toner to the photoreceptor. The Coulomb force itself increases as the particles get smaller. So in the transfer step the force created by an electric field must overcome these effects. The net result is that usually no more than 95% *transfer efficiency* is achieved; values as low as 85% have been reported. The last transfer problem is that the medium that will carry the final image, paper in particular, is not always well-defined electrically and that the electric impedance properties vary as the humidity changes. This means that performance in transfer depends on the local environment and on the paper type. If transfer efficiency is low, additional noise is introduced into the image, which degrades the print quality.

Figure 4.13 shows the first method employed to create an electric field: another corona device. It is essentially the same scheme as used for charging (Fig. 4.3). The

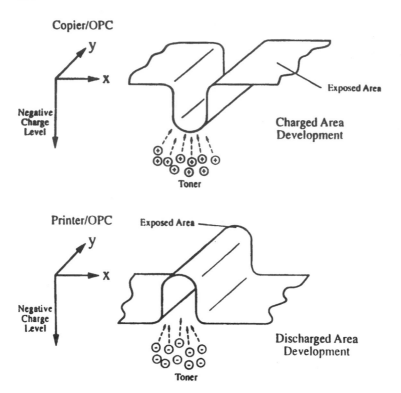

Figure 4.12   This is an illustration and summary of the two methods of creating and developing a latent electrostatic image in conjunction with an OPC. Both cases show the development of the image of a line. The upper drawing shows the situation for a copier wherein the OPC was initially uniformly negatively charged and the image area remains charged after exposure. This image area has electrostatic fringe fields that attract positively charged toner to it. The lower sketch is for a printer wherein the OPC was initially uniformly negatively charged and the image area is discharged after exposure. This image area has electrostatic fringe fields such that negatively charged toner is attracted to it.

charges deposited on the back side of the media for DAD systems using organic photoreceptors are positive, which creates an electric field directed so as to attract the negatively charged toner to the underside of the medium. This field has to be strong enough to attract much of the toner, but yet not so large as to create air breakdown in the small air gaps, which are ever-present due to the roughness of the paper. In practice this means electric fields in the range of 25–35 volts/μm have to be created. The force of attraction between the charges on top of the paper and the ground plane of the photoreceptor are such that considerable pressure is created: this pressure helps to enhance the transfer efficiency and to prevent image disturbances. The other place where such disturbances can occur is in the act of removing the paper carrying the toned image from the photoreceptor. The electric field strengths are such that air breakdown can be encountered as the air gap during this stripping process increases. To counteract this, another corona device is often used in order to reduce the electric field strength during paper stripping from the photoreceptor.

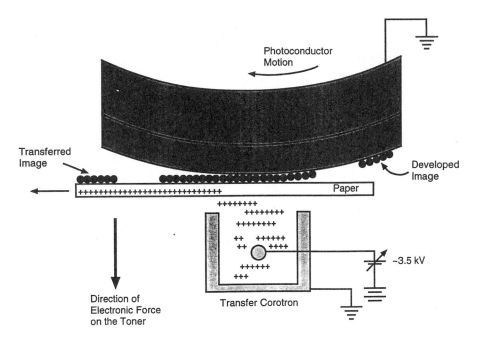

Figure 4.13 This is an example of a transfer charging scheme. It shows a charging device similar in concept to that in Fig. 4.3, which is used to create an electric field across the paper/toner/photoreceptor sandwich. The photoreceptor substrates provide the ground reference potential. In this example, the developed toner is negatively charged, and the charges deposited on the back of the paper are positive so as to create an electric field that creates an attractive force in the direction shown.

The other method for achieving transfer is similar to the biased roll charging scheme (Fig. 4.4). The same considerations as for scorotron transfer apply. In this case, though, additional pressure can be controlled by the loading on the *biased transfer roller*. The reasons for using this scheme are again compactness and less generation of oxides of nitrogen and ozone, plus somewhat better mechanical control over the media.

4.2.6 Fusing

The last step in the imaging path is to fix the toner permanently to the media. The toner resin is typically a thermoplastic. Moderate amounts of heat will cause it to flow and melt. If the respective surface energies of the molten toner and the media are correctly matched, the toner will flow into the fiber structure of the paper or simply adhere to smooth media like transparencies. So the function of the *fuser* is to melt and coalesce the toner, to cause it to flatten so that its final surface roughness profile matches the desired outcome (e.g., rough for a matte finish and smooth for a glossy look), and to create minimum image disturbance (e.g., no loss of resolution). As for the other subsystems, there are several ways to fix toner to the paper to achieve the required final print appearance. The principal methods are the contacting or the noncontacting application of heat. The noncontacting method is done either by flash fusing or by radiant fusing (e.g., as shown schematically on the right-hand side of Fig. 4.2, step 7). The contacting methods either use only pressure to squeeze the toner flat and into the paper, or employ a combination of heat and pressure,

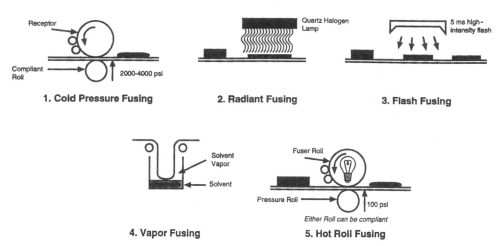

Figure 4.14   This figure shows sketches of the various kinds of fusing process that can be found in the marketplace. The selection of the fusing process depends upon such things as power consumption requirements, process speed, toner gloss requirements, and cost.

or exposure to solvent vapor to achieve coalescence. Only hot roll fusing will be discussed; the references give additional details about all methods.

Figure 4.14 shows a summary of the various methods. The lower right-hand picture shows schematically the standard hot roll method. Both rolls consist of at least two layers. There is a metal core covered by a conformable elastomer in order to be able to create a nip. In the case of the heated roll, it must have good heat-conducting properties. The nip width and the roll temperature are adjusted so that the toner is heated hot enough and for long enough to achieve the desired end. There is often another layer added to achieve good life and to achieve good toner or paper release properties for both of the rolls, since duplex operation is often a necessity. In some cases, the fuser rolls are lightly treated with an oil to aid in release; in other cases the toner has a waxlike release agent designed into it. This last is mostly true for low-speed black-and-white printers. The surface roughness of the heated roll, the nip pressure, the paper thickness and mechanical structure, and the toner pile height and melt rheology properties all combine to yield the degree of gloss and the fix level in the final image. From the foregoing, it can be realized that this is not a simple process. The toner is the intermediate agent between development and fusing. These two subsystems impose some stringent boundary conditions on the toner designer.

Figure 4.15 shows a typical time–temperature profile as the toner pile representing an unfused image passes through the fuser. As can be seen, the process of fixing the toner to the paper proceeds in three seemingly distinct steps: melting, coalescing, and flowing. In reality, since the toner is heated from above, the processes merge into one another throughout the toner pile. If the total quantity of heat applied is too low, the image fix is poor, and some toner adheres to the fuser roll because its internal cohesion is too low due to a lack of full coalescence (*cold offset*). If the fuser roll is too hot, the toner becomes too liquid and again sticks to the fuser because its internal cohesion is again too low (*hot offset*). The desired state is that the toner becomes hot enough to melt and coalesce and flow, yet not so hot as to have low internal cohesion. The adhesion to the medium must be greater than that to the fuser roll surface under these conditions, of course.

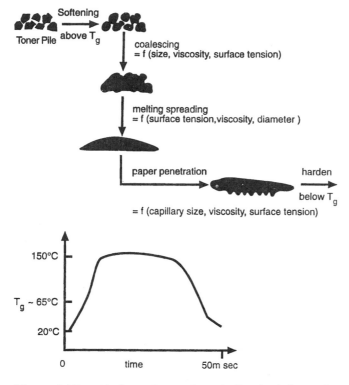

Figure 4.15 This figure shows schematically what is happening to the dry powder toner pile as it goes through the fusing nip. The graph at the left shows a typical time–temperature profile. As the toner softens, coalesces, and melts, various physical properties of the toner are controlling these processes, as shown at each step. These properties are all temperature dependent. The smoothness of the final toner layer also depends on the fuser roll surface properties. The final step is to cool the toner as it exits the fuser, by which time it has penetrated the paper fibers. The cooling step is by unforced convection and conduction.

4.2.7 Cleaning and Erasing

The final process step before the whole imaging cycle begins anew is to refresh the photo-receptor by removing any memory of the previous image; that is, to restore the photorecep-tor to as close to a pristine state as possible. This means eradicating both the remaining electrostatic image and the traces of toner that were not transferred or that randomly found their way into the background areas. The electrostatic image is typically eliminated by flooding the photoreceptor with uniform illumination (*erase*), but it can also be done by recharging the photoreceptor uniformly. The former eliminates any possibility of a "ghost" image; the latter preferably requires the use of an alternating voltage corotron in order to create both positive and negative charge. The biased charging roll already uses both AC and DC voltages, so the erase lamp is usually omitted in this case.

The remnants of toner are removed either by mechanical means alone or with some form of reverse charging assistance that lowers the charge on the toner and consequently the adhesion force to the photoreceptor. The mechanical systems in use are brushes or blades or combinations of these. One such scheme is shown in Fig. 4.16 for a *blade cleaner*. It will be noticed that the waste or used toner is collected in a sump. In some

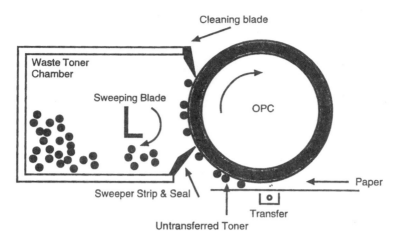

Figure 4.16 This figure illustrates perhaps the simplest cleaning system for removing toner from the photoreceptor. The blade simply scrapes off the toner into a waste hopper. Naturally, there is an art to designing the shape of the blade. Its effectiveness depends on this shape and the pressure exerted on it. Both the blade and the photoreceptor become damaged and worn during long usage.

systems, especially small machines, this toner accumulates until the sump is full. It must then be emptied or disposed of. In several larger systems, this toner is recycled within the system. The reason for this is understandable when one considers that if the transfer efficiency is 90%, 10% of the toner is being wasted.

4.3 APPLICATIONS TO COLOR

The first commercial color copier using the electrophotographic process was introduced by Xerox in 1973. This was a light lens device, not a digital printer. It took another 10 years or more before the digital versions appeared. This had as much to do with progress in computer technology as it did with progress in electrophotography. The method by which color pages are produced essentially mimics any printing process. The color image is broken down into four color planes, and each is printed separately and sequentially on the same piece of paper so that when the eye integrates the image it sees a full-color print. For information on the color process itself, see Refs. 17, 18, 19, and 21–29. These references are mostly general in nature and the additional criteria that they teach about color per se has been applied. The references used earlier for monochrome printing are still valid. Here we shall only address implications for the electrophotographic process. The four colors in question are cyan, magenta, yellow, and black (*CMYK*) since we deal with the so-called subtractive color process. What this means is that the whole electrophotographic process must be repeated four times within the confines of the same piece of hardware under the control of a microprocessor. The colors available from the process are determined by the specific pigments used in the toners, the media printed upon, and the method of digitally imaging and overlaying the four-color planes.

The same basic electrophotographic steps must still be accomplished. But first let us consider color in the abstract as compared to black-and-white documents. Primarily, black-and-white documents consist of text with some use of graphics; they are meant to convey information or communicate in mostly a verbal or arithmetic manner. Thus the images must be above about 1.0 *optical density* with a matte finish, edges must be crisp, serifs or other small features must be resolved, the background must be clean, solid areas must be uniform, and graphics half-tones must be free from noise. In the United States, the typical black-and-white page is 8.5 × 11″, is often duplex, is on plain (uncoated, nonglossy) paper, and has a toner area coverage between 3 and 8%.

Color documents, on the other hand, expand beyond many of these requirements and in other cases impose much stricter quality bounds. The primary changes are, first, the mechanical ones associated with large paper sizes and the handling of glossy or coated stock: these changes do not greatly affect the electrophotographic process. The major problem encountered is an engineering one of "stretching" all the subsystems to deal with a wider imaging path. Second, the requirement is to deliver uniformly glossy images at much greater toner area coverages, often up to 30% per color: these changes affect the design of the development subsystem, the toner, and the fuser. The changes to the fusing subsystem and development subsystem are often modifications to existing designs. Last, the issues related to print quality have to do with the fact that two or more CMYK toners are required to achieve a full *color gamut*: to achieve a uniform solid area of green, say, the yellow and cyan toners must be perfectly registered everywhere on the page; to achieve good definition in both highlight and shadow areas, single pixels need to be developed and properly fused; to achieve a sufficient number of gray levels with reasonable resolution to reduce graininess, at least 600 dpi is required; to get flat documents with no perceptible toner piles in dark areas, low developed toner mass is required; and to get a uniform level of gloss for different gray levels on glossy or coated stock, toner and fuser designs must be changed. The imaging system design is also impacted by these considerations, and the transfer subsystem often is as well. The systems that are largely left unchanged (but not unchallenged) are the charging, photoreceptor, and cleaning-and-erase subsystems. Usually, the first two are required to deliver more spatially uniform performance than for the monochrome case.

Precisely how this is accomplished is again to some degree a matter of choice. If the emphasis is to be on compactness and cost at the expense of a loss of speed and some degradation of print quality compared, say, to offset prints, then the same photoreceptor is used to image all four color planes. This results in a *multipass* system. But further choices are still to be made for the method where each color plane is stored before the final color image is complete, as illustrated in Fig. 4.17. The three choices are (1) on the photoreceptor as exemplified by the HP Color LaserJet®; (2) on the paper as exemplified by the Canon CLC® series and the Xerox Majestik/Regal®; and (3) on an intermediate as exemplified by the Tektronix Phaser 540/550® and the Xerox Xprint® series. There are further nuances within these boundaries depending on whether the CMYK developer stations are fixed or movable, and whether the photoreceptor is a belt or drum geometry.

If, on the other hand, the emphasis is on speed and productivity with excellent print quality but size and cost are secondary considerations, then a *single-pass* system is preferred. The choices in this case are illustrated in Fig. 4.18. The choices remain largely the same except that now the CMYK developer stations are always fixed. The currently preferred method is to use a separate electrophotographic engine for each color plane as exemplified by the Ricoh 8015®, the Xeikon DCP/32D®, the Xerox DocuColor 40®, and the Canon CLC1000® and their successor products.

In order to fix ideas more clearly, two examples only will be discussed—one each of multipass and single-pass architectures. Figure 4.19 shows the multipass examples. Most of the useful subsystem elements have been left out of the schematic to illustrate

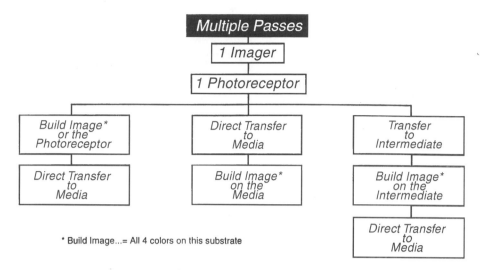

Figure 4.17   This figure shows the various pathways for implementing a multipass color electro-photographic process. There are many options, and the particular choice is dictated more by business, market, and patent access considerations than by technological ones. Examples of each path exist in the marketplace.

the features. The first is that the development stations have to be rotated to come into coincidence with their respective electrostatic latent images. This causes the development system to be the largest machine component. The second feature to note is that although the paper paths can be very different, the cut sheet of paper itself is stored on another drum (which is also larger in size than the photoreceptor) during the whole imaging process and

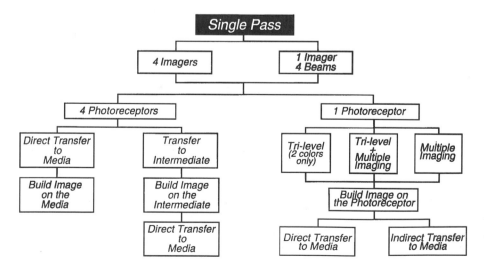

Figure 4.18   This figure shows the various pathways for implementing a single-pass color electro-photographic process. There are many options and the particular choice is dictated more by business, market, and patent access considerations than by technological ones.

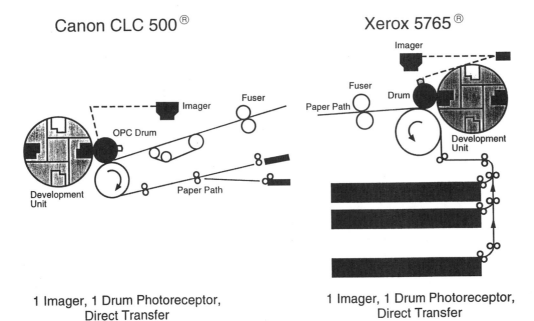

1 Imager, 1 Drum Photoreceptor, Direct Transfer

1 Imager, 1 Drum Photoreceptor, Direct Transfer

Figure 4.19 This figure shows two similar examples of a multipass color system. In both cases the paper is stored on a drum. Each color plane is imaged in sequence. As it is processed, it is transferred from the OPC to the paper on the drum. To develop the next color electrostatic latent image, the developer unit is rotated 90° to bring the next color into contact with the OPC. Both systems use laser diodes and spinning polygons for imaging, and hot roll fusing. The differences are in the details of the paper path and the image processing algorithms. Both units have built-in scanners for copying (see Fig. 1). Duplex printing or copying requires moving the media through the system a second time.

accumulates the image color plane by color plane. So compared to black-and-white printing, the complexity introduced by the need to keep four operations synchronous is considerable. This sort of scheme is capable of printing at the rate of 10 color pages per minute with OPCs no larger than 100 mm in diameter. To go faster, the size typically must increase somewhat and the development stations must be switched more quickly. The empirical limit to this scheme appears to be about 10 ppm.

Figure 4.20 shows the single-pass examples; again many of the details of the standard subsystems have been omitted from the sketches. The previous two examples were capable of duplex printing by having a convoluted paper path, since the image needs to be fused to each side separately. The single-pass example on the left suffers from this same difficulty; the one of the right overcomes this problem by having eight separate electrophotographic engines contained within it. There are other major differences related to choices the designers have made. The slower speed machine on the left uses laser diodes, in a rather complex imager, and relatively small OPCs, each of which is imaged with a separate laser beam. The system on the right is faster and uses larger OPCs, each of which is imaged by its own LED image bar. There remain two other distinctive differences. The machine on the left uses heated roll fusing and handles only cut sheet media. That on the right uses a radiant fuser and handles only a web of paper or other media.

Ricoh Artage 8015®   Xeikon DCP-1®

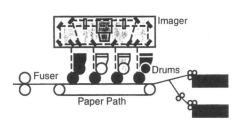

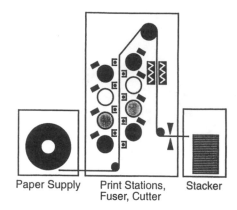

1 Imager, 4 Beams,
4 Drum Photoreceptors,
Direct Transfer

Paper Supply   Print Stations,   Stacker
                Fuser, Cutter

8 Imagers, 8 Drum
Photoreceptors,
Direct Transfer, Duplex

Figure 4.20   On the left of this figure is an example of a single-pass color system employing a single spinning polygon with optics to create four beams that image on four separate OPCs. Development, transfer, and fusing are all standard. Cut-sheet paper is "escorted" past each OPC on a transport belt before being delivered to the fuser, which is a heated pressure roll. Duplex printing requires the paper to be sent through the process yet again, so that duplex pages are produced at approximately half the speed of simplex pages. On the right of the figure is a more complex, higher speed design, which uses eight separate LED bars to image onto eight separate and somewhat larger OPCs. The medium in this case is fed from a roll and preconditioned before entering the process. Printing is simplex or duplex in the same pass through the process; the fusing in this case is by radiant fuser.

In both cases, however, the medium acquires each color plane in sequence by electrostatic transfer as it passes by each photoreceptor.

There are many more commercial variants on these themes. At present, all the color copiers and printers in production use digital imaging on organic photoreceptors with discharge area development and a CMYK toner set with an average size of about 7 μm. Nearly all use hot roll fusing and corotron transfer.

4.4   MAJOR ALTERNATIVES

This section will address briefly the three major alternatives to the xerographic version of electrophotography. Each of them utilizes several aspects of the basic process, so only the differences will be covered. Each of the systems discussed is in commercial use; for a more detailed discussion of them, the reader is again directed to the references. These brief discussions also illustrate the point made at the beginning of the chapter, that there are many ways to practice print-on-demand from a printing engine viewpoint (Refs. 5, 14, 15, 16, 19, and 20).

4.4.1   Ionography

The major difference in principle between ionography and electrophotography is that the optical imaging system of the latter is replaced by a direct charge writing scheme that is imagewise capable; i.e., it can be turned on and off at the pixel level to write individual pixels. It has also been called "electron beam imaging." Since no light is involved, no photoreceptor is required. This item is replaced by a simple dielectric member of suitable thickness and electrical properties; e.g., it must retain charges only on its surface with no leakage into the bulk, or image ghosting will result. The charge writing system still consists of a glow discharge in air, but now much more closely spaced multiple electrodes with apertures that control the pixel size are used to switch on and off the flow of current to create the images. The drawback to this scheme is that as the optical imaging techniques have moved on to 1200 dpi, this technology has found it harder and harder to keep pace. As practiced by Delphax, the development system is single component and the fusing has been done by simultaneous transfer and cold-pressure fixing. There is no intrinsic reason why this need be so: two component systems, standard transfer, and hot-roll fusing can be used, which opens up the door to color printing using this approach. This technology is fully capable of monochrome speeds above 300 ppm.

4.4.2 Magnetography

The major difference in principle between this method and electrophotography is that the optical imaging system is replaced by a magnetic writing head and the photoreceptor by a magnetizable receiver. The scheme bears some resemblance to tape recording in this regard. The system inevitably must use magnetic toner that is attracted to the magnetic latent image. This requirement makes this technology unsuited to color printing. As practiced by Nipson, transfer and fusing are done by using magnetic force and pressure, respectively. This technology is fully capable of monochrome speeds above 300 ppm.

4.4.3 Liquid Toner Development

The major difference in principle between this method and electrophotography is that the dry powder development system is replaced by a liquid ink system. The liquid "ink" consists of fine toner particles in a low molecular weight hydrocarbon solvent or carrier that also contains some additives that ensure that the toner particles acquire a net charge. Managing the carryout and mass balance of the carrier and ensuring a good fix level to most popular printing papers are two of the challenges for this technology. As practised by Indigo, an intermediate belt is used to prepare the image for transfer and fixing. The principal advantages of this technology are the low toner mass on the media, that toners can be mixed on line, and that exotic toners such as metallics can be fairly readily adapted to the system. This technology is fully capable of monochrome speeds above 200 ppm.

4.4.4 Electrography

This system is a combination of two of the above, ionography and liquid toner development, but with the simplification that the image is written directly onto the media by an array of niblike electrodes that may be switched on or off at the pixel level. As practised by Xerox ColorGrafx, this technology is quite slow and confined to wide format systems.

4.5 SUMMARY

This brief introduction has been intended to introduce the reader to the electrophotographic process as a system where all the serial process steps have to perform well in the context of both the ultimate print quality and machine reliability, and the requirements set by the preceding and proceeding subsystems. It has deemphasized the copying function and focused on digital printing, because this is the direction the technology has taken and this is the path to color. It has attempted to point out the changes required of the basic monochrome process as a result of this shift. The references are not intended to be comprehensive, merely suitably more detailed for the reader who wishes to pursue the subject to find out more of the specifics related to variant ways of performing each process step, or to find out more about how to think about color. An attempt has been made in the figures to illustrate how the process works in principle rather than to indicate the best way to design each subsystem. There has been no attempt to be exhaustively technical; electrophotography is such a multidisciplined process that only teams can successfully design a robust, reliable, and effective printer. The skills of many are required, and the team members learn best by doing and teaching each other. An introduction such as this can at best be only a preface.

Given the number of papers that are presented and the numbers of attendees from around the world at the IS&T's Non-Impacting Printing (NIP) Congresses (Ref. 15), the subject is a long way from being exhausted. Electrophotography has just begun to move out of the office into the commercial printing markets represented by offset presses and wide format printing, whether as a complete press or as part of some hybrid system. The miniaturization that occurred in the monochrome electrophotographic process and allowed printers on the desktop has now begun to occur in the area of color printing. The combination of multiple functions in a single machine is also now well established.

This brief introduction has also been meant as context setting for the chapters on toners, developers, and photoreceptors, and other materials that follow. In this regard, some allusions to the more important system level properties of these items have been made. It should be remembered that the design of these materials is done in conjunction with the subsystems, and even systems, in which they will be used. So, just as there are many different embodiments of the various process elements, so there are perhaps an even greater variety of materials used in conjunction with these process elements.

ACKNOWLEDGMENTS

This brief introduction is an outgrowth of a tutorial that I have presented at the Toner and Photoreceptor Conferences, organized by Diamond Research Corp.: I thank Art Diamond of that company for the resulting opportunity to have developed a broader knowledge of electrophotography. I also need to thank my many colleagues at Xerox Corp. who have answered my questions with patience and provided an atmosphere conducive to continuous learning.

REFERENCES

1. Carlson, C. F. *Electrophotography*. U.S. Patent 2,297,691, 1942.
2. Dessauer, J. H. *My Years with Xerox; The Billions Nobody Wanted*. Doubleday, Garden City NJ, 1971.
3. Mort, J. *The Anatomy of Xerography, Its Invention and Evolution*. McFarland, Jefferson NC, 1989.
4. *SpecCheck Copier, Printer and Facsimile Guides*. DataQuest, San Jose CA, 1996.
5. *Color Printer Data Base*. InterQuest, Charlottesville VA, 1996.

6. Dessauer, J. H., and Clark, H. E. eds. *Xerography and Related Processes*. Focal Press, New York, 1965.
7. Durbeck, R. C., and Sherr, S. *Output Hard Copy Devices*. Academic Press, New York, 1988.
8. Johnson, J. L. *Principles of Non-Impact Printing*. Palatino, Irvine CA, 1986.
9. Mees, C. E. K., and James, T. H. *The Theory of the Electrophotographic Process*. Macmillan, New York, 1966.
10. Schaffert, R. M. *Electrophotography*. Focal Press, London, 1975.
11. Scharfe, M. E. *Electrophotography Principles and Optimization*. Research Studies, Letchworth UK, 1984.
12. Schein, L. B. *Electrophotography and Development Physics*. Springer-Verlag, New York, 1988.
13. Williams, E. M. *The Physics and Technology of Xerographic Processes*. Wiley, New York, 1984.
14. Levy, A. U., and Biscos, G. *NonImpact Electronic Printing*. InterQuest, Charlottesville VA, 1993.
15. *Proceedings of the 1st–16th International Congress On Non-Impact Printing Technologies*. IS&T, Springfield, VA, 1982–2000.
16. Shimuzu, K. I., ed. *Hard Copy and Printing Technologies*. *SPIE Proceedings Vol.* 1252. SPIE, Bellingham WA, 1990.
17. *Proceedings of the 1st–17th Toner and Photoreceptor Conference*. Diamond Research, Ventura CA, 1983–2000.
18. *Proceedings of the 1st–7th Laser Printing Conference*. IMI, Kingfield ME, 1989–1996.
19. *Proceedings of the 1st–7th Digital Electronic Printing Presses Conference*. IMI, Kingfield ME, 1993–1999.
20. Pai, D. M., and Springett, B. E. *Reviews of Modern Physics* 65, 163 (1993).
21. Yule, J. A. C. *Principles of Color Reproduction*. John Wiley, New York, 1967.
22. Kang, H. R. *Color Technology for Electronic Imaging Devices*. SPIE Press, 1996.
23. Pierce, P. E. and Marcus, R. T. *Color and Appearance*. Federation of Societies for Coatings Technology, Blue Bell PA, 1994.
24. Hunter, P. S. *The Basics of Color Appearance*. Hunter Associates Lab., Reston VA, 1992.
25. *Precise Color Communication: Color Control from Feeling to Instrumentation*. Minolta Corp., Ramsey NJ, 1993.
26. *Understanding Color*. 3M Printing and Publishing Systems, St. Paul MN, 1994.
27. *The Color Guide and Glossary*. X-Rite, Grandville, MI, 1996.
28. *An Introduction to Digital Color Printing*. Agfa Educational Publishing, Randolph MA, 1996.
29. *A Look Inside the Color Laser Writer*. Apple Computer, Cupertino CA, 1995.

5

Dry Toner Technology

PAUL C. JULIEN and ROBERT J. GRUBER

Xerox Corporation, Webster, New York

5.1 INTRODUCTION

5.1.1 History of Xerography

The growth of xerography has been a classic example of the successful commercialization of an invention (Dessauer, 1971). Chester Carlson first used a photoconductive material to make an image on October 22, 1938. At the time Carlson was working with patent applications and was frustrated with the enormous labor involved in making copies of documents. The invention was the direct response to a perceived problem. Nonetheless it was almost 10 years before the Haloid Corporation saw the technology at the Battelle Memorial Institute and decided to commercialize it, and it was more than 10 years from that time before Haloid (now called Xerox) introduced the Xerox 914 automatic copier in 1959.

From this time the growth of Xerox was explosive, reaching $1 billion in sales by 1968. Although Carlson had used sulfur as the first photoactive material, selenium metal was found to be much more practical, and this was the basis for Xerox's technology in the 1960s. The first commercial "dry inks" used with selenium were based on styrene methacrylate copolymers and had a negative electrical charge.

Other companies were eager to participate in the profitable business of xerography. Thus during the 1970s IBM and Kodak developed and introduced copiers based on organic photoactive materials and positive charging toners. With the 1075 copier, Xerox introduced its own organic photoreceptor and positive toner, while continuing to introduce products based on improved selenium photoreceptors and negative toners.

During the 1980s, Japanese manufacturers such as Canon and Minolta started introducing low-speed copiers based on selenium and cadmium sulfide photoreceptors and again using negative toner. They also introduced dry toner copiers using single-component development, eliminating the use of carrier beads.

Since the early 1980s, many combinations of single- and two-component development and positive and negative toner have been used in the industry. Presently companies are improving toner performance within the design options that they have chosen. None of the options has decisively shown itself to be superior in all applications.

The 1990s have been distinguished by the emergence of color copiers and printers as a major trend in the industry. This has been driven by computers in two ways: first, through creating the need for color output by the creation of color images at computer workstations, and second, through enabling sophisticated correction algorithms to achieve proper color in a given machine.

5.1.2 The Xerographic Process

Several books and articles on the xerographic process and its individual steps have been written over the years (Schaffert, 1980; Scharfe, 1984; Williams, 1984; Schein, 1996; Hays, 1998). The following summary places the role of toner in the context of the entire process. Each step in the process has been the recipient of considerable ingenuity over the years and has a multiplicity of options available. Typical realizations of a particular step are mentioned, but those given are by no means exhaustive.

While the xerographic process comprises many steps involving different materials technologies, the key phenomenon that enabled Chester Carlson to invent his copying process is the existence of materials that will hold a charge in darkness but conduct electricity when exposed to light. These electrically resistive substances, called photoconductors, typically work best when charged to a particular polarity. The photoreceptors (light-sensitive devices) based on the metal selenium used in the earliest xerographic machines are typically sensitized with a blanket positive charge, while later organic materials are negatively charged. These differences normally lead to a change in sign throughout the xerographic process when lenses are used to cast the image of a document on the platen onto the photoreceptor. In particular, they determine the sign of the dry toner used to develop the latent image. On the other hand, laser imaging allows us to discharge the photoreceptor either where we want the image to appear after development (discharged area development) or where we want blank areas after development (charged area development, as in light lens based copiers). The choice of either of the two options will change the charge polarity of the toner required. The choice is typically determined by many factors such as print quality or the availability of a suitable toner of the correct charge.

A practical copier based on photoconductivity goes through six basic steps in reproducing a document (Fig. 5.1): charging, exposure, development, transfer, fusing, and cleaning. A seventh step (erasure) is discussed below but not shown in Fig. 5.1. All the steps are repeated for each additional copy.

In the charging step, the photoreceptor is covered with ions of the appropriate polarity through the use of a wire or grid biased to high voltage.

In the exposure step, an optical system forms an image of the document on the photoreceptor. For a light lens copier, where the document is white, there is sufficient light to cause the photoreceptor to conduct charge and neutralize the image, but the dark lines of text leave the charge undisturbed when imaged on the photoreceptor. This process forms a latent image of charge, duplicating the original document.

In the development step, toner of the appropriate polarity is typically brought into contact with this image. As discussed above, this will vary with the choice of photoreceptor and imaging method. There is now an image of the document on the photoreceptor

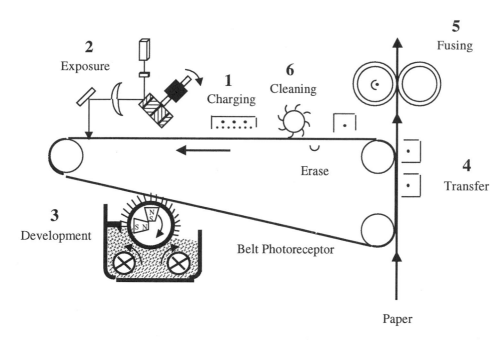

Figure 5.1   The basic steps in the xerographic process: (1) charging the photoreceptor, (2) exposing to form the latent image, (3) developing the latent image into a real image, (4) transferring the image to paper, (5) fusing the image to paper, and (6) cleaning residual toner from the photoconductor.

formed by individual toner particles held to the photoreceptor primarily by electrostatic forces.

In the transfer step, a piece of paper is brought into contact with the photoreceptor, and the back side of the paper is charged with ions opposite in polarity from the toner. This attaches the great majority of the toner particles to the paper. The paper is then removed from the photoreceptor and passed through the fuser.

In the fusing step, the toner is melted onto the surface of the paper. This is done either by running the sheet of paper between two rolls, at least one of which is heated, or by exposing it to radiant heat from a lamp. In an alternative arrangement called cold pressure fixing, high pressure is used without heat to force the toner into the paper.

In the cleaning step, the small amount of toner remaining on the photoreceptor after transfer typically should be removed if the photoreceptor is to be reused in its original condition. The toner is swept off the photoreceptor with a brush of furlike material, a brush of xerographic carrier beads electrically biased to remove toner, or a conformable rubber blade.

In the erase step, the photoreceptor charge is reduced to zero by exposing it to a lamp, which causes the entire width of the photoreceptor to conduct electricity. This erases any remnants of the latent image.

The entire process is then repeated as the photoconductive drum or belt returns to its starting point.

The steps that are most important to dry toner design are the development and fusing steps. The latter to a large extent determines the polymers that can be used in toner fabrication, since the toner should melt in the fuser and adhere to the paper without contaminating the fuser itself. Similarly, cold pressure fixing has its special requirements for toner design (Bhateja and Gilbert, 1986).

Once a toner polymer has been selected and developed, most of the work of toner design goes into assuring that all the toner particles have the proper charge level to give sufficient development without background. This charge level will vary with the machine design.

In addition to these primary considerations, the effects of the toner design on the other steps in the process should be considered. The bulk of the toner should leave the photoreceptor during the transfer step; any remaining toner is typically removed during cleaning. Any microscopic toner constituents that are not removed by the cleaner should not degrade the charging properties of the photoreceptor, because this would affect subsequent charging, exposure, and erase steps.

5.1.3 Two-Component and Single-Component Developers

There are two primary methods of charging the dry ink and presenting it to the charge pattern on the photoreceptor to develop the latent image. The first, two-component development, was universally used in the early commercial applications of xerography. The charge is generated on the toner particles by mixing them with much larger beads chosen so that there is sufficient difference in the electrical nature of the toner and carrier materials to generate a charge of the desired magnitude on the toner. The mixture is then brought in contact with the latent image on the photoreceptor. A combination of impact and electrostatic forces strips the toner particles from the carrier beads, and the fields produced by the latent image attach them to the photoreceptor (Schein, 1996).

The earliest machines used nonmagnetic carrier beads and poured the developer over the latent image to bring the toner into contact in a process called cascade development (Schein, 1996).

Modern designs almost invariably have magnetic carrier beads and use magnets to control the flow of the developer and bring it into contact with the image. Since the magnetic fields in the contact zone with the latent image are typically designed to form a brush of carrier beads, this form of development is called "magnetic brush development" (Schein, 1996). Figures 5.2 and 5.3, respectively, show a typical two-component developer housing and a strand of the magnetic brush between the developer roll and the photoreceptor. Electrostatic forces drive the toner to image on the photoreceptor, while magnetic forces hold the carrier beads on the roll. Section 5.4 discusses toner properties for this application in more detail.

The use of carrier beads is not essential to xerography. The possibility of charging toner and then bringing it into contact with the latent image without using carrier has been mastered by many companies, with the introduction of successful products incorporating what is called "single-component technology" (Schein, 1996; Bares, 1993). A typical housing utilizing this technology is shown in Fig. 5.4 (Takahashi et al., 1982). Section 5.5 discusses toner designs targeted for this type of application.

At present both technologies play an important role in the marketplace. Single-component designs tend to be used mostly in smaller, slower machines where compactness, simplicity, and low cost are prime requirements, while two-component designs still dominate the high-speed machines, where clean copy quality and stability over long periods of time are most important. While the same toner technologies can be used for both types of application, the details of the toner design may be significantly different.

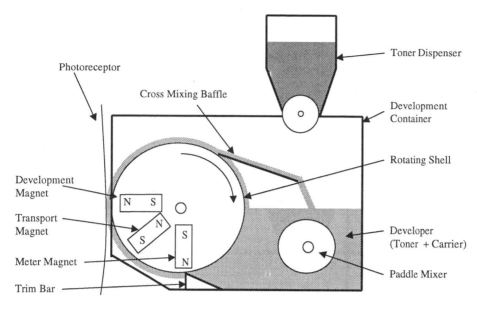

Figure 5.2 Structure of a typical two-component housing.

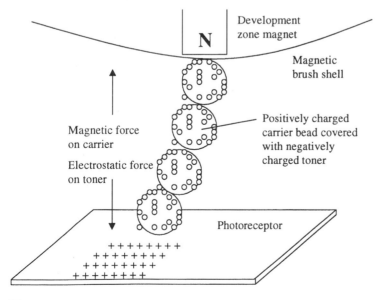

Figure 5.3 Schematic development, indicating direction of electrostatic force on toner and magnetic force on carrier above an image. (Adapted from Scharfe et al., 1989.)

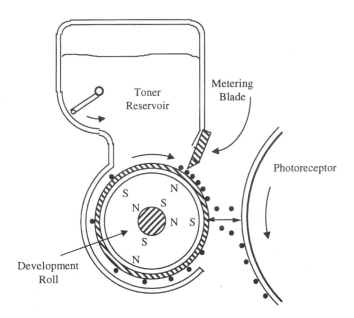

Figure 5.4   Single-component development unit with magnetic toner. (Adapted from Hays, 1998.)

5.2 TONER COMPONENTS

Dry xerographic toners consist of a colorant in a binder resin. Beyond these essential ingredients, a particular toner design may contain charge control additives to control the charge level, surface additives to control flow and cleaning properties, magnetic additives to aid in toner control, and waxes to promote toner release from the fuser roll. These components are introduced in the subsections that follow and are treated in more detail when the toner function they affect is discussed.

5.2.1 Resin

Several different families of resins (polymers) have found frequent application in xerography depending on the fixing technique selected (Table 5.1). The role of the resin in a toner is to bind the pigment to the paper or transparency material to form a permanent image. This is typically done by selecting a polymer that will melt at a reasonable temperature when heat is applied in any of a number of ways or one that can be forced into the paper fibers at high pressure without additional heat.

The materials for the last application are typically lower molecular weight polypropylenes, polyethylenes, ethylene–vinyl acetate copolymers, and mixtures of these materials. These cold pressure fix materials have the advantage of requiring low power in operation and no standby power (Bhateja and Gilbert, 1985a, b, 1986). They commonly have the disadvantage of producing high-gloss images that can be easily damaged by rubbing. However, they are perfectly acceptable for some applications, most notably computer printing (Rumsey and Bennewitz, 1986).

In other applications, a continuous radiant source of heat such as a quartz lamp or heated coil is used to melt the toner into the paper fibers. The viscosity of the toner usually reaches quite low values in flowing into the paper, but the time allowed for heating can

Table 5.1 Common Toner Resins

Polymer name or class	Morphology	Melt properties T_g or T_m (°C)	Molecular weight range ($\times 10^{-3}$)	Application
Polystyrene *n*-butyl methacrylate	Amorphous	T_g 50–60	50–60	Roll and flash fusing
Polystyrene *n*-butyl acrylate	Amorphous	T_g 50–60	50–60	Roll and flash fusing
Polyester	Amorphous	T_g 50–60	8–30	Radiant and roll fusing
Epoxy	Amorphous	T_g 60–100	1–10	Roll and flash fusing
Polyethylene	Crystalline	T_m 86–130	0.5–15	Cold pressure and roll fusing
Polypropylene	Crystalline	T_m 130–165	3–15	Cold pressure, roll fusing, and release agent[a]

[a] Added to provide release between fuser roll and toner surface.
(*Source*: Adapted from Gruber et al., 1989.)

be up to 500 milliseconds. Here polyesters and epoxies are often used. In this application the molecular weight of these materials ranges from 5,000 to 50,000 and the glass transition temperature Tg, from 50 to 60°C (Palermiti and Chatterji, 1971). A notable contemporary use of radiant fusing is the simultaneous duplex digital press developed by Xeikon (Tavernier et al., 1995).

In flash fusing, the toner is melted into the paper by a very short high-intensity flash of light lasting less than 5 milliseconds. Toner temperatures typically exceed 200°C in attaining the low viscosities required, and the thermal decomposition of the toner polymer is a significant problem. Styrene copolymers, epoxies, and copolycarbonates have all been used (Gruber et al., 1982; Narusawa et al., 1985).

The great majority of copier designs use hot roll fusers for fixing the image (Kuo, 1984; Prime, 1983; Lee, 1975). The paper with the unfused toner passes through a nip formed by a heated roll and a backup roll forced against the heated roll at fairly high pressures. This combination of temperature and pressure gives the best overall performance for most applications. Styrene copolymers such as styrene acrylates, methacrylates, and butadienes are used. Molecular weights range from 30,000 to 100,000, and glass transition temperatures range from 50 to 65°C (Nelson, 1984). Where lower melting temperatures Tm, are desired, polyester resins have been used (Fukumoto et al., 1985). When multiple layers of toner are fused, the low melt characteristics of polyester toners are especially important, and most contemporary full-color copiers and printers use polyester toners.

The above-named polymers can be specially modified for a particular application by incorporating monomers or side chains to accomplish such functions as charge level modification (Gibson, 1984). However, the primary role of the polymer resin is to fix the image, and polymer designs are developed with their fusing characteristics as the primary consideration.

Toners are typically manufactured by the attrition of the blend of the ingredients. Here the ease with which the polymer fractures has a very significant influence on the rate at which one can produce toner. At the same time, the toner should not break up when subjected to the stresses within the developer housing (Ahuja, 1976). These mechanical characteristics are another essential consideration in polymer design.

Since usually 90% or more of the toner is polymer, its cost is very important in determining the cost of the final product. Polyester resins are typically more expensive than styrene acrylate resins and thus tend to be limited to color toners or to black toners intended to be used with the color toners. It is also possible to form hybrids of polyester and styrene acrylate, and these may have improved properties (Kawaji et al., 1995; Aoki et al., 1996).

Recently, chemical methods of producing toner have received increased attention from several manufacturers. These are discussed in more detail in Section 5.6.6. Polymers that are typically produced by suspension or emulsion polymerization methods have an advantage here, and much of the published literature is concerned with styrene acrylate type resins rather than those typically produced by bulk polymerization such as polyesters.

5.2.2 Colorants

The most common colorant for xerographic toners is carbon black (Julien, 1993). Most manufacturers offer a range of blacks that differ in such properties as tinting strength and acidity. Important properties of carbon blacks for xerographic applications are their dispersibility in the resin in hot melt mixing and their tendency to charge either positive or negative. Carbon blacks are usually used in toner at a loading of 5 to 15% by weight.

Besides carbon black there are several other materials that can be used to make black toners. Magnetite is often used in toners to allow for magnetic control of the toner. The substance is typically black and is seldom used as a pigment per se, but often the loadings for magnetic properties are sufficiently high that additional pigment is not necessary.

Some charge control additives such as nigrosine are good black pigments, and their use in a toner can lead to the reduction or elimination of the carbon black.

Pigments other than black are increasingly playing a role in xerography in two applications. The first is a color to be used in addition to black when there is a desire to highlight certain information. Typical colors used for this application are red, blue, green, and brown, made from either a single pigment or a blend of pigments.

The other major application is in the creation of full-color documents. Here the subtractive set of pigments, cyan, magenta, and yellow, is used. These pigments are chosen for colorimetric properties such as spectral purity and their ability to generate as broad a gamut of colors as possible when blended together. To give permanence to the color image, a degree of lightfastness is useful (Blaszak et al., 1994). This suggests that pigments are more useful than dyes. Usually organic pigments are used. Copper phthalocyanines are often used for cyans and blues, azo pigments for yellows, and quinacridones or rhodamines for magentas and reds (Bauer and Macholdt, 1995). Figure 5.5 shows graphically the color range accessible by the various chemical classes with typical pigments within each class. Color pigments are chemically active materials, and this can affect both their xerographic properties and their compatibility with existing chemical laws (Macholdt and Bauer, 1996).

5.2.3 Charge Control Additives

Charge control additives are often added to a toner when the pigment (chosen for its color) blended into the polymer (chosen for its fusing performance) does not give an adequate

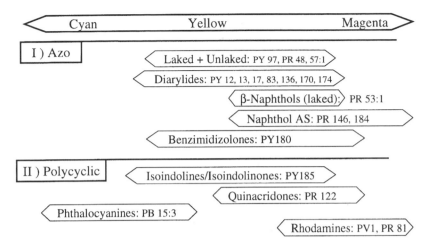

Figure 5.5 Pigments for full-color copying and printing; PY = Pigment Yellow, PR = Pigment Red, PB = Pigment Blue. (Adapted from Macholdt and Baur, 1996.)

charge level or rate of charging. This can occur in both positive and negative charging applications.

For positive applications, one family of charge control additives is the quaternary ammonium salts (Lu, 1981). These compounds are mostly colorless, allowing their use in color applications.

Nigrosine is an organic pigment that is effective in imparting a positive charge to toner (Guay et al., 1992). Because it is black, however, it is unsuitable for most color applications.

For negative applications, acidified carbon blacks have been quite successful as charge control agents in addition to their pigment qualities (Julien, 1993). In applications calling for transparent internal negative charge control, metal complexes have been found to be effective (Inoue et al., 1985; Birkett and Gregory, 1986). However, the oxide surface additives described in the next section have become increasingly important as charge control agents.

5.2.4 Surface Additives

The multifunctional benefits of surface additives such as silicas and titanias have received increased attention in recent years. When materials such as fumed silica are added to the surface of a toner, the flow properties often improve dramatically (Veregin and Bartha, 1998). The silicas can also improve transfer from the photoreceptor to paper by lowering the adhesion of the toner to the photoreceptor surface (Akagi et al., 1993) while improving the charge stability of the toner and carrier mixture (Nash and Bickmore, 1988; Stuebbe, 1991). The chemical treatment of silica surfaces has been found to have a profound influence on the properties of toners made with them (Julien et al., 1993; Veregin et al., 1997). In particular, amine treatments convert silicas from negative to positive charging materials (Takenouchi, 1986; Heinemann and Epping, 1993).

For blade cleaning, surfactant materials such as zinc stearate are often blended with the toner to lubricate the blade passing over the photoreceptor (Weigl, 1982). Fumed silicas may be used with the stearate to control the buildup of material on the photoreceptor.

5.2.5 Magnetic Additives

Magnetite is primarily added to toner in single-component applications, where it enables the transport of the toner through the developer housing and against the latent image under magnetic control (Button and Edberg, 1985). The additives are typically a few tenths of a micrometer in size.

It has been found that even in two-component development, where magnetite is not necessary for developer transport, this material offers advantages in controlling machine dirt (Knapp and Gruber, 1985). Here the typical loadings are about 15 to 20% as opposed to the 60 to 70% necessary for toner transport. Without magnetite, low charge toner often gives severe dirt and dusting problems.

If sufficient magnetic remanence is present within a toner formulation, the toner can be used for magnetic ink character recognition (MICR), a special xerographic application for check sorting, discussed in Section 5.4.3. Typically, a high-remanence magnetite is used at a loading in excess of 20% by weight.

5.2.6 Other Additives

Fuser rolls typically require the use of a release agent such as silicone oil to prevent the adhesion of the toner to the hot roll surface (Seanor, 1978). Hardware design is simplified if this release agent management system can be eliminated. This both lowers the cost of the machine and eliminates the need for the customer to handle fuser oil. It is possible to do this by incorporating a low molecular weight polyethylene or polypropylene wax into the toner itself (Gruber et al., 1986). This flows very readily at temperatures sufficient for toner fusing and fills the role of the silicone oil. Partial cross-linking of the polymer in the toner also helps prevent adhesion to the fuser roll (Inoue et al., 1985). This technology is now being used for color toners (Kawaji et al., 1997).

5.3 TONER REQUIREMENTS AND CHARACTERIZATION TECHNIQUES

The development of a toner involves the choice of the components described above, starting with the choice of polymer and proceeding through the selection of each component to fit the particular application. Techniques have been developed for each of the requirements to facilitate selection, and these are often unique to the xerographic application.

5.3.1 Rheology

There are three or four xerographically significant temperatures necessary to characterize a toner for xerography. The most obvious is the temperature at which the image is fixed to the paper. This will vary with the degree of fix required, but for an adequate fix level it is called the minimum fix temperature. Above this is a temperature at which the toner is so fluid that it simply splits apart when the paper leaves the fuser roll, leaving traces of the image on the fuser roll to contaminate the next sheet. This is called the hot offset temperature. Next, the toner should not sinter when left on a loading dock or in the machine toner hopper, even at an elevated temperature. The temperature at which significant sintering occurs is called the blocking temperature (Weigl, 1982).

For color printing, the gloss of the image is also very important (Dalal et al., 1991), and it is strongly affected by fusing conditions (Nakamura, 1994). Thus there can be a fourth critical temperature for color toners, that is, the fuser temperature at which a toner reaches the desired gloss level, called the gloss temperature. Ideally, a toner should provide the desired gloss level at the minimum fix temperature, but this is not always true.

Adequate fix is a subjective judgment and can vary with the application. Measurements of fix level generally involve abrading an image on paper in a controlled way and measuring the amount of toner lost from the paper or the change in the density of the image (Bhateja and Gilbert, 1985a; Hayakawa and Ochiai, 1993).

The minimum blocking temperature allowable is determined by the highest temperature one expects the toner to see outside the fuser. It is often asked to be about 115°F. To a large extent this is a characteristic of the toner resin. However, it can be modified by the other ingredients of a toner. For example, carbon black can act as a reinforcing agent, raising all the characteristic temperatures including the blocking temperature (Ahuja, 1980c; Fox, 1982). Similarly, surface additives such as fumed silica will typically prevent sintering and hence raise the blocking temperature (Barby, 1976).

Even though additives can modify the results, the primary means for controlling the fusing properties of a toner is through the resin. The lowest possible minimum fix temperature is desirable, since this will typically minimize power consumption and maximize fuser life. For a given minimum fix temperature, the hot offset temperature should be as high as possible to maximize fuser latitude. The latter two temperatures in particular should be measured in the fuser of interest. More general rheological measurements can serve as an aid to polymer design, although they are not sufficient for final approval.

A commonly measured characteristic of a polymer is its glass transition temperature, T_g, where the polymer changes from a hard glass to a rubbery state. This is measured in a differential scanning calorimeter, which looks for the change in heat capacity at the transition. For adequate blocking, toners generally should have T_g values above 50°C.

Another relevant characteristic of the polymer is its melt viscosity, that is, its ability to flow at a given temperature. The rate at which this falls at and above T_g determines the temperature above the blocking temperature at which the toner flows into the paper and the temperature above this at which the toner splits, leaving residue on the fuser roll as well as on the paper.

These characteristics can be controlled by varying the composition and molecular weight of copolymers. This can be illustrated by examining one of the most common copolymers, styrene-butyl methacrylate. For a random copolymer, the T_g is given by

$$T_g = W_1 T_{g1} \pm W_2 T_{g2} \pm k W_1 W_2 \tag{5.1}$$

where W_1 and W_2 are the weight fractions of constituents 1 and 2 and T_{g1} and T_{g2} are the transition temperatures of the homopolymers (Manson and Sperling, 1976). The constant k allows for deviations from the ideal case. However, the transition temperature of the copolymer also depends on its molecular weight for lower molecular weights. This approaches an asymptote for high molecular weights due to chain entanglement. Figure 5.6 shows that the transition temperature for the copolymer decreases as the amount of *n*-butyl methacrylate increases (Scharfe et al., 1989). For any copolymer, the ratios are adjusted to obtain a T_g that corresponds to an acceptable blocking temperature.

The effect of molecular weight is shown in Figure 5.7 (Scharfe et al., 1989). T_g appears to reach a plateau at a molecular weight of about 23K. For a polystyrene homopolymer, the plateau is reached at about 34K. The plateau is reached earlier for the copolymer because its chains are more flexible, allowing earlier entanglement.

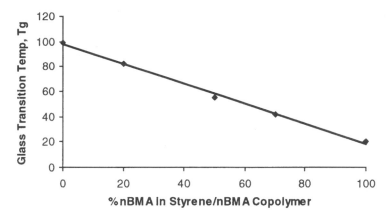

Figure 5.6 Effect of copolymer on glass transition temperature T_g of styrene-n-butyl methacrylate copolymers. (Adapted from Scharfe et al., 1989.)

The viscoelastic properties of polymers are measured in a viscometer. The viscosity of polymers typically depends on the frequency of the driving force as well as the temperature. It is found that time (the inverse of frequency) and temperature are interchangeable for typical (linear) polymers. Data often are expressed as viscosity as a function of reduced frequency $a_T\omega$, where ω is frequency, and a_T is the temperature-dependent shift factor.

Figure 5.8 shows the viscosity of styrene-n-butyl methacrylate toners as a function of reduced frequency (Ahuja, 1980b). Clearly the 399K molecular weight polymer has a higher viscosity, as well as a greater dependence of viscosity on temperature and time (which could be time in the fusing nip). When random copolymers of two different molecular weights (e.g., A = 29K and B = 371K) are blended, the mixture will have a viscoelastic response that is between that of the two original polymers and is monotonically dependent on the amounts of the two polymers, as shown in Figure 5.9 (Ahuja, 1980b). In particular, the addition of even 10% of the higher molecular weight material significantly raises the viscosity at higher times and temperatures. Blends of different molecular weights

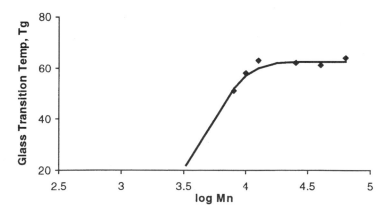

Figure 5.7 Effect of molecular weight on glass transition temperature T_g of styrene-n-butyl methacrylate copolymers. (Adapted from Scharfe et al., 1989.)

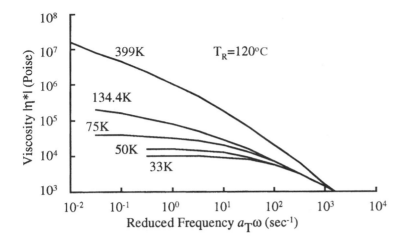

Figure 5.8 Viscosity as a function of reduced frequency $a_T\omega$ for various styrene-*n*-butyl methacrylate copolymers at reference temperature T_R in °C. (Adapted from Ahuja, 1980b.)

are a common method of modifying the fusing properties of a toner. For example, a broad molecular weight distribution or a blend of low and high molecular weight resins has been shown to allow high hot offset temperatures while retaining low minimum fix temperatures (Hayakawa and Ochiai, 1993).

The effect of the addition of carbon black to a particular polymer is shown in Figure 5.10 (Ahuja, 1980c). This effect is related to the reinforcement properties of the pigment, a property varying with the carbon black but controlled and characterized by the carbon black industry, particularly for rubber applications. In general, high surface area carbon blacks tend to raise viscosity more than low surface area carbon blacks. Similarly, high loadings of any color pigment can reinforce the pigment–polymer matrix, leading to higher fusing temperatures.

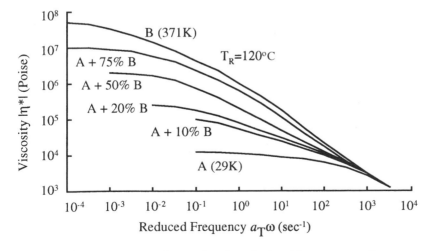

Figure 5.9 Complex viscosity as a function of reduced frequency $a_T\omega$ for styrene-*n*-butyl methacrylate blends. (Adapted from Scharfe et al., 1989.)

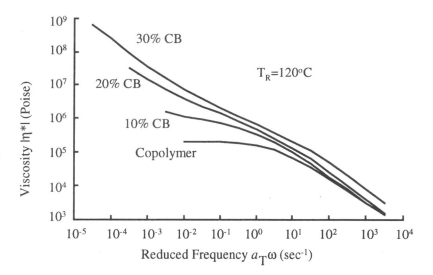

Figure 5.10   Effect of carbon black (CB) concentration on melt viscosity as a function of reduced frequency. (From Ahuja, 1980c.)

Molecular weight and copolymer ratio are also important in toner processing. Toners made from a molecular weight above a value called the brittle–ductile transition are very ductile and difficult to micronize. Similarly, copolymers such as styrene-butadiene with a large amount of butadiene are well known to be rubbers, hence are also very difficult to micronize efficiently. On the other hand, the more ductile polymers are less susceptible to fracture in a developer housing. In practice, polymer compositions are selected to effect a compromise between toughness and processibility. As discussed below, chemical toner preparation processes eliminate the need for micronization and thus may enable new or different classes of polymers to be used for toner application.

5.3.2   Colorimetrics

For black toners, the primary consideration is to generate sufficiently high optical densities with practical developed masses. In practice this has meant that the use of 5% or more of almost any of the available carbon blacks is quite sufficient.

For highlight color toners there is again a requirement to develop sufficient optical density of the color of interest. The amount of pigment necessary to do this will differ with the color strength of the pigment, which can vary quite significantly. Beyond the strength or chroma of the color, a hue that will be pleasing to the largest number of customers should be chosen. For example, the red for highlight color applications can be chosen from an infinite number of possibilities.

For process color developers the goal is to generate as wide a color gamut as possible from a particular set of cyan, magenta, and yellow toners. The range of colors possible depends on the detailed spectral absorption of the various pigments. As a result, a wavelength-scanning spectral densitometer is necessary. With this tool, the range of colors possible for a given pigment set can be calculated. Various combinations of pigments can then be evaluated to find the optimum set.

It is extremely difficult to find subtractive pigments that perfectly match any red, green, and blue filters used for optical color separation. To compensate for this, the image is completely converted to electronic information, adjustments are made to the strength of each color dot depending on the complete color signal, and the image is then reformed on the photoreceptor using a laser printer (Schein, 1996). In another application of electronic color correction, black toner can be added to the image to reduce the amount of more expensive color toner necessary.

Each of the candidate pigments is typically evaluated for lightfastness (Blaszak et al., 1994). If one of the pigments fades quickly, especially under office lights, this will distort the initial colors produced by the copier (Fujii et al., 1988). The degree of lightfastness necessary will depend on the application; however, since one wants to use a limited number of pigments, the goal is usually to obtain values sufficient for almost any possible application.

Pigment evaluation can only predict the largest gamut possible. The real color gamut a given set of toners will produce in a particular color copier is also affected by the smoothness of the image formed by the fused toner and the general quality of the image formed by the system (Nakaya et al., 1986).

5.3.3 Particle Size

Toner particle sizes have been slowly decreasing over the years and are now generally in the range of about 7 to 12 μm in diameter. Particle sizes significantly larger than this usually produce ragged lines and dots and thus degrade copy quality. As a result, smaller sizes have been found to be superior for color reproduction (Chiba and Inoue, 1988) and for noise reduction in general (Shigehiro et al., 1993). However, for a given resin the smaller sizes require longer grinding times in manufacturing and hence are more expensive to produce. Also, smaller sizes tend to produce more dirt at a given charge-to-mass ratio and to cause more rapid developer degradation (Nash and Bickmore, 1988).

Even if the average size is reasonable, a broad particle size distribution will introduce significant amounts of the small and large toner particles that cause dirt and copy quality problems. As a result, toner processing strives for the narrowest practical size distribution. This is typically done with well-designed micronization equipment followed by air classification and possibly sieving.

These limitations on the practical size of conventional dry toner are one of the attractions of the new chemical methods for making dry toner (Takezawa et al., 1997; Edwards et al., 1998).

The size distribution is measured by the use of the Coulter Counter (Beckman Coulter, Inc., Fullerton, CA) or similar instrumentation. This places the toner particles in a conducting fluid and examines changes in signal as the particles are passed through an orifice. The instrument is sensitive to the volume of liquid displaced by each toner particle and hence gives an accurate measurement of the volume size distribution of the toner. The distribution can be characterized in various ways, most typically by the geometric (logarithmic) standard deviation, the number fraction of particles below a certain size, and the volume fraction of particles above a certain size.

An optical image analyzer can also be used for toner size analysis if the particles are spread out on a microscope slide. This method looks at the projection of the particles in the object plane and can be used for linear or areal measurements. This method is often used in conjunction with a charge spectrograph to make simultaneous measurements of toner size and charge as discussed in Section 5.3.4.

Since the electrolytic displacement and optical methods measure the size in quite different ways, there can be a significant displacement between measurements from the two techniques, and comparisons should be made using the same method (Julien, 1996).

5.3.4 Charging

To control toner in the electric field of the photoreceptor, the particles are given a well-defined charge. Much of the progress in the creation of clean, crisp copies and prints by means of xerography has come through increased control over the toner charge distribution. This in turn has been facilitated by increased sophistication in charge characterization methods.

The primary method of characterization of the charge on toner has been the measurement of the charge-to-mass ratio (Schein, 1996). For two-component development, this is typically done by putting the developer in a metal cage with screens on each end; the screens are large enough to allow toner but not carrier beads to escape. The mass of the developer in the cage is determined, and the cage is connected to an electrometer. The toner particles are then blown off, and the charge and mass differences that result are measured. The ratio of these differences gives the charge-to-mass ratio for the toner blown off. This quantity is called the tribo or blowoff tribo for that developer.

The same quantity can be determined for toner on a photoreceptor or single-component donor roll by drawing the toner into a chamber connected to an electrometer and weighing the chamber before and after. Because of size selectivity in development, the blowoff tribo measured for toner developed onto a photoreceptor will not necessarily be equal to that for the same toner when still mixed with carrier.

Another method of charge characterization that has become increasingly common in the industry is the charge spectrograph. In the embodiment of this instrument used at Xerox (Lewis et al., 1981a,b; 1983), the toner is drawn into an airflow while simultaneously exposed to a perpendicular electric field. The combination of viscous drag and electrical forces determines where each individual toner particle falls on a collection filter. Rather than the charge-to-mass ratio (q/m) that the previous technique measures, the charge spectrograph measures the charge-to-diameter ratio (q/d). The average charge-to-diameter ratio can be determined by eye from the trace on the filter very quickly, but the instrument also allows the simultaneous determination of the charge-to-diameter ratios and the diameters for all the individual toner particles. For the latter application, computer analysis of the collection filters is necessary.

Other, commercially available, realizations of charge spectra characterization are described in the literature (Tsujimoto et al., 1991; Epping, 1988).

It is important to realize that charge spectrograph measurements can be significantly different from blowoff tribo measurements. Most theories of toner charging predict that the charge will be proportional to the toner area (Gutman and Hartmann, 1995), and this is indeed the general result (Nakamura et al., 1997). This implies that q/d^2 is a constant as the toner size changes, and this in turn leads to the result that q/m (proportional to q/d^3) increases while q/d decreases as the toner size decreases (Julien, 1996). For different toner sizes with the same composition, one method of charge characterization will give higher charge levels while the other will give lower levels. Thus it is always important to keep toner size and characterization technique in mind when interpreting charging results.

Knowledge of the charge on an individual toner particle may still be insufficient to predict the behavior of the particles. Van der Waals forces can introduce irregularity into the adhesion of the toner particle to a carrier bead or to the photoreceptor (Schmidlin, 1976), but a nonuniform distribution of charge on the toner particle can lead to even larger variations in toner release characteristics (Hays, 1978; Lee and Ayala, 1985). Considerable work has been done in the last decade to investigate toner adhesion (Ott et al., 1996), and the results have generally been consistent with the nonuniform distribution of charge (Feng and Hays, 1998). The inability of the toner designer to control charge at this level contributes to the empirical nature of much toner formulation work.

For the 8 μm toner particles typical of modern color printers, the useful range of charge, using charge-to-mass ratios, is from 15 to 40 μC/g. Toner particles with higher charge are difficult to strip from the carrier and deposit relatively little mass for a given amount of charge neutralization on the photoreceptor. Since the amount of mass deposited on the photoreceptor for a given latent image voltage is a prime measure of development efficiency, one would like the largest mass for a given amount of charge on the toner particle and hence the lowest q/m. However, values below 15 μC/g generally cause both dirt in the machine and background on the copy. As with so much else in xerography, the value used is a compromise between maximum development and minimum background for a particular hardware configuration.

Having the average charge level in the correct range is merely the first charging requirement that a toner faces. It should also reach this equilibrium charge level as rapidly as possible when added to the existing developer in a machine (Gutman and Matteson, 1998). Otherwise, it may still have little or no charge when the developer flow brings it into contact with the photoreceptor, and this can lead to background even though the bulk of the toner has adequate charge.

The charge on the toner should also be maintained as thousands of prints are made on the device. A design with initially excellent charging characteristics can rapidly lose its level in a period of time much shorter than the design intent of the copier or printer. Extensive work has been done on the aging mechanisms of two-component developers, particularly through the filming of the carrier by toner (Nash and Bickmore, 1988).

Furthermore, a toner or developer design is expected to retain its charge level through different environmental conditions. Moist air can have a dramatic effect on charge levels; very high levels such as 80 or 90% relative humidity can lower the charge, while very low levels such as 10 or 20% RH raise the charge. Hydroscopic additives in the toner can lead to very low charge levels at high humidity, but high charge levels at low humidity, and the resulting low developability can also be a problem. As a result, toners and developers should be tested early in a machine program on a bench scale and later in a full machine configuration at high, low, and moderate humidity levels.

5.3.5 Toner Flow and Adhesion

Toner flow and adhesion are related yet somewhat distinct properties. Toner flow describes how toner behaves in contact with itself (cohesion), while toner adhesion describes how toner behaves in contact with other materials, such as a donor roll or photoreceptor. Often, good toner flow is cited as an important toner property when the relevant process, such as transfer, really requires low toner adhesion (Tavernier et al., 1995; Sata et al., 1997). However, since low cohesion typically implies low adhesion, powder flow measurements are often a good guide for identifying toners with improved transfer.

There are many ways of measuring powder flow properties; most of them are static surrogate measurements for the dynamic properties of the toner. The angle of repose of a pile of toner measures how easily the toner slides, the bulk density or compressibility measures how easily it packs, and cohesivity measures how easily it passes through screens of various sizes. Hosokawa Micron manufactures a Powder Tester that can perform each of these measurements.

Toner adhesion as a unique property has received increased attention over the last decade. This is due to the increased presence of full-color copiers, where excellent transfer properties for all colors are essential to optimum copy quality, and to the increased presence of single component donor roll development, where low toner adhesion is required for high development efficiency. It is of course especially important for full-color non-magnetic single-component systems.

There are three common techniques for investigating toner adhesion (Ott et al., 1996). On a microscopic scale the atomic force microscope can be used to investigate the adhesion of individual toner particles to either other toner particles or to a substrate of interest. On a macroscopic scale the amount of toner transferred from one parallel plate to another as a function of the electric field can be monitored. Finally, centrifugal detachment can be used to measure the inertial forces required to separate a toner particle from the substrate upon which it is resting. The two macroscopic techniques are probably more common (Fukuchi and Takeuchi, 1998). The electric field detachment method is the closest surrogate to the actual development or transfer system itself.

Toner flow and adhesion are typically controlled by surface additives (Akagi et al., 1993; Sata et al., 1997; Veregin and Bartha, 1998).

5.3.6 Compatibility with Other Subsystems

The primary requirements for a toner concern the control of charge for development performance and the control of rheology for fusing performance. However, a toner should also be compatible with the other steps in the xerographic process, at times in ways that are not completely obvious.

The first subsystem the toner contacts after leaving the development zone is the photoreceptor. The toner should neither chemically alter the photoreceptor nor coat it with a thin layer of toner or toner constituents. A problem in this area shows up as a defect in the electrostatic image. For example, if the photoreceptor either cannot hold a charge or cannot be erased in an area where an image being used in testing contains a solid area patch, the root of the difficulty usually turns out to be a toner-photoreceptor interaction.

Ideally, the contact between the toner and the photoreceptor should be as gentle as possible to minimize the possibility of contamination. However, a typical cleaning system, whether a blade, a "fur" brush, or a carrier brush, effectively scrubs toner along the photoreceptor and hence has a tendency to smear it over the surface. With blade cleaning in particular, a lubricating film is often necessary for smooth blade action (Weigl, 1982). A soaplike molecule such as zinc stearate is often used for this purpose. This film should not alter the electrical performance of the photoreceptor.

In fusing, the obvious requirement is that the toner fuse to the paper without offset to the fuser roll. At the same time, the toner should not chemically attack the roll at the high temperatures encountered. Fuser rolls are often expected to last for hundreds of thousands of copies, and testing for interactions that can degrade ultimate performance is a very tedious but necessary part of any machine development program.

Once specific problems have been identified, either with the photoreceptor or with the fuser, quicker, more convenient bench tests usually can be implemented to assist in problem solving. Problems usually are first detected in full-scale systems tests, however, and any solution must eventually be verified in the same way.

5.3.7 Safety

The past 40 years, a period that has seen the xerographic industry reach maturity, has also seen greatly increased awareness of safety hazards and the corresponding development of means to detect them. This has in turn strongly affected the way materials are developed. In particular, as the concept of cancer-causing agents has become better known, the use of the Ames and related assays for mutagenicity has become an integral part of the toner design process.

There is nothing inherent in xerography that would require the use of dangerous materials, and the thrust in the industry is to make the materials as safe as any material a user would come in contact with. The safety of the various possible components that could go into a toner is considered before the research and development process is started. Often development of a particular compound is avoided even if mutagenicity tests have not been performed if related compounds have some form of toxicity. In the absence of detailed knowledge, xerographic evaluation is performed with reasonable precautions for personnel working with a comparatively unknown material.

Once a material has been identified as having promise, mutagenicity assays are performed, and any positive signal will disqualify the material. Acute toxicity tests and tests for skin and eye irritation are typically also performed. Assuming that the individual components of a toner pass all the safety requirements, the final toner design is again tested to assure that toner processing does not introduce new compounds that may be hazardous even though the initial ingredients were harmless. The results of the tests on the toners are summarized under the Toxicology and Health Information heading in Material Safety Data Sheets.

In addition to looking at the toner as a substance independent of its application, one can also carefully evaluate conditions (such as fusing) that can lead to the creation of hazardous substances from an otherwise benign toner. Any situation indicating a quantifiable (even if low-level) hazard can be examined to see whether there is a way of reducing the levels and hence minimizing the possibility of higher problem levels in the future.

It is prudent to monitor routinely even a material that has passed all safety requirements and has been introduced into the marketplace. The reason for this is that the source of an undesired substance is often a contaminated raw material. The use of a new processing technique or a new source of raw materials by a vendor of toner constituents can introduce mutagenic materials into the toner. The aim is to catch these changes as soon as they become detectable and eliminate the problem before it becomes hazardous.

5.4 TWO-COMPONENT DEVELOPERS

5.4.1 The Role of Carrier

The carrier in two-component systems provides two functions for the toner: charge generation and transport through the developer housing. First, the rubbing of the carrier against the toner generates the desired magnitude and sign of charge on the toner and a corresponding countercharge on the carrier bead itself, leaving a net neutral developer. Second, toner particles attach themselves to a carrier bead through electrostatic forces and can be moved through the machine by the action of magnets on the magnetic carrier core material.

The ability of a carrier to control toner charge is very powerful, primarily because it is possible to coat the core with a selected polymer. In contrast to the toner polymer, which is typically chosen primarily for its rheology, the carrier polymer can be chosen primarily for its charge generation properties. An important requirement is that one must be able to coat it on the carrier bead, either from a solution or a suspension or as a dry powder that can then be fused onto the surface. It then should have sufficient integrity on the carrier surface to ensure that the resultant toner–carrier pair has long life in the machine. In practice, different carrier polymers can often be chosen to drive the charge level on a given toner from beyond the range of useful charge in the positive direction through useful charge levels both positive and negative to beyond the range of useful charge in the negative direction. Thus a methacrylate-coated carrier can charge a toner to -40 μc/g, while a fluoropolymer-coated carrier will drive the same toner to $+50$ μc/g. While this flexibility does not guarantee that the charging rate will be adequate for both signs, it is still a powerful tool.

The transport function of carrier is also useful. Almost invariably either rotating magnets or a rotating shell over stationary magnets are used to transport developer through the development zone and to create a "magnetic brush" to transfer toner onto the latent image. Thus magnets are used to control toner even though the toner itself is nonmagnetic. Most single-component systems retain the use of magnets for toner transport, but this requires the use of magnetic toner, which either is black or has a very limited set of magnetic color pigments. Through the use of carrier beads, magnetic transport is possible with all color toners.

5.4.2 Two-Component Charging

Toner charge is primarily determined by chemical composition. Since the typical toner charges involve only 10^{-4} or 10^{-5} of the surface atoms, charging can be extremely sensitive to impurities. Nevertheless, the industry has been able to design toners and implement development based largely on empirical knowledge of the charging properties of the toner components.

The most basic description of our empirical understanding of the charging of different materials is to rank them according to which gains a positive charge and which gains a negative charge when two materials are rubbed together. This generates a triboelectric series, with materials that acquire a positive charge at one end of the series and those that acquire a negative charge at the other. Isolated cases of a particular polymer, metal, or other substance may violate this ranking, presumably as a result of impurities or surface treatment, but the tribo series does provide general guidelines for developing materials for xerographic application.

The extensive experimental and theoretical work on charge exchange between polymers and between metals and polymers has been reviewed by Harper (1967), Seanor (1972), and Lowell and Rose-Innes (1980). Models of toner charging in particular have been discussed over the years by Anderson (1994), Gutman (Gutman and Mattison, 1998), Nash (Nash et al., 1998), Schein (1998), and Takahashi (Lee and Takahashi, 1998). The references cited are only the most recent presentations at "Non-Impact Printing" conferences conducted by the Society for Imaging Science and Technology (IS&T).

Duke and Fabish (1978) modeled the electronic structure of polymers based on donor and acceptor states capable of forming localized molecular ions upon charge transfer. However, the spectroscopic techniques on which their work is based are still controversial (Lowell and Rose-Innes, 1980). More limited attempts to correlate specific chemical structures to charging tendency have led to general guidelines (Cressman et al., 1974; Gibson, 1984). In particular Gibson (1975) quantitatively correlated tribo charging with the Hammett constant, a measure of electron-withdrawing power, for a set of substituents for benzene. As a result of these and other studies, it has been established that highly halogenated polymers, such as polytetrafluoroethylene and polychlorotrifluoroethylene, because of the electron affinity of the halogens, are strongly negative charging. At the opposite end of the triboelectric series, polymers that have nucleophilic sites such as nitrogen or oxygen atoms either within the polymer backbone or as a side group tend to be strongly positive. Examples are polyamides, polyamines, and polyacrylates.

Copolymers have a triboelectric nature dependent on their constituents. A copolymer of styrene, a relatively negative polymer, and *n*-butyl methacrylate, a relatively positive polymer, has a triboelectric nature intermediate between the two, with copolymers containing more methacrylate being more positive than those containing more styrene. Thus any change that modifies the rheological properties of a resin can also modify its charging characteristics.

The polymers most useful for toner because of their rheological characteristics tend not to be highly nitrogenated or halogenated and thus are intermediate in charging characteristics, although they can differ significantly among themselves. Nitrogenous compounds rather than polymers are often added to toner to improve their positive charging, as discussed later, while methacrylates like polymethylmethacrylate are often used as positively charging carrier coatings to drive the corresponding toner more negative. Similarly, the highly halogenated polymers are often used as electronegative carrier coatings to drive toner positive. The rheological properties of a polymer to be used as a carrier coating are not as important as its toughness and resistance to toner filming.

When a pigment is added to the toner polymer, the charging of the toner depends on the pigment also. One of the most significant pigments for xerographic toners is carbon black (Table 5.2). As a result, the charging properties of carbon black in polymers have been extensively studied; this work was recently reviewed by Julien (1993). Carbon blacks without surface functional groups are found to be relatively neutral, lying near the styrene-acrylate resins commonly used in toners. The addition of these blacks to such resins has a relatively small effect on charging properties. On the other hand, there is a large class of acidified carbon blacks that are more electronegative because of the electrophilic nature of the acidic surface groups. Their position in the triboelectric series is close to that of pure polystyrene (Fig. 5.11). As a result, the choice of carbon black can significantly alter toner charging properties.

Carbon blacks can also affect the rate of charging because of their conductivity (Julien et al., 1992). If the charge on a toner particle resides on its carbon black, it is readily available for "charge sharing" when this toner particle collides with a recently added uncharged toner particle. On the other hand, if the charge resides on the insulating polymer, it is not readily available for exchange between toner particles. The location of the charge on a toner particle is determined in part by the difference in energy levels between the charge exchange sites. For negative charging, the carbon blacks lie at the extreme in a triboelectric series of carrier polymer, toner resin, and carbon black; as a

Table 5.2 Typical Carbon Black Properties[a]

Type[b]	Surface area (m^2/g)	Volatile content (%)	pH	Contact potential[c] (V)
Regal 330	94	1	8.5	0.00
Raven 1020	82	1	7.0	0.00
Raven 420	28	0.4	9.0	0.00
CSX99	520	5.1	8.3	−0.05
Raven 8000	935	9.6	2.4	−0.25
Black Pearls L	138	5.0	3.4	−0.40
CSX137	290	9.5	3.2	−0.50

[a] Except for contact potential, the values are those provided by manufacturer.
[b] The Raven blacks are produced by the Columbian Chemical Company. All the others are produced by the Cabot Corporation.
[c] Relative to unoxidized (low volatile content) blacks.
(*Source*: Adapted from Julien, 1982.)

result, they acquire the bulk of the negative charge. This charge is then available for charge exchange. On the other hand, if the same toner is driven positive through the use of a halogenated carrier polymer, the toner resin now is the extreme component, and it gains the bulk of the charge. This is not readily available for charge exchange, and charging processes are slow. Thus the same toner can exhibit very different charging rates depending on the sign of charge it is given.

The charging properties of color pigments are equally important in determining the properties of the associated toners. Macholdt and Sieber (1988) have made toners incorporating pigments in styrene-methacrylate copolymers and charged them against carrier

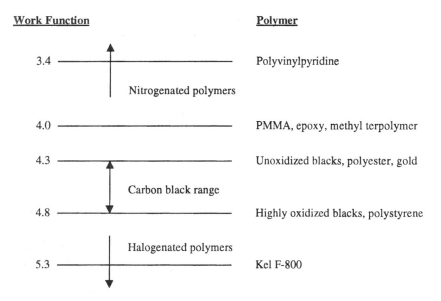

Figure 5.11 Carbon black in a triboelectric series; approximate work function values in volts. (From Julien, 1982.)

coated with the same polymer. They found that the copper phthalocyanines typically used for cyan and blue toner are relatively neutral, while the quinacridones used as magenta pigments are more negative, and the azo pigments used for yellow are more negative still (Fig. 5.12). However, the same authors demonstrate that treatment of the pigment surface can alter charging characteristics dramatically. Treatments include the inclusion of electrophilic or nucleophilic side groups, either deliberately or as a side effect of pigment finishing.

To relate Figs. 5.11 and 5.12, a styrene-methacrylate copolymer would probably lie between polyester and epoxy in work function in Figure 5.11, and the carbon blacks would probably look like the quinacridones in Fig. 5.12.

Of course, a pigment is chosen primarily for its visual characteristics. Since these are often highly critical, it is at times necessary to use pigments whose charging properties in the polymer of choice do not fall within the desired charge range. As a result, certain compounds called charge control agents (CCAs) are often added to toner for the specific purpose of modifying their charge. With black toners it is possible to use colored dyes as CCAs, e.g., nigrosine for positive charging (Guay et al., 1992) and metal complexes for negative charging (Fig. 5.13, Birkett and Gregory, 1984). With the metal complexes the nature of the positive counterion is extremely important and can change the triboelectric character of the material from negative to positive depending on whether the counterion or the metal complex is the more labile (Matsuura et al., 1993).

For general application with color toner, on the other hand, it is necessary for the charge control agent to be essentially colorless. For positive charging there are many nitrogenous compounds that are colorless. The most useful for many applications are the general class of quaternary ammonium salts (Fig. 5.14). Long chain aliphatic groups attached to the quaternary nitrogen tend to make the treated toner less hydroscopic and

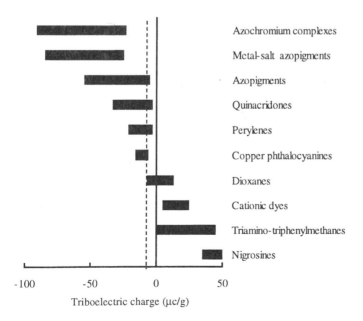

Figure 5.12 Effect of the incorporated pigments on the triboelectric charge of the test toner; broken line indicates the charge of the pure resin. (Adapted from Macholdt and Sieber, 1988a.)

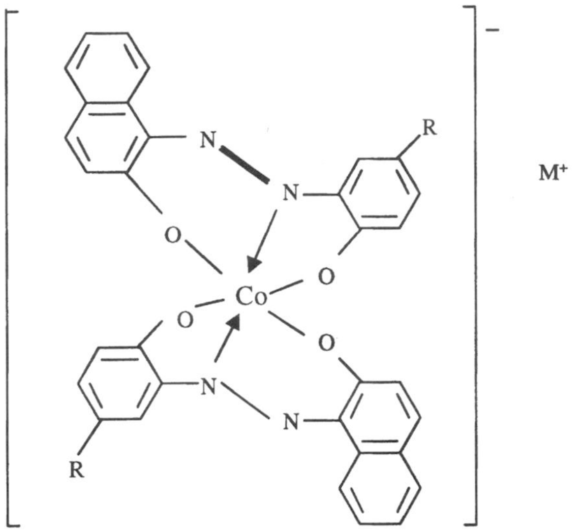

Figure 5.13 Typical metal complex for negative charge control. R can be SO_2, NH_2, Cl, etc.; M^+ can be Na^+, K^+, NH_4^+, etc. (From Birkett and Gregory, 1984.)

therefore less likely to experience changes in charge level with changes in humidity. Many variations on the aliphatic chain and counterion are typically examined to find the compound giving the best overall performance in a given formulation.

Compounds related to the quaternary ammonium salts but based on phosphorus or sulfur rather than nitrogen have also been examined for charging performance (Yourd et al., 1985). The concept remains that a large positively charged molecule with at least one aliphatic chain to couple it to the host polymer is combined with a small labile negative ion. One hypothesis for the charging mechanism of these materials is that some of the negative ions migrate away from the toner particle, leaving a net positive charge.

For negative charging black toners, carbon black can often provide all the necessary charging characteristics. On the other hand, color pigments are typically neither conductive nor at the negative extreme in triboelectric nature of the various toner components. Materials intended for use as negative CCAs typically follow the same design concepts as those

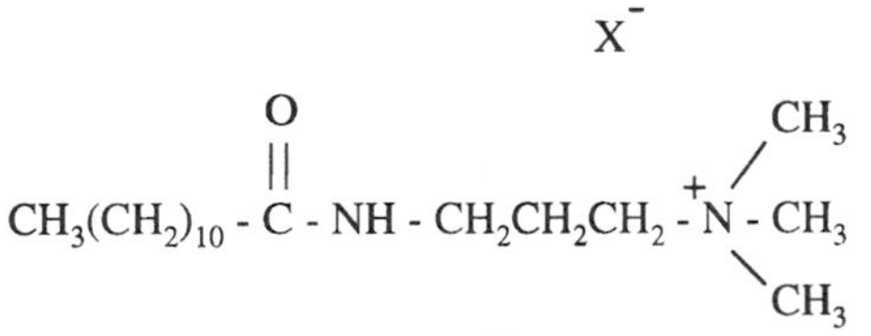

Anion X^-

Halogen e.g. Cl, Br, I
Fluoroborate BF_4
Sulfate $R\text{-}SO_4$
Sulfonate $R\text{-}OSO_2$

Figure 5.14 Typical example of a quaternary ammonium salt used in positive charging toners. The counter ion, X^-, is commonly chloride, bromide, or sulfate. (From Scharfe et al., 1989.)

for positive charge control agents but with opposite sign. Thus a large immobile molecule carrying a negative charge is paired with a smaller positively charged ion. This positive ion can be as simple as the proton itself. Gruber et al. (1983) found that the charging ability of various acids was directly related to their pKa value (i.e., their tendency to give up a proton).

Metals can also be used as labile positive ions. However, many metal complexes with large organic molecules are colored, hence of limited utility for general application to colored toner.

Surface additives such as fumed silicas used for flow properties can also be effective negative charge control agents (Schein, 1996).

The strong negative charging tendency of fumed silicas becomes a problem if one desires to improve the flow of positive toners. Sufficient silica to improve flow can reduce the positive charge on the toner to an unacceptably low level. Takenouchi (1986) discloses that the treatment of the silica with an aminosilane or other material can give the silica positive charging properties but renders it humidity sensitive. He found that additional treatment with a hydrophobic agent such as hexamethyldisilazane may be necessary. Degussa-Hüls (2000) and Wacker-Chemie (1998) have developed amino-silane treated silicas for positive toner applications.

There are so many requirements on charge control agents in terms of charge level and rate, aging and environmental stability, and interactions with the other systems in a machine that no additive performs ideally in all respects. As a result, there is continual research for better variations.

5.4.3 Special Magnetic Applications

While magnetite in the toner is unnecessary for developer flow when carrier beads are used, it is still useful to have. Inertial forces in the development nip tend to knock the larger toner particles off the carrier beads, generating either background on the copy or dirt in the machine. A small amount of magnetite in the toner generates an additional holding force, holding the particle against the developer roll. By adjusting the electric fields in the nip, one can obtain development as efficient as with nonmagnetic toner but with less toner released into the machine environment. This is especially true if there is a significant amount of low-charge toner in the developer.

An application in which magnetite in the toner is essential is in magnetic ink character recognition printing, abbreviated as MICR (Gruber et al., 1985). Here the characters at the bottom of a check (among other applications) can be read magnetically at very high speed by generating specific waveforms in specialized readers. This is usually done with lithographic inks containing high levels of magnetite. Through the use of special magnetic toners with higher levels of magnetite chosen to retain its magnetization longer than the grades usually used in xerography, printers such as the Xerox 9700 can generate images capable of being read in existing reading devices. A recent discussion of the special requirements of MICR toners has been given by Neilson (1998).

5.5 SINGLE-COMPONENT DEVELOPERS

5.5.1 Advantages and Disadvantages

There are several possible advantages that single-component technology may have over two-component developers.

The first of these concerns the volume occupied by the carrier. There is a constant desire to make things smaller in copiers and printers, to give the customer capabilities that earlier were available only in larger machines. This has put pressure on design engineers to reduce the development subsystem. By eliminating carrier beads, it may be possible to reduce the volume of the housing by an equivalent amount.

At the same time, two-component housings are becoming increasingly smaller. It is quite possible that a two-component housing could be made as small as a single-component housing and still deliver the same performance.

The charge level of the toner on the carrier is dependent on the amount of toner in the developer mix. To keep the toner concentration in a range where development is adequate without background, this concentration is typically continuously monitored and adjusted. This requires additional sensors, logic, and means of controlling toner dispensing. Thus a single-component system could be simpler, which should reduce cost and improve reliability.

Here again two-component technology is continually improving. Sensors and the associated electronics are becoming cheaper and may become an insignificant fraction of the cost of a housing. On the other hand, single-component development typically requires much closer tolerances in the development nip, among other places. A simpler but more precise housing may yet cost more.

Another consideration is development efficiency. Lower toner charge-to-mass ratios deliver more mass and hence density onto the photoreceptor for a given degree of charge neutralization. However, very low charge levels are difficult to work with in two-component xerography because carrier bead collisions can shake the toner off and allow it to fly around the housing, generating dirt and background. Single-component systems do not have to contend with the relatively heavy carrier beads and hence can operate at charge levels below 10 µc/g.

Most early single-component development systems relied on high magnetic loadings to control the movement of the toner through the housing. Over the last few years, several companies such as Canon and Lexmark have introduced nonmagnetic single-component systems for full color applications, but there has been little discussion of the toner requirements for these systems in the literature.

5.5.2 Single-Component Charging

The great majority of single-component systems still rely on bringing charged toner into the development nip and using the charge on the toner to control the development process (Fig. 5.4) Without the carrier beads to generate a charge on the toner, other means must be used. Typically the donor roll materials are selected to generate a charge of the right polarity on the toner when the latter is brought in contact with the roll. The toner layer formed on the donor roll either by electrostatic or magnetic forces is then passed through a metering/charging zone before entering the development zone. This zone is formed with either a blade or a pad brought into close proximity with the roll. The pressure or the position of the metering device is controlled to produce a toner layer of the desired thickness on the roll as it enters the development zone. In the case of a metal blade, an applied voltage can be used to modify the charge on the toner.

The design of the donor roll and the metering/charging nip can be at least as challenging as the design of the carrier, and all the questions about the contamination and stability of the carrier surface carry over to the donor roll and metering/charging device surfaces.

If the bulk of single-component charging is done in a small metering/charging zone, there will typically be only a fraction of a second for charging to occur. The result is that toner charging in single-component systems is usually not as complete as is possible for the same toners with appropriate carriers in two-component systems. The charge levels are not as high, leading to more low-charge toner. This low-charge toner can be accommodated by the jumping development process (Takahashi et al., 1982). An AC field is added to the DC development and cleaning fields to agitate the toner in a gap left between the donor roll and the latent image on the photoreceptor. Without the AC agitation, no toner reaches the latent image, and there is no development and no background. The AC field primarily agitates the charged toner; hence only the controlled portion of the distribution actually has a chance to move over to the photoreceptor. Only if the level of low-charge toner is quite large is it pulled along in the toner cloud in numbers sufficient to cause high background.

To satisfy the flow requirements for single-component toner, surface additives are almost always used. These will always have a dramatic effect on charging, because of their location on the surface of each toner particle. Fumed silica is by far the most common surface additive, and it can be treated to modify the charging properties with amines or titanates if positive charging is necessary (Takenouchi, 1986).

5.5.3 Toner Transport and Flow

One of the major challenges of single-component development is to move the toner through the developer housing. The attritted plastic that toner consists of is intrinsically a poor flowing material. In two-component development raw toner is moved from the toner bottle to the developer. This is typically done with foam rolls, brushes against screens, or augers. The last is probably the most practical method when the overall design requires that the toner bottle be well separated from the developer.

In single-component development, the lack of a carrier for transport raises the requirement for good flow characteristics. In practice, essentially all single-component toners have surface additives such as fumed silicas to improve powder flow.

5.5.4 Inductively Charged Toner

For some applications it is possible to charge toner in the development nip itself (Rumsey and Bennewitz. 1986). When toner of an appropriate conductivity is exposed to the development field, charge will flow from the conducting donor roll onto the toner. If the photoreceptor or other device holding the electrostatic image is insulating, the charge remains on the particle as it leaves the development zone. It is thus attached to the latent image by electrostatic attraction and remains on the photoreceptor to form the image.

There are several limitations to this scheme. Magnets are used to move the uncharged conductive toner through the development zone, requiring the toner to be magnetic, hence in almost all cases black. Since there is no intrinsic polarity to the charging/development phenomenon, reversed-bias cleaning fields cannot be used to control background, and only magnetic forces can be used for this control. Electrostatic transfer of conductive toner onto plain paper typically runs into problems because of charge leakage, and either coated paper or an alternative means of transfer such as pressure transfix is often used. However, for applications that fall within the boundaries posed by these limitations, inductive charging can be very effective, since it eliminates many of the problems of charging level, rate, and degradation encountered in triboelectric charging.

Recently, Océ developed a technology enabling bright colors with magnetic conductive color toner (Geraedts and Lenczowski, 1997). The toners are opaque, requiring the use of a seven-color system of cyan, magenta, yellow, red, green, blue, and black to develop a useful color gamut with toner monolayers.

5.6 TONER FABRICATION

5.6.1 Introduction

As stated earlier, dry xerographic toners are pigmented bits of plastic about 10 μm in diameter. Pigment and additive dispersion and particle size and size distribution are parameters that determine the quality of the resultant image; These parameters are in turn influenced by manufacturing techniques. The challenge to manufacturing is to produce the highest quality materials at the lowest possible price.

The great bulk of toner is manufactured by a multistep process consisting of melt mixing pigments and internal additives with the base toner polymer, breaking the pigmented polymer into particles of approximately the desired size, removing unwanted sizes from the size distribution, and blending in any external additives that may be necessary (Figure 5.15). Each of these steps generally requires at least one dedicated piece of equipment, as described in the sections that follow.

At the same time there has been a continuous effort to eliminate many of these steps, and chemical processes for simplified toner fabrication are described in Section 5.6.6.

5.6.2 Melt Mixing

The initial major step in toner manufacture is the blending or kneading of pigments and other internal additives with the polymer binder. The polymer is raised to a temperature at which it flows but with a relatively high viscosity. Higher temperatures at which the

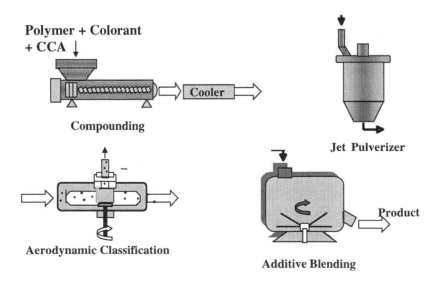

Figure 5.15 Conventional manufacturing process basic steps.

polymer flows more readily will not disperse the pigment as well, while lower temperatures require too much power to knead the pigment–polymer blend. Operating blend temperatures are generally not high compared to polymer processing in general, since the polymers have been designed to melt at as low a temperature as possible to accommodate the fusing of the toner to the paper. Temperatures from 300 to 400°F are typically adequate.

The mixing can be done in either a batch process or a continuous process. One implementation of the batch process uses a mixer such as the Brabender or Banbury. The materials to be melt mixed are premeasured and possibly premixed in a dry powder mixer such as those made by Henschel, Lodige, and others. These are then fed into the preheated melt mixing unit for a given length of time. Generally this is sufficient to form a polymer matrix, but component dispersion may be poor. If necessary, the output of the mixer can be passed repeatedly through a rubber mill until the desired dispersion is obtained.

In general, the intent is to obtain dispersions closely approximating perfectly distributed ingredients; however, it is possible that an ultimate dispersion is not the most useful. A well-documented nonxerographic example of this is the use of chained carbon blacks to give conductivity to a polymer (Sommers, 1984). If the mixing is too vigorous or too long, the chains will be broken, leading to a loss in conductivity. The most effective dispersion of many xerographic additives is not well understood, and often a range of dispersions is produced and evaluated in the course of the engineering studies. It is also true that the best dispersion typically requires the longest blend times and hence the highest processing costs; this also may lead to the use of a less-than-complete dispersion. In these cases it becomes a challenge to manufacturing to maintain the desired dispersion.

The continuous method of melt blending can be done on an extruder. This device allows higher toner production rates and hence lower processing costs than the batch processes. Here a smaller amount of polymer is in the extruder at any time, but the processing time that a given segment of polymer sees is also smaller. As a result, the forces to which a polymer is exposed may be required to be higher in the extrusion methods than in the batch processes.

Extruders offer the option of adding different components at different positions along the mixing barrel and hence at different stages in the melt history of the polymer. This offers more possibilities for process optimization, but also more stringent requirements on reproducibility.

5.6.3 Attrition

The manufacturing scale equipment used for batch melt mixing produces a slab of blended plastic weighing generally in excess of 50 pounds. The output of an extruder is more convenient in size, consisting of pellets roughly an eighth of an inch in diameter. Both these outputs are ground down until particles about 100 μm in size are produced. These are then fed into an air jet mill (jet pulverizer, or micronizer), which entrains the particles in a high velocity airstream in such a way that they collide against one another with sufficient velocity to cause fracture. The geometry of the airstream is designed so that particles below a certain size are carried out of the attrition zone, while larger particles are retained and subjected to further collisions. In this way 100 μm powder is fed in and smaller powder is continuously removed from the apparatus.

As stated in Section 5.2.1, on toner resin, the ability to use this method of toner manufacture is an important attribute of polymer design, independent of the fusing temperature of the resin. For example, a rubbery polymer may have acceptable fusing performance but poor jetting performance. Lowering the molecular weight of a polymer will in general lead to faster jetting rates by decreasing the mechanical strength of the material.

5.6.4 Classification

Attrition as it is currently practiced is a violent process. While the average particle size can be closely controlled, many particles are produced that are significantly smaller than desired, and a significant number exit the attrition zone even though they still have a fairly large size. The small particles can have undesirably low charge levels, leading to high background, while the large particles can resist fusing and form spots on the copy. These should be controlled for optimum copy quality.

The small and large particles can be removed by air entrainment technology similar to that in the attrition step. Particles are carried along by an airstream that flows around sharp corners. Inertia tends to carry the larger particles to the outside of the bend, and a knife edge can then separate the large from the small particles. The velocity of the airstream can be changed to adjust the diameter of separation. The cut point between the high and low sizes is generally not very sharp, and often more than one pass through the classifier (or a pass through more than one classifier) is necessary to remove the amount of large or small toner desired without taking away a significant portion of the desired toner size. Material that is rejected by the classifier can often be recycled through the melt mixing stage, but even this represents a loss in throughput capability. The goal is to adjust the attrition parameters so that a minimum amount of material is rejected by classification.

Since toner comes out of the attrition step entrained in an airstream, it is possible to direct it immediately into a classifier, thus producing a continuous process with no loss in productivity. A second stage of classification, if required, would necessitate an additional classifier, trading capital cost for throughput capability.

5.6.5 Finishing

At times an additional step, the addition of external additives, is needed in toner processing. This can be done by injecting the additives into the airstream carrying the toner as it leaves the classifier. Under these conditions, however, it is difficult to obtain good mixing of the often extremely light additives with the toner. More often, the additives are blended in a batch process, using any of a number of dry powder mixers. When extremely small surface additives such as the 10 nm fumed silicas are blended with 10 to 20 μm toner, often highly intensive blending is necessary to break down the additive to the smallest unit and distribute it uniformly over the surface of the toner particle.

The equipment type is typically selected at the time of the design of the plant, but the choice of the best mixing speed and time typically varies with each toner design. With some designs, a plot of the desired property such as flow will reach its final value sooner than others, and unnecessary mixing time can be avoided. Others will actually show a degradation in the desired property with prolonged blending.

5.6.6 Chemical Processes for Toner Fabrication

The polymers that are subjected to all the foregoing processing steps are typically made by suspension or emulsion polymerization processes that produce spherical particles with a fairly uniform size. It would obviously save a great deal of work if the pigment and charge control additives could be incorporated into the polymer during this stage, resulting in pigmented particles in the proper size range. This would eliminate melt mixing, attrition, and classification.

Polymer beads are typically made to be about a millimeter in size, but they can be produced in sizes down to a fraction of a micrometer. A more fundamental problem is the incorporation of pigment into the polymerization process. The pigment is usually chemically active and can interfere with the delicate reactions that must be repeated hundreds of times without interruption to form a polymer of sufficient molecular weight to be useful. However, several authors have described processes for producing "suspension polymerized" or "dispersion polymerized" toner (Takezawa et al., 1997; Fukuda et al., 1996; Totsuda et al., 1998). Also, Nippon Zeon has been producing commercial chemical black toner based on suspension polymerization for several years (Yanagida, 1998), and Canon and Hewlett Packard introduced products using color chemical toner in 1998.

The problem of polymerization in the presence of pigments can be avoided by starting the toner preparation process with a submicron polymer latex and aggregating it with a pigment dispersion. These "latex aggregation" processes have also been discussed at recent conferences (Edwards et al., 1998; Koyama et al., 1994; Nakamura et al., 1991), and Nippon Carbide has been making black aggregation toner on a commercial basis.

Both processes have the ability to produce small particles with a narrow size distribution. The suspension polymerization techniques form intrinsically spherical particles, although the surface texture can be controlled, while the latex aggregation techniques allow a range of morphologies from "potato" to sphere. The spherical morphology helps transfer (Yanagida, 1998), but it may have more problems with blade cleaning (Koyama et al., 1994). The suspension polymerization technique is well suited to building a layered structure into the toner (Fukuda et al., 1996), while the latex aggregation technique readily allows the blending of different molecular weight latices (Edwards et al., 1998). Many variations within or combining chemical processes are possible, and considerable activity is expected over the next few years.

5.7 CONCLUSIONS

5.7.1 Summary

Dry toner technology is relatively mature, with an enormous base of raw materials, characterization techniques, working knowledge, and processing methods that has been built up over the years. There are an enormous number of options for the toner as part of the overall machine design. Few of the options have been shown to be unworkable or uncompetitive over the years. The sign of the toner charge, the size of the toner particle, and the use of single- or two-component development are matters that have not been decided firmly.

5.7.2 Future Directions

Probably the most obvious future direction in xerography will be the presence of even more color in the marketplace. Personal computers are giving people the ability to command color on their display screen, and color ink-jet personal printers have become the norm rather than the exception. There will then be a need to make copies of these prints. Eventually it will penetrate all volume bands with a mix of color capabilities from single highlight color options to full process color. All copier hardware will require color options. This will put an enormous burden on the toner designer either to deal more effectively with color in new toner designs or to provide more generic designs that can be used in a variety of hardware.

A multiplicity of color toner designs will also stress the manufacturing capabilities of the various vendors. Individual designs probably will have insufficient volume to justify unique processing equipment, and a sensible manufacturing strategy will be very important. Chemical toner fabrication offers the possibility of efficiently making high quality toner at any scale and should become increasingly important as the technology is mastered.

Lower melting temperatures with adequate blocking characteristics are desirable for improved fuser performance and will become more prevalent as their performance justifies their cost.

ACKNOWLEDGMENT

We acknowledge the contribution of S. K. Ahuja to the discussion on rheology.

REFERENCES

Many of the references below are taken from the proceedings of the International Conferences on Digital Printing Technologies sponsored by IS&T, the Society for Imaging Science and Technology. Recent proceedings are available from the Society at 7003 Kilworth Lane, Springfield, Virginia 22151.

Ahuja, S. K. (1976). *J. Colloid Interface Sci.*, *57*: 438.

Ahuja, S. K. (1980a). *Rheol. Acta*, *19*: 307.

Ahuja, S. K. (1980b). *Rheol. Acta*, *19*: 299.

Ahuja, S. K. (1980c). *Rheology*, *2*: 469.

Akagi, H., Takayama, H., Sugizaki, Y., and Moriya, H. (1993). Application of small particle size toner to color xerography. *Ninth International Conference on Advances in Non-Impact Printing*, Yokohama, Japan.

Anderson, J. (1994). Surface state models of tribocharging of insulators. *Tenth International Conference on Advances in Non-Impact Printing*, New Orleans, LA.

Aoki, K., Kawaji, H., and Kawabe, K. (1996). A Hybrid Resin for Toner II. NIP12: *International Conference on Digital Printing Technologies*, San Antonio, TX.

Barby, D. (1976). In *Characterization of Powder Surfaces* (G. D. Parfitt and R. S. W. Sing, eds.). Academic Press, New York.

Bares, J. (1993). Single component development—a review. *Ninth International Conference on Advances in Non-Impact Printing*, Yokohama, Japan.

Baur, R., and Macholdt, H. T. (1995). Organic pigments for digital color printing. *Eleventh International Conference on Advances in Non-Impact Printing*, Hilton Head, SC.

Bhateja, S. K., and Gilbert, S. K. (1985a). *J. Imaging Technol. 11*: 267.

Bhateja, S. K., and Gilbert, S. K. (1985b). *J. Imaging Technol. 11*: 273.

Bhateja, S. K., and Gilbert, S. K. (1986). *J. Imaging Technol. 12*: 156.

Birkett, K. L., and Gregory, P. (1986). *Dyes Pigments*, *7*: 341.

Blaszak, S., Dalal, E., Natale, K., and Swanton, P. (1994). Lightfastness in xerography and competitive technologies. *Tenth International Conference on Advances in Non-Impact Printing*, New Orleans, LA.

Button, A. C., and Edberg, R. C. (1985). *J. Imaging Technol.*, *11*:261.

Chiba, S., and Inoue, S. (1988). Toner requirements for digital color printer. *Fourth International Conference on Advances in Nonimpact Printing Technology*, New Orleans.

Cressman, P. T., Hartman, G. C., Kuder, J. E., Saeva, F. D., and Wychick, D. I. (1974). *J. Chem. Phys.*, *61*: 2740.

Dalal, E., Blaszak, S., and Swanton, P. (1991). Gloss Measurement of Xerographic Images. *Seventh International Conference on Advances in Non-Impact Printing*, Portland.

Degussa-Hüls (2000). Special Hydrophobic Aerosil (SHA) for Toners. Bulletin TI1222.

Demizu, H., Saito, T., and Aoki, K. (1986). Development properties of the mono-component non-magnetic development system. *Third International Conference on Advances in Nonimpact Printing Technology*, San Francisco.

Dessauer, J. H. (1971). *My Years with Xerox*. Doubleday, Garden City, NY.

Duke, C. B., and Fabish, T. J. (1978). *J. Appl. Phys.*, *49*: 315.

Edwards, E., Ellis, G. Morris, D., Ormesher, N., and Nevin, B. (1998). Chemical toners from a latex aggregation process. *IS&T's NIP14: International Conference on Digital Printing Technologies*, Toronto.

Epping, R. H. (1988). Lifetime simulation and charge related parameters of two-component developers. *Fourth International Conference on Advances in Nonimpact Printing Technology*, New Orleans.

Feng, J., and Hays, D. (1998). Theory of Electric Field Detachment of Charged Toner Particles. *IS&T's NIP14: International Conference on Digital Printing Technologies*, Toronto.

Fox, L. P. (1982). In: *Carbon Black—Polymer Composites* (E. K. Sichel, ed.). Marcel Dekker, New York.

Fujii, E., Fujii, H., and Hisanaga, T. (1988). *J. Photogr. Sci.*, *36*: 87.

Fukuchi, Y., and Takeuchi, M. (1998). A comparative study on toner adhesion force measurements by toner jumping and centrifugal measurements. *IS&T's NIP14: International Conference on Digital Printing Technologies*, Toronto, Ontario, Canada.

Fukuda, M., Takezawa, S., Watanuki, T., and Sawatari, N. (1996). Toner with gradated resin composition made by suspension polymerization technique. *NIP12: International Conference on Digital Printing Technologies*, San Antonio, TX.

Fukumoto, H., Inoue, S., Sasakawa, M., and Doi, S. (1985). U.S. Patent 4,533,614.

Geraedts, J., and Lenczowski, S. (1997). Oce's productive colour solution based on the Digital Imaging Technology. *IS&T's NIP13: International Conference on Digital Printing Technologies*, Seattle.

Gibson, H. W. (1975). *J. Am. Chem. Soc.*, *97*: 3832.

Gibson, H. W. (1984). *Polymer*, *25*: 3.

Gruber, R. J., Pacansky, T. J., and Knapp, J. F. (1982). U.S. Patent 4,318,947.

Gruber, R. J., Bolte, S. B., Koehler, R. F., and Connors, E. W. (1983). U.S. Patent 4,378,420.

Gruber, R. J., Knapp, I. F., and Bolte, S. B. (1985a). U.S. Patent 4,517,268.

Gruber, R. J., Koehler, R. F., Knapp, J. F., and Bolte, S. B. (1985b). Generating magnetically encoded images using a laser printer. *Proceedings of the International Electronic Imaging Exposition and Conference*, Boston.

Gruber, R. J., Koch, R. J., and Knapp, J. F. (1986). U.S. Patent 4,578,338.

Gruber, R. J., Ahuja, S., and Seanor, D. (1989). In: *Encyclopedia of Polymer Science and Engineering* (H. F. Mark, N. M. Bikales, C. G. Overberger, and G. Menges, eds.). Wiley-Interscience, New York.

Guay, J., Miller, J. L., Nguyen, H., and Diaz, A. F. (1992). Charging properties and characterization of nigrosine blends. *Eighth International Conference on Advances in Non-Impact Printing*, Williamsburg.

Gutman, E., and Hartmann, G. (1992). *J. Imaging Sci. Technol.*, *36*: 335.

Gutman, E., and Hartmann, G. (1995). *J. Imaging Sci. Technol.*, *39*: 285

Gutman, E., and Mattison, D. (1998). A model for the charging of fresh toner added to a two-component charged developer. *IS&T's NIP14: International Conference on Digital Printing Technologies*, Toronto.

Harper, W. R. (1967). *Contact and Frictional Electrification*. Oxford University Press, Oxford.

Hayakawa, N., and Ochiai, S. (1993). The relations between the toner properties and the viscoelasticity and molecular weight distribution of toner resins. *Ninth International Conference on Advances in Non-Impact Printing*, Yokohama, Japan.

Hays, D. (1978). *J. Photogr. Sci. 22*: 232.

Hays, D. (1998). Xerography *Encyclopedia of Applied Physics, Vol. 23* (G. Trigg, ed.). Wiley-VCH, Weinheim and New York.

Heinemann, M, and Epping, R. (1993). Free flow characteristics and charge parameters of monocomponent toners with positive polarity. *Ninth International Conference on Advances in Non-Impact Printing*, Yokohama, Japan.

Inoue, S., Sasakawa, M., Fukumoto, H., and Doi, S. (1985). U.S. Patent 4,535,048.

Julien, P. C. (1982). In: *Carbon Black—Polymer Composites* (E. K. Sichel, ed.) Marcel Dekker, New York.

Julien, P., Koehler, R., Connors, E., and Lewis, R. (1992). Charge exchange among toner particles. *Eighth International Conference on Advances in Non-Impact Printing*, Williamsburg.

Julien, P. (1993). In: *Carbon Black—2nd Edition* (J.-B. Donnet, R. Bansal, and M.-J. Wang, eds.). Marcel Dekker, New York.

Julien, P., Koehler, R., and Connors, E. (1993). The relationship between size and charge in xerographic developers. *Ninth International Conference on Advances in Non-Impact Printing*, Yokohama, Japan.

Julien, P. (1996). The relationship between blowoff tribo and charge spectrograph measurements. *NIP12: International Conference on Digital Printing Technologies*, San Antonio, TX.

Kawaji, H., Aoki, K., and Kawabe, K. (1995). A hybrid resin for toner. *Eleventh International Conference on Advances in Non-Impact Printing*, Hilton Head, SC.

Kawaji, H., Shimizu, J., and Omatsu, S. (1997). Full color toner for oil free fuser. *IS&T's NIP13: International Conference on Digital Printing Technologies*, Seattle, WA.

Knapp, J. F., and Gruber, R. J. (1985). U.S. Patent 4,520,092.

Koyama, M., Hayashi, K., Kikuchi, T., and Tsujita, K. (1994). Synthesis and characteristics of non-spherical toner by polymerization method. *Tenth International Conference on Advances in Non-Impact Printing*, New Orleans, LA.

Kuo, Y. (1984). *Polym. Eng. Sci., 24*: 9.

Lee, L. H. (1975). In: *Adhesion Science and Technology* (L. H. Lee, ed.) Plenum Press, New York, p. 831.

Lee, M. H., and Ayala, I. (1985). *J. Imaging Technol., 11*: 279.

Lee, W.-S., and Takahashi, Y. (1998). Comparison and evaluation of various tribo-charging models for two-component developer. *IS&T's NIP14: International Conference on Digital Printing Technologies*, Toronto, Ontario, Canada.

Lewis, R. B., Connors, E. W., and Koehler, R. F. (1981a). A spectrograph for charge distributions on xerographic toner. *Fourth International Conference on Electrophotography*, Washington, DC.

Lewis, R. B., Connors, E. W., and Koehier, R. F. (1981b). U.S. Patent 4,375,673.

Lewis, R. B., Connors, E. W., and Koehler, R. F. (1983). *Denshi Shashin Gakkaishi (Electrophotography), 22*: 85.

Lewis, R. B., Julien, P. C., Gruber, R. I., and Koehler, R. F. (1984). U.S. Patent 4,426,436.

Lowell, J., and Rose-Innes, A. C. (1980). *Adv. Phys. 29*: 947.

Lu, C. (1981). U.S. Patent 4,298,672.

Macholdt, H.-T., and Sieber, A. (1988). *J. Imaging Technol., 14*: 89.

Macholdt, H.-T., and Baur, R. (1996). Recent trends in organic color pigments. *NIP12: International Conference on Digital Printing Technologies*, San Antonio, TX.

Manson, J. M., and Sperling, L. H. (1976). *Polymer Blends and Composites*. Plenum Press, New York.

Matsuura, Y., Anzai, M., and Mkudai, O. (1993). Function of charge control agent. *Ninth International Conference on Advances in Non-Impact Printing*, Yokohama, Japan.

Miyabe, K. (1994). LED page printer by using a polymerizing toner. *Tenth International Conference on Advances in Non-Impact Printing*, New Orleans.

Nakamura, M. (1994). The effect of toner rheological properties on fusing performance. *Tenth International Conference on Advances in Non-Impact Printing*, New Orleans, LA.

Nakamura, Y., Takezawa, S., Katagiri, Y., and Sawatari, N. (1991). Polymerization toner techniques in mono-component non-magnetic development. *Seventh International Conference on Advances in Non-Impact Printing*, Portland.

Nakamura, Y., Moroboshi, Y., Terao, Y., Suzuki, Y., Tanabe, J., Sekine, T., Sasabe, S., Yokoyama, T., and Mazumdar, M. (1997). Tribocharging of toner particles in two-component developer and its dependence on the particle size. *IS&T's NIP13: International Conference on Digital Printing Technologies*, Seattle, WA.

Nakaya, F., Kita, S., Takeuchi, K., and Tanaka, T. (1986). *J. Imaging Technol.*, *12*: 304.

Narusawa, T., Sawatari, N., and Okuyama, H. (1985). *J. Imaging Technol.*, *11*: 284.

Nash, R. J., and Bickmore, J. (1988). Toner impaction and triboelectric aging. *Fourth International Conference on Advances in Nonimpact Printing Technology*, New Orleans.

Nash, R., Grande, M., and Muller, R. (1998). The effect of toner and carrier composition on the average and distributed toner charge values. *IS&T's NIP14: International Conference on Digital Printing Technologies*, Toronto, Ontario, Canada.

Neilson, I. (1998). Magnetic and thermo-mechanical properties of raw materials for high speed magnetic ink character recognition (MICR) toners. *IS&T's NIP14: International Conference on Digital Printing Technologies*, Toronto, Ontario, Canada.

Nelson, R. A. (1984). U.S. Patent 4,469,770.

Ott, M., Eklund, E., Mizes, H., and Hays, D. (1996). Small particle adhesion: measurement and control. *NIP12: International Conference on Digital Printing Technologies*, San Antonio.

Palermiti, F., and Chatterji, A. (1971). U.S. Patent 3,590,000.

Prime, R. B. (1983). *Photogr. Sci. Eng.*, *27*: 1.

Rumsey, J. R., and Bennewitz, D. (1986). *J. Imaging Technol.*, *12*: 144.

Sata, S., Shirai, E., Shimizu, J., and Maruta, M. (1997). Study on the surface properties of polyester color toner. *IS&T's NIP13: International Conference on Digital Printing Technologies*, Seattle, WA.

Schaffert, R. M. (1980). *Electrophotography*. Focal Press, New York.

Scharfe, M. E. (1984). *Electrophotography, Principles and Optimization*. Research Studies Press, Letchworth, England.

Scharfe, M. E., Pai, D. M., and Gruber, R. J. (1989). In: *Imaging Processes and Materials; Neblette's Eighth Edition* (J. M. Sturge, V. K. Walworth, and A. Shepp, (Eds.). Van Nostrand Rienhold, New York.

Schein, L. B. (1996). *Electrophotography and Development Physics, Rev. 2ⁿᵈ Edition*. Laplacian Press, Morgan Hill, CA.

Schein, L. (1998). Recent advances in our understanding of toner charging. *IS&T's NIP14: International Conference on Digital Printing Technologies*, Toronto, Ontario, Canada.

Schmidlin, F. W. (1976). In: *Photoconductivity and Related Phenomena* (I. Mort and D. M. Pai, eds.). Elsevier, New York.

Seanor, D. A. (1972). In: *Electrical Properties of Polymers* (K. Frisch and A. V. Patsis, eds.). Technomic Press, Westport, CT.

Seanor, D. A. (1978). *Photogr. Sci. Eng.*, *22*: 240.

Shigehiro, K., Arai, K., Machida, Y., Fukuhara, T., Hirose, Y., Takiguchi, K. (1993). The effects of toner particle size and image structure on the image quality in electrophotography. *Ninth International Conference on Advances in Non-Impact Printing*, Yokohama, Japan.

Sommers, D. I. (1984). *Polym-Plast. Technol. Eng.*, *23*: 83.

Stuebbe, A. (1991). Pyrogenic oxides for improving the charge stability of two component dry toners. *Seventh International Conference on Advances in Non-Impact Printing*, Portland.

Takahashi, T., Hosomo, N., Kanbe, J., and Toymono, T. (1982). *Photogr. Sci. Eng.*, *26*: 5.

Takenouchi, M. (1986). U.S. Patent 4,618,556.

Takezawa, S., Fukuda, M., Watanuki, T., and Sawatari, N. (1997). Effect of particle size distribution on the triboelectric charge of toners. *IS&T's NIP13: International Conference on Digital Printing Technologies*, Seattle, WA.

Tavernier, S., Alaerts, L., and Debie, H. (1995). Offset quality "Short Run Colour Printing" using a dry 4-colour bi-component electrophotographic process. *Eleventh International Conference on Advances in Non-Impact Printing*, Hilton Head, SC.

Totsuda, H., Maeda, M., Suzuki, Y., Ozawa, J., and Nagare, T. (1998). Development of polymerized toner. *IS&T's NIP14: International Conference on Digital Printing Technologies*, Toronto.

Tsujimoto, H., Kaya, N., Huang, C.-C., Yamamoto, H., and Mazumdar, M. K. (1991). Electrostatic characterization of toners measured by E-spart analyzer. *Seventh International Conference on Advances in Non-Impact Printing*, Portland.

Veregin, R., Powell, D., Tripp, C., McDougall, M., and Mahon, M. (1997). Kelvin potential measurement of insulative particles. Mechanism of metal oxide triboelectric charging and relative humidity sensitivity. *IS&T's NIP13: International Conference on Digital Printing Technologies*, Seattle, WA.

Veregin, R., and Bartha, R. (1998). Metal oxide surface additives for xerographic toner: adhesive forces and powder flow. *IS&T's NIP14: International Conference on Digital Printing Technologies*, Toronto, Ontario, Canada.

Wacker-Chemie (1998). Wacker HDK Fumed Silica for Toners and Developers. Bulletin 5479e.

Weigl, J. W. (1982). In: *Colloids and Surfaces in Reprographic Technology* (M. Hair and M. D. Croucher, eds.). American Chemical Society, Washington, DC, p. 139.

Williams, E. M. (1984). *The Physics and Technology of Xerographic Processes*, Wiley Interscience, New York.

Yanagida, N. (1998). Polymerized toner: its feature and future. *Diamond Research Conference on Toner & Photoreceptors '98*, Santa Barbara, CA.

Yourd, R. A., III, Majumdar, D., and Gruber, R. I. (1985). U.S. Patent 4,537,848.

6

Carrier Materials for Imaging

LEWIS O. JONES*

Consultant, Ontario, New York

INTRODUCTION

Since the publication of the first edition of this book there has been considerable activity in the field of xerography, with much attention focused on the carriers used in two-component development. These developments have not revolutionized our understanding of how carriers work or why they are needed. There has been enough activity to tell us that they are still considered important and are necessary, especially in the most rapidly growing sector of the market: digital color imaging. Many attempts have been made to eliminate the need for carriers, an objective that might be achieved sometime in the future, but for the present these unique imaging materials seem firmly entrenched in the world of electrophotography.

During the past 10 years, more than 150 United States patents have been issued in the field of xerographic carriers and carrier manufacturing processes. Of these about 10% have been aimed directly at the color copying or printing portion of the market. Much of the effort in color has been directed at noncontact development, processes in which the carrier approaches, but never touches, the developed image. At least 37 patents have this idea as their subject matter. The details of these and other recent developments in xerographic carrier materials will be covered at the end of this chapter, but for now the previously published information stands with few changes or updates.

6.1 CARRIER DEFINITION AND FUNCTION

Carrier is a general term used in xerography to describe the toner transporting species in a two-component dry developer. In such a mixture, the smaller toner particles are carried

---

* Retired

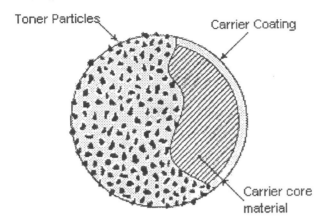

Figure 6.1 Cross section view of a coated spherical developer.

on the surface of the larger carrier granules, as depicted in Fig. 6.1. Here, a coated, spherical carrier is shown in cross section. In practice, however, the carrier may be coated or uncoated, spherical or irregular in shape.

Toner is the "ink" that actually makes the mark on the paper (1). It generally consists of a low melting polymer compounded with approximately 10% carbon black and ground to a particle size ranging from about 12 microns in older products to about 6 microns in the more recent "microfine" toner powders.

The carrier is also a granular material which ranges from 3 to 50 times the average diameter of the toner particle. This component of the developer mix is the subject of the remainder of this chapter.

Xerographic developer mixtures typically contain from 1 to 10% by weight of toner. The toner is transported by the carrier and brought into close proximity with an invisible, electrostatically charged image on a photosensitive surface in a copier, laser printer, fax machine, or multifunctional imaging device. At this point, toner particles abandon the carrier surface and deposit either in charged areas on the latent image, or are repelled into neutral image areas, depending upon the relative charge polarity of image and toner.

One of the most important functions of the carrier is to impart a static charge to the toner particles. This is accomplished by frictional surface contact with the toner during mixing, a phenomenon known as *triboelectrification*. This term comes from a combination of Greek words meaning to charge by rubbing and is likened to the familiar experiment of charging a glass rod by rubbing it with cat's fur to attract and hold small bits of paper. Another example of tribocharging is the unpleasant experience of walking on a carpet on a dry day and being shocked by touching a doorknob or other grounded element.

In the case of xerographic developers the combination of toner and carrier properties must be chosen to produce the correct level and polarity of electrostatic charge on the toner. This will ensure that the desired amount is attracted to the oppositely charged image area. It is important that the charge on the toner be neither so high that it cannot be stripped from the carrier nor so low that it is not held tightly to the carrier. Loose toner can float around the imaging device, depositing on optical and other machine components, producing dust in the environment, and settling in nonimage areas of the print as unwanted level of background density.

The methods and materials used to accomplish these design tasks are the subject of the remainder of this chapter.

6.2 MATERIALS (HISTORICAL)

6.2.1 Xerox Corporation

Any discussion of carrier materials must start with those substances used initially by the Haloid Corporation, the cradle of the electrophotographic process, which later was renamed the Xerox Corporation. In an early demonstration of xerography by its inventor Chester Carlson in 1938, the carrier used was uncoated iron powder and the image was formed by cascading the developer over the image until enough toner had been deposited to produce a visible mark or readable character (2). In the course of reducing Carlson's imaging method to a commercial process, carriers were chosen from powders available in the market place; most of the design work was empirical because little was known about the triboelectric charging effect.

In 1950 the Haloid Corporation introduced a manually operated offset platemaker called the Model D, in which the carrier was washed and screened sand. This carrier, measuring approximately 600 microns in diameter, bore a lacquer coating to obtain the desired charging effect. From that beginning, many different powdered materials have evolved (see 6.2.2) as the industry expanded to encompass other equipment manufacturers and to suit a wide range of customer requirements. Carrier technology become more sophisticated, enabling output image quality to progress from something barely readable to razor sharp prints with rich black solids and smooth halftones at ever increasing speeds.

Subsystems have progressed from cascade (Fig. 6.2) to magnetic brush development (Fig. 6.3) for delivering the toner to the image. The design of carrier materials has kept pace with these advances, moving much of the early empiricism into the realm of science. As machine performance and copy quality have progressed, so have the developer packages and naturally the carriers. During the early improvements in machine design, there was little carrier development. Design engineers and materials scientists were faced with

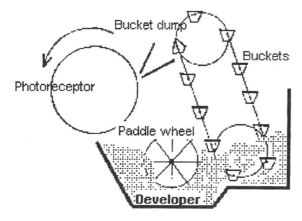

Figure 6.2 Cascade development process.

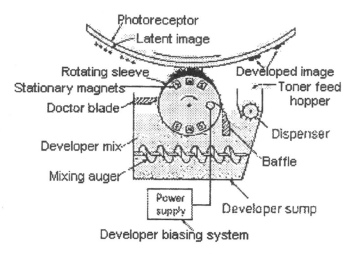

Figure 6.3   Magnetic brush development.

an exploding demand for copiers and duplicators. Further, almost any toner–tribo pair would work as long as the two dissimilar granules rubbed together exchanged charge and could sustain their performance as a developer mix. Toner was consumable and habit forming. The market expansion got somewhat ahead of the technology, a reaffirmation of the dictum, ''the art comes before the science.''

Existing powders that were available for other applications were also utilized. Some were specially screened distributions obtained from abrasive powder suppliers. Originally these materials were designed for shot blasting or scarfing applications such as cleaning plastic or metal parts. Others were taken from the powdered metals designed for use in pressing parts by compaction and sintering. Generally copier manufacturers used tailings or fines from these sources rather than develop a process for customized copier powders. As the industry grew, competition and customer awareness of copy quality forced increasingly stringent specifications to be placed on the carriers as well as the toners.

6.2.2   Specific Carriers Used

Table 6.1 is an historical summary of the various powder materials that have been used since the beginning of electrostatic imaging. It can be seen that the industry has progressed from dusting the image with dark powder to delivering the toners with sand, glass, aluminum, iron, steel, nickel, magnetite, and ferrites. All these carrier materials can be used in cascade development. Indeed, the latest magnetic roll systems are based mainly on the last five materials. Starting in the early 1960s, development programs were launched to design xerographic carriers. To date, there have been a few basic carriers generated that were designed directly for specific machines. Some of these are:

1. Spherical iron powders processed by the rotating electrode process by Nuclear Metals Inc. (see Fig. 6.4)
2. Spherical ferrite (magnetic ceramic) powders produced by spray drying and high-temperature sintering (see Fig. 6.5)
3. Magnetite spheroidized by plasma energy
4. Irregular iron powder produced by the reduction of screened magnetite ore

Table 6.1 History of Electrostatic Copier Powders

Year	Event	Copier powder
1777	Lichtenberg demonstrates electrostatic image development	Lycopodium powder
1920	Selenyi develops a facsimile recording process using an electrified stylus, writing on an insulating surface, and powder development	Lycopodium powder
1938	Chester Carlson invents xerography, an electrostatic imaging process for plain paper copying	Uncoated iron carrier and toner
1950	Haloid introduces manually operated xerographic equipment for offset platemaking: the Model D	600 μm lacquered sand carrier and toner
1954	RCA researchers Grieg, Giaimo, and Young invent Electrofax, an electrostatic imaging process for coated paper copying using a magnetic brush	Uncoated irregular iron carrier and toner
1959	Haloid-Xerox introduces the first automated plain paper copier to use Carlson's xerographic technology: the Xerox 914, a 7 page/min (ppm) machine	600 μm lacquered sand carrier and toner
1961	American Photocopy Equipment Co. markets the first plain paper copier to use RCA's Electrofax technology: the Dristat	Uncoated irregular iron carrier and toner
1964	Xerox announces the first copier-duplicator to operate at 40 ppm: the Xerox 2400	600 μm lacquered sand carrier and toner
1969	Xerox demonstrates a major advance in xerographic copier quality using a conductive steel bead carrier and cascade development: the Xerox 3600-III, a 61 ppm machine	450 μm lacquered spherical steel carrier and toner
1970	Xerox unveils a compact high-speed plain paper copier: the Xerox 4000, a 43 ppm machine	250 μm lacquered spherical steel carrier and toner
1971	IBM enters the office copying field with a 10 copy/min machine: the Copier I	170 μm teflon coated spherical steel carrier and toner
1972	Xerox markets a plain paper copier based on the finest spherical steel carrier ever used for xerography: the Xerox 3100, a 12 ppm machine	100 μm lacquered spherical steel carrier and toner
1972	3M Company revolutionizes electrostatic image development with the VQC series of copiers, the first to use single-component magnetic toner	Magnetic toner, 20 μm carrier
1973	Xerox introduces the first nonimpact computer printer based on xerography: the Xerox 1200	450 μm lacquered spherical steel carrier and toner

Table 6.1   (Continued)

Year	Event	Copier powder
1973	Xerox rolls out the first xerographic color copier: the Xerox 6500	2 toner colors with lacquered 100 μm spherical steel carrier and the third color with 100 μm nickel carrier
1975	Xerox demonstrates the fastest xerographic copier-duplicator using mag brush development, operating at 120 ppm: the Xerox 9200	80 μm lacquered spherical ferrite carrier and classified toner
1975	Eastman Kodak enters the office copying field with equipment using nonspherical iron carrier: the Ektaprint 100/150	170 μm laquered irregular iron carrier and toner
1976	IBM commercializes the first laser beam xerographic computer printer: the IBM 3800 operates at more than 150 ppm	170 μm Teflon-coated spherical iron carrier and toner
1979	Minolta markets a copier using ''microtoning'' technology, the first electrostatic developer to use a fine particle carrier: the EP-310	40 μm magnetic carrier and toner
1979	Canon introduces Ion Projection Development in the first practical plain paper copier to use one: component toner, the Canon NP-200	10–15 μm magnetic carrier and magnetic toner
1980	Xerox offers the first copier in its line to use irregular iron carrier: the Xerox 3300	100 μm lacquered irregular iron carrier and toner
1983	Xerox debuts its first office copier to use an organic photoreceptor: the Xerox 1075	130 μm lacquered irregular iron carrier and toner

There are many variations on these, such as partially reduced magnetite or oxidation of irregular iron for conductivity variation and control. The important powder properties will be discussed later, but it can be noted that the trends are toward smaller size distribution for higher surface area, lower density, lower resistivity, and better flow characteristics.

6.3   MATERIAL PROGRESS

6.3.1   Copier Evolution Effects

Of the two commercially available electrostatic copying methods: Electrofax, the direct, coated paper copier (CPC) process has all but disappeared from the market, while xerography, the transfer, or plain paper copier (PPC) process has flourished. Both methods employ a photoconductive member that can be charged uniformly and discharged in the nonimage areas by reflected light from the original document. In the CPC process, the paper is coated with a photoconductive zinc oxide pigment in a resin binder. This layer is charged and exposed to a light image yielding a latent electrostatic charge pattern corresponding to light and dark areas on the original document. In the development step the charged image attracts charged toner particles from the carrier rendering the image visible. The toner deposit is fixed or fused onto the sheet by heat, pressure, or a combination of heat and pressure. Thus the photoconductor is used but one time, becoming the designed copy.

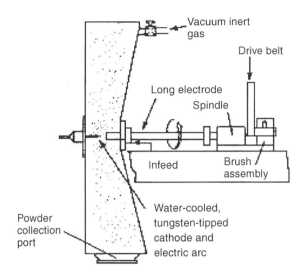

Figure 6.4   The rotating electrode process.

In the transfer process a reusable drum or belt, coated with a photosensitive material such as selenium, is charged and exposed in the same manner as the CPC process, but the toned image is then transferred to a plain paper or film susbstrate upon which it is permanently adhered by heat or pressure fixing.

Four methods of toner development are known: cascade (Fig. 6.2), magnetic brush (Fig. 6.3), powder cloud, and touchdown. Since cascade and magnetic brush are the most widely used, and the only ones of importance in two-component development, they will be discussed in detail here. Powder cloud is used in xeroradiography, while touchdown development remains experimental and has not found commercial applications.

Cascade development generally uses more spherically shaped beads to obtain better flow characteristics. The machines that use cascade development resemble a Ferris wheel with buckets on a chain drive to pick up the developer from a sump or reservoir. The developer mix is dumped into a trough which allows the material to cascade down over the exposed photoreceptor and the return to the sump.

Toner particles leave the carrier surface upon being attracted to highly charged image areas as the beads roll and bounce over the photoreceptor. In many cases the impact dislodges poorly charged toner particles causing them to become airborne and to deposit in nonimage (background) areas. The result is dust and dirt in and around the machine and a reduction in the quality of the final print. This drawback of the cascade process is one of the reasons that the magnetic brush system is preferred in most of the copiers, duplicators, and printers being manufactured today. This may not be a completely fair statement, because the cascade systems of the past would probably deliver better performance if they were to take advantage of today's developer technology.

Another limitation of the old cascade process is the poor development of solid areas on the copy. This is particularly bad when insulating materials such as sand and glass are used. Again, today's materials would likely improve solid area fill, but magnetic brush development has become the accepted technique over the last 15 years.

Magnetic brush development was invented by Giaimo and Young (3–5). This process uses a mixture of magnetic carrier beads with the toner to make up the developer. The size of the carrier ranges from 30 to 450 microns. Van Engeland gives an excellent

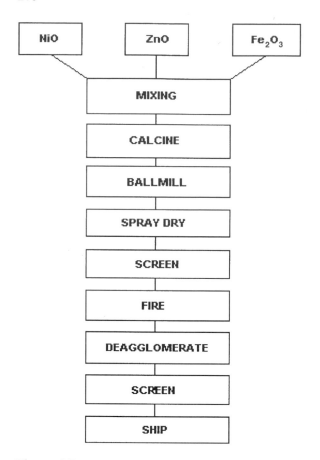

Figure 6.5 Ferrite process flowsheet.

review of this process (6). The physics of cascade and magnetic brush development are treated by L. B. Schein along with other dry and liquid toner development processes (58).

Figure 6.3 illustrates a schematic of a magnetic brush development subsystem. The magnetic developer is attracted to the surface of a nonmagnetic rotating roller by stationary magnets positioned inside the roller. The quantity of material transported is metered by a trimmer bar or doctor blade, which controls the pile height of the brush bristles. The carrier material aligns itself in the field of the magnets and forms a bristle or brush that can make contact with the photoreceptor surface, enabling the toner to develop the charged image.

As the carrier moves past the development zone, the field decreases with distance from the internal magnets. Gravity and centrifugal force return the partially spent developer to the sump. Fresh toner is added automatically to the mixture from the toner hopper to replenish the toner lean developer mix. A mixing auger stirs the material to assure that the fresh toner is charged and uniformly distributed in the developer mix.

The toner concentration of the developer can be monitored in several ways. Most machines have sensors that detect changes in resistivity, magnetics, or reflectance as a function of toner concentration. These sensors can determine when the toner concentration is low and signal the toner hopper to replenish the spent mix by dispensing toner to the developer housing.

The characteristics of the brush in the development zone are important to determine the supply of toner available to the image. In any given housing the magnetic properties of the carrier determine how hard or soft the brush will be. The resistivity of the developer mix is a major factor in the development of solid areas.

Spherical carriers have better flow properties that improve mixing efficiency and reduce the torque required to drive the housing; irregular carriers can be designed to provide controlled resistivity, but this feature comes at the expense of wear to the photoreceptor, somewhat inferior flow properties, and wear on the carrier itself.

6.3.2 Magnetic Materials Required

As stated previously, all the materials mentioned can be adapted to cascade development. What is required is a mechanism that will move and mix the developer. In the last 20 years, magnetic brush development has restricted the choices to "soft" magnetic materials. This term means that the material is magnetically soft or has very low loss and low remanence. Once the material has been in a magnetic field, it retains very little polarization when removed from the field. If the remanence is high, each bead becomes a tiny permanent magnet, and all of the beads attract each other; this severely restricts the flow properties of the carrier returned to the sump for remixing. The "soft" magnetic materials used are iron, low-carbon steels, annealed nickel, magnetite, and ferrites. Recently, Eastman Kodak broke new ground by introducing "hard" ferrite carrier into some of its equipment. This material has some of the problems of flow after magnetization, but Kodak scientists introduced some rather ingenious developer housing designs to solve that problem. The "hard" materials have considerable remanence and are generally used in permanent magnet applications such as the novelty refrigerator magnets.

The need for cleaner, sharper, high-quality copies with rich solid areas for color reproduction and graphics has prompted many improvements in carrier materials. The ultimate goal is to have copies that compare to lithography for resolution, solid fill, and uniformity. This requires a controlled and consistent resistivity of the developer, generally lower than the insulating sand or glass yet higher than bare iron or steel core materials. Therefore most of the iron powders are oxidized to control the resistivity and partially coated to control electrostatic charging. Irregular iron powders have been used in the last few years, and the theory is basically that the high points are oxidized (15) and poorly coated to supply the required resistivity while the valleys are better coated to supply the charging effect required.

Many of the Japanese copiers used irregular iron such as flake or filings. In recent years most of the two-component copiers have shifted to ferrites because of the semiconducting properties they possess. These materials have resistivities in the range desired because they are transition metal oxides and magnetic ceramic materials that can in some applications be used without partial coating for toner charging.

6.4 CURRENT CARRIER POWDERS

6.4.1 Steel (Spherical)

One of the first uses of 450 micron spherical steel came in 1969 with the introduction of a 60 ppm machine by Xerox Corporation called the 3600-III. The process for these types of materials is described in patents by R. Hagenbach and R. Forgensi for classifying and manufacturing by a two-wire process (7,8).

In the same time frame, development of a rotating electrode process was going on at Nuclear Metals Inc. This process is depicted schematically in Fig. 6.4 and was one of the first materials designed specifically for carrier powders. Another highly successful machine, the Model 3100 copier, was introduced by Xerox in 1972. This 12 ppm machine utilized 100 micron spherical steel as the carrier in a magnetic brush developer housing. The success of the 3100 led to a series of similar machines all using the same developer, many of them are still in use today.

In 1971 IBM joined the copier field with the introduction of its Copier I which used a 170 micron Teflon coated spherical steel carrier. IBM continued to use this material in most of its subsequent machines. Since all these copiers used coated carrier, the developer was quite insulating, and consequently solid area reproduction was poor. In addition, background toning and machine dirt became problems.

6.4.2 Iron (Irregular)

Irregular iron was first demonstrated by Greig, Giaimo, and Young with the invention of Electrofax at RCA for a coated paper copier. This became a commercial reality in 1961 when it was marketed by the American Photocopy Equipment Company as "DriStat." The next important entry of irregular iron came in 1975 when Eastman Kodak entered the copier field with the Ektaprint 100/150 (17). This machine employed a coated 170 micron carrier and made use of resistivity and charge controls by oxidation (15) and by proprietary carrier coating methods as described previously. This material is a reduced, screened magnetite that provides a very rough bead of nearly pure iron.

The same type of material is utilized in the Xerox 1075, which came on the market in 1983. The carrier size is slightly smaller at 130 microns, but the same resistivity and charge control processes were applied. A different 100 micron irregular iron made by water atomization was introduced in 1980 by Xerox Corporation in a machine called the Model 3300. This carrier is oxidized and coated similarly for the same reasons.

In the late 1980s, a partially reduced magnetite was used in some Japanese copiers as described in a joint patent between Kanto Denka Kogyo in Japan and Hoeganas AB in Sweden (9). All of these machines reproduce solid areas quite well and deliver excellent copy quality.

6.4.3 Soft Ferrites (Spherical)

The simple ferrite process flowsheet of Fig. 6.5 illustrates how the spherical powder is formed in the spray drying step. Many ferrite formulations may be produced by this method, which has been adapted to the manufacture of carriers for xerography. The first mention of ferrites to be used in magnetic brush development was by Joseph Wilson of Haloid Xerox Inc. in 1958 (19). The first commercial use came in 1975. This family of materials has grown steadily in the last 15 years from the first introduction in a high-speed machine, the Xerox 9200 (10–14), operating at 120 ppm. Since that time ferrites have been used in over 20 different Xerox machine models.

The use of ferrite carriers spread radidly to Japanese original equipment makers (OEMs). In 1986 alone they introduced 38 new copiers and printers of which 34 utilized ferrite carrier. These materials provide several important advantages in magnetic brush

Table 6.2 Comparison of Bare, Uncoated Carrier Core Properties

Property	Iron powders	Magnetite	Ferrite	Hybrids	Composites
Magnetic saturation (emu/g)	160–190	88–98	30–96	50–65	30–65
Resistivity @ 100 volts (ohm-cm)	10^3–10^8	10^6–10^8	10^3–10^8	10^{37}–10^9	$>10^{12}$
Apparent density (g/cc)	2.4–2.5	2.0–3.0	1.5–3.0	1.5–2.5	1.0–1.5
Susceptibility to oxidation	Moderate	High	Low	Low	None
Cost (1–10) Low–high	1–5	4–8	5–10	5–10	8–10
Use in full color applications	Very difficult	Difficult	Preferred	Difficult	Moderate
Versatility of design in systems	Moderate	Low	High	Moderate	Low

Source: Courtesy of Powdertech Corporation.

development. Primarily, they allow the carrier to be tailored to the development housing design and to the charging specifications required for any given machine.

In addition, the magnetic saturation (magnetic saturation moment) can be varied by formulation and process from about 30 to 96 emu per gram, which relates to the developer pickup on the brush and the stiffness of the brush. Too high a moment will result in a stiff brush that might scratch the image, causing nonuniformities in the solids and ragged edges on the line copy or letters.

A comparison of bare carrier core properties (Table 6.2) for iron powders, magnetite, ferrites, hybrids, and composites was recently presented by W. R. Hutcheson of Powdertech Corporation (104). In the case of iron or steel carriers, magnetic saturation moments range from 160 to 190 emu per gram and cannot be varied much outside these limits.

Ferrites are lower density materials and therefore require lower torque to drive the developer housing; this translates into lower energy consumption and less heat generation. Ferrites are natural semiconductors with an electrical resistivity that can be varied across the useful range of carriers, generally from 10^6 to 10^{12} ohm-centimeters. They have been used in some cases without the requirement of partial coating. The surface microroughness can be controlled so that the material is consistently coatable in any given coating process, should it be required. The size range is variable from 10 to 120 microns for the spherical ferrites and can be extended down to 2 microns for irregular ferrites, a property that has begun to raise some interest in the industry (16).

6.4.4 Hard Ferrites (Spherical)

The use of a hard or permanent magnet type of ferrite entered the marketplace with the introduction by Eastman Kodak of the Coloredge machine in 1988. It uses a smaller carrier of a type described in the patent issued to E. Miskinis and T. Jadwin (18). Coloredge is a 23 ppm full color copier, which can run highlight color at higher speeds and black or monochrome copies at 70 cpm. The machine has multiple developer housings to accommodate the primary colors and black for a full spectrum output.

6.4.5 Hybrids and Composites

The Xerox 6500, introduced in 1973, was another color copier that used multiple developer housings to achieve full color reproduction. Some industry analysts believe it entered the market before its time, as there were not enough color originals for it to copy. Now that full color printers are flourishing, there is an explosion of interest in this area.

The 6500 is included in the hybrid category because it used two different carriers. One color utilized 100 micron ''nickel-berry'' beads, electrolytically grown nodules of high-purity nickel. Other colors used coated steel carriers. Because nickel is a costly metal that has undergone wild fluctuations in price, and in view of the military priorities placed on this material, it has not been used in other applications to this writer's knowledge.

Another hybrid ferrite composition was developed in 1986 by Hitachi Metals, Ltd., of Japan. It uses two different ferrite crystalline structures mixed together in the standard ferrite process to make a copier powder (20). The composition, used in some of the Japanese machines, is a mixture of a hexagonal ''hard'' ferrite and a spinel ''soft'' ferrite. Although it has not been disclosed, the resultant carrier core probably has a two-phase structure. The hard phase would be little affected in the low magnetic fields experienced in the magnetic roll of the machine and has little magnetization effect. However, even in low-gauss fields, flow characteristics and torque requirements differ from those of soft ferrite carriers.

Still another hybrid is a mixture of two or more of the materials already mentioned. An example of this is a mixture of spherical ferrite and iron flake noted in at least one of the Panasonic copiers. There is also a patent by J. Cooper and A. Goldstein which describes a similar mixture (21).

A ''composite carrier'' as defined by Hutcheson (104), contains magnetic powder dispersed in a nonmagnetic polymer or resin matrix. Composites were developed for their extremely stable triboelectric properties and low bulk density, but they typically exhibit low magnetic moment and a very high resistivity.

6.5 CARRIER MANUFACTURING PROCESS

This section is included to describe some of the methods used to convert the raw powder into carrier. Many carriers are coated, or more likely partially coated, for two main reasons. The first is to enhance toner charging, because polymer surfaces give better charge exchange and less charge recombination than inorganic surfaces.

The second reason is that adhesive force or tackiness is lower with polymer surfaces; thus less toner is adhered to the carrier surface. Such impacted toner impedes development and shortens the useful life of the carrier. When toner is adhered or too highly charged it uses up carrier area until there is not enough charging surface left to support the toner required to produce the desired level of image density. Any further toner additions result in low or zero charging, increased background deposits, and machine dirt. Generally, this is a criterion for ''developer failure,'' which means a costly maintenance call to replace the developer and service the machine.

There is a triboelectric series that has been generated by several investigators by rubbing dissimilar surfaces together and determining the charge exchange between them. The series is arranged to show that a material will acquire a positive charge from the material just below it in the series and a negative charge from the material just above it. The following short list is an example (22).

Air
Glass
Nylon, wool, and silk

Aluminum
Cellulose
Cellulose acetate
Polymethylmethacrylate
Iron
Polyester
Polyurethane
Polystyrene
Polyethylene
Polypropylene
Polytetrafluoroethylene (Teflon®)

Many variables affect the charging properties of a triboelectric pair, including moisture, impurities, and surface irregularities. This may account for differences in the triboelectric series as reported by different sources (22). The list of polymer coatings is too long to include here but most of the carrier patents listed in the references for this chapter provide a short history of this technology and an extensive list of materials that have been used.

The majority of the carriers in production today are coated by one of two methods, which will be discussed next. Another minority group uses uncoated carrier or raw powder. Still another small percentage uses an oxidized iron carrier core material, and this type will also be briefly covered.

6.5.1 Solvent Coating

One of the most common methods of coating carrier employs a resin dissolved in a solvent or solvent blend. As an example, polymethylmethacrylate (PMMA) is commonly applied to carrier core materials from a solution in toluene.

There are several types of carrier coating equipment in use today to perform this operation; they range from the very simple to the highly sophisticated. One method uses a vibratory tub with a cover connected to a suction fan to pull the solvent vapors through a cooling unit that condenses and recovers the toluene. Carrier powders, or beads, are normally preheated before being charged to the unit. Some vibratory coaters are constructed with a steam heated jacket to elevate the carrier core temperature. After the powder is heated to a specific temperature, the solution of PMMA in toluene, generally containing 10 to 15 percent PMMA by weight, is added slowly by spraying onto the powder. The combination of heat and the vacuum created causes rapid solvent evaporation, leaving the dried polymer on the surface of the beads.

Most vibratory tubs have a curved bottom such that the constant vibration keeps the powder flowing in one direction and turning over on itself. Generally a stirring bar is required to break up agglomerates that form when the coating goes through a tacky phase. The resin solution is added until the precalculated coating weight is achieved. Agitation and heating are continued until the bed of powder starts to rise in temperature, indicating that the solvent is gone. The agitation continues until the temperature drops to a level for safe removal of the coated product.

A more sophisticated coater is a twin-cone (or twin-shell) vacuum drier equipped with a heated shell and metered solution input. This unit is connected to a vacuum pump, which pulls the vapors through a condenser for solvent recovery. Many varieties of this apparatus exist, some with feedback controls that vary the vacuum pumpdown and solution feed rates to control the temperature throughout the coating cycle.

With such a double-cone blender/drier, the bare powder is charged to the unit at ambient temperature and heated to the initiation temperature. At this point, addition of the resin solution begins and proceeds at a controlled rate. The rest of the process is similar to the vibratory tub cycle. Programmed twin-shell blenders generally provide a more consistent coating, but it is important to note that most carrier powders require only a partial coating, and different methods will yield different toner charging properties even at the same coat weight. Several other solvent coating devices are in use, but the principles are the same, and the results must be tuned to the ultimate application of the carrier in a specific imaging system.

6.5.2 Powder Coating

A second extensively utilized coating method is powder coating. Here, the resin coating is applied as a fine powder that is fused to the carrier bead by heat. Polyvinylidene fluoride (PVF) resin is often used in carrier coating applications. The particle size of the coating powder is less than 5 microns. PVF is added to the core at the desired coating weight and tumbled in a twin-shell or double-cone blender until it has coated the core either by adhesive or electrostatic forces. The material is then subjected to a temperature high enough to melt the polymer and spread it over the core surface.

This operation is generally performed in a rotary tube furnace so that the powder rolls over itself and progresses down the inclined tube. While the output is a coated carrier, it should be noted that the process parameters must be tuned to yield the desired toner charging results for the specific application. This type of coating is mainly used for irregular carrier powders and is effective in controlling resistivity and charging response.

6.5.3 Oxidation

Surface oxidation applies mainly to iron and steel powders, as the ferrites are already oxidized. The process forms a ferritelike oxide surface and raises the resistivity of the base material. Oxidation also prepares the carrier surface for coating, providing better adhesion just as most metal surfaces are primed for painting or coating.

One method of oxidation is described in the patent noted in Ref. 15. In this case the powder is fluidized in a cylindrical vessel. The fluidizing air is heated to a specific temperature for a controlled length of time and then cooled in a controlled atmosphere to produce the fine-grained type of surface oxide required. Often this process is controlled to give a specific color, because the various oxides of iron range from red to yellow to dark blue depending on the time and temperature of the treatment. Of course, the qualifying parameter is ultimately the resistivity of the powder. It may or may not be necessary to coat this product afterwards, depending upon the application.

The fluidized bed operation just described is a batch process. Many carrier makers prefer to use a continuous process, such as a rotary tube furnace, to oxidize their powder. This can be tricky, because the temperature required is generally just below one of the phase transition temperatures of iron oxide, a phase change that is exothermic. If exceeded, there can be a runaway reaction with localized melting. At best, this stops the process; at worst, it can cause damage to the furnace.

6.6 CARRIER PROPERTIES

6.6.1 Magnetic Properties and Testing

The magnetic characterization of carrier powders is important, as are the physical, chemical and electrical properties, in order to obtain a controlled product that will deliver consis-

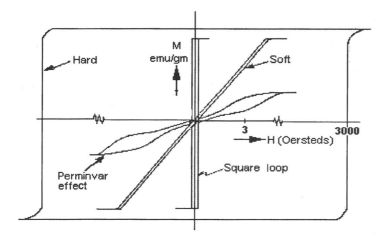

Figure 6.6 Generic hysteresis loops for hard and soft magnetic materials.

tent performance. Most developer manufacturers require two and sometimes three points on the direct current "hysteresis loop" curve to ensure that the carrier meets the specifications set for this product.

The hysteresis loop is plotted by a powder magnetometer that subjects the carrier beads to a slowly increasing magnetic field and measures the magnetization of the sample in that field. The magnetic field is increased to saturate the material and then reduced to zero and reversed to saturate the sample in the opposite sense; the field is then reduced to zero again. The sample magnetization curve is plotted as a function of the magnetic field applied, generating a hysteresis loop that characterizes the sample. Some stylized loops are illustrated in Fig. 6.6. Note especially the soft and hard type of magnetic materials that have been discussed.

Equipment for obtaining this data is illustrated in Fig. 6.7 and is called a "vibrating sample magnetometer" (VSM). This device was developed by S. Foner in the early 1960s and sold commercially since that time by Princeton Applied Research. The principle of operation is simply that the sample is vibrated in a vertical mode while the magnetic field of the electromagnet is applied orthogonally to that motion. Since the magnetization of the sample is the only parameter varying with time, it is picked up by the sensing coils, integrated electronically, and displayed as a function of the applied field.

The VSM is calibrated using 99.9999% pure nickel, which has a saturation moment of 55.01 emu per gram. This calibrates the y-axis for electromagnetic units (emu), and the sample value is divided by its weight, resulting in a traceable, precise, and accurate measurement. The voltage on the coils is

$$E = N\frac{d\Phi}{dt} \tag{1}$$

where N = the number of turns, Φ = the flux from the sample, and t = the time. Integration with respect to time gives

$$E(\Phi) = \frac{Et}{N} \quad (\text{V*s}) \tag{2}$$

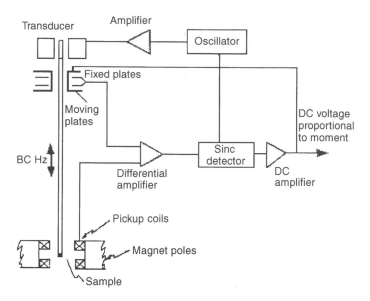

Figure 6.7   Vibrating sample magnetometer schematic.

but this output voltage is calibrated with the standard in emu so that the sample can be read directly in those units. A complete hysteresis loop can be displayed on an x–y plotter giving the type of curve shown in Fig. 6.6.

   Recently a more compact and less expensive unit that performs the same task has been offered by Princeton Measurements Corporation (23). It is called an "alternating-gradient magnetometer" (AGM). In addition, there are single point instruments which measure either the saturation magnetization or the remanence magnetization or both. From the xerographer's point of view these values are useful. The saturation moment, M_s, can be correlated to the stiffness of the brush in any given machine. The remanence moment, M_r, can be correlated to the flow of material in the sump during mixing. The coercive force, Hc, or the magnetic field required to reduce the remanence field to zero, is a measure of the difficulty in magnetizing the material or the "softness." The higher the Hc, the harder the material magnetically.

6.6.2   Electrical Properties and Testing

Measuring the electrical properties of a powder is not a simple task, as the material is solid but behaves like a liquid. A myriad of methods have been developed to obtain a number representing the resistivity of these materials. Most xerographic engineers like to measure the powder in a cell that simulates the action of a magnetic brush. Many OEMs have constructed cells that resemble the schematic design in Fig. 6.8. It consists of a miniature magnet roll and power supply and is fitted with the appropriate meters. Some measure electrical properties with the brush moving while others form the brush and make the measurements after it has been stopped. Different voltages are used, and the calculations of resistivity all assume some brush cross-sectional area.

   There several problems with this method. First, it is difficult to build two identical cells that would enable operators at different locations to obtain identical results. This is not the best situation for vendor–customer correlation.

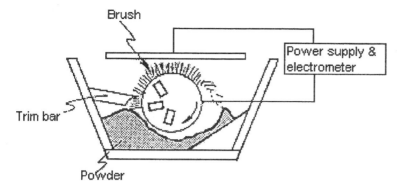

Figure 6.8   Generic dynamic cell.

Second, because there are no standards for powder resistivity, the parties involved must choose variations in the material, establish measurement standards, and negotiate an agreement on the correlation of their measured values. Unfortunately, there is no standard industry cell available; these devices are usually built by the parties engaged in the manufacture and use of the carrier powders.

Although there is no standard material for reference, a more precise resistivity measurement, one that *can* be correlated, is possible with a static cell such as the one illustrated in Fig. 6.9. This guarded electrode cell is normally used to determine the resistivity and dielectric properties of liquids; it is being used, however, to evaluate copier powders, requiring about 1 pound of material for the measurement. The cell is easily cleaned, and the micrometer adjustment gives accurate spacing.

The powder is assumed to be a liquid, and a voltage is applied across the sample while measuring current flow. The area of the inner electrode is used in the resistivity calculation. The measurement is repeatable, and correlation is good between different locations. This cell is an old standard in the industry and is available to everyone. Another

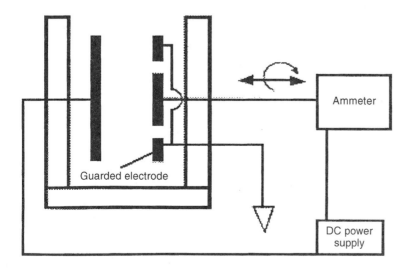

Figure 6.9   Static resistivity cell.

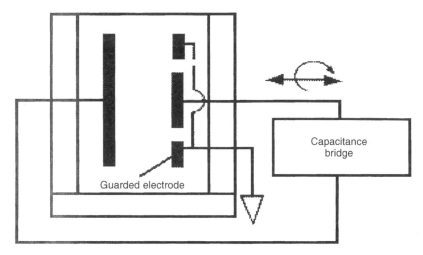

Determine -capacitance & dissipation factor
Derive -K' & K'' complex permitivity
-Loss tangent tan delta = K''/K'

Figure 6.10   Dielectric measurement.

way of applying this cell is actually to measure the dielectric constant and loss factor of
the material, especially in the case of ferrites, as shown in Fig. 6.10. Metals are so high
in dielectric constant and losses that it would apply to coated materials only. This is done
with a capacitance-dissipation factor bridge. Several types are available, some featuring
automatic balancing and digital readout.

6.6.3   Physical Properties and Testing

Several physical properties will be discussed in this section, and here is a list for reference:

    Sieve analysis and calculated specific surface area
    BET gas absorption surface area and roughness
    Bulk density and flow properties
    Triboelectric properties (toner charging)

Sieve analysis of carrier materials is generally performed according to ASTM B 214-
66 (American Society for Testing and Materials). Often the parties involved agree to a
modification of this standard to eliminate screens or use a different weight of sample. The
equipment is a Ro-Tap sieve shaker with a timer. The screens used are U.S. Standard
testing sieves covering the size distribution required. As an example, for a 90 micron
material, the screen mesh sizes would be #120, 140, 170, 200, 230, 270, and 325 with a
pan and cover. The Ro-Tap is started and timed for 15 minutes. Then the weight of material
is recorded from each screen, and each is calculated as a percentage of the total weight of
the sample. Even this procedure is sometimes difficult to correlate between two locations.
Matching data to a master screen set by both parties and continual rotation of the screens
as they wear are necessary practices to ensure good correlation.

    The calculated surface area is just a calculation from the screen analysis that makes
the assumption that all the particles are spheres and all particles on any given screen are

exactly the average diameter of the two screen meshes for that fraction. The individual bead density is used to calculate the surface area per gram that a toner particle could see or touch or with which it can exchange charge.

BET surface area is derived by a procedure which uses gas adsorption as a measure of the area that a krypton atom will find by detecting all the crevices, surface porosity, grain boundaries, etc. It is basically the surface that a solvent coating solution could contact. The most common equipment in use is the Quantasorb Surface Area Analyzer from Quanta Chrome Corp. In the unit, Kr is adsorbed on the surface when a Kr/He mixture is passed through the sample at liquid nitrogen temperature. The Kr is desorbed when the sample is removed from the liquid nitrogen, and the amount desorbed is proportional to the surface area of the sample. The amount of Kr is determined by comparison to a known amount of nitrogen that has been correlated to the signal from Kr gas. The BET value divided by the calculated surface area (CSA) results in a quality factor that relates to the coatability of the powder. As an example, a powder with a very high roughness factor requires much more coating weight than a lower roughness material to achieve the same triboelectric charging potential.

Bulk density and powder flow rate are both determined by the ASTM B-212(1) procedure, which establishes a standard funnel arrangement. The rate of flow is timed for a sample of known weight and reported in grams per second. The material from the funnel drops freely into a brass cylinder of known volume (50 cc). The overflow is scraped off and the remaining material is weighed. This weight (in grams) divided by 50 is the bulk density in grams per cubic centimeter. This value depends not only on the true density of each bead but also on the packing factor—a function of the shape of the size distribution. As an example, a very narrow size distribution will result in a lower packing factor and therefore a lower bulk density.

The triboelectric properties of a carrier are best described in terms of a standard toner. This may be the toner to be used in a particular machine that will use this developer, or it may be compared to other carriers by using a common toner for all. In this procedure, the toner is blended with the carrier in the desired ratio (e.g., 1% by weight of toner) in a clean dry container (generally glass) for a predetermined time (typically 30 minutes) on a roll mill.

A small sample of this mixed developer (1–2 grams) is then placed in a Faraday cage, a stainless steel cylinder fitted with appropriate screens at each end. The screen mesh is chosen so that the toner can pass through it, but the carrier cannot. The Faraday cage is clipped to a well insulated electrode that is connected to an electrometer. The toner is then blown out of the cage with dry air or nitrogen, taking its charge with it. The carrier is left with an equal and opposite charge, which is read on the electrometer.

Using the tare weight of the cage, the weight of the cage with developer, and the final weight of the cage, the weight of toner actually blown off is determined. The charge reading is divided by the weight of toner and the tribo results reported in microcoulombs per gram. This value is called the "total blow off tribo" of that developer at that toner concentration. Samples of developer taken from operating copiers are also tested to determine the tribo and toner concentration as a function of the age of the developer expressed in number of copies produced.

6.7 TYPICAL CARRIER PROPERTIES

Most of the carrier powders in current use have been discussed in this chapter. Those properties of special interest today and in the near future will likely center on the electrical

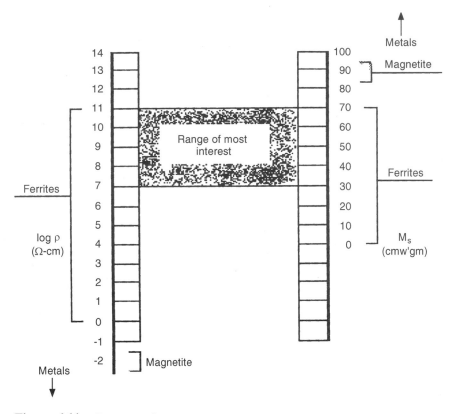

Figure 6.11   Current carrier property targets.

resistivity and magnetics of these unique materials. An illustration of this is shown graphically in Fig. 6.11. Here, two vertical axes represent the logarithm of resistivity, $\log \rho$, and the saturation magnetization, M_s, respectively. The scales are positioned so that the range of current interest matches on each scale. The resistivity is determined by the type of static cell illustrated in Fig. 6.9.

Note that ferrites and oxidized iron both fall nicely within the interest range for resistivity. Unoxidized iron and steel must be partially coated to have these effective resistivities, and they are in many cases successful. Ferrites also fall neatly into the interest range for magnetic properties, but iron powders must be very porous and irregular to bring the magnetization down to these levels. Developer housing design can compensate for these high magnetic moment materials, which are useful in many applications. Spheroidized magnetite has not gained much interest in the imaging industry, but partially reduced, irregular magnetite is gaining ground in Japan.

6.8   NEW PATENT AND LITERATURE ACTIVITY

Now, as mentioned in the first section of this chapter, the carrier activity of the last ten years should be discussed. This section will cover what is relevant in the carrier patent and publication field since the first edition of this book was published. The anticipated demise of two-component developer systems and the carriers involved is still far in the future. Since 1990 over 230 U.S. patents have been issued which cover four basic categories pertaining to xerographic carrier manufacturing, processing, and end use. These are:

1. Carrier cores and carrier coating processes
2. Color-specific carrier applications
3. Scavengeless image-on-image (IOI) systems, mainly for color reproduction
4. Ferrite-specific based carriers

Note that only U.S. patents that have been issued since 1990 were surveyed for this update.

6.8.1 Carriers and Coating

Nearly 160 patents fall into this category, with most involving additives for improved flow or charge control or methods of mixing coating materials to stabilize charging toner characteristics, or new combinations of coating materials to improve performance. At least 20 assignees (companies) are included in this patent progress. Following is a table illustrating the important features of this progress.

Company	Number	Ref. No.	Comments
Canon KK	15	26	Ten of these describe properties and surface treatments of carriers
Eastman Kodak Co.	22	27	Most of these refer to hard ferrites (Ba or Sr), some mixed with soft ferrite. Five describe two- or three-phase ferrite composites. Coatings of silicone or PVDF are included
Fuji Xerox	21	28	Three of these are variations in ferrite composition. Seven involve coatings of fluoro resins, PVDF, vinyl, or acrylic polymers
Hitachi Metals Ltd.	11	29	Ten cover magnetic carriers of various magnetic and conductivity properties, high surface area for long life; and the most recent is a Li ferrite
Konica Corporation	13	30	Five discuss powder coating, some with fluoro resins, Si compounds, or polyolefin. Three relate to Mg surface atoms and negatively charging developer. Most concern ferrite carriers
Kyocera Corporation	4	31	All describe two-component developer
Matsushita Electric	1	32	Two-component developer properties
Minolta Camera KK	9	33	Three cover composite magnetic carriers, two involve 1.5 component developer (very small consumable carrier), one covers powder coating and two pertain to size
Mitsubishi KK	1	34	Carrier with vinylidene chloride coating and a charge control agent

Company	Number	Ref. No.	Comments
Mita Industrial Co., Ltd.	12	35	Seven cover various coatings, two involve resistivity and carrier relaxation time
Ricoh Co., Ltd.	10	36	Three refer to forming microfields, one covers size and charge, one covers a Si modified acrylic coating, one has a mixture of charge control agents, and one has 2 carriers with one cleaning the other
Tomoegawa Paper Co.	2	37	One covers a ferrite composition and one covers magnetic properties of carrier
Xerox Corporation	22	38	Five involve carrier conductivity, eight cover two-polymer coatings, four cover coating methods, one includes additives, and one refers to fly ash

6.8.2 Color Specific Carrier Patents

Company	Number	Ref. No.	Comments
Canon KK	4	39	All cover ferrite carrier, two for yellow toner, one for four colors, one with 2-resin mix coating
Eastman Kodak Co.	1	40	Refers to charge area or discharged area development
Kao Corporation	1	41	Fluoro resin for positive charge for color
Minolta Camera KK	2	42	One covers coating molecular weights and one involves cross-linked styrene acrylic copolymer and melamine resin
Xerox Corporation	4	43	One covers variations in percentage of bare core in developer, two describe preferred properties, and one has charge control and flow additives

6.8.3 Scavengeless Development

Company	Number	Ref. No.	Comments
Eastman Kodak Co.	2	44	One covers method and one involves spacing to keep the brush from touching the previous image developed
Hitachi Metals Ltd.	1	45	Controls the magnetic brush height to prevent touching
Xerox Corporation	33	46	All involve image-on-image development with various methods of supplying an intermediate step for the two-component developer to charge and transfer toner without touching the previous image developed

6.8.4 Ferrite Core Specific

Company	Number	Ref. No.	Comments
BASF	1	47	Ferrite or oxidized iron with controlled resistivity
Canon KK	4	48	Two refer to yellow toner, one involves 4-color development, and one coating of two resins
Eastman Kodak Co.	9	49	Three involve 2- or 3-phase ferrite compositions, one refers to a mix of hard and soft ferrite, one gives a method of producing a ferrite, one covers silicone resin coating, and one provides lower dusting
Fuji Xerox	3	50	All refer to ferrite compositions
Hitachi Metals Ltd.	1	51	Lithium ferrite composition
Konica Corporation	2	52	One refers to a ferrite composition and one covers coating by impacting powder
Minolta Camera KK	1	53	Refers to composite ferrite/polymer carrier
Mita Industrial Co., Ltd.	1	54	Coating ferrite with acrylates to specific electrical and magnetic properties
Powdertech Co., Ltd.	7	55	Three provides lithium ferrites, one offers a mixed hard and soft ferrite, one refers to a small iron carrier (25–40 microns), and one covers coating by microwave heating
Steward Manufacturing	1	56	Liquid phase sintering to make ferrite with no nickel, zinc, or copper
Xerox Corporation	2	57	One coating to specific resistivity and one refers to 2-polymer coating for tribo control

6.8.5 Literature Survey

Over 60 articles have been published on this subject during the 1990s. They can be divided among eight basic categories:

1. General theory
2. Contact charging
3. Two-component developer
4. Coating methods and materials
5. Conductivity (electrical)
6. Triboelectric properties
7. Adhesion
8. Future

No attempt has been made to elaborate on or judge the relative merits of these papers; they were chosen as they represent the most informed workers in this field. The reader interested in pursuing is referred to these references for more in-depth study.

General Theory

This group includes L. B. Schein's book on electrophotography (58) as well as his toner charging theory (59). Surface state models of tribocharging (60) and comparison to experimental data for two-component developer (61) are presented by J. H. Anderson. Veresh-chagin and Krivov offer an analysis of the behavior of material particles in an electric field (62), while Jeyadev and Stark (63) describe a transfer function for development to complete field neutralization.

Contact Charging

Castle and Schein (64) present a model for sphere–sphere insulator contact electrification. Impact charging of insulators is also the subject discussed by Masui and Murata (72). A. Diaz et al. discuss an ion transfer model for contact charging and the effects of ionomer ions or dissociated ions (65–67). L. H. Lee presents a dual mechanism for contact electri-fication by metal–polymer contact (68). Veregin et al. discuss the relation of Kelvin poten-tial to charging (69) and effects of relative humidity on charging for metal oxides (70). Itakura et al. relate tribocharge to contact potential differences of the surfaces (71). J. Q. Feng et al. present arguments concerning electric field effects on nonuniformly charged spheres on dielectric surfaces (73). The contact charging of polymers and the penetration depth of the charge are presented by Watson and Zhao-Zhi (74).

Two-Component Developers

Anzai et al. offer some considerations on development efficiency for two component de-velopers in a magnetic brush system (75). The correlation of low-charge toner to back-ground development, mentioned previously in this chapter, was justified in a paper by Gutman and Hollenbaugh in 1997 (76). P. C. Julien related the relationship of size and charge in developers (77) and the effect of toner clouding from two-component developer in a magnetic brush system (78). Aging of two-component developers as related to toner concentration and how conductive developer aging affects the xerographic response are subjects of papers by R. J. Nash et al. (79,80). Nash et al. also have a paper describing toner charge instability (81), and Nash and Muller offer a paper about the effect of toner and carrier composition on tribocharge and toner concentration (82).

Coating Methods

In 1997 K. Adamiak published a numerical modeling scheme of tribocharging for powder coating systems (83). Muzumder et al. studied the influence of powder properties on the electrostatic coating process (84), and Stotzel et al. described adhesion measurements for this process (85).

Carrier Conductivity Considerations

E. Gutman and G. Hartmann presented in 1996 a study of the conductive properties of two-component developer materials (93). The effect of toner impaction on conductivity aging of a developer was reported in 1992 by R. Nash and J. Bickmore (94), and in 1996 Nash et al. describe how conductivity aging relates to xerographic response (95). N. Felici discusses interfacial effects and offers a criticism of the conduction model (96).

Triboelectric Properties

E. Gutman and G. Hartmann collaborated on two papers that discuss the triboelectric properties of two-component developers (86) and describe the role of an electric field in tribocharging two-component systems (87). P. Julien et al. (88) in 1993, Takezawa et al. (89) in 1997, and Nakamura et al. (90) in 1997 discuss the effect of particle size and size distribution on the charging of toners. K.Y. Law et al. offer a tribocharging mechanism in a model toner (91), while Shijo et al. (92) report general charging characteristics in two-component developers. In 1997, Lee and Takahashi described the dependence of triboelectric charging on external additives in two-component developers (101).

Adhesion Properties

In 1994, J. C. Maher (97) offered a phenomenological model for describing toner adhesion to carrier and D. Hays (98) gave a general discussion of toner adhesion. In the same year, E. Eklund et al. also presented a paper (99) on toner adhesion physics and measurements of the contact area of toner to substrate surfaces, while M. Ott (100) described the effect of relative humidity on toner adhesion in the presence of surface additives.

6.9 FUTURE TRENDS

Indications are that the future will see carrier design moving further in the directions already implied in this chapter. More irregular powders will likely be seen, probably in smaller size distributions. Indeed, W. R. Hutcheson's study of trends in carrier particle size over the past 30 years, shown graphically in Figure 6.12 (104), supports this direction. The shift toward smaller carrier beads has several advantages: it offers more surface area for transporting toner, it generally results in longer developer life, and for a given toner concentration, the resulting developer mix is lower in resistivity as it offers more core-to-core contact. The latest machines incorporate designs that are better able to contain these fine materials in the developer housing. In addition, with finer carrier materials it is possible to raise the level of copy quality. From a statistical standpoint, more toned carrier surface is offered to the photoreceptor in the development zone.

The industry also continues to move to lower density powders (1) to minimize the torque required to transport the carrier material, (2) to reduce the weight of any individual machine developer charge, and (3) to form a softer brush in the development zone.

All carrier materials are purchased by the pound and used by the cubic inch. As an example, a small copier that might require 2.5 pounds of steel would only require 1.6 pounds of ferrite to achieve the same total carrier surface area for the toner.

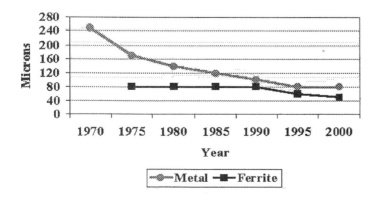

Figure 6.12 Trends in carrier size. (Courtesy of Powdertech Corporation.)

Greater interest is being shown in the various ferrites for most of the reasons given above as the manufacturers move to reduce cost by using less expensive raw materials and more efficient quality and process control. Fine particle ferrites are already in use or on the drawing board with 50, 30, 12, and 5 micron materials now in the field.

The aim has always been to emulate lithographic resolution, density uniformity, and zero background especially in the high volume printers and copiers. A continued increase is expected in color copier and printer applications as the generation of color originals expands from ink-jet, thermal, and toner-based printers. Emphasis will undoubtedly be placed on faster, lower cost color applications using two-component development. Of course, many other reproduction methods are increasing and will continue to increase in importance.

Trends in carrier consumption are illustrated in Table 6.3 for the U.S. imaging industry (in tons). Spherical ferrites exhibited the largest increase, about 8.5% compound annual growth rate (CAGR) over the past five years (1995–2000), with spherical carriers advancing at about 3.9% CAGR. The use of spherical steel powders in xerographic imaging systems has declined at an average annual rate of about 10%, while irregular iron and steel powders demonstrated about 6.0% CAGR. Spherical nickel and irregular sand and glass carriers have all but disappeared from use in U.S. office copiers and duplicators.

Liquid toner development has great capabilities but presents an environmental problem associated with the volatile organic compounds (VOCs) employed. Two-component developers in magnetic brush systems should hold steady for some time to come because the toner charge distribution required for optimum marking can be designed into the system. Monocomponent toner applications have problems in this charge distribution requirement as currently practiced, but the future holds the possibility of a corona charged fluidized bed recharge-expose-and-develop (REaD) as a very strong contender. Recent patent activity indicates research in this area at Xerox Corporation and Moore Business Forms (103).

L. B. Schein provides an excellent review of recent advances in this technology (102), giving the reader an expert's persepective on future trends. For more information about the xerographic process, the reader is referred to books by Schaffert (2) and Dessauer and Clark (24).

Table 6.3 Estimated U.S. Consumption of Carrier Materials (1985 to 2000) (Tons)

	1985[a]	1990[a]	1995[a]	2000	CAGR, %
Spherical Carriers					
Steel	4,430	4,311	4,069	2,400	−10.0
Ferrite	3,415	5,838	8,546	12,850	+8.5
Magnetite	235	348	544	650	+3.6
Nickel	68	23	0	0	—
Subtotal	8,147	10,520	13,159	15,900	+3.9
Irregular Carriers					
Iron and steel	3,252	5,039	8,142	10,900	+6.0
Sand and glass	250	85	16	0	—
Subtotal	3,502	5,124	8,158	10,900	+6.0
TOTAL	11,648	15,644	21,317	26,800	+4.7

[a] *Source*: Ref. 25, p. 582.

ACKNOWLEDGMENTS

I would like to express my appreciation to Arthur S. Diamond for many hours of information exchange and efforts in generating this and previous publications (25). Also, many thanks to Robert J. Hagenbach for his help in many ways along these same lines.

REFERENCES

1. Diamond, A. S., et al. Reprographic chemicals—USA. Specialty chemicals—strategies for success. SRI International, Menlo Park, CA, Oct 1982.
2. Schaffert, R. M. Electrophotography yesterday, today, and tomorrow. Photo. Sci. Eng., 22, May/June 1978, pp. 149–153.
3. Young, C. J. Electrophotographic developing apparatus. U.S. Patent 2,786,439, Mar. 26, 1957.
4. Giaimo, E. Electrophotographic developing apparatus. U.S. Patent 2,786,440, Mar. 26, 1957.
5. Young, C. J. Apparatus for applying electrostatic developer powder by means of a magnetic brush. U.S. Patent 2,786,441, Mar. 26, 1957.
6. Van Engeland, J. Special characteristics of magnetic brush development: a review. Photo. Sci. Eng., 23, March/April 1979, pp. 86–92.
7. Hagenbach, R. Highly shape-classified oxidized steel carrier particles. U.S. Patent 3,849,182, Nov. 19, 1974.
8. Hagenbach, R., and Forgensi, R. Xerographic carriers by the two wire spheroidization. U.S. Patent 4,018,601, April 19, 1977.
9. Katayama, M., et al. Carrier for use in electrophotographic developers. U.S. Patent 4,732,835, Mar. 22, 1988.
10. Jones, L. O. Stoichiometric ferrite carriers. U.S. Patent 4,042,518, Aug. 16, 1977.
11. Jones, L. O. High surface area carrier. U.S. Patent 4,040,969, Aug. 9, 1977.
12. Jones, L. O. Humidity insensitive ferrite developer materials. U.S. Patent 3,996,392, Dec. 7, 1976.
13. Jones, L. O. Stoichiometric ferrite carriers. U.S. Patent 3,929,657, Dec. 30, 1975.
14. Berg, A. C., et al. Production of ferrite electrostatographic carrier materials having improved properties. U.S. Patent 4,075,391, Feb. 21, 1978.
15. McCabe, J. M., et al. Method for producing oxide coated iron powder of controlled resistance. U.S. Patent 3,767,477, Oct. 23, 1973.
16. Tosaka, H., et al. Magnetic developer for developing latent electrostatic images. U.S. Patent 4,670,368, June 2, 1987.
17. Drexler and Altmann. Apparatus for development of electrostatic images. U.S. Patent 3,543,720.
18. Miskinis, E. T., et al. Two-component, dry electrographic developer compositions containing hard magnetic particles. U.S. Patent 4,546,060, Oct. 8, 1985.
19. Wilson, J. C. Method of developing electrostatic images. U.S. Patent 2,846,333, Aug. 5, 1958.
20. Iimura, T., and Chinju, M. Spherical EPG magnetoplumbite-type hexagonal ferrite carrier powder. U.S. Patent 4,623,603, Nov. 18, 1986.
21. Cooper, J., and Goldstein, A. Dry process electrostatic developer, round magnetic carrier and flake type carrier. U.S. Patent 4,683,187, Jul. 28, 1987.
22. Winkelmann, D. Electrostatic aspects of electrophotography. J. Electrostatics, 4, 1977/1978, pp. 193–213.
23. Flanders, P. J. An alternating-gradient magnetometer. J. Appl. Phys., 63(8), Apr. 15, 1988.
24. Dessauer, J. H., and Clark, H. E. Xerography and Related Processes. Focal Press, New York, 1965.

25. Diamond, A. S., and Jones, L. O. Copier Powders, Metals Handbook, 9th ed., Vol. 7, Powder Metallurgy. ASM, Metals Park, OH, June 1984, pp. 580–588.

26. U.S. Patents 5,187,523, 5,215,848, 5,340,677, 5,439,771, 5,447,815, 5,455,666, 5,459,559, D0,366,898, 5,494,770, 5,624,778, 5,670,288, 5,712,069, 5,766,814, 5,795,693, 5,885,742.

27. U.S. Patents 4,990,876, 5,061,586, 5,096,797, 5,104,761, 5,106,714, 5,124,223, 5,190,841, 5,190,842, 5,217,804, 5,235,388, 5,255,057, 5,268,249, 5,272,039, 5,306,592, 5,316,882, 5,332,645, 5,364,725, 5,381,219, 5,411,832, 5,500,320, 5,512,404, 5,516,615, 5,705,307, 5,709,975.

28. U.S. Patents 4,898,801, 4,912,004, 4,929,528, 5,085,963, 5,110,703, 5,202,210, 5,256,511, 5,275,902, 5,288,578, 5,362,596, 5,482,806, 5,629,120, 5,634,181, 5,665,507, 5,672,455, 5,693,444, 5,752,139, 5,783,345, 5,783,350, 5,897,477.

29. U.S. Patents 5,422,219, 5,429,900, 5,483,329, 5,496,673, 5,516,613, 5,547,795, 5,731,121, 5,733,699, 5,786,120, 5,790,929, 5,876,893.

30. US Patents 4,882,258, 5,104,762, 5,182,181, 5,194,360, 5,200,287, 5,272,038, 5,350,656, 5,376,488, 5,441,839, 5,478,687, 5,486,901, 5,637,431, 5,643,704, 5,795,691, 5,932,388.

31. U.S. Patents 5,256,513, 5,395,717, 5,633,107, 5,824,445.

32. U.S. Patents 5,593,806.

33. U.S. Patents 4,996,126, 5,204,204, 5,206,109, 5,275,901, 5,285,801, 5,391,451, 5,663,027, 5,688,622, 5,689,781, 5,736,287.

34. U.S. Patent 5,360,691.

35. U.S. Patents 4,963,454, 5,079,124, 5,085,964, 5,212,034, 5,212,038, 5,217,835, 5,232,806, 5,232,807, 5,240,804, 5,258,253, 5,360,690, 5,514,509, 5,634,174, 5,683,846.

36. U.S. Patents 5,225,302, 5,315,061, 5,403,690, 5,424,814, 5,451,713, 5,638,159, 5,652,079, 5,666,625, 5,674,408, 5,678,125.

37. U.S. Patents 5,290,652, 5,484,676.

38. U.S. Patents 4,894,305, 4,935,326, 4,937,166, 5,002,846, 5,015,550, 5,071,726, 5,087,545, 5,100,753, 5,102,769, 5,162,187, 5,171,653, 5,194,357, 5,213,936, 5,223,368, 5,230,980, 5,236,629, 5,238,770, 5,304,449, 5,324,613, 5,330,874, 5,332,638, 5,395,450, 5,401,601, 5,424,160, 5,451,481, 5,484,681, 5,496,675, 5,506,083, 5,510,220, 5,516,612, 5,516,614, 5,518,855, 5,595,851, 5,882,834.

39. U.S. Patents 5,1167,111, 5,149,610, 5,164,275, 5,256,512.

40. U.S. Patents 5,748,218.

41. U.S. Patents 5,173,387.

42. U.S. Patents 5,212,039, 5,260,159.

43. U.S. Patents 4,920,023, 5,021,838, 5,336,579, 5,536,608.

44. U.S. Patents 5,409,791, 5,489,975.

45. U.S. Patents 5,430,528.

46. U.S. Patents 5,053,824, 5,128,723, 5,144,371, 5,153,648, 5,172,170, 5,245,392, 5,253,016, 5,311,258, 5,322,970, 5,338,893, 5,359,399, 5,360,940, 5,404,208, 5,409,791, 5,420,672, 5,422,709, 5,430,528, 5,473,418, 5,489,975, 5,504,563, 5,521,677, 5,537,198, 5,539,505, 5,557,393, 5,572,302, 5,579,100, 5,592,271, 5,600,418, 5,600,430, 5,640,657, 5,666,612, 5,666,619, 5,729,807, 5,734,954, 5,794,106.

47. U.S. Patent 4,925,762.

48. U.S. Patents 5,116,711, 5,149,610, 5,164,275, 5,256,512.

49. U.S. Patents 5,096,797, 5,104,761, 5,106,714, 5,190,842, 5,268,249, 5,316,882, 5,332,645, 5,500,320, 5,709,975.

50. U.S. Patents 4,898,801, 5,629,120, 5,693,444.

51. U.S. Patents 5,876,893.

52. U.S. Patents 5,350,656, 5,637,431.

53. U.S. Patent 5,663,027.

54. U.S. Patent 5,212,034 (Mita Industrial Co., Ltd. and TDK Electronics). This is a joint patent of a copy machine manufacturing with a carrier core vendor.

55. U.S. Patents 5,204,204 (Minolta Camera KK with Powdertech Co, another joint patent), 5,419,994, 5,466,552, 5,491,042, 5,518,849, 5,595,850, 5,798,198.
56. U.S. Patent 5,422,216.
57. U.S. Patents 5,162,187, 5,516,614.
58. Schein, L. B. Electrophotography and Development Physics. Revised second edition. Laplacian Press, Morgan Hill, CA, 1996.
59. Schein, L. B. Theory of toner charging. J. Imaging Sci. Technol., 37, 1, 1993.
60. J. H. Anderson. Surface state models of tribocharging of insulators. IS&T's 10th Int'l Congr. on Adv. in NIP Technol., 111, 1994.
61. Anderson, J. H. A comparison of experimental data and model predictions for tribocharging of two-component electrophotographic developers. J. Imaging Sci. Technol., 378, 1994.
62. Vereshchagin, and Krivov. Analysis of the material particles behavior on the surface in the electric field. J. Electrostatics, 40, 363, 1997.
63. Jeyadev, S., and Stark, H. M. Modulation transfer function for development to complete field neutralization. J. Imaging Sci. Technol., 40, 4, 369, 1996.
64. Castle, G. S. P., and Schein, L. B. General model of sphere–sphere insulator contact electrification. J. Electrostatics, 36, 165, 1995.
65. Diaz, A., and Alexander, D. F. An ion transfer model for contact charging. Langmuir, 9, 1009, 1993.
66. Diaz, A., et al. Effect of ionomer ion aggregation on contact charging, J. Polymer Sci., B, 29, 1559, 1991.
67. Diaz, A., et al. Importance of dissociated ions in contact charging. Langmuir, 8, 2698, 1992.
68. Lee, L. H. Dual mechanism for metal–polymer contact electrification. J. Electrostatics, 32, 1, 1994.
69. Veregin, Powell, Tripp, McDougall, and Mahon. Kelvin potential measurement of insulative particles. Mechanism of metal oxide triboelectric charging and RH sensitivity. IS&T's 13th Int'l Congr. on Adv. in NIP Technol., 133, 1997.
70. Veregin, R. P. N., et al. The role of water in the triboelectric charging of alkylchlorosilane treated silicas as toner surface additives. J. Imaging Sci. Technol., 39, 5, 429, 1995.
71. Itakura, T., et al. The contact potential difference of powder and the tribo charge. J. Electrostatics, 38, 213, 1996.
72. Masui, N., and Murata, Y. Impact charging of insulators, J. Electrostatics, 32, 31, 1994.
73. Feng, J. Q., Eklund, E. A., and Hays, D. A. Electric field detachment of a nonuniformly charged sphere on a dielectric coated electrode. J. Electrostatics 40, 289, 1997.
74. Watson, P. K., and Zhao-Zhi, Y. The contact electrification of polymers and the depth of charge penetration. J. Electrostatics, 40, 67, 1997.
75. Anzai, et al. Some consideration on developing efficiency for dual component magnetic brush development. IS&T's 13th Int'l Congr. on Adv. in NIP Technol., 89, 1997.
76. Gutman, E. J., and Hollenbaugh, W. Background development and low charge toner in the charge distribution of two-component xerographic developers. IS&T's 13th Int'l Congr. on Adv. in NIP Technol., 41, 1997.
77. Julien, P. C. Toner clouding from a two-component magnetic brush. IS&T's 10th Int'l Congr. on Adv. in NIP Technol., 160, 1994.
78. Julien, P. C. The relationship between size and charge in xerographic developers. IS&T's 6th Int'l Congr. on Adv. in NIP Technol., 1990.
79. Nash, R. J., and Bickmore, J. T. The influence of toner concentration on the triboelectric aging of CCA-containing xerographic toners. IS&T's 9th Int'l Congr. on Adv. in NIP Technol., 68, 1993.
80. Nash, R. J., Bickmore, J. T., Hollenbaugh, W. H., and Wohaska, C. L. Xerographic response of an aging conductive developer. IS&T's 11th Int'l Congr. on Adv. in NIP Technol., 40, 4, 347, 1996.

81. Nash, R. J., Silence, S. M., and Muller, R. N. Toner charge instability. IS&T's 10th Int'l Congr. on Adv. in NIP Technol., 95, 1994.

82. Nash, R. J., and Muller, R. N. The effect of toner and carrier composition on the relationship between the toner charge to mass ratio and toner concentration. IS&T's 13th Int'l Congr. on Adv. in NIP Technol., 112, 1997.

83. Adamiak, K. Numerical modelling of tribo-charge powder coating systems. J. Electrostatics, 40, 395, 1997.

84. Mazumder, et al. Influence of powder properties on the performance of electrostatic coating process. J. Electrostatics, 40, 369, 1997.

85. Stotzel, et al. Adhesion measurements for electrostatic powder coatings. J. Electrostatics, 40, 253, 1997.

86. Gutman, E. J., and Hartmann, G. C. Triboelectric properties of two-component developers for xerography. J. Imaging Sci. & Technol., 36, 4, 335, 1992.

87. Gutman, E. J., and Hartmann, G. C. The role of the electric field in triboelectric charging of two-component xerographic developers. J. Imaging Sci. Technol., 39, 4, 285, 1995.

88. Julien, P. C., Koehler, R. F., and Conners, E. W. The relationship between size and charge in xerographic developers. IS&T's 9th Int'l Congr. on Adv. in NIP Technol., 1993.

89. Takezawa, T., et al. Effect of particle size distribution on the triboelectric charge of toners. IS&T's 13th Int'l Congr. on Adv. in NIP Technol., 1997.

90. Nakamura, et al. Tribocharging of toner particles in two-component developer and its dependence on the particle size. IS&T's 13th Int'l Congr. on Adv. in NIP Technol., 173, 1997.

91. Law, K. Y., Tarnawskyi, I. W., Salamida, D., and Debies, T. Tribocharging mechanism in a model xerographic toner. IS&T's 10th Int'l Congr. on Adv. in NIP Technol., 122, 1994.

92. Shinjo, et al. Study of tribo-charging characteristics between toner and carrier. IS&T's 13th Int'l Congr. on Adv. in NIP Technol., 123, 1997.

93. Gutman, E. J., and Hartmann, G. C. Study of the conductive properties of two-component xerographic developer materials. J. Imaging Sci. Technol., 40, 4, 334, 1996.

94. Nash, R. J., and Bickmore, J. T. Toner impaction and conductivity aging. IS&T's 8th Int'l Congr. on Adv. in NIP Technol., 131, 1992.

95. Nash, R. J., Bickmore, J. T., Hollenbaugh, W. H., and Wohaska, C. L. Xerographic response of an aging conductive developer. J. Imaging Sci. Technol., 40, 4, 347, 1996.

96. Felici, N. J. Interfacial effects and electrorheological forces: criticism of the conduction model. J. Electrostatics, 40, 567, 1997.

97. Maher, James C. Characterization of toner adhesion to carrier: a phenomenological model. IS&T's 10th Int'l Congr. on Adv. in NIP Technol., 156, 1994.

98. Hays, D. A. Toner adhesion. Proceedings of 17th Ann. Mtg. Symp. on Particle Adhesion, 91, 1994.

99. Eklund, E. A., Wayman, W. H., Brillson, L. J., and Hays, D. A. Toner adhesion physics: measurements of toner/substrate contact area. IS&T's 10th Int'l Congr. on Adv. in NIP Technol., 142, 1994.

100. Ott, M. L. Humidity sensitivity of the adhesion of pigmented polymer particles treated with surface modified surface additives. IS&T's 10th Int'l Congr. on Adv. in NIP Technol. 142, 1994.

101. Lee, W.-S., and Takahashi, Y. Dependence of triboelectric charging characteristics of two-component developers on external additives. IS&T's 13th Int'l Congr. on Adv. in NIP Technol., 144, 1997.

102. Schein, L. B. Electrostatic marking technologies—recent advances and future outlook. IS&T's 10th Int'l Congr. on Adv. in NIP Technol., 30, 1994.

103. U.S. Patents 5,532,100, 5,862,440, 5,899,608, 5,926,674, 5,953,571.

104. Hutcheson, W. R. Color Imaging and the Future for Carrier Core Materials. Toners and Photoreceptors 2000. Diamond Research Corporation, Santa Barbara, CA, June 5, 2000.

7

Liquid Toner Materials

JAMES R. LARSON and GEORGE A. GIBSON

Xerox Corporation, Webster, New York

STEVEN P. SCHMIDT

Dade Behring, Glasgow, Delaware

7.1 INTRODUCTION

Liquid toners are charged, colored particles suspended in a nonconductive liquid, used to develop electrostatic images. Liquid toner based imaging systems incorporate features of electrostatic imaging similar to those of dry toner based systems. However, liquid toner particles are significantly smaller than dry toner particles. Because of their small particle size, ranging from 3 microns to submicron, liquid toners are capable of producing very-high-resolution toned images. This high-resolution capability, and the capability of liquid toners to produce pure, transparent color, has led to their use of liquid toners in high-quality color printing and related applications.

7.2 LIQUID TONER ELECTROGRAPHIC AND ELECTROPHOTOGRAPHIC PROCESSES

Liquid development of electrostatic images was first demonstrated independently by Metcalfe (1955) and by Mayer (1957). Since then a large number of liquid toner based electrostatographic processes have been developed and turned into products. The electrographic and electrophotographic processes and their individual steps have been well described in several books, including those of Schaffert (1975), Williams (1984), and Schein (1988).

Self Fixing Processes

Printers, platemakers for short-run offset presses, radiographic systems, and copiers were all made exploiting this basic process, and a number of such systems are still commercially

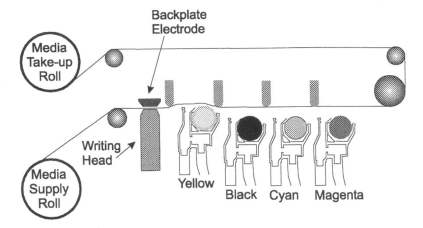

Figure 7.1 Electrographic process. A schematic diagram of the Xerox ColorgrafX Systems 8900 family of wide-format graphics printers. This system is exemplary of a product using the self fixing type of liquid toner process. An electrostatic image is applied to a dielectric medium, which may be paper or film, by means of metal styli addressed with high voltage.

important. These systems are called self fixing because they require no additional heat to fuse the image to its final receiver. For the most part they employ engineered substrates as both the latent image bearing member and as the image receiving substrate. The most common copier embodiment involves a ZnO dispersion coated paper which serves as both the photoreceptor and the final substrate. This substrate is corona (or similarly) charged, and the latent image is generated by standard reflective optics. Development could be quite crude with immersion in the liquid toner and (by today's standards) crude development electrodes. Air knives or rollers are generally used to remove excess toner, and the image is fixed by evaporation of the volatile liquid dispersant. This process is capable of generating quite high-resolution images, although the solid areas are often poor.

Another version of this process uses a conductive paper covered with a dielectric coating. The latent electrostatic image is generated by a direct writing stylus array. Figure 7.1 presents a schematic of the Xerox ColorgrafX Systems 8900 Digital Color Printer, a family of wide-format graphics printers that employ this variation. An electrostatic image corresponding to one separation of the CMYK image is written on the paper by the styli array. This image is developed by flooding the paper with the toner and metering the excess off with a counter-rotating development electrode. The developed image is then dried by passing it through an airstream. The paper is then rewound, and the other separations are produced in turn. Because of the image-on-image nature of this architecture, the toner is required not only to develop sufficient mass (per unit area) to give appropriate color but also to develop sufficiently close to charge neutrality that the preceding image will not be color shifted after passing through the next developer. Simplicity is the primary strength of this family of techniques. Applications where environmental constraints do not dominate and those where the properties (and cost) of an engineered substrate are not barriers will be most suitable for such processes.

Liquid Toner Transfer (LTT) Processes

The requirement for copiers to use plain paper led to the creation of the LTT process and its derivatives. Developed by Savin Corporation and a large number of collaborators

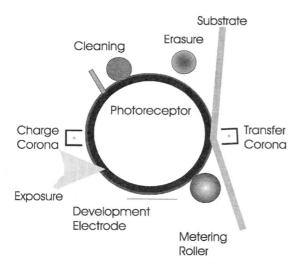

Figure 7.2 Electrophotographic process. A schematic diagram of the liquid toner transfer (LTT) process used in a number of Savin Corporation copiers.

(Jacobson and Hillkirk, 1986) this process enables liquid toner to mark uncoated papers. Figure 7.2 presents the schematic of such a system based on the Savin 870 copier. Conceptually centered around a photoreceptor, the process begins with corona charging followed by exposure, development, metering, transfer, erasure, and cleaning. The imaged paper is then fused by the application of heat (with or without pressure).

A variety of photoreceptors have been used for liquid toner based devices including ZnO, CdS, Se, As_2Se_3, and, most recently, organics. The combination of the optimum charging polarity of the photoreceptors with the ability to practice either charged area development (CAD) or discharged area development (DAD) allows the system designer to make the choice of toner charge sign on system level considerations (environmental issues, exposure source issues, etc.).

The latent electrostatic image is generated by photodischarge of the uniformly charged photoreceptor. In the copier embodiments this is accomplished with reflective scanning optics. However, in later printer embodiments like the AM Graphics Electropress®, diode arrays are also used. In principle, any analog or digital source whose wavelength and intensity are suitable for the particular photoreceptor and speed can be used.

In some embodiments (Riesenfeld et al., 1988; Tam et al., 1988) the photoreceptive element is permanently imaged to form areas of differential conductivity. Uniform electrostatic charging followed by differential discharge of the imaged element creates a latent electrostatic image. These elements are called electrographic or xeroprinting masters, as they can be repeatedly charged and developed after a single imaging exposure.

For development with liquid toners, the liquid developer is brought into direct contact with the electrostatic image. Usually a flowing liquid is employed, to ensure that sufficient toner particles are available for development. The field created by the electrostatic image causes the charged particles, suspended in a nonconductive liquid, to move by electrophoresis. The charge of the latent electrostatic image is thus partially neutralized by the oppositely charged particles. The theory and physics of electrophoretic development with liquid toners are well described by Schaffert (1975), Schein (1988), Chen (1988, 1995), and Chen et al. (1996a, b).

If a reimageable photoreceptor or an electrographic master is used, the toned image is transferred to paper (or other substrate). Most generally, the back side of the substrate to be imaged is charged by a corona, having a polarity opposite to that of the toner particles. This causes the developed image to transfer to the substrate.

Finally, the toned image is fixed to the paper. Heat is applied by the use of plates, rollers, or ovens, wholly analogous to a graphic arts drier. For heat-fusible toners, thermoplastic polymers are used as part of the particle. Heating both removes residual liquid and fixes the toner to the paper. Interaction of the carrier liquid with the toner resin allows the temperature at which fusing occurs to be relatively low compared to that required for dry toners.

Most emissions from the liquid dispersant generated by the process occur during fusing. Remediation techniques include simple venting, incineration, and recovery. The technique selected is application dependent and is responsive to the particular environment (industrial or office, for example).

Completing the cycle, the residual electrostatic charge is erased from the photoreceptor by either blanket exposure or the use of an AC corotron, and any residual toner is removed.

A limitation of the LTT process is its narrow substrate range. If the paper is smooth and the developed image is relatively thick, then the paper can actually cause material flow in a direction opposite to the process direction. This causes image smear, which can be alleviated by decreasing the thickness of the developed image. For smooth papers, this is an acceptable solution. When rough papers are used, however, the thin image does not contain enough toner to cover the paper irregularities, and low-density images are produced. A recent report discusses liquid toner image transfer and paper properties (Simms et al., 1992).

In an effort to address this limitation, Landa developed toners that contained a population of rigid particles such as microspheres, that are larger than the developed image thickness. The paper is uniformly gapped by these separate particles. Toner transfer was purported to occur by the formation of pseudopodia, which carry toner to the paper (Landa, 1983, 1984).

Landa later described a liquid toner called ElectroInk® (Landa, 1986; Landa et al., 1988) designed to give higher resolution and higher density images across a wider range of substrates than were achievable with previous toners. ElectroInk is designed to afford a rigid, cohesive mass upon development, a mass that resists the aforementioned deformation or squashing under pressure of transfer of relatively thick images. Landa's mechanistic model is based on a morphological structure in which particles have tentacles that interlock when brought together, forming the rigid mass. ElectroInk is used in the Indigo E-Print 1000, a digital offset color press. (Niv, 1994).

7.3 LIQUID TONER COMPOSITION

Liquid electrostatic toners are composed of a colloidal dispersion of pigmented or dyed resin particles suspended in an insulating liquid dispersant with added charge control agents that impart an electrostatic charge on the particles. This section describes typical materials that are useful for the preparation of liquid toners.

7.3.1 Dispersants

A liquid used as a toner dispersant must meet a number of demanding requirements. First, the liquid must be essentially nonconductive, to avoid discharging the latent electrostatic

Table 7.1 Isoparaffinic Hydrocarbons Used as Liquid Toner Dispersants

Material	Producer	Predominant carbon structure	Boiling range (°C)	Flash point (°C)
Isopar G	Exxon	C10–C11	155–176	40
Isopar H	Exxon	C11–C12	169–193	49
Isopar L	Exxon	C11–C13	185–206	60
Soltrol 100	Phillips Petroleum	C9–C11	157–173	41
Soltrol 130	Phillips Petroleum	C10–C13	176–208	56
Shell-Sol 71	Shell	C9–C12	179–202	52

Source: Mullin et al. (1990).

image. Resistivities of dispersants are typically greater than 10^{10} ohm cm. The liquid must be chemically inert with respect to other materials or equipment used in the electrographic or electrophotographic processes, including photoreceptors. Low-viscosity liquids are preferred, to allow rapid movement of the charged particles during development. The liquid must be volatile enough to allow its removal from the final imaged substrate, but volatility also should be low enough to minimize evaporation from the developer. Finally, the dispersant must be safe, in terms of physical properties, including those relating to flammability, toxicology, and human health. Moreover, the dispersant must be environmentally acceptable to manufacturers, regulators, government authorities, end-users, and the public at large.

A number of classes of organic liquids meet some or many of the requirements outlined. A variety of aliphatic and aromatic hydrocarbons, certain chlorofluorocarbons, and siloxanes have all been used. With increasingly demanding requirements for safety and environmental suitability, a class of aliphatic hydrocarbons, the isoparaffins, emerged as preferred materials. The isoparaffinic hydrocarbons are highly branched alkanes; those employed as dispersants have carbon skeletons ranging from C10 to C15. They are available from Exxon, Shell Oil, and Phillips Petroleum under the trade names Isopar®, Shell-Sol®, and Soltrol®, respectively. They are offered with various boiling ranges and volatilities (Table 7.1). The Isopars®, in particular, have enjoyed widespread use. Details of environmental aspects of liquid toners and isoparaffinic hydrocarbons are described in Section 7.7.

7.3.2 Resins

Liquid toner resins serve as the vehicle for the dispersed pigments or dyes, provide colloidal stability, and aid in fixing of the final image. The resin must also contain charging sites or be able to incorporate materials that have charging sites. Multifunctional resins that serve to impart a number of desired characteristics are often prepared specifically for use as liquid electrostatic toner resins (Santilli, 1977; Myers et al., 1987; Tavernier, 1988; Elmasry and Kidnie, 1990). Some commercially available resins that have been found effective are homopolymers such as polyethylene, polypropylene, polystyrene, polyesters, polyacrylates, polymethacrylates; ethylene vinylacetate copolymers (Elvax® resins, E. I. du Pont de Nemours, Wilmington, DE); ethylene acrylic acid or methacrylic acid copolymers (Nucrel® resins, E. I. du Pont de Nemours, Wilmington, DE; Primacor® resins, Dow Chemical Company, Midland, MI); ionomers of ethylene acrylic acid or methacrylic acid (Suryln® ionomer resin available from E. I. du Pont de Nemours, Wilmington, DE);

acrylic copolymers and terpolymers (Elvacite® resins, E. I. du Pont de Nemours, Wilmington, DE); styrene or vinyltoluene coplymers with butadiene or alkylacrylate (Pliotone® Resins, Goodyear Tire and Rubber Company, Akron, OH).

It has been recently reported that liquid toners were prepared as dispersions of colored pigments in hydrocarbon solvent using novel polymeric dispersants containing reactive functional groups such as thermally cross-linkable peroxy groups. The overprinted toner film layers could be thermally cross-linked at 140°C to improve mechanical durability (Rao et al., 1996).

7.3.3 Charge Control Agents (Charge Directors)

A charge control agent (CCA) is added to the toner to impart a charge to the toner particles. Ionic surfactants or metal soaps that form inverse micelles in the liquid dispersant are often used as charge control agents. Toner particles can obtain either positive or negative charge depending on the combination of particle material and charge control agent used.

Suitable charge control agents include alkylated aryl sulfonates, which can be obtained with excess calcium or barium carbonate, resulting in basic forms if desired. Typical examples are Basic Barium Petronate®, Neutral Barium Petronate®, Calcium Petronate® from Witco Inc., neutral barium dinonylnaphthalene sulfonate, basic barium dinonylnaphthalene sulfonate, neutral calcium dinonylnaphthalene sulfonate, basic calcium dinonylnaphthalene sulfonate from R. T. Vanderbilt Inc., and dodecylbenzenesulfonic acid sodium salt from Aldrich Inc. Another class of basic charge control agent is the polyisobutylene succinimides such as Chevron's Oloa® 1200 (El-Sayed and Taggi, 1987; Fowkes et al., 1990).

Soy lecithin, mixtures of soy lecithin with *N*-vinyl pyrrolidone polymers, and copolymers of lecithin with ethylenic comonomers have been noted to be particularly effective charge control agents for negative toners (Gibson, 1990; Gibson et al., 1992). Another class of charge control agents are the sodium salts of phosphated mono- and diglycerides with saturated and unsaturated acid substituents such as Witco's Emphos® D70-30-C and Emphos® F27-85. AB diblock copolymers of (1) polymers of 2-(*N,N*) dimethylaminoethylmethacrylate quaternized with methyl-*p*-toluene sulfonate and (2) poly-2-ethylhexylmethacrylate were shown to be excellent charge directors (Page and El-Sayed, 1990). They enable charge director properties to be customized by selecting the molecular weight selection (Chen et al., 1996a,b).

Divalent and trivalent metal carboxylates are excellent charge control agents for positive toners. Examples of these materials are aluminum tristearate, barium stearate, chromium strearate, magnesium octoate, calcium stearate, iron naphthenate, and zinc naphthenate (Croucher et al., 1984b).

7.3.4 Colorants

The pigment or dye used will depend largely on the color of the image required. However, other critical pigment characteristics are particle size, solvent and light fastness, dispersibility, and insolubility in the toner dispersant. The pigment often impacts a number of toner properties including particle charging, so the choice of pigment will also depend on the combination of toner resin and charge director used. Some suitable pigments are listed in Table 7.2. Sublimable dyes are effective colorants for processes involving the transfer of a colored design from a print on paper to a textile or film substrate.

Table 7.2 Typical Pigments for Liquid Electrostatic Toners

Pigment brand name	Registered trade mark	Manufacturer	Pigment Color Index
Permanent Yellow DHG		Hoechst	Yellow 12
Permanent Yellow GR		Hoechst	Yellow 13
Permanent Yellow G		Hoechst	Yellow 14
Permanent Yellow NCG-71		Hoechst	Yellow 16
Permanent Yellow GG		Hoechst	Yellow 17
Hansa Yellow RA		Hoechst	Yellow 73
Hansa Brilliant Yellow 5GX-02		Hoechst	Yellow 74
Dalmar Yellow YT-858-D	√	Heubach	Yellow 74
Hansa Yellow X		Hoechst	Yellow 75
Novoperm Yellow HR	√	Hoechst	Yellow 83
Chromophtal Yellow 3G	√	Ciba-Geigy	Yellow 93
Chromophtal Yellow GR	√	Ciba-Geigy	Yellow 95
Novoperm Yellow FGL	√	Hoechst	Yellow 97
Hansa Brilliant Yellow 10GX		Hoechst	Yellow 98
Lumogen Light Yellow	√	BASF	Yellow 110
Permanent Yellow G3R-01		Hoechst	Yellow 114
Chromophtal Yellow 8G	√	Ciba-Geigy	Yellow 128
Irgazin Yellow 5GT	√	Ciba-Geigy	Yellow 129
Hostaperm Yellow H4G	√	Hoechst	Yellow 151
Hostaperm Yellow H3G	√	Hoechst	Yellow 154
L74-1357 Yellow		Sun Chemicals	
L75-1331 Yellow		Sun Chemicals	
L75-2377 Yellow		Sun Chemicals	
Hostaperm Orange GR	√	Hoechst	Orange 43
Paliogen Orange	√	BASF	Orange 51
Irgalite Rubine 4BL	√	Ciba-Geigy	Red 57:1
Quindo Magenta	√	Mobay	Red 122
Indofast Brilliant Scarlet	√	Mobay	Red 123
Hostaperm Scarlet GO	√	Hoechst	Red 168
Permanent Rubine F6B		Hoechst	Red 184
Monastral Magenta	√	Ciba-Geigy	Red 202
Monastral Scarlet	√	Ciba-Geigy	Red 207
Hehogen Blue D 7072 DD	√	BASF	Prussian Blue 15:3
Hehogen Blue L 6901F	√	BASF	Blue 15:2
Hehogen Blue NBD 7010	√	BASF	
Hehogen Blue K 7090	√	BASF	Blue 15:3
Hehogen Blue L 7101F	√	BASF	Blue 15:4
Heucophthal Blue G XBT 583D		Heubach	
Pahogen Blue L 6470	√	BASF	Blue 60
Hehogen Green K 8683	√	BASF	Green 7
Hehogen Green L 9140	√	BASF	Green 36
Eupolen Blue 70-8001	√	BASF	Prussian Blue 15:3
Monastral Violet R	√	Ciba-Geigy	Violet 19
Monastral Red B	√	Ciba-Geigy	Violet 19
Quindo Red R6700	√	Mobay	
Quindo Red R6713	√	Mobay	
Indofast Violet	√		Violet 23
Monastral Violet Maroon B	√	Ciba-Geigy	Violet 42
Sterling NS Black	√	Cabot	Black 7
Sterling NSX 76	√	Cabot	
Tipure R-101	√	Du Pont	
Mogul L Carbon Black	√	Cabot	
BK 8200 Black Toner		Paul Uhlich	

7.4   LIQUID TONER PREPARATION

Liquid toners are dispersions of pigments in a resin binder. Because paints and inks are also based on pigment dispersions, it is not surprising that the traditional production equipment used in the manufacture of liquid toners emerged from the paint and ink industries. Batch milling processes are used predominately with equipment such as steel and pebble mills. Milling of pigment with resin and liquid dispersant is typically carried out until a dispersion of a specific particle size has been achieved. Steel mills provide a finer grind (which may or may not be desirable) but also introduce metal contamination, thus requiring the use of a magnetic filter. High shear equipment, such as colloid mills and attritors, are used when the materials comprising the liquid toner formulation are not readily compatible and difficult to disperse. Toners are typically produced in concentrated form and subsequently diluted with additional liquid dispersant to afford a solids concentration of ~0.1 to 2% by weight for use in imaging devices. Charge control agents are typically added after completion of milling.

An alternate approach to the preparation of liquid toner particles was developed at Xerox Research Centre of Canada (Croucher, 1987; Croucher et al., 1984a, 1985, 1988; Duff et al., 1987). Particles are made by nonaqueous dispersion polymerization in the hydrocarbon solvent and subsequently colored with pigments or dyes. The attraction of this approach is that particles can be made with well-controlled particle size and size distributions.

To carry out the nonaqueous dispersion polymerization method, a monomer, which is soluble in the hydrocarbon medium, is polymerized in the presence of an amphipathic polymer (a polymer that has two distinct ends, one essentially insoluble in the hydrocarbon medium, the other soluble). When the polymerization has generated a certain chain length, the polymer precipitates from solution, and a core particle is formed. The amphipathic copolymer is adsorbed onto this nucleus, which then grows as a discrete particle. Coloring of the particle can be accomplished by ball-milling pigments into the particles. Alternatively, the particles can be colored by dying. Charge control agents may then be added as required.

7.5   MEASUREMENT OF LIQUID TONER CRITICAL PROPERTIES

The characterization of a liquid toner requires the measurement of a number of toner physical properties and image quality testing in an electrostatic imaging device (Novotny, 1981; Schein, 1988). Toner particle size is critical, as it in part determines the ultimate resolution capability of the toner. This capability may or may not be realized in the imaging hardware. Sufficient toner particle charge and electrophoretic mobility are required for development of the latent electrostatic image and if required the electrostatic transfer of the developed image to the final substrate. Other critical properties include but are not limited to conductivity, colloidal stability, viscosity, morphology, surface tension and wetting characteristics, fusing or fixing characteristics, and color and transparency.

7.5.1   Particle Size

No perfect system for sizing liquid toners has yet been demonstrated. Each of the techniques used for the sizing of particles has been applied to liquid toners, and each has its set of advantages. Each, however, also needs to be interpreted in light of the differences between the native state of the liquid toner and the conditions under which the analysis is performed. In most cases, several techniques are used in a complementary fashion to ensure that a comprehensive picture of the effective size distribution is to be obtained.

A general question that arises in the interpretation of all particle size data concerns choosing the moment of the distribution that is most highly correlated with the functional property of interest. Care must be taken in proper selection of the moments and in investigation of the algorithms used to generate them. It is common in some techniques to make assumptions about the nature of the distribution (e.g., it is monomodal and Gaussian) that do not accurately reflect the sample.

Particle shape can also be an important consideration. In techniques that do not reveal the actual shape of the particles, the equivalent spherical radii that tend to be reported may not correspond to any physical dimension of the particle in question. One of the other common shape-induced pitfalls concerns particles with aspect ratios much different from one. In these cases, care must be taken to understand the effects of the technique selected on the orientation of the particles, and what effect that orientation may have on the analysis. This is demonstrated by the comparison in which a single toner dispersion was measured with three instruments. The results are given in Table 7.3.

Microscopy

Direct measurement of particle size can be obtained by microscopy, but several factors must be weighed in the interpretation of such data. Many liquid toners are of submicron diameter, and ordinary light-microscopic techniques are not applicable for the study of the fundamental particles. Often even liquid toners that have submicron fundamental particles, however, have aggregate structures that are important in their performance that are well within the reach of light microscopy. Liquid toners are also quite optically dense, so often the use of either dilute toner or a thin sample is necessary to obtain sufficient light transmission for analysis. Dilution can alter the charge distribution within the toner, either by changing the charging equilibrium or by effectively changing the ''ionic strength'' of the medium, thus altering the aggregate stability.

Both scanning electron microscopy (SEM) and transmission electron microscopy (TEM) have been used for liquid toner size distribution and morphology determination. Sample preparation artifacts and contrast ratio concerns are important in the interpretation of the results. For SEM, the largest complication is effects induced by removing the carrier liquid. Since many liquid toner resins interact with the carrier liquid, removal of that liquid may be expected to change the size and structure of the particle.

Table 7.3 Particle Size Determination: Comparison of Measurements on a Single Toner by Three Instruments[a]

Number	Area	Volume
Brinkman Particle Size Analyzer (median µm)		
1.95 (0.05)	7.00 (0.10)	7.33 (0.40)
Horiba Capa 700 (median µm)		
0.52 (0.07)	2.22 (0.26)	3.15 (0.32)
Malvern Particle Sizer 3600E (median µm—volume)		
10% less than	50% less than	90% less than
3.2 (0.1)	6.7 (0.1)	13.0 (0.7)

[a] Standard deviations in parentheses.

In TEM both vacuum and contrast pose problems. The contrast problems can be addressed by the use of contrast agents including many of the heavy metal salts used commonly as liquid toner charge directors.

Light Scattering

These techniques offer many advantages for particle sizing. There is typically little sample preparation required, and the devices are reasonably automated, leading to short analysis times. Most, however, require that the sample be optically dilute to avoid multiple scattering.

There are two types of optical scattering instruments: Fraunhofer diffraction devices and autocorrelation devices. These instruments are used for both dry and wet particle size analysis of powders, suspensions, emulsions, and sprays. The attainment of appropriate resolution and reproducibility is dependent on suitable sample handling. Again, sample preparation is key, and the best results are only achieved when the sample is suitably dispersed before the analysis.

Representative of the commercial equipment is the Coulter® LS Series (Beckman Coulter Corporation, Miami, FL). The LS devices use both Fraunhofer and Mie theories to deduce the particle size distribution from the spatial scattering distribution. The LS 230 uses multiwavelength analysis to give a range from 0.04 μm in a single scan using 116 size channels. The Malvern Particle Sizer 3600E (Malvern Instruments Ltd, Malvern, Worcestershire, UK) uses the principle of laser diffraction to assign particle size. Laser light is scattered and detected over all scattering angles. Large particles scatter light at small angles and small particles scatter light at large angles.

Dynamic light scattering, also called quasi-elastic light scattering or photon-correlation spectroscopy, is applicable to particles suspended in a liquid, which are translating due to Brownian motion (i.e., particles generally of 2–3 μm diameter and smaller). The speed of the translation is inversely proportional to the particle size, and the velocity distribution can be deduced by analyzing the time dependence of the light intensity fluctuations (measurement of the autocorrelation function). Accuracy and information about particle shape and polydispersity of the particle size distribution can be obtained by making scattering measurements at several angles as most commercial instruments do. Some commercial instruments are capable of sizing particle ranges from 0.003 to 3 μm in diameter. The small end of this range is determined by measurement wavelength, incident power, and sample concentration; the high end by gravitational stability.

Representative commercial instruments are the Coulter Nano-Sizer® N4 and N4 Plus. These systems detect particles within the range of 3 μm down to 40 nm analyzing the autocorrelation function by the method of cummulants for determination of polydispersity of an assumed unimodal distribution.

Sedimentation

Based on the Stokes–Einstein equation, the size distribution of a particulate sample can be deduced by the settling velocity (Allen, 1974). The oldest of such techniques uses settling tubes in which the particle suspension was allowed to stand. While this process only worked well for particles relatively large with respect to most liquid toners, it has the advantage of being useful even in quite concentrated suspensions as long as proper account is made for viscosity. A variety of light sources have been employed including x-rays. Modern instruments employ automated observation and data analysis techniques to improve accuracy and productivity.

Representative devices include the Micrometrics SediGraph 5100 Particle Size Analysis System. This instrument is designed for completely automatic operation and employs a moving, temperature-controlled cell and a fixed-position x-ray source/detector. Most commonly used for inorganic powders, the device has a typical range of 300 to 0.1 μm.

Using a centrifuge to increase the settling rate allows faster analysis times and a broader analysis range, especially when the centrifuge is used to change speed during the analysis. Representative equipment of this type includes the Horiba Capa 500–700 (Horiba Instruments Inc., Irving, CA). In this method, the Stokes sedimentation equation is combined with the proportional relationship between absorbency and particle concentration (Bohren and Huffman, 1983).

Volume Determination

Electrical volume determination is another technique that has been developed to determine particle size. The prototype instrument of this class is the Coulter counter (Coulter Corporation, Miami, FL). The Coulter method of sizing and counting particles is based on the changes in electrical resistance produced by nonconductive particles suspended in an electrolyte as they are pumped through a calibrated aperture between electrodes. As particles displace the supporting electrolyte, a voltage pulse is recorded whose height is proportional to the volume of the particle. The manufacturer quotes the measurement range as 0.4–1200 μm.

Particle size can also be determined by optical volume measurement. An example of this type of device is the Brinkman Particle Size Analyzer (Brinkman Instruments, Inc., Westbury, NY). This instrument utilizes a laser based time of transition analysis system that presents statistical information including particle size, area and volume distribution, and sample concentrations. Time of transmission analysis determines particle diameter by sensing the time required for the particle to pass through the laser beam path.

In conclusion, care is necessary in comparing the absolute particle sizes of liquid toners assigned by different instruments. However, a good correlation between particle sizes determined over a broad range by two instruments that utilize very different techniques has been demonstrated. The median particle size of 67 toners was measured with the Horiba and Malvern instruments described above with the correlation shown in Table 7.4. The expected range of Horiba values (median by area) was determined using linear regression at a confidence level of 95% (Trout and Larson, 1988).

Table 7.4 Correlation Between Malvern 3600 E and Horiba Capa 500 Particle Size Instruments

Median particle size (μm) determined by Malvern 3600E Particle Sizer	Expected median particle size (μm) range for Horiba Capa-500
30	9.9 ± 3.4
20	6.4 ± 1.9
15	4.6 ± 1.3
10	2.8 ± 0.8
5	1.0 ± 0.5
3	0.2 ± 0.6

7.5.2 Toner Particle Charge and Electrophoretic Mobility Measurement

Charging of liquid toners is often studied by calculating toner charge-to-mass ratio (Q/M) from the DC current that flows as toner mass deposits onto an electrode in the presence of an applied electric field (Dahlquist and Brodie, 1969; Novotny and Hair, 1979). These measurements are frequently complicated by electrical noise and interference from conductive ingredients in the toner. Electrophoretic mobility measurements provide similar charging information (Schaffert, 1975):

$$\text{Particle mobility} = \frac{6\pi(\text{Particle radius})(\text{Fluid viscosity})}{\text{Particle charge}} \tag{1}$$

Direct measurement of a toner velocity distribution can be made with laser Doppler electrophoresis (LDE). Knowledge of the velocity distribution, electric field, and particle size allows calculation of the zeta potential distribution. These relationships are well known for polar liquids and especially cases where the interparticle separation is larger than the Debye length. However, they have only recently appeared for cases germane to liquid toner (Chen et al., 1996a,b). At the fields and conductivities characteristic of liquid toners, hydrodynamic instabilities can disrupt measurement of mobility (Rhodes et al., 1989), so care must be taken to collect data from within the stationary layer of the electrophoretic cell.

Two commercial instruments are available from Beckman Coulter and Malvern. The Coulter Delsa 440SX uses LDE measured simultaneously at four angles. This zeta potential instrument can determine the relative sizes of particles at different mobilities. The Coulter Delsa contains four 256-channel autocorrelators to discriminate between particles from four optimized angles and detect very small mobility or zeta potential differences. The manufacturer says that this results in an increased accuracy that cannot be achieved by repeating single-angle analyses. The instrument is sensitive over a wide range of particle sizes, from 10 nm to greater than 30 μm.

The Malvern Zetasizer 3 (Malvern Instruments Ltd, Malvern, Worcestershire, UK) uses a patented cell design for measurements in low dielectric dispersants. Intersecting laser beams create an interference pattern at the stationary layer in a flat cell. The beams enter the cell through a transparent electrode. A brief DC electric field (continuously variable up to 0.4 V/μm) is applied to the cell. Toner particles move through the interference pattern, and the resulting scattered light is collected by a photomultiplier. The mobility of the particles is calculated from the Doppler shift in the frequency spectrum. This is checked by reversing the polarity on the electrodes, sampling the scattered light, and recalculating the mobility (McNeil-Watson and Pedro, 1987; Degiorgio, 1982).

In the past few years, mobility measurements have been developed based on the Debye colloid potential that can measure the zeta potential of particles in concentrated dispersions (Oja et al., 1985; Niv et al., 1986; McNeil-Watson and Pedro, 1987; Larson, 1993). Representative commercial equipment includes the Matec Electrokinetic Sonic Analysis (ESA) (Matec Instruments, Inc., Hopkinton, MA). The ESA applies an oscillating electric field to the sample and measures the sound wave produced by the motion of the double layers around the particles (Oja et al., 1985). The ESA measurement is directly proportional to the AC electrophoretic mobility of the particles and represents the inverse of the Debye colloid potential. The measurement is effective at low fields and has been found useful in diverse systems (Isaacs et al., 1990). Mobilities determined by the Matec

ESA have been demonstrated to correlate with the image quality obtained in an electrostatic imaging device (Larson et al., 1989; Felder et al., 1990).

Direct measurement of the Debye colloid potential can be accomplished by the Pen Kem System 7000 (University of Maine, Orono, ME). This device applies an ultrasonic disturbance to a colloidal sample and measures the electric field produced.

A comparison of mobilities for a series of toners obtained with the Malvern Zetasizer and Matec ESA instruments is shown in Fig. 7.3 (Felder et al., 1990). Similar studies have been reported more recently (Caruthers et al., 1994).

The Indigo Mobility Analyzer (Indigo Ltd, Rehovot, Israel) also measures particle density. This instrument uses a DC field applied across parallel plates to cause toner particles in suspension between the plates to move to one of the electrodes. As the particles deposit onto the electrode, light transmission through the cell increases. The rate of increase of light transmission is related to mobility. This method is useful for determining electrophoretic mobilities at high fields up to 2.0 V/μm (Niv et al., 1986).

Staples and Koop (1991) describe a technique by which the toner charge and mass are measured simultaneously in a deposition cell. Electrical field flow fractionation is an analytical method based on the differential migration of particles in a flowing stream

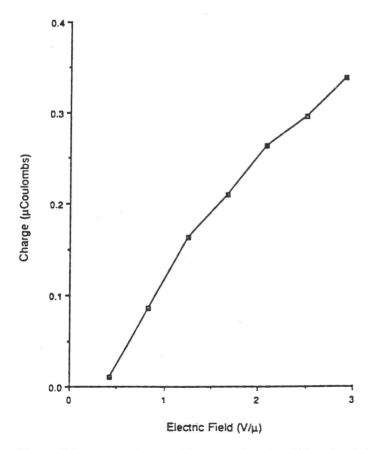

Figure 7.3 Matec ESA vs. Malvern Zetasizer 3 mobility. Correlation coefficient $R^2 = 0.89$. (From Felder et al., 1990.)

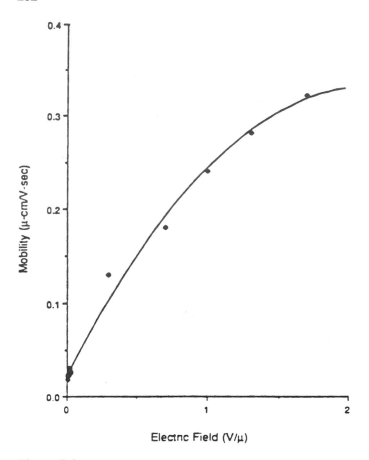

Figure 7.4   Toner charge vs. electric field. Toner charge was calculated from the charge-to-mass ratio, which was determined by placing the toner between two electrodes with a 1 mm separation and applying an electric field until all of the toner was developed. (From Felder et al., 1990.)

that passes through an externally applied electric field. It was reported to be attractive in determining the properties of charged particles in a nonpolar medium. This off-line method has similarities to liquid toner development processes in that charged particles, dispersed in a nonpolar fluid, are passed through an electric field orthogonal to the fluid flow (Russell, 1993).

Characterizing a toner by electrophoretic mobility measurements over a range of electric fields is useful, because as the electric field strength increases, the ionization process required for toner particle charging becomes increasingly favorable. This leads to increased charge on the toner particle and hence increased toner particle mobility (Stotz, 1978; Niv et al., 1986; Felder et al., 1990). This effect is demonstrated in Figs. 7.4 and 7.5. The charge on the toner particles is also impacted by the polarity of the liquid dispersant (Mitchell, 1987).

7.5.3   Conductivity

The deposition behavior of liquid toners makes it difficult to measure the conductivity of such systems. In general, to be effective, liquid toners must have AC conductivities as

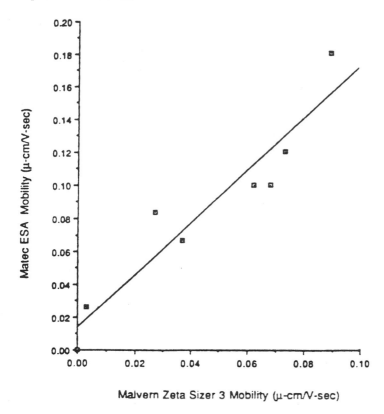

Figure 7.5 Toner particle mobility vs. electric field. Low field values were obtained with the Malvern Zeta Sizer 3, and high field values were obtained with the Indigo Mobility Analyzer. (From Felder et al., 1990.)

much as three orders of magnitude below the sensitivity of most commercial equipment. The device most commonly used for conductivity studies is the Scientifica Model 623 (Scientifica, Princeton, New Jersey) conductivity meter. It is available in both dip and flow-through probe configurations.

7.5.4 Rheology

The rheological behavior of liquid toners varies greatly. While most liquid toners display Newtonian behavior in working developers, there is sometimes a dramatic departure from ideality at concentrations characteristic of developed images. This rheological behavior is shown, in some instances, to map on imaging performance (Gibson et al., 1992). A great variety of commercial instruments exists for viscosity measurement and care must be taken in instrument selection. Measurement geometry and shear rate employed in the measurement should closely correspond to that which the liquid toner experiences in use. Representative commercial equipment includes the SR-2000 and SR-5000 controlled stress rheometers (Rheometric Scientific, Inc. Piscataway, NJ).

7.5.5 Color and Color Strength

The color of a liquid toner can be measured in a number of ways. Most commonly a dried toner film is measured with one of the wide variety of commercially available instruments.

Care must be taken in such measurements to account for the mass per unit area of the toner film and of the optical properties of the substrate upon which the film is deposited. Three types of instruments are generally distinguished: densitometers, colorimeters, and spectrophotometers.

Densitometers measure the transmission or reflection density of a sample viewed through a filter. Differences in filter sets and aperture distinguish among instruments and the various standards. Representative of this type of equipment is the Macbeth 1200 Series densitometers (Macbeth Division of Kollmorgen Instruments Corporation, New Windsor, NY). Such instrumentation is generally rugged and is designed to be used in production environments and for routine quality control.

Colorimeters are typically more sophisticated than densitometers, often acquiring a full reflectance or transmission spectrum of the sample under observation. This data is analyzed and color is reported in one of the common color spaces (L*, a*, b* for example).

Spectrophotometers like the UltraScan XE, LabScan II, and ColorQUEST from HunterLab (Hunter Associates Laboratory, Reston, VA) offer greater flexibility when measuring a sample's color or color difference as it would appear under different lighting conditions. Flexibility within illuminant, sample form, and data reduction characterize these instruments. While the most flexible and sophisticated of these instruments are best suited to a laboratory environment, several companies offer microspectrophotometers, which combine some of the robustness of the densitometers with some of the sophistication of the spectrophotometers.

7.6 LIQUID TONER CHARGING

Particles dispersed in a hydrocarbon or other nonconductive liquid become charged by the addition of an ionic surfactant (charge control agent or charge director), which forms inverse micelles in low dielectric media. The spontaneous separation of charge between the particle and micelle phases can be accounted for by a number of mechanisms including (1) differences of electron affinity of the micelle and particle; (2) physical trapping of nonmobile charge in one phase; (3) difference of cation or anion affinity of the particle and micelle; and (4) surface group ionization (Hunter, 1981). Although these mechanisms may play a role alone or in combination, mechanisms 3 and 4 are probably predominant. Toner charging can often be accounted for by the acid–base or donor–acceptor properties of the toner materials.

Acid–base chemistry between the particles and the ionic surfactant micelles is believed to result in charging of the particles. According to this model, the formation of a negatively charged particle is enhanced by proton or cation exchange from the particles to the micelles, and formation of positively charged particles is enhanced by proton or cation exchange from the micelles to the particles (Fowkes et al., 1984; Croucher et al., 1984b). This process results in the formation of a diffuse double layer with zeta potentials of over 100 mV in cases of strong acid–base interactions (Ross and Morrison, 1988). A model for liquid toner particle charging and charge director ionization based on a series of reversible equilibria has been proposed and shown to compare well against a limited set of experimental data (Larson et al., 1995).

7.6.1 Acid–Base Chemistry and Positive Particle Charging

$$(\text{PARTICLE}) + (\text{MICELLE}){:}-\text{H} \longrightarrow (\text{PARTICLE})-\text{H}^+ + (\text{MICELLE}){:}^-$$

| Basic particle | Acidic micelle | Positive particle | Negative micelle |

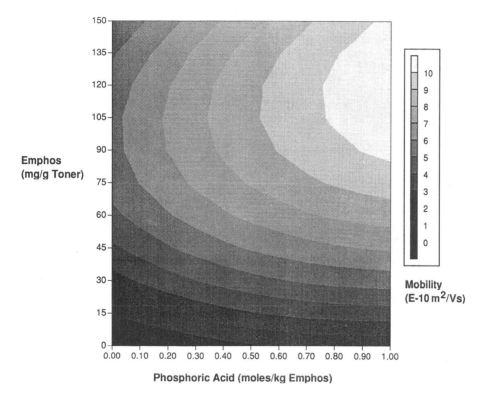

Figure 7.6 Particle mobility vs. Emphos® D70-30C and dichloroacetic acid concentration. Particle mobility determined with Matec ESA.

It has been demonstrated that the addition of micelle soluble acidic materials to the toner dispersion enhances particle positive charging. The acids can be incorporated into the particle initially and then leached into the micelles (El-Sayed and Trout, 1990; Chan and Trout, 1990) or added directly to the micelles (Pearlstine et al., 1990; Pearlstine and Swanson, 1992). Suitable acids include *p*-toluenesulfonic acid, *p*-nitrobenzoic acid, *p*-chlorobenzoic acid, phosphoric acid, dichloroacetic acid, and dodecylphosphonic acid. The impact of acid and micelle forming ionic surfactant concentration on toner charge as indicated by changes in toner particle mobility is shown in Fig. 7.6. The corresponding impact on toner conductivity is shown in Fig. 7.7. Examination of Figs. 7.6 and 7.7 shows that toner particle mobility is primarily a function of acid concentration, while conductivity is primarily a function of ionic surfactant concentration.

Divalent and trivalent metal carboxylates and sulfonates are very effective charge control agents for positive toners (Croucher et al., 1984b). It has been proposed that this is due to the acidity of the micellar cores resulting from the strong acidity of the cations (Fowkes et al., 1990).

7.6.2 Acid–Base Chemistry and Negative Particle Charging

$$(PARTICLE)-H + (MICELLE): \longrightarrow (PARTICLE)^- + (MICELLE)-H^+$$

Acidic	Basic	Negative	Positive
particle	micelle	particle	micelle

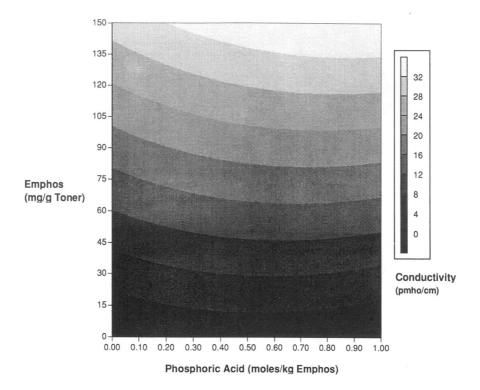

Figure 7.7 Toner conductivity vs. Emphos® D70-30C and dichloroacetic acid concentration. Conductivity determined at 5 Hz and 5 V.

It was demonstrated that copolymers of ethylene or styrene with sulfonic acid containing monomers were better for the preparation of negative liquid toners compared to the homopolymers or copolymers of ethylene with carboxylic acid containing monomers (Larson and Trout, 1987; Larson, 1988a). Suitable sulfonic acid containing monomers include 2-acrylamido-2-methyl-1-propanesulfonic acid and 2-sulfoethylmethacrylate. Similarly, it has been shown that the blending of highly acidic polymers into the toner particle and that the adsorption of acids onto the surface of the toner particle enhance negative charging of the particle (Larson and Trout, 1988, Gibson, 1990).

It has been noted that effective charge control agents for negative toners include basic barium petroleum sulfonates and polyisobutylene succinamiides with basic amine groups (Fowkes et al., 1990). The addition of micelle-soluble bases to the toner has also been shown to increase the negative charge on the particles. Effective bases include organic soluble hydroxides like tetabutylammonium hydroxide and tetraethylammonium hydroxide (El-Sayed and Larson, 1988), aliphatic amines such as tributylamine, 1-hexylamine, dibutylamine, and 1-dodecylamine (El-Sayed et al., 1990), and organic diamines including ethylenediamine, 1,2-diaminobenzene, and 1,2-diaminocyclohexane (Larson, 1988b). Aminoalcohol compounds, including triethanolamine, triisopropanolamine, and ethanolamine, added to a liquid toner, are very effective at enhancing the electrodeposition of negative toner particles (Larson, 1987; Larson et al., 1989).

A number of charge additives are effective even though they do not obviously follow the acid–base charging mechanism as outlined above. It was demonstrated that the amino-

Table 7.5 Effect of Piperidine and Pyridine Derivatives on
the Mobility of Cyan Toner Particles

Additive	Mobility[a] $(10^{-10}$ m^2 V · s)	pK$_a$[b]
Control	1.6	na
Piperidine	3.5	11.2
2-Piperidinemethanol	5.0	
3-Piperidinemethanol	3.7	
Pyridine	1.7	5.2
2-Pyridinemethanol	4.6	4.9
3-Pyridinemethanol	7.8	4.9
3-Pyridinepropanol	7.0	5.5

[a] ESA (1.2 MHz) Mobility for cyan toner 1.5% solids in Isopar L. 50 mg/g
toner solids of basic barium petronate. 40 mg/g toner solids of additives.
[b] Data from D. D. Perrin. *Dissociation Constants of Organic Bases in Aqueous Solution.* Butterworths. London. 1965. For further experimental details,
see Larson et al (1989).

alcohol was much more effective than the corresponding unsubstituted amine, even though
their basicities are similar. This effect was assumed to be due to increased particle charging
but is not anticipated by the acid–base model because there was no apparent reason for
increased proton transfer from the particle to the micelles with aminoalcohols as charge
additives relative to the corresponding amines, as shown in Table 7.5. Aminoalcohols and
diamines have also been noted to enhance the charge stability of liquid toner compositions
(Larson, 1987; Larson, 1988b). Similar charge stabilizing effects have been observed for
polyoxyethylene alcohols (Almog and Gutfarb, 1994). Metal salts of beta-diketones such
as aluminum acetylacetonate have been found to be very effective negative charging agents
and do not follow the pattern expected by an acid–base charging mechanism. It was proposed that the charging in these systems is due to the partitioning of surface ions into
charge director micelles (Lane, 1990).

7.7 ENVIRONMENTAL HEALTH AND SAFETY OF LIQUID TONERS

Recent years have brought an increasing awareness of the impact of materials and activities
on the environment and on human health and safety. Environmental issues of liquid toners
are largely those of the liquid dispersant. The requirements for and selection of liquid
toner dispersants have been described above. The widely used isoparaffinic hydrocarbons
are synthetic materials, purified by fractional distillation. As high-purity aliphatic hydrocarbons these materials have inherently low odor and low toxicity, which is reflected by
their acceptance for other consumer applications. Some of these materials, for example,
are used in cosmetics. Also, the isoparaffinic hydrocarbons enjoy U.S. Food and Drug
Administration clearances for use in a variety of direct and indirect food applications.

7.7.1 Physical Classifications

The physical classification of liquid toners reflects the physical classification of the dispersant. Flammability is the key parameter. It is normally quantified by flash point, the lowest
temperature at which a liquid gives off enough vapor to form an ignitable mixture with

air and produce a flame when an ignition source is present. The isoparaffinic hydrocarbons used in liquid toners have been selected in part because of their relatively high flash points. With flash points above 40°C, all isoparaffinic hydrocarbons of Table 7.1 are classified as combustible materials, rather than flammable, by U.S. Department of Transportation criteria. For air and international shipments, materials are considered combustible if flash points are above 60.5°C. The flash point also bears on classification of liquid toner waste under the U.S. Environmental Protection Agency.

7.7.2 Toxicology

Potential human exposure during manufacture and use of liquid toners exist by inhalation of volatilized dispersant or by skin contact with liquid dispersant. Toxicological considerations thus are focused on the dispersant. The isoparaffinic hydrocarbons used widely as liquid toner dispersants have been studied extensively, and a comprehensive summary of the toxicity studies and human exposure data has been published (Mullin et al., 1990).

The isoparaffinic hydrocarbons are practically nontoxic in acute exposures (single high-level doses) by oral, dermal, and inhalation routes. In numerous long-term exposure studies with animals, no chronic toxicity issues with relevance to humans have been found. In developmental and genetic evaluations the isoparaffins have also been found to be nontoxic.

More specifically, in dermal tests, isoparaffinic hydrocarbons have been found to be nonirritating when evaporation is allowed to occur freely, conditions which realistically simulate human workplace exposure. When evaporation is not allowed to occur, these materials can be slightly irritating, most likely as a result of their ability to solubilize skin oils and defat the skin. Dermal safety of other toner components is considered during formulation of liquid toners, and skin irritation tests are usually performed on the final toner design.

LC_{50}, the conventional measure of acute inhalation toxicity derived from rat testing, could not even be determined for Isopars G and L; this was because the LC_{50} was higher than concentrations that could be generated for the tests (above 2000 ppm for Isopar G). At least eight other acute and chronic inhalation exposure tests with a variety of animal species have been reported, all supporting the inherent low toxicity of the isoparaffins. Isopar® G has also been tested for respiratory irritation in mice and was found not to be irritating even at ~400 ppm (Mullin et al., 1990).

7.7.3 Occupational Exposure Guidelines

The most widely used guidelines for occupational exposures are the Threshold Limit Values (TLVs) established by the American Conference of Governmental Industrial Hygienists. TLVs apply to time-weighted averages for 8-hour-day, 40-hour-week exposures. However, no TLVs have been set for occupational exposure to isoparaffinic hydrocarbons. Likewise no U.S. OSHA Permissible Exposure Limits, German MAKs, or other governmental limits have been created specifically for the isoparaffinic hydrocarbons (American Conference of Governmental Industrial Hygienists, 1990). However, occupational exposure limits of 100–400 ppm have been established by several manufacturers of isoparaffinic hydrocarbons or of liquid toners (Greenwood, 1990). The isoparaffinic hydrocarbon occupational exposure levels, based on the toxicology data summarized above, are judged to be those levels to which workers may be repeatedly exposed without adverse health effects. Liquid toner processes and equipment thus are designed to ensure that worker exposures are below these levels.

7.7.4 Indoor Air Quality

Indoor air quality, a growing area of interest, concerns comfort and health issues in homes, schools, public buildings, and offices. Factors beyond the toxicology of dispersants and toners must be considered when using liquid toner processes in such nonoccupational environments.

Scores of volatile organic compounds (VOCs), encompassing a wide range of chemical compositions, are typically present in indoor environments, being emitted from building materials, household and office products, and human activities. Such VOCs, particularly when present as complex mixtures, have been associated even at low-level exposures with a variety of subjective human responses including sensory irritation, the perception of odor, and the perception of poor air quality (Mølhave et al., 1986; Kjaergaard et al., 1989; Hudnell et al., 1990). The nature of these responses and the roles of VOC composition and concentration in those responses are only beginning to be determined. Still, as isoparaffinic hydrocarbons can be considered part of the broad class of VOCs, and have been identified as components of VOCs in indoor air of buildings containing liquid toner process equipment (Tsuchiya et al., 1988), their potential role in such subjective responses needs to be considered.

Several studies have addressed human sensory response to isoparaffinic hydrocarbons and to liquid toner process emissions. In a Danish study, human volunteers, exposed to 100 ppm of an isoparaffinic hydrocarbon for 6 hours, were given a questionnaire to evaluate sensory irritation, fatigue, and other responses. No symptoms associated with solvent exposure were reported (Pederson and Cohr, 1984). In a study reported by Ricoh Corporation, a human panel, exposed to emissions from a liquid process photocopier, ranked odor and discomfort. Odor was almost imperceptible at 22 ppm with slight discomfort noted at 54 ppm (Mullin et al., 1990). Thus, while the nature of human responses to low level VOC exposures continues to be studied, it appears, based on these studies and other workplace experience (Mullin et al., 1990), that acceptability of liquid toner processes in indoor environments, where odor and comfort are considered, is achievable if isoparaffinc hydrocarbon concentrations are kept below ~20–50 ppm.

Standards for human exposures in indoor environments do not exist separate from the occupational exposure guidelines such as threshold limit values (TLVs) previously discussed. However, the suggestion has been made that TLVs are not sufficiently stringent for indoor environments, recognizing that exposures in the indoor environment may be to a complex mixture of materials, for more than a 40-hour week, and may include a broad range of population. It has been noted that it is customary to use as a guideline that a concentration of 0.1 TLV would not produce complaints in a nonindustrial population in an indoor environment (American Society of Heating, Refrigerating and Air-Conditioning Engineers, 1989).

In assuring that airborne concentrations of isoparaffinic hydrocarbons do not exceed recommended guidelines, both process emission and room ventilation rates must be considered. Design of liquid toner imaging equipment and processes will impact source emission rates, which can be determined through experiments in controlled environments. Required minimum ventilation rates for the field placement can then be used to assure that intended guidelines are achieved.

7.7.5 Total Emissions

Another area of current awareness is the emission of volatile organic compounds (VOCs) to the outside environment. VOCs, including hydrocarbons, have been implicated in complex lower atmosphere chemistry involving nitrogen oxides and sunlight in the production of ozone pollution. U.S. Environmental Protection Agency guidelines require the use of emission control equipment for sources emitting more than 100 tons of VOCs per year, and some regional air authorities have imposed more stringent limits. However, with numerous regions of the United States remaining in noncompliance with federal ozone standards, the EPA and regional air authorities are considering more stringent limits on VOC emissions, and are increasingly turning to so-called small generators in an effort to reduce overall emissions. In particular, the printing industry has received considerable attention from these agencies (Jones, 1990).

Total emission for liquid toner processes is an issue only for high-volume applications, where potential emissions to the outside environment are in the tons of hydrocarbons per year range. Even in these high-volume applications, control of hydrocarbon emissions is achievable.

7.7.6 Containment Concepts

Several approaches to reducing or controlling hydrocarbon emissions from liquid toner processes have been described recently. A solvent recovery system can be used to condense hydrocarbon vapors that are generated in the fusing process (Howe and Hsu, 1988; Szlucha et al., 1988). Alternatively, hydrocarbon vapors generated in fusing can be oxidized catalytically to carbon dioxide and water (Landa and Sagiv, 1985).

In an innovative technique to reduce the overall volume of liquid toner used, liquid dispersant can be recycled for reuse in toning after electrophoretic separation and deposition of solids from excess toner (Day, 1989, 1990).

7.8 RECENT ADVANCES

The main body of this chapter was written in 1996. Here we provide references to more recent work on liquid toner technology.

General: Status of liquid toner technology and problems to be solved have been reviewed by Omodani. et al. (1998). The history of liquid toner innovation from 1953–1997 was recently summarized by Case et al. (1998).

Liquid toner materials: Liquid toners were designed using silver flake as a component to enable printing of conducted traces on printed circuit boards and was reported by Kydd et al. (1998). The relationship between paper properties and image defects observed in a liquid toner printing system was reported by Caruthers et al. (1999). Particle mobility in a model liquid toner system and CuPc particles charged with $ZrO(Oct)_2$ have been measured. These experiments were obtained by a toner charging mechanism based on a site-bonding model as reported by Keir et al. (1999). Performance of liquid carriers for liquid toner systems have been reviewed (Larson, 1999) and a demonstration of optimized design of liquid toners using perfluorinated solvents as liquid carriers was presented in a recent report by Rao et al. (2000).

Liquid toner process: A liquid toner printing process was modeled as a dynamic analog electronic circuit by A. Kotz (1998), and using a similar technique the impact of "pixel-pixel cross talk" was demonstrated and modeled by A. Kotz (1999). "Reverse charge printing" liquid toner process was demonstrated to give results similar to offset printing (Liu, 1999) and there have been demonstrations of the use of highly concentrated liquid toner to develop high-quality, noise-free images as reported by Mori et al. (1999), Matsumoto et al. (2000), and S. Horii et al. (1997). Fluid removal from liquid toner images by the use of a vacuum-assisted blotter roll was shown to be very effective in a report by Chang et al. (1999). Coordinated theoretical and experimental study of the impact of lateral conduction on liquid toner images was reported (Chen et al., 1996). Optimization of liquid toner composition for use in a process that utilizes an electronic print plate consisting of a two-dimensional array of individually controlled electrodes was presented in a recent report by G. Bartscher et al. (1996). Image on image process using liquid toners was described in a recent report by H. Yagi et al. (1997). Hydrodynamics of reverse roll metering was modeled and demonstrated by Wang et al. (1997).

Liquid toner characterization: The electrophoretic mobility of a model liquid toner system and silica particles in hydrocarbon carrier charged with AOT, was measured using an electrophoretic light-scattering apparatus. The electric field dependence of the mobility was demonstrated (Jin et al., 1998). Measurement of liquid toner layer cohesion was measured using a visualization cell. It was demonstrated that the integrity of liquid toner images are understood through the plastic behavior of granular materials, namely tensile strength and consolidation stress, as reported by Chang et al. (1998). Charging characterization of liquid toner materials using a plate-out cell was described. Charge-transport models were used to rationalize the data in a report by Wang et al. (1999). Charge density in liquid toner images was determined by a series capacitor technique in a report by Chen (1997). Liquid toner electrophoretic mobility was measured with electrokinetic techniques and reported by Russell et al. (1997). Charged-micelle mobilities were determined in a lecithin Isopar system by measuring time-dependent conductivities and reported by Davis (1997).

Liquid toner control processes: Control of liquid toner properties to enable consistent print system output by a replenishment system was modeled and demonstrated by Gibson et al. (1998). Electronic control of the Indigo Electronik® technology was demonstrated to meet the demands of digital production printing in a recent report by Levy et al. (1997).

7.9   OUTLOOK

The best opportunities for liquid toner processes are in areas that take full advantage of the high resolution and high-quality color capabilities inherent in liquid toners. Products based on liquid toners are expanding into application areas that challenge commercial and industrial printing processes, including short-run on-demand printing, variable information printing, and printing on nonpaper substrates.

Recent progress has furthered the mechanistic foundation of liquid toner particle charging. Also, new charge additives continue to be identified. Both areas strengthen the framework used in predicting and controlling charging properties of new liquid toners being designed and developed.

Much attention has been given to assuring the environmental suitability and acceptability of liquid toner processes. Continued interest in this area will likely prompt more advanced methods of controlling hydrocarbon emissions, work which will only further enhance the future of liquid toner based imaging processes.

REFERENCES

Allen, T. (1974). Particle Size Measurement. John Wiley, New York.

Almog, Y., and Gutfarb, J. (1994). IS &T Tenth International Congress on Advances in Non-Impact Printing Technologies Proceedings, 199–200.

American Conference of Governmental Industrial Hygienists (1990). Guide to Occupational Exposure Values— 1990. Cincinnati, OH.

American Society of Heating, Refrigerating and Air-Conditioning Engineers (1989). ASHRAE Standard 62-1989, Ventilation for Acceptable Indoor Air Quality. Atlanta, GA.

Bartscher, G., Breithaupt, J., and Hill, B. (1996). IS &T Twelfth International Congress on Advances in Digital Printing Technologies, 349–353.

Bohren, C. F., and Huffman, D. R. (1983). Absorption and Scattering of Light by Small Particles. John Wiley, New York.

Caruthers, E., and Zhao, W. (1999). IS &T Fifteenth International Congress on Advances in Digital Printing Technologies, 642–645.

Caruthers, E. B., Gibson, G. A., Larson, J. R., Morrison, I. D., and Viturro, E. R. (1994). IS &T Tenth International Congress on Advances in Non-Impact Printing Technologies Proceedings, 210–214.

Case, C. (1998). IS &T Fourteenth International Congress on Advances in Digital Printing Technologies, 226–230.

Chan, D. M. T., and Trout, T. J. (1990). U.S. Patent 4,917,986.

Chang, S., Ramesh, P., LeStrange, J., Domoto, J., and Knapp, J. (1996–1999). IS &T 12th–15th International Congress on Advances in Digital Printing Technologies, 638–641.

Chen, I. (1988). J. Imaging Sci., *32*, 201.

Chen, I. (1995). J. Imaging Sci. Technol., *39*, 473.

Chen, I., Mort, J., Machonkin, M. A., and Larson, J. R. (1996a). J. Imaging Sci. Technol., *40*, 431–435.

Chen, I., Mort, J., Machonkin, M. A., Larson, J. R., and Bonsignore, F. (1996b). J. Appl. Phys., *80*, 6796.

Croucher, M. D. (1987). Surfactants in Emerging Technology (M. J. Rosen, ed.). Marcel Dekker, New York, pp. 1–30.

Croucher, M. D., Duff, J. M., Hair, M. L., Lok, K. P., and Wong, R. W. (1984a). U.S. Patent 4,476,210.

Croucher, M. D., Drappel, S., Duff, J., Lok, K. and Wong, R. W. (1984b). Colloids and Surfaces, *11*, 303–322.

Croucher, M. D., Lok, K. P., Wong, R. W., Drappel, S., Duff, J. M., Pundsack, A., and Hair, M. L. (1985). J. Appl. Poly. Sci., *29*, 593.

Croucher, M. D., Wong, R. W., Ober, C. K., and Hair, M. L. (1988). U.S. Patent 4,789,616.

Dahlquist, J. A., and Brodie, I. (1969). Applied Physics, *40*, 3020.

Davis, T. (1997). IS &T Thirteenth International Congress on Advances in Digital Printing Technologies, 352–356.

Day, G. F. (1989). U.S. Patent 4,799,452.

Day, G. F. (1990). U.S. Patent 4,895,103.

Degiorgio, V. (1982). The Application of Laser Light Scattering to the Study of Biological Motion (J. C. Earnshaw and N. Steer, eds.). Plenum Press, New York.

Duff, J. M., Wong, J. M., and Croucher, M. D. (1987). Surface and Colloid Science in Computer Technology (K. L. Mittal, ed.). Plenum Press, New York, pp. 385–397.

Elmasry, M. A., and Kidnie, K. M. (1990). U.S. Patent 4,925,766.

El-Sayed, L. M., and Larson, J. R. (1988). U.S. Patent 4,783,388.

El-Sayed, L. M., and Taggi, A. J. (1987). U.S. Patent 4,702,984.

El-Sayed, L. M., and Trout, T. J. (1990). U.S. Patent 4,917,985.

El-Sayed, L. M., Larson, J. R., and Trout, T. J. (1990). U.S. Patent 4,935,328.

Felder, T. C., Marcus, S. M., and Pearlstine, K. A. (1990). International Symposium on Surface Charge Characterization, 21st Annual Meeting, Fine Particle Society.

Fowkes, F. M., Jinnai, H., Mostafa, M. A., Anderson, F. W., and Moore, R. J. (1984). Mechanism of electric charging of particles in non-aqueous liquids. A.C.S. Symp. Ser., *200*, 282–306.

Fowkes, F. M., Lloyd, T. B., Chen, W.-J., and Heebner, G. W. (1990). Proceedings—Hard Copy and Printing Materials, Media, and Processes. International Society for Optical Engineering, 52–62.

Gibson, G., Caruthers, E., Pan, D., and McGrath, R. (1998). IS &T Fourteenth International Congress on Advances in Digital Printing Technologies, 214–217.

Gibson, G. A. (1990). U.S. Patent 4,891,286.

Gibson, G. A., Caruther, E., Chow, J., Harrington, R., Luebbe, R., Simms, R., Thomas, T., and Wen, J. (1992). IS &T Eighth International Congress on Advances in Non-Impact Printing Technologies Proceedings, 209–211.

Greenwood, M. (1990). Indoor Air '90; Proceedings of the 5th International Conference on Indoor Air Quality and Climate, *5*, 169–173, Toronto.

Horii, S., and Horii, T. (1997). IS &T Thirteenth International Congress on Advances in Digital Printing Technologies, 344–347.

Howe, W. C., and Hsu, T. C. (1988). U.S. Patent 4,731,636.

Hudnell, H. K., Otto, D. A., House, D. E., and Mølhave, L. (1990). Indoor Air '90; Proceedings of the 5th International Conference on Indoor Air Quality and Climate, *1*, 263–268, Toronto.

Hunter, R. J. (1981). Zeta Potential in Colloid Science. Academic Press, London.

Isaacs, E. E., Haung, H., Babchin, A. J., and Chow, R. S. (1990). Colloids and Surfaces, *46*, 177–192.

Jacobson, G., and Hillkirk, J. (1986). Xerox: American Samurai. Macmillan, New York.

Jin, F., Davis, T., and Fennell Evans, D. IS &T Fourteenth International Congress on Advances in Digital Printing Technologies, 206–209.

Jones, G. A. (1990). GATF World, *2*(4), 29–36.

Keir, R., Quinn, A., Jenkins, P., Thomas, J. C., and Ralston, J. (1999). IS &T Fifteenth International Congress on Advances in Digital Printing Technologies, 611–614.

Kjærgaard, S., Mølhave, L., and Pedersen, O. F. (1989). Environ. Int., *15*, 473–482.

Kotz, A., and Ender, D. A. (1998). IS &T Fourteenth International Congress on Advances in Digital Printing Technologies, 231–238.

Kotz, A. R. (1999). IS &T Fifteenth International Congress on Advances in Digital Printing Technologies, 619–622.

Kydd, P. H., and Richard, D. (1998). IS &T Fourteenth International Congress on Advances in Digital Printing Technologies, 222–225.

Landa, B. (1983). U.S. Patent 4,413,048.

Landa, B. (1984). U.S. Patent 4,454,215.

Landa, B. (1986). Third International Congress on Advances in Non-Impact Printing Technologies, SPSE, Springfield, VA, 307–309.

Landa, B., and Sagiv, O. (1985). U.S. Patent 4,538,899.

Landa, B., Ben-Avraham, P., Hall, J., and Gibson, G. (1988). U.S. Patent 4,794,651.

Lane, G. (1990). Hardcopy and Printing Materials, Media, and Process. SPIE, *1253*, 29–36.

Larson, T. M., and Jarnot, B. M. (1999). IS &T Fifteenth International Congress on Advances in Digital Printing Technologies, 631–633.

Larson, J. R. (1987). U.S. Patent 4,702,985.

Larson, J. R. (1988a). In: Annette Jaffe, ed. Fourth International Congress on Advances in Non-Impact Printing Technologies. Society for Imaging Science and Technology, 142–145.

Larson, J. R. (1988b). U.S. Patent 4,780,388.

Larson, J. R., and Trout, T. J. (1987). U.S. Patent 4,681,831.

Larson, J. R., and Trout, T. J. (1988). U.S. Patent 4,772,528.

Larson, J. R., Lane, G. A., Swanson, J. R., Trout, T. J., and El-Sayed, L. M. (1989). Fifth International Congress on Advances in Non-Impact Printing Technologies. Society for Imaging Science and Technology.

Larson, J. R., Caruthers, E. B., Gibson, G. A. (1995). Society of Electrophotography of Japan Journal, *34*, 415–419.

Larson, J. R. (1993). *Electroacoustics for Characterization of Particles and Suspensions*, NIST Special Publication 856, 301–314.

Levy, D., and Freminger, J. (1997). IS &T Thirteenth International Congress on Advances in Digital Printing Technologies, 363–369.

Liu, C., and Zhao, W. (1999). IS &T Fifteenth International Congress on Advances in Digital Printing Technologies, 627–630.

Matsumoto, S., Mori, A., Matsuno, J., Sasaki, A., Akasaki, T., and Kamio, K. (1998). IS &T Fourteenth International Congress on Advances in Digital Printing Technologies, 239–242.

Matsumoto, S., Satou, K., Matsuno, J., Sasaki, A., Akasaki, T., and Kamio, K. (2000). IS &T Sixteenth International Congress on Advances in Digital Printing Technologies, 251–254.

Mayer, E. F. (1957). U.S. Patent 2,877,133.

McNeil-Watson, F. K., and Pedro, C. D. N. (1987). UK Patent Application GB 2194112A.

Metcalfe, K. A. (1955). J. Sci. Instrum., *32*, 74.

Mitchell, R. D. (1987). U.S. Patent 4,663,264.

Mølhave, L., Bach, B., and Pedersen, O. F. (1986). Environ. Int., *12*, 167–175.

Mori, A., Matsumoto, S., Matsuno, J., and Kamio, K. (1999). IS &T Fifteenth International Congress on Advances in Digital Printing Technologies, 634–637.

Mullin, L. S., Ader, A. W., Daughtrey, W. C., Frost, D. Z., and Greenwood, M. R. (1990). J. Appl. Toxicology, *10*, 135–142.

Myers, D. Y., Jr., Alexandrovich, P. S., Pearce, G. T., Santilli, D., Sreekumar, C., Berwick, M. A., Upson, D. A., and Goebel, W. K. (1987). U.S. Patent 4,708,923.

Niv, Y. (1994). Proceeding of the Tenth International Congress on Advances in Non-Impact Printing. Society for Imaging Science and Technology, 196.

Niv, Y., Adam, Y., and Krumberg, Y. (1986). Presentation at the Third International Congress on Advances in Non-Impact Printing. Society for Imaging Science and Technology.

Novotny, V. (1981). Colloids and Surfaces, *2*, 373–385.

Novotny, V., and Hair, M. L. (1979). J. Colloid Interface Science, *71*, 273–282.

Oja, T., Petersen, G. L., and Cannon, D. W. (1985). U.S. Patent 4,497,208.

Omodani, M., Lee, W., and Takahashi, Y. (1998). IS &T Fourteenth International Congress on Advances in Digital Printing Technologies, 210–213.

Page, L. A., and El-Sayed (1990). Proceedings—Hard Copy and Printing Materials, Media, and Processes. International Society for Optical Engineering, 37–39.

Pearlstine, K. A., and Swanson, J. R. (1992). Journal of Colloid and Interface Science, *151*, 343–350.

Pearlstine, K. A., Swanson, J. R., and Page, L. A. (1990). SPSE 43rd Annual Conference.

Pederson, L. M., and Cohr, K. (1984). Acta Pharmacol. Toxicol., *55*, 317–324.

Rao, S. P., Hitzman, C. J., and Pathre, S. V. (2000). IS &T Sixteenth International Congress on Advances in Digital Printing Technologies, 242–245.

Rao, S. P., Mikelsons, V., and Ruta, A. G. (1996). IS &T 49th Annual Conference Proceedings, 545–548.

Rhodes, P. H., Synder, R. S., and Roberts, G. O. (1989). Journal of Colloid and Interface Science, *129*, 78.

Riesenfeld, J., Bindloss, W., Blanchet, G., Dessauer, R., and Dubin, A. S. (1988). U.S. Patent 4,732,831.

Ross, S., and Morrison, I. D. (1988). Colloidal Systems and Interfaces. John Wiley, New York.

Russell, D., Hargrove, D., and Trent, J. (1997). IS &T Thirteenth International Congress on Advances in Digital Printing Technologies, 348–351.

Russell, D. D. (1993). Proceeding of the Ninth International Congress on Advances in Non-Impact Printing. Society for Imaging Science and Technology, 397–400.

Santilli, D. (1977). U.S. Patent 4,052,325.

Schaffert, R. M. (1975). Electrophotography. Focal Press, London.

Schein, L. B. (1988). Electrophotography and Development Physics. Springer-Verlag, Berlin.

Simms, R. M., Dreyfuss, D. D., Caruthers, E. B., and Wen, J. (1992). IS &T Eighth International Congress on Advances in Non-Impact Printing Technologies Proceedings, 212–214.

Staples, P. E., and Koop, S. A. (1991). IS &T Seventh International Congress on Advances in Non-Impact Printing Technologies Proceedings, 561–569.

Stotz, S. (1978). Journal of Colloid and Interface Science, *65*, 118.

Szlucha, T. F., Dyer, D. A., and Langdon, M. J. (1988). U.S. Patent 4,731,636.

Tam, M. C., Pundsack, A. L., Gundlach, R. W., Vincent, P. S., Kovacs, K. J., Jennings, C. A., and Loutfy, R. O. (1988). J. Imag. Sci., *32*, 247–254.

Tavernier, S. M. F. (1988). Proceedings of the Fourth International Congress on Advances in Non-Impact Printing Technologies. Society for Imaging Science and Technology, 101.

Trout, T. J., and Larson, J. R. (1988). U.S. Patent 4,783,389.

Tsuchiya, Y., Clermont, M. J., and Walkinshaw, D. S. (1988). Environ. Tox. Chem., *7*, 15–18.

Wang, F. J., Domoto, G. A., Till, H. R., and Knapp, J. F. (1999). IS &T Fifteenth International Congress on Advances in Digital Printing Technologies, 615–662.

Wang, F. J., Morehouse, P., Knapp, J. F., and Domoto, G. A. (1997). IS &T Thirteenth International Congress on Advances in Digital Printing Technologies, 357–362.

Williams, E. M. (1984). The Physics and Technology of Xerographic Processes. John Wiley, New York.

Yagi, H., Shinjo, Y., Oh-oka, H., Saito, M., Ishii, K., Takasu, I., and Hosoya, M. (2000). IS &T Sixteenth International Congress on Advances in Digital Printing Technologies, 246–250.

8

Dielectric Papers and Films

LUBO MICHAYLOV

Worldwide Images, Carmel Valley, California

DENE H. TAYLOR

Specialty Papers & Films, New Hope, Pennsylvania

8.1 INTRODUCTION

The first publication of this chapter in 1991 (1) was based upon material compiled and submitted for publication in 1990, just when many industry observers were confidently predicting the death of the electrostatic plotter market. Right then, the industry was jolted by the introduction of the Scotchprint® Imaging Process from the Commercial Graphics Division of 3M Corporation (St. Paul, Minnesota). At about the same time, Harry Bowers (Bowers Associates, Berkeley, CA), using proprietary algorithms, demonstrated to the world that electrostatic plotters were capable of producing large-format graphics of quality not seen before. This was the beginning of the rebirth of the electrostatic plotter and media markets. What followed was an exciting period.

The growth witnessed during the 1990s would have been difficult, if not impossible, without the close alliances forged between hardware manufacturers, system integrators, and media and toner manufacturers. For instance, the close cooperation between 3M, Synergy Computer Graphics (Sunnyvale, CA), James River Graphics (now Rexam Graphics, South Hadley, MA), and Hilord Chemical Corp. (Hauppauge, NY) led to the success of the Scotchprint graphics fabrication process and its acceptance by the graphic arts and printing industry worldwide.

An alliance between Specialty Toner Corporation (Fairfield, NJ) and Harry Bowers led to the formation of Cactus (Chino, CA), a system integrator and a major supplier of printing systems for the wide format graphics market. Others, such as Onyx (Salt Lake City, UT), Visual Edge (South San Francisco, CA—formed when Bowers Associates restructured), and Colossal Graphics (San Francisco, CA), soon followed, and all set their eyes not just on traditional printing but also on photographic reproduction. At James River Graphics a separate business unit, Display Media, was formed specifically to support these markets.

During the same period there was also a considerable restructuring in the hardware business. Synergy, who manufactured the fast single-pass Color Writer® 410 electrostatic plotter, was acquired by Nippon Steel Co. (Japan) and shortly thereafter closed its doors for good. Precision Image Corp. (Redwood City, CA), a manufacturer of drum electrostatic plotters, saw a similar fate. The company was acquired by Graphtec (Japan) and ceased to exist. Hewlett-Packard Co. (Palo Alto, CA) discontinued the sales and marketing of their line of electrostatic plotters, and CalComp (a subsidiary of Lockheed Corp., Sunnyvale, CA) began a slow retreat from the electrostatic equipment market.

We witnessed the formation of Phoenix Precision Graphics Corp. (Sunnyvale, CA) and the success of Raster Graphics Inc. (San Jose, CA). RGI, having developed novel 22-, 24-, and 36-inch wide plotters, introduced the first 54-inch wide high productivity printer, the Digital Color Station® (DCS) DCS 5400. Xerox (Stamford, CT) restructured the San Jose, California, electrographic arm of Xerox Engineering Systems to form Xerox ColorgrafX Systems and introduced its version of a super-wide multi-pass color electrostatic plotter, the 8954.

Perhaps the most dramatic event during this period was the introduction of the Scotchprint graphic fabrication system by 3M. Initially, this was a turnkey operation, consisting of a Sun computer workstation, proprietary software, a scanner, a Synergy Color Writer 410 color electrostatic printer, and a Pro-Tech® laminator capable of transferring the image from the special dielectric Scotchprint Transfer Media onto self-adhesive vinyl. Later on, the Xerox 8954 and the Raster Graphics DCS 5400 printers were qualified and packaged as Scotchprint systems.

The fact is that Bowers Associates, 3M, Cactus, Onyx, Visual Edge, and the other value-added resellers (VARs) generated a whole new market in the graphics industry, the short-run graphic! In doing so, they created a strong impetus for the media manufacturers to satisfy a growing demand for specialty, high-performance products. Their response to this demand, and their creation of new markets by producing novel papers, forms the basis for the latter part of this chapter.

Today, no one in the industry talks about *plotters*; they are called *printers*. Indeed, the latest generation of these imaging machines has been so productive they may well be referred to as *presses*! This shift in terminology did not occur overnight. It evolved from the gradual penetration of wide format electrostatic devices in the printing industry. It might appear superficial to the casual observer, but the name change has its roots in the primary markets served by these devices. During the 1980s the primary use for them was in the CAD/CAM (computer aided drafting/computer aided manufacturing) market where they produced computer-generated drawings or plots. During the 1990s, on the other hand, they went predominantly into the graphic arts, where they are used to produce computer generated, short-run prints and graphics for advertisements, signage, and display. Thus in this chapter they are printers.

An imaging device, however, would be worthless without the appropriate imaging media. In our opinion, it is the availability of properly designed media for specific markets that facilitates the sales of hardware, not the other way around. Numerous disasters occurred in the early days of electrostatic plotter development, when a manufacturer rushed to introduce a new hardware model to market before the supplies development process was complete. Availability of properly designed supplies for each model of hardware usually trailed after the new model was shipped to the customer. Strict secrecy about any new hardware development would prevent any meaningful cooperation with the suppliers during the design phase. Therefore the advantages of the new hardware would be poorly exploited.

A superb example of how important a joint effort is among the hardware, software, and media and toner suppliers, can be found in the Scotchprint imaging process introduced by 3M in 1991. Its success in the marketplace was assured by a perfect match between hardware design, software design, a specially designed toner transfer dielectric paper (Scotchprint Transfer Media), a variety of self-adhesive vinyls with a toner receptor layer, an assortment of laminating films, and a toner with good ultraviolet (UV) light stability and color saturation. This system of interactive components, known as the 3M Matched Component System or MCS®, played a fundamental role in the highly successful marketing campaign by the manufacturer.

Another fruitful joint effort was the introduction of materials for outdoor advertising, specifically Rexam's Outdoor Poster Grade and Specialty Toner Corp's Weather Durable Toner. These developments were initiated by Cactus, who brought together the technical and marketing groups of these two companies with Gannett Outdoor, a highly respected leader in the outdoor signage (billboard) market.

It is fair to say that the introduction of the Scotchprint graphic fabrication process opened up a whole new market for imaging supplies manufacturers, i.e., the short-run end of the screen printing industry. We have seen the introduction of new and exciting media in the market, such as Rexam's Wear Coat®, a dielectric paper with a special coating which, upon imaging and transfer onto a substrate, produces an image that does not require an overlamination film. The dry transfer image is imbedded in a hard polymeric matrix that protects the toner particles from abrasion, vandalism, and ultraviolet radiation to some extent. Additionally, Wear Coat media can transfer the image under heat and pressure to a large variety of substrates (plastic films and sheets of almost any type, coated canvas, fabric, and many others). Rexam has been hailed as a pioneer for this and other exciting products for the short-run graphic fabricators.

Another similar transfer technique relies on the easy release of the dielectric layer from wet dielectric paper. This so-called *wet transfer* process is a simple, although messy, means to get an image onto a wide range of substrates. The success and phenomenal growth of the short-run graphic market since the introduction of the Scotchprint process provided a strong stimulus to ink jet printer and media manufacturers as well.

Today, the end user will find a wide spectrum of dielectric media that will satisfy almost any commercial application. If one includes transfer media and the associated substrates, as well as the dye sublimation process, the selection of media and the versatility of applications is indeed impressive. During those early days of color electrostatic printing (1980 to 1990), the images, when exposed outdoors, lasted only a few days. Today, outdoor display materials (banners, signs, vehicle graphics) are guaranteed to last up to five years if certain conditions are met.

Shortly after the first edition of this handbook was published, Raster Graphics introduced a novel family of printers for the smaller format market: the printers took sheets cut from a roll and mounted these on a stainless steel belt. The most impressive development from Raster Graphics was full-width writing heads with individually driven writing nibs. The printers had a substantial speed edge and were free of many of the constraints of the prior technology. Raster Graphics demonstrated leadership again in 1993, introducing the 54-inch DCS 5400 printer. It has proven the mainstay of several market segments, especially those, such as the out-of-door industry, where high speed production factors into success. The most recent version, the DCS 5442, can print at 10 inches per second (ips)!

The renaissance in the technology stimulated a newcomer in the hardware area, i.e.,

Phoenix Precision Graphics Corp. (Sunnyvale, CA). A transplant from the now extinct Precision Image Corp., the team at Phoenix Precision Graphics first designed a 36-inch color electrostatic printer that recycles the toner internally, thus minimizing the need for toner waste disposal and making the printer self-sufficient in terms of toner supply. On occasion, the operator need only add dispersant to compensate for evaporation and carry-out by the paper during printer operation. More recently a 60-inch wide version was announced.

The continued growth of the short-run graphic market convinced 3M to develop the most productive yet single-pass, color electrostatic printer, the Scotchprint 2000. At a speed of 2 ips, this monster can produce the graphics for a 40 foot truck trailer in about 18 minutes. No doubt, the 2000 will further erode the low-end runs of the screen printers and convince those who are still hesitating to jump on the digital wide format printing bandwagon.

Color electrostatic printers are still relatively expensive devices. For instance, the 54-inch wide printers sold by Xerox CGS and Raster Graphics cost about $100,000. That is why we find the greatest population of printers in commercial photographic laboratories, screen printers, service bureaus, and to a lesser extent the in-house printing and CAD/CAM departments of large firms. They are also being adopted by other industries such as the outdoor advertisers. The printers are best suited for businesses where their high production speed and image durability are necessities.

Another recent major business is imaging woven fabric materials (primarily polyester) by dye sublimation (also, *sublimatic*) printing. The key to success was the introduction of special toners containing sublimation dyes. Synergy and Hilord were the pioneers of this technology in this country. In Japan, Nippon Steel Chemical (a subsidiary of Nippon Steel Co.) was the leading force.

One reason the process was not implemented on a large scale until recently was that color consistency was almost impossible to achieve in long production runs, and the process was too slow to be economical. Both problems have now been overcome by the efforts of Cactus, PowerPrint Technologies (Sunnyvale, CA), Paedia Corp. (San Francisco, CA), Hilord, Specialty Toner, and others. Sublimatic printing (see Section 8.6.7) is being adopted for many other materials and being used for snowboards, skis, furniture, and art. Xerox entered the market directly with the 8900 ''DS'' series, while Raster Graphics had its own variant specifically tailored for this application.

The five toner stations of the latter printer have been used by Specialty Toner Corp. to address one of the undesirable artifacts of sublimatic printing: The dye diffusion transfer process results in high dot gain. Dots can double in size. The few dark pixels needed to produce correct light colors, are often visible to the naked eye. STC's solution, ''V® toners'' dramatically reduces this phenomenon and produces graphics that withstand close scrutiny.

Although the rate of development in electrographic technology has slowed recently, new applications are continually being found, and the sales of consumables remains strong—a testament to the power of this versatile technology.

8.2 THE IMAGING PROCESS

8.2.1 Principles of the Electrographic Process

Electrography can be defined as an imaging process in which a latent electrostatic charge image is formed on the surface of a dielectric medium, and made visible by applying oppositely charged toner particles. The essential process elements of electrostatic writing are as follows:

An electrical potential, typically 400–600 volts (V), is applied across a dielectric layer with a conductive backing, by a writing head and a set of counter electrodes. When the potential across the air gap exceeds the threshold for air breakdown (circa 380 V), a discharge occurs and a net electrostatic charge is deposited on the dielectric layer. By selectively energizing a set of styli in the writing head to levels above the threshold voltage and by moving a charge receptive sheet or web past that head, a latent image is generated. The resulting latent image is then developed by applying toner particles carrying the opposite charge. Electrostatic attractive forces are the primary factor responsible for image development.

To be functional, the dielectric medium used to receive the charge must satisfy certain basic requirements, such as:

The dielectric layer must have high capacitance.

The charge decay rate of the dielectric must be low enough to ensure that the latent image reaches the toning station before any significant charge loss occurs.

The substrate carrying the dielectric must have a short electrical time constant, to ensure that the charge deposition will reach its maximum value during the duration of the writing pulse.

The essential design of the substrate for the process is simple: a dielectric coating on a conductive base (Fig. 8.1). However, the technology in the substrate, required to provide the proper functionality, is complicated and extremely challenging. In particular, electrography is the only commonplace printing process where the substrate is an integral part of the generation of the latent image—in no other is it a major part of the electronic circuitry involved in the placement of colorant! Thus the dielectric imaging element must have, in addition to all the needs for the end use, electrical properties befitting a sophisticated electronic component. That this is a difficult task to accomplish is often reiterated in this chapter.

8.2.2 Evolution of the Electrographic Process

The history of the evolution of experiments and observations with electrostatic charge, from the ancient Greeks through to the early generations of color plotters, is well described in the first edition of this book. The applications covered are now either obsolete or in rapid decline. It is the evolution of color printing that has been the focus of the industry over the last 5 years, and it is that which is addressed here.

Dielectric Layer

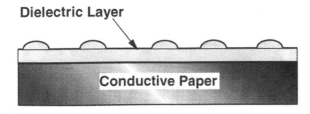

Figure 8.1   Schematic structure of dielectric paper.

Color Electrostatic Printing Technology

In 1983, the first commercial Electrostatic Color Plotter was introduced by Versatec (Xerox). Since then, a succession of companies have entered the market; most have dropped out, but there remains a strong cadre of players that ensures its viability and future.

The Versatec ECP 42 was a multipass 42-inch wide machine. Like all Xerox printers it featured through back-grounding (through grounding) and multiplexed writing. This printer not only revolutionized CAD but also sowed the seeds for the graphic arts industry. It was replaced by the CE 3000 Series, 200 and 400 dots per inch (dpi) machines in 1986, and these in turn were superseded by the 8900 family beginning in 1989. Xerox responded to Raster Graphics superwide printer introduction with the 8954 in 1992. After the ECP 42, the next revolutionary introduction was by Benson in 1987 with a 48-inch wide single-pass printer also using through grounding and multiplexed writing. Only a limited number of Benson printers were produced before the venture halted.

CalComp followed in 1987 with the successful 5700 series, which was a multipass printer with front grounding and multiplexed writing. The subsequent color models were less tolerant and therefore not widely adopted. CalComp departed the market in 1995.

Synergy's single-pass entry, the ColorWriter 400 of 1989, was another strong technological player; its adoption by 3M for the Scotchprint process ensured its place in the market but did not guarantee Synergy's survival. Nippon Steel purchased the company and closed the Sunnyvale offices in 1992.

Precision Image Corporation, founded in 1985, developed and marketed a unique drum plotting method using cut sheet media. The C448 was marketed from 1986 to 1988. It had a ceramic writing head and a unique toning system.

Hewlett Packard filled the gap between pen plotting and inkjet with a robust front-grounded printer, built by Matsushita of Japan, which was sold from 1988 to 1992.

The major contribution Raster Graphics gave the industry was the silicon writing bar, a full-width writing head with individually driven nibs. The first machine, the 22-inch wide Color Station® was sold in 1991. A series of 200 and 400 dpi 24- and 36-inch printers followed. This product line was discontinued by 1995 as the company focused on wide format high-speed printers. The first of these was the highly productive DCS 5400, which was followed, in 1996, by the DCS 5442. Oddly, no one else has adopted writing technology similar to the silicon writing bar.

Phoenix Precision Graphics introduced its Model 360, a 36-inch wide multipass printer. Instead of utilizing a 36-inch wide writing head, the PPGC device employs a head several inches wide that shuttles along and across the paper web during writing. This allows significant savings in production cost, and enabled the manufacturer to offer the machine at a lower price than other 36-inch electrostatic printers. In 1999 PPGC unveiled a 60-inch printer, but disappointing sales led the firm to close its doors in the year 2000.

The Scotchprint 2000, from 3M Commercial Graphics, has been the sole important hardware introduction after 1996. A 54-inch wide single-pass machine that runs at 2 ips, it is a veritable media eater. Like all the other new printing systems introduced since 1994, it was designed to satisfy the rigid requirements of the graphic arts market for speed, image quality, and operator control.

8.2.3 Implementation

There are three basic commercial printer configurations: through-grounded multiplexed multipass from Xerox CGS; through-grounded, individually driven multipass from Raster

Graphics and PPGC; and front-grounded, multiplexed, single-pass from 3M. The three configurations are shown in Fig. 8.2. The basic printer components are

The web handling system, which includes the media supply and take-up spools, the drive roll, and the turning bars or rollers

The writing system, which consists of the imaging electronics, the printhead, and the electrodes

The developing system, which includes the toner supplies, the pumps, hoses, fountains, rollers and vacuum bars, as well as the dryer fans

The printing and control hardware, firmware, and software

All have added features that are intended to satisfy the requirements of the service bureaus producing advertising materials, i.e., they are faster, offer better image quality, and offer more flexibility in setting up printing parameters. The issues of easy maintenance and color consistency have also been addressed.

The 54-inch wide Xerox 8954 printer offers a choice between single-layer and double-layer toning. It incorporates a humidifier and measures the temperature and percentage relative humidity (RH) inside the printer housing. There are many other useful features available from the control panel or via software connections from the operator or designer's computer. For example, the writing voltage of each color can be adjusted

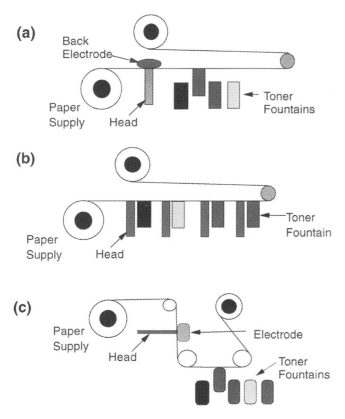

Figure 8.2 Commercial printer configurations: (a) 3M single pass (optional 5th fountain not shown), (b) Raster Graphics multipass, and (c) Xerox multipass.

independently. Further, each color can be written and toned multiple times, or written once and toned twice. Versions with 200, 300 and 400 dpi resolution were offered, with print speeds of 4, 3, and 1.5 ips, respectively.

Xerox also has two models differentiated specifically for sublimatic printing, the 8954 Series III and the 8954DS. The printing process used by Xerox is especially designed for full surface coverage. This, coupled with the 300 and 400 dpi resolution, gave this machine family the edge for photorealistic printing. Xerox also benefited from its world-wide customer and technical support organization.

The Raster Graphics DCS 5442, also a 54-inch wide printer, is a much improved version compared with the earlier 22-, 24- and 36-inch devices. Registration, speed, and paper handling, in addition to other features, have been optimized. The printer has a num-ber of advanced options in addition to the full-width imaging bar with individually driven nibs. These include very high print speeds: 10 ips in draft and 5 ips in premium quality mode. Although the nominal registration is 200 dpi, the length of the dot can be selected as either 200 or 400 dpi for extra fine resolution. There is an optional 5th toner station which can be used for spot colors, or for ''Digital Varnish,'' an overlacquer that improves print appearance and durability. The DCS 5442 comes ready to run standard or sublimatic toners. In the last configuration, it is the basis for the high-resolution V toner system from STC.

The Scotchprint 2000, like the earlier 36-inch 3M 9512 printer, will lay down 4 colors at 400 dpi in a single pass. The big difference is speed—it can output 2,400 sq. ft. of printed media per hour. Some industry observers predict that the 2000 machine will further penetrate the low end of the screen printing business. This rate means producing 400 posters, each 2 ft. $\times$ 3 ft., in one hour, complete vinyl graphics for a 48-ft. semitrailer in 19 minutes, or an 8-ft. $\times$ 10-ft. mural in two minutes. The printer accepts a variety of transfer media, such as Rexam's Wear Coat, standard interior and exterior imaging papers, and the new direct print Scotchprint Vinyl Media. Several options further expand the capability of this device, such as a high-capacity toner module that will allow an 8 hour uninterrupted printing cycle, a fifth toning station that permits the application of accent colors or UV protective dispersions in line with the printing, and a wind/unwind module that facilitates the loading and unloading of the voluminous supply rolls.

The Phoenix Precision Graphics printers have a single head, approximately 2.5 inches wide, containing 1,024 nibs. This head shuttles across the paper (the X direction), then the paper steps forward (the Y direction), in a mode similar to ink jet cartridges in ink jet printers. It offers two print resolutions: 200 dpi (draft mode) and 400 dpi (premium). The print speed varies as a function of the print mode, that is, 1.0 ips for premium, 1.5 ips for standard, and 2.0 ips in draft mode. The printer can accept paper weights from 75 to over 150 grams per square meter (gsm) and film thicknesses up to 5 mils. A remarkable feature of the PPGC printer is the recycling and reuse of the toner. The depleted toner is plated out in a high-voltage cell; the solid residue is then redispersed in Isopar® (Exxon Chemical's isoparaffinic hydrocarbon carrier fluid) and returned to the bottle of working premix. There are several patents covering the recycling operation.

8.2.4 System Interactions

The quality of the printed article, and to some extent the speed at which it is produced, is what the customer sees and pays for. Once the decision has been made to acquire an imaging device, be it an electrostatic printer, an ink jet printer, or a laser printer, the end user's main concern is print quality and how consistent that quality is over time. Other considerations also come into play, such as hardware reliability, frequency of repairs, maintenance cost, cost of disposal, and availability of local technical support, depending on the application and the production environment.

Service bureaus in the graphic arts market are the most demanding customers from the point of view of quality, consistency, and fast technical support. They operate on tight, rigid schedules and cannot afford to waste time waiting for a service engineer to show up, or to clean the device, or tweak the parameters to get rid of some artifacts.

In comparison with other nonimpact printing technologies, electrostatic printers are highly reliable. There are few moving parts in the system—toner pumps, the differential drive roller, and a few pulleys. It is not uncommon to find printers in the field that have been operating for five or more years without a major failure. Extending the MTBF (mean time between failure) depends largely upon the type of media used in the printing process, the relative humidity in the printer room, and the preventative maintenance performed on the hardware. The writing heads in an electrostatic printer, depending on usage, can last five or more years. In our experience, head damage usually results from metal particles imbedded in dielectric media, although large chunks of conductive agglomerates sticking on the surface of the dielectric papers and films can have the same detrimental effect on the writing head.

Printer design features are only one factor in the image quality equation. The type of media and toner used are equally important. The reason is that the printer, media, and toner form a highly interactive system. Different hardware implementations, and even different models made by the same manufacturer, have their specific requirements with respect to properties and performance. Moreover, each printer manufacturer considers certain system design features to be of a sensitive nature and is unwilling to share this information with the toner and media producers. Only firms that meet certain criteria established by the original equipment manufacturer (OEM) and whose products have been qualified for use in their hardware have access to the inside know-how of system parameters.

If an OEM decides to accept a new supplies vendor, then an extensive qualification process is launched that may require anywhere from 6 to 12 months of product submissions and performance testing. This is a well-justified and necessary ritual. It is a common practice that printer OEMs sell the consumables under their own brand name for two reasons: first, the sale of supplies is a highly profitable business, and second, the end user is assured of high performance and quality, thanks to the OEM's qualification and continuing quality acceptance programs. Today, strong alliances exist between equipment and supplies manufacturers to ensure optimal performance of the printers in the field. The industry has matured.

8.2.5 Strengths and Weaknesses of Electrography

Every printing technology has its limitations. Our perception of the strengths and weaknesses of electrography, as we know it today, depends fundamentally on the applications. As shown in the previous sections, each application has its peculiarities and demands on print quality. The following characterization, therefore, is based on the performance of electrostatic printers in the graphic arts industry, as that is where most of the new printer installations can be found. In the background, we are comparing electrography with ink jet technology.

Strengths
    Speed and productivity
    Reliability
    Resolution (up to 400 dpi)
    Long-term outdoor durability (up to 7 years with overlaminate)
Weaknesses
    High equipment cost
    Highly specialized media
    Flammable toner dispersant liquids
    Susceptibility to numerous image defects
    Requirement for highly skilled machine operators

The main reason that electrostatic printers are still popular in the graphics printing industry is their speed, high resolution, and durability of the printed image when certain conditions are met.

8.3 IMAGE GENERATION

There are three distinct phases to the creation of an electrographic image: deposition of charge to form the latent image, image development, and toner fixing. The first step is the most demanding of the substrate, again because it must function dynamically as a component of the electrical circuit that determines image formation. The subsequent demands are comparable to those of other digital imaging media. Understanding the process is aided by considering the electrical circuitry involved within the substrate, and by a separate description of the events. This is the approach adopted in this section.

8.3.1 The Electrical Circuit in the Substrate

The fundamental approach to explaining the development of the voltage gradient, and hence the charging process, is helped by studying the electrical circuits involved. (Fig. 8.3). [A thorough description is also provided by Johnson (2)]. The conductive materials are semiconductors, while other parts are not conductors but function as capacitors. The components of the circuit for a back-grounded printer are the power supplies for the nibs and the electrodes, the nibs (conductors) and electrodes themselves (conductors or semiconductors), the air gap (a capacitor), the dielectric layer (another capacitor), and the conductive substrate (a resistor and capacitor in parallel connecting two resistive layers—the conductive coatings).

The description of the circuit for a front-grounded printer is somewhat different (2) because the back electrode is capacitively coupled to the front side conductive layer. However, the differences mean little for the general discussion of the process that follows. Similarly, nibs energized by either multiplexed drivers or individual drivers respond equivalently.

When a voltage is applied to the electrodes, the potential initially divides between the air gap, the dielectric layer, and the conductive base, so the potential across the air gap is defined by the relationship

$$V_a = V\left(1 + \frac{C_a}{C_b} + \frac{C_a}{C_d}\right) \tag{1}$$

where the terms are as defined in the caption to Fig. 8.3.

(a) Through coupling

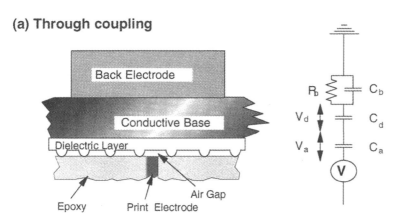

(b) Front coupling

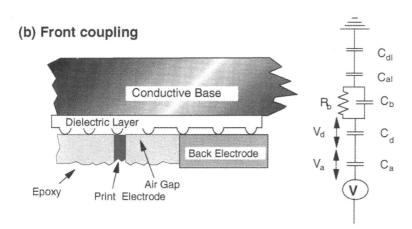

Figure 8.3 Equivalent electric circuit for (a) through coupling and (b) front coupling electrographic printers. V = potential applied to the nib; V_a = potential across the gap; V_d = potential across the dielectric layer; R_b = lumped resistance of the conductive base; C_a = capacitance of the air gap under the nib; C_d = lumped capacitance of the dielectric under the nib; C_b = lumped capacitance of the base, C_{dl} = capacitance of the dielectric under the electrode; C_{al} = capacitance of the gap between the dielectric and the electrode.

Initially no charge is deposited as the potential is divided among the three capacitors, C_a, C_b, and C_d. The resistance R_b of the conductive base controls the rate of transfer of the mobile charge and hence the redistribution of the voltage. The resistance must be kept low enough so that the relaxation time of the substrate, τ_b, defined by

$$\tau_b = R_b C_b \tag{2}$$

is low enough for the short write pulses to get maximum charging efficiency. Capacity is defined by

$$C = \frac{\xi A}{\ell} \tag{3}$$

where ξ is the permittivity, A is the area of the capacitor, and ℓ is the thickness of the layer. Thus relative values of capacitance per unit area, C', in the air gap, the dielectric layer, and the base paper, are estimated from

$$C' = \frac{C}{A} = \frac{\xi}{\ell} \tag{4}$$

Some measured capacitances for dielectric papers are $C'_d = 740$ pF/cm^2 and $C'_d = 60$ pF/cm^2. If the total applied potential is 550 V, the air gap is about 6–15 microns (μm), the surface electrical resistivity (SER) of the base is 10 MΩ/$\square$ (megohms per square), and the volume electrical resistivity (VER) is about 50 M$\Omega \cdot$ cm, an applied surface voltage (ASV) of about 120 V should be obtained.

8.3.2 Latent Image Generation—The Charging Process

In an electrographic printer, a charge is deposited imagewise on the surface of the dielectric paper or film medium by applying a sufficiently high voltage to the nib and across the dielectric layer. The high nib potential causes a spontaneous emission of electrons, and an avalanche of ion flow, as the air in the gap between the nib and the dielectric surface breaks down.

The conditions necessary for the first step in charging, that is, the emission of the electrons from a nib, include a high field gradient and a suitable emissivity of the nib itself. High field strength is found where there is a high potential gradient, expressed in volts per micron (V/μm) and a sharp surface. Thus it is found at the edges of the nibs, and especially when these are freshly polished or abraded. Emission is also easier from clean metals than oxide surfaces and polymers. Again, areas of fresh metal will be better emitters and more likely sites for discharge to begin. Figures 8.4 attempts to illustrate conditions in the gap between the nib and the conductive layer of constant field strength.

The current from electron emission is small—too small to be useful for printing. It must be enhanced somehow. If there is a sufficiently high voltage gradient across an air

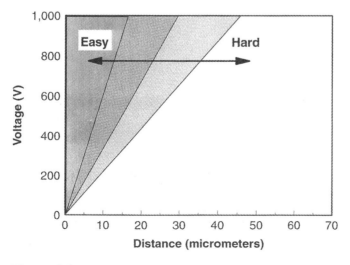

Figure 8.4 Conditions in the gap where nibs of different emissivity will spontaneously emit electrons.

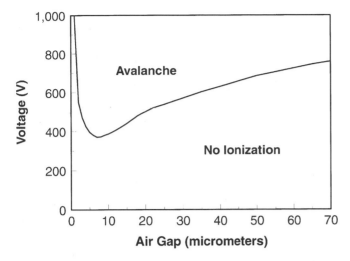

Figure 8.5 The ideal Paschen curve.

gap, electrons in that gap can be accelerated to the velocity that causes them to ionize the nitrogen molecules in the air that they collide with. This phenomenon is described by the Paschen Curve (Fig. 8.5). For a 10 μm gap the potential must be at least 300 V, or 30 V/μm. Once such ionization occurs, there is a cascade; ions are accelerated to high velocity and stimulate further breakdown in the air gap.

The electrographic printing system requires first, suitable conditions for electron emission, and second, a sufficient gap between the nib and the surface for breakdown to occur (Fig. 8.6). The means to achieve and control this gap and the emissivity of the nib are described below.

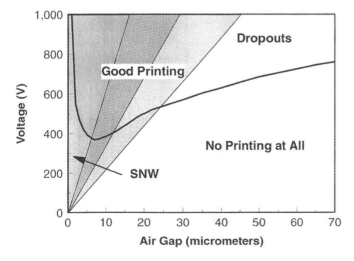

Figure 8.6 Combined Paschen curve and emissivity states showing effect on writing (SNW = spurious nib writing).

The first step in writing is the application of about 400 volts between the back electrode and the nib. This immediately causes a potential gradient between these two elements of about 4 V/μm with a 100 gsm paper, not sufficient to cause either emission or breakdown. Immediately upon establishment of this field there is a migration of charge, both within the conductive layers and between the front and the back conductive layers, to ground which, if the paper has sufficient Z-directional conductivity, rearranges the field so it is dominantly between the top of the front conductive layer and the nib. Thus across the dielectric layer and the air gap, a combined distance of about 10 μm, there is now a voltage gradient of about 40 V/μm. This is usually more than sufficient to initiate emission, which leads to avalanche and reproduction of the footprint of the nib as a dot of charge on the dielectric.

The discharge has several effects both on the nib and on the paper. The plasmalike conditions erode the nib surface especially where the field gradient is high. This rounds out sharp edges and oxidizes pure metal. Additionally, there is probably deposition of detritus after the discharge, because of the plasma vaporizing some of the dielectric layer. The nib is therefore in a less active state than before the discharge and so will require a higher voltage to discharge again if the surface is not rejuvenated.

The surface charge on the paper is balanced by a charge separation in the conductive layer as electroneutrality is sought. A surplus of negative charges seek ground. Because these charges are moving under a lower voltage than the applied potential, it will take them longer to be neutralized. If the front conductive layer is not well grounded this charge will accumulate with successive firings, especially if the path to ground has a high resistance. Eventually the accumulated charge will be sufficient to interfere with subsequent printing steps. Examples of the problems caused are described in Section 8.5.3 on image artifacts.

There are two distinctly different nib geometries, which produce dissimilar dot shapes (Fig. 8.7). There are also two different methods of depositing the charge that forms a dot. With solid shape nibs of large two-dimensional footprint, such as wires, the dot is produced by a pulse of short duration—approximately 60 to 120 microseconds (μs). During that time the area under the nib fills in with charge starting with the areas near the

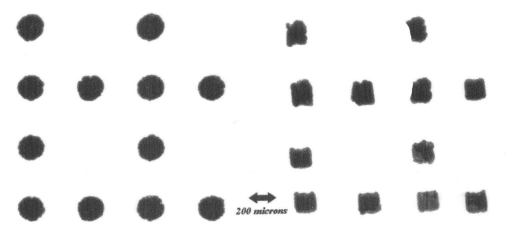

Figure 8.7 Typical high-quality image dots. Left: wire-wound nibs used by Xerox, 3M, and others. Right: ribbon nibs used by Raster Graphics Inc. (Courtesy of T. McBride.)

electrode edge, and then filling in until the voltage gradient in the gap is dropped below the critical value by the buildup of charge on the dielectric. At this point the voltage gradient exists substantially across the dielectric layer. If there is a significant build up of the interlayer voltage from inadequate charge depletion to ground, then the solid area will be incompletely charged. The dot will appear as a donut, or in extreme cases as a crescent (Fig. 8.8).

Printers with individual drivers have narrow, ribbonlike nibs. Dots from these printers are made by a ''paintbrush'' approach—the nib is energized when the paper at the beginning of the dot passes under the nib, and deenergized when the dot should end. The time for this is about 1 to 5 μs. The charge is actually deposited in a series of bursts, or flashes not unlike lightning strikes. Such dots appear square (or rectangular when the 200 × 400 dpi feature is being used.)

Lines from the two writing methods have distinctly different appearances. Wire nibs produce lines that are a series of circles, while the paintbrush heads give lines with greater edge definition (Fig. 8.9). Edge quality is related to the surface structure. The area that gets charged is the area determined by the shadow cast by the nib. When the surface is rough, the nib is further from the surface, and so it subtends a greater area. In the paintbrush mode, the line width clearly follows the topography of the paper surface.

8.3.3 Image Development—The Toning Process

The paper bearing the charge is advanced through a toning station. Here it is brought into contact with the toner, a dispersion of pigment finely ground in a resin binder, and a charge director (or, charge control agent) in an insulating liquid. The ideal carrier for liquid toner is odorless mineral spirits, sold commercially by Exxon under the Isopar® tradename, by Phillips Petroleum as Soltrol®, and by Shell Chemical as ShellSol®. These names represent a series of highly refined petroleum distillates, having different flash points and containing at most trace amounts of water and sulfur compounds.

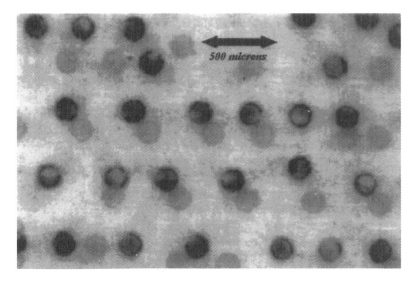

Figure 8.8 Incomplete dots deposited by wire nibs. Note halo/donut structure. (Courtesy of T. McBride.)

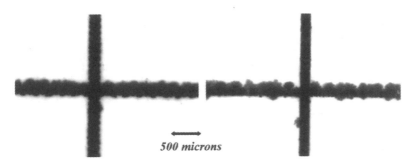

Figure 8.9 Typical electrographic line structure. Left: line from a wire nib with pulse-mode writing. Right: line from a ribbon nib written by paintbrush mode. Note flare on dot leading edge, or nib trailing edge—see Image Artifacts, Section 8.5.3. (Courtesy of T. McBride.)

The toner station is composed of three separate components: an applicator, which ensures that the paper surface is flooded with fresh, active toner; a high-shear mixing zone to impact constantly fresh toner particles with the dielectric in order to neutralize the surface charge completely; and a decoating or doctoring process to remove excess liquid toner. The first station is typically a fountain.

The mixing and liquid removal are both done with a reverse rotation metal roller. On the fountain side this brings about a region of turbulence and mixing as the dielectric is coated with a liquid toner film. On the decoating side, the reverse roller acts as a doctoring member by removing excess liquid from the paper. After this there may be a vacuum channel to complete the removal of the surplus toner and to aid in drying the wet sheet.

There are at least three demands on the dielectric surface: most importantly, it must not dissolve in the mineral spirits dispersant; next, it must protect the weakly bound toner from the agitation of the roller, and finally, it must not be a trap for toner, lest excessive background deposit occur.

8.3.4 Toner Fixing

The toners used in electrography most typically are fixed by drying. These toners are resin-coated particles usually ground in a steel ball mill to a submicron size. Because it is only partially soluble in the dispersant, the polymer is somewhat swollen, or "solvated," by the carrier fluid giving it a larger volume in suspension than when dry. During the development process the particle is drawn to the surface by electrostatic forces, but it becomes bound thereafter by van der Waals forces. These increase on drying.

The resin is also chosen from a number of polymers for its ability to wet the dielectric—i.e., the dielectric will preferentially adsorb the polymer from the solvent. Thus the soft layer will have conformed to the dielectric surface. On drying this bond is enhanced, increasing the adhesion of the toner to the surface. Cohesion is also critical—it is necessary that not just the first layer be well bound to the surface. Generally, if there is good adhesion to a dielectric there is good cohesion. If not then the choice of that particular resin binder is likely to be the problem. Cohesion between different layers is best if the wettability (surface energy) of the toner polymers for the different toners increases with sequence.

Toner fixing is also aided by heat. However, heat is not designed into any of the current commercial printers; they rely upon room temperature impingement air driven by fans.

8.4 FUNDAMENTAL DESIGN FEATURES

8.4.1 Critical Features

There are three fundamental components of media for electrographic printing, regardless of the printer design, operation, or end use. These are the dielectric layer, the conductive agents, and the substrate. Each must satisfy the functional requirements of the physics of the imaging process, the most critical of which is the deposition of charge on the dielectric surface during, and only during, deliberate writing. The difficulty of design is compounded by the use of liquid toners, printer mechanical design, and especially the end use. It is even further complicated by the contradictory nature of many of the requirements.

The most important requirement for the creation of an individual dot of image density is obtaining a high charge level on the dielectric surface. As will be shown in the following discussion, this is dependent upon the voltage gradient across the gap, the time the voltage is applied, the thickness of the air gap, the composition of the gas in the gap, and the composition of the dielectric layer itself. Each of these is in turn dependent upon a large number of factors including in each case contributions from the paper. All of the following media factors impact charge deposition, and hence image density, in some way: media thickness, uniformity of media thickness, media density, paper porosity, paper surface roughness, conductive material content, conductive material distribution, conductive material chemical composition, dielectric layer chemical composition (dielectric constant), dielectric layer purity (conductivity), dielectric layer distribution (penetration into the base), dielectric layer pigment type, size distribution, and content, dielectric layer roughness, and the content of water in all layers.

Coupled to this are the printer factors which include applied voltage, write time, electrode and nib composition, wrap angle of the media over the writing head, web tension, pressure of any back electrode on the paper, contact area with the back electrode, and speed of movement of the printer. For color printing the dryness of previously applied layers (the quantity of dispersant remaining in the toned layer) is an additional parameter. Another is the nature of the image being printed—there are usually unintended differences in the densities of dots in solid fill areas versus individual dots spaced widely apart.

8.4.2 The Dielectric Layer

The dielectric layer functions, first, as a capacitor and electron acceptor during the generation of the latent image, second, as an abrader of the nibs, to keep them clean and in an electron emitting state, third, as a toner receptive surface during image development, and fourth, as may be required in a particular end use application. The capacitance of the layer is determined by the dielectric constant and its thickness. These in turn are determined by the composition of the materials used, the manner in which they are coated, and the surface of the conductive substrate to which they are applied.

Three different approaches have been employed in applying the dielectric layer to the substrate: aqueous coating, solvent coating, and hot melt extrusion. Solvent coating dominates the industry and has yet to be matched for reliability of performance and for image quality. Therefore it is the main focus of this section.

Aqueous coating formulations have been used (3) but lack the performance of solvent coatings, usually because of difficulty in preserving the dielectric layer as a contiguous insulating film at the surface of the substrate. Aqueous coatings tend to penetrate the base, and the coating methods employed, whether rod, roll, or blade, promote intermixing of the dielectric composition with the water-soluble conductive layer. Double coating can address this (3) but adds cost and lowers the thickness uniformity of the thin layer.

Hot melt extrusion was investigated thoroughly in the late 1970s at Crown Zellerbach. This is an attractive technology for applying dielectric coatings because the layer is discrete and lies atop the conductive substrate. When properly formulated, it need neither penetrate that base nor have an open structure. The extruded layer had three components as shown in Fig. 8.10. The drawback at the time—difficulty of obtaining a suitably thin coextrudate of the polymers and polymer/pigment mixture—prevented it from being commercialized. Substrate adhesion was achieved by first extruding a thin layer of a copolymer containing carboxylic functional groups. Simple studies in the late 1980s to assess some new materials produced satisfactory imaging on some polymers but were confounded by poor adhesion to the base paper. The result, a clear but weakly supported film with an image on it, was a step towards the development of the image transfer technology described later in this chapter (4).

Solvent coating is used for all but a minor amount of dielectric paper and film products. Toluene with cosolvents such as acetone or ethanol prevails. The three primary components of the dry layer are resin, pigment, and plasticizer. There may be small amounts of other materials such as optical brighteners to whiten the media, viscosity modifiers, or tracers for coat weight measurement and control.

One of the most valuable tools available to the media designer is the microscope. Typical high-performing dielectric surfaces are shown in Fig. 8.11.

Resins

Resins with dielectric constants of about 4 to 5 are most desirable. Typical of these are polyacrylates, most especially Rohm and Haas Desograph® E 342. Polyvinylbutryal (e.g., Monsanto's Butvar® B-76) and polystryrene acrylonitrile are also employed. These are soluble in blends of toluene, ethanol, and acetone for preparation of the coating mixture. The properties endowed by these resins include good film forming (the ability to give an intact, relatively uniform coating layer with limited penetration of the base paper); high dielectric constant, charge acceptance, and charge retention; and surface pH control. The functional groups included in the resin are important for the last factor. The use of amido groups to reduce background has been patented (5).

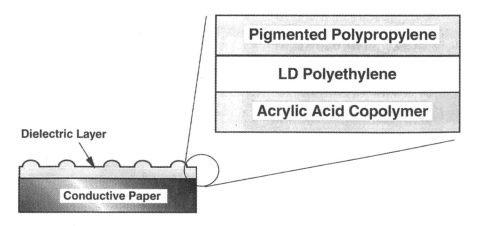

Figure 8.10 Triple-layer structure of hot melt extrusion dielectric coating.

(a)

(b)

Figure 8.11 Scanning electron micrographs at 500×: (a) dielectric paper; (b) dielectric film. (Courtesy of L. Kondik.)

Resins suitable for aqueous dielectric formulations are largely emulsions of polymers similar to those used in solvent coatings. These include various acrylics, carboxylated polyvinylacetates, and polyvinylbutyral. Cross-linking reduces charge decay in an aqueous system; perhaps the mechanism involves a reduction in the mobility of the conductive agent that inevitably pollutes aqueous dielectrics. The most advanced aqueous coating was developed by du Pont (3).

Plasticizer

Like many resins used in the dielectric formulation, Desograph E 342 is brittle. On drying it will crack unless it is plasticized with, for example, oligomers of polystyrene or other materials. These tend to be more readily extracted into toner than other components, no doubt because of the low molecular weight, but this is not generally a problem with current media nor printers.

Pigments

There are three functions of pigment in the dielectric coating. First, it controls the air gap between the nib and the surface—the so-called *spacing effect* (6,7) for which particles of about 5 to 12 μm diameter are preferred.

Second, it abrades the nibs, keeping them clean and in the proper emitting state. The ideal pigment scrapes the nib to a depth of only a few nanometers, like the finest emery polishing cloths.

Third, it protects the wet toner layer until it dries and becomes affixed to the substrate by the toner resin binder—an effect similar to providing the spacing for the air gap. It is especially critical for the dielectric transfer papers using transfer toners.

Ground calcium carbonate dominates as the pigment of choice for dielectric paper coating. It is low in cost, easily dispersed, readily available in a choice of narrow particle size ranges, and delivers consistent image quality. Additionally, it has an intermediate level of hardness so that under mild conditions it can maintain good, although not ideal, abrasion of the writing nibs.

By comparison, ground silica has a broad particle size distribution. Thus, in a formulation with an appropriate proportion of particles in the 5 to 12 μm range there can be a large number, on an areal basis, of oversize silica particles. Some of the larger ones will move down into the fibrous mat of the sheet, but too many protrude and gouge the soft metal of the nib, leaving it prone to spurious nib writing, or flare.

Precipitated calcium carbonate is available in more discrete particle size ranges and surface treatments. However, it has always been associated with either low image density or high background stain. Further efforts with surface treatments similar to those used for amorphous silica (8) may ultimately resolve this deficiency.

Amorphous silica is also popular because it is a mild abrasive agent, because it is translucent, and because it is available in controlled particle sizes. It is commonly used in this regard as a spacer. When used over highly mobile water-soluble conductive layers, it is necessary to change the surface chemistry from hydrophilic, which promotes wicking of the salts through the particle pores, to hydrophobic (8). However, by itself it has usually not been sufficiently abrasive, and hence it has needed to be blended with a harder material. Silica, unlike ground calcium carbonate, is readily sorted by air classification.

Crystalline silica (9) has been added to dielectric coating formulations to ensure high abrasion and avoid missing dots. Powders having controlled particle size a little smaller than the primary spacer can be included at low concentrations in formulations with amorphous silica to achieve this. The downside is that the abrasions themselves are large and so cause substantial nib writing and flare. A balance can be made by combining soft spacer particles with hard abrasive ones (9), but this is a difficult approach with invariable inconsistencies in performance. Crystalline silica dust is also a health hazard, and so it is unlikely to be incorporated in any new formulations.

Blending both spacer and abrasive particles can be approached also by having a fine abrasive pigment at a relatively high concentration in the coating, mixed with less abrasive spacer particles. The abrasion is then obtained from the small particles in the coating that remain on the spacer after drying.

On the other hand, a particle will protrude from a coating layer at low pigment: binder (P:B) ratios as long as its diameter is greater than the layer thickness. This means that the layer of coating over the spacer particles can have abrasive components, provided these are at a sufficient volume concentration in the formulation. It is necessary to remove the volume of the spacers themselves from the calculation—they effectively act as a discrete phase. Finely ground calcium carbonate and titanium dioxide act in this manner in formulations with spacers. This approach can be used with amorphous silica, starch, and beads of polymers such as polyethylene and polystyrene or polyacrylonitrile.

A substantial effort has gone into establishing the proper concentration of particles in the dielectric layer on film (10–12). These particles are either the abrasive elements themselves or carriers for the abrasive particles. Their surface concentration determines the ratio of nib firing opportunities to nib abrasions or emission regeneration opportunities. If the abrasive event is large, the nib can write multiple times (typically, about 10) without regeneration, although the first firing will undoubtedly have flare. The frequency of particles need only be such as to ensure that contact is most probable in an area covered by the nib during the imaging of multiple (in this case, 10) rows.

On the other hand, if the abrasive event can only sustain one discharge, the nib needs to be given a regeneration event each time the media moves forward one row. The surface concentration of the contacting particles must be high, most especially for a 400 dpi printer. The distribution of the particles on the surface is not regular, so in calculating the preferred concentration a randomness factor is required. A further effect is the size distribution of the spacers. If the abrasiveness is such that a nib needs regeneration after each firing, then the height of the spacers must be similar, otherwise large particles will bear the load for all those around them.

Frequent contact of the abrasive with the nibs is typical for premium monochrome paper grades (13). It is also characteristic of Japanese papers, where high image acuity is essential to reproduce the subtle strokes present in kanji characters. The drawback from having a high frequency is that the capacity of the surface to hold toner is decreased, so densities drop off, marks from the impressions of the drive (pinch) rollers become obvious, and toner is scraped from secondary and tertiary colors.

The search continues for an ideal surface structure, one which has proper spacing, a high frequency of nib regeneration events (all of which are gently abrasive), and the capacity to hold all the toner.

8.4.3 Conductive Materials

The conductive substrate is the embodiment of the active electrical component in the paper. It, too, is critical to the printer's ability to create a complete dot. There are a number of chemical compositions that can give conductivity suitable for electrography. They are of two basic categories, ionic or electronic. Ionic conductors include inorganic salts, organic salts, and conductive clays. As the charge carrier in these materials is an ion, and salts in air are typically hydrated, conductivity is dependent upon the water content of the material (14).

Electronic conductors include semiconductive materials such as polyanilines, tin oxide doped with antimony, indium, or copper, copper iodide, and small concentrations of good conductors such as carbon black, titanate whiskers, or metallic fibers. These materials function independently of water content; they are not affected by the humidity of their environment.

There is a substantial difference in cost among conductive treatments for papers and films. Simple economics have ensured that the dominant conductive materials for electrography are quaternary ammonium polymer salts, often augmented with simple inorganic salts. Unfortunately, as these salts are hygroscopic, the ambient relative humidity determines the moisture content of the substrate, and therefore its conductivity. Although this is a limitation, it has not prevented the universal adoption of ionic conductive technology.

Inorganic Salt Conductors

The simplest ionic conductors are common salts such as sodium chloride (table salt), sodium nitrate, and potassium chloride. Although sodium chloride is the cheapest, sodium nitrate is less corrosive. As stated earlier, because all these salts are hygroscopic, their conductivity depends upon moisture content. This is determined by the amount of moisture in the surrounding air from which an equilibrium moisture content is established.

Simple inorganic salts have a characteristic relative humidity above which they pass from solid to liquid. This property renders them unsuitable as conductors in the form of continuous films. When used with paper, however, they are applied with humectants and become part of the thin film of the water-soluble/water swellable material that coats the fibers. Conductivity (resistivity) changes continuously with relative humidity as shown in Fig. 8.12. Because the response is steep, extremely tight RH control of the printing environment is needed to enable salts to function reliably—too tight, unfortunately, for commercial air handling systems that are normally installed in graphic arts establishments.

Organic Salt Conductors

Three quaternary ammonium polymers have dominated commercial electrographic papers: polyvinyl benzyl trimethyl ammonium chloride (15), polydiallyl dimethyl ammonium chloride, and polymethyl methacrylate trimethyl ammonium chloride. The first, as Dow Chemical's ECR 34, was the most common material in use in the United States until it was withdrawn because one of the precursors is carcinogenic. A family of these organic salts is made in Japan by Sanyo Chemicals and sold worldwide under the tradename Chemistat®. It was used for one of the first-generation electrographic films (8).

Three properties set this family apart from other common polymer quaternary salts: (1) it is a good film former, (2) it is conductive without being sticky at relative humidity levels below 70%, and (3) it shows a lower variation of conductivity with changes in relative humidity.

Among its shortcomings: it is relatively expensive and tends to be used only for high-performance CAD and graphic arts media, and it may have a slightly fishy odor.

Polydiallyl dimethyl ammonium chloride is a common flocculant used in water treatment systems; thus it is readily available in high purity. While being highly conductive and relatively low in cost, it is known to create problems based upon its tackiness at preferred printing room relative humidities. Papers made with this conductive agent are almost invariably high in defects, are prone to blocking, and feel sticky to the touch. In

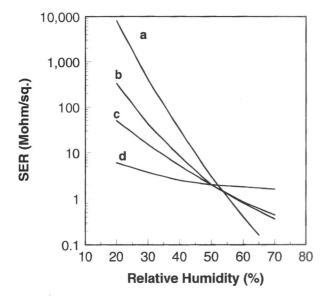

Figure 8.12 Effect of relative humidity on surface electrical resistivity of paper coated with different conductive materials, at equilibrium: (a) simple ionic salts; (b) anionic polymer; (c) cationic polymer; (d) electronic conducting pigment.

spite of these drawbacks, it is used extensively in CAD and low end graphic arts papers. Commercially it is available as Calgon's Conductive Polymer 261® or as Agestat 41T® from CPS Chemical Company.

Polymethyl methacrylate ammonium chloride is intermediate between the other two polymers in all properties: film forming, tackiness, and cost. It is found in papers of all types, especially in Europe, where it is available as Induquat ECR 69L® (formerly Makrovil) from Indulor Chemie. CPS Chemical Company also provides a similar acrylic designated Agestat 1401®. Other quaternary ammonium compounds have been used experimentally but have had limited commercial application (16–18).

Polystyrene sulfonic acid and its sodium and potassium salts (products of National Starch and Chemical) have also been used for dielectric papers and films. Because this acid is highly conductive, only thin coatings are required; but as a strong acid, it is reactive. Additionally, the humidity response of anionic salts is so steep as to make it nearly impossible to maintain adequate control of moisture content.

Because quaternary ammonium polymers are the dominant conductive treatments for dielectric papers and films, ambient relative humidity is an extremely critical aspect of electrographic printing. The equilibrium resistivity of the paper varies exponentially with relative humidity in the 10% to 55% range. This relationship is a strong characteristic, a fingerprint that serves as a means to evaluate different conductive treatments. For commercial papers using polydiallyl dimethyl ammonium chloride, it has been found that surface resistivity, R, and paper moisture content, $[H_2O]$ (in percent by weight), are related by the empirical formula

$$R = [H_2O]^{-4} \qquad (5)$$

Thus a 2% decrease in moisture content can raise paper resistivity by a factor of 5!

Interestingly, surface and volume resistivity can have different dependencies upon moisture content. This can be the cause of otherwise unexplainable leading edge fog.

Paper absorbs or loses water rapidly to the surrounding air stream. Figure 8.13 shows the rate of pickup for an oven-dried paper in still air. The rate of change is much faster when the air is blown. Needless to say, in only seconds paper can lose a percent of moisture in a dry room and hence see a substantial change in performance. Given the narrow range for optimum performance for graphic arts printing, it is essential that printer room humidities be controlled tightly. A range of not more than 5% is recommended. For example, presentation media should be printed in rooms at 50% to 55% RH. This is much tighter than the norm for CAD rooms.

The rate at which paper can exchange moisture with the ambient is a function of the driving force, i.e., the difference between ambient moisture content and the equilibrium moisture content of the sheet. High-quality papers are properly and consistently moisturized within a range of no more than $+1.0-0.5\%$ upon leaving the paper mill.

There are two well-known approaches to controlling the moisture content of coated papers. First, when water-based coatings are used, the coating and paper are dried to the target. Second, when solvent coatings are used, water is added back to the paper after most of the solvent has been removed.

A device such as a Dahlgren LAS® thin film water applicator, or a steam foil, is normally used. Thus a dielectric paper that meets a conductivity specification within 0.5% is possible. Before this refinement was implemented, the surface resistivity of the base paper would fluctuate anywhere in the 20 to 75 MΩ/□ range, leading to large quantities of out-of-spec product. After the moisturizing step, the resistivity of the final product could be controlled within the 1–10 MΩ/□ range, a great improvement for CAD materials.

All the ionic conductive agents are water soluble. If the paper or film becomes wet, the dielectric coating will detach from the substrate. There have been many attempts to develop coatings that retain their integrity when wet, especially for use as poster and

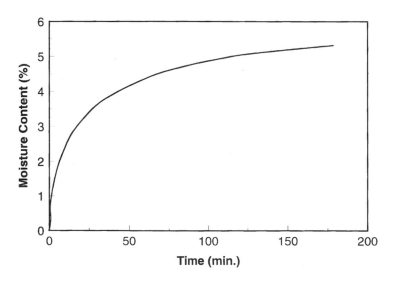

Figure 8.13   Rate of moisture pickup by oven-dried dielectric paper in still air.

billboard papers. Two approaches have been tried in dealing with this issue. The first is to include in the polymer unsaturated groups that can be cross-linked after the coating has been applied to the substrate. This has proven impractical, as it is difficult to drive to completion the necessary chemical reactions given the limitations of the dryer sections of most coating machines. Even when cross-linked, these polymers remain hygroscopic; the bond to the dielectric after wetting becomes very weak.

The second approach has been to form a copolymer of a quaternary monomer with other compounds so that the product is solvent soluble but not water soluble (19).

Conductive paper coating formulations with organic polymer salts generally contain a latex binder and a filler, to ensure that the conductive coating fills the pores between fibers and builds a good platform onto which the dielectric layer can be formed. These coatings give both structural and electrical properties to the paper. The binder selection is difficult, because most latices are anionically stabilized and immediately coagulate when mixed with a cationic quaternary ammonium polymer. Nonionic, or cationic, stabilized latices are more suitable. While most fillers are stabilized anionically, they can be dispersed in a cationic polymer solution if they are "swamped" with the polymer so that there is no chance for them to bridge between the pigment particles. Ground calcium carbonate or kaolin clays can be used.

Better solvent holdout (the ability to keep the dielectric as a discrete layer on the paper surface) can be achieved if hydrophilic polymers such as polyvinyl alcohol, alginates, or starch are added (20,21). Fluorocarbon surfactants can also be used for this purpose (22).

A further factor in applying conductive coatings is the placement of the conductive agent in the paper, and its effect on curl. Using alcohol as a cosolvent can modify the final position of the conductive agent and also allow unbalanced coatings to be applied with minimal curl. This option is becoming more difficult to execute with the rising pressure to reduce emissions of volatile organic compounds (VOCs). Coating monomer blends and curing them by UV radiation is a more environmentally friendly solution. The formulation can be adjusted to optimize penetration, conductivity, and holdout of the dielectric. This approach is now being practiced by Rexam (23,24).

Ionically Conductive Pigments

The montmorillonite clays have expanding lattices in which the hydrated ions that counterbalance the structural charges are able to move. This suggests their use as conductive agents, especially as they are pigments and water insoluble. Various attempts have been made to use natural clays in dielectric papers and films (25,26), but inability to obtain a reliable supply of a consistent product has handicapped most efforts.

Laponite®, a synthetic hectorite clay from Laporte, has good conductivity (27–29) and is available in a variety of forms. One that is best suited for paper coating has a component, neighbourite, which cannot be bound and causes dusting, a problem that has been solved by centrifuging (30,31). Laponite sees use in wet strength papers (32) and premium graphic arts papers. Although it is still affected by ambient moisture, Laponite exhibits the least RH dependence of all the common ionic conductive agents.

Conductive clays must be formulated with binders so that they have physical integrity. Zeolites have pore structures that can also act as channels for hydrated ions but have not found commercial application for dielectric papers (33).

Tin Oxides

Doped tin oxides are electronic semiconductors. They are also water insoluble, making them highly desirable for special dielectric media applications. Invariably, these applications have been with film or filmlike media, as they are too costly for dielectric paper coating. It is interesting to note that those end users who cannot afford to install ambient humidity control are also not prepared to pay the premium price for tin oxide doped media. Ironically, those electrographic printing establishments that produce high-value-added prints and can justify a higher media cost already have the humidity controlled environment that will enable less costly conductive treatments to be used effectively!

Dopants used to enhance the conductivity of tin oxide include indium (34,35), antimony (3,9,10,36), and copper (37). Fluorine is a new dopant for commercial grades.

The dominant application driving the use of doped tin oxide conductive layers was film for CAD. Clarity is necessary because the film is used either as a diazo master or as an overlay. Achieving this high level of transparency with a naturally white or off-white pigment necessitates grinding it until its particle size is too small to cause light diffraction, i.e., until the average particle size is about 50 nanometers (9,10). When combined with antimony doped tin oxide, this approach has been the most successful for electronic ground planes in electrography.

The same pigments can be used with much less grinding for opaque films such as direct write, pressure-sensitive vinyls. This promises to be a major growth area for these materials.

Formulating with these materials is largely a matter of adding the small amount of binder that is needed to give the coated layer physical integrity and flexibility without sacrificing the conductivity. With too little binder, the coating will crack if bent and will show poor adherence to the substrate. On the other hand, with too much binder the coating will lose conductivity because the particles will not be in ohmic contact. A further compounding factor is that solvent dielectric coatings often interfere with the electrical characteristics of this layer.

An alternative approach to obtaining a thin clear film depended upon sputtering a mixture of tin and copper in a controlled atmosphere of oxygen (37). Much of the formulation expertise lies in preparing the metal electrode for sputtering. This technology failed for lack of exploitation during a window of opportunity that closed when pigment doped tin oxides were commercialized.

Polyaniline and Other Electronic Conductors

The first commercial electrographic film, Kodak's product introduced in the early 1980s, was rendered conductive with polyaniline, an electronically conducting organic polymer (38,39). It was humidity independent, water insoluble, and tough. It was, however, colored and expensive, properties that have been the bane of most electronic conductors in this technology.

Potassium titanate whiskers (19,40) and ground copper iodide (41,42) have also been used for conductive media but with little market success.

8.4.4   The Substrate

Commercial electrographic printing substrates fit into two classes, papers and films. As films are usually dimensionally stable, durable insulators, the focus in this section is on the electrical properties of the base sheet and other factors that affect the dielectric layer.

In the imaging process, the properties of the base paper must be such as to allow complete discharge of all the nibs. There are three other important considerations: strength, printability, and esthetic properties.

Base Paper Electrical Properties

A most important electrical property of the paper substrate is the relaxation time between the two sources of the voltage gradient. For a given charging pulse length this time must be significantly shorter to ensure maximum charge deposition. A ratio of about 10:1 to 15:1 is a good guide in developing conductive substrates. For write times as short as 40 μs, the optimum base relaxation time is 4 to 6 μs. The relationship between relaxation time and write pulse length is illustrated in Fig. 8.14, where the total write voltage is plotted against the write time for two values of base relaxation time, 7 and 64 μs. This plot, which was constructed from experimental data, shows that the maximum potential across the dielectric at 40 μs write time can be achieved if the base relaxation time is 7 μs or less, while at 64 μs relaxation time the effective potential is significantly reduced.

The conductivity of the base paper can vary significantly depending on the placement of the conductive material. In general, conductivity is applied by coating processes which tend to leave the bulk of the material near the surface. Even size press applications tend to be surface oriented because the formulations are pigmented and of moderate viscosity. The result is papers that have good surface conductivity but low volume conductivity. Achieving high volume conductivity generally requires large quantities of conductive agent—an expensive addition to the unit manufacturing cost.

Requirements for the earlier CAD multiplexed plotters with back-grounding generally performed well at surface resistivities of 5 to 10 MΩ/□ and volume resistivities of about 1 to 50 MΩ · cm at 50% RH. Front grounding CAD printers perform best with a highly conductive layer under the dielectric, usually near 1 MΩ/□, while a minimum volume resistivity of about 50 MΩ · cm is sufficient to ensure the countercharge is grounded.

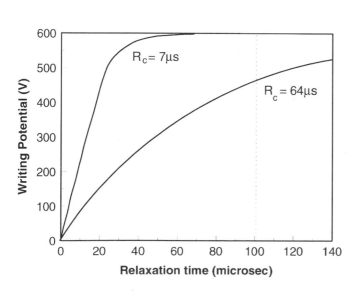

Figure 8.14 Effect of base relaxation time on charging efficiency.

Graphic arts printers now perform at higher speeds and with larger solid fill areas than a decade ago. In all cases these require SERs of 1 to 3 $M\Omega/\square$ and VERs of less than 1 $M\Omega \cdot cm$. There remains with these printers a narrow sweet spot where the complications of multiplexing between the region where conductivity is too low (low image density, many unpleasant artifacts, and low-density striations) and too high (reverse image striations and strong overtoning) do not exist.

Printers with individually driven nibs are much more tolerant, especially when using a paintbrush mode to write dots. Because the write time is long, even poorly conductive papers will write acceptably for line work. But should the paper be used for printing heavy solids, it becomes necessary to provide sufficient volume conductivity to ensure that the countercharge is properly dispersed. Generally speaking, the SER should be lower than 5 $M\Omega/\square$ and the VER below 0.5 $M\Omega \cdot cm$. As described above, control of the amount of conductive agent, its placement, the paper moisture content as packaged, and the relative humidity of the print room are all critical factors in maintaining proper paper electrical characteristics.

Properties that Influence the Dielectric Layer

There are substantial interactions between the base paper and the dielectric coating that can dramatically influence the image. The base properties should be such that the dielectric layer is of uniform thickness (uniform capacitance) and is uniformly distant from the nibs (uniform emission and discharge); and the high points of the layer should be in uniform contact with the nibs (uniform abrasion).

Minimizing grain requires that the penetration of the coating into the base paper must be minimized by uniform solvent holdout. The Kubelka–Munk equation describes the penetration of a fluid into a pore:

$$L^2 = \frac{(R_p \gamma \cos \theta)t}{2\mu} \tag{6}$$

where L is the distance of penetration that occurs in time, t, R_p is the pore radius of the substrate, γ is the surface tension of the fluid, and θ is its contact angle, while μ is the fluid viscosity. Keeping the dielectric out of the paper is then aided by having small holes, high viscosity, low surface tension, poor wettability, and short times. The ideal paper will have not only optimized values of these but also consistent values.

Consistency of pore structure has two components. The first is time dependent. The papermaker must make the same product minute-by-minute, hour-by-hour, day-by-day, and month-by-month. Good process control, consistency of supplies, and regular preventive maintenance are the hallmarks of good practice and will show up in the statistical control charts for air porosity, or solvent holdout using the Hercules Size Tester.

The second pore consistency determinant is spatial. Paper has short-term mass variations from the flocculation that occurs during sheet formation on the wire. The flocs have more matter and are densified to a greater degree during calendering. Pores between the fibers are then much smaller than those in the areas between the flocs—the interflocs. If the formation is poor and holdout is marginal, it is likely that the paper will be grain prone. Poorly made paper has streaks of greater or lesser mass originating from the head box. These too can reflect grain to differing degrees across the image.

Poor formation on a sheet with high solvent holdout will still show poor imaging, but in this case it is more from the dielectric surface being at different distances from the nibs—it is closer on top of the flocs and further away in the interflocs. The result is mottle—density differences that follow the paper formation. Mottle and grain often go together, although there may be a grainlike deficiency if the paper roughness is the main contributor to the gap. Smooth dielectrics will have low density on the tops of flocs.

The printer design will either exaggerate formation-related artifacts or relieve them. Wrapping the paper around the writing head reduces mottle and grain tendencies (paper will distort so that the inside surface will follow the contour). Longer write times, higher voltages, and good grounding also lower mottle and grain levels.

Paper Strength Properties

Strength properties include tear, tensile, elongation, burst, fold endurance, stiffness, and internal bond, factors that are important for the processing and handling of the material during manufacture, printing, and end use. In general, if the paper has sufficient strength to be coated, it can be printed. Most of the emphasis on these properties is for end use applications.

Paper Printability Properties

Printability refers to those properties needed to ensure that the material can be properly processed in high yield in the printing establishment. Although normal paper strength factors are included, freedom from tears, holes, and wrinkles are more critical. Additionally, flatness and straightness are required. Papers that are not flat, papers with large edge flutes or puckers, may not be held flat going across toner fountains, rollers, or vacuum channels.

There is an interaction between flatness and stiffness in various printers. In those with vacuum channels, nonflat paper that is flaccid will be pulled down so that there is a good vacuum seal. If, however, the paper is stiff, the vacuum seal may be broken and (unless the printer has a low vacuum detector) liquid toner may pass into other parts of the machine. In printers where either the head or elements of the toner system form hard nips with the backing electrode, nonflat flaccid paper will tend to fold and crease, whereas stiffer paper will transport smoothly. The solution to both issues is flat paper.

A related issue is nonstraight paper. If the paper is bowed, unless its elasticity allows the printer to pull out the bow, creases and wrinkles are probable. Roisum (43) has defined both flatness and straightness and given guidelines for good runnability in web handling equipment. Additionally, bow can give registration differences between the two sides for multicolor prints.

A factor that has a much greater influence on registration, though, is dimensional change during printing. Paper shrinks or expands depending upon its water content. Some papers will grow by 1% when equilibrated to a 10% increase in RH. If the paper is not at equilibrium in the environment that it is being printed in, then it will either be absorbing or losing water as the printing proceeds. This can have annoying effects for precisely defined images such as dither patterns and overlaid lines. The effect can readily be visualized by printing four color vertical and horizontal lines spaced by one line. Dimensional changes of as little as a 20 μm will show up as color shifts or color bands. This problem is minimized if paper is provided with a moisture content equivalent to what it would have at 50% RH and 70°F, and the printer is kept in the same environment. This is typical practice for good offset printing houses, whose behavior digital printers need to emulate. The hygroscopic expansivity of a typical dielectric paper is shown in Fig. 8.15.

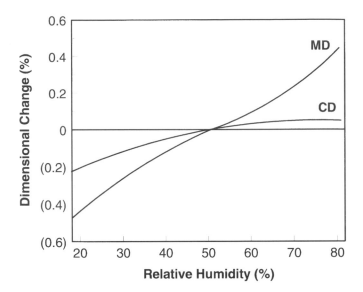

Figure 8.15   Hygroscopic expansivity of dielectric paper initially equilibrated at 50% RH, where MD = machine direction, and CD = cross-direction.

Gross misregistration is generally attributed to a mismatch between the coefficient of friction of the back side of the paper with the rubber tread of the drive roll. Unless there is a good grip, the roll may turn at a speed different from the paper. A consistent level of roughness is recommended. This may come from the paper fiber structure or from the pigmentation and binders in the coatings. In designing these coatings it is critical that they do not come off the paper and accumulate on any part of the printer they might contact. All printers have stationary elements of some type, whether this is a turn bar, a back electrode, or a grounding element. Once material begins to stick there is a snowball effect which continues until the effect is sufficiently gross that the operator stops the printer and cleans up, or the defect eventually tears the web and aborts the run.

Two printer designs that had little issue with registration were the Precision Image Corp. helical scan printer and the Raster Graphics 22-, 24-, and 36-inch printers. All held paper in sheet form firmly onto metal drums or belts. The dot placement was then quite precise (44).

Paper Esthetic Properties

Esthetic properties include color, opacity, whiteness, gloss, brightness, and freedom from dirt, holes, and pinholes. These influence the appearance of direct image prints, especially in background (white or lightly colored) areas. Although whiteness, for example, is generally considered an advantage, overall these properties are most properly defined by the end use.

8.4.5   Overall Design Considerations

The design of a new medium for electrography is nowadays a major task. Development scientists need access not just to good laboratories and pilot equipment but also to a typical end user facility where the vagaries of the printers are understood in a way that focuses upon application output. Additionally, there are few opportunities to develop a paper for a single printer that functions with only one toner. There are at least two sources of toner and also different variants from the OEM to be considered. Additionally, papers are expected to function in earlier, obsolete, printers, and often, because many printing shops have machines from two OEMs, the expectation crosses platforms.

8.5 IMAGE QUALITY

At first appearance, images produced electrographically by skilled printers can rival those from offset, ink jet, and even photography. Only on close inspection can the limitations be seen. On the other hand, it is a difficult technology to practice, and from the media manufacturer's perspective, it offers endless possibilities to introduce artifacts and imperfections. In this section, the basic nature of the image is described, the causes of artifacts and imperfections are discussed, and comparisons are made to the images available from other technologies.

8.5.1 Image Composition

The most commonly referred to image attributes in the digital printing world relate to printing fine detail, bright colors (solid area), and matching targets (process colors), all without deficiencies.

Resolution

Electrography is a binary process—a dot is either there or it is not. Although there have been attempts to modulate dot size and dot intensity, the complications of the writing process, especially during the generation of the latent image, have thwarted these efforts. This immediately limits resolution, and as the dots are larger than 100 μm the best image appearance requires that it be viewed from at least 4 feet. At this distance, the dots are barely resolved by human eye; they merge, creating the pleasing illusion of a continuous tone image.

The effect of resolution is most obvious with process colors, which are generated, of course, by printing different concentrations of dots of the four primary colors. When light colors are needed, the individual dots are far apart and again quite readily visible, especially if a small amount of black or magenta is being applied to a white or pale yellow background. Intense colors are less affected by dot size.

STC's solution to hiding the visible dots in light colors, most obvious in dye sublimation printing, is its V color process, which uses the three subtractive primaries—cyan, magenta, and yellow—in full strength, along with two additional toners: light cyan and light magenta. Black is generated using three process color. Dark colors are built from the standard-intensity tones, while the lights are made with yellow and the low-intensity cyan and magenta.

Solid Colors

Solid colors have all of the media surface colored with image, often saturated or near saturated with one of the primaries. While this might be seen as relatively straightforward to produce, considering that there are no worries about resolution nor dot drop out, clean, pure, uniform, and consistent solid colors are the most difficult demand for the system to fulfill.

Process Colors

The critical issue in developing a good color palette for a binary process is to obtain the proper concentration of dots of the different colors in a way that is acceptable for the end use. Colors in most CAD software was with pixels, or matrices of dots, often 3×3 or 4×4. These are prone to patterns from the regular arrangements of the dots, which is actually desired in CAD but unacceptable for most graphic arts. Similarly, except for long distance viewing, multidot pixels, which for a 200 dpi printer could be 50 dpi equivalent, are unwanted in imaging. The coarseness of texture is visually displeasing. A substantial breakthrough for the technology was the generation of software that introduced error diffusion. A recent refinement, well received in graphic arts, is stochastic dot placement. Essentially all graphic arts work is now done with dithering that randomizes dot placement and gives process colors with little organized structure.

8.5.2   Typical Image Properties

Dots and Lines

On good media individual dots made with round wire nibs have a high degree of circularity and uniformity (45). 200 dpi printers have dots about 220 μm in diameter, while 400 dpi printers produce roughly 160 μm diameter dots on the same media. In each case the dot is substantially larger than the nib—the discharge obviously spreads laterally about 45 to 50 μm. The degree of spreading is controlled by the electrical characteristics of the paper, especially those factors that determine the amount of charge that can be deposited, and also by the paper surface roughness. Rougher paper produces larger dots. The resolution of the technology is limited by the amount of dot spreading (dot gain)—the complexity of using closer nib spacing is not repaid with better resolution, as the dots will be at least 100 μm from even the finest nibs. (Spurious nib writing shows that finer resolution would be possible with a process that relies only upon emission without the avalanche.)

Line quality reflects the nature of the nib, whether rectangular or circular, and the charging sequence, whether pulse or duration. Circular nibs give lumpy lines in all directions, while rectangular nibs have high line quality MD and CD but show steps for lines at angles to the paper direction. Paper surface roughness also affects line width uniformity—the line width corresponds to the microroughness.

Solid Areas

Solid areas are more difficult to print than dots or lines because of the interference by residual charges in adjacent areas. With the single pulse printers the dots formed are more likely to be donut shaped and have lesser density in the center—a consequence of the field never reaching its peak. Overall, the density is reduced. The "paintbrush" printers show a different response—they have more difficulty in fully imaging the total surface and tend to leave the higher points without coverage. It is surmised that these printers deposit charge in a multitude of small discharges, confined to small locations rather like lightning strikes. The charge is then highly directed. It is perhaps this factor that distinguishes the two in terms of ultimate quality.

The major media characteristic that influences density is dielectric coating thickness—higher coat weight means less density as predicted from capacitance values. Figure 8.16 is a normalized example. There is a maximum density because at low coat weights the dielectric is too thin to hold the charge—it will pass straight through. Note the steeper

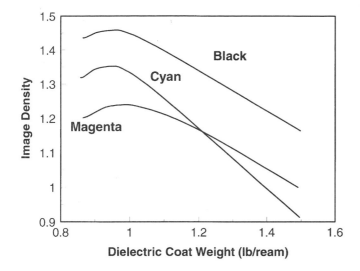

Figure 8.16 Typical relationship between dielectric coat weight and solid area image density.

response of the cyan toner—some cyan toners have distinctly different resins and so behave uniquely. The actual density that is obtained in any system is as much a function of the toner, the printer, and the mode of operation, as it is of the media. The stronger systems tend to give black density of at least 1.50, cyan of 1.45, magenta of 1.35, and yellow of 1.15. This allows large gamuts to be achieved directly, with a substantial further boost from overlamination.

Consistency of density is a major determinant of quality. High quality offset printing maintains image density variations of less than 0.05 units within any sheet and within a run. To maintain that degree of consistency it is essential that variations in the dielectric coat weight are held below 0.06 lbs./ream, or less than about 5%.

The matte nature of the image detracts from its initial visual impact—the apparent image density is low compared to other printing processes especially on glossy material. The porous pigment structure of the toner layer means that light is scattered within the colorant. Intensity can be enhanced when the air is displaced from the toner layer. This can be done by heating the toned image to cause individual particles to coalesce. A hot roll laminator is particularly effective for this purpose. Apparent density can also be enhanced by overlaminating with either hot or cold laminates, or by applying an overlacquer. Image transfer paper takes advantage of this factor too (45).

8.5.3 Image Artifacts

Electrography has its own lexicon describing image artifacts, so many of these are unique. The most common two that interfere with line and dot quality are flare and drop outs (missing dots). Voids, glitches (zippers), striations, reverse striations, overtoning (overplating), mottle, grain, bleeding (tailing), pinch roll marks, and toner scraping all detract from solids. Background areas and light colors can show spurious nib writing, ghosting, and stain. Furthermore, the image can be spoiled by scratches caused by material from the dielectric building up on the writing head, or from the backside accumulating on rolls, electrodes, and stationary elements.

Flare

Flare manifests itself as large "explosions" of charge, occurring somewhat randomly around otherwise well-formed dots (Fig. 8.17). They are typically small, rounded, and on the leading edge of the dot (the trailing edge of the writing nib). They are more problematic with 400 dpi than with lower resolution printers. In severe cases they may double the area of the dot. The flare area, however, is of lower density than the dot proper. In graphic arts, flare is a problem with both light tones and halftones—in the latter case by making the leading edge of a color much more intense than the bulk.

The extent of flare is dependent upon the work function of the nib metal (the ease at which it emits electrons), the roughness of the media, the abrasiveness of the media surface, the relative humidity, and the writing voltage. Electron microscopy has been used to investigate media surfaces directly before large flares. Typically, a substantial abrasive event is found. Metal fragments, like turnings from a lathe, have also been seen nearby on strongly flare-prone media. It is now believed that flares are explosive events that occur during the plasma discharge of conductive shavings abraded from the nib surface. These shavings will accumulate at the trailing edge of the nib. They will not repeat—once burnt off they must be regenerated. They will become visible when nibs have had substantial exposure to abrasion without writing. They can also be seen after cleaning the surface of the writing head by lapping, that is, removal of undesired deposits using abrasive materials.

Although printer manufacturers have developed certain techniques to reduce flare (46,47), it remains as one of the unwanted compromises in electrographic printing technology.

Dropouts

Missing dots are called drop outs whether in a halftone area or in a line. Typically they are measured in line patterns because they are much more visible to the naked eye. The effect in CAD is loss of data, which is unacceptable. In graphic arts, dropout is undesirable but unacceptable only in gross cases.

Dropouts are seen when nibs do not discharge as programmed. The most common cause is insufficient regeneration of frequently firing nibs. After a nib has fired, the field gradient needed to cause spontaneous electron emission is raised. When there are a large number of firings in sequence, eventually the nib will not fire, unless it is regenerated.

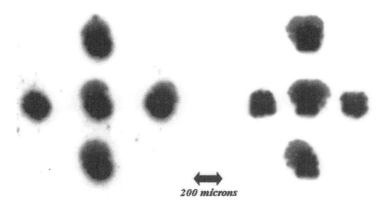

200 microns

Figure 8.17   Typical examples of flare. Left: wire nib. Right: ribbon nib. (Courtesy of T. McBride.)

These methods are effective: increasing the abrasiveness of the media, increasing the pressure of media onto the nib by raising the force from the back electrode, and raising the applied voltage. Using pigments of greater hardness, of greater size, or of greater frequency in the dielectric is the typical media design response to inadequate abrasion. Alternatives are harder dielectric resins, less residual solvent in the layer, and harder conductive substrates.

Dropouts are seen only when nibs are fired in isolation. When adjacent nibs are fired there is a synergistic effect from charge spreading that ensures the discharge, as one nib triggers the discharge of its neighbors.

Voids and Starvoids

Voids are areas in dense solid colors where the image is missing. They are most commonly caused by lumps of matter that increase the distance between the media surface and the nib beyond that at which discharge can occur. They may also be formed by large holes in the dielectric or by the absence of conductive agent in the substrate.

Starvoids are voids in only one color of a multicolor image—the void-causing event being present for only some, but not all, of the printing passes. Particles as small as 50 μm can cause visible defects. Obviously, coating lumps, dust, dirt, dandruff, and paper fibers can be the problem.

A common cause is pickoff of conductive matter from the web that is transferred to the dielectric surface in the roll. The pickoff may happen in the printer but is more commonly an artifact introduced during the paper manufacturing process. With modern production operations in place, void levels can be directly related to dust levels in printer rooms. In general, printers that produce consistently clean prints have clean rooms with wet-mopped floors, low airborne dust levels, and no trash. Multipass multiplexed printers are the most prone to these problems—single-pass machines offer little opportunity for web contamination.

Glitches

There is an artifact that is unique to solid prints in multiplexed pulsed printers. Glitches, also called ''zippers'' because of their characteristic appearance (see Fig. 8.18), are formed when a conductive pathway through the dielectric shorts out a nib. This means that the conductive layer near the nib will not hold charge, and the voltage gradient between it and the neighboring nibs will be too low to produce a proper discharge. When the printer moves forward, the short circuit stops and printing resumes. Glitches commonly persist for only one or two line widths.

Commonly glitches come from paper fibers, holes from bubbles in the coating or pinholes in the base, lumps of conductive material, or pickouts of dielectric pigment particles. They also may be formed by conductive polymer wicking through the pores of hydrophilic spacer particles (8). Rashes of glitches are often seen when the dielectric coat weight is too low.

Striations

Multiplexed printers also suffer from striations, or low-density bands associated with the boundaries of the elements of the backplate. They occur when printing solid colors. When groups of nibs fire, there is a residual charge left in the conductive layer under the dielectric. If this charge is not quickly dissipated, the intensity of the adjacent image will be reduced. Xerox introduced ping-pong writing sequences so that the adjacent groups were

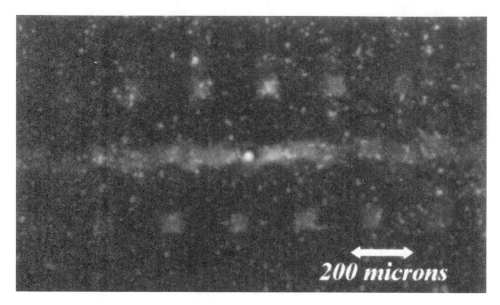

Figure 8.18   Glitches in a solid image. Note the hole in the center. (Courtesy of T. McBride.)

not written sequentially (48). This helped significantly until other speed-limiting factors were overcome and the issue reappeared in less than optimum situations.

Striations are seen when the relaxation time for the media is long compared to the time between adjacent nib groups printing. The characteristic media deficiency is insufficient conductivity in the base and especially to ground. Often it is seen with media that has been allowed to dry out, is being printed at high speed in a room of lower than optimum relative humidity, or was perhaps insufficiently conductive for the application. Ways to reduce the problem include raising printer room humidity, increasing printer writing voltage, and running the printer at a slower speed.

Reverse Striations

Reverse striations—high density bands between the backplate segments—can occur if the media is too conductive. This is often seen with front-grounded printers when media moisture content is relatively high. Paper should be provided with a narrow range of moisture content—it should be within 0.5% of that it will have if equilibrated at 50% RH. The protective packaging used by most manufacturers should keep it at that level for extended periods.

Overtoning

If the charge deposited on a dielectric sheet is not neutralized by toner it can persist for long periods. On any subsequent pass this charge will attract toner. Thus cyan can become mauve and magenta red. The phenomenon is called *overtoning or overplating*. Likely causes are dielectric coat weights that are too low, writing voltages that are too high, insufficiently charged toner, or a development time that is too short.

Overtoning can also occur on a small scale and reflect paper base formation or fiber structure. The relationship between coat weight and degree of overtoning is well known

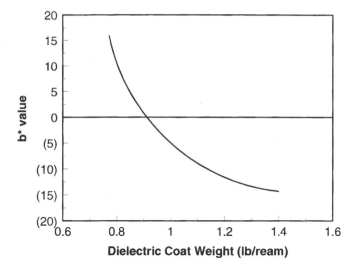

Figure 8.19 Effect of dielectric coat weight on overtoning of magenta with yellow. Yellow shift is shown as $b*$.

(Fig. 8.19). Maintaining consistent color with heavy solid areas requires close attention to dielectric coat weight on both a macro (roll-to-roll) and a micro (point-to-point) scale.

Mottle

Cloudy density patterns in the image of the order of several millimeters are described as mottle. Inevitably, they follow the formation of the base paper (see for example Ref. 44). In these areas the paper thickness and density vary so that the conductive coating, dielectric layer, or gap may vary in thickness. The mottle pattern is often visible as either overplating or grain.

Grain

Fine structures that are improperly toned are referred to as grain. Most often they correlate with high paper roughness in which poorly toned areas mirror gaps between fibers. In these areas the dielectric surface is too far from the nib. Poor solvent holdout exacerbates this imperfection.

Topographic analysis is useful for studying grain. The average surface roughness for good paper should be not more than 2 μm (13). The surface itself should be about 5 to 10 μm below the plane of contact of the spacer particles. Pits deeper than 10 μm will not be properly toned.

Mottle and grain can both be measured by image analysis (49).

Pinch Roller Marks

There are several artifacts that arise when media is too smooth. Pinch roller marks appear when these drive elements come in contact with the wet toner and leave an impression. A Sheffield roughness of about 100 mL/min prevents this. The toner layer must itself be thick, thus the problem is associated with solid secondary or tertiary colors where more than one toner is laid down.

Toner Scraping

Toner scraping is also symptomatic of a media surface that is so smooth that it cannot protect the incompletely dried, liquid toner layer. Contact with stationary elements scrapes off the weak toner layer. This is a particular problem in single-pass machines where the dry time is limited, and again is seen in the high-density areas of secondary or tertiary colors. The thickness of a normal single-toned layer, in comparison to the roughness of the dielectric, is shown in Fig. 8.20.

Bleeding

Bleeding and tailing are terms used to describe fuzzy outlines around printed images. The usual cause is under charged toner.

Ticks

Sometimes on low-grade papers at high humidity, little check marks in the machine direction are seen at the leading and trailing edges of features such as lines. These repeat each time a particular nib fires. Ticks (also extradata) arise from conductive material transferring by electrophoresis from the paper surface to the writing head; they extend the region of nib discharge.

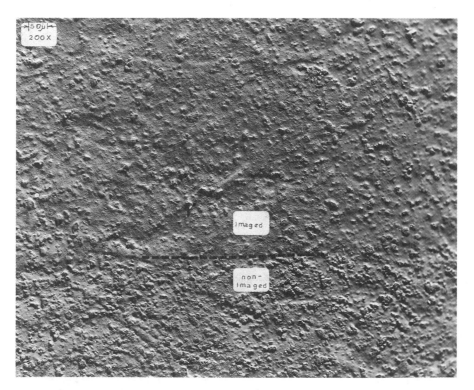

Figure 8.20   Photomicrograph at 200× of imaged (top) and nonimaged (bottom) dielectric papers. (Photo courtesy of D. Quackenbush.)

Spurious Nib Writing

Like flare, *spurious nib writing* (the characteristic trail of dots is shown in Fig. 8.21) is another artifact unique to electrography. It has essentially the same origin: excessively large abrasive events that produce metal spurs on the trailing edges of the writing nibs. When pulsing the printhead in the nonwriting mode, the field strength at the spur may be sufficient to cause spontaneous electron emission (see Ref. 50 and Section 8.4.2). Close microscopic inspection generally shows that it starts immediately after a large pigment particle or clump of poorly dispersed particles passes across the printhead.

Ghosting and Leading Edge Fog

Reflection of the image in nonimage areas is called *ghosting*. The most common type is leading edge fog which is seen when the countercharge is unable to move freely to ground, but can ground capacitively through the toner fountain. This charge then attracts toner into unwanted areas.

Stain

Any other accumulation of toner in otherwise unimaged areas is often classified as *stain*. This may be caused by entrainment of toner in the media surface crevices, by stray electrical fields, or by electrochemical attraction between the dielectric surface and the toner particles. Resin choice and pigment treatment have a strong effect on stain.

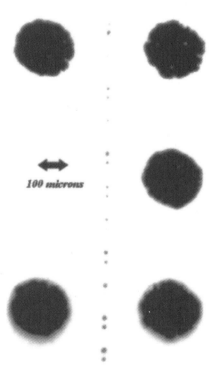

Figure 8.21 Spurious nib writing among regular halftone dots. (Courtesy of T. McBride.)

Other Defects

In addition to the above artifacts, the image can be spoiled by scratches caused by material torn from the dielectric layer that accumulates on the writing head. Similarly, residue from the backside of the dielectric paper or film material can accumulate on rolls, electrodes, and stationary elements causing image defects. There is also a myriad of defects that can be introduced by careless workers responsible for handling the media. Such damage is beyond the scope of this chapter.

Overall Quality

The manufacture of dielectric papers and films involves complicated operations and processes that require absolute attention to detail, rigorous quality systems, and thorough, frequent review of the quality control data, to ensure that good-quality imaging can be obtained reliably. Consistently obtaining high-quality imaging is also the responsibility of the printer operator, who must be well skilled at balancing the various factors under his or her control.

8.5.4 Comparison to Other Wide Format Printing Technologies

The quality of electrographic printing can be quite high when all parts of the system are optimized and the image being printed has been well chosen. With overlaminating it is possible to have images that appear flawless when viewed from a moderate distance. Thus electrographic images find extensive use in point-of-purchase displays, posters, billboards, truck signage, and other wide format reproduction applications. It is when the image is inspected closely or when compromises must be made that the flaws inherent in the technology become apparent. For closely viewed images electrography also lacks the resolution and fine structure of halftone technologies such as offset printing and continuous ink jet. However, it is much better suited to short runs than offset and of course has productivity benefits over this type of ink jet.

Screen printing is well suited for solid designs with extensive use of spot colors and highlights. However, producing process colors with this method requires extensive preparation of the screens and has long lead times. Electrography, while doing an adequate job of solid prints, executes process colors well, making it a strong contender for wide format imagery.

8.6 PRESENT TYPES OF DIELECTRIC PAPERS

Today there exists a broad range of dielectric papers available for the operators of electrostatic printers. This includes products for the conventional electrostatic writing process, as well as a choice of papers for transfer printing. In the former category are distinctions by image quality, basis weight, and transparency to ultraviolet light. Among the latter, we recognize three product types: toner transfer, image transfer, and sublimatic printing. This section deals with the construction and use of the dominant commercially available dielectric papers.

8.6.1 CAD Papers

Although electrography is rapidly losing favor to ink jet technology in the CAD market, it is still the workhorse for large corporate engineering organizations and service bureaus where high productivity and fast turnaround are essential. Monochrome printers have been extensively replaced by ink jet, direct thermal, and laser printers, especially those in the 22-, 24-, and 36-inch widths. New color electrostatic printers retain an edge for productivity, but the ease of operation and low cost make the transition to color ink jet hard to resist. Even so, substantial quantities of dielectric paper are still being made specifically for this market.

There are four types of dielectric paper for CAD: Report (Opaque), Translucent, Vellum and Semitranslucent. Each is distinct and designed to meet those market demands that evolved during the transition of CAD from pen plotting to electrography.

Report

Report papers are the electrographic equivalent to bond or plain papers. They are the low end opaque grades with typical basis weights of about 65 to 75 gsm (40 to 45 lb./sq.ft.) designed for either monochrome or color printing.

Monochrome papers have the simplest construction and the least sophisticated coating. As the only critical requirement is good contrast between the black lines and the unprinted area, the tolerances for performance fall within a rather broad range. Low-cost components tend to be used, the manufacturing processes are run at relatively high speeds, and only modest levels of production control are needed to make a print of satisfactory quality. Cost considerations dictate a low level of conductive treatment based at least upon some simple salts that deliver poor image quality at low relative humidities (below 40% RH).

Because color report papers are used for color CAD drawings, they must be capable of reproducing not only lines but also solid fill areas. Seismographs and microelectronic circuits (integrated circuit chip design, printed circuit board design, etc.) are usually printed on these grades. To meet the higher image and durability requirements, color papers start from better formed base stocks with greater solvent holdout, tighter porosity, greater smoothness, and more uniform and higher volume and surface conductivity. These parameters are controlled much more tightly than in the case of monochrome grades. Additionally, the dielectric coating is formulated with blends of several polymers and pigments—often with twice the number of ingredients as for monochrome papers. The greater complexity is reflected in the price of these dielectric grades.

Translucent

Translucent electrographic papers are designed for use as masters for diazo reproduction. Therefore, their design and price points are focused on achieving a suitable image quality in the reproduction process while allowing for easy generation of multiple copies. A secondary use is for overlays and tracing, where images must be seen through several— usually four to five—sheets of paper. Accordingly, the critical parameters of translucent papers are optical transmission to UV and visible light, good line quality, and physical strength. They are generally about 80 gsm in weight and have high density but low tear strength.

Because these papers are typically very brittle, their useful life as a diazo master depends upon the care taken in handling. For example, creasing and folding will lead to stress fractures, which propagate tears. Creases also scatter UV light, forming defects in the diazo copy.

Translucent dielectric papers are made in the same manner as tracing papers, i.e., by using highly beaten pulps, large quantities of starch and synthetic polymer, and extensive calendering to ensure they are free of any opacifying component or artifact. As a consequence, the base sheet is quite sensitive to changes in relative humidity, and can show a dimensional change of up to 3% with ambient changes from 20% to 80% RH. Prone to curl and cockle, they are unsuitable for drawings with precise dimensions.

The dielectric layer for translucent papers must be relatively transparent to UV light and thus is formulated with a low pigment content (P:B ratios of between 0.5:1 to 0.8:1) and without opacifying agents. Calcium carbonate is the primary pigment, but low concentrations of silica and talc are sometimes used. As the primary use is monochrome line printing, image quality expectations are relatively low.

Vellum

Vellums used in reprography are made from cotton (rag or linter) fibers. These are much more durable than wood fibers. Drawings made on vellum can be stored for very long periods, hence it is the preferred substrate for archival keeping. Other unique properties include high folding endurance, excellent tensile strength (a measure of toughness), good dimensional stability, and high resistance against crease formation. With these characteristics it is preferred as a master in the diazo process when many copies are to be produced.

Vellum derives its translucency from a transparentizing process wherein the raw paper is saturated with oils like low molecular weight polystyrene or polyalphamethyl styrene. Traditional transparentizing uses solvents (e.g., xylene) to transport the oil into the paper on a squeeze roll coater. A drawback is that oil present in the base sheet raises solvent retention, giving this dielectric grade the tendency to exhibit a stronger solvent odor. One maker uses an aqueous, hence solvent-free, transparentizing treatment that enables the addition of conductive agent early in the process. This gives the product additional volume conductivity.

Dielectric formulations for vellums differ little from those used for translucents. As might be expected, imaging performance is similar, although the slightly higher level of volume conductivity delivers higher image densities.

Semitranslucent Paper

The high cost of vellum and translucent papers created an opportunity in the market for a sheet based on chemical pulp that could be priced closer to opaque papers yet feature only minor compromises in overall performance. These semitranslucent papers are made from moderately beaten wood fiber. The degree of defibrillation is higher than standard opaque papers but much less than for translucents. It is made without any opacifying pigment to keep the UV light opacity at moderate levels yet sufficiently opaque to provide adequate contrast after imaging.

Conductivizing is achieved to a greater degree when performed on the paper machine because the sheet then has a modest level of porosity, reducing the demand for a heavy conductive coating on the surface of the sheet. The dielectric coating composition is similar to that used for translucent paper.

Overall imaging performance of the semiopaque papers is very good for CAD work, but one drawback is slower diazo reproduction speed. Nevertheless, the lower cost of this grade resulted in its widespread adoption.

8.6.2 Lightweight Color Papers

Early developments in electrography focused on lightweight papers for color CAD. At the time, they were called *premium* or *color report* grades, and they functioned as the preferred material when purity of color, whiteness of background, and opacity sufficient for direct viewing were all called for—a role they still fill for CAD. Color electrography was also adopted quickly by the seismic industry. This use placed two additional requirements on the paper that were instrumental in making it suitable for graphic arts. The first was freedom from defects. Voids and starvoids could lead to misinterpretation of the seismographs alerting plant personnel to surface contaminants and pickouts at all phases of the manufacturing process.

The second was adjusting the paper to the stop–start mode of the printer. In these early days, the computers storing and transmitting the data could not match the speed of the printer. Thus the printer would print in a stop–start mode. While stopped the toner particles on the paper needed to stay in place and not be "washed" away by the continuing flow of toner fluid. Additionally, soluble materials in the dielectric could be extracted into the toner and poison it. A solution for this problem came through dielectric coating formulation and toner modifications.

Color report grades are now the everyday papers of the graphic arts segment of the electrographic industry—the paper used when bond or "plain" paper would be called for in other modes of printing. Dominant uses in this market are volume low end graphics (in-store advertising), and for checking images prior to printing on more expensive grades. They may also be used for the wet transfer process—see Section 8.6.6.

8.6.3 Premium Color (Presentation) Papers

Soon after the first color photographs were reproduced by electrography the deficiencies of the existing dielectric papers were recognized. There arose a demand for high-performance papers comparable to the premium offset printing grades.

Presentation grade papers were specifically developed to meet this need. They were introduced at different basis weights, first 100 gsm, then 80 gsm, later 150 gsm (100 lb), and most recently, 120 gsm. They are characterized by high whiteness, exceptional formation, high consistency, low artifact levels, and premium image quality, properties obtained from their heritage of photographic paper making. Typical properties for these papers are shown in Table 8.1. Presentation papers are the image quality leaders. With them, printing is most consistent, gamuts are broadest, mottle and grain are least observed, and other visual nonuniformities such as striations are minimized.

Table 8.1 Typical Properties of Presentation Grade Papers

Presentation grade	80	100	120	150
Basis weight (gsm)	80	100	120	150
Thickness (μm)	73	88	110	125
Density	1.10	1.15	1.10	1.2
Burst	180	250	300	400
Stiffness				
MD	20	40	90	150
CD	35	80	150	250
Opacity (%)	80	85	87	92
Brightness (%)	92	94	88	94
Moisture content (%) (at 50% RH)	7.0	7.0	6.8	6.8
Suppliers to North American market	Rexam	Rexam	Rexam	Rexam
	Sihl	Sihl	Sihl	

Overlaminating adds greatly to image intensity and snap—it is normal for photore-productions. Stiffness is a significant aid to laminating. Yield losses from attempting to laminate lightweight papers can often greatly exceed the anticipated cost savings that might be realized by using the lighter paper. The heaviest grade is preferred for applications where lighter grades must be laminated, not for esthetic reasons, but merely for stiffness and durability. However, this stiffness is also a drawback because it makes runnability more difficult in some printers.

Higher basis weight papers have some water-resistance, retaining sufficient strength to be actually handled when wet, especially if the time of exposure is short. Thus some use of these grades for wallpapers, murals, and even outdoor posters has developed, although extra special care is required for these applications.

A wide range of applications has developed with these papers. Any use requiring true color photoreproduction is best handled with this family. Examples are point-of-purchase advertising, standees, and backlit advertising (with two images mounted together in perfect register to obtain the additional image density).

8.6.4   Wet Strength Papers

There are three basic markets for digital color printing that require wet strength papers: billboards, maps, and wall coverings.

Billboards

Electrography lends itself to billboards and other outdoor advertising applications. It has good resolution, has good color, and can produce customized prints for short runs—even single images or "one offs"—cost effectively. This application was envisioned soon after the first graphic arts capable systems had been assembled. However, the papers and toners available at the time were totally unsuitable. The color report grade papers had no wet strength, and the color toners faded and bled or ran when wet. If the opportunity was to be realized, a paper that, once printed, could be immediately used by the existing installation industry was needed. Simultaneously, toners suitable for the outdoors had to be introduced.

The issues were solved together in a manner characteristic to the industry: Cactus, a VAR, brought together the materials manufacturers (Specialty Toner Corp., Chartham Paper Mill, and James River Graphics) and a potential end-user (Gannett Outdoor). The result was both a paper, namely, Rexam Graphics' Outdoor Poster Grade (27), and a toner (Specialty Toner Corp.'s Weather Durable®) that met the standards of the billboard industry.

The critical properties for a billboard grade include good imaging performance, good wet strength, controlled wet expansion, and some level of paste absorption. Obtaining wet strength in paper is well known—a chemical that will provide covalently cross-linked bridges between the fibers is added at the wet end or size press. Styrene maleic anhydride and Kymene® (a polyamide-polyamine epichlorohydrin adduct wet strength resin offered by Hercules) are two commonly used reagents. These react with the cellulose when heated during drying and so provide strength whenever water loosens the hydrogen bonds that normally give paper its strength. The minimum wet burst of paper that can withstand the common practice of prepasting is around 100 lbs.

Changes in the manufacture of the base paper are needed to obtain paper with such performance. It is typical to cut refining back, but this can be at the expense of solvent holdout and hence loss of image quality relative to presentation papers. A modest degree of paste absorption ensures that the paper will remain bonded to the surface it is applied to when rained upon. Nonabsorbent papers tend to slide off when put up with the potato starch pastes most commonly used in outdoor advertising. This is actually more important than having wet strength once the board is installed.

Opacity must be high to hide the colored material remaining on the billboard face from prior advertisements, when a new face is being installed. The requirement is not just for dry opacity, which is comparatively easy to obtain, but for wet opacity, which is much more difficult; the opacity should be at least 90%.

Maps

Cartographic applications have somewhat different requirements for a dielectric paper substrate. The attraction of electrography becomes apparent when it is realized that geographical databases are always in a state of flux. Because new information is always being generated, maps can go out of date even before the proofs are prepared. In some instances, such as military events, obsolescence can occur in hours. Additionally, these databases are now digital. Map printing directly from such databases is an ideal that is attainable provided the toner, the paper, and the final image can all stand up to the rigors of the end use.

Traditionally, quality maps are printed on paper with a moderate degree of wet strength so they can be read in the rain. Outdoor poster grade has some suitability for mapping, but the existing 36-inch printers that have been adopted by the cartographic industry do not handle it. Nor does it have the fine surface structure needed to reproduce fine lettering. Some success has been possible with a wet strength 100 gsm presentation grade. However, most electrographic maps are printed on presentation imaging grades and laminated for extreme service. There remains open a good opportunity for a dielectric paper specially designed for map making.

Wallpaper

Wallpapers require wet strength for installation only, unless washability is essential. A grade specifically designed for wallpaper has been marketed. It supplements outdoor poster and heavyweight presentation grades that have been used with moderate success. These papers, protected by a layer of Rexam's Fluorex® film (see Section 6.6) have demonstrated high functionality as wall coverings in kitchens and bathrooms.

8.6.5 Toner Transfer Paper

Toner transfer is an operation in which the toned image alone is transported, intact, from one substrate to another. The first electrographic transfer process was the Scotchprint® vinyl printing system. It is deceptively simple: an electrographic paper that weakly binds toner and is printed in mirror-image form, mated with a suitable substrate (usually a vinyl), and subjected to heat and pressure enabling the toner to transfer to the substrate. The vinyl print is then overlaminated (Fig. 8.22). The image is thus highly protected, trapped between the substrate and the overlaminate. The process as developed is well-defined in the patent literature (51). It has been extremely successful.

As in many cases, what appears simple in concept has many complications in the

(a)

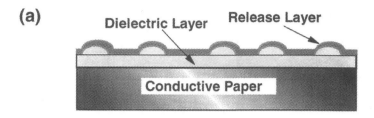

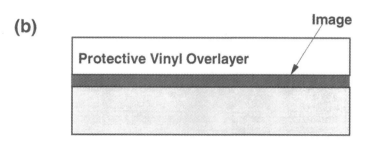

(b)

Figure 8.22 Schematic representation of 3M Scotchprint materials: (a) transfer media with toner release coating over dielectric layer; (b) final product with the image encapsulated between the substrate and the overlaminate.

details. The critical factors in the toner transfer process are balancing opposing forces at each step. The first of these is to have a surface that will hold toner firmly during the multiple writing and toner steps, yet will release the toner completely in the transfer process. There are two elements to this—chemical and physical. The surface of the dielectric needs to have a low surface energy, but not so low that it has no affinity for the toner. This is achieved by coating a thin layer of a release agent, such as Syl-Off® (51) or silicone urea block copolymers (52,53), over a dielectric paper with good charge retention properties—the surface energy being controlled by the ratio of nonpolar dimethyl siloxane groups to polar urea elements. The toner properties are such that it does not fix upon drying but remains weakly bound. It is the combination of both the surface layer of the dielectric and the properties of the toner that ensures this response.

The release surface must remain with the dielectric coating during both the toning step and the transfer process. Thus it must be insoluble in the toner carrier fluid, and it must be well-fixed or integral to the dielectric layer. The silicone urea block copolymer satisfies these requirements, whereas commercially available release agents, when used in this application, became contaminated with residual silicone, which poisons the toner.

The weak adhesion of the toner to the paper raises the need for physical protection to be provided by the dielectric paper to ensure that the image is not degraded by scraping from machine elements encountered along the paper path. The spacer particles in the dielectric layer serve this purpose. The test is most severe for the thickest toner layers— not just secondary colors but tertiaries bearing the richest hues. Making the particles large enough to ensure that the thickest layers are not in jeopardy, however, makes the surface excessively rough—too rough for the best imaging, and too rough for the total transfer to the substrate.

The ability to print without scratches varies among the different printer designs. The original single-pass machine had this capability with little modification. The Xerox 8900 family has also proven reliable. Both have the paper wrapping the writing head with pressure from a compliant back element. The Raster Graphics printers have been prone to scratches on all materials, but most especially for toner transfer, because there is no compliant element at the writing head—both the head and the back element are comparatively rigid, so that the paper cannot distort when a large particle is to be passed. This has been improved with the DCS 5442, which has the web wrapping the head, while a thin stainless steel back electrode serves as the compliant member.

After printing, the image on the release paper is fragile and must be handled carefully. It is preferred to leave the paper on the wind up core in the printer, although sometimes it may be unwound and reviewed for manual touch-up of imperfections.

Transfer of the toner to the receptor substrate is effected with both heat and pressure. Originally this was done with a flat-bed hot laminator (51), but commercial operations now use a hot roll laminator. The transfer takes place in the nip between two heated rolls. One is covered with silicone rubber, while the other may either be rubber-covered or chrome steel. A relatively high pressure is applied by air cylinders at each end of the roll pair. Electrical elements inside the rollers provide the heat source.

One of the difficulties has been maintaining a comparatively constant temperature across the roll face; it is normal to use a roll at least 8 to 10 inches wider than the material to be transferred to avoid the end zone where the temperature drops off. There are separate supply shafts for the imaged paper and the substrate. Twin take-off shafts wind up the substrate with the image and the used release paper.

The most fundamental requirement of the receptor is that it makes good contact with the toner layer in the laminator. Extremely high pressures must be used to ensure 100% surface contact, even with a compliant layer. This paper is not recommended for toned image transfer onto a rigid substrate such as polycarbonate, but compliant materials, such as polyvinyl chloride, are suggested. A suitable coating on the receptor sheet can also improve transfer efficiency.

The substrate itself requires a receptive surface with both proper surface energy and glass transition temperature to bond the toner firmly. It also must have the ability for at least the surface layer to flow when hot so that the toner layer becomes one with the new substrate. Standard calendered or cast polyvinyl chloride has generally been unsuitable and not given sufficient adhesion. Special thermoplastic receptors have been necessary. Coatings of polymers of ethylene acrylic acid, methylmethacrylate, or butylmethacrylate can give the proper qualities, while vinyls can be modified with acrylics or vinylacetates to do the same. Both approaches have been utilized.

The thickness of this coating is adjusted to compensate for the rigidity of the receptor substrate—pressure-sensitive vinyls have thin layers, while more rigid materials have much thicker layers, sometimes as much as one-third the thickness of the paper itself. A good description is provided in the patent literature (54).

After transfer the image is still susceptible to damage from abrasion, making an overcoat or overlaminate essential. Overlamination is the predominant process. Clear vinyl films, with either heat seal or pressure-sensitive adhesives, can be applied by the same laminator used for the transfer step. When pressure-sensitive laminates are selected, the transfer and overlamination can be done simultaneously in one process step. The overlaminate is chosen to suit the application.

When the image is to be in a gentle environment, for example in point-of-purchase advertising, a light 3-mil clear vinyl is appropriate. However, for heavy duty work such as fleet or floor graphics, 7- or even 10-mil thick laminates may be taken. The durability of the laminate has been matched by the light fastness of the toner.

Overlaminated prints on self-adhesive vinyl have found numerous applications in the advertising world, such as murals, trade show displays, corporate logos, sporting events, museum exhibits, trucks, vans, trains, airplanes, and many others. A new material has been designed for use on windows whereby the graphic is seen from outside while from inside the view is essentially unrestricted. When applied on transportation vehicles, with proper material selection, the images can last for five years without appreciable fading, cracking, or other deterioration caused by environmental factors. The concept of wrapping buses with printed vinyls, first developed by and commercialized by SuperGraphics, Inc. (Sunnyvale, CA), has been especially successful. By renting the bus space to advertisers, public transit agencies are deriving hefty revenues (up to $8,000 per month) which help the agencies keep the bus fare low. It is a situation where everyone wins, including the bus passenger.

8.6.6 Image Transfer Papers

Image transfer is the term used to describe methods where both the image and the dielectric coating that supports it are transferred to a new substrate. Two approaches are used—the simple dry transfer process and the more complicated wet method.

Dry Transfer

Dry transfer printing (55,56) is similar to toner transfer in that the image is printed in mirror form on a special paper—3M's Image Transfer or Rexam's Wear Coat. The image is transferred to a substrate with a laminator, but both the dielectric and the image are transferred. The image is then sandwiched between the substrate surface and the dielectric (Fig. 8.23). The product from this one-step process is then ready for use in environments where the dielectric layer provides sufficient protection for the image.

Dry transfer paper has the same basic structure of dielectric media—a conductive substrate carrying an insulating coating—but there is a release layer under the dielectric.

The release layer is a critical element of the structure. Its surface energy has been carefully matched with the dielectric so that the bond is weak. It has the integrity to hold the dielectric in place during manufacturing and printing operations, but it will separate if the dielectric is bound to another surface. There have been two variants of the release layer. The first generation of image transfer used a thin, high surface energy, silicone release coating. While this was highly functional, it also acted as a dielectric material and therefore reduced the image density achievable with useful thicknesses of the dielectric layer.

The second generation of dry transfer paper uses a proprietary combined conductive release coating that greatly increases the apparent density of the imaged product. Resolution is not changed—the transferred dot remains essentially the same size on transfer (45). Grain and other microscale nonuniformities can be reduced.

Further, the conductive release layer can be applied at much higher coat weights; it can be laid down with many different textures, which are then reproduced in the surface of the wear coat on transfer. Glossy, sheen, and flat surfaces can be produced (23,24).

(a)

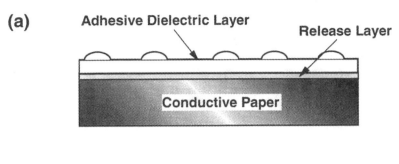

(b)

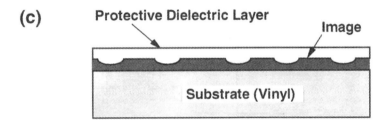

(c)

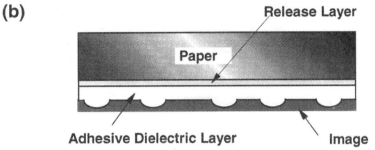

Figure 8.23 Schematic representation of the Rexam Dry Transfer dielectric paper, showing (a) the special adhesive dielectric and release layer, (b) imaged paper ready for transfer, and (c) the final product after the carrier sheet is removed. The image is encapsulated between the substrate and the thin transparent dielectric coating.

The transferred dielectric layer functions to protect the image for short periods, of approximately 30 days duration. Image durability is good and particularly useful for promotional applications, such as banners, signs, and advertising. When protected from intense sunlight and harsh abrasion, it has shown little loss of integrity even after outdoor exposure for more than two years.

Nevertheless, regular dielectric coatings, especially if formulated to act as hot-melt adhesives, are not very durable. They degrade under UV light, can harbor mildew in damp environments, and are prone to scratching from even mild abrasion. They also tend to offset if folded or rolled—a fact that induced most banner producers to overlaminate dry transfer prints.

A solution to this was introduced by Rexam (57). In the new product, a thin layer of Rexam's Fluorex, a durable polymer alloy, is applied on the release coating, and this in turn is coated over with a thin dielectric layer. The combined layer has the thickness

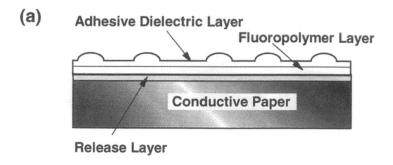

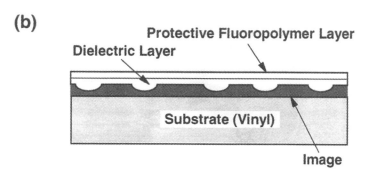

Figure 8.24   Schematic representation of the Rexam Dry Transfer Paper (a) with Fluorex layer on top of the conductive layer, and (b) with the protective Fluorex layer exposed.

of a standard dielectric. When transferred, it is now the Fluorex layer that is exposed to the elements (Fig. 8.24). This layer raises the durability of the coating dramatically—abrasion resistance increases tenfold, UV degradation is cut to less than 1/10th, chemical resistance is greatly improved, nonsolvent graffiti removers can be used to remove paint and other materials, and the coating is mildew resistant.

Wet Transfer

We transfer takes advantage of the fact that the common quaternary ammonium polymers used to conductivize most dielectric papers are water soluble. When wet, these cause failure of the adhesive bond between the dielectric layer and the base paper. The typical process starts with direct imaging onto the substrate. An overlaminating film is then applied to the image surface in one pass through a laminator. This may be either a hot or a cold laminating film. The paper is then passed through a water saturating station, such as a bath, which weakens the dielectric bond. The paper can then be stripped off and wound up on a separate core. The overlaminate/image/dielectric sandwich is wound up separately. Usually, there is residue from the paper on the back side of the dielectric that must be wiped off continually. In a second step, the dielectric is bonded to a substrate. The whole process is shown in Fig. 8.25.

Wet transfer has the advantage of using off-the-shelf direct imaging papers, producing the same image as the direct write paper and less expensive imaging paper. On the other hand, in the most common approach it is two step process, requires an overlaminate, and is typically messy.

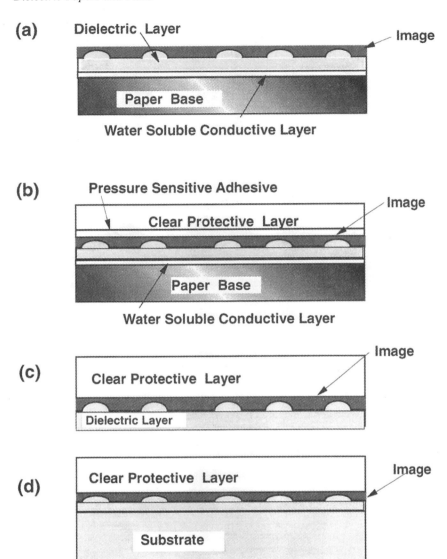

Figure 8.25 Schematic representation of the steps of the wet image transfer process, showing (a) imaged dielectric paper, (b) imaged dielectric paper overlaminated with protective film, (c) image and protective laminate remaining after wetted paper is stripped off, and (d) final product mounted on substrate. Note that the image is encapsulated between substrate and overlaminate.

8.6.7 Sublimation Transfer Printing Papers

Printing fabrics by dye sublimation transfer is a traditional process for fine silks, limited edition blouses, white goods, and other items. Dyes that will sublime or diffuse at high temperature are printed onto paper by either offset or serigraphy (silk screen) as a mirror image. The paper and fabric are mated and exposed to about 350° to 400°F for about 30 seconds in a transfer press. The paper is peeled off and the fabric is thereby imaged. The technology is, of course, limited to combinations of fabrics and dyes that will function this way. Polyester is the staple of the industry. This plastic softens at a suitably low temperature and can absorb the dyes that are mobile, again at that temperature. The dye, in vapor form, penetrates, or becomes dissolved in the polyester fiber in a process called imbibition printing. Once trapped inside the PET strand, the dye cannot be attacked, making it highly resistant to laundering and dry-cleaning.

This is another market area that is fertile ground for on-demand, digital color printing—the textile industry. In this highly fashion-oriented market, styles and colors change rapidly. The long lead times required for prepress or makeready operations and the extended run lengths needed for these methods to be economically feasible make electrography an attractive alternative.

The appeal of a fully digital process is obvious. Electrography has been adopted readily because it has appropriate image quality for many applications and good productivity; and the dyes can be incorporated into liquid toners. There have been a number of systems in Japan, and a handful in the U.S., supported first by Synergy and then by Nippon Steel (58), printing this way since the early 1990s. Recently, the technology has been picked up by value-added resellers, and about 30 systems have been placed with either Xerox or Raster Graphics printers. They have been concentrating on flags, banners, proofing, and uniforms, and other applications where polyester textiles are functional or already in use.

It is not textiles alone that can receive the dye. Certain polyester-based coatings are available, such as STC's Soltex™, that enable the transfer of images by dye sublimation onto everything from wood, vinyl, and wallpaper to ceramics, textiles, and metals.

Electrographic sublimatic technology is just what is expected by printing with electrography and transferring traditionally. A summary of the process is outlined here.

Printing

Software used for apparel design and fabric prints has been adapted to electrography. The patterns are printed in a regular manner although the optimum printer operation can be quite different from processes using pigment toners—the image as produced is muddy-looking, and true colors are not apparent until the transfer is made. Color matching has a distinct art component, although advances in software are simplifying this step.

Printers

Resolution is important to most fabric printers, as fabric is usually seen at close proximity. Therefore 400 dpi has become the norm, although 200 dpi is entirely satisfactory for banners, flags and furniture.

Toners

The toners are similar in structure to other electrographic toners, i.e., dispersions of pigments with polymers and charge directors in odorless mineral spirits. The pigments, of course, are the same as or similar to those dyes used for traditional sublimatic printing.

Transfer Presses

The two types of transfer press are roll fed and sheet fed. The sheet-fed units have a stationary hot press and a fabric carrier on which the imaged paper and the fabric are placed. The carrier is pushed into the press, which closes automatically for the preset time. It is then pulled out, the treated sample is removed, and a fresh combination is put down. The primary problems with this type of press are paper curl before the press closes, and image blurring when the image shifts across the fabric surface just when the press is opened, but before the temperature has dropped sufficiently to condense the gaseous dye.

Roll-fed presses are normal for large runs of printed material. Curl is not an issue. However, this process rapidly raises the paper temperature so that there will be a propensity for blistering with heavier paper, especially if it has a modest water content. Hence, lighter weight papers have not only a cost advantage, but also a performance advantage for roll-to-roll transferring. However, superior imaging on the fabric has given a preference for 100 gsm presentation grade. The blistering problem can be overcome with a simple radiant heater in the web path between the printed paper unwind and the hot roll. This can bring the paper moisture content down to from 4 to 5%.

The two process steps that especially stress the imaging paper are image generation and hot transfer. The transfer process requires that the paper not curl uncontrollably on entering the press nor blister in it. It should not shrink independently of the fabric, as the image will appear diffuse. Excessive adhesion of the dielectric coating to the fabric must be avoided for obvious reasons, and finally the paper should not transfer objectionable odors or chemicals in the press. In fabric printing, background from any cause is problematic: "White must be white." Thus low stain is a requirement, and a total absence of nib writing is an ideal. Both of these are functions of the dielectric coating, the toner, and the printer writing voltage. Low dot gain is desired also—it is achieved by avoiding conditions that promote flare and by keeping writing voltage low.

The ideal paper is one with low moisture content, to avoid blistering and shrinkage; low refining, to give good dimensional stability; good conductivity, both surface and volume; a thin dielectric layer having a uniform roughness level; and a low-temperature adhesive. This ideal leads to many conflicts and compromises: dimensional stability, for example, is better for coarser fiber papers that are weakly refined, but such papers have poor coating holdout and uneven surfaces, so that they produce images with poor small-scale uniformity.

8.6.8 Self-Adhesive Papers

Direct imaging, self-adhesive papers have utility for posters and advertisements. They are simple in construction (Fig. 8.26) with a direct imaging paper laminated to a pressure-

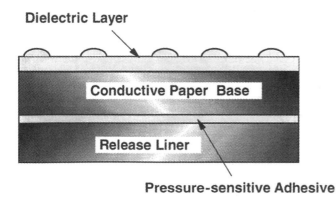

Figure 8.26 Schematic diagram of self-adhesive dielectric paper. Removing the release liner exposes the adhesive on the paper.

sensitive adhesive coated onto a release liner. After printing, the image is trimmed to size, the release liner removed, and the print installed. Repositionable adhesives are preferred, as the paper must be applied dry and will tear if the bond is too tight. The compromise in printing has been leading edge fog, as release liners and pressure-sensitive adhesives are not good conductors. This is solved by separating the writing and toning passes.

The size of the market has yet to be determined. The competitive methods include laminating liner and adhesive to imaged stock after printing—most of the laminate manufacturers offer such goods.

8.7 PRESENT TYPES OF DIELECTRIC FILMS AND FABRICS

The dielectric paper substrates described thus far have a conductive agent dispersed or diffused completely through it, so that the conductive layer under the dielectric coating is in electrical contact with electrodes on the back side. This section describes the structures of dielectric films and fabrics and the challenges that have been met in adapting them to electrographic imaging. The main focus is on polyester film, because that technology has been fully described in the patent literature.

8.7.1 Polyester Films

Historical Overview and Markets

Polyester films (polyester terephthalate, or PET), when compared to paper, offer clarity, dimensional stability, and durability. Additionally, they have sufficient tensile strength to be transported through the printer without distortion—a limitation for many otherwise useful plastics. These are properties desired by the seismic industry, where huge amounts of data are obtained and which were being analyzed by visual comparison of massive seismographs, laid one atop another. This market sector provided a major impetus for the development of an electrographic film product. Development began seriously in 1979, with most of the work in the United States occurring at GAF (59,60) and Eastman Kodak (38). At the same time, a plotter modification was necessary to handle and print film successfully. Versatec succeeded with the introduction of a suitable printer-plotter in 1980.

The major printing obstacle was the large electrical barrier of the film in the backgrounding configured printers—there can be no electrical path through the film because it is an insulator. There have been efforts to make a film with suitable volume conductivity, but these have failed because the full range of properties has not been obtained, while the compromises have been unacceptable.

However, modeling studies showed that fast capacitive coupling could be effected if the footprint of the back electrode was much larger, and higher write voltages were used. Once the relative ratios of the electrodes was determined and the higher write voltages needed were established, the program moved from ''bootleg'' to ''sanctioned,'' and a working electrostatic film plotter appeared shortly thereafter. It had an enlarged, flexible backplate electrode, high writing potential (up to 750 V), and new electronic architecture (write time and pulse amplitude), features that have not only permitted the use of electrographic film but also given printers the capability to image nonstandard paper grades. Today, a separate ''film'' mode is a standard option on electrographic printers.

Front-grounded plotters were able to accommodate film more readily because they relied upon capacitive coupling through the dielectric. However, the interaction of the dielectric pigment with the electrodes was an issue demanding careful design.

In addition to the seismic market, electrographic film found wide usage in the reprographic industry where high speed led to substantial displacement of pen plotters as CAD placements grew. Both clear and matte films were provided—the matte type was preferred because it could be altered or overdrawn by hand, but both could serve as diazo masters.

The bulk of the film used for CAD was made with water-soluble conductive polymers. Unfortunately, there were a few incidents in which film plots—thought to be archival—were destroyed by water from floods, pipe breaks, or sprinkler accidents. The media manufacturers then invested in electronic conductor layers, so that the current films are all archival.

A special application for film arose in the aerospace industry, where film templates are used as masters to control large cutting tools. High precision drawings are made on polyester film. The traditional process with pen plotters was very slow—electrographic plotters offered a massive time saving, but with two important criteria: media that is flat and straight so that highly accurate drawings could be made on it; and plotters that would guide the film and image on it with the desired accuracy. Both requirements taxed the plotter manufacturer, the media suppliers, and the polyester film manufacturers. Nevertheless, a moderate degree of success was obtained, and the approach was implemented.

The final area that film attempted to enter was full-color printing for graphic arts applications. Some spectacular prints were produced on film made with white opaque base.

Polyester Film Structure

Polyester electrographic films share the same five structural elements (Fig. 8.27):

1. A PET substrate
2. A conductive layer with surface electrical resistivity from 1 to 5 MΩ/$\square$
3. A dielectric layer over the conductive layer
4. Conductive grounding stripes at each edge
5. An antistatic backcoat

Substrate

The polyester is typically 3 to 4 mils thick, although 7 mil film is used for precision plotting. It provides the high-temperature stability, low moisture absorption, durability, high tensile strength, resistance to tear, and dimensional stability required for film applica-

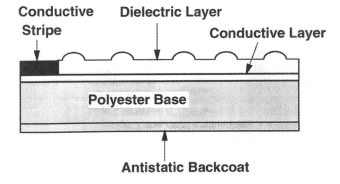

Figure 8.27 Schematic diagram showing a dielectric film structure.

tions. Optical clarity is a requirement for clear films but not for matte. Adhesion promoting bond coats are provided on each side to ensure the coatings are properly bound. Suitable films are offered by DuPont Teijin Films, Kimoto, Mitsubishi, and Toray.

Conductive Layer

The surface resistivity of the conductive layer typically falls into a 3 MΩ/$\square$ wide band somewhere in the region of 1 to 10 MΩ/$\square$. It must be situated under the dielectric layer for the process to function. However, because it is there it must have other properties as well: it should be colorless, essentially transparent for clear film, ensure the dielectric layer remains firmly in place (good adhesion with the base and the dielectric), and it must not detract from the other performance attributes of the medium. Of these, surface resistivity has been the most difficult property to control.

Six chemical families have been used to lend conductivity to dielectric films: quaternary ammonium polymers, polystyrene sulfonic acid or its salts, polyanilines, indium doped tin oxide (ITO), copper doped tin oxide, and antimony doped tin oxide (ATO).

Early on in dielectric film development, quaternary ammonium polymers were already in use with conductive papers. Polystyrenesulfonic acid and its sodium salt were likewise in use by the film coating industry for antistatic purposes. Thus the extension of both to conductive films was a natural step. There are many issues, though, in manufacturing quality imaging materials with water-soluble materials, not the least of which is their unacceptable behavior when wet (the dielectric layer and hence image can slide off the film!).

The conductivity of these layers is dependent upon the inherent conductivity of the chemical mix, its thickness (coat weight), the content of water in the layer, and the uniformity of the coating layer itself. The first is a simple design issue, but the loss of conductivity when binders are added was a major impediment to providing an archival cost-effective film. Coat weight is a normal manufacturing parameter, but an exceptionally high level of uniformity is necessary—these are thin layers in which small differences will show up as imaging artifacts.

The most difficult aspect of achieving and maintaining proper conductivity has been ensuring consistent moisture content. Unlike paper, polyester film is neither a source of water nor is it permeable to water vapor. Thus the required moisture level must either remain in the layer after coating (generally beyond the capability of most film coaters) or be added to a separate operation by passing the film through a steamer. Unfortunately, moisture gain varies with changes in web speed, steam level, temperature, and dielectric thickness. The goal was never adequately met. Additionally, the film could pick up or lose moisture at the end user site if not properly stored and handled.

Coupled with these factors is the propensity of the conductive polymers to induce image defects, for example, by migrating through hydrophilic pathways in the dielectric layer (8), or from gels in the coating process. Quaternary ammonium polymers and other water-soluble conductives were thus quickly abandoned when better and more cost-effective technology became available.

Kodak introduced electrographic film with a thin polyaniline conductive treatment (38). It produced good images and performed reliably, especially as it is an electronic conductor and thereby independent of water content and relative humidity. This Kodak product was the first truly RH-independent electrographic imaging medium. The primary drawback of the polyanilines is their green/grey color. That fact, coupled with high cost, limited the film to a small sector of the market.

A film with similar properties was made using sputter coated, copper doped tin oxide (37). This dielectric film had an attractive golden tint, but its commercial life was short due to the introduction of films coated with a slurry of antimony doped tin oxide (ATO).

Ref. 3 describes film made with ITO-coated silica. This chemical is a reliable semi-conductive material with few factors other than composition affecting its electrical properties. It is water insoluble and compatible with many binders. The primary impediment to its use in electrographic films was its particle size as supplied. The powder is opaque and contains particles that can protrude through the dielectric layer. Thus special grinding and dispersion techniques are needed to reduce the material to a much finer powder. Particle diameters must be less than 5 µm to avoid penetrating the dielectric layer, and even smaller to prevent light diffraction in the layer. For this, the particle size must be under 100 nm. Both Atheron (9) and Katsen (10) describe methods for size reduction.

ATO is now the conductive treatment of choice, having been commercialized by the three dominant manufacturers: Kimoto, Arkwright (9), and Rexam Graphics (10).

Dielectric Coatings

Critical requirements of the dielectric layer are the same as any paper substrate: high charge acceptance, low charge decay, proper spacing, proper nib cleaning without excessive generation of artifacts, low stain, toner adhesion, protection of the wet toner layer, and adhesion to the conductive base. On top of this, there may be special needs to satisfy the end use application. For example, CAD films must accept common drawing office pens and pencils, be erasable, and also be tolerant of desk accidents such as spilled coffee. The formulations developed over the years have been continually refined to meet these challenges, and they have by and large succeeded.

Dielectric coating resins (often with plasticizer) are essentially the same for films as those used in paper coating. They generally comprise 80 to 90% of the coating mass, as a low P:B ratio is essential to good clarity. In addition to accepting and holding the charge, the resins serve to bind the toner.

A mixture of pigments—such as precipitated or ground calcium carbonate and amorphous or crystalline silica—is normal. Plastic pigments may also be used for clear films (12) because, like the silica, their refractive index is similar to the resins, so light scattering is minimal. Blends are chosen to give the proper relationship between abrasion and spacing. Although profilometry is the most precise measurement of smoothness and pigment concentration, Sheffield smoothness is still the common roughness test. Typically a roughness of 100 to 120 mL/min is optimum; it ensures a 6 to 10 µm gap for nib discharge during writing. It also protects the wet toner leaving the toning station. As stated earlier, there should be sufficient abrasion to maintain the nibs in a good firing state while at the same time minimizing the frequency of highly abrasive events from oversize or excessively hard particles, which result in flare and spurious nib writing.

Different approaches to this balance are described in the patent literature (9–11,54). The Atherton (9) approach tends to have the abrasive event leave the nib with good emissivity for several firings, because it relies on a moderate concentration of hard materials. The Katsen (10) approach prefers to have a greater number of less abrasive events, each capable of ensuring emissivity for only one or two firings. Dot size analysis demonstrates the lower frequency of flaring and hence the better image quality. The drawback with plastic pigments is their low abrasiveness. Dot dropout is a common issue. A sufficient content of fine amorphous silica in the coating left on top of the larger pigments can provide the necessary microabrasion to prevent line dropout.

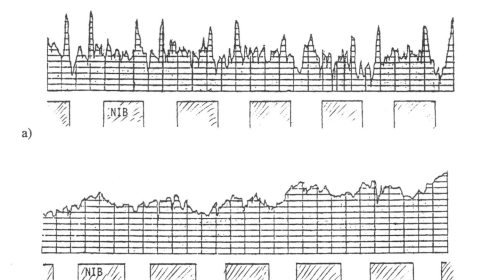

a)

b)

Figure 8.28 Effect of high and low abrasion pigments on the writing head: (a) new film with high-density silica; (b) old film.

Edge Grounding Stripes

The insulating nature of polyester film blocks direct grounding of the countercharge that forms in the conductive layer on deposition of the latent image. Therefore, to prevent capacitive grounding via the toner fountain and the generation of leading edge fog, an alternative grounding method is needed. All electrographic films require edge stripes of highly conductive ink that can provide a pathway between the conductive layer and the printer ground. The usual stripe composition contains carbon black dispersed in a dielectric resin binder. The SER of this layer is typically 104 to 105 $\Omega/\square$. A grounding surface is provided in the printer to contact the image side of the film. Printers with drums perform better if the stripe is also on the film end (called *wraparound striping*), so that contact between the conductive layer and the back is assured. Prior to the introduction of ping-pong writing it was advantageous to have an unbalanced relationship between the stripes on the side where image writing began and where it ended (61).

The grounding stripes themselves are only effective if they can remove the charge at a rate comparable to that at which it is formed without the potential over the toner fountain rising by more than a few volts. There are two factors that determine the effectiveness of stripes: the relative distance from the center of the film to the edge compared with that to the toner station, and the speed of the printer. With the introduction of wider printers, the problem grew worse; higher printer speed also exacerbated the problem. Xerox solved this in the 8500 Series printers by applying a controlled positive potential to the stripes, the so-called *ghostbuster* technology (62).

For the Raster Graphics DCS 5400 series it is customary to minimize fog by separating the writing pass from the toning pass. This gives sufficient time for the surplus charge to bleed off so that good quality images can be obtained. A similar practice is embodied in the high-speed Xerox printers but is possible only with materials that exhibit good charge retention.

Single-pass printers such as the Synergy machines do not have the same ability to bleed off charge via edge stripes, simply because they were never designed to run dielectric film. When attempts were made to write on film, leading edge fog appeared to a degree proportional to image area coverage. Images with large solids invariably showed heavy leading edge fog. The problem could be eliminated by installing bias plates at each end of the writing head and applying about +10 to +20 volts DC. Several printers were so adapted for specific customers.

Back Side Coatings

Polyester films, like most plastics, generate static when transported across many materials. This static will stay with the film for extended periods of time—as much as months. When the film is passed through the toner development system of the plotter, toner will be attracted to the areas where the static resides, causing an objectionable background pattern. More familiarly, polyester film also draws dust and dirt, while in large diazo duplicating machines it may cause jams, because the film master and the copy paper fail to separate. To eliminate static buildup it is customary to apply an antistatic coating to the reverse side (60). This is the case not only for clear and opaque films but also for those with a matte coating. Provided the SER is about 108 to 1010 $M\Omega/\square$ at 50% RH, static is usually not an issue. The quaternary ammonium nitrate compound Cyastat SN® (American Cyanamid) is a common antistatic agent. ATO can also be used (10).

Prevention of blocking and improvement in transport requires a degree of roughness that the drive rolls can grip. Amorphous silica provides this in many antistatic formulations that use a resin binder to bond particles to the film. Most film coaters have antistatic antiblocking coatings suitable for clear and opaque electrographic film.

Matte coatings can be applied to the back for drafting applications. The coatings are usually derived from drafting film compositions and modified only to accommodate different antistatic levels, opacity, or transport requirements.

8.7.2 Pressure-Sensitive Vinyl Films

Vinyl films are ubiquitous—they are used for advertising, decoration, protection, and direction. Direct imaging on pressure-sensitive adhesive vinyl has been an ideal; developments with ATO conductive materials have solved some of the critical performance factors allowing such materials to be introduced. Penetration of the transfer technologies into signage, flags and banners, fleet graphics, and other wide-format markets brought the market to a scale sufficient to make direct-write vinyl films affordable.

Structure

The structure of pressure-sensitive vinyls is similar to electrographic film except that a composite of vinyl film, pressure-sensitive adhesive, and release liner replace the polyester. The structure comprises the following elements (see Fig. 8.29):

1. The dielectric layer and edge stripes
2. The conductive layer
3. The vinyl film
4. The pressure-sensitive adhesive
5. The release liner

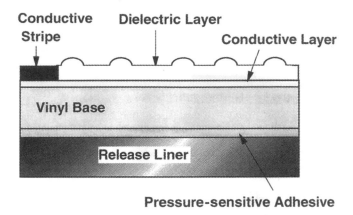

Figure 8.29   Schematic diagram showing the structure of direct-write, self-adhesive dielectric vinyl.

The structure is typically assembled on the release liner. A pressure-sensitive adhesive is first coated onto the liner, and in the same operation the vinyl film is laminated to the liner using a pressure-sensitive adhesive. Because the vinyl has a high surface energy, the adhesive bonds preferentially to it. Next, the conductive and dielectric coatings are applied. Conductive striping, slitting, and rewinding into finished rolls complete the process.

Release Liner

Release liners are generally paper materials of about 50 to 70 lb with a release layer on one side. The release layer is a low surface energy silicone polymer often coated from monomer and cured by radiation. The silicone layer may be applied directly to the paper or on top of another coating. The paper may be coated on both sides with polyethylene coatings to make it waterproof and hence suitable for wet installing. The cost is related to the quality of materials used and whether the liner is water resistant.

Adhesive

There exist a wide variety of pressure-sensitive adhesives that serve many different demands. The adhesive may be "repositionable" so that the vinyl can be moved around during installation until the position is proper. Controltac®, a product of 3M, features weak adhesion until the position is finalized; then permanent bonding is achieved using heavy pressure. Options exist for removable or permanent bonding, as well as for special surfaces such as glass. One of the dilemmas facing the manufacturer of electrographic vinyl is the choice of adhesive.

Vinyl Layer

A white cast vinyl is the preferred substrate (the value added by conversion to an electrographic medium does not warrant the use of calendered vinyl). This material is made by dissolving polyvinyl chloride in a solvent such as MEK (methylethyl ketone) and casting it (coating) onto a heavy release coated paper. This material has a low tensile strength and is elastic, so that it does not lend itself to traditional solution or dispersion coating. Instead, the paper-backed vinyl is brought together with the adhesive-coated release liner, and the backing upon which the vinyl is cast is then removed.

Conductive Layer

The vinyl is opaque, permitting the use of a water-insensitive conductive layer based upon a pigment such as doped tin oxide. It is not necessary to obtain a haze-free coating as for clear polyester films. An average pigment particle size of about 0.5 μm is preferred for such films versus the 0.1 μm of transparent coatings. Without the requirement for opacity the formulator can also run to higher coat weights, giving the benefit of higher binder contents and tougher coatings.

Dielectric Coating

The opacity of these vinyl films affords some latitude to the formulator of the dielectric layer. These can be similar to paper coatings with higher pigment loadings and easier topological control.

Conductive Grounding Stripes

Conductive striping is similar to those used on polyester film.

Performance

The use of direct-write vinyl films has never blossomed. Image quality is generally useful, although spurious nib writing and flare are issues. Leading edge fog is often seen, and although it can be dealt with by separating the writing and toning sequences, this solution doubles the printing time. Only a limited range is available. Transfer techniques are still needed to obtain electrographic images on most vinyl types and for special applications such as window film.

Manufacturer

Kimoto and 3M (with a product developed by Azon) (63) have entries in the direct-write pressure-sensitive vinyl market. A similar product, using opaque polyester, has also been discussed (64).

8.7.3 Direct Imaging Dielectric Fabrics and Nonwovens

A direct-write polyester fabric was brought to market by Xerox but later withdrawn. This material was intended for banners and flags but (1) lacked sufficient image quality, (2) was generally considered to be priced too high, and (3) various transfer techniques proved a better alternative.

du Pont's popular Tyvek® can also be made into a direct-write material (65).

8.8 FUTURE OUTLOOK

Liquid-toner-based electrographic printing, as portrayed in this chapter, remains an important imaging process among competing technologies. Compared to ink jet printing, for example, it is faster, has better image permanence, and generally costs less in terms of imaging material (ink jet ink versus toner). Keeping this in mind, and despite the demise of certain printer OEMs (e.g., CalComp, Phoenix Precision Graphics, Precision Image, and Synergy) there is no doublet that this technology will continue to serve many applications in the present decade.

To its advantages we must add the availability of a wide selection of dielectric papers and films and the growth of sublimation transfer applications. Today's polyester-based receptor coatings have made it possible to place near-photographic quality images on textiles, garments, drums, tambourines, floor coverings, ceramic tiles, shower enclosures, pools and spas, and other architectural surfaces. Image permanence is a proven fact, as electrographic prints withstand sunlight exposure, weather, and swimming pool and spa chemicals.

Admittedly, ink jet printers are available at a fraction of the cost of electrographic machines, but their low productivity is a drawback for many print shops and service bureaus. The impending solution for ink jet—the pagewide array printhead—will certainly increase imaging speed but will also bring a quantum leap in hardware cost. Thus the economics of wide-format printing will likely continue to favor electrography.

Wide-format, digital color printing continues to exhibit rapid growth. It dominates a few key applications and promises to capture many others where it has yet to be tried. Electrography is still being used creatively, although it has become a mature technology.

One cannot ignore the existing installed base of electrographic printers, a vast population that requires large quantities of consumables. This business volume is sufficient to sustain innovation and support ventures into novel applications. Indeed, it will be a long time before the last electrographic printer is retired.

ACKNOWLEDGMENTS

The knowledge we are sharing here was accumulated over the years we have spent with electrography. Much of it came from informal, open, and frank discussions with our colleagues from throughout the industry. It is too large a task to recognize all of these people directly. We trust they will accept our thanks this way.

In the preparation of the chapter we can recognize people who have been especially cooperative and who have ensured that their organizations have been supportive: Vern Rylander at 3M Commercial Graphics, Frank Perry and Frank Shah of Xerox ColorgrafX Systems, Rak Kumar of Raster Graphics Inc., Don Balbinder of Hilord Chemical, Romit Bhattacharya of Specialty Toner Corp., Boyd Jones and Gene Day of Phoenix Precision Graphics, and Rich Himmelwright of Rexam Graphics.

Dene Taylor wishes especially to recognize Don Brault, his colleague at Rexam Graphics, who foresaw the market that exists today, who knew it could not happen without media, and who drove for the provision of them. The industry would not be the same without his vision.

REFERENCES

1. L. Michaylov. *Handbook of Imaging Materials* (A. S. Diamond, ed.). Marcel Dekker, 1991.
2. J. L. Johnson. *Principles of Nonimpact Printing*. 2d ed., Palatino Press, 1992.
3. R. A. Work, III, et al. USP 5,192,613, 1993. Electrographic recording element with reduced humidity sensitivity.
4. D. A. Brault et al. USP 5,601,959, 1997. Direct transfer electrographic imaging element and process.
5. F. J. Ragas et al. USP 4,339,505, 1982. Electrographic coatings containing acrylamide polymers.
6. A. D. Brown, J. Blumenthal. USP 3,657,005, 1972. Electrographic record medium.
7. A. D. Brown, J. Blumenthal. USP 3,711,859, 1973. Electrographic record system having a self spacing medium.

8. R. P. Lubianez, E. Bennett. USP 4,656,087, 1987. Dielectric imaging sheet through elimination of moisture induced image defects.

9. D. Atherton, S. K. Gadodia, M. E. Gager. USP 5,126,763, 1992. Film composite for electrostatic recording.

10. B. J. Katsen et al. USP 5,399,413, 1995. High performance composite and conductive ground plane for electrostatic recording of information.

11. T. Maekawa et al. USP 4,795,676, 1989. Electrostatic recording material.

12. T. Sugimori et al. USP 4,752,522, 1988. Electrostatic recording material.

13. D. H. Taylor et al. Effects of media surface topography on the effects of images printed by the electrographic process. Proceedings of IS&T's Tenth Non-Impact Printing Congress, New Orleans, 1994.

14. W. C. Meyer. Tappi 57, 86 1975. Understanding of electrical conductivity–humidity relationships of electroconductive resins.

15. L. H. Silvernail, M. W. Zembal. USP 3,011,918, 1961. Electroconductive coated paper and method of making the same.

16. C. H. Lu, J. S. Chow. USP 5,130,177, 1992. Conductive coating compositions.

17. D. M. MacDonald, L. H. Deed. USP 4,024,311, 1977. Electroconductive paper coating.

18. T. Ohmae et al. USP 4,919,757, 1990. Aqueous dispersion of cationic polymer.

19. K. Iwamoto et al. USP 5,234,746, 1993. Conductive substrate and printing media using the same.

20. R. H. Windhager. Tappi 57, 75, 1974. The importance of barrier coatings in conductive base stock manufacture.

21. R. H. Jansma, D. W. Holty. Tappi 58, 96 1975. Binder selection for conductive coatings.

22. R. H. Windhager. Tappi 64, 91, 1981. One pass conductive coating colors for reprographic grades.

23. D. H. Taylor et al. USP 5,759, 636, 1998. Electrographic imaging element.

24. D. A. Cahill et al. USP 5,869,179, 1999. Imaging element having a conductive polymer layer.

25. H. J. Bixler. USP 3,639,162, 1972. Electroconductive coating.

26. R. M. Levy et al. USP 3,653,894, 1972. Electroconductive paper, electrographic recording paper and method of making same.

27. A. N. Fellows. USP 4,336,306, 1982. Electrostatic imaging sheet.

28. P. Wacher. USP 5,240,777, 1993. Electrostatic recording media.

29. P. Wacher. USP 5,360,643, 1994. Electrostatic recording media.

30. K. W. Barr, D. V. Royston. USP 4,739,003, 1988. Aqueous conductivising composition for conductivizing sheet material.

31. W. K. Barr, D. V. Royston. USP 4,868,048, 1989. Conductive sheet material having an aqueous conductive composition.

32. C. V. Willetts et al. USP 5,385,771, 1995. Outdoor poster grade electrographic paper.

33. E. W. Sawyer, F. J. Dzierzanowski. USP 3,9694,202, 1972. Paper containing electroconductive pigment and use thereof.

34. H. Mikoshiba et al. USP 5,225,273, 1993. Transparent electroconductive laminate.

35. Matsushita News. MEP-79-8. March 15, 1979.

36. H. R. Linton. USP 5,236,737, 1993. Electroconductive composition and process of preparation.

37. J. B. Fenn, Jr. *Vacuum Deposited Films for Reprographics: An Update*. Ninth Annual Specialty Papers and Films Conference and Tutorial, Diamond Research Corp., Santa Barbara, CA, 1991.

38. D. A. Upson, D. J. Steklenski. USP 4,237,194, 1980. Conductive polyaniline salt-latex compositions, elements and processes.

39. W. K. Goebel, D. M. Rakov. USP 4,920,356, 1990. Electrographic recording receiver.

40. T. Oki, K. Iwamoto. USP 5,384,180, 1995. Electrostatic recording medium.

41. B. J. Katsen. USP 5,158,849, 1992. Process for preparing stable dispersions useful in transparent coatings.

42. B. J. Katsen. USP 5,210,114, 1993. Process for preparing stable dispersions useful in transparent coatings.
43. D. R. Roisum. Tappi 79 (10), 217, 1996. The mechanics of wrinkling.
44. G. F. Day. *Ultra-Precise Dot Placement: A Breakthrough in Electrostatic Imaging.* Third Color Imaging Conference, Diamond Research Corp., Santa Barbara, CA, 1989.
45. D. H. Taylor. *Relative Image Quality of Direct and Transferred Full Color Images Printed by the Electrographic Process.* Proceedings of IS&T's Eleventh Non-Impact Printing Conference, Hilton Head Island, 1995.
46. J. J. Bakewell. USP 4,415,403, 1983. Method of fabricating an electrostatic print head.
47. P. A. O'Connell. USP 4,766,450, 1988. Charging deposition control in electrographic thin film writing head.
48. L. K. Hansen et al. USP 5,061,948, 1991. Electrographic marking with modified addressing to eliminate striations.
49. T. McBride, R. S. Himmelwright. Image uniformity for electrographic images. Proceedings of IS&T's Twelfth Non-Impact Conference on Non-Impact Printing, San Antonio, 1996.
50. R. J. Gable. USP 5,200,770, 1993. Background from an electrographic printer through modulated off states.
51. H. Chou et al. USP 5,262,259, 1993. Toner developed electrostatic imaging process for outdoor signs.
52. P. J. A. Brandt et al. USP 5,045,391, 1991. Release coatings for dielectric substrates.
53. P. J. Wang et al. USP 5,106,710, 1992. Release coatings for dielectric substrates.
54. R. S. Steelman et al. USP 5,852,121, 1998. Electrostatic toner receptor layer of rubber modified thermoplastic.
55. D. A. Cahill et al. USP 5,483,321, 1996. Electrographic element having a combined dielectric/ adhesive layer and process for use in making an image.
56. D. A. Cahill et al. USP 5,488,455, 1996. Electrographically produced imaged article.
57. T. M. Chagnon et al. USP 5,688,581, 1997. Electrographic image transfer element having a protective layer.
58. H. Soga et al. USP 5,159,356, 1992. Web printing apparatus.
59. H. Burwasser, J. B. Wyhof. USP 4,112,172, 1978. Dielectric imaging member.
60. H. Burwasser. USP 4,287,286, 1981. Toner repellant coating for dielectric film.
61. H. Yamauchi et al. EPA 439 177 A2, 1991. Electrostatic recording material.
62. L. K. Hansen et al. USP 5,055,862, 1991. Film ghost removal in electrographic plotters by voltage bias of the plotter fountain or film edge strip.
63. T. L. Morris, W. A. Neithardt. USP 5,736,228, 1998. Direct print film and method for making same.
64. K. Furugawa. JPO Appln. HEI 3[1991]-69960, 1991. Electrostatic image recording adhesive sheet.
65. Anonymous. Research Disclosure, May 1990. Coating compositions for dielectric printing.

9

Photoreceptors: The Chalcogenides

S. O. KASAP

University of Saskatchewan, Saskatoon, Saskatchewan, Canada

9.1 INTRODUCTION

The commercial importance of amorphous selenium (a-Se) and its various alloys at present lies in their use as xerographic photoreceptor materials (e.g., Se–Te alloys and As_2Se_3) and more recently as x-ray photoconductors in x-ray imaging though, in the past, crystalline selenium has had successful applications in photocells, solar cells, and rectifier diodes. In a much smaller quantity, amorphous Se–Te–As alloys are also used in Hitachi's Saticon TV pickup tubes (Goto et al., 1974; Maruyama, 1982). The xerographic photoreceptors over the last decade have been progressively using more organic photoconductors rather than selenium alloys, and this trend is expected to continue (Schein, 1988; Springett, 1989, 1994). Some large-volume copying applications still use a-Se alloys since they provide many copies per drum. Another challenge to the chalcogenide photoreceptor comes from a-Si:H photoreceptors, which have good sensitivity in the red and IR regions and exceptionally long machine lifetimes, as is discussed by Mort (Chapter 16) and Joslyn (Chapter 11) in this handbook. Recent research on x-ray imaging systems utilizing the x-ray sensitivity of a-Se photoconductors, however, suggests that the x-ray photoconductor usage is likely to experience substantial growth, as will be discussed later in this chapter. There are currently a number of potential applications for selenium-based amorphous semiconductors in high-sensitivity TV pickup tubes, called the HARPICON (Tanioka et al., 1988), in large area x-ray sensitive vidicons for medical imaging, called the X-icon (Luhta and Rowlands, 1991), in ELIC (electrophotographic light-to-image converter) imaging devices (Kempter et al., 1983), in optical storage (Koshino et al., 1985; Matsushita et al., 1987), in IR fiber optics (Klocek et al., 1987) and in optical recording of images via selective photodeposition of a-Se films from a colloid (the Selor process), as demonstrated by Peled and coworkers (Peled and Dror, 1988; Peled et al., 1992).

The xerographic process and the basic photoreceptor requirements for xerography are summarized in this handbook by Borsenberger and Weiss as well as by numerous authors in the past (see, e.g., Tabak et al., 1973; Mort and Chen, 1975; Schaffert, 1975; Weigl, 1977; Berger et al., 1979; Mort, 1984; Scharfe, 1984; Williams, 1984; Pai and Melnyk, 1986; Burland and Schein, 1986; Mort, 1989; Pai and Springett, 1993). There is also an extensive literature on the physics and technology of selenium based photoreceptors (Berkes, 1974; Cheung et al., 1982; Springett, 1984, 1988). In its simplest form, a xerographic photoreceptor consists of an amorphous selenium (a-Se) alloy film deposited by vacuum evaporation techniques onto an aluminum drum. Vitreous selenium alloys are normally thermally evaporated from long stainless steel boats onto heated cylindrical aluminum substrates, which are rotated during deposition. Typically the a-Se film is 50–70 μm in thickness and 100 cm^2 in surface area, though areas as large as 1 m^2 are used in the largest machines. The fabrication of many photoreceptor drums requires special vacuum coaters that can accommodate a large number of drums (e.g., 50 or more) and can achieve the required film composition and xerographic characteristics with high yield.

Many aspects of selenium alloy photoreceptor fabrication are poorly understood, especially in the case of Se–Te and Se–As based alloys, where fractionation effects dominate (e.g., Berkes, 1974; Schottmiller, 1975; Springett, 1988). There are still a number of photocopying and printing machines that use either a-As$_2$Se$_3$ or a-Se$_{1-x}$Te$_x$ alloy films. Amorphous Se$_{1-x}$Te$_x$ photoconductors contain various amounts of Te alloying and are either in monolayer form or in multilayer geometries. The major reason for alloying with Te is to shift the spectral response of a-Se toward the red region of the spectrum to match the photoreceptor spectral sensitivity with the efficiency of the light source used in the copier or printer. In laser printer applications, the laser light wavelength is invariably beyond the visible red region and demands appreciable Te alloying to bring the photosensitivity of the photoreceptor into the source wavelength band. With as high as ~25% Te alloying, many additional problems arise in the xerographic performance of the photoreceptor, such as rapid dark discharge and high residual potentials, which must be overcome. The solution has been to utilize multilayer photoreceptor geometries incorporating a selenium-rich charge transport layer (CTL) on the Al substrate, a Se–Te alloy based photogeneration layer (PGL) on the CTL, and a Se–As alloy type protection or overcoating layer (OL) on the PGL. In addition, various amounts of halogenation are used to attain specific xerographic criteria. These multilayer geometries have interesting xerographic properties that have been investigated during the 1980s, as is discussed below (Kiyota et al., 1980; Taniguchi et al., 1981; Cheung et al., 1982; Melnyk et al., 1982; Tateishi and Hoshino, 1984). The advantage of a-As$_2$Se$_3$ photoreceptors is that they have a wide spectral photosensitivity extending from the blue region to ~700 nm.

9.2 PROPERTIES OF SELENIUM AS AN IMAGING MATERIAL

9.2.1 Structure of Selenium

The structure of amorphous selenium has been the subject of many discussions in the literature, inasmuch as for a long time it was believed that the amorphous phase consisted of selenium chain, Se$_n$, and 8-ring, Se$_8$, and perhaps 6-ring, Se$_6$, structures mixed together.

This model essentially arose from the fact that in the crystalline phase, selenium can exist in two forms, α-monoclinic Se (α-Se) and trigonal Se (γ-Se). The former has Se_8 rings and the latter Se_n chains, and therefore it was quite natural to seek a structure for the amorphous phase based on a mixture of ring and chain members. Recent structural studies on selenium and its alloys however favor a "random chain model" in which all the atoms are in twofold coordinated chain structure and the dihedral angle ϕ is constant in magnitude but changes in sign randomly (Lucovsky, 1979; Lucovsky and Galeener, 1980; Feltz, 1993).

The dihedral angle, as illustrated in Fig. 9.1, is defined as the angle between two adjacent bonding planes. Its definition therefore involves four atoms, say 1, 2, 3, and 4, so that it is observed as shown in Fig. 9.1 by looking down the bond connecting atoms 2 and 3. In the crystal, the positions of all the atoms are fixed by the symmetry and the bond length r and the bond angle θ; consequently the magnitude of ϕ is constrained as a function of bond length and angle. In the trigonal form, the dihedral angle rotates in the same sense in moving along a chain to give a spiral pitch of three atoms. In the Se_8 molecular unit, however, its sign alternates in moving around the ring. Thus in a-Se, the change in the sign of the dihedral angle ϕ leads to regions that are ringlike or to regions that are chainlike, depending on a particular sequence of ϕ. If + or − is used to indicate the relative phase of the dihedral angles between adjacent bonding planes, then a sequence of the type $+ - + -$ has been termed a ringlike and a sequence of $+ + +$ or $- - -$ chainlike by Lucovsky (1979).

The local order shown in Fig. 9.2, for example, can be characterized as $+ + + - + - + - - -$. This model, which assumes only local molecular order within a selenium chain, has been used successfully to explain the vibrational spectra of a-Se to account for the presence of various Se_8-like spectral features in the infrared absorption and Raman scattering spectra without invoking a mixture of Se_n and Se_8 members for the structure. Other structural studies of a-Se, in particular those by Meck (1976), Robertson (1976), and Long et al. (1976), generally support the random chain model.

An important common feature of nearly all the chalcogenide glasses is the fact that these materials contain thermodynamically derived charged structural defects, called valence alternation pairs (VAP), which correspond to some of the chalcogen atoms being

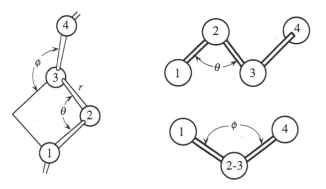

Figure 9.1 Selenium chain molecule and definition of the dihedral angle ϕ. The definition involves an angle between planes and thus four atoms labeled 1, 2, 3, and 4. It is observed looking down the bond joining atoms 2 and 3.

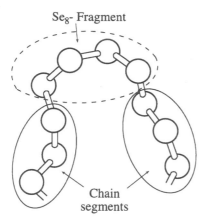

Se₈- Fragment

Chain
segments

Figure 9.2 Local molecular order in a selenium chain in which there are segments characterized by repetition of the same dihedral angle (''chainlike'' in the sense of trigonal Se) and segments characterized by alternating dihedral angles (''ringlike'' in the sense of Se_8 molecules). (From Lucovsky, 1979.)

under- and overcoordinated (Kastner et al., 1976; Adler, 1977; Fritzsche, 1977; Kastner, 1977, 1978; Feltz, 1993). The absence of electron spin resonance (ESR) signal in chalcogenide glasses (Abkowitz, 1967; Agarwal, 1973) has indicated that the lowest energy defect in these glasses does not have a dangling bond. The lowest energy defects are charged centers in the structure that have their electrons paired. For example, in the a-Se structure, the Se_2 is the lowest energy configuration for the Se atom and represents the normal bonding configuration. The lowest energy structural defect, however, is not a singly bonded neutral Se atom, Se_1^0, or a triply bonded neutral atom, Se_3^0, but a pair of charged centers of the type Se_1^- and Se_3^+. If the atoms of the pair are in close proximity, they will form an intimate valence alternation pair (IVAP).

The VAP model is essentially based on the fact that it is energetically more favorable to form a diamagnetic pair of charged over- and undercoordinated chalcogenide centers, Se_1^-, Se_3^+, than to form paramagnetic singly or triply coordinated defects, Se_1^0 or Se_3^0. The latter are unstable. For example, a dangling bond, Se_1^0, can lower its energy by approaching the lone pair on the normally coordinated Se_2 atom and generate an IVAP. The diffusion of the resulting species can further reduce the Gibbs free energy. Thus the reaction $Se_1^0 + Se_2^0 \rightarrow Se_1^- + Se_3^+$ is exothermic because lone pair electrons have been absorbed into dative bonding. Figure 9.3, a schematic representation of a typical a-Se structure with VAP centers, illustrates the nature of lowest energy defects in chalcogen glasses.

Many photoelectric properties of a-Se and its alloys can be at least qualitatively explained by using concepts based on VAP or IVAP centers and interconversions between the diamagnetic charged centers and the paramagnetic defects. The physics of such processes has been extensively discussed in the literature (see, e.g., Mott and Davis, 1979; Elliott, 1984, 1986). Their existence and the possible defect reactions that can occur in the structure have led to many important predictions and much insight into the behavior of chalcogenide semiconductors. For example, the linear dependence of the steady state photoconductivity on the light intensity for a-Se has been interpreted via photoinduced IVAP-type centers (Carles et al., 1984).

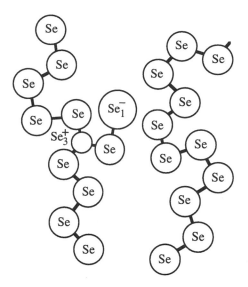

Figure 9.3 Schematic illustration of the amorphous selenium structure showing valence alternation pair (VAP) defects; Se_3^+, Se_1^-.

9.2.2 Optical and Electrical Properties

During the 1960s and 1970s the optical and electrical properties of a-Se films were extensively studied by many authors. The experimental research at that time coincided with the lack of proper understanding of the physics of amorphous semiconductors, and many of the papers contained models and interpretations that were not fully applicable. Nonetheless, the wealth of experimental data was later put into perspective by various authors (Davis, 1970; Mott and Davis, 1979; Elliott, 1984). The optical absorption coefficient of a-Se exhibits an Urbach edge, of the form (Hartke and Regensburger, 1965) $\alpha = 7.35 \times 10^{-12} \exp[h\nu/0.058 \text{ eV}]$ (cm^{-1}), whereas at high photon energies the absorption coefficient has been found (Davis, 1970; Mott and Davis, 1979; Al-Ani and Hogarth, 1984) to obey $(\alpha h\nu) \sim (h\nu - E_0)$, where $E_0 \approx 2.05$ eV is the optical "band gap" at room temperature. The latter behavior has been attributed to a sharp rise of the density of states at the band edges. A dependence following Tauc's law $(\alpha h\nu) \sim (h\nu - E_0)^2$ has been also found with an optical bandgap E_0 of about 1.9 eV (Adachi and Kao, 1980; Chaudhuri et al., 1983; Nagels et al., 1994, 1996). Although the absorption coefficient indicates considerable absorption at photon energies above 2 eV, the quantum efficiency (defined as the number of electron–hole pairs collected per absorbed photon) has been found to evince a strong field and photon energy dependence even above the fundamental band edge. Figure 9.4 shows the dependence of the absorption coefficient α and the quantum efficiency η on the photon energy $h\nu$, where it is clear that the quantum efficiency reaches a xerographically acceptable value only at high electric fields and photon energies above the fundamental edge.

The mechanism for the field-dependent quantum efficiency observed for a-Se is common to other molecular solids and can be explained by the Onsager theory for the dissociation of an electron–hole pair (Pai and Enck, 1975). In essence, the Onsager theory calculates the probability that an electron-hole pair will diffuse apart under an electric

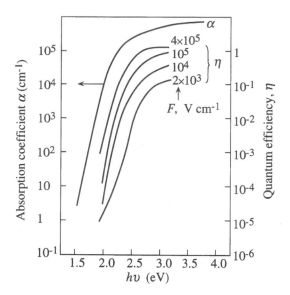

Figure 9.4 Absorption coefficient α and quantum efficiency η, as a function of incident photon energy $h\nu$ at various applied fields F. (Data from Hartke and Regensburger, 1965; Ing and Neyhart, 1972; Pai and Enck, 1975.)

field. The quantum efficiency depends on the electric field F, the temperature T, and the initial separation of the photogenerated electron–hole pair r_0, the thermalization length. The quantum efficiency η is thus given by

$$\eta = \eta_0 f(F, T, r_0) \tag{1}$$

where $f(F, T, r_0)$ is the probability that the electron–hole pair will separate and $\eta_0(h\nu)$ is the quantum efficiency of the intrinsic photogeneration processes. The significance of the field and photon energy dependence of the quantum efficiency is that the xerographic design has to consider the charging voltage together with the spectrum of the illumination to obtain optimal performance from the photoreceptor. The quantum efficiency affects the nature of photoinduced xerographic discharge and thus the contrast potential. For example, the transit time of photogenerated holes across the thickness of an a-Se based photoreceptor is invariably much shorter than the exposure time, which means that the photoinduced discharge mechanism is emission limited (Mort and Chen, 1975). As the field in the photoreceptor decays, the photogeneration rate diminishes, by virtue of the field dependence of the quantum efficiency. Consequently the total number of photogenerated charge carriers is not simply proportional to the exposure, which makes gray scale images more difficult to replicate.

It is instructive to mention that recently Moses (1992, 1996) has challenged the Onsager interpretation of the quantum efficiency in a-Se. He carried out subnanosecond transient photoconductivity experiments using a matched microwave stripline technique and a very short laser pulse (25 ps at 1.8–2.29 eV photon energy). The quantum efficiency as inferred from the peak photocurrent in the subnanosecond time scale shows no field or temperature dependence, in contrast to that inferred from TOF transient photoconductivity and xerographic photoinduced discharge experiments (i.e., as that shown in Fig. 9.5) in the time scale of microseconds and above. The quantum efficiency interpretation in the latter, according to Moses, is a carrier supply yield rather than the intrinsic quantum efficiency. More careful experiments are now needed to clarify the present controversy.

Charge transport and trapping in a-Se and its alloys has been a subject of much interest and research inasmuch as it is the product of the charge carrier drift mobility and the trapping time (or lifetime), $\mu\tau$, termed the range of the carriers, which determines the xerographic performance of a photoreceptor. The nature of charge transport in a-Se alloys has been extensively studied by the time-of-flight (TOF) transient photoconductivity (TP) technique. The principle of the technique is illustrated in Fig. 9.5. The sample is sandwiched between two electrodes, typically Au and Al, the former being semitransparent. Since the external resistance R is much less than the sample resistance, the applied bias V appears across the thickness L of the specimen. A short light pulse of appropriate wavelength photogenerates a packet of electron–hole pairs near the surface of the specimen. The absorption depth $\delta(\lambda) = 1/\alpha(\lambda)$ is chosen to be much shorter than the thickness L. Electrons become neutralized almost immediately by reaching the top electrode, whereas holes drift toward the substrate, generating a transient current in the external circuit R. The circuit time constant $C_{sample}R$ is maintained much shorter than the transit time of the charge packet, $T_t = L^2/\mu V$, across the film, to ensure that the voltage signal across R is proportional to the photocurrent in the specimen. While the holes are drifting in the specimen, there is a photocurrent $i(t)$ in the sampling resistor R, and the shape of the photocur-

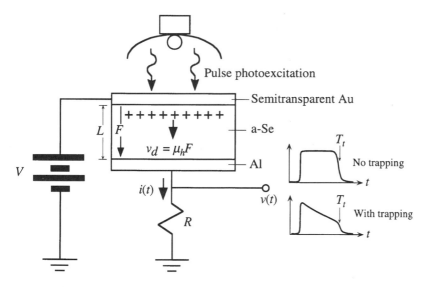

Figure 9.5 Schematic diagram illustrating the principle of time-of-flight (TOF) measurements. The top electrode is semitransparent. Following pulse photoexcitation, electron-hole pairs are generated within an absorption depth $1/\alpha(\lambda) \ll L$. As the holes drift across the specimen, they generate an external photocurrent $i(t)$. The shape of the photocurrent $i(t)$ depends on the nature of trapping within the solid. T_t is the transit time of the photoinjected holes across the sample thickness, L, and is given by $T_t = L/v_d$ where v_d is the drift velocity, $v_d = \mu_h F$, μ_h is the hole drift mobility, and F is the electric field (V/L).

rent waveform at any time represents not only the average concentration of mobile holes but also the mean velocity of the carriers at that instant, that is,

$$i(t) = eAp(t)v_d(t) \tag{2}$$

where $p(t)$ is the mean concentration of mobile holes in the specimen at time t, which is obtained by averaging the hole concentration $p(x, t)$ over the thickness L, and $v_d(t)$ is the drift velocity at time t.

Injected charge is normally much less than that on the electrodes, so that the electric field in the sample remains relatively unperturbed. Consequently, charge trapping and release events involving various distributions of traps in the material result in the transient photocurrent decaying with time. In the extreme case, when the trapping time is much shorter than the transit time and there is insignificant release from the traps over a time scale of the order of T_t, the observed photocurrent is an exponential-looking rapid decay without a transit time. The latter is termed trap-limited response. Many recent papers in the literature extract the distribution of localized states in the mobility gap from the shape of the TOF transient photocurrent. Some of these theories have been successfully applied to characterize the distribution of localized states in plasma-deposited amorphous silicon (a-Si:H) films, which is of interest in a-Si:H photoreceptor design. The virtue of the TOF measurement lies in its ability to monitor the motion of charge carriers across a photorecep-tor film and so to provide a direct evidence of whether the photoinjected charges are making it across the sample.

Furthermore, by using an interrupted field time-of-flight (IFTOF) measurement, it is also possible to study the nature of trapping and release kinetics in the material at any location (Kasap et al., 1988, 1990c). This technique involves interrupting the electric field for a duration t_i during the flight of the photoinjected carriers. The field is removed when the charge carrier packet is at one location and is reapplied at the end of the interruption time t_i to extract the remaining carriers. The fractional change in the recovered photocur-rent represents the trapped charge and allows the carrier lifetime, τ, to be evaluated at that location. Indeed, one can readily determine the variation of the carrier trapping time across a photoreceptor film and correlate this with the composition across the film (Kasap and Polischuk, 1995).

Figure 9.6 shows the temperature dependence of the drift mobility in various phases of selenium obtained mainly from transient photoconductivity measurements; the only exception consists of the drift mobility versus temperature data on trigonal Se (γ-Se), which was measured by the acoustoelectric current as well as the magnetoresistance meth-ods (Mort, 1967; Mell and Stuke, 1967). It is immediately apparent from Figure 9.6 that, except in γ-Se, the drift mobilities at low temperatures in the liquid, monoclinic, and amorphous phases are thermally activated. In particular, the hole drift mobility activation energy in amorphous and α-monoclinic (α-) Se are comparable, though they encompass different temperature ranges. The basic interpretation of drift mobility–temperature data for both a-Se and α-Se has been a shallow trap–controlled transport mechanism in which the hole TOF drift mobility (or the effective drift mobility) is given by

$$\mu_h(T) = \mu_0(T) \frac{\tau_c}{\tau_c + \tau_r} = \mu_0(T) \left[1 + \frac{N_t}{N_v} \exp\left(\frac{E_t}{kT}\right) \right]^{-1} \tag{3}$$

where μ_0 is the microscopic (conductivity) mobility, τ_c and τ_r are the mean capture and the mean release times, respectively, N_t is the shallow trap concentration, N_v is the density

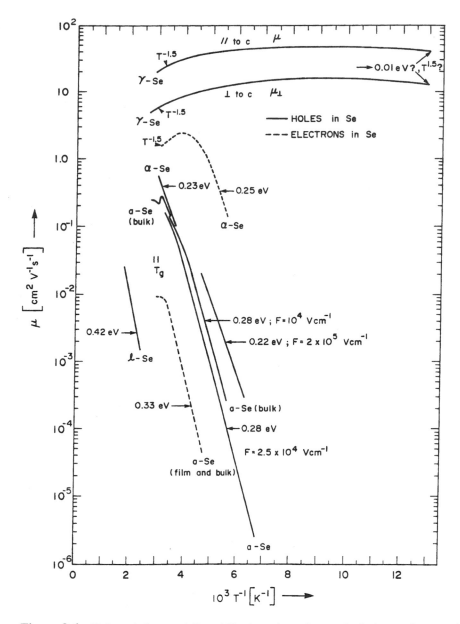

Figure 9.6 Hole and electron drift mobility in various phases of selenium: a-Se, amorphous Se; α-Se, α-monoclinic Se; γ-Se, trigonal Se; *l*-Se, liquid Se. (Data collected from Mort, 1967; Marshall et al., 1974; Spear, 1961; Juska et al., 1974; Marshall and Owen, 1972; Pfister, 1976; Marshall et al., 1974; Schottmiller et al., 1970; Juska and Vengris, 1974; and Abkowitz, 1979.

of states at the valence band mobility edge E_v, and E_t is the energy depth of the shallow traps from E_v.

The nature of the basic microscopic conduction process, whether by extended state transport or by hopping—for example, in the localized tail states—has not been conclusively established. The lack of pressure dependence in the thermally activated drift mobility down to ~230 K has been used by various authors as evidence that the measured drift mobility does not represent a purely hopping transport process (Dolezalek and Spear, 1970). The temperature dependence of the microscopic mobility seems to follow a $\mu_o \sim T^n$ type of behavior, where n is an almost diffusive type of index ($n \approx 1$) (Kasap and Juhasz, 1985).

Figure 9.7 represents the presently accepted model for the electronic density of states for a-Se as developed mainly by Abkowitz and coworkers through various transient photoconductivity and electrophotographic measurements of cycled-up residual and dark discharge (Abkowitz, 1984a–c, 1985, 1987, 1988). There is a wealth of experimental evidence that the localized states, both shallow and deep, in the mobility gap are due to various structural defects that are thermodynamically stable at room temperature (Abkowitz, 1981, 1984a–c; Abkowitz and Markovics, 1984). Almost exponentially decaying shallow trap densities with discrete manifolds at certain energies near the transport bands have been determined from picosecond-resolution transient photoconductivity experiments using microwave stripline techniques (Orlowski and Abkowitz, 1986). The high concentration of traps at the relatively discrete energies ~0.29 eV above E_v and ~0.35 eV below E_c essentially control hole and electron drift mobility as originally proposed by Spear (1957, 1960) and Hartke (1962). Although these traps are known to be native defects, their exact nature has not been conclusively determined. Lucovsky and coworkers (Wong, Lucovsky, and Bernholc, 1985) proposed that they may be due to dihedral angle distortions in the random structure of a-Se in which the lone pair orbitals on adjacent Se atoms approach parallel alignment. The energy distribution of the deep localized hole states with

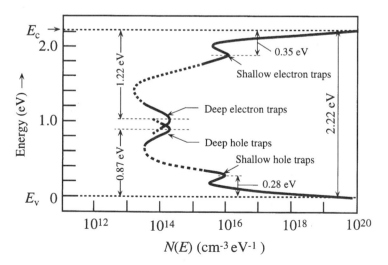

Figure 9.7 Density of slates function $N(E)$ for a-Se derived from various optical, TOF, and xerographic measurements. (Adapted from Abkowitz, 1988.)

a peak around 0.85 eV has been also measured by cycled time-of-flight experiments as described by Veres and Juhasz (1993).

Early TOF measurements were quite successful in identifying the drastic effects of various impurities and alloying elements on the nature of charge transport in a-Se (Kolomiets and Lebedev, 1966; Schottmiller et al., 1970; Tabak and Hillegas, 1972; Pai, 1974; Takahashi, 1979; Kasap and Juhasz, 1982; Takasaki et al., 1983; Baranovskii and Lebedev, 1985; Oda et al., 1986). It was shown, for example, that additions of halogens in small amounts to a-Se completely destroy electron transport. Halogens even in the parts per million range introduce sufficient concentration of electron traps to diminish electron transport. On the other hand, halogenation has been reported to improve the hole lifetime. The impurity effects in a-Se have been reasonably well documented. By appropriate chemical modification, it is possible to obtain an optimum xerographic performance from an a-Se based photoreceptor (Abkowitz and Jansen, 1983; Abkowitz et al., 1985b, 1986; Badesha et al., 1986).

9.2.3 Xerographic Properties

An optimal photoreceptor design will require, among many other factors, high charge acceptance, slow dark discharge, low first and cycled-up (saturated) residual voltages, and long charge carrier ranges ($\mu\tau$). The latter factor has been addressed above. There are essentially three important types of xerographic behavior, generally termed the dark discharge, first cycle residual, and the cycled-up residual voltage, which must be considered in evaluating the electrophotographic properties of a-Se and its alloys. The three xerographic properties are illustrated in Fig. 9.8.

Dark discharge rate must be sufficiently low to maintain ample amount of charge on the photoreceptor during the exposure and development steps. A high dark decay rate will limit the available contrast potential. The residual potential remaining after the xerographic cycle must be small enough not to impair the quality of the electrostatic image in the next cycle. Over many cycles, the cycled up residual potential should also be small, to avoid deterioration in the copy quality after many cycles. In the case of a-Se, these xerographic properties have been extensively studied (Abkowitz and Enck, 1980, 1982, 1983). In addition to the magnitude of the saturated residual voltage, the rate of decay and the temperature dependence of the cycled-up residual potential are important considerations, since they determine the time required for the photoreceptor to regain its first cycle xerographic properties.

Figure 9.9 displays the simplest experimental setup for xerographic measurements. The rotating photoreceptor drum is charged at station A by a corotron-type device. The surface potential is measured at B, and the photoreceptor is then exposed to a controlled wavelength and intensity illumination at station C, following which its surface potential is measured again at station D. In some systems, the surface potential is also monitored during exposure at C via a transparent electrometer probe to study the photoinduced discharge characteristics (PIDC). Normally, the charging voltage, speed of rotation, and exposure parameters (energy and wavelength) are user-adjustable.

Figure 9.10 shows typical positive and negative dark discharge curves for pure a-Se films prepared under different conditions (Schaffert, 1975), where it can be seen that the dark discharge rate depends on the substrate temperature. The presently accepted model for the dark decay in a-Se films is that which involves substrate injection as well as bulk

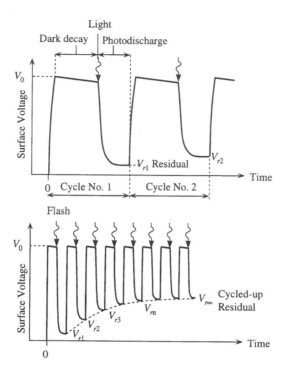

Figure 9.8 Typical photoreceptor behavior through xerographic cycles showing dark decay, first cycle residual potential V_{r1}, and cycled-up residual potential $V_{r\infty}$, after many cycles.

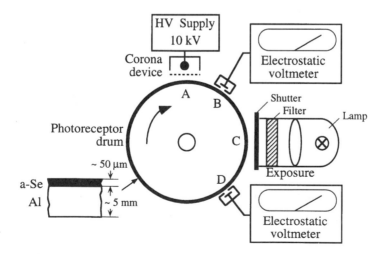

Figure 9.9 Simplified schematic diagram of a xerographic measurement. The photoreceptor is charged at A and exposed at C. Its surface potential is measured before and after exposure at B and D.

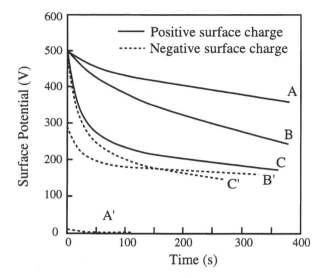

Figure 9.10 Dark discharge of surface potential on a-Se layers. A, B, and C involve a-Se deposition under different substrate temperature (T_{sb}) conditions: A and A′ at T_{sb} = 75°C; B and B′ at T_{sb} = 50–60°C; C and C′ at T_{sb} = 25–50°C and uncontrolled. (From Schaffert, 1975.)

thermal generation and depletion. The latter process involves thermal generation of holes in the bulk and their sweepout from the sample by the internal field, leaving behind a bulk negative space charge (see Section 9.3.2). With thick films and a good blocking contact between a-Se and the preoxidized aluminum substrate, the latter phenomenon dominates. The main reasons for a-Se possessing good dark decay characteristics are (1) there are not many deep localized states in the mobility gap of a-Se; (2) the energy location of these localized states is deep in the mobility gap, so that the thermal generation process of holes (or electrons) from these centers is slow; (3) injection from the substrate can be reduced substantially by using oxidized Al substrates (Zhang and Champness, 1991).

Figure 9.11a displays the buildup of the residual voltage V_{rn} on an a-Se film with the number of xerographic cycles n. If blue light is used for the discharge process, then the absorption is very close to the charged surface, and one can assume that the discharge process involves the transport of photogenerated holes through the bulk. Trapping of these holes in the bulk then results in the observed first residual potential, V_{r1}. In the case of a-Se photoreceptor films it has been found that V_{r1} is well predicted by the simple Warter expression (Kasap et al., 1991a, b)

$$V_{r1} = \frac{L^2}{2\,\mu_h\tau_h} \tag{4}$$

where L is the film thickness, μ_h is the hole drift mobility, τ_h is the hole lifetime, and $\mu_h\tau_h$ is the hole range.

It can be seen from Fig. 9.11a (and also from Fig. 9.8) that as the xerographic cycle is repeated many times at a constant repetition frequency, the residual voltage rises and eventually saturates. The saturated residual voltage $V_{r\infty}$ is much larger than the first cycle residual potential V_{r1}. Both the first residual and the cycled-up saturated residual potential, V_{r1} and $V_{r\infty}$, are sensitive to preillumination as well as to impurities and alloying. For

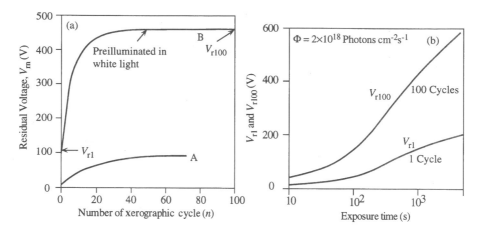

Figure 9.11 (a) Residual potential V_m as a function of the xerographic cycle n for pure a-Se films of comparable thickness ($L = 50$–55 μm). A: dark rested specimen; B: specimen preilluminated with white light. (Data from Abkowitz and Enck, 1982, 1983.) (b) Dependence of the first and hundredth cycle residual potentials V_{r1} and V_{r100} on the exposure time under white light illumination. (From Abkowitz and Enck, 1983.)

example, when a-Se films are preilluminated with white light, the buildup of the residual potential occurs more rapidly toward a much higher saturated residual potential, as indicated by curve B in Fig. 9.11a. Furthermore, the residual potentials V_{r1} and $V_{r\infty}$ increase with the exposure time (Fig. 9.11b). It is clear from Fig. 9.11b that exposure to white light generates an appreciable concentration of deep hole traps. There are essentially two ways of accounting for the saturation of the residual voltage in Fig. 9.11a. First, the observed saturation may be due to the dynamic balance between trapping and release of charge carriers as the xerographic cycle is repeated. Alternatively, it may be due to the filling of the deep trap population so that the saturated residual potential is given by

$$V_{r\infty} = \frac{L^2 e N_t}{2\varepsilon_0 \varepsilon_r} \tag{5}$$

where N_t is the concentration of deep traps and ε_0 and ε_r are, respectively, the absolute permittivity and relative permittivity of the photoreceptor material.

The rate of decay and the temperature dependence of the saturated voltage can be used to obtain the concentration and energy distribution of the deep traps responsible for the residual voltage. Thus $V_{r\infty}$ provides a useful means of studying the nature of deep traps in amorphous semiconductors and has been successfully used to derive the energy distribution of deep localized states in the mobility gap of both a-Se and a-Si:H (Abkowitz and Enck, 1982; Abkowitz and Markovics, 1984; Imagawa et al., 1986).

Figure 9.12 shows the decay of the saturated residual potential on an a-Se film at the end of a large number cycles. As thermal release proceeds, holes are emitted and swept out from the specimen, resulting in the decrease of the measured surface potential. The decay rate of the saturated potential is strongly temperature dependent due to thermal release from deep mobility gap centers, which are ~0.9 eV above E_v for holes. The discharge of the saturated potential due to electron trapping occurs much more slowly, as indicated in Fig. 9.12. The reason is that the energy depth of electron traps from E_c is about ~1.2 eV, which is greater than that of hole traps from E_v.

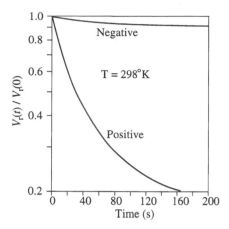

Figure 9.12 The decay of the saturated residual potential $V_{r\infty}$ with time following the cessation of xerographic cycling for positive charging and negative charging. The residual potential at time t is normalized to that at $t = 0$ marking the end of xerographic cycling.

9.2.4 Thermal Properties and Aging (Structural Relaxation)

Inasmuch as a-Se is basically an inorganic polymeric glass, it exhibits many of the properties of glassy polymers. Its glass transformation and crystallization behaviors have been extensively studied by numerous authors. Most thermal studies on a-Se and its alloys have used the differential scanning calorimeter (DSC) type of differential thermal analysis (DTA) measurements, which involves monitoring the rate of heat flow into the specimen as a function of sample temperature while the sample is heated at a constant rate.

Figure 9.13 shows a typical DSC thermogram on a pure a-Se film at a heating rate of $10°C$ min^{-1} where the glass transformation, crystallization, and melting phenomena are clearly visible as endothermic, exothermic, and endothermic peaks, respectively. The glass transformation temperature T_g, the crystallization onset temperature T_o, the maxi-

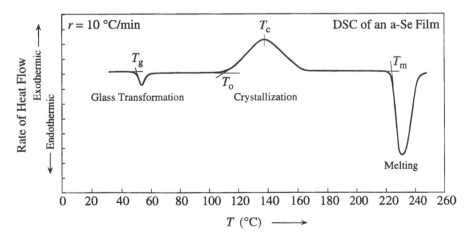

Figure 9.13 A typical DSC thermogram for an a-Se film showing the glass transition region, crystallization, and melting transitions. Heating rate, $r = 10°C/min$. (From Kasap and Juhasz, 1986.)

mum crystallization rate temperature T_c, and the melting temperature T_m are defined in the figure.

The dependence of the glass transition temperature T_g on the heating rate r and the aging time t_A has been investigated by Stephens (1976, 1978) and Larmagnac et al. (1981) and can be readily explained by the enthalpy $H(T)$ vs. temperature behavior of a typical glass-forming material shown in Fig. 9.14. Once the glassy state G has been reached—by, for example, cooling from the melt—and the glass is at temperature T_A, the enthalpy will relax via structural relaxation toward the enthalpy of the metastable equilibrium state $H_E(T_A)$ at A. In the case of a-Se and in a number of other glassy polymers, several days of annealing at the room temperature brings the glass enthalpy close to $H_E(T_A)$, as evidenced by the endothermic peaks in the DSC thermograms at T_g. There is much evidence from a variety of experiments under thermal cycling conditions to indicate that following prolonged annealing (e.g., ~1000 hours) at room temperature, the thermodynamic state of a-Se corresponds almost to that of the supercooled metastable liquid at A (Abkowitz, 1981, 1984a–c, 1985, 1987).

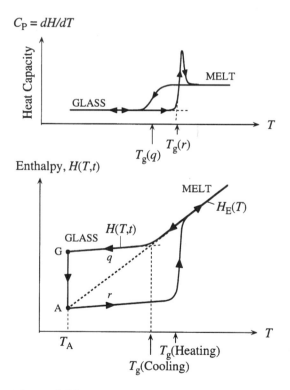

Figure 9.14   Enthalpy H versus temperature T for a typical glass-forming liquid. As the melt is cooled at a rate q, eventually the structure goes through a glass transformation at $T_g(q)$. The enthalpy of the glass is higher than the equilibrium liquidlike enthalpy $H_E(T)$. When the glass at point G at T_A is annealed, the structure relaxes toward the equilibriumlike metastable state at A. On reheating at a rate r, the glass enthalpy is retarded and recovers toward H_E at $T = T_g$ (heating) or $T_g(r)$. The upper curve shows the heat capacity versus temperature behavior observed during heating and cooling.

When the sample at $H(T_A)$ is heated at a constant rate r, the enthalpy of the system, as a result of long relaxation times at these low temperatures, follows the glass H–T curve until the structural relaxation rate is sufficiently rapid to allow the system to recover toward $H_E(T)$. This sharp change in $H(T)$ leads to the glass transformation endotherm in the DSC heating scan. The observed glass transition temperature T_g during heating depends not only on the heating rate but also on the initial state $H(T_A)$. On the other hand, during cooling (at a rate q) from the melt down to low temperatures one observes only a change in the base line, i.e., a change in the heat capacity. T_g during cooling depends only on the cooling rate q (for a-Se, see, for example, Kasap et al., 1990a).

The significance of the glass transformation behavior of a-Se is that as T_g is approached, many of the physical properties of the film exhibit sharp changes as a result of relaxation phenomena. For example, the hole range $\mu\tau$ evinces a sharp drop in the T_g region, which reflects as a deterioration of the xerographic performance. When the photoreceptor is returned to its normal operating temperature, the recovery of the various physical properties occurs over a time scale determined by the relaxation time at that temperature. The mean relaxation time τ for structural relaxations has a Vogel $\tau \sim \exp[A/(T - T_0)]$ type strong temperature dependence (Kasap et al., 1990a, 1990b). Relaxations at room temperature occur over a time scale ~ 1000 hours. The Vogel type of relaxation phenomenon exhibited by a-Se is typical of many organic polymers and is also consistent with the Williams–Landel–Ferry behavior observed for the dielectric relaxation in this material (Abkowitz et al., 1980). Nearly all the physical properties of a-Se evince an aging behavior, which means that the property changes with time as the film is left to anneal isothermally. For example, immediately after deposition, the Vickers microhardness H_V of an a-Se film may typically be about 25 kgf · mm^{-2}, but H_V increases with aging, and after aging for several hundred hours H_V stabilizes at around 35 kgf · mm^{-2}. The basic principle of the aging process can be explained by enthalpy–temperature or volume–temperature diagrams as illustrated in Fig. 9.14, which show structural relaxation toward the metastable liquidlike state $H_E(T_A)$ at A. Aging behavior has been reported for various properties of a-Se, e.g., density, heat capacity, density of structural defects, etc. (e.g., Das et al., 1972; Stephens, 1976, 1978; Larmagnac, et al., 1981; Abkowitz, 1985) and is a fundamental property of all glasses.

Crystallization of a-Se in bulk and film form has been extensively examined (see, e.g., Cooper and Westbury, 1974). It is well known that pure a-Se is sensitive to crystallization and invariably crystallizes at a rate determined by the nucleation process, the morphology of growth, and the temperature. Isothermal crystallization rate has been found to be well described by the Avrami equation. Crystallization data from a variety of experiments suggest that the growth rate during the crystallization process is inversely proportional to the melt viscosity and that the latter can be adequately described over a temperature range by a Vogel type of temperature dependence as originally proposed by Felty (1967). The effects of various impurities on the crystallization kinetics have been reported by many authors, as reviewed by Cooper and Westbury (1974).

Alloying a-Se with As has been found to be very effective in retarding the crystallization rate. In fact, once the As content has reached a few percent, the alloy is almost totally resistant to crystallization (Nemilov and Petrovskii, 1963a, b). During the 1960s it was found, essentially by experiments, that 0.3–1% As addition to a-Se is sufficient to diminish the crystallization rate but has the adverse effect of creating hole traps. The latter xerographic disadvantage was overcome by adding Cl in ppm amounts to compensate for the As-induced traps (Tabak and Hillegas, 1972; Schottmiller, 1975). It should be mentioned that with As alloying the glass transformation temperature increases, which is a distinct advantage, and the films become mechanically more stable (Kasap and Juhasz, 1987). Although most a-Se based modern photoreceptors are essentially Se–Te alloys or a-As$_2$Se$_3$, chlorinated Se:0.5% As alloy is still in use as x-ray photoconductors in x-ray imaging applications, as will be discussed later, and often forms the overcoat layer in multilayer Se–Te based photoreceptors. Chlorinated Se:0.5% As alloy is typically referred as *stabilized* a-Se.

9.3 PROPERTIES OF SELENIUM ALLOYS

9.3.1 Selenium–Arsenic

It has been mentioned that alloying selenium with arsenic results in improvements in the thermal and mechanical properties. There are essentially two composition ranges for Se–As alloys that are useful for electrophotographic applications; the low As content end (As content <1 at%) and the near stoichiometric composition (As content 35–38 at%), which is simply termed As$_2$Se$_3$. The physical properties of the Se–As alloy system over a wide range of compositions have been extensively studied, as reviewed by Barisova (1981) and Feltz (1993), for example. There is also much literature on the electrophotographic properties of a-As$_2$Se$_3$ that examines its optical, charge transport, and xerographic properties (Tabak et al., 1973; Pfister and Morgan, 1975, 1980; Pfister et al., 1977; Pfister, 1979; Scharfe, 1984; Pinsler et al., 1986).

The major advantages of a-As$_2$Se$_3$ photoreceptors over those of a-Se:0.5% As and a-Se$_{1-x}$Te$_x$ alloys are superior thermal and mechanical properties, as manifested in a higher glass transformation temperature, almost absent crystallization, and a considerably higher microhardness, as summarized in Table 9.1. The latter advantages naturally lead to a long operational lifetime for a-As$_2$Se$_3$ photoreceptors in large-volume copying applications. Interband absorption in a-As$_2$Se$_3$ starts at a photon energy of ~1.8 eV, which makes the photoreceptor almost panchromatically sensitive to the visible spectrum. Unity xerographic photosensitivity, for example, occurs around 500 nm for a-Se but at about 720 nm for a-As$_2$Se$_3$. Only holes are mobile in a-As$_2$Se$_3$, with an effective drift mobility (as determined by TOF type experiments) that is highly field dependent and nearly 3 to 4 orders of magnitude smaller than in a-Se. Consequently, the photoinduced discharge in a-As$_2$Se$_3$ photoreceptors is controlled mainly by bulk transport. Generally a-As$_2$Se$_3$ used as a photoreceptor material is also typically halogen doped (e.g., 1000 ppm Br) to increase the hole drift mobility (Pfister et al., 1977). It should be mentioned that the transport of photoinjected holes through an a-As$_2$Se$_3$ film is based on multiple trapping and release events involving a distribution of localized states in the mobility gap and cannot be simply described in terms of a constant drift mobility as in the case of a-Se. The dark decay of the surface potential is controlled by a depletion discharge process (Melnyk, 1980) and has many similarities to that found for a-Se$_{1-x}$Te$_x$ alloys, as is discussed in detail in the next section.

The xerographic properties of a-As$_2$Se$_3$ in terms of dark discharge and residual potential are generally worse than those for a-Se, essentially as a result of a high population of deep mobility gap states in a-As$_2$Se$_3$. Another shortcoming of a-As$_2$Se$_3$ is that in a monolayer form it is not generally suitable for laser printer applications requiring long

Table 9.1 Physical Properties of a-Se and a-As$_2$Se$_3$

Property	a-Se	a-As$_2$Se$_3$
Density, g · cm^{-3}	4.29	4.55
Glass transformation temperature (DSC at 5°C/min)	50°C	~180°C
Crystallization temperature (DSC at 5°C/min)	100°C	~300°C
Melting temperature, °C	218	~360
Microhardness (Vickers hardness number kgf/mm^2)	40	~150
"Photoconduction band gap" (eV)	2.5	1.8
Wavelength for unity photosensitivity, nm	500	720
Dark resistivity (Ω · cm)	10^{14}	10^{12}
Hole drift mobility, cm^2/V · s (F is the electric field in V · cm^{-1})	0.16	~2 × 10^{-10} F
Electron drift mobility, cm^2/V · s	7 × 10^{-3}	—
Quantum efficiency	Field dependent	Field dependent
Xerographic dark discharge	Depletion discharge	Depletion discharge

Sources: Barisova (1981, Ch. 1), Kasap and Juhasz (1987), Lutz (1987), Lutz and Reimer (1982), Scharfe (1984).

wavelength responsivity, since bulk absorption results in trapped electrons. By using a double-layer photoreceptor consisting of a thin As$_2$Se$_{3-x}$Te$_x$(x < 0.5) layer for photogeneration and a thick As$_2$Se$_3$ layer for charge transport, the photosensitivity can be shifted further into the long wavelengths (Pinsler et al., 1986).

The cyclic buildup of the residual potential in As$_2$Se$_3$ photoreceptors can be stabilized to ensure reproducible copy quality by employing light sources of various wavelengths for prefatiguing or erasure as described by Pinsler et al. (1986). In addition, the photoreceptor can be operated at a stable temperature, above the room temperature, to reduce variations in the xerographic performance due to temperature changes in the environment. Both dark discharge and residual potential depend on the population of deep localized states which are thermally generated structural defects. Thus the dark discharge and residual potential are highly temperature sensitive (Ing and Neyhart, 1972; Melnyk, 1980).

Figure 9.15 shows the effect of adding 0.5% As to a-Se on the charge transport and trapping properties from TOF measurements. Although the mobility is relatively unaffected, the trapping time becomes shorter, which leads to, for example, high residual potentials. With ppm amounts of Cl addition, however, the hole lifetime is restored. It was the success of this combinational doping to normalize the properties of a-Se that led to the continued use of a-Se as an imaging material. The physical process that restores the hole transport can be qualitatively accounted for by various possible defect reactions in the structure (Juhasz and Kasap, 1985); the compensation effects of As and Cl in a-Se is still a subject of topical interest (Pai, 1997), given the recent importance of this material as an x-ray photoconductor. The improvement in the hole lifetime when a-Se is doped with small amount of Cl is also apparent in xerographic measurements in which the residual potential, V_{r1} in Eq. (4), falls sharply with small amounts of Cl addition (<10 ppm) (Wang and Champness, 1995).

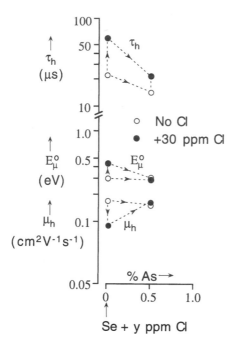

Figure 9.15 Hole transport parameters, namely hole drift mobility (μ_h), mobility activation energy (E_μ^0), and hole lifetime (τ_h) in a-Se$_{1-x}$ As$_x$ + y ppm Cl; solid points contain Cl. (From Juhasz and Kasap, 1985.)

9.3.2 Selenium–Tellurium

The central reason for alloying amorphous selenium with tellurium is the shift in the spectral response toward the red region with Te content, which allows the photosensitivity of the photoreceptor to be matched with the exposure spectrum; hence an improvement in the overall xerographic efficiency. Figure 9.16 shows that the optical band gap decreases almost linearly with Te addition, which leads to an appreciable increase in the quantum efficiency. For example, for ~20 at% Te alloy, the quantum efficiency is almost 3 orders of magnitude higher at a wavelength of 600 nm. The photosensitivity is also enhanced with Te alloying, as illustrated in Fig. 9.17. From both these figures it clear that there is a distinct advantage of alloying Se with Te, since the spectral response of the photoreceptor can be tailored to specific exposure needs.

Although a-Se$_{1-x}$Te$_x$ alloys have desirable spectral photosensitivity, their electrophotographic properties, especially when the Te content is high, are not suitable for monolayer photoreceptor applications. Monolayer a-Se$_{1-x}$Te$_x$ photoreceptors exhibit rapid dark decay and high residual potentials.

The decay of surface potential on a-Se$_{1-x}$Te$_x$ alloy films has been found to be controlled by the depletion discharge process (Abkowitz et al., 1985a; Baxendale and Juhasz, 1990; Kasap et al., 1991c). In essence, the xerographic depletion discharge model is based on bulk thermal generation involving the ionization of a deep mobility gap center to produce a mobile charge carrier of the same sign as the surface charge, and an oppositely charged ionic center. Assuming, as above, positive charging, a mobile hole would be thermally generated and the ionized center would be negative. As thermally generated

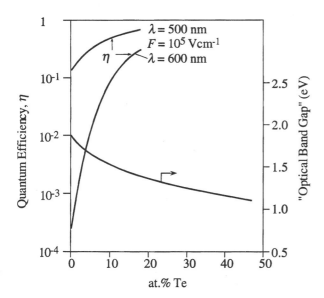

Figure 9.16   Quantum efficiency, $\eta(\lambda, F)$, at two wavelengths and "optical band gap" E_0 in a-Se$_{1-x}$Te$_x$ alloys as a function of Te content. Quantum efficiency data at a field F of 10^5 Vcm^{-1}. (From Hagen and Derks, 1984.) Optical band gap refers to E_0 in $(h\nu\alpha)^{1/2} \sim (h\nu - E_0)$. (As determined by Adachi and Kao, 1980.)

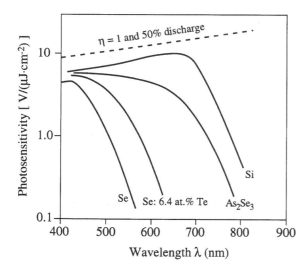

Figure 9.17   Photosensitivity versus wavelength for pure a-Se, a-Se$_{1-x}$Te$_x$, a-As$_2$Se$_3$, and a-Si: H photoreceptors. "Photosensitivity" is simply the amount of surface potential discharge per unit incident radiation energy, i.e., V/(J · cm^{-2}). It is widely determined by measuring the required flux of radiation for 50% discharge. The absolute values of photosensitivity, however, can vary considerably depending on the experimental conditions used (e.g., initial charging voltage). The broken line is the theoretical photosensitivity for unity quantum efficiency and 50% surface potential discharge. (Data mainly from Lutz, 1987; Lutz and Reimer, 1982; and Nakayama et al., 1982.)

holes are swept out by the electric field, a negative bulk space charge builds up with time in the specimen, causing the surface potential to decay with time as shown in Fig. 9.18. If the buildup in the bulk negative charge density is spatially uniform, the internal electric field F falls linearly with distance from the top surface. At a certain time called the depletion time t_d, the electric field F at the grounded end of the sample becomes zero. From that time onward the field will be zero at a distance $X(t) < L$, the sample thickness, and consequently there will be a neutral region from X to L inasmuch as holes generated in $0 < x < X$ and arriving into $X < x < L$ will not be swept out. The shrinkage of the depleted volume with time $t > t_d$ means that t_d marks a functional change in the dark decay rate, as indicated in Fig. 9.18, and therefore is readily obtainable from dark discharge experiments.

Under low charging voltages the depletion time indicates the time required for the surface potential to decay to half its original value. Under high charging voltages, however, field-enhanced emission from the deep mobility gap centers also plays an important role, and the surface potential initially decays at a much faster rate so that at the depletion time the surface potential is in fact less than half the initial value (Kasap et al., 1991c). Figure 9.19 shows the dependence of the depletion time t_d and the half-time $t_{1/2}$ on the charging voltage V_0, where it can be seen that at the highest charging voltages there is no improvement in $t_{1/2}$ with further increase in the charging voltage V_0. Inasmuch as the dark decay in a-Se$_{1-x}$Te$_x$ alloys is a bulk process, the rate of discharge increases with the square of thickness ($dV/dt \sim L^2$) and can be reduced only by using thin a-Se$_{1-x}$Te$_x$ layers. The latter concept leads naturally to the design of multilayer photoreceptor structures.

The origin of the deep localized states in the mobility gap that control the dark decay has been attributed to structural native thermodynamic defects (Abkowitz, 1984a–c, 1985). Thermal cycling experiments of Abkowitz and coworkers show that the response of the depletion time to temperature steps is retarded, as would be expected when the structure relaxes toward its metastable liquidlike equilibrium state.

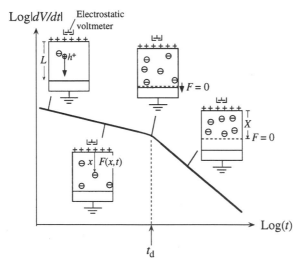

Figure 9.18 Typical log–log plot of the dark discharge rate versus time for an a-Se$_{1-x}$Te$_x$ film. The break point identifies the depletion time t_d. Various stages in the depletion discharge model are also illustrated. F is the electric field. (From Kasap, 1989a.)

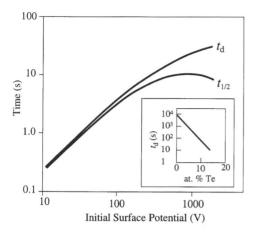

Figure 9.19 Log–log plots of the depletion time t_d and time for the surface potential to decay to its half value $t_{1/2}$ versus charging voltage V_0 for an a-Se:13 wt% Te photoreceptor film of thickness 70 μm. (From Kasap, 1989a.) The inset shows the dependence of the depletion time t_d on the Te content. (From Abkowitz et al., 1985b.)

Figure 9.20 illustrates the relaxation effect on the depletion time t_d as an a-Se:Te film is temperature stepped. When the temperature is stepped from 22°C to 35°C, the depletion time instantaneously drops from (a) to (b) to a value determined by the isostructural temperature dependence. Then as the structure relaxes toward the equilibrium state, t_d decreases further toward (c) until the structure has equilibrated. Stepping the temperature back to 22°C again results in an instantaneous change in t_d from (c) to (d), followed by a gradual increase toward its original value as the structure equilibrates. The only possible inference is that t_d must be controlled by structure-related thermodynamic defects. The generation of such defects is therefore thermally activated. It should also be noticed that a change of 13°C in the temperature has changed the depletion time, and hence the dark discharge time, by more than an order of magnitude: (a) to (c). We should note that since the depletion discharge mechanism involves the thermal emission of carriers from deep localized states, it is strongly temperature dependent. For example, t_d increases in an approximate Arrhenian fashion with decreasing temperature (Baxendale and Juhasz, 1990).

In addition to the deterioration of the dark decay, there is an increase in the residual potential associated with a-Se$_{1-x}$Te$_x$ alloys. Figure 9.21 displays the $\mu\tau$ product for holes and electrons (Abkowitz and Markovics, 1982) determined from the xerographic residual potential in a-Se$_{1-x}$Te$_x$ monolayer films, where it can be seen that even with very little Te alloying there is a considerable rise in both hole and electron deep traps. The relationship between the trapping time τ and the residual potential has been evaluated by Kanazawa and Batra (1972) and Kasap (1992). It can be seen that once the Te concentration exceeds 12 wt% Te, the residual potential is more than an order of magnitude larger than typical values for a-Se.

The relatively large residual potentials in a-Se$_{1-x}$Te$_x$ alloys can be effectively reduced by doping the alloy with chlorine. Addition of chlorine even in the ppm range has been found to reduce drastically the residual potential, as shown in Fig. 9.22. On the other hand, both the charge acceptance, as indicated by the initial charging voltage V_0, and the dark discharge, as indicated by the surface potential 1 second later, V_{01}, decrease with the

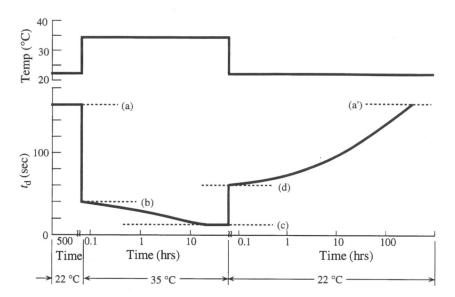

Figure 9.20   Time evolution of the depletion time t_d for an a-Se : 12 wt% Te film 50 μm thick after temperature steps. The initial charging voltage was 100 V. Imposed thermal history is represented in the upper half of the figure. Almost instantaneous changes (a)–(b) and (c)–(d) in t_d are due to the isostructural dependence of t_d on the temperature. The much slower response (b)–(c) and (d)–(a′) reflect the structural relaxation induced changes in t_d. (From Abkowitz, 1984a, b, 1985).

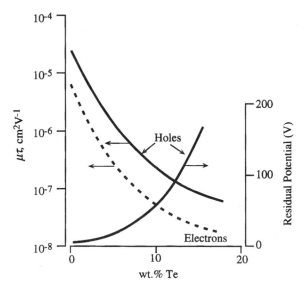

Figure 9.21   Hole and electron drift mobility lifetime product μτ and residual potential versus Te content in a-Se$_{1-x}$ Te$_x$ films. (The μτ product was xerographically measured by Abkowitz and Markovics, 1982; residual potential data from Onozuka et al., 1987.)

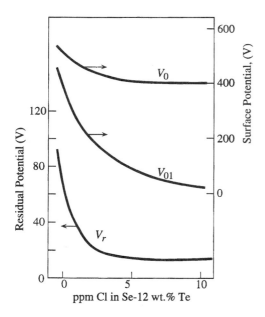

Figure 9.22 Effect of Cl doping on the residual potential, the initial voltage V_0, and the dark surface potential after 1 second for a-Se:12 wt% Te films; $V_0 - V_{01}$ represents dark discharge in one second. (Data from Onozuka et al., 1987.)

Cl content. The dark decay, as gauged by $V_0 - V_{01}$ in Fig. 9.22, therefore increases sharply with chlorination in accordance with the effects of Cl addition to pure a-Se (Abkowitz et al., 1985b). The general xerographic effects of Cl addition to a-Se:12% Te films highlighted in Fig. 9.22 are also similarly observed for Cl addition to a-Se films (Wang and Champness, 1994, 1995).

Charge transport in a-Se$_{1-x}$Te$_x$ alloys has been studied extensively by the TOF technique. Typical hole and electron TOF photocurrent waveforms are shown in Fig. 9.23. Compared with the a-Se case, where the TOF waveform is almost ideal, the photocurrents in Fig. 9.23 evince noticeable dispersion. Figure 9.24 summarizes the dependence of the hole and electron drift mobility and their activation energies in the alloy system a-Se$_{1-x}$Te$_x$ with various amounts of chlorination in the ppm range. The electron drift mobility contin-

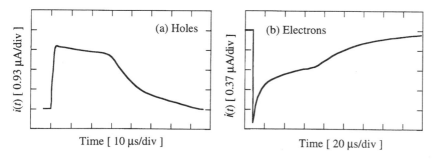

Figure 9.23 Time-of-flight hole and electron transient photocurrents in an a-Se$_{0.966}$Te$_{0.034}$ alloy photoreceptor film. (From Kasap, 1989b; Kasap et al., 1991c.)

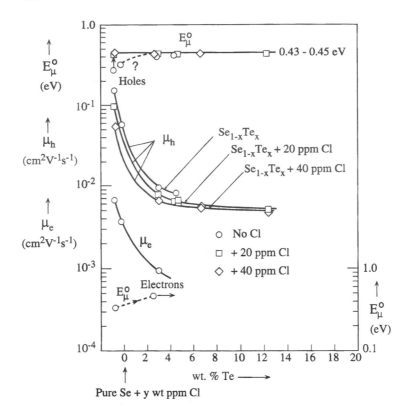

Figure 9.24   Charge transport parameters, namely drift mobility μ and zero field activation energy E_μ^0, in chlorinated a-Se$_{1-x}$Te$_x$ alloys. (From Kasap, 1989b.)

ues to fall with Te addition. But the hole drift mobility, after an initial sharp decrease, saturates at a low value determined by the extent and concentration of localized states. As discussed by the present author (Kasap, 1989b), the charge transport mechanism in halogenated a-Se$_{1-x}$Te$_x$ films is most probably trap-controlled hopping in the band tail states similar to doped As$_2$Se$_3$ (Pfister and Morgan, 1980). The sharp drop in the drift mobility coupled with the rise in the population of deep traps as inferred from Fig. 9.21 mean that the charge carrier range in a-Se$_{1-x}$Te$_x$ alloys is shorter than that in pure a-Se and leads to the rapid buildup of the saturated potential. The effect of deep traps can be reduced, though not completely eliminated, by doping with Cl, since it is well known that Cl addition enhances hole transport.

   The exact way in which charge transport and trapping are controlled in a-Se$_{1-x}$Te$_x$ is not well understood, but there is no doubt that under- and overcoordinated charged structural defects, VAP-type centers, must play a key role. Recently, by considering the possible defect reactions in the a-Se$_{1-x}$Te$_x$ structure, it was shown, for example, that as the Te content is increased, there is an initial rapid rise in the population of localized states, which eventually saturates as the Te content becomes appreciable (Springett, 1990). The latter behavior is in qualitative agreement with the observed rapid fall and saturation of the hole drift mobility.

   The alloying of Se with Te does not result in changes in the glass transformation and crystallization behavior as drastic as those observed for the Se–As glass system. The

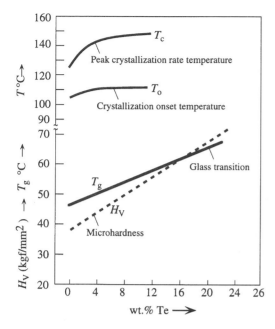

Figure 9.25 Thermal and mechanical properties of a-Se$_{1-x}$Te$_x$ alloy films from DSC and micro-hardness measurements. DSC data at a heating rate of 5°C/min; T_g = glass transition temperature; T_o = crystallization onset temperature, T_c = maximum crystallization rate temperature (defined in Fig. 9.13), H_v = Vickers microhardness. (From Kasap and Juhasz, 1986.)

glass transformation temperature increases almost linearly with the Te content, whereas the crystallization onset temperature has been reported to increase only slightly (Kasap and Juhasz, 1987).

Figure 9.25 summarizes the thermal and mechanical properties of a-Se$_{1-x}$Te$_x$ alloys over the xerographically useful alloy range. The improvements in the glass transition temperature T_g and the microhardness are a consequence of a-Se$_{1-x}$Te$_x$ having stronger secondary bonds between the Se–Te chains and the increase in the average mass of the chains with Te inclusion.

9.4 PHOTORECEPTOR DESIGNS

The basic electrophotographic requirements are high charge acceptance, slow dark decay, and low first and cycled-up residual potentials, assuming that the spectral response of the photoreceptor has been matched to that of the exposure lamp. With high Te content in the a-Se$_{1-x}$Te$_x$ film, both the dark decay rate and the residual potential are high, and the only practical alternative is to design multilayer photoreceptor structures in which the charge generation and charge transport functions have been separated, as is illustrated in Fig. 9.26, where single-, double-, and triple-layer photoreceptors are shown with typical structures that would be expected for copying and printing applications. The photogeneration layer (PGL) has a high Te composition and is made as thin as allowed by the penetration depth, $1/\alpha(\lambda)$, of the exposing radiation. Typically this layer is a few micrometers thick and has a Te content of 10–15% for copying applications and ~25% for laser printer

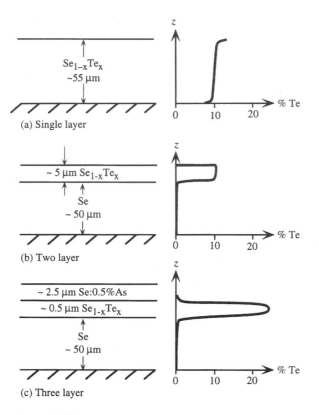

Figure 9.26   Typical single- and multilayer photoreceptor structures for copying and printing applications. (a) Single-layer photoreceptor for copying. (b) Double-layer photoreceptor for copying. (c) Triple-layer photoreceptor for printing. (From Cheung et al., 1982.)

applications. The printer-type photoreceptor also needs an overcoating layer (OL) to attain the required xerographic charge acceptance, to prevent surface injection and conduction at these high Te content levels. The overcoating layer also serves as a protection layer, since it often has some As added to enhance the thermal and mechanical stability.

With the high Te containing PGL region confined to a limited thickness, both the dark decay rate and the residual potential are reduced. The photosensitivity versus wavelength behavior, however, depends on whether the photoreceptor structure is a double- or a triple-layer type, and on the thickness of the photogeneration and overcoating layers. Figure 9.27 shows the xerographic photosensitivity of a typical three-layer photoreceptor as a function of wavelength for various overcoating thicknesses; it is obvious that the photosensitivity has been extended to the wavelength region of AlGaAs solid state lasers. It should be remarked that the photosensitivity axis in Fig. 9.27 cannot be directly compared with that in Fig. 9.17 inasmuch as the definition used in Fig. 9.27 is based on the quadratic-field-dependent generation model for the photoinduced discharge characteristics (Scharfe, 1984). It is interesting to mention that because of a large difference between the Te contents of the PGL and the OL, the photosensitivity exhibits a ''gap'' around the ~560 nm region when OL becomes ''thick.'' This gap is actually attributable to the failure of the 500–600 nm photons absorbed by the overcoating layer to result in photoconduction,

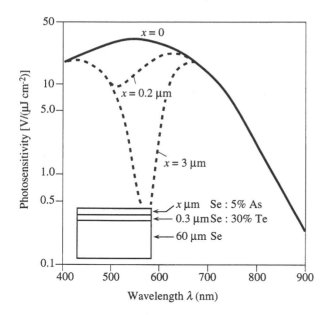

Figure 9.27 Photosensitivity of a triple-layer a-Se:Te:As-type photoreceptor as a function of wavelength for various thicknesses (x μm) of Se:5%As overcoating layer. The definition of photosensitivity differs from that shown in Fig. 9.16 and is based on the quadratic-field-dependent generation model. (From Melnyk et al., 1982.)

as was discussed in Section 9.2.2. Since the prime application in this case was for a printer utilizing an 825 nm solid state laser source, such a wavelength gap in the photosensitivity has no adverse effect on the xerographic printing performance of the photoreceptor.

The multilayer photoreceptor structures shown in Fig. 9.26 give the impression that the Te concentration profile changes sharply from one layer to another. This is not actually the case, however, since during the fabrication process there are fractionation and interdiffusion effects that tend to prevent sharp Te concentration changes. Nonetheless, there is the possibility that the interface between the layers may be electronically active: it can thermally generate and capture charges simply because strain in this region, as a result of some atomic mismatch due to the density change, can lead to structural defects. The properties of such interfaces between the various layers have not been studied in sufficient detail to permit the formulation of a general multilayer photoreceptor theory. It seems that a sharp change in the Te concentration is not desirable inasmuch as it will lead to more interfacial strain and thus to more interface localized states. Pinsler (1988) reported a xerographic TOF study of 15 wt% Te:Se/Se double-layer photoreceptor to conclude that the nature of charge transport in the CTL can be fully accounted by trapping and release kinetics and dispersion in the PGL without any evidence of interfacial traps. In other experiments, however, some evidence for trapping at the interface has been reported from the saturated residual potential on double-layer photoreceptors (Kasap et al., 1991c).

The main drawback of such multilayer PR designs is that they add to the complexity of the fabrication process, eliminating the simplicity of evaporating the alloy from a single boat with only basic source temperature and deposition controls. Not only must additional boats be used, but they must be properly shuttered or need careful control of the evaporation process to achieve the required Te profile across the film. Given unavoidable fractionation effects, which are not well characterized, the established evaporation methods are essentially empirical.

9.5 PHOTORECEPTOR FABRICATION

The requirement of a ~60 μm thick photoreceptor coating for xerographic applications implies that the most efficient method of depositing the photoreceptor films is by thermal evaporation. Vitreous Se alloy pellets are simply loaded into stainless steel boats and evaporated at temperatures around 300–350°C onto drum substrates held above the typical glass transformation temperature of the alloy. Figure 9.28a illustrates schematically the principle of thermal evaporation for depositing a-Se alloy photoreceptor films. Typical deposition rates reported in the literature have been around 2 μm/min, which indicates an evaporation process lasting about 30 minutes. During this time the material is continuously evaporated from the boat, and there are many changes taking place not only in the source material but also in the vapor composition and in the film deposited. There have not been any detailed studies of the evaporation kinetics of Se–Te alloys, although the Se–As sys-

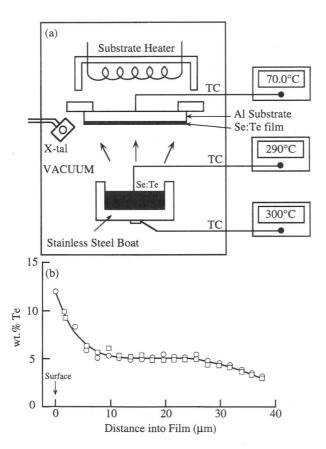

Figure 9.28 (a) Schematic sketch of a-Se$_{1-x}$Te$_x$ photoreceptor film fabrication by vacuum deposition. Evaporation is often from a directly heated open boat. The substrates are normally Al drums that are heated and rotated during evaporation. TC means a thermocouple. (b) Fractionation in a-Se$_{1-x}$Te$_x$ alloy photoreceptor films; plot shows a typical Te content across the film thickness.

tem has been examined by Sigai (1975) and Hordon (1989). Sigai, for example, was able to explain at least qualitatively the As composition variation across the photoreceptor film in terms of the evaporation process occurring in the boat. The Se–Te case however is poorly documented, and very little information is available on fractionation effects.

Figure 9.28b shows the Te composition of a $Se_{1-x}Te_x$ photoreceptor film deposited by a typical evaporation process from an open stainless steel boat. It should be noticed that fractionation effects cause the Te concentration to vary significantly across the film thickness, being Se-rich near the substrate and having a Te concentration 2 to 3 times as much on the surface. The latter concentration has been found to depend on the deposition rate. The composition of the condensed material at any time during the deposition depends on the partial pressures of Se and Te in the vapor. Deviations from Raoult's law result in condensed layers having different compositions (i.e., fractionation effects). The vapor pressure of Te over the $Se_{1-x}Te_x$ alloy is lower than its Raoult partial pressure because the latter occurs in an ideal solution in which all the bonds, Se — Se, Se — Te, and Te — Te, are assumed to be alike. Inasmuch as these bond energies are given by $E_{Se-Se} = 1.91$ eV, $E_{Te-Te} = 1.43$ eV, and $E_{Se-Te} = 1.8$ eV, the Te atoms in the $Se_{1-x}Te_x$ are more tightly bound than those in the parent (pure Te) material; consequently, their escape tendency is reduced in the $Se_{1-x}Te_x$ alloy, resulting in a lower Te partial pressure than the Raoult value. Fractionation effects are therefore a thermodynamic necessity of the evaporation process.

The actual fractionation that occurs during deposition depends in a much more complex way on the evaporation process, since temperature nonuniformities in the boat, finite material thermal conductivity, and limited diffusion in the source material all contribute to fractionation. As the evaporation proceeds, the Se-rich initial vapor leaves the surface region of the source rich in Te, which then results in an inhomogeneous source material. It may be that the constant Te concentration in the mid-bulk region of the photoreceptor film in Fig. 9.28b is a consequence of a dynamic equilibrium between the evaporation process and the diffusion process between the bulk of the source material and its Te-rich surface region.

It is apparent from the discussions above that fractionation effects result in inhomogeneous photoreceptor films. Inasmuch as many physical properties (e.g., density, drift mobility, deep trap concentration, glass transformation) depend strongly on Te concentration, the resulting film will have these properties varying along the film thickness. Very high Te concentrations on the surface can reduce the charge acceptance and accelerate the dark decay. Furthermore, the surface and substrate regions of the photoreceptor will structurally relax at different rates in response to a temperature change, resulting in unpredictable behavior. It should be remarked that although fractionation and noncongruent evaporation effects can be overcome by using flash evaporation techniques (Jansen 1982), such methods do not transfer directly to volume fabrication of photoreceptor drums.

9.6 TV PICKUP TUBES

A well known and successful application of amorphous Se–As–Te alloys is in Hitachi's Saticon (Goto et al., 1974), which is a commercially available TV pickup tube. Figure 9.29 displays the structure of a typical a-Se Saticon, which utilizes the high panchromatic photosensitivity of the Se–Te alloy and the relatively fast hole drift mobility of a-Se. Layers of CeO_2 and Sb_2S_3 act as hole and electron blocking contacts, respectively. Electrons injected by the scanning electron beam become trapped in the Sb_2S_3 layer, forming

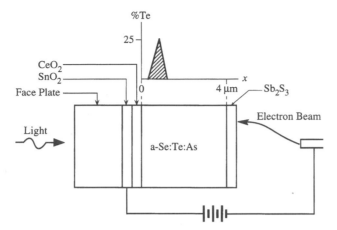

Figure 9.29 Schematic diagram of a Saticon TV image pickup tube utilizing an a-Se : Te : As alloy. The Te composition is concentrated in a narrow region, whereas the bulk is mainly a-Se : As. (From Maruyama, 1982.)

a negative space charge in this layer. Photogenerated holes in a-Se, from exposure to the image, transit across toward Sb_2S_3 to recombine with the electrons trapped in Sb_2S_3. The photogeneration is contained in a region of high Te concentration. The crystallization of a-Se is inhibited by the addition of $\sim 1\%$ As to a-Se. Inasmuch as Saticon uses an amorphous material (i.e., grainless and uniform material), it exhibits high resolution. Further details on the fabrication and operation of the Saticon may be found in Maruyama (1982).

Recently Tanioka and coworkers (Tanioka et al., 1987, 1988; Takasaki et al., 1988; Tsuji et al., 1991) have developed a super-sensitive photoconductive target called the HARP ("high-gain avalanche rushing photoconductor") for use in HDTV (high-definition television) camera pickup tubes. The vidicon using the HARP target has been called the HARPICON or a-Se avalanche vidicon. The basic structure of the HARP target and the principle of operation are schematically illustrated in Fig. 9.30. The entire target is typically about 2 μm thick though it may be thicker in ultrasensitive targets. The transparent signal electrode (SnO_2) is biased positively with respect to the cathode. The CeO_2 and

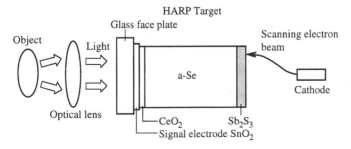

Figure 9.30 Schematic illustration of the structure of the HARP target and the principle of operation of the HARPICON. Avalanche multiplication occurs in the a-Se layer where the electric field exceeds 80 V/μm and causes hole multiplication by impact ionization. (Adapted from Tanioka et al., 1988.)

SbS_3 layers, as in the Saticon, act as blocking contacts for hole and electron injection, respectively. The incident light from the object is absorbed mainly in the a-Se layer. The electron–hole pairs photogenerated in the a-Se layer are then drifted by the applied electric field and constitute the signal current. As holes drift through the a-Se layer towards the back electrode, as a result of the large applied electric field (greater than 8×10^5 V/cm or 80 V μm), they experience avalanche multiplication and hence yield a quantum efficiency greater than unity. The effective quantum efficiency resulting from avalanche multiplication depends on the fields as well as the photoconductor thickness. For example, in the 2 μm thick HARP target the quantum efficiency is about 10 at a field of 120 V/μm, whereas the quantum efficiency is about 1000 in an a-Se target of thickness 24.8 μm at a field of 100V/μm operating at a wavelength of 400 nm (Tsuji et al., 1991). TV pickup tubes using such HARP targets clearly have far superior sensitivity than conventional TV pickup tubes and accordingly constitute ultrahigh sensitive image pickup tubes.

9.7 X-RAY IMAGING

The use and properties of stabilized a-Se (a-Se:0.2–0.5% As doped with 5–20 ppm Cl) layers on Al substrates for x-ray imaging by xeroradiography is well documented (Boag, 1973; Leiga, 1990), but this system suffers from the difficulties and noise associated with the powder development technique. Xeroradiography is no longer competitive because of the toner readout method, not the underlying properties of the a-Se photoconductor. By replacing the toner readout with an electronic readout, a-Se has again become the basis of a clinical imaging system, and the commercial interest in a-Se has recently been revived. Recent research at the Sunnybrook Health Science Centre (University of Toronto) by Rowlands and coworkers (Rowlands et al., 1991, 1992; Zhao et al., 1995; Que and Rowlands, 1995; Yaffe and Rowlands, 1997), at Philips in Germany (Schiebel et al., 1986; Hillen et al., 1988), at Dupont de Nemours in the U.S. (Lee et al., 1995), and at Thompson CSF in France (De Monts and Beaumont, 1989) has shown that an x-ray imaging system based on the x-ray sensitivity of a-Se:As photoconductors has enormous potential for digital radiographic applications in medical diagnosis. Further, within the last few years commercial x-ray medical diagnostic imaging systems have been introduced into the market that are based on using the x-ray sensitivity of a-Se:As photoconductive layers (Neitzel et al., 1994) (e.g., Philips). In addition, two significant patents (e.g. U.S. Patent 5,396,072, May 1995, and 5,319,206, June 1994) have been filed by major corporate laboratories that use an a-Se:As layer vacuum coated onto a thin film transistor (TFT) active matrix array (AMA) and use this plate as an x-ray image detector. Images obtainable have been potentially superior to film based radiology, and the technique renders itself inherently to digital processing and storage.

Very recent research has shown clearly that one of the most promising digital radiographic systems is based on using a large area TFT matrix array (as used in flat panel displays for example) with an electroded x-ray photoconductor (e.g., Zhao et al., 1995; Zhao and Rowlands, 1995; Lee et al., 1995; Rowlands and Kasap, 1997). Figure 9.31 illustrates how a TFT-AMA can be used to read the amount of charge on each pixel electrode. Each pixel electrode carries an amount of charge that is proportional to the amount of incident x-ray radiation. All FETs in a row have their gates connected, whereas all FETs in a column have their sources connected. When gate line i is activated, all FETs in that row are turned ON and N data lines from $j = 1$ to N then read the charges on the pixel electrodes in row i. The parallel data are multiplexed into a series data and then

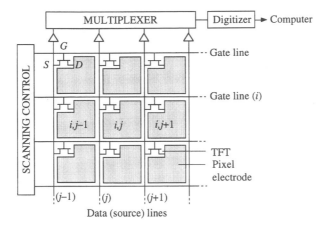

Figure 9.31 Thin film transistor (TFT) active matrix array (AMA) for use in x-ray image detectors with self-scanned electronic readout. (After Zhao et al., 1995.)

digitized and fed into a computer for imaging. The scanning control then activates the next row, $i + 1$, and all the pixel charges in this row are then read and multiplexed so on until the whole matrix has been read from the first to the last row.

The basic structure of an a-Se based TFT-AMA digital x-ray image detector is shown schematically as a cross section in Fig. 9.32. The a-Se layer is vacuum coated onto the TFT-AMA and carries a top electrode (A). Each pixel (i, j) carries a charge collection electrode, B, connected to a signal storage capacitor, C_{ij}, whose charge can be read through properly addressing the TFT (i, j) via the gate (i) and drain (j) lines. An external readout electronics and software, by proper self-scanning, converts the charges read on the C_{ij} to a digital image. The electron–hole pairs (EHPs) that are generated in the photoconductor by the absorption of an x-ray photon travel *along* the field lines. Holes accumulate on the storage capacitor C_{ij} and thereby provide a charge-signal q_{ij} that can be read during self-scanning. In this type of a-Se coated TFT-AMA image detector shown in Fig. 9.32, the resolution is determined by the pixel size, which in present experimental image detectors

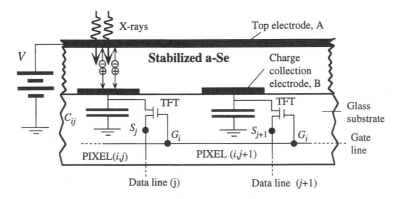

Figure 9.32 A simplified schematic diagram illustrating two neighboring pixels (i, j) and $(i, j + 1)$ and the formation of a charge image in pixel i, j.

is typically ~150 µm but is expected to be smaller than 50 µm in future detectors to achieve a better resolution than the current film based systems for mammography.

Ideally the photoconductive layer should possess the following material properties: (1) There should be no dark current, which means that contact A should be hole injection blocking and contact B should be electron injection blocking. (2) There should be no deep trapping of EHPs, which means that for both electrons and holes the Schubweg $\mu\tau F \gg L$ holds, where μ is the drift mobility, τ is the deep trapping time (lifetime), F is the electric field, and L is the photoconductor thickness. (3) There should be no recombination, since EHPs are generated in the bulk of the a-Se layer. (4) EHP x-ray photogeneration efficiency, as gauged by the reciprocal of the energy required to create a free EHP (E_{EHP}) should be at a maximum. (5) The above should not change or deteriorate with time as a consequence of repeated exposure to x-rays, i.e., x-ray fatigue and x-ray damage should be zero. (6) The photoconductor should be easily coated onto the AMA panel, for example, by conventional vacuum techniques. Special processes are generally more expensive.

Provided that both holes and electrons have long ranges (i.e., $\mu\tau F \gg L$; L is the a-Se layer thickness), which will be the case for good quality a-Se material, the x-ray sensitivity of a given thickness photoconductor layer is then determined by the energy required to create a free electron–hole pair, E_{EHP}. This energy decreases with the electric field because at high fields more EHPs can escape recombination. At a field of 10 V/µm, E_{EHP} is about 50 eV. Experiments indicate that at sufficiently high fields, E_{EHP} may be expected to saturate at about 4 to 6 eV, which represents operation at maximum efficiency (Kasap et al., 1998).

There have been no systematic and fundamental studies in the literature that address the material properties of a-Se related to medical x-ray imaging requirements, even though all experiments, including commercialized systems and patents, indicate that it is an excellent x-ray photoconductor. It is quite likely that x-ray imaging will be one of the major uses of a-Se in the near future (Rowlands and Kasap, 1997).

ADDED IN PROOF

Recent work (Song, H. Z., et al., *Phys. Rev. B, 59.* 10610) examining the post-transit transient photocurrent in time-of-flight photoconductivity experiments has suggested a distinctly different density of states (DOS) distribution than that shown in Figure 9.7. This recent DOS model shows that there are two peaks at 0.40 eV above E_v and 0.55 eV below E_c. The most interesting feature in this model is the fact that these peaks are roughly close to the theoretical expectations for valence alternation pair defects at $\frac{1}{4}E_g$ ($= 0.55$ eV) and $\frac{1}{3}E_g$ ($= 0.74$ eV) from band edges as further discussed in Kasap, S.O. and Rowlands, J. A., 2000, *J. Mater. Sci: Elec. 11*; 179.

ACKNOWLEDGMENTS

The author is grateful to NSERC for their continuing support of his research program in amorphous semiconductor materials and devices since 1986.

REFERENCES

Abkowitz, M. (1967). *J. Chem. Phys. 46*:4537.
Abkowitz, M. (1981). *Ann. N. Y. Acad. Sci. 371*:171.

Abkowitz, M. (1984a). *J. Non-Cryst. Solids 66*:315.

Abkowitz, M. (1984b). *Polym. Eng. Sei. 24*:1149.

Abkowitz, M. (1984c). *Physics of Disordered Materials*. (D. Adler, H. Fritzsche, S. R. Ovshinsky, eds.). Plenum Press, New York, p. 483.

Abkowitz, M. (1985). *J. Non-Cryst. Solids 77–78*:1191.

Abkowitz, M. (1987). *J. Non-Cryst. Solids 97–98*:1163.

Abkowitz, M. (1988). *Philos. Mag. Lett. 58*:53.

Abkowitz, M., Enck, R. C. (1980). *J. Non-Cryst. Solids 35–36*:831.

Abkowitz, M., Enck, R. C. (1982). *Phys. Rev. B. 25*:2567.

Abkowitz, M., Enck, R. C. (1983). *Phys. Rev. B. 27*:7402.

Abkowitz, M., Jansen, F. (1983). *J. Non-Cryst. Solids 59–60*:953.

Abkowitz, M., Maitra, S. (1987). *J. Appl. Phys. 61*:1038.

Abkowitz, M., Markovics, J. M. (1982). *Solid State Common. 44*:1431.

Abkowitz, M., Markovics, I. M. (1984). *Philos. Mag. B. 49*:L31–L36.

Abkowitz, M., Pai, D. M. (1978). *Phys. Rev. B. 18*:1741.

Abkowitz, M., Pochan, D. F., Pochan, J. M. (1980). *J. Appl. Phys. 51*:1539.

Abkowitz, M., Foley, G. M. T., Markovics, I. M., Palumbo A. C. (1984). Metastable photoenhanced thermal generation in a-Se-Te alloys. In: *Optical Effects in Amorphous Semiconductors*, AIP Conference Proceedings No. 120, p. 117.

Abkowitz, M., Foley, G. M. T., Markovics, J. M., Palumbo, A. C. (1985a). *Appl. Phys. Lett. 46*: 393.

Abkowitz, M., Jansen, F., Melnyk, A. R. (1985b). *Philos. Mag. B. 51*:405.

Abkowitz, M., Badesha, S. S., Knier. F. E. (1986). *Solid State Common. 57*:579.

Adachi, H., Kao, K. C. (1980). *J. Appl. Phys. 51*:6326.

Adler, D. (1977). *Sci. Am. 236*:36.

Agarwal, S. C. (1973). *Phys. Rev. B. 7*:685.

Al-Ani, S. K. J., Hogarth, C. (1984). *J. Non-Cryst. Solids 69*: 167.

Anderson, J. C. (1986). *J. Vac. Sci. Technol. A4*:610.

Badesha, S. S., Abkowitz, M. A., Knier. F. E. (1986). *J. Mater. Res. 1*:10.

Baranovskii, S. D., Lebedev, E. A. (1985). *Sov. Phys. Semicond. 19*:635.

Barisova. Z. (1981). *Glassy Semiconductors*. Plenum Press, New York, Chs. 1 and 2.

Baxendale, M., Juhasz, C. (1990). *SPIE Proc. 1253*:212.

Berger, S. B., Enck, R. C., Scharfe, M. E., Springett, B. E. (1979). In: *Physics of Selenium and Tellurium* (E. Gerlach and P. Grosse, eds.). Springer-Verlag, New York.

Berkes, J. S. (1974). In: *Electrophotography: Second International Conference* (D. R. White, ed.). Society of Photographic Scientists and Engineers, Springfield, VA, p. 137.

Boag J. W. (1973). *Phys. Med. Biol. 18*:3 and references therein.

Burland, D. M., Schein. L. B. (1986). *Physics Today 39*:46.

Carles, D., Lefrancois, J. P., Larmagnac, J. P. (1984). *J. Phys. Lett. 45*:L901.

Chaudhuri, S., Biswas, S. K., Choudhury, A., Goswami, K. (1983). *J. Non-Cryst. Solids 54*:179.

Chen, I., Mort, J. (1972). *J. Appl. Phys. 43*:1164.

Cheung, L., Foley, G. M. T., Fournia, P., Springett, B. E. (1982). *Photogr. Sci. Eng. 26*:245.

Cooper, W. C., Westbury, R. A. (1974). In: *Selenium* (R. A. Zingaro, W. C. Cooper, eds.). Van Nostrand, New York, Ch. 3.

Cukierman, M., Uhlmann, D. R. (1973). *J. Non-Cryst. Solids 12*:199.

Das, G. C., Bever, M. B., Uhlmann, D. R. (1972). *J. Non-Cryst. Solids 7*:251.

Davis, E. A. (1970). *J. Non-Cryst. Solids 4*:107.

Davis, J., et al. (1989). In: *Proceedings of the Fourth International Symposium on Uses of Selenium and Tellurium*. Selenium-Tellurium Development Association, Darien, CT, p. 267.

Dawar, A. L., Joshi, J. C., Narain, L. (1981). *Thin Solid Films 76*:113.

Dembovsky, S. A., Chechetkina, E. A. (1986). *Philos. Mag. B. 53*:367.

De Monts H., Beaumont, F. (1989) *Med. Phys. 16*:105.

Dolezalek, F. K., Spear, W. E. (1970). *J. Non-Cryst. Solids 4*:97.

Elliott, S. R. (1984). *Physics of Amorphous Materials.* Longman, New York, Chs. 5 and 6.

Elliott, S. R. (1986). *J. Non-Cryst. Solids 81*:71.

Felty, E. J. (1967). Reprographie: 2-Bericht über den 2. International Congress on Reprography. Helwich Verlag, Darmstadt, West Germany, 1967.

Feltz, A. *Amorphous Inorganic Materials and Glasses.* VCH, Weinheim, 1993.

Fritzschc, H. (1977). *Chinese J. Phys. 15*:73.

Goto, N., Isozaki, Y., Shidara, K., Maruyama, E., Hirai, T., Fujita, T. (1974). *IEEE Trans. Electron. Dev. ED-21*:662.

Hagen, S. H., Derks, P. J. A. (1984). *J. Non-Cryst. Solids 65*:241.

Hartke, J. L. (1962). *Phys. Rev. 125*:1177.

Hartke, J. L., Regensburger, P. J. (1965). *J. Phys. Rev. 139*:A970.

Hillen, W., Schiebel, U., Zaengel, T. (1988). In: *Medical Imaging II* (R. H. Schneider, S. J. Dwyer, eds.). *SPIE 91*:253.

Hordon, M. (1989). In: *Proceedings of the Fourth International Symposium on Uses of Selenium and Tellurium*, Banff, May 1989. Selenium-Tellurium Development Association, Brimbergen, Belgium, pp. 156–171.

Imagawa, O., Iwanishi, M., Yokoyama, S. (1986). *J. Appl. Phys. 60*:3176.

Ing, S. W., Neyhart, J. H. (1972). *J. Appl. Phys. 43*:2671.

Ishiwata, T., Fujimaki, Y., Shimizu, L., Kokado, H. (1980). *J. Appl. Phys. 51*:444.

Jansen, F. (1982). *J. Vac. Sci. Technol. 21*:106.

Jeromin, L. S., Klynn, L. M. (1979). *J. Appl. Photogr. Eng. 5*:183.

Juhasz, C., Kasap, S. O. (1985). *J. Phys. D: Appl. Phys. 18:* 7′1.

Juhasz, C., Vaezi-Nejad, M., Kasap, S. O. (1987). *J. Mater. Sci. 22*:2569.

Juska, G., Vengris, S., Viscakas, J. (1974). In *Amorphous and Liquid Semiconductors* (J. Stuke, W. Brenig, eds.). *Proceedings of the Fifth Internatiorial Conference on Amorphous and Liquid Semiconductors*, September 1973. GarmischPartenkirchen, West Germany, Taylor and Francis, London, p. 363.

Kanazawa, K. K., Batra. I. P. (1972). *J. Appl. Phys. 43*:1845.

Kasap, S. O. (1988). *J. Appl. Phys. 64*:450.

Kasap, S. O. (1989a). *J. Electrostat. 22*:69.

Kasap, S. O. (1989b). *Can. J. Phys. 67*:1053.

Kasap, S. O. (1992). *J. Phys. D. 25*:83.

Kasap, S. O., Juhasz, C. (1982). *Photogr. Sci. Eng. 26*:239.

Kasap, S. O., Juhasz, C. (1985). *J. Phys. D: Appl. Phys. 18*:703.

Kasap, S. O., Juhasz, C. (1986). *J. Mater. Sci. 21*:1329.

Kasap, S. O., Juhasz, C. (1987). *J. Mater. Sci. Lett. 6*:397.

Kasap, S. O., Polischuk, B (1995). *Cnd. J. Phys. 73*:96.

Kasap, S. O., Thakur, R. P. S., Dodds, D. (1988). *J. Phys. E: Sci. Instrum. 21*:1195.

Kasap, S. O., Aiyah, V., Yannacopoulos, S. (1990a). *J. Phys. D: Appl. Phys. 23*:553.

Kasap, S. O., Polischuk, B., Aiyah, V., Yannacopoulos, S. (1990b). *J. Appl. Phys. 67*:1918.

Kasap, S. O., Polischuk, B., Dodds, D. (1990c). *Rev. Sci. Instrum. 61*:2081.

Kasap, S. O., Aiyah, V., Polischuk, B., Liang, Z., Bekirov, A. (1991a) *J. Non-Cryst. Solids 137–138*:1329.

Kasap, S. O., Aiyah, V., Polischuk, B., Bhattarcharyya, A., Liang, Z. (1991b). *Phys. Rev. B. 43*: 6691.

Kasap, S. O., Baxendale, M., Juhasz, C. (1991c). *IEEE Trans. Indust. Appl. 27*:620.

Kasap, S. O., Aiyah, V., Polischuk, B., Baillie, A. (1998), in press.

Kastner, M. (1977). In: *Amorphous and Liuid Semiconductors.* Seventh International Conference on Amorphous and Liquid Semiconductors (W. E. Spear, ed.). University of Edinburgh, Edinburgh, 1977, p. 504.

Kastner, M. (1978). *J. Non-Cryst. Solids 31*:223.

Kastner, M., Adler, D. Fritzsche, H. (1976). *Phys. Rev. Lett. 37*:1504.

Kempter, K., Kiendl, A., Muller, W., Voit, H. (1983). *J. Non-Cryst. Solids 59–60*:1219.

Kiyota, K., Teshima, A., Tanaka, M. (1980). *Photogr. Sci. Eng. 24*:289.

Klocek, P., Roth, M., Rock, R. D. (1987). *Opt. Eng. 26*:88.

Kolomiets, B. T. (1964). *Phys. Stat. Solidi 7*:713.

Kolomiets, B. T., Lebedev, E. A. (1966). *Sov. Phys. Solid State 8*:905.

Koshino, N., Maeda, M., Goto, Y., Itoh, K., Ogawa, S. (1985). *Off. Eq. Prod.* June, p. 64.

Larmagnac, J. P., Grenet, J., Michon, P. (1981). *J. Non-Cryst. Solids 45*:157.

Lee, D. L., Cheung, L. K., Jeromin, L. (1995). *SPIE Proc. 2432*:237.

Leiga, A. G. (1990). In: *Proceedings of the Fourth International Symposium on Uses of Selenium and Tellurium*. Selenium-Tellurium Development Association (STDA, Grimbergen, Belgium, 1989), pp. 249–256.

Long, M., Gallison, P., Alben, R., Connell. G. (1976). *Phys. Rev. B. 13*:1821.

Lucovsky, G. (1979). In: *Physics of Selenium and Tellurium* (E. Gerlach, P. Grosse, eds.). Springer-Verlag, New York.

Lucovsky, G., Galeener, F. L. (1980). *J. Non-Cryst. Solids 35–36*:1209.

Luhta, R., Rowlands, J. A. (1991). In: *Proc. Conf. Photoelectric Image Devices*. London, 1991.

Lutz, M. (1987). *SPIE Proc. 759*:35.

Lutz, M., Reimer, B. (1982). In: *Advances in Non-Impact Printing Technologies for Computer and Office Applications*. Proceedings of the First International Congress, June 1981. Venice, Italy (J. Gaynor, ed.). Van Nostrand Reinhold, New York, p. 446.

Marshall, J. M., Owen, E. A. (1972). *Phys. Stat. Solidi 12*:181.

Marshall, J. M., Fisher, F. D., Owen, E. A. (1974). *Phys. Stat. Solidi 25*:419.

Matsushita, T., Suziki, A., Nakau, T., Okuda, M., Rhee, J. C., Naito, H. (1987). *Jpn. J. Appl. Phys. 26*:L62.

Maruyama, E. (1982). *Jpn. J. Appl. Phys. 21*:213.

Meek, P. E. (1976). *Philos. Mag. 34*:767.

Mell, H., Stuke, J. (1967). *Phys. Stat. Solidi A 24*:183.

Melnyk, A. R. (1980). *J. Non-Cryst. Solids, 35–36*:837.

Melnyk, A. R., Berkes, J. S., Schein, L. B. (1982). In: *Advances in Non-Impact Printing Technologies for Computer and Office Applications*. Proceedings of the First International Congress, June 1981, Venice, Italy (J. Gaynor, ed.). Van Nostrand Reinhold, New York, p. 503.

Mort, J. (1967). *Phys. Rev. Lett. 18*:540.

Mort, J. (1984). *J. Vac. Sci. Technol. B. 2*:823.

Mort J. (1989). *Anatomy of Xerography: Its Invention and Evolution*. McFarland, Jefferson, NC.

Mort, J., Chen, I. (1975). *Appl. Solid State Sci. 5*:69.

Moses, D. (1992). *Philos. Mag. 66*:1.

Moses, D. (1996). *Phys. Rev. B 53*:4462.

Mott, N. F., Davis, E. A. (1979). *Electronic Processes in Non-Crystalline Materials*. 2d ed. Oxford University Press, Oxford, Chs. 6 and 10.

Nagels, P., Sleeckx, E., Callaerts, R., Tichy, L. (1994). In: *Proc. of the Fifth Int. Symp. on Uses of Se and Te*, 8–10 May 1994 (S. C. Carapella, J. E. Oldfield, Y. Palmieri, eds.). Brussels, Belgium, pp. 215–220.

Nagels, P., Sleeckx, E., Callaerts, R., Marquez, E. (1996). In *Proc. of the Ninth International School in Condensed Matter Physics, Varna, September 9–13, 1996*. J. M. Marshall, (ed.). World Scientific, 1997.

Nakayama, Y., Sugimura, A., Nakano, M., Kawamura, T. (1982). *Photogr. Sci. Eng. 26*:188.

Nakayama, Y., Akita, S., Kawamura, T. (1988a). *Jpn. J. Appl. Phys. 27*:L320.

Nakayama, Y., Akita, S., Tomomatsu, Y., Takahashi, T., Kawamura, T. (1988b). Measurement of the density of deep emission states in amorphous silicon alloys by depletion-discharge transient spectroscopy. *Proceedings of the 30th Anniversary Conference of Japan, May 1988*, Nohkyo Hall, Tokyo (Japan Hardcopy 88), pp. 349–352.

Nakayama, Y., Kita, H., Takahashi, T., Akita, S. and Kawamura. T. (1988c). *J. Non-Cryst. Solids* *97–98*:743.

Neitzel U., Maack, I., Guenther-Kohlfahl, S. (1994). *Med. Phys. 21*:509.

Nemilov, S. V., Petrovskii, G. T. (1963a). *J. Appl. Chem. (USSR) 36*:932.

Nemilov, S. V., Petrovskii, G. T. (1963b). *J. Appl. Chem. (USSR) 36*:1853.

Oda, O., Onozuka, A., Tsuboya, I. (1986). *J. Non-Cryst. Solids 83*:49.

Okamoto, Y., Nakamura, K. (1979). *Electophotography (Japan) 17*:60.

Onozuka, A., Oda, O., Tsuboya. I. (1987). *Thin Solid Films 149*:9.

Orlowski, T. E., Abkowitz, M. (1986). *Solid State Commun. 59*:665.

Pai, D. M. (1974). In: *Amorphous and Liquid Semiconductors, Proceedings of the Fifth International Conference on Amorphous and Liquid Semiconductors, September 1973, Garmisch-Partenkirchen, West Germany* (J. Stuke, W. Brenig, eds.). Taylor and Francis, London, p. 355.

Pai, D. M. (1997). *J. Imaging Sci. Tech. 41*:135.

Pai, D. M., Enck, R. C. (1975). *Phys. Rev. B. 11*:5163.

Pai, D. M., Ing, S. W. (1968). *Phys. Rev. 173*:729.

Pai, D. M., Melnyk, A. R. (1986). *SPIE Proc. 617*:82.

Pai, D. M., Springett B. E. (1993). *Rev. Modern Phys. 65*:163 and references therein.

Papin, P. I., Huang, H. K. (1987). *Med. Phys. 14*:322.

Peled, A., Dror, Y. (1992). *Thin Solid Films 218*:201.

Peled, A., Friesem, A. A., Vinokur, K. (1988). *Opt. Engin. 27*:482.

Pfister, G. (1976). *Pass. Ret. Lett 36*:271.

Pfister, G. (1979). *Contemp. Phys. 20*:449.

Pfister, G., Morgan, M. (1975). *Philos. Mag. 32*:1341.

Pfister, G., Morgan, M. (1980). *Philos. Mag. B. 41*:209.

Pfister, G., Melnyk, A. R., Scharfe, M. E. (1977). *Solid State Common. 21*:907.

Pinsler, H. (1988). *J. Imaging Sci. 32*:5.

Pinsler, H., Walsdorfer, H., Lutz, M. (1986). In: *Proceedings of the Third International Congress on Advances in Non-impact Printing Technologies* (J. Gaynor, ed.). Springfield, VA, p. 7.

Que W., Rowlands, J. (1995). *Med. Phys. 22*:365.

Robertson, I. (1976). *Philos. Mag. 34*:13.

Rowlands, J. A., Hunter D. M., Araj, N. (1991). *Med. Phys. 18*:421.

Rowlands, J. A., DeCrescenzo, G., Araj, N. (1992). *Med. Phys. 19*:1065.

Rowlands, J. A., Kasap, S. O. (1997). *Physics Today*, November 1997, p. 24.

Schaffert, R. M. (1975). *Electrophotography*. Society of Photographic Scientists and Engineers. Focal Press, London, Chs. 1 and 2.

Scharfe, M. E. (1984). *Electrophotography. Principles and Optimization*. Research Studies Press Letchworth, Hertforshire, U.K., Chs. 1 and 4.

Schein, L. B. (1974). *Phys. Rev. B. 10*:3451.

Schein, L. B. (1988). *Electrophotography and Development Physics*. Springer-Verlag, New York, Chs. 1 and 2.

Schiebel, U., Hillen, W., Zaengel, T. (1986) *SPIE Proc. 626*:176.

Schottmiller, I. C. (1975). *J. Vac. Sci. Technol. 12*:807.

Schottmiller, I. C., Tabak, M., Lucovsky, G., Ward, A. (1970). *J. Non-Cryst. Solids 4*:80.

Sigai, A. G. (1975). *J. Vac. Sci. Technol. 12*:958.

Spear, W. E. (1957). *Proc. Phys. Soc. (London) B70*:669.

Spear, W. E. (1960). *Proc. Phys. Soc. (London) B76*:826.

Spear, W. E. (1961). *J. Phys. Chem. Solids 21*:110.

Spear, W. E. (1969). *J. Non-Cryst. Solids 1*:197.

Springett, B. E. (1984). *Proceedings of the Third International Symposium on Industrial Cases of Selenium and Tellurium*. Selenium-Tellurium Development Association, Grimbergen, Belgium, p. 285.

Springett, B. E. (1988). *Phosphorus Sulphur 38*:341.

Springett, B. E. (1989). *Proceedings of the Fourth International Symposium on Uses of Selenium and Tellurium*. Selenium-Tellurium Development Association, Grimbergen, Belgium, p. 126.

Springett, B. E. (1994). *Proceedings of the Fifth International Symposium on Uses of Selenium and Tellurium*. Selenium-Tellurium Development Association, Grimbergen, Belgium, p. 187.

Stephens, R. B. (1976). *J. Non-Cryst. Solids 20*:75.

Stephens, R. B. (1978). *J. Appl. Phys. 49*:5855.

Tabak, M. D. (1970). *Phys. Rev. B 2*:2104.

Tabak, M. D., Hillegas, W. J. (1972). *J. Vac. Sci. Technol 9*:387.

Tabak, M. D., Ing, S. W., Scharfe, M. E. (1973). *IEEE Trans. Electron. Dev. ED-20*:132.

Takahashi, T. (1979). *J. Non-Cryst. Solids 34*:307.

Takasaki, Y., Maruyama, E., Uda, Y., Hirai, T. (1983). *J. Non-Cryst. Solids 59–60*:949.

Takasaki, Y., Tsuji, K., Hirai, T., Maruyama, E., Tanioka, K., Yamazaki, J., Shidara, K., Taketoshi, K. (1988). *Mater. Res. Soc. Symp. Proc. 118*:387.

Taniguchi, Y., Yamamoto, H., Horigome, S., Saito, S., Maruyama, E. (1981). *J. Appl. Phys. 52*: 7261.

Tanioka, K., Yamazaki, J., Shidara, K., Taketoshi, K., Kawamura, T., Ishioka, S., Takasaki, Y. (1987). *IEEE Elec. Dev. Letts. EDL-8*:392.

Tanioka, K., Yamazaki, J., Shidara, K., Taketoshi, K., Hirai, T., Takasaki, Y. (1988). *Adv. Electron. Electron Phys. 74*:379.

Tateishi, K., Hoshino, Y. (1984). *IEEE Trans. Electron. Dev. ED-31*:793.

Tsuji, K., Ohshima, T., Hirai, T., Gotoh, N., Tanioka, K. Shidara, K. (1991). *Mater. Res. Soc. Symp. Proc. 219*:505.

Veres, J., Juhasz, C. (1993). *J. Non-Cryst. Solids 164–166*:407.

Wang Y., Champness, C. H. (1994). *Proceedings of the Fifth International Symposium on Uses of Selenium and Tellurium*. Selenium-Tellurium Development Association, Grimbergen, Belgium, p. 175.

Wang Y., Champness C. H. (1995). *J. Appl. Phys. 77*:722.

Weigl, J. W. (1977). *Agnew. Chem. Int. Ed. Engl. 16*:374.

Williams, E. M. (1984). *The Physics and Technology of Xerographic Processes*. John Wiley, New York, Chs. 1 and 2.

Wong, C. K., Lukovsky, G., Bernholc, J. (1985). *J. Non-Cryst. Solids 97–98*:1171.

Yaffe, M. J., Rowlands, J. (1997). *Phys. Med. Biol. 42*:1.

Zhang, Q. H., Champness, C. H. (1991). *Cnd. J. Phys. 69*:278.

Zhao, W., Rowlands, J. A. (1995). *Med. Phys. 22*:1595.

Zhao, W., Rowlands, J. A., Germann, S., Waechter, D., Huang, Z. (1995). *SPIE Proc. 2432*:250.

10

Photoreceptors: Organic Photoconductors

PAUL M. BORSENBERGER[†]

Eastman Kodak Company, Rochester, New York

DAVID S. WEISS

Heidelberg Digital L.L.C., Rochester, New York

10.1   INTRODUCTION

The research and development of organic photoreceptors for xerography has received considerable emphasis during the past two decades. The interest in these materials is primarily due to three considerations. First, these can be readily fabricated in large areas on a web or drum substrate; as a result, there is flexibility in designing the copier or printer architecture. Second, the electronic properties, particularly the dark resistivity and photoconductivity, are well suited for xerography. Third, relative to competitive technologies such as α-Si and the chalcogenide glasses, organic materials offer significant cost and environmental advantages. In the past two decades, organic materials have been increasingly employed and are currently used in most applications. At present, the annual revenues of these materials are approximately $5 billion. This is the major photoelectronic application of organic materials.

The first organic photoreceptor used in a copier was based on the charge-transfer complex formed between poly(N-vinylcarbazole) (PVK) and 2,3,7-trinitro-9-fluorenone (TNF), introduced in the IBM Copier I in 1970. In 1975, Eastman Kodak Company introduced the dye–polymer aggregate photoreceptor in the Kodak Ektaprint 100 copier. Both the PVK:TNF and the aggregate photoreceptors were coated on a flexible web substrate

---

[†] *Deceased*

and then fabricated into a loop configuration. The PVK:TNF and dye–polymer aggregate technologies demonstrated the suitability of organic photoreceptors for high-volume, high-quality xerographic applications.

The next significant development was the introduction of the dual-layer photoreceptor configuration. In this format, the charge generation and transport functions are separated into two adjacent layers. These are described as the generation and transport layers. Dual-layer photoreceptors offer the advantages of higher sensitivity and extended process lifetimes. This configuration was introduced by Kalle in 1976. In the late 1970s, IBM replaced the PVK:TNF photoreceptor with a dual-layer photoreceptor containing a generation layer based on the bisazo pigment Chlorodiane Blue. In 1980, Eastman Kodak Company introduced a dual-layer photoreceptor containing a dye–polymer aggregate generation layer. By the late 1980s, most organic photoreceptors were prepared in the dual-layer configuration.

During the 1980s, there was increasing emphasis on the development of photoreceptors based on pigment-containing generation layers. These can be provided in large quantities, with acceptable levels of purity at low cost. Further, many of these materials show high sensitivity in the near infrared. Pigment-based generation layers are usually prepared by dispersion coating techniques. These materials are widely used for both copier and printer applications.

This chapter discusses the use of organic photoreceptors for xerography. The general photoreceptor requirements are described. A brief review of the photoelectronic properties of organic materials is given. Fabrication techniques are discussed. The xerographic properties are reviewed with emphasis on the materials used for generation and transport layers.

For further reviews of organic photoreceptors, see Pai and Melnyk (1986), Abkowitz and Stolka (1988), Melnyk and Pai (1990), Pai (1991), Borsenberger and Weiss (1993, 1998), Law (1993), Pai and Springett (1993), Stolka and Mort (1994), and Stolka (1995). For reviews of xerography, see Schaffert (1975), Williams (1984), Burland and Schein (1986), and Schein (1995).

10.2   PHOTORECEPTOR REQUIREMENTS FOR XEROGRAPHY

10.2.1   The Xerographic Process

The xerographic process involves the formation of an electrostatic latent image on the surface of a photoconductive insulator. The latent image is made visible by toner particles, transferred to a receiver, and then made permanent by a fusing process. The overall process can involve as many as seven steps, as illustrated in Fig. 10.1. In step 1, a uniform electrostatic charge is deposited on the photoreceptor surface. This can be accomplished by a corotron or a roller charging device. In the second step, the photoreceptor is exposed with a reflected or digitally rendered image. This selectively dissipates the surface charge in the exposed regions and creates a latent image in the form of an electrostatic charge pattern. In step 3, electrostatically charged toner particles are brought into contact with the latent image. The toner particles are transferred to a paper receiver in step 4 and then fused in step 5. Fusing is normally accomplished by passing the paper receiver through a set of heated rollers. In step 6, the remaining toner particles are removed from the photoreceptor surface, usually by means of a rotating brush. Finally, in step 7, the photoreceptor is uniformly exposed to remove any remaining surface charges. Following step 7, the process can be repeated.

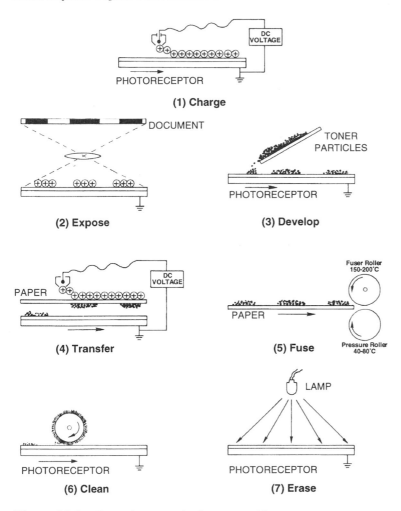

Figure 10.1 The various steps in the xerographic process.

The processes outlined in Fig. 10.1 can be carried out by either charged-area-development (CAD) (Fig. 10.2) or discharged-area-development (DAD) (Fig. 10.3) processes. In CAD, the toner particles are charged to the opposite polarity of the photoreceptor surface. The toner particles are thus attracted to the charged, or unexposed, regions of the photoreceptor. CAD processes are widely used for optical copiers. CAD exposures are usually obtained from a Xe-filled lamp for flash exposures or a quartz–halogen or fluorescent lamp for continuous or scan exposures. In DAD, the toner particles are of the same polarity as the photoreceptor surface. By means of a biased development electrode, the toner particles are attracted to the discharged regions of the photoreceptor. DAD processes are commonly used for printers. Exposures for DAD processes are usually derived from a laser or an array of light-emitting diodes. Figures 10.2 and 10.3 illustrate the CAD and DAD processes. In CAD, the photoreceptor areas that are exposed correspond to the background areas of the image that is to be reproduced. In DAD, the exposed regions correspond to the image areas. The development potential is the potential difference be-

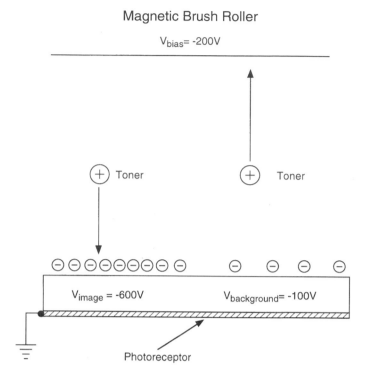

Figure 10.2   The charged-area-development (CAD) process.

tween the image and bias potentials. Photoreceptor requirements that are specific to DAD processes have been discussed by Chen (1990), Lutz (1995), Mort (1995), Jeyadev and Pai (1996), and Pai (1995).

The xerographic process places several fundamental requirements on the photoreceptor. The key requirements are that (1) the rate of thermal generation of free carriers must be extremely low, (2) the photoreceptor must have high sensitivity throughout the appropriate region of the spectrum, (3) the charge-transport processes must occur in the absence of deep trapping over an extended range of fields, and (4) the electronic properties must be stable to highly oxidizing corona atmospheres under high field and exposure conditions. Finally, the photoreceptor must have good mechanical properties, high abrasion resistance, and be amendable to large-area, low-cost manufacturing processes. These are separately discussed in the sections that follow.

10.2.2   Dark Discharge

The initial step in the xerographic process involves the deposition of electrostatic charges on the photoreceptor surface. To sustain the latent image for a period of time sufficient for the desired process development, the rate of dark discharge of the surface potential must be extremely low.

There have been many dark discharge models described in the literature. For a review, see Borsenberger and Weiss (1993, 1998). Most are premised on the assumption that the thermal generation process creates a free and deeply trapped carrier of opposite sign. The discharge that occurs under these conditions is described as depletion discharge and has been treated by Abkowitz and coworkers (Abkowitz and Markovics, 1982, 1984;

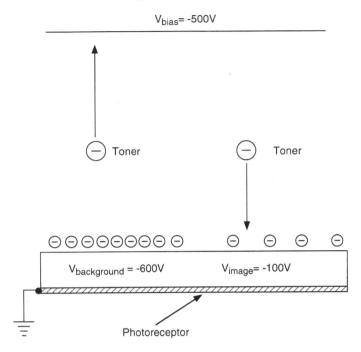

Figure 10.3 The discharged-area-development (DAD) process.

Abkowitz and Jansen, 1983; Abkowitz et al., 1985; Abkowitz, 1987, 1987a; Abkowitz and Maitra, 1987). The results are described in two time zones, $t \lessgtr t_d$. Here t_d is a depletion time, defined as the time when the thermally generated bulk space charge is equal to one-half of the charge initially deposited on the surface. For $t < t_d$,

$$\frac{dV}{dt} = -\frac{L^2}{2\varepsilon \varepsilon_0} apt^{p-1} \tag{1}$$

For $t > t_d$,

$$\frac{dV}{dt} = -\frac{\varepsilon \varepsilon_0}{2a} \left(\frac{V_0}{L}\right)^2 pt^{-p-1} \tag{2}$$

At $t = t_d$,

$$\rho(t_d) = \frac{\varepsilon \varepsilon_0 V_0}{L^2} \tag{3}$$

and

$$t_d = \left(\frac{2V_d \varepsilon \varepsilon_0}{aL^2}\right)^{1/p} = \left(\frac{\varepsilon \varepsilon_0 V_0}{aL^2}\right)^{1/p} \tag{4}$$

In Eqs. (1) to (4), L is the thickness, ε the dielectric constant, ε_0 the permittivity of free space, V_0 the initial potential, and a and p are constants. For an exponential distribution of energies, p is inversely proportional to the width of the distribution. For a discrete energy, $p = 1$. The key assumption of the model is a field-independent thermal generation process that creates a free and deeply trapped carrier of opposite sign.

For other treatments of dark discharge, see Kasap et al. (1987, 1992), Kasap (1989), and Scott and Lo (1990). For treatments of the charge acceptance, see Pai (1988), Jeyadev and Pai (1995), and Mishra and Pai (1996).

10.2.3 Photogeneration

To form an electrostatic latent image, a photoreceptor is charged to an initial potential and then exposed to a pattern of radiation that corresponds to the image that is to be reproduced. The absorption of the image exposure creates bound electron–hole pairs, a fraction of which separate and migrate to the appropriate surfaces. The exposure required to create the latent image is determined by the efficiency by which free electron–hole pairs are created in conjunction with any loss mechanisms, such as recombination or trapping. A key parameter is the photogeneration efficiency, the ratio of the number of free electron–hole pairs created to the number of photons absorbed. The models that have been most widely used to describe photogeneration phenomena of organic materials are based on surface-enhanced exciton dissociation or geminate recombination. For reviews, see Pope and Swenberg (1982, 1984), Pope (1989), Silinsh (1989), and Silinsh and Capek (1994).

Surface-enhanced exciton dissociation arguments are premised on the assumption that the absorption of a photon creates an exciton that diffuses to the surface where it either recombines or dissociates into a free electron–hole pair, or a free and a deeply trapped carrier of opposite sign. These arguments were first proposed in the early 1960s. Many early studies of anthracene were described by surface dissociation models. These were usually based on the assumption that the dissociation occurred at surface sites occupied by O_2 that act as deep electron traps. Most studies of the phthalocyanines have been explained by surface-enhanced exciton dissociation arguments, and usually attributed to the presence of O_2. For a review of exciton dissociation processes in the phthalocyanines, see Mizuguchi (1987). Photogeneration has also been described by surface-enhanced exciton dissociation arguments involving charge transport molecules (Umeda and Hashimoto, 1992; Umeda et al., 1993; Umeda, 1994, 1998, 1999; Umeda and Niimi, 1994, 1994a; O'Regan et al., 1995, 1996; Molaire et al., 1997; Umeda and Yokoyama, 1997; Popovic et al., 1999).

The limitation of exciton dissociation models is that they are based solely on exciton diffusion length considerations and provide little further insight into the physical processes involved in photogeneration. Further, these do not address the field or temperature dependencies of the photogeneration process, which are of central relevance to xerography.

Geminate recombination is the recombination of an electron with its parent cation. Geminate recombination arguments are based on the assumption that the formation of a free electron–hole pair involves the dissociation of an intermediate charge-transfer state. There are many references in the literature to geminate recombination in organic solids, as well as organic liquids, α-Se, and α-Si. The most widely cited are based on theories due to Onsager (1934, 1938).

The Onsager theories are derived from the Smoluchowski (1916) equations that describe the escape probability of an electron from its parent cation in the presence of a field. In models based on the Onsager theories, free carriers are assumed to be created by a two-step process. The first involves photon absorption and the creation of a charge-transfer state. The probability of creating the state is described by a primary quantum yield η_0. In the second step, the state can either dissociate into a free electron and a free hole, or recombine. The photogeneration efficiency η is then given as the product of the efficiency of creating the state η_0 and the dissociation probability.

The Onsager expression for the photogeneration efficiency is

$$\eta = \eta_0 \left(1 - \frac{kT}{eEr_0} \sum_{j=0}^{\infty} I_j \frac{eEr_0}{kT} \right) \tag{5}$$

where

$$I_j(x) = I_{j-1}(x) - \frac{\exp(-x)x^j}{j!} \tag{6}$$

$$I_0(x) = 1 - \exp(-x) \tag{7}$$

and E is the field and r_0 the electron–hole separation distance of the charge-transfer state. The derivation of Eq. (5) is based on the assumption that the distribution of electron–hole separation distances can be described by an isotropic δ-function. In the literature, r_0 is usually described as a thermalization distance. The Onsager formalism leads to strongly field-dependent photogeneration efficiencies that approach nonzero values as the field goes to zero. At high fields, the efficiencies approach a limiting value. Field-dependent photogeneration was first described in the mid 1960s and has since been observed in virtually all organic materials described in the literature.

In the past two decades, most photogeneration studies of organic solids have been described by the Onsager theories. The limitations of the formalism are that the Onsager solutions to the Smoluchowski equations are based on a somewhat arbitrary set of boundary conditions and the absence of a theory to describe the thermalization process. Most models based on the Onsager theories are premised on the assumption that the thermalization distance is independent of field and temperature. These are highly questionable assumptions. For alternative treatments of geminate recombination, see Rackovsky and Scher (1984, 1988), Silinsh and Jurgis (1985), Ries and Bässler (1987), Berlin et al. (1990), Racovsky (1991), Arkhipov and Nikitenko (1993), Scher (1993), and Albrecht and Bässler (1995). For a review, see Pai (1985).

10.2.4 Charge Transport

Xerographic photoreceptors can be prepared in the single- or dual-layer configuration, as shown in Figs. 10.4 and 10.5. For either configuration, the photoinduced discharge must occur in the absence of trapping and within a time commensurate with the process development requirements. The photoreceptor property that determines the discharge time is the mobility, the ratio of the carrier drift velocity to the field. For complete discharge in a range of 10 ms, mobilities in excess of 10^{-5} cm^2/Vs are typically required.

In the past decade, most studies of organic materials have been described by a formalism based on disorder, due to Bässler and coworkers (Bässler, 1993, 1994, 1994a, 1994b; Hartenstein and Bässler, 1995; Hartenstein et al., 1996; Borsenberger et al., 1998; Wolf et al., 1997; Visser et al., 1999). The disorder formalism is premised on the argument that charge transport occurs by hopping through a manifold of localized states that are distributed in energy and distance. The principal assumptions are that (1) the distributions of site energies and distances are Gaussian, (2) hops to higher site energies are scaled by a Boltzmann probability, while hops to lower energies occur with a probability of unity, (3) polaronic effects can be neglected, and (4) the process is incoherent and phase memory

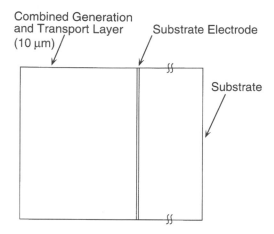

Figure 10.4 The single-layer photoreceptor configuration. Thickness of single-layer photoreceptors are typically 10 to 15 µm. The substrate serves as the electrode in drum photoreceptors. In this
configuration, the photoreceptor can be discharged with either a positive or a negative potential.

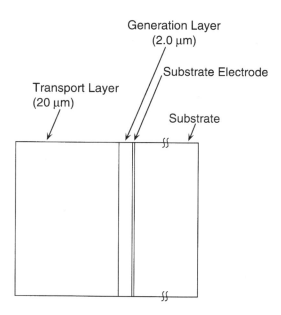

Figure 10.5 The dual-layer photoreceptor configuration. Transport layer thicknesses are typically
15 to 30 µm. Generation layers are usually between 0.5 and 5.0 µm. The substrate serves as the
electrode in drum photoreceptors. In the configuration illustrated, the polarity of the surface potential
must be opposite to the conductivity type of the transport layer. Hole transport layers require negative
surface potentials and vice versa.

is lost after every jump. Of these, the key assumption is the formalism for computing the hopping rates. The validity of this approach has been reviewed by Kenkre and Dunlap (1992) and Dunlap (1995). Due to the asymmetry in the hopping rates, any analytic treatment becomes difficult, particularly for the case of a Gaussian distribution of site energies. For this reason, predictions of the formalism have been largely derived from Monte Carlo simulations. For a discussion of the simulation procedures, see Bässler (1993). At high fields, the simulations give (Borsenberger et al., 1991)

$$\mu(\hat{\sigma}, \Sigma, E) = \mu_0 \exp\left[-\left(\frac{2\hat{\sigma}}{3}\right)^2\right] \exp[C(\hat{\sigma}^2 - \Sigma^2)E^{1/2}] \tag{8}$$

where μ is the mobility, σ the energy width of the hopping site manifold, $\hat{\sigma} = \sigma/kT$, Σ the degree of positional disorder, E the field, μ_0 a prefactor mobility, C an empirical constant of 2.9×10^{-4} $(cm/V)^{1/2}$, and kT has its usual meaning. Equation (10.8) is valid only for fields above a few multiples of 10^5 V/cm and $T_g > T > T_c$, where T_g is the glass transition temperature and T_c the nondispersive-to-dispersive transition temperature. At low fields, the mobility is field independent or, in the case of large Σ, increases as the field is further reduced. The materials parameters of the formalism are σ, μ_0, and Σ.

From Equation (8), the basic predictions of the formalism are the field and temperature dependencies of the mobility and the temperature dependence of the field dependencies of the mobility. These agree with experimental results reported for a wide range of molecularly doped polymers, main chain and pendant group polymers, as well as vapor-deposited molecular glasses (Borsenberger et al., 1993).

For other treatments of transport, see Sahyun (1984), Movaghar (1987, 1991), Kenkre and Dunlap (1992), Schein (1992), Dunlap and Kenkre (1993), Novikov and Vannikov (1993, 1994, 1994a), Bässler et al. (1994), Gartstein and Conwell (1994, 1994a, 1995), Dunlap (1995a, 1996, 1997), Parris (1995, 1996, 1997), Gartstein et al. (1995), Dunlap et al. (1996), Parris et al. (1997), and Novikov et al. (1998).

10.2.5 Recombination

Electron–hole recombination depends on the product of the electron and hole concentrations and thus may become important for high-intensity exposures. In low mobility materials, recombination is usually explained by a theory due to Langevin (1903). Langevin recombination arguments are based on the assumption that the recombination process is diffusion controlled. This requires that the electron and hole mean free paths be small compared to the Coulomb radius.

Recombination-limited discharge models have been described by Collins (1974), Fox (1974), Kerr and Rokos (1977), Chen (1978), and Mey et al. (1979). Recombination plays an important role in determining the photodischarge characteristics with high-intensity exposures such as those used in certain digital applications.

10.2.6 Photoreceptor Evaluation

Latent image formation involves the creation of free electron–hole pairs by the absorption of an image exposure and the displacement of these carriers by a field. These processes are determined by the photogeneration efficiency and mobility. Experimental techniques for their characterization are thus of considerable importance to xerography.

Photogeneration efficiencies are usually derived from potential discharge measurements. These can be made with continuous or flash exposures. The distinction is based on the transit time. Continuous exposures are long compared to the transit time, while flash exposures are short compared to the transit time. Continuous exposures are usually monochromatic. The results can be described by the xerographic gain. The gain is defined as the ratio of the rate of decrease of the surface charge density to the incident photon flux,

$$G_x = -\frac{\varepsilon \varepsilon_0}{LeI}\left(\frac{dV}{dt}\right) \tag{9}$$

where I is the photon flux and dV/dt the rate of photoinduced discharge. As defined in Eq. (9), the maximum gain is unity. In the absence of trapping and recombination, the gain is equal to the photogeneration efficiency η when normalized to the absorbed photon flux. The gain is commonly used to characterize the spectral dependence of the sensitivity.

Flash exposures are usually made with constant initial potentials and multiple exposures of different intensities. The surface potentials are sampled at a time well in excess of the transit time, typically 1 to 2 s. Normally, the sample is erased after each exposure. The results are usually described by the exposure required to discharge the photoreceptor from an initial potential V_0 to an arbitrary potential, typically $V_0/2$, at a specified time after the exposure. The exposures are usually expressed in erg/cm^2 or μJ/cm^2. The wavelength, the exposure time, the time at which the potential is sampled, and the erase conditions must be specified. Results of these measurements can also be used to determine the photogeneration efficiency, although the analysis is somewhat more complex than for continuous exposures. The sensitivity is a useful metric for the evaluation of photoreceptors. Comparisons are difficult, however, when measurements are made with different conditions. For practical evaluations, it is usually necessary to evaluate the gain or sensitivity with cycling. Commercial instrumentation has been developed for this purpose (Tse, 1994, 1996).

Potential discharge measurements are also used for image evaluation. By dividing the overall xerographic process into the different subsystems and then describing each subsystem with a transfer function, the output image density can be related to the input density through a four-quadrant plot (Paxton, 1978; Pai and Melnyk, 1986). In this manner, the effects of changes in the various subsystems on the output density can be determined.

The standard method for measuring mobilities in materials with long dielectric relaxation times is the time-of-flight photocurrent technique, first described by Haynes and Shockley (1951) and Lawrance and Gibson (1952). These techniques were developed in considerable detail during the 1950s and 1960s by Brown (1955), Spear (1957, 1960), Kepler (1960), Le Blanc (1960), and others who successfully applied these techniques to a wide range of both crystalline and amorphous solids. By this method, the time for a sheet of carriers generated by a short flash of radiation to transit a sample of known thickness L is measured. The mobility can then be determined from the expression

$$\mu = \frac{L^2}{t_0 V} \tag{10}$$

where t_0 is the transit time of the carrier sheet.

For a review of potential discharge and photocurrent transient techniques, see Melnyk and Pai (1993).

Digital xerography has placed new requirements on photoreceptor materials. Because most digital engines use DAD processes, any microscopic defects in unexposed areas will have lowered surface potentials, and small spots will be developed. These are sometimes called breakdown spots because of the inability of these regions to maintain the surface charge. The need to examine photoreceptors for such defects has led to new measurement techniques. The evaluation of microscopic electrical defects has been treated by Popovic et al. (1991), Lin et al. (1993), Prichard and Uehara (1993), and Lin and Nozaki (1995). Technologies have also been developed to determine the presence of such defects by examining the photoreceptor toned image (Tse, 1994, 1996).

10.3 FABRICATION

Photoreceptor fabrication involves the sequential application of one or more layers onto a substrate. The coating methods depend on whether the substrate is a flexible web or a metal drum. Flexible webs, such as poly(ethylene terephthalate), are usually metallized; Ni or Al are common. The substrate conductivity requirements have been discussed by Chen (1993). Polymeric electrodes (Katsen et al., 1994) have been used in some applications. With metal drums, the drum itself serves as the electrode. In either case, the substrate sometimes receives a very thin preliminary polymer coating as an interlayer between the substrate and the photoconductive layers. The interlayer has several purposes. It may serve as a blocking layer to prevent charge injection, which would otherwise give rise to excessive dark discharge and/or localized breakdown. The interlayer may also serve as an adhesive to ensure adequate bonding of the subsequently applied photoreceptor layers. Finally, the interlayer may serve to smooth a rough substrate surface, or alternatively it may provide a roughened surface to prevent the formation of interference patterns. Because of these many features, the choice of interlayer materials depends on the characteristics desired. Polymers such as nylons, polyvinyl chlorides, and sarans have been used. An overcoat layer may also be provided to protect the photoreceptor from chemical and/or mechanical damage. Finally, to fabricate an image loop, the ends of the appropriate length sections must be joined. This can be accomplished by an adhesive splice or by ultrasonic bonding techniques. The seam region is positioned outside the image area to prevent the production of an image artifact. As long as the seam does not interfere with belt alignment or the development and cleaning subsystems, it can be neglected.

10.3.1 Solvent Coating Techniques

For research purposes, solutions or pigment dispersions can be readily coated by a doctor blade. For commercial applications, materials must be produced under more carefully controlled conditions. This necessitates the use of coating machines that permit the precise control of the many coating variables. Among the more important are web stability and velocity, solution delivery rate and geometry, and drying conditions. Solvent emission control is also an important component of the coating process. Because organic materials are generally water insoluble, organic solvents are commonly used. The factors influencing the choice of solvent are solubility, solvent evaporation rate, solution surface tension, and toxicity. Ketones (acetone and methyl ethyl ketone), alcohols (methanol), chlorinated hydrocarbons (dichloromethane or 1,2,2-trichloroethane), aromatics (toluene), ethers (tetrahydrofuran), esters (methyl acetate), and acetonitrile are commonly used. The *Federation Series on Coatings Technology* comprises relevent reviews including *Organic Pigments* by Lewis (1988), *Solvents* by Ellis (1986), *Coating Film Defects* by Pierce and Schoff (1988), and *Film Formation* by Wicks (1986). In addition, monographs such as *Coating and Drying Defects* by Gutoff and Cohen (1995) and *Organic Coatings: Science and Technology, Volumes 1 and 2* by Wicks et al. (1992, 1994) are excellent sources of practical information.

While there are several coating methods available, a common technique is to pump the coating solution or dispersion from a hopper through a slot and onto the web. The design of the hopper assembly, the shape of the hopper lip, and the relative position and distance from the web are all critical. The optimum conditions will be specific to the particular coating under investigation. There are many coating defects to be avoided. These include surface patterns, thickness variations, voids, cracks, streaks, blush, and mottle. These can frequently be eliminated by modifications of the coating conditions, such as the elimination of vibrations, the prevention of static discharge in the roller nip, adjustments of hopper geometry, and changes in coating speed and drying conditions. Other defects are best controlled by modifying the coating solution.

10.3.2 Dispersion Methods

Most photoreceptors have generation layers composed of pigment dispersions. These are fabricated by solvent coating methods. Techniques for the large-scale preparation of such dispersions are well known, especially for applications in the paint industry (Coulson and Richardson, 1983). The pigment is usually received as a powder cake or as crystal nuggets and must be crushed to a small size before milling. Mills of various types can reduce the pigment to submicron size. Devices commonly used are ball or vibrating mills, where small balls of a hard material such as stainless steel or a ceramic are placed in a slurry of the pigment in a nonsolvent. The action of rolling, or vibration, imparts sufficient energy to break the pigment into small primary particles. Roll mills are common in the paint industry. In these devices, two (or more) rollers move at different speeds in opposite directions. The pigment is circulated between the rollers until the desired combination of uniformity and particle size is attained. Other methods use centrifugal and sand or salt milling techniques. It is usually necessary to treat the milled pigment further to prepare the final dispersion solution. In the paint industry, the techniques are specific to each pigment due to the need to control pigment crystal structure, morphology, and surface characteristics as well as the extent of aggregation (McKay, 1989). Ball milling is the most commonly used technique for the preparation of photoreceptor dispersions. For environmental reasons, it is desirable to coat the generation layer from aqueous dispersions. Hoshino et al. (1990) discuss one such application.

10.3.3 Vapor Deposition Techniques

In some cases, generation layers can be prepared by vapor deposition techniques. Examples of materials that have been prepared by these methods include various phthalocyanine and perylene derivatives. The major advantage of these techniques is that a very thin layer can be deposited in a controlled manner with excellent uniformity. The principal disadvantage is that vapor-deposited pigment films are usually amorphous. To achieve optimum sensitivity, the material must be converted to a crystalline state. This requires a subsequent thermal or solvent conversion process. A further disadvantage is that vacuum deposition methods are slow and expensive compared to solvent coating techniques.

10.3.4 Drum Techniques

In principle, drum coating can be carried out in a manner similar to that of web coating. The simplest method is dip coating, where the drum is withdrawn from a coating solution.

The film thickness increases with increasing viscosity and withdrawal speed and decreases with increasing surface tension. An alternative technique is ring coating. In this method, the drum is slowly withdrawn through an annulus, which determines the coating thickness. In addition, various spray coating methods may be used. The preferred method must be selected for each application. Virtually all desktop printers use drums with diameters of approximately 30 mm. Larger diameter drums have been produced, however, such as the 140 mm drum used in the Xeikon Digital Color Press (De Schamphelaere et al., 1994).

10.3.5 Overcoat Layers

Organic photoreceptors are relatively soft and easily deformed, making them susceptible to physical damage and wear. Further, they are sensitive to the deleterious effects of corona exposures. Many overcoat technologies have been investigated in order to improve the photoreceptor stability and process life. The overcoat conductivity must be carefully controlled. If it is too insulating, a residual potential will develop during cycling. If it is too conductive, the latent image will spread. Therefore overcoats must be formulated for an optimized conductivity. Several approaches have been described and some commercialized.

Polymeric materials are one approach. Sato et al. (1992) described an overcoat composed of a polyurethane and a silicone. The best image quality was obtained by the use of a hydrophobic silica in an overcoat with a high urethane content. Yamamoto et al. (1982) described an overcoat composed of a polyester doped with 1,1′-dimethylferrocene. The overcoat was applied by spray coating and had a conductivity of approximately 10^{-10} $(\Omega\ \text{cm})^{-1}$. The photoreceptor had satisfactory image quality at elevated humidities.

Another approach involves the use of the polysiloxanes (Weiss et al., 1999). These materials are extremely tough and wear resistant. The conductivity must be carefully controlled by doping. An example of this technology is UltraShield™ from Optical Technologies Corp. (Cornelius, 1994). This material has been successfully used in both drum and web applications. A disadvantage of this material is that the dark conductivity is sensitive to environmental conditions, which can lead to latent image degradation. Polysiloxane overcoats are prepared from aqueous solutions using standard solvent coating technologies. These must be thermally cured to achieve the desired physical properties.

Another technique is to utilize a very thin layer of a refractory material such as SiC, SiN, or diamondlike carbon (DLC). These materials are usually prepared by plasma-enhanced chemical-vapor-deposition processes. Sato and Hisada (1996) described the use of a pulsed molecular beam process to prepare photoreceptor drums with DLC overcoats. DLC has been commercialized as Diamond 4™ (HDS Corp.) for drum applications. The dark conductivity of this layer must be controlled to minimize environmental effects. Kochelev et al. (1996) have compared DLC and polysiloxane overcoats.

10.4 XEROGRAPHIC PHOTORECEPTORS

10.4.1 Configuration

Organic photoreceptors are between 10 and 50 μm in thickness and can be prepared in a single- or a dual-layer configuration. Single-layer photoreceptors are composed of a pigment in a polymer doped with either a donor or an acceptor molecule and are typically between 10 and 15 μm in thickness. In the dual-layer configuration, the photoreceptor is prepared with two separate layers. The generation layer contains a pigment and is usually coated adjacent to the substrate with the transport layer uppermost. Transport layers contain either a donor or an acceptor molecule in a polymer. Thicknesses of dual-layer photoreceptors are usually between 15 and 50 μm. Most transport layers are doped with donor molecules and thus transport only holes. For this reason, dual-layer photoreceptors are usually charged negatively. When it is desired to charge the photoreceptor positively, the layers may be inverted, so that the transport layer is coated on the substrate and the charge generation layer is uppermost.

10.4.2 Charge Generation Materials

Generation materials are generally classified according to their chemical structure. The physical forms may be molecular complexes or pigments. In either the single- or the dual-layer arrangement, both electrons and holes must transit the generation layer thickness in the absence of trapping to affect complete photodischarge. For this reason, generation layers usually contain high pigment concentrations and are fabricated as thin as possible while maintaining the desired absorption characteristics. Molecular complexes were widely used in early applications. More recently, pigments have received considerable attention. Both vapor-deposited and dispersion layers have been used.

Because of the long history of the synthesis and commercialization of pigments, the desired physical and chemical characteristics can be largely predicted and achieved. Most pigments exhibit polymorphism (Dunitz and Bernstein, 1995; Gavezzotti and Filippini, 1995), and the crystal form can be influenced by various finishing processes (Hunger and Merkle, 1983). Typical of these are solvent treatments, the introduction of surfactants, and thermal treatments. Because of the low solubilities of pigments, specialized purification technologies such as solvent extraction, acid pasting, and train sublimation are often used. Crystal modification is known to affect color, tinctorial strength, shade, and lightfastness. Other important factors are crystal morphology, degree of crystallinity, particle size distribution, and surface conditioning. The advent of semiconductor lasers and light-emitting diodes has necessitated generation layers with sensitivity in the near infrared. Thus there has been interest in predicting and producing crystal modifications with the desired absorption characteristics.

The properties of generation materials that are of principal interest are the useful spectral range and the sensitivity. The former is determined by the absorption spectrum and the latter by the efficiencies of carrier generation and injection into the transport layer. In the section that follows, we discuss different generation materials. These are classified as molecular complexes (donor–acceptor charge transfer and aggregates), pigments (poly-azos, phthalocyanines, squaraines), polycyclic aromatics, and other materials.

Molecular Complexes

Charge-transfer complexes have a long history as generation materials. Charge-transfer complexes of various electron acceptors with poly(N-vinylcarbazole) (PVK) have been extensively studied. The complex formed between PVK and 2,4,7-trinitro-9-fluorenone (TNF) (Shattuck and Vahtra, 1969; Schaffert, 1971; Melz, 1972; Vahtra and Wolter, 1978) was the first organic photoreceptor to be used in a copier. The field and spectral dependencies of the xerographic gain have been described by the Onsager formalism (Melz, 1972; Kato et al., 1983; Andre et al., 1989). Values of the thermalization distance and primary quantum yield were approximately 35 Å and 0.20. The use of the fullerenes (C_{60} and C_{70}) as electron acceptors in sensitizing the photoconductivity of PVK (Wang, 1992; Chen et al., 1996) demonstrates the continuing interest in this class of materials.

Dye–polymer aggregate photoreceptors are a class of two-phase materials that contain a light-sensitive crystalline phase dispersed in a polymer. Usually a donor molecule is added to enhance the sensitivity. Perlstein and Borsenberger (1982) have reviewed the photoelectronic properties of these materials. The dye–polymer aggregate phase is a highly colored filamentary structure. The dye can be any of several aryl-substituted pyrylium or thiapyrylium salts, while the polymer is bisphenol-A polycarbonate. The dye–polymer aggregate can be prepared by solvent fuming a homogeneous coating of a pyrylium or thiapyrylium dye and the polycarbonate. Aggregation shifts the absorption maximum from typically 580 to 680 nm. Substitution of the seleno- or telluropyrylium dye for the thiapyrylium shifts the absorption maximum to longer wavelengths. Dye–polymer aggregate photoreceptors can be prepared in the single- or dual-layer configuration (Nguyen and Weiss, 1988, 1989). In the dual-layer format, the aggregate-containing layer comprises the generation layer.

Figure 10.6 shows the spectral dependencies of the xerographic gain of a mixture of 4-(4-dimethylaminophenyl)-2,6-diphenylthiapyrylium perchlorate, the polycarbonate, and the donor molecule 4,4′-bis-(diethylamino)-2,2′-dimethyltriphenylmethane before and after aggregation by fuming with dichloromethane. Aggregation results in a considerable increase in the gain. Both electrons and holes are mobile in the aggregate phase, while only holes are mobile in the homogeneous precursor. In the absence of carrier range limitations, the xerographic gain is symbatic with the absorption spectrum. Similar results are obtained for positive or negative potentials. The field dependence of the xerographic gain has been described by the Onsager theory with thermalization distances between 30 and 60 Å and a primary quantum yield of approximately 0.50. The thermalization distance

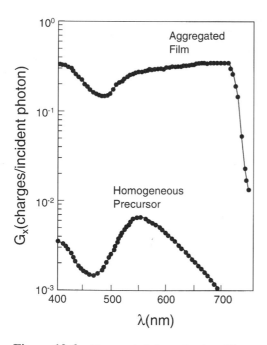

Figure 10.6   The spectral dependencies of the xerographic gain of a mixture of 4-(4-dimethylaminophenyl)-2,6-diphenylthiapyrylium perchlorate, bisphenol-A polycarbonate, and the donor molecule 4,4′-bis-(diethylamino)-2,2′-dimethyltriphenylmethane before and after solvent fuming with dichloromethane. The thickness was 20 μm, the temperature 296 K, and the field 4.0×10^5 V/cm. The measurements were made with a positive surface potential.

increases with increasing donor concentration. The photogeneration is believed to occur via a singlet exciton that diffuses to the interface between the crystalline and homogeneous phases and then dissociates into a free electron and a free hole via an interaction with the donor molecule (O'Regan et al., 1995, 1996; Molaire et al., 1997).

Reciprocity phenomena in aggregate materials have been described by Mey et al. (1979). A loss in sensitivity of 0.14 logε was observed and explained by Langevin recombination.

Polyazo Materials

Azo dyes and pigments have been important commercial products for the past century. Following the discovery of diazonium compounds in 1858, and the first azobenzene in 1861, this class of compounds was rapidly exploited for the production of dyes and pigments. One of the reasons for the success of this class is their relative ease of synthesis. Thus large numbers of compounds can be readily prepared in pure form. The basic synthetic steps, diazotization and coupling, are outlined in Fig. 10.7. The starting materials are primary aromatic amines, of which aniline (shown) is the simplest example, plus a coupling compound. For photoreceptor applications, the latter is typically an anilide of an *o*-hydroxyaromatic carboxylic acid. The example shown uses 2-hydroxy-3-naphthanilide. This compound is commercially available as Naphthol AS. The synthesis commences with the preparation of the diazonium salt from the amine. This is accomplished by reaction of the amine with cold aqueous nitrous acid. The coupling reaction is a classical electrophilic aromatic substitution. Being a very weak electrophile, the diazonium ion is most reactive toward activated (electron-rich) aromatics. Hydroxyl substituents are common. Depending on steric effects, such substituents direct the coupling to the *o* or *p* positions. In Fig. 10.7, only the *o* position is available. The final product is readily isolated from the reaction mixture in crystalline form. Subsequent treatments, such as stirring with hot dimethylformamide followed by water washing (Cort et al., 1983), yield the α-form. This pigment form is readily milled to produce stable dispersions.

Azo compounds are classified according to the number of azo groups: mono, bis or dis, tris, and tetrakis. These are prepared from the appropriate mono-, di-, tri-, and tetraamino-substituted starting material. To describe the complex structures of the azo dyes used in generation layers, it is helpful to locate the portions of the molecule that were the original hydroxyaromatic coupling moiety, and the aromatic amine (I and II, respectively, in Fig. 10.7). In virtually all the literature on azo pigments as generation materials, the molecular structures are drawn as azo ($-N=N-$) compounds. Law et al. (1993), however, have shown that azo compounds prepared from 2-hydroxy-3-naphthanilide exist exclusively as the hydrazone tautomer. Pacansky and Waltman (1992) studied Chlorodiane Blue and came to the same conclusion. Ono et al. (1991) carried out similar studies on a series of bisazo pigments prepared from 2,5-bis(4-aminophenyl)-1,3,4-oxadiazole and a variety of couplers. Thus the azo tautomers are not always accurate structural representations of these materials, although the molecules continue to be drawn in this manner because it is easier to identify the structural units. Figure 10.7 shows the hydroxyazo to keto-hydrazone tautomerism.

Figure 10.8 shows representative azo pigments that have been used as generation layers (Nakanishi, 1987). Bisazo compounds dominate, although tris and tetrakis compounds are being increasingly used for applications that require sensitivity in the near infrared. The first use of an azo pigment as a photoreceptor was in Chlorodiane Blue. This material was used in early IBM copiers and printers. Both single- and dual-layer

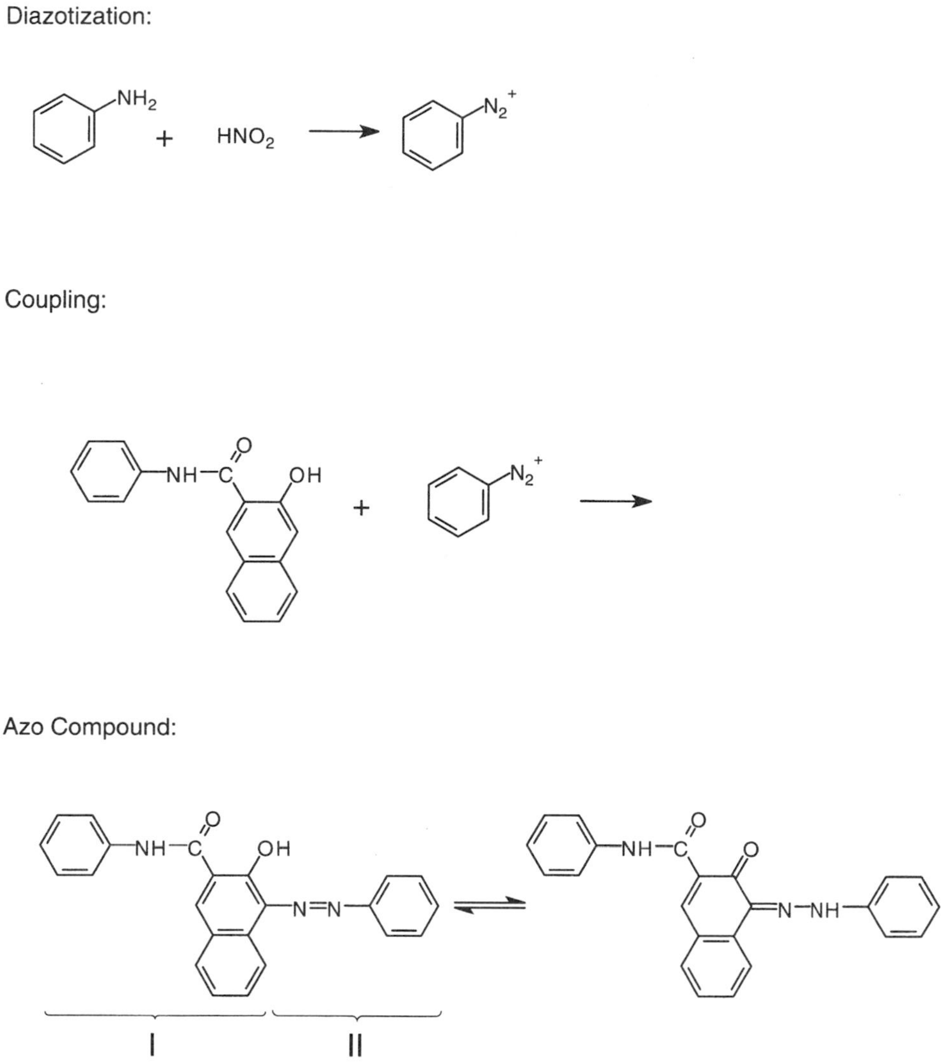

Diazotization:

Coupling:

Azo Compound:

Figure 10.7 Azo compound synthesis and the molecular structure of the hydroxy-azo tautomer showing the positions of the coupler (I) and the amine (II) components. The keto-hydrazone tautomer is shown in equilibrium with the hydroxy-azo tautomer.

structures were used. The dual-layer configuration was preferred because of a considerable increase in lifetime and sensitivity (McMurty et al., 1984). Permanent fatigue occurred with excessive exposure to room light.

Single- and dual-layer photoreceptors using Chlorodiane Blue have been described by Khe et al. (1984). The sensitivity was improved by the addition of an indoline electron donor. Kakuta et al. (1981) showed that the sensitivity depends on the ionization potential of the donor molecule of the transport layer. Pacansky and Waltman (1992) reported that the pigment in the generation layer is in the azo tautomer, as initially coated from a solvent mixture of ethylenediamine and tetrahydrofuran, but it converts to the hydrazone tautomer

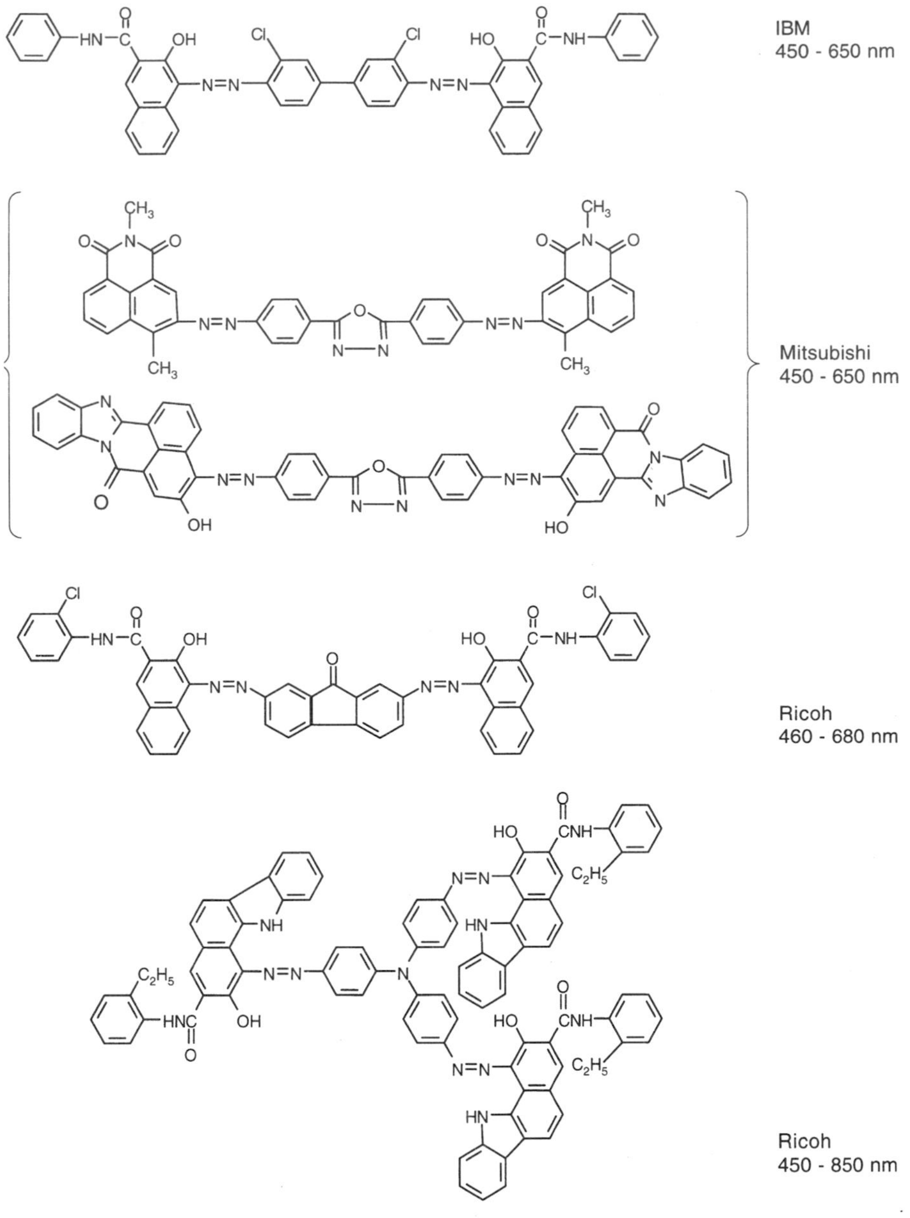

Figure 10.8 Polyazo compounds used in pigment dispersion charge generation layers.

during the subsequent transport layer overcoating process, which uses a solvent mixture of tetrahydrofuran and toluene.

The photoreceptors shown in Fig. 10.8 have been described by Nishijima (1985) and Ohta (1986). The trisazo material was used in a photoreceptor developed for printer applications (Seki et al., 1989). Umeda and Hashimoto (1992) and Umeda et al. (1993) have carried out extensive studies on the photogeneration mechanisms for this pigment. The more recently disclosed Mitsubishi materials have been developed for higher sensitiv-

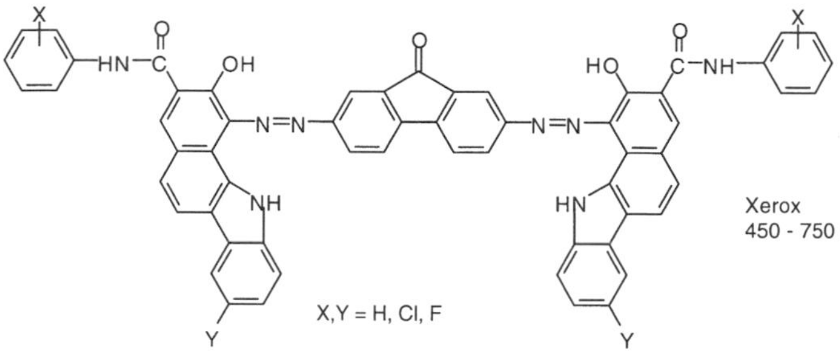

Xerox
450 - 750 nm

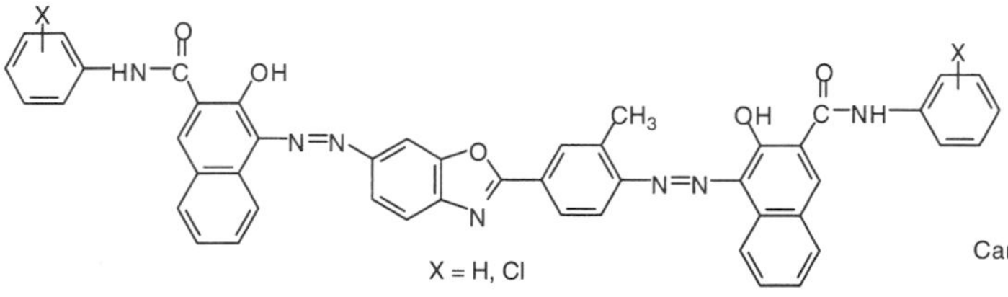

X,Y = H, Cl, F

X = H, Cl

Canon

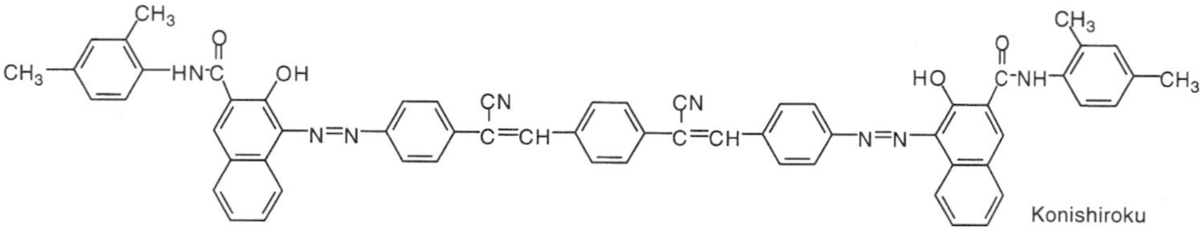

Konishiroku

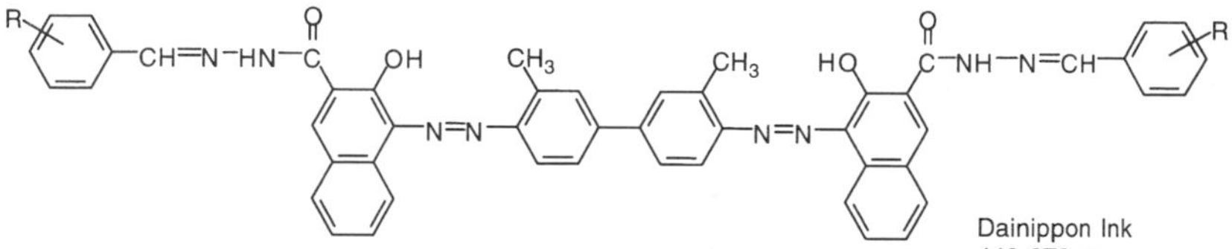

Dainippon Ink
440-670 nm

ity and increased thermal and fatigue resistance (Otsuka et al., 1986). By using a coupling component that is a mixture of isomers, the perinone bisazo pigment is a mixture of three isomers. The x-ray diffraction peaks were very broad, indicative of low crystallinity and small particle size. Various treatments were unsuccessful in modifying these crystal properties. Dual-layer photoreceptors were prepared using generation layer dispersions. The transport layers contained mixtures of different hydrazones. These materials have significant blue absorption; thus the sensitivity decreased sharply for wavelengths below 470 nm. The xerographic gain was much less field dependent than for Chlorodiane Blue.

There are a few studies on the effects of different substituents with a given parent pigment and the importance of the crystal modification. In a study by Hashimoto (1986), several bisazo pigments based on diaminofluorenone were synthesized and characterized. Table 10.1 shows the results for pigments related to the bisazo Ricoh entry in Fig. 10.8. Here X refers to the Naphthol AS phenyl ring substituents. The long wavelength absorption edge and absorption maximum were not systematically altered by the phenyl substitu-

Table 10.1 The Effects of X Substituents on the Absorption and Sensitivity of a Series of Dual-Layer Photoreceptors Using Fluorenone-Based Bisazo Pigments in the Generation Layer, Fig. 8. The Sensitivity is the Energy Required to Discharge the Photoreceptor to $V_0/2$

X	Absorption edge (nm)	Absorption maximum (nm)	$V_0(V)$	$\varepsilon(erg/cm^2)$
H	673	610	−1064	15.2
4-OCH$_3$	678	620	−1246	24.4
2-CH$_3$	665	620	−1090	6.4
3-CH$_3$	682	620	−990	12.0
4-CH$_3$	674	620	−1258	20.4
2-Cl	650	590	−690	1.9
3-Cl	655	590	−440	2.0
4-Cl	668	640	−844	8.6
2-NO$_2$	684	640	−1192	4.9

Source: After Hashimoto, 1986. Reprinted from *Denshi Shashin (Electrophotography)*, *25*:10. Copyright 1986, Japanese Electrophotography Society.

ents. There were, however, large differences in sensitivity. The generation layers were used with transport layers containing a hydrazone derivative. A Hammett free energy analysis showed a clear correlation between the sensitivity and increasing electronegativity of the anilide ring substituent. Interestingly, intramolecular hydrogen bonding also became stronger with substituent electronegativity. The compound with the 2-chloro substituent has recently received considerable attention in a series of papers by Niimi and Umeda (1993, 1994), Umeda and Niimi (1994, 1994a), and Umeda (1999).

There has been considerable activity in the preparation and study of new azo pigments. Law and Tarnawskyj (1993, 1995, 1995a) described pigments prepared with a variety of novel aromatic diamines, such as 2-nitro-4-4′-diaminodiphenylamine, 2,7-diaminofluorenone, and different 2-hydroxy-11(H)-benzo(a)carbazole-3-carboxanilide couplers, which extend the sensitivity into the near infrared. The sensitometry of many of these bisazo pigments have been described by DiPaola-Baranyi et al. (1990). Attempts have been made to understand the results in terms of structure–activity relationships (Law et al., 1994).

Phthalocyanines

From their discovery in Great Britain in 1927 to the present day, the phthalocyanines, Fig. 10.9, have been of significant commercial relevance. These materials have exceptional colorant characteristics, as well as stability to heat and light. Present applications extend from the traditional inks and paints (McKay, 1989) to xerographic photoreceptors (Gregory, 1991). Due to their importance as colorants, synthetic methods have been developed for their preparation on a large scale (Moser and Thomas, 1963, 1983; Thomas, 1990). Because the phthalocyanines are insoluble in most solvents, specialized isolation and purification methods have been developed. Phthalocyanines with many different central metal ions have been prepared and studied as generation materials. Since phthalocyanines containing ring substituents or polyaromatic rings are known, the number of possible compounds appears to be virtually limitless (Gregory, 1988).

The phthalocyanines can be used as vapor-deposited or dispersion layers. Of particular interest are those with strong absorption in the near infrared. Considerable effort has

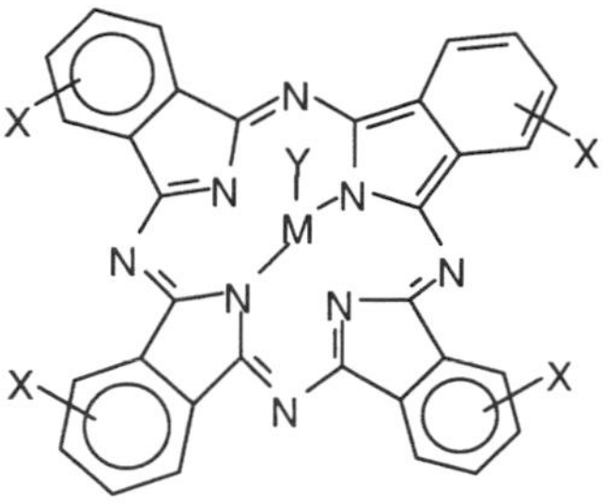

Figure 10.9 General molecular structure of a phthalocyanine (YM)(X_4-P_c).

gone into the synthesis of new compounds and the development of methods for the conversion of known phthalocyanines into morphologies that are infrared sensitive. This has been difficult, because the absorption spectra and generation characteristics depend on the chemical and crystal structure, particle size and morphology, and the presence of absorbed surface species (Sappok, 1978; Whitlock et al., 1992; Kubiak et al., 1995). Synthetic techniques and methods of photoreceptor fabrication must thus be designed to ensure that the desired characteristics are retained and/or induced (Mayo et al., 1994; Yao et al., 1995).

Because most phthalocyanines are insoluble in organic solvents, specialized methods of purification have been developed (Liebermann et al., 1988). Commonly used techniques are solvent extraction, acid pasting, and train sublimation. Solvent extraction is useful in removing trace organic contaminants. Acid pasting involves dissolving the material in concentrated mineral acid, typically sulfuric, followed by precipitation by the addition of cold water (Mayo, 1993). A limitation of this method is that some phthalocyanines (e.g., H_2Pc) may decompose during the acid pasting process. In addition, acid pasting may result in increased contamination by heavy metals such as Fe (Loutfy and Hsiao, 1979). A novel variation of acid pasting, carried out by dissolving the pigment in a solution of a Lewis acid and nitromethane, has been described by Hsieh and Melnyk (1996, 1998). Train sublimation is done by placing the pigment in one end of a quartz tube. The tube is placed in a furnace and the end containing the pigment is heated to the sublimation temperature. A flow of an inert gas, such as N_2 or Ar, is maintained from the high- to the low-temperature ends of the tube. The purified material is deposited in the cooler zones. This procedure has been described in detail by Wagner et al. (1982) and Kitamura et al. (1988).

Copper phthalocyanine (CuPC) was the first of this pigment class to be studied as a photoreceptor. Purity, crystal structure, morphology, and dispersion preparation conditions are important factors in the sensitometry. Effects of the polymer on the sensitometry of α- and β-CuPc dispersions have been described by Kishi et al. (1984). Other polymorphs, such as x-CuPc (Sharp and Abkowitz, 1973) and σ-CuPc (Enokida and Hirohashi, 1991), have also been reported. Single-layer photoreceptors containing CuPc are of interest for positive-charging applications. A single-layer drum photoreceptor containing α-CuPc has been introduced by Kentek for applications that require near-infrared exposures (Decker et al., 1991). Fujimori et al. (1992) described the preparation of a single-layer photoreceptor with excellent cycling characteristics when α-CuPc is doped with nitrated CuPc. Dual-layer photoreceptors using α-CuPc, β-CuPc, ε-CuPc, and partly nitrated α-CuPc in the generation layer have been described by Enokida and Hirohashi (1992). The partly nitrated α-CuPc gave the best overall performance. Similarly substituted CuPc has also been bonded to polymers such as PVK (Chen et al., 1993; Wang et al., 1993, 1996).

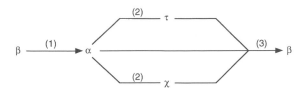

Figure 10.10   Polymorphic interconversions of H_2Pc.

Metal-free phthalocyanine (H_2Pc) has been prepared in several polymorphic forms: α (tetragonal), β (monoclinic), x (hexagonal), and τ. Figure 10.10 shows the interconversions of these materials. For example, treatment of Monolite Fast Blue GS with hot dimethylformamide followed by ball milling in isopropyl alcohol yields the β-form as needles. Acid pasting produces the α-form as needles and flakes (Loutfy, 1981). Milling under different conditions yields the τ-form as needles or flakes (Takano et al., 1984; Kakuta et al., 1985) or the x-form as needles (Sharp and Lardon, 1968; Hackett, 1971). The τ-form as rods has been produced directly by carrying out the synthesis in the presence of 1,8-diazabicyclo[5.4.0]-undec-7-ene and seed crystals of the τ-form (Enokida and Ehashi, 1988). All of these morphologies convert to the stable β-form under equilibrium thermal, mechanical, or solvent conditions. The polymorphic forms are commonly identified by x-ray powder diffraction patterns and infrared absorption spectra. Enokida et al. (1991b) reported that cross-polarization magic-angle spinning ^{13}C NMR spectroscopy can be used to identify the various forms. Assignments have been problematic. Kubiak et al. (1995) argued that the x-form is actually β-H_2Pc and that the α-forms of divalent metal phthalocyanines, such as MgPc, are complexes with N and O. This confusion is a recurring feature of phthalocyanine technology. Figures 10.11 and 10.12 show the absorption spectra of the four morphologies. Whatever their nature, the long-wavelength absorptions of the τ- and x-forms are such that they are of interest for near-infrared applications. A dual-layer photoreceptor based on τ-H_2Pc has been described by Kakuta et al. (1985) and Shimada et al. (1987). The generation layer contained the τ-form in a polysiloxane or polycarbonate. The transport layer contained an oxazole derivative in the same polymers. Further refinements of the τ-form have been reported by Enokida et al. (1991a) and Endo et al. (1994).

Positively charged photoreceptors are of considerable interest for printer applications. Tsuchiya et al. (1995) described a single-layer photoreceptor prepared with x-H_2Pc. Nakatani et al. (1985) reported that doping β-H_2Pc with electron acceptors, such as 2,4,5,7-tetranitrofluorenone, substantially enhanced the sensitivity. Similar observations were later reported by Kobayashi et al. (1996). A structure with the generation layer uppermost has been described by Takahashi and Yamamoto (1995). The generation layer was composed of x-H_2Pc dispersed in a TiO_2 glass.

Generation layers composed of titanyl phthalocyanine (TiOPc) dispersions have received considerable recent attention. A procedure for photoreceptor fabrication described by Ohaku et al. (1988) involved first converting the as-synthesized pigment to the α-form by extended ball milling. A dispersion of the pigment and a polyester was used to prepare the generation layer. The transport layer contained an indoline derivative in a polycarbonate. The photoreceptor showed satisfactory charge acceptance and high sensitivity in the 520 to 900 nm spectral range. A similar photoreceptor prepared with β-TiOPc showed considerably higher dark discharge and lower sensitivity. Recent work has led to the identification of several new polymorphs, some of which have had commercial application.

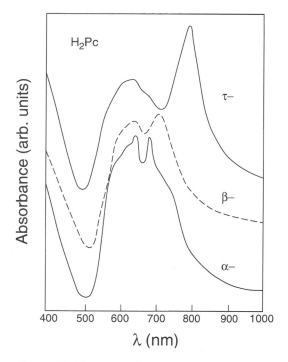

Figure 10.11 Absorption spectra for α-, β-, and τ-H₂Pc. (After Kakuta et al., 1985. Reprinted from *J. Imaging Technol. 11*:7. Copyright 1985, Society for Imaging Science and Technology.)

Unfortunately, different names are frequently given to what appear to be the same materials. Some of these may be small molecule complexes and not polymorphs. Enokida et al. (1990) described five forms of TiOPc (amorphous, α, β, m, and γ). Miyazaki et al. (1990) described the I-, M-, E-, G-, and H-forms. Takano et al. (1991) reported on Phases I, II, and III. Fujimaki (1991) and Fujimaki et al. (1991) described the A-, B-, D-, and Y-forms. Effects of ring substituents were described by Watanabe et al. (1993). Martin et al. (1995) studied Types I, II, III, IV, and X. Oka and Okada (1993) examined the relationship between crystal structure and sensitivities of Phases I and II and the C- and Y-forms. Hung et al. (1987) developed a ring-fluorinated TiOPc for use in dispersion-based generation layers with high sensitivity at 830 nm. Other TiOPc derivatives have been described by Suzuki et al. (1988).

Many other phthalocyanines have been investigated. Most undergo morphology changes upon solvent treatment, giving rise to new or enhanced absorptions (Hor and Loutfy, 1983) and increased sensitivity between 800 and 900 nm (Loutfy et al., 1985). Exposure to methylene chloride (fuming, ball milling, or overcoating) is typical, but other solvents such as acetone, toluene, tetrahydrofuran, and isopropyl alcohol will also induce polymorphic transformations. Table 10.2 lists the absorption maxima of vapor-deposited phthalocyanine films before and after exposure to methylene chloride. Tables 10.3 and 10.4 compare the sensitivities of dual-layer photoreceptors with dispersion and vapor-deposited generation layers. The transport layers contained N,N′-diphenyl-N,N′-bis(3-methylphenyl)-(1,1′-biphenyl)-4,4′-diamine (TPD) in a polycarbonate. The enhanced long-wavelength absorption and sensitivity were believed to be related to solvent-induced

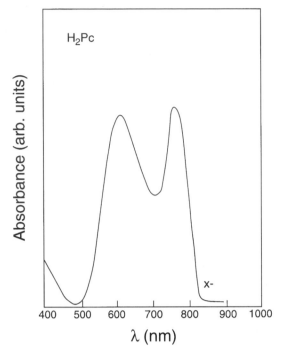

Figure 10.12 Absorption spectra of the x-H$_2$Pc. (After Loutfy, 1981. Reprinted from *Can. J. Chem. 59*:549. Copyright, Canadian Chemical Society, 1981.)

Table 10.2 The Absorption Maxima of a Series of 500 Å Vapor-Deposited Phthalocyanine Films Before and After Exposure to Methylene Chloride

	As deposited		After exposure to methylene chloride	
Phthalocyanine	λ_{max} (nm)	Optical density	λ_{max} (nm)	Optical density
MgPc	625	0.72	620	0.36
	695	1.00	700	0.46
			832	1.00
ZnPc	625	1.00	610	0.86
	700	0.83	700	1.00
			820	0.86
VOPc	630	0.56	630	0.52
	740	1.00	740	0.74
	830	0.77	830	1.00
ClAlPc	660	0.60	650	0.89
	730	1.00	720	0.96
			810	1.00

Source: After Loutfy et al., 1985. Reprinted with permission from *J. Imaging Sci. 29*:116. Copyright 1985, The Society for Imaging Science and Technology.

Table 10.3 The Sensitivity and Dark Discharge for a Series of Dual-Layer Photoreceptors with Phthalocyanine Dispersion-Based Generation Layers. The Sensitivity is the Energy Required to Discharge the Photoreceptor to $V_0/2$. The Wavelength Was 830 nm, and the Initial Potential -830 V. The Transport Layer Contained N-N'-Diphenyl-N-N'-bis(3-Methylphenyl)-[1,1'-Biphenyl]-4,4'-Diamine in a Polycarbonate

Phthalocyanine	Generation layer thickness (μm)	Dark discharge (V/s)	$\varepsilon_{1/2}$ (erg/cm^2)
VOPc	0.40	30	4.1
τ-H$_2$Pc	0.30	20	5.0
TiOPc	0.30	30	5.6
HO$_2$GePc	0.50	60	7.0
ClInPc	1.40	60	8.0
ClAlPc	0.45	35	10.0
ClAlPcCl	0.45	25	11.0
MgPc	0.10	80	30.0

Source: After Loutfy et al., 1988. Reprinted from *Proceedings of the Fourth International Congress on Advances in Non-Impact Printing Technologies* (A. Jaffe, ed.). SPSE, Springfield, VA, p. 52. Copyright 1988, The Society for Imaging Science and Technology.

Table 10.4 The Xerographic Sensitivity and Dark Discharge for a Series of Dual-Layer Photoreceptors with 0.10 μm Vapor-Deposited Phthalocyanine Generation Layers. The Sensitivity is the Energy Required to Discharge the Photoreceptor to $V_0/2$. The Transport Layers Contained N-N'-Diphenyl-N-N'-bis(3-Methylphenyl)-[1,1'-Biphenyl]-4,4'-Diamine in a Polycarbonate. The Wavelength was 830 nm and the Initial Potential -830 V

Phthalocyanine	Dark discharge (V/s)	$\varepsilon_{1/2}$ (erg/cm^2)
ClInPc	45	2.6
ClAlPc	40	6.0
PbPc	130	10.0
VOPc	30	16.0
Cl$_2$GePc	100	>100
FCrPc	26	>100
Cl$_2$SnPc	60	>100
SnPc	50	>100

Source: From Loutfy et al., 1988. Reprinted from *Proceedings of the Fourth International Congress on Advances in Non-Impact Printing Technologies* (A. Jaffe, ed.). SPSE, Springfield, VA, p. 52. Copyright 1988, The Society for Imaging Science and Technology.

changes in molecular stacking and spacing. Thus, α-MgPc has a 13.2 Å interplanar spacing and a 3.8 Å intermetal spacing along the phthalocyanine stacking axis. These distances are 11.6 and 3.5 Å after methylene chloride exposure. Similar changes occur with ClAlPc. Loutfy et al. (1985) used the term aggregation to describe these solvent-induced morphology changes. Generation layers prepared with vapor-deposited layers, Table 10.4, were converted to the infrared-sensitive morphology by overcoating with a methylene chloride solution of the transport layer (Loutfy et al., 1987).

Haloindium phthalocyanines have been used for dispersion (Hung et al., 1987a) and vapor-deposited (Kato et al., 1988) generation layers. Chloroindium phthalocyanine (ClInPc) shows near infrared absorption as synthesized. This was unchanged upon purification by train sublimation or ball milling in halogenated solvents or tetrahydrofuran (Loutfy et al., 1985a). The x-ray powder diffraction spectra likewise indicated that these materials were the same polymorph. Dispersion generation layers used with transport layers containing TPD in a polycarbonate showed increased 830 nm sensitivity and increased dark discharge as the thickness of the generation layer was increased. The results are summarized in Table 10.4.

Chloroindium phthalocyanine with ring chlorination, ClIn(Cl-Pc), has been described by Kato et al. (1985). The pigment was a mixture of the parent and the ring-chlorinated materials. Exposure to tetrahydrofuran vapors induced a modest absorption enhancement at longer wavelengths. Interestingly, the x-ray powder patterns for the as-synthesized material, and the vapor-deposited layers before and after tetrahydrofuran exposure, were identical. Thus ClIn(Cl-Pc) exists in one stable morphology.

In contrast to ClIn(Cl-Pc), the long-wavelength absorption maximum of vapor-deposited ClAl(Cl-Pc) undergoes a shift from 740 to 825 nm on immersion in tetrahydrofuran or acetone (Arishima et al., 1982). Generation layers of ClAl(Cl-Pc) of specific composition and x-ray powder patterns have been described by Nogami et al. (1988). These results, however, must now be viewed in light of the observation that ClAlPc forms a hydrate with spectral characteristics similar to those previously ascribed to polymorphism (Whitlock et al., 1992).

Vanadyl phthalocyanine (VOPc) exists in two crystal phases as characterized by absorption and emission spectroscopy (Huang and Sharp, 1982). Phase I, obtained by vapor deposition, has a long-wavelength absorption maximum at 725 nm. Acid pasting, milling in methylene chloride, and thermal treatments favor the formation of Phase II, which has a new absorption at 840 nm. A dual-layer photoreceptor with the generation layer uppermost showed high sensitivity from 400 to 900 nm (Grammatica and Mort, 1981). A soluble phthalocyanine, $VO(t\text{-}Bu_{1.4}Pc)$, was fabricated into a single-layer photoreceptor (Law, 1988). The absorption spectrum had maxima at 650 and 694 nm, with a shoulder at 810 nm. Exposure to ethyl acetate vapor resulted in increased crystallinity, as evidenced in x-ray powder diffraction patterns and absorption spectra. Further, the intensity of the absorption at 810 nm was substantially increased. These changes were identified with Phases I and II as in VOPc. The Phase II film had a sensitivity at 600 nm in excess of 300 times that of Phase I. This was attributed to the crystalline nature of Phase II and the presence of a particular stacking structure with a close approach of neighboring molecules.

Hydroxygallium phthalocyanine (HOGaPc), with and without ring halogenation, has been described by Kato et al. (1986). The preparation and polymorphic transformations have been described by Mayo et al. (1994) and Daimon et al. (1996). Type I is formed during acid pasting and is converted to the more desirable Type V by subsequent milling in N,N-dimethylformamide. Photoreceptors prepared with the latter showed high near-infrared sensitivity and good cyclic and environmental stability (Hsiao et al., 1994). A photoreceptor composed of a generation layer of HOGaPC in poly(vinyl butyral) and a transport layer of TPD in a polycarbonate had high sensitivity between 650 and 850 nm.

Naphthalocyanines have also been used as generation layers. Nikles et al. (1992) used an *n*-butanol soluble derivative, $[RO(CH_3)_2SiO]_2$ SiNc, and prepared the generation layer by solution coating. Hayashi et al. (1992) described generation layers prepared from $[C_3H_7\text{-}SiO]_2 SiNc$ as dispersions and as vapor-deposited layers with different transport layers. The best results were obtained with a vapor-deposited generation layer and a transport layer containing 1,1-bis(*p*-diethylaminophenyl-4,4-diphenyl)-1,3-butadiene.

Molaire et al. (1997) described dual-layer photoreceptors prepared with tetrafluorinated titanylphthalocyanine $TiO(F_4\text{-}Pc)$ generation layers. The transport layers contained tri-*p*-tolylamine doped into a polycarbonate. High sensitivity in the near-infrared and low dark discharge were reported. The spectral and field dependencies of the xerographic gain are shown in Figs. 10.13 and 10.14.

Finally, photoreceptors using mixed phthalocyanines (Wang et al., 1991, 1993; Itami et al., 1991; Hayashida et al., 1994) and phthalocyanines doped with C_{60} (Chen et al., 1995; Narushima et al., 1996) have been described.

Squaraines

Squaraines can be readily synthesized via the reaction of squaric acid (Treibs and Jacob, 1965) or its diester (Law and Bailey, 1986) with aromatic amines. Figure 10.15 shows

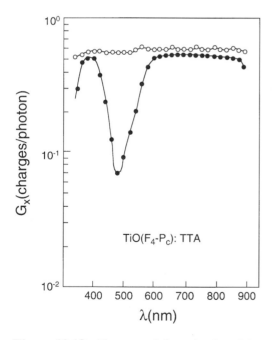

Figure 10.13 The spectral dependencies of the xerographic gain of a dual-layer photoreceptor computed on an absorbed (open circles) and incident (solid circles) photon basis. The generation layer contains a dispersion of a tetrafluorophthalocyanine in poly(vinyl butyral). The transport layer contains a mixture of 40% tri-*p*-tolylamine (TTA) and a polycarbonate (Molaire et al., 1997). The thickness was 20 μm, the field 4.0×10^5 V/cm, and the temperature 296 K.

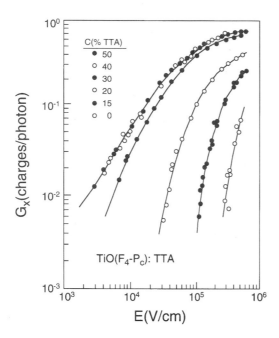

Figure 10.14 The field dependencies of the xerographic gain of a dual-layer photoreceptor. The generation layer contains a dispersion of a tetrafluorophthalocyanine in poly(vinyl butyral). The transport layer contains different concentrations of a mixture of tri-*p*-tolylamine (TTA) and a polycarbonate (Molaire et al., 1997). The thickness was 20 μm, the wavelength 680 nm, and the temperature 296 K.

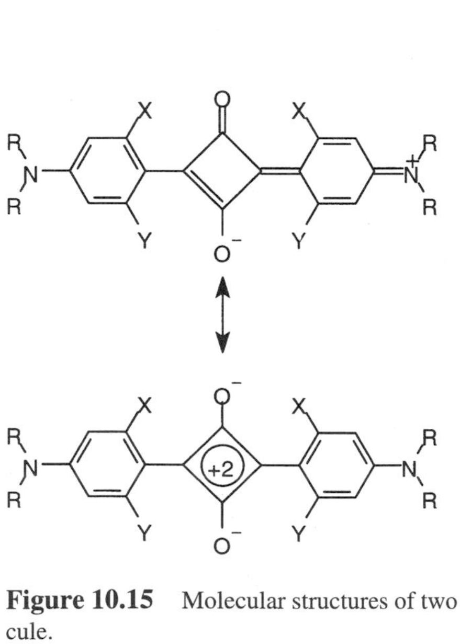

Figure 10.15 Molecular structures of two of the resonance forms of the parent squarylium molecule.

the molecular structure. Law and Bailey (1987) reported that squaraines prepared via the diester give enhanced sensitometry. This has been ascribed to the formation of a different crystal modification with a lower concentration of impurities. Other purification methods have been reported (Lin and Dudek, 1986). The synthesis of squaraines as generation layer materials has been widely studied during the past decade (Law and Bailey, 1988, 1991, 1992, 1993).

In the late 1970s, researchers at IBM developed dual-layer photoreceptors using squaraine dispersions as generation layers. The transport layers contained different pyrazoline and hydrazone doped polymers (Champ and Shattuck, 1974; Melz et al., 1977; Tam, 1980). A photoreceptor for laser printers has been described in some detail by Wingard (1982) and Champ (1987). An Al-coated polyester is the substrate. This was overcoated with a thin polyamide layer that serves as an adhesive for the generation layer as well as a blocking layer to prevent charge injection from the Al. The generation layer is an aggregated hydroxy-squarylium ($R = CH_3$, $X = OH$, $Y = H$). The transport layer is composed of p-diethylaminobenzaldehyde diphenylhydrazone (DEH) in a blend of a polycarbonate and a polyester plus polydimethylsiloxane, a surfactant for coating quality, and a dye to absorb fluorescent room light. The latter is required because the photoreceptor undergoes fatigue on exposure to room illumination. The action spectrum shows high sensitivity from 500 to 800 nm. The residual potential increases with cycling. The rate of increase is such that the useful lifetime was approximately 50,000 cycles. Schwartz (1991) has described a drum photoreceptor with similar composition that has been developed for use in a printer.

The photogeneration efficiency of the squaraines is related to their ability to form aggregates that have specific short-range intermolecular charge-transfer interactions (Law, 1988). Squaraines in solution exhibit narrow absorption bands with maxima between 500 and 600 nm. Materials that are useful in generation layers undergo aggregation, giving rise to very broad absorption spectra in the 400 to 1000 nm region. Law has reported several studies that correlated the sensitometry with solution electrochemistry (Law et al., 1990), fluorescence emission (Law, 1990), and state of aggregation (Law, 1992).

The effects of generation layer fabrication variables on the sensitivity of dual-layer photoreceptors prepared with bis(4-dimethylaminophenyl) squaraine ($R = CH_3$, $X = Y = H$) have been extensively investigated (Law, 1987). The dispersions were coated under humidity-controlled conditions. The transport layer contained TPD in a polycarbonate. The sensitivity was dependent on the generation layer polymer, milling conditions, pigment concentration, layer thicknesses, and the ambient humidity. The quality of the generation layer appeared to be of major importance. Some polymers produced unstable dispersions that precipitated large pigment agglomerates. Photoreceptors prepared with these polymers usually showed low charge acceptance and high residual potentials. Layers with the best dispersion quality, however, did not necessarily produce the highest sensitivity. Increased pigment concentration, or generation layer thickness, resulted in decreased charge acceptance, increased dark discharge, and increased sensitivity. Under high-humidity conditions, the dark discharge and charge acceptance were degraded. These effects were reversible. It was suggested that water absorption on the pigment surface gives rise to increased dark charge injection.

Table 10.5 compares several squaraines in a dual-layer configuration (Loutfy et al., 1988a). The generation layers were coated on an Al substrate prepared with a siloxane blocking layer. The transport layers were composed of a mixture of TPD in a polycarbonate. Increased squaraine solubility in the polymer (e.g., $R = Bu$) had a detrimental effect

Table 10.5 The Xerographic Sensitivity and Dark Discharge of a Series of Dual-Layer Photoreceptors with Different Squaraine Generation Layers. The Sensitivity is the Energy Required to Discharge the Photoreceptor to $V_0/2$. The Wavelength was 830 nm and the Initial Potential -830 V

X	Y	R	Dark discharge (V/s)	$\varepsilon^{1/2}$ (erg/cm^2)
HO	H	CH$_3$	35	3.0
H	H	CH$_3$	42	3.5
F	H	CH$_3$	70	3.5
F	CH$_3$	CH$_3$	35	5.5
HO	CH$_3$	CH$_3$	15	10.0
H	CH$_3$	CH$_3$	20	13.0
H	SCH$_3$	CH$_3$	40	33.0
HO	HO	CH$_3$		Not photosensitive
H	OCH$_3$	CH$_3$		Not photosensitive
HO	H	Et		Not photosensitive
H	H	Et		Not photosensitive
HO	H	Bu		Not photosensitive
HO	CH$_3$	Bu		Not photosensitive
HO	H	Julolidine	15	50.0
HO	HO	Julolidine	20	52.0

Source: After Loutfy et al. 1988a. Reprinted from *Pure Appl. Chem. 60*:1047. Copyright 1988, International Union of Pure and Applied Chemistry.

on the sensitivity and residual potential. Optimization of the photoreceptor with R = CH$_3$, X = F, Y = H (Kazmaier et al., 1988) used poly(vinyl butyral) as the generation layer polymer. This material exhibited high sensitivity in the 500 to 900 nm spectral range, as well as good stability to temperature, humidity, and cycling.

The presence of certain phenols, such as 3,5-dihydroxytoluene (orcinol), during the synthesis of a squaraine corresponding to R = CH$_3$, X = OH, Y = CH$_3$ led to enhanced sensitivity (DiPaola-Baranyi et al., 1988). Thus a photoreceptor similar to those described in Table 10.5 showed a 6 times increase in sensitivity on doping with >20 mol% orcinol). Neither the spectral sensitivity nor the dark discharge was affected. Doping of the pigment dispersion prior to coating had no effect. This was ascribed to the isomorphic substitution of orcinol in the squaraine crystal lattice.

Unsymmetrical squaraines (Yanus and Limberg, 1986; Kazmaier et al., 1988a) prepared from two different aromatic amines have been compared to the corresponding symmetrical compounds (Kazmaier et al., 1988b). In the same study, a mixture of two aromatic amines was used in the reaction, resulting in a composite product mixture consisting of two symmetrical squaraines and the nonsymmetrical compound. Most of these materials showed high dark discharge and moderate near-infrared sensitivity. However, an unsymmetrical squaraine has been used in the development of a photoreceptor with excellent sensitometry (Law, 1992a). The pigment is the unsymmetrical fluorinated 3,4-dimethoxyphenyl-2'-fluoro-4'-(dimethylamino)phenylsquaraine. The generation layer polymer was poly(vinyl formal). The transport layer contained TPD in a polycarbonate. The photoreceptor showed low dark decay and high sensitivity from the visible into the near-infrared.

A squaraine prepared from N-chlorobenzyl-N-methylaniline and squaric acid has been used as a generation layer in a dual-layer photoreceptor with an inverted structure for positive charging applications (Yamamoto et al., 1986). The generation layer polymer was poly(vinyl acetate). The photoreceptor was prepared with a transport layer of TPD in a polycarbonate, an interface layer of a hydrolytically cured silane and zirconium, and an overcoat composed of tin oxide in a thermally cured polyurethane. The action spectrum was essentially constant from 400 to 900 nm. The photoreceptor exhibited full process stability with cycling.

There are many literature references to photoreceptors with generation layers based on dispersions of novel squaraine pigments. See, for example, Kin et al. (1986), Champ and Vollmer (1987), and Tanaka et al. (1987).

Polycyclic Aromatics and Other Compounds

Diimides of perylene-3,4,9,10-tetracarboxylic acid, Fig. 10.16, are frequently referred to as perylenes. These compounds are readily available, mainly because of their use in automotive finishes. As is the case with many pigments, the solid state absorption is red-shifted and broadened with respect to the solution absorption. This is believed to be due to specific intermolecular interactions in the crystal. Thus small changes in the structure of the R substituent give rise to absorption changes by influencing the crystal structure.

Several perylene pigments have been used in generation layer dispersions (Hor and Loutfy, 1986; Wiedemann et al., 1987; Kazmaier et al., 1988c; Staudenmayer and Regan, 1988). In one example, R = R' = 2-phenethyl, conversion of the amorphous material occurs on overcoating with the transport layer. This results in a generation layer with enhanced absorption around 620 nm that shows high sensitivity and panchromatic response (Staudenmayer and Regan, 1988). A dual-layer photoreceptor with R = R' = CH$_3$ has been described by Schlosser (1978). The transport layer was 2,5-bis(4-diethyl-aminophenyl)-1,3,4-oxadiazole in a mixture of a polyester and poly(vinyl chloride).

A positively charged single-layer photoreceptor prepared with a dispersion of R = R' = 3,5-dimethylphenyl in PVK has been described (Khe et al., 1984a). The as-synthesized pigment was observed to undergo a change in morphology with various milling techniques as well as exposure to organic solvent vapors. Good cyclic stability was reported. This pigment is used in single-layer drum applications (Matsumoto and Nakazawa, 1988). The effects of perylene structural modification on the sensitometry of vapor-deposited generation layers have been investigated by Loutfy et al. (1989) and Duff et al. (1990, 1991).

A near-infrared-sensitive photoreceptor based on 1,4-dithioketo-3,6-diphenyl-pyrrolo[3,4-c]pyrrole, Fig. 10.17, has been described by Mizuguchi and Rochat (1988). This material exists in three crystal forms: α, β, and Γ (Arita et al., 1991). Treatment of a sublimed film with solvent vapors, or pigment milling, results in conversion to the β-

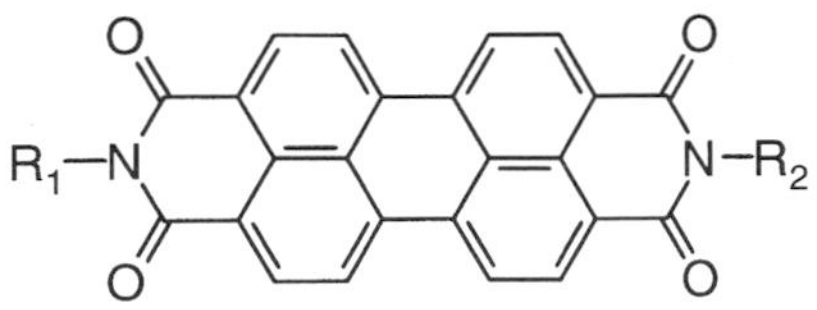

Figure 10.16 The molecular structure of perylene tetracarboxylic diimide.

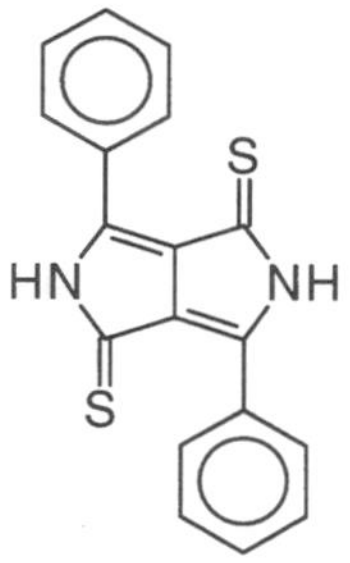

Figure 10.17   The molecular structure of 1,4-dithioketo-3,6-diphenylpyrrolo[3,4-c]pyrrole.

form with a shift in the absorption spectrum to longer wavelengths and the appearance of a new 830 nm absorption. Many solvents were successful, but acetone was preferred. The conversion also occurred upon ball milling in a mixture of xylene and ethylene glycol monomethylether. Both the converted and the nonconverted pigment forms were stable upon heating to 250°C. From transmission electron microscopy and x-ray diffraction patterns, the authors speculate that the conversion is due to a decrease in the interplanar separation in the crystal. It was further suggested that this stacking change permits enhanced intermolecular charge-transfer interactions, explaining the increase in the sensitivity of photoreceptors prepared with the converted material. Dual-layer photoreceptors with both vapor-deposited and dispersion-generation layers were prepared. The transport layers contained a mixture of DEH and a polycarbonate. The photoreceptors prepared with dispersion-generation layers showed a residual potential that increased with the generation layer thickness.

Takenouchi et al. (1988) described a dual-layer photoreceptor composed of a generation layer containing 2,7-dibromoanthanthrone, Fig. 10.18, dispersed in a polycarbonate. The transport layer contained a styryl triphenylamine derivative in a polycarbonate. The transport layer was degraded on exposure to ozone during corona charging. This resulted in a rising residual potential with cycling. Effects due to different halogen substituents on the pigment have also been studied (Allen et al., 1989). The unsubstituted parent compound had the highest sensitivity and the bromo- and iodo-substituted compounds the lowest. The results, however, depended on the donor component of the transport layer. The decreased sensitivity with bromine and iodine substitutents was ascribed to a heavy atom effect, giving rise to enhanced rates of intersystem crossing from the singlet charge generating state to the triplet.

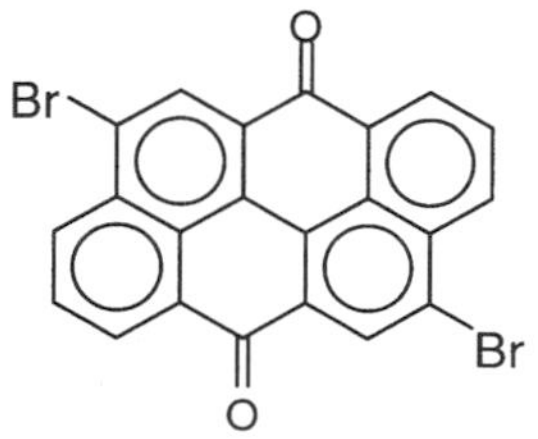

Figure 10.18   The molecular structure of 2,7-dibromoanthanthrone.

Generation layers composed of pigment mixtures have also been described. Nakazawa et al. (1994) reported that mixtures of phthalocyanines (H_2Pc, CuPc, TiOPc) with perylene, anthanthrone, or bisazo pigments offer enhanced sensitivity. Nishino et al. (1995) described a single-layer photoreceptor with TiOPc and a perylene pigment. Cyanine dyes, aromatic polyenes, fullerenes (Mort et al., 1992; Hosoya et al., 1995), and similar large conjugated organic molecules have also been used.

10.4.3 Charge Transport Materials

All photoreceptors contain donor or acceptor molecules to enhance the transport of holes or electrons. The donor or acceptor molecules can be doped into a polymer or chemically attached to the polymer. The former procedure is more common because it permits the greatest flexibility in the use of mixtures for optimum performance.

The basic requirement is the same for single- and for dual-layer photoreceptors. Charge transport must occur in the absence of trapping without a mobility limitation. The latter places requirements on both the transport material and the xerographic process (Pai and Yanus, 1983). The key requirement is that there be negligible residual surface potential. A residual potential can be due to either a mobility limitation or trapping. A mobility limitation arises when the transit time of electrons or holes created in the image exposure becomes comparable to the time between the exposure and development steps. There are several approaches to decreasing a mobility limitation. Increased donor or acceptor concentrations or reducing the photoreceptor thickness will reduce the residual potential. The same result can be obtained by increasing the time between the exposure and development processes or increasing the field. Traps may be present as a result of insufficient purification, chemical instability of the donor or acceptor molecule, or chemical instability induced by radiation or exposure to the chemicals associated with the corona discharge. When the potential discharge is trap-limited, the residual potential usually increases with cycling.

Hole Transport

Figure 10.19 shows examples of different classes of donor molecules. In the literature, these are often referred to by their compound class: arylamines, enamines, hydrazones, oxadiazoles, triphenylmethanes, etc. This classification is convenient but misleading in that (with the exception of the polysilanes and polygermanes) it obscures the one basic feature common to all donor molecules, the presence of any arylamine substituent. That is, all these materials can be considered to be substituted aromatic amines. The presence of a nonbonding electron pair on a N atom gives rise to both a low oxidation potential, with the production of a chemically stable radical cation, and the potential for effective overlap of the nonbonding molecular orbitals of adjacent molecules. Both are important for efficient charge transport.

Some basic principles have been proposed in the design of molecules for hole transport. One is the presence of multiple conjugated arylamine moieties. This permits delocalization of the radical cation via resonance, thereby maximizing the probability for electron exchange with a neighboring donor molecule. To this end, there have been several studies on the effects of molecular orbital distributions on transport (Sugiuchi et al., 1990; Aratani et al., 1991, 1996; Kitamura and Yokoyama, 1991; Hirose et al., 1992; Okada, 1992; Hirano et al., 1995). The molecule should also be designed to prevent intramolecular ground state and/or excited state (excimer) dimer formation but at the same time maximize

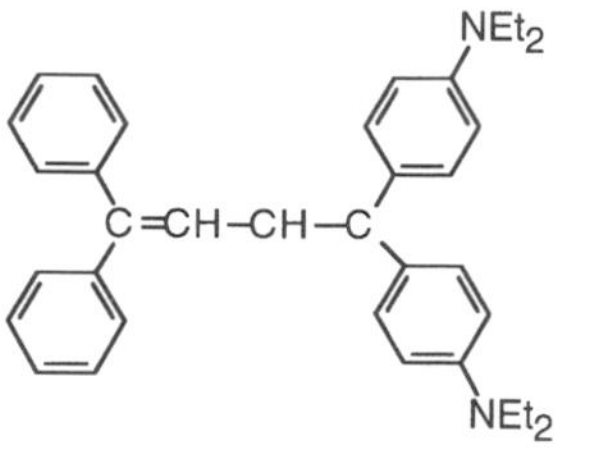

Butadienes

1,1-bis(4-diethylaminophenyl)-4,4-
diphenyl)-1,3-butadiene
Aratani et al. (1990)

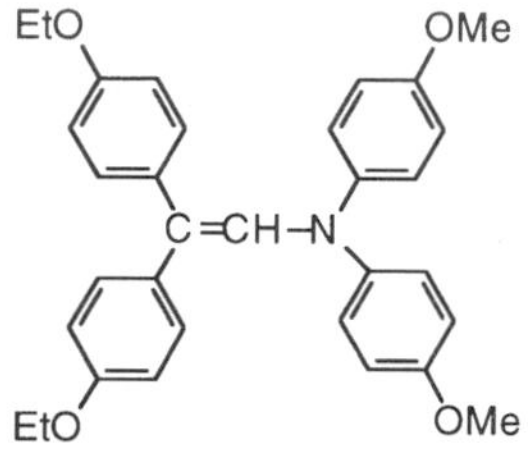

Enamines

bis(p-ethoxyphenyl)acetaldehyde
di-p-methoxyphenylamine enamine
Rice et al. (1985)

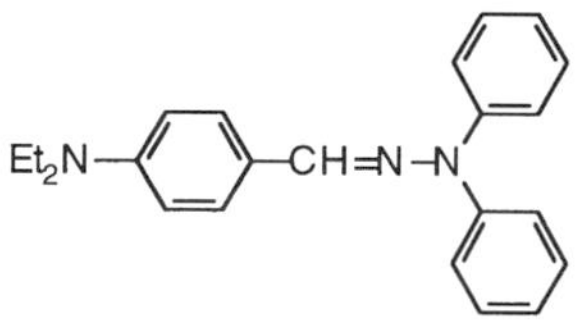

Hydrazones

p-diethylaminobenzylaldehyde
diphenylhydrazone (DEH)*
Schein et al. (1986)

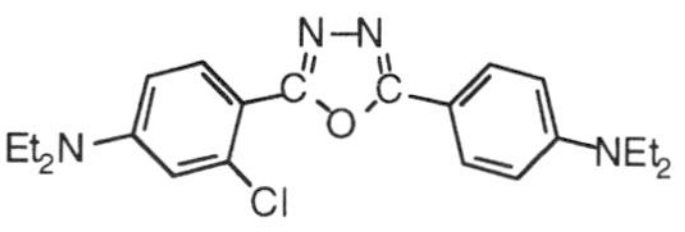

Oxadiazoles

2,5-bis(4-N,N'-diethylaminophenyl)-
1,3,4-oxadiazole
Schlosser et al. (1978)

Figure 10.19   The molecular structures, chemical names, and common acronyms of various classes of hole transport molecules.

Oxazoles

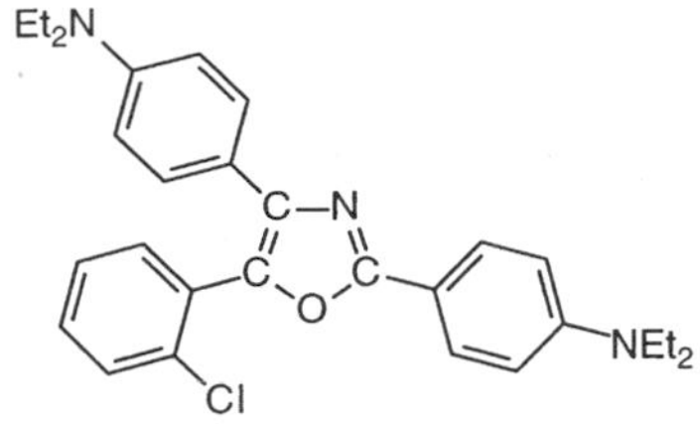

2,5-bis(4-diethylaminophenyl)-4-
(2-chlorophenyl)oxazole
Kakuta et al. (1981)

Pyrazolines

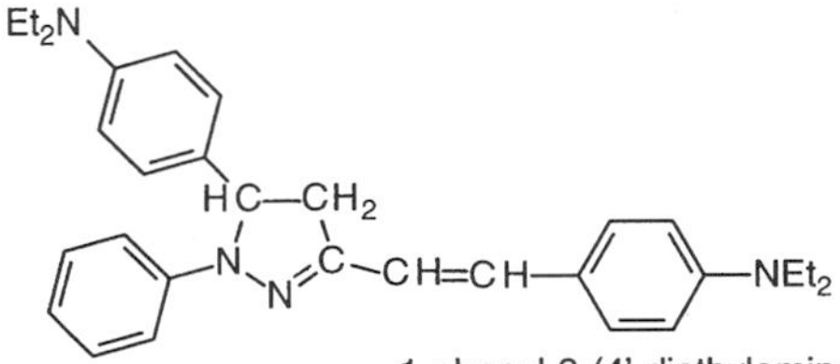

1-phenyl-3-(4'-diethylaminostyryl)-5-
(4"-diethylaminostyryl)pyrazolene (DEASP)*
Meitz et al. (1977)

Triarylamines

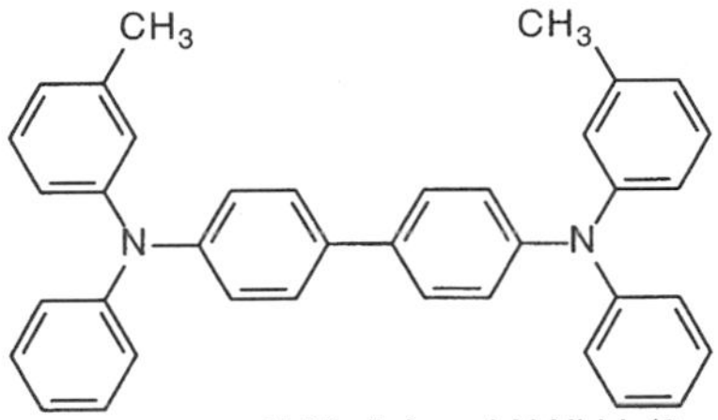

N,N'-diphenyl-N,N'-bis(3-methylphenyl)-
[1,1-biphenyl]-4,4'-diamine (TPD)*
Stolka et al. (1984)
Pai et al. (1984)

Triarylmethanes

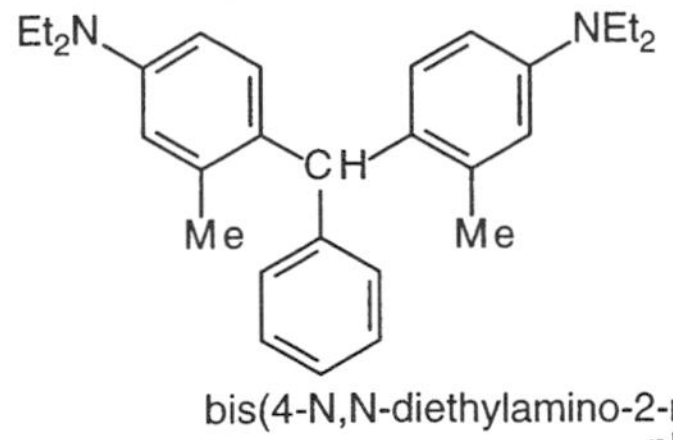

bis(4-N,N-diethylamino-2-methylphenyl)
phenylmethane
Borsenberger et al. (1978)

intermolecular orbital overlap (Murayama et al., 1988). Dimer sites are believed to be hole traps in polyfunctional and polymeric charge transport materials such as PVK. Molecules designed according to these principles have been found to have improved transport properties. Examples are the styryl or stilbene classes that have been described by Makino et al. (1988), Sasaki (1988), and Ueda (1988).

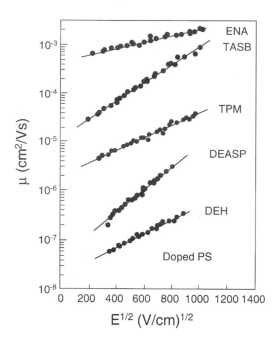

Figure 10.20   The field dependencies of the room temperature hole mobilities for a series of triphenylmethane (TPM), arylamine (TASB), enamine (ENA), pyrazoline (DEASP), and hydrazone (DEH) derivatives doped into poly(styrene) (PS). The donor concentrations were 40%.

Figure 10.20 shows the field dependencies of room temperature mobilities of different donor molecules doped into poly(styrene): N,N′-bis(2,2-diphenylvinyl)-N, N′-diphenylbenzidine (ENA), bis(ditolylaminostyryl)benzene (TASB), bis(4-N,N-diethylamino-2-methylphenyl)(4-phenylphenyl)methane (TPM), 5-(p-diethylaminophenyl)-1-phenyl-3-(p-diethylaminostyryl)-pyrazoline (DEASP), and p-diethylaminobenzaldehyde diphenylhydrazone (DEH). In virtually all doped polymers described in the literature, the mobilities are field dependent, varying approximately as $\mu \propto \exp{(\beta E^{1/2})}$, where β is a temperature-dependent constant. The mobilities are very low, strongly field and temperature dependent, as well as dependent on the dopant molecule, the dopant concentration, and the polymer host. Figure 10.21 shows the effects of dopant concentration for 1-phenyl-3-((diethylamino)styryl)-5-(p-(diethylamino)phenyl)pyrazoline (DEASP) doped polycarbonate. The mobilities are strongly concentration dependent, increasing with increasing concentration. Describing the results by the disorder formalism yields values of σ between 0.07 and 0.16 eV. In most materials, σ, the energy width of the hopping site manifold, is independent of the dopant concentration. Values of the positional disorder parameter Σ are between 1.0 and 5.0, increasing with increasing dilution.

Recent studies have also shown that the presence of large dipole moment functionalities has a deleterious effect on transport. The dipoles can be associated with the donor or acceptor molecules, the polymer repeat unit, or polar addenda. The presence of highly polar groups results in an increase in σ. This has been described by a model of dipolar disorder, due to Borsenberger and Bässler (1991), and more recently by Dieckmann et al. (1993), Sugiuchi and Nishizawa (1993), Young (1995), Parris (1996), and Hirao and

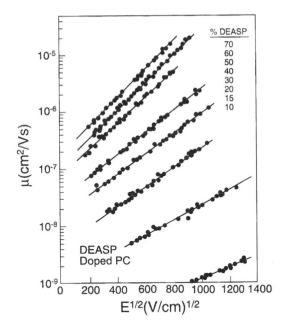

Figure 10.21 The field dependencies of the room temperature hole mobilities of 1-phenyl-3-((diethylamino)styryl)-5-(p-(diethylamino)phenyl)pyrazoline (DEASP) doped polycarbonate for different DEASP concentrations.

Nishizawa (1997). For a review of dipolar effects on transport, see Young and Fitzgerald (1995) and Young et al. (1995).

For many years, the polymer was thought to have little influence on the transport properties. Recent evidence, however, shows that the polymer can have a very considerable effect on transport (Sasakawa et al., 1989; Kanemitsu and Einami, 1990; Aratani et al., 1990; Kanemitsu, 1992; Yuh and Pai, 1992; Hirao et al., 1993). Depending on the polymer, the mobilities vary by as much as three orders of magnitude. Figure 10.22 shows the hole mobilities of N,N′-bis(2,2-diphenylvinyl)-N,N′-diphenylbenzidene (ENA) doped into different polymers: poly(styrene) (P-1), bisphenol-A polycarbonate (P-2), poly(4,4′-isopropylidene bisphenylene-co-4,4′-hexafluoroisopropylidene bisphenylene (50/50) terephthalate-co-azelate (65/35) (P-3), poly-(4,4′-(2-norborylidene)diphenylene terephthalate-co-azelate (40/60)) (P-4), a phosgene-based polyester carbonate (P-5), and poly(vinyl butyral) (P-6). In part, these results are due to dipole moments associated with the polymer repeat unit. Apolar polymers, such as poly(styrene), result in low values of σ, while highly polar polymers, such as polycarbonates or polyesters, give high values of σ. Subtle effects such as ground state charge-transfer complex formation between the transport molecule and the polymer may also be important (Scott et al., 1990).

In addition to efficiently transporting charge, the transport layer must also accept charge from the generation layer with high efficiency (Kanemitsu and Imamura, 1987, 1989). Thus a correlation is predicted between the injection efficiency and the oxidation (reduction) potential of the donor (acceptor) molecule of the transport layer with low potentials favoring injection. Correlations between the oxidation potentials of the donor

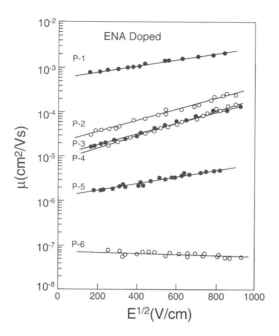

Figure 10.22   The field dependencies of the room temperature hole mobilities of N,N′-bis(2,2-diphenylvinyl)-N,N′-diphenylbenzidine (ENA) doped into poly(styrene) (P1), bisphenol-A polycarbonate (P2), a fluorinated polyester (P3), a polyester (P4), a fluorinated polyester-copolycarbonate (P5), and poly(vinyl butyral) (P6). The ENA concentration was 40%.

molecules and the injection efficiencies have been reported by Melz et al. (1977), Kakuta et al. (1981), and Umeda et al. (1993). For photoreceptors with generation layers composed of an azo pigment and transport layers containing different enamine derivatives, the sensitivity was the highest for enamine derivatives with the lowest oxidation potentials (Rice et al., 1985). Others have noted that within a given class of compounds, correlations between photoinjection efficiency and oxidation potential are complex and depend on the generation material (Murayama et al., 1988). Kubo et al. (1988) reported that dark decay caused by hole injection from polypyrrole into various transport layers correlated with the oxidation potential of the donor molecule of the transport layer.

Polysilanes $(R_1 R_2 Si)_n$ (West, 1986; Miller and Michl, 1989) are a class of materials that have high hole mobilities. These have been used as transport layers with generation layers of H_2Pc and TiOPc dispersed in poly(vinyl butyral) (Yokoyama and Yokoyama, 1989). A poly(phenylmethylsilane) transport layer used with a TiOPc generation layer performed satisfactorily in a laser printer. For a variety of reasons, including perhaps synthetic complexity, physical characteristics, and photoinstability, these materials have not been widely utilized as hole transport materials.

Transport active polymers, in which the polymer and the donor molecule are combined, are an alternative to molecularly doped polymers. The hole mobilities of these materials are significantly lower than related molecularly doped polymers, typically by a factor of 10 to 100. The cause, at least in the case of PVK, is believed to be hole trapping at carbazole dimer sites whose signature is excimer emission. Considerations of the importance of molecular rotation in permitting overlap of the molecular orbitals of neighboring transport molecules (Slowik and Chen, 1983) has led to the synthesis of carbazole-substituted methacrylate and acrylate polymers. It was thought that the carbazole moiety would be far enough removed from the polymer backbone to be relatively free to rotate at temperatures below the polymer glass transition. Some of these polymers did exhibit enhanced hole mobilities relative to PVK (Oshima et al., 1985, 1985a). Polymers that do not show excimer emission have also been prepared (Ito et al., 1985), and some of these show improved hole mobilities relative to PVK (Hu et al., 1988). Polymeric transport materials continue to be investigated (Domes et al., 1989; Frechet et al., 1989; Murti et al., 1991).

In addition to their relevance to transport layers, transport phenomena in generation layers are also important. Most studies have been made of various phthalocyanines. Both vapor-deposited and dispersion layers have been investigated.

As normally prepared, CuPc is p-type, although n-type conductivity has been observed following various heat treatments (Delacote et al., 1964). Hole mobilities have been reported by Sussman (1967), Hamann (1968), Mycielski et al. (1982), Ziólkowska-Pawlak et al. (1983), and Gould (1985). The values were between 10^{-4} and 10^{-2} cm^2/Vs. Usov and Benderskii (1970) reported a mobility of 0.40 cm^2/Vs for H$_2$Pc. Nespurek and Lyons (1981) reported the hole range in H$_2$Pc as 2×10^{-10} cm^2/V. Ahmad and Collins (1991) reported a mobility of 6×10^{-6} cm^2/Vs for triclinic PbPc. Kontani et al. (1995) measured hole mobilities in vapor-deposited TiOPc. A key result of the work of Kontani et al. is that the mobilities are strongly dependent on the substrate temperature during the vapor-deposition process. The mobilities were between 6.0×10^{-6} and 8.0×10^{-5} cm^2/Vs. The differences were attributed to an increase in film crystallinity as the substrate temperature was increased. Hole mobilities of vapor-deposited ClAlPc were measured by Ioannidis et al. (1993, 1994). The mobilities were between 2 and 6×10^{-4} cm^2/Vs.

Kitamura and Yoshimura (1992) measured hole mobilities of dispersions of TiOPc in a polyester and a polycarbonate. The mobilities were in the range of 10^{-7} to 10^{-6} cm^2/Vs. Enokida et al. (1993) measured hole mobilities of TiOPc dispersions in different polymers. The results were dependent on the specific form of the pigment as well as the polymer. Valerian and Nespurek (1993) measured the hole range of vapor-deposited α-H$_2$Pc. The values were between 4×10^{-11} and 2×10^{-9} cm^2/V. The results indicated that the layers were highly inhomogeneous with the distribution of trapping centers distributed predominately in the vicinity of the substrate. Omote et al. (1995) measured hole mobilities in dispersions composed of x-H$_2$Pc in poly(vinyl butyral) and poly(styrene). At room temperature, the mobilities were in the range of 10^{-6} to 10^{-5} cm^2/Vs.

Electron Transport

The approach taken for the search for suitable acceptor molecules is guided by several principles. First, the molecule must be nontoxic. The mutagenicity of TNF suggests that caution must be used with this and related materials. Second, it must be readily synthesized and purified. Third, the molecule should have a low reduction potential (high electron affinity). Since molecular oxygen, a potential electron trap, is always present in relatively high concentrations, the reduction potential of the acceptor molecule must be lower than that of oxygen. Reduction potentials determined in solution can be correlated with gas phase electron affinities (Loutfy et al., 1984). Fourth, the acceptor molecule must be weakly polar. Finally, the acceptor molecule should be highly soluble so that high-concentration films can be prepared.

The above requirements have made the search for electron transport materials difficult. To achieve the desired reduction potential, it is necessary to attach electron-withdrawing substituents, such as nitro or dicyanomethylene functionalities, on a parent aromatic ring such as a fluorene (Kuder et al., 1978). Unfortunately, many of these molecules are highly polar. Further, these frequently show irreversible redox chemistry, with correlated electron trapping. Also, the introduction of electron-withdrawing substituents reduces the solubility so that high-concentration layers often cannot be prepared without crystallization. For the above reasons, the development of electron transport materials has been slow. This is somewhat surprising because among the earliest organic photoreceptors were IBM's PVK:TNF charge-transfer complex and the dye–polymer aggregate developed at Eastman Kodak Company. In both materials, electron transport occurs over considerable distances in the absence of trapping. The mobilities, however, are too low for current applications.

Figure 10.23 shows structures of representative acceptor molecules. Unlike donor molecules, there are no molecular classes by which these molecules can be described. Figure 10.24 shows the field dependencies of the room temperature mobilities of different acceptor molecules doped into poly(styrene): 3,5'-dimethyl-3',5-di-*t*-butyl-diphenoquinone (DPQ), N,N'-bis(1,2-dimethylpropyl)-1,4,5,8-naphthalenetetracarboxylic diimide (NTDI), 1,1-dioxo-2-(4-methylphenyl)-6-phenyl-4-(dicyanomethylidene)thiopyran (PTS), 2-*t*-butyl-9,10-N,N'-dicyanoanthraquinonediimine (DCAQ), and 2-methyl-2-pentyl-1,3-bis(dicyanomethylene)indane (IND). Electron mobilities of molecularly doped polymers are typically some one to two orders of magnitude lower than hole mobilities. The field, temperature, and concentration dependencies, however, are similar. The results of most recent studies have been described by the disorder formalism. The results yield values of σ between 0.09 to 0.16 eV, considerably higher than values for donor doped polymers.

Aklyl-substituted nitrated fluorene-9-ones were synthesized with the expectation of improved solubility and the hope for lowered toxicity (Loutfy and Ong, 1984; Loutfy et al., 1984; Ong et al., 1985). These compounds showed improved solubility. Compared to TNF, the mobilities were somewhat improved. Studies have been carried out with dual-layer photoreceptors using transport layers of (4-*n*-butoxycarbonyl-9-fluorenylidene) malonitrile and generation layers of a dispersed squaraine (Murti et al., 1987) and vapor-deposited VOPc (Loutfy et al., 1985b). Electron trapping was found to be important in both materials. The electron mobilities of this molecule in a polyester have been reported by Borsenberger and Bässler (1991a). An acceptor doped polymer, with this acceptor functionality pendant, has been prepared and used as the transport layer in a photoreceptor with a TiOPc generation layer (Sim et al., 1996). The discharge was highly efficient, although there was a significant residual potential.

More recent activities have centered on the synthesis of new classes of materials. Thus far, however, none of these materials have mobilities comparable to donor doped polymers. When these materials have been used in a dual-layer configuration, the thicknesses of the transport layers are considerably less than in photoreceptors with hole transport layers.

Some of the more recently investigated classes of compounds include 4*H*-dicyanomethylene-thiopyran-dioxides (Scozzafava et al., 1985; Detty et al., 1995; Borsenberger et al., 1995a), N-(nitrofluorenylidene)anilines (Matsui et al., 1993), 6,13-diazanaphtho[2, 3-b]fluorene-5-arylimino-7,12-diones (Mizuta et al., 1996), 9,10-N,N'-dicyanoanthraquinone-diimines (Borsenberger et al., 1994, 1995), 4,4'-diphenoquinones (Yamaguchi and Yokoyama, 1991), and naphthyl-substituted oxadiazoles (Tokuhisa et al., 1995). The latter is interesting because oxadiazoles are usually considered to be donor molecules.

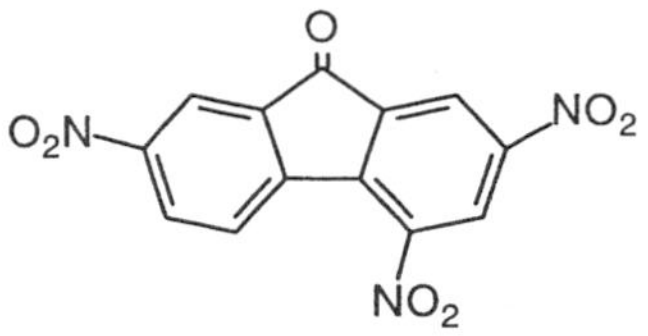

2,4,7-trinitro-9-fluorenone (TNF)
Gill (1972)

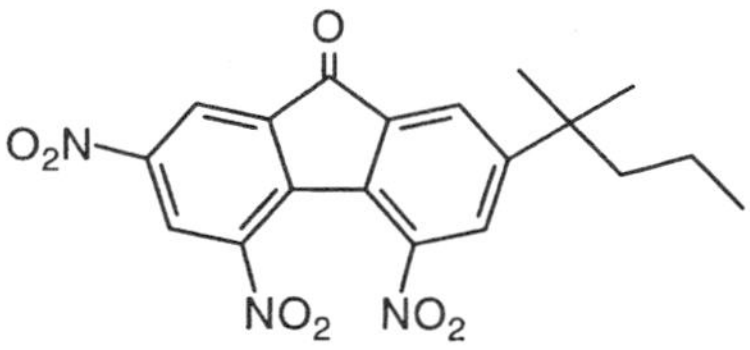

2-(1,1-dimethylbutyl)-4,5,7-trinitro-9-fluorenone
Ong et al. (1985)

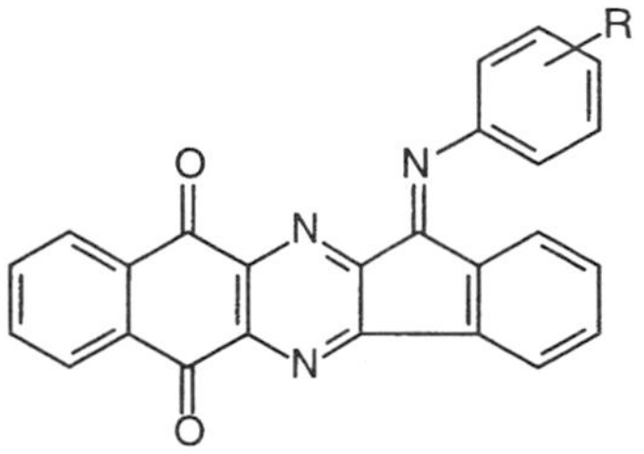

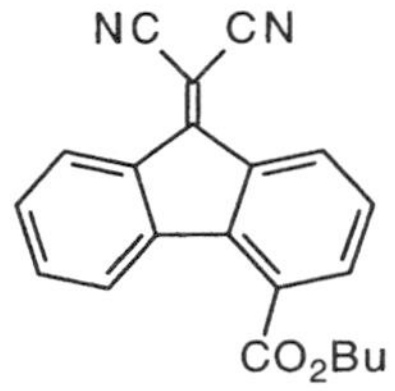

(4-butoxycarbonyl-9-fluorenylidene) malononitrile
Murti et al. (1987)

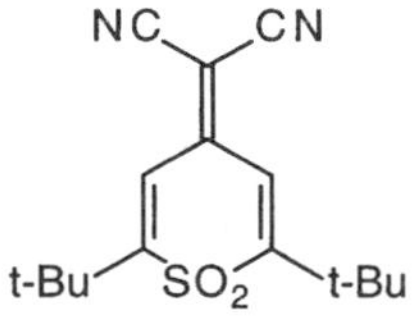

6,13-diazanaphtho[2,3-b]fluorene-5-arylimino-7,12-diones
Mizuta et al. (1996)

2,6-di-tert-butyl-4-dicyanomethylene-4-H-thiopyran-1,1-dioxide
Scozzafava et al. (1985)

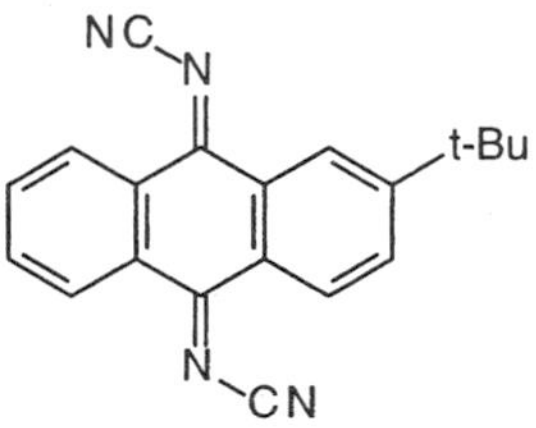

2-*tert*-butyl-9,10-N,N'-dicyanoanthraquinonediimine
Borsenberger et al. (1994, 1995b)

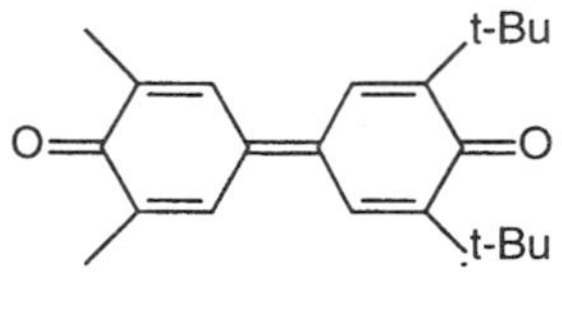

3,5-dimethyl-3',5'-di-*tert*-butyl-4,4'-diphenoquinone
Yamaguchi et al. (1990)

Figure 10.23 The molecular structures, chemical names, and common acronyms of typical classes of acceptor molecules.

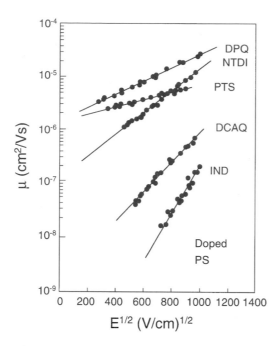

Figure 10.24 The field dependencies of the room temperature electron mobilities of a series of dephenoquinone (DPQ), naphthalene diimide (NTDI), sulfone (PTS), dicyanoanthraquinone (DCAQ), and indane (IND) derivatives doped into poly(styrene). The acceptor concentrations were 40%.

Magin and Borsenberger (1993) measured electron mobilities in vapor-deposited films of N,N′-bis(2-phenethyl)-perylene-3,4:9,10-bis(dicarboximide). The field dependencies of the mobility showed the $\ln \mu \propto \beta E^{1/2}$ relationships commonly observed in doped polymers. The temperature dependencies, however, could not be described by either an Arrhenius relationship or a $\ln \mu \propto -(T_0/T)^2$ relationship. The width of the hopping site manifold σ was determined as 0.080 eV. Consistency with predictions of the disorder formalism, however, required an additional source of activation. The source of the activation was suggested to be either polaron formation or trapping.

Ioannidis and Dodelet (1997, 1997a, 1997b) measured electron mobilities of vapor-deposited ClAlPc. The mobilities increased with increasing substrate temperature and decreasing deposition rate. This was attributed to an increase in order. The mobilities were significantly degraded by the presence of air. The electron and hole mobilities were of comparable magnitude.

Thus far, positive-charging photoreceptors have been prepared either in the single-layer configuration or in a dual-layer configuration with the generation layer uppermost. For DAD applications, however, there is a need for conventional dual-layer photoreceptors that can be charged positively. Thus there is a need for electron transport materials. Suitable acceptor molecules, however, must have a combination of a low reduction potential, a high solubility, and a low dipole moment, a combination that has thus far proved elusive.

Bipolar Transport

An interesting recent development is that polymers doped with mixtures of donor and acceptor molecules show bipolar transport in which the mobilities of each are unaffected

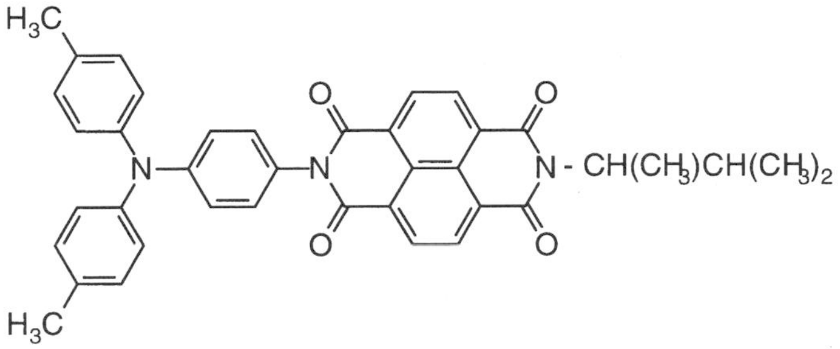

Figure 10.25   A bipolar charge transport molecule.

by the presence of the other. Thus Yamaguchi et al. (1990, 1991) described single-layer photoreceptors containing mixtures of 3,5-dimethyl-3′,5′-di-*t*-butyl-4,4′-diphenoquinone and N,N,N′,N′-tetrakis(*m*-methylphenyl)-1,3-diaminobenzene in poly(4,4′-cyclohexyli-denediphenylcarbonate) with TiOPc. Gruenbaum et al. (1996) described electron mobili-ties of transport layers composed of 4*H*-1,1-dioxo-4-dicyanomethylidene-2-*p*-tolyl-6-phe-nylthiopyran and tri-*p*-tolylamine in a polyester with a TiOPc generation layer. Lin et al. (1996) described electron and hole mobilities of a vapor-deposited bipolar transport layer composed of N,N′-bis(1,2-dimethylpropyl)-1,4,5,8-naphthalenetetracarboxylic diimide and tri-*p*-tolylamine. A bifunctional molecule N-[*p*-(di-*p*-tolylamino)phenyl]-N′-(1,2-di-methylpropyl)-1,4,5,8-naphthalenetetracarboxylic diimide was described by Kaeding et al. (1996). The structure is shown in Fig. 10.25. In this material, electron and hole mobili-ties are comparable with similar field and temperature dependencies.

10.4.4   Fatigue

Fatigue is the occurrence of changes in electrical properties with use. Changes in charge-ability, dark discharge, sensitivity, and residual potential are examples. These may occur gradually with extended cycling under nominal conditions or rapidly after unintended exposures to environmental changes. Fatigue may necessitate the use of complex process compensation to avoid image quality degradation. Thus a considerable effort has been devoted to achieving fatigue resistance, either by the appropriate choice of materials or by the addition of additives.

Fatigue, manifested as an increase in the residual potential with use, commonly results from range limitations due to trapping (Okuda et al., 1982; Pai and Yanus, 1983; Murti et al., 1991a; Kasap et al., 1992). Trapping at the interface between the generation and transport layers has also been reported (Kanemitsu and Imamura, 1989, 1989a, 1990; Kanemitsu et al., 1990; Kanemitsu and Funda, 1991). Effects of environmental factors such as seasonal changes in humidity have been described by Law (1987).

Radiation-induced fatigue has recently received considerable attention. Unfiltered cool-white fluorescent light has a significant ultraviolet component that can be absorbed by the transport layer and lead to photochemical damage. The IBM photoreceptor using the bisazo pigment Chlorodiane Blue in the generation layer is permanently damaged by exposure to room illumination (McMurty et al., 1984). The transport layer thus includes a dye to absorb the ultraviolet component (Champ and Vollmer, 1987). Recently, the details of the fatigue were elucidated by relating these effects to the photochemistry of the DEH donor molecule (Pacansky et al. 1987, 1991). DEH undergoes the photocycliza-

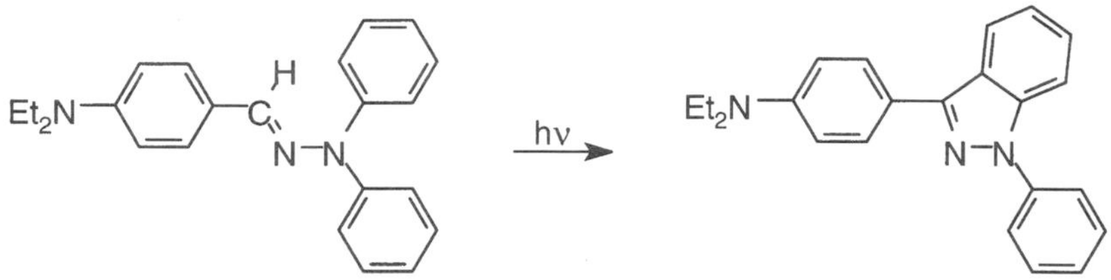

Figure 10.26 The solid-state photocyclization reaction of p-diethylaminobenzaldehyde diphenyl-hydrazone (DEH), a donor molecule commonly used in transport layers. (After Pacansky et al., 1987. Reprinted from *J. Photochem. 37*:293. Copyright 1987, Elsevier Science Publishers.)

tion, shown in Fig. 10.26, with a very high quantum efficiency. The effects of ultraviolet exposures on hole transport of DEH doped polycarbonate have been described by Stasiak et al. (1995) and Stasiak and Storch (1996, 1996a, 1996b). Mizuta et al. (1993) reported that single-layer photoreceptors containing 9-isopropylcarbazole N,N-diphenylhydrazone as the donor molecule are fatigued on exposure to radiation absorbed by the transport layer. The results were interpreted by a mechanism in which ultraviolet radiation induces isomerization of the anti- to the syn-isomer. The latter has a lower ionization potential and is thus a hole trap. In the single-layer configuration, hole trapping near the free surface results in the creation of a space charge that causes decreased sensitivity. Weiss and Chen (1995) have shown that transport layers containing tri-*p*-tolylamine undergo ultraviolet-induced photochemical reactions that cause fatigue in the form of an increasing residual potential with use. Enokida et al. (1991) reported that ultraviolet-radiation-induced fatigue of a photoreceptor containing a poly(methylphenylsilylene) hole transport layer could be explained in terms of transport layer photodecomposition.

In addition to changes in the transport properties, exposure of some transport layers to ultraviolet radiation leads to increased dark discharge (Kanemitsu and Imamura, 1988; Kanemitsu et al., 1989; Nabeta et al., 1990; Hirao et al., 1994). Although the mechanistic details are unclear, one likely explanation is the photoinduced electron transfer from the donor molecule to an acceptor molecule, creating a radical cation (Limburg et al., 1983). Since donor molecules are designed to have low oxidation potentials, even weak acceptors will suffice. This effect can be enhanced and such films used as electrophotographic masters (Ogawa et al., 1993).

The patent literature has many examples of addenda that can act as radical traps and quenchers of excited states, singlet oxygen, and superoxide to prevent transport layer photooxidation. See, for example, Limburg and Pai (1982) and Limburg and Renfer (1986). Surprisingly, ultraviolet radiation and electron beam curing have been investigated for the preparation of dual-layer photoreceptors (Pacansky et al., 1987a). Preliminary results indicated feasibility for the generation layer, but high dark discharge rates occurred when the transport layer was radiation cured. Radiation curing is a novel approach that avoids the use of the solvents necessary for solution coating technologies.

It is well known that ozone, a by-product of the corona discharge process, is very reactive toward most organic materials (Nashimoto, 1988). It has been shown that a styryl donor molecule undergoes ozonolysis, yielding carbonyl-containing products that are hole traps (Takenouchi et al., 1988). Interestingly, it was found that for this family of donor molecules, photoreceptors containing those with lower ionization potentials showed decreased ozone sensitivity. Other by-products of the corona discharge process are nitrogen oxides (Haridoss et al., 1982; Goldman and Sigmond, 1985; Nashimoto, 1988). These are potent oxidizing agents. Reactions of transport layers with nitrogen oxides can lead to degraded sensitometry (Weiss, 1990). In addition, acids formed by a combination of these nitrogen oxides with water can lead to image degradation by increasing the conductivity of the photoreceptor surface (Yarmchuk and Keefe, 1989; Weiss et al., 1996).

10.5 FUTURE DIRECTIONS

In the past two decades, organic photoreceptors have played an increasingly important role in xerography. Because of their photoelectronic properties, their low cost, and the ability to prepare these materials in large areas as flexible layers, or on drum substrates, organic materials are now used for a wide range of both copier and printer applications. Based on recent developments, we expect this trend to continue.

We anticipate that increased sensitivity, longer process lifetimes, higher fatigue resistance, and higher mobilities will be primary objectives. In addition, there will be increasing emphasis on materials for DAD applications. Here, enhanced near infrared sensitivity, electron transport layers, and lower dark discharge are primary requirements. Finally, we anticipate increasing emphasis on materials that can better withstand the chemical and physical environment of xerographic engines.

REFERENCES

Abkowitz, M. (1987). *SPIE 763*:88.

Abkowitz, M. A. (1987a). *J. Non-Cryst. Solids 97–98*:1163.

Abkowitz, M. A., Jansen, F. (1983). *J. Non-Cryst. Solids 59/60*:953.

Abkowitz, M., Maitra, S. (1987). *J. Appl. Phys. 61*:1038.

Abkowitz, M. A., Markovics, J. M. (1982). *Solid State Commun. 44*:1431.

Abkowitz, M. A., Markovics, J. M. (1984). *Philos. Mag. B. 49*:L31.

Abkowitz, M. A., Stolka, M. (1988). In *Proceedings of the International Symposium on Polymers for Advanced Technologies* (M. Lewin, ed.). VCH Publishers, New York, p. 225.

Abkowitz, M., Jansen, F., Melnyk, A. R. (1985). *Philos. Mag. B. 51*:405.

Ahmad, A., Collins, R. A. (1991). *Phys. Status Solidi (a) 123*:201.

Albrecht, U., Bässler, H. (1995). *Chem. Phys. Lett. 235*:389.

Allen, N. S., Robinson, E. T., Scott, C. M., Thompson, F. (1989). *Dyes Pigm. 10*:183.

Andre, B., Lever, R., Moisan, J. Y. (1989). *Chem. Phys. 137*:281.

Aratani, S., Saito, T., Kawanishi, T., Kinjo, N. (1990). *Jpn. J. Appl. Phys., 29*:L1682.

Aratani, S., Kawanishi, T., Kakuta, A. (1991). *Jpn. J. Appl. Phys. 30*:L1656.

Aratani, S., Kawanishi, T., Kakuta, A. (1996). *Jpn. J. Appl. Phys. 35*:2184.

Arishima, K., Hiratsuka, H., Tate, A., Okada, T. (1982). *Appl. Phys. Lett. 40*:279.

Arita, M., Homma, S., Fujushima, K., Yamamoto, H., Kura, H., Okamura, M. (1991). In: *Proceedings of the Sixth International Congress on Advances in Non-Impact Printing Technologies* (R. J. Nash, ed.). IS&T, Springfield, VA, p. 321.

Arkhipov, V. I., Nikitenko, V. R. (1993). *J. Non-Cryst. Solids 164/166*:587.

Bässler, H. (1993). *Phys. Status Solidi (b) 175*:15.

Bässler, H. (1994). *Int. J. Mod. Phys. 8*:847.

Bässler, H. (1994a). In: *Disorder Effects on Relaxation Processes* (R. Richert, A. Blumen, eds.). Springer-Verlag, Berlin, p. 485.

Bässler, H. (1994b). *Mol. Cryst. Liq. Cryst. 252*:11.

Bässler, H., Borsenberger, P. M., Perry, R. J. (1994). *J. Polym. Sci. Part B: Polym. Phys. 32*:1677.

Berlin, Yu. A., Chekunaev, N. I., Goldanskii, V. I. (1990). *J. Chem. Phys. 92*:7540.

Borsenberger, P. M., Bässler, H. (1991). *J. Chem. Phys. 95*:5327.

Borsenberger, P. M., Bässler, H. (1991a). *J. Imaging Sci. 35*:79.

Borsenberger, P. M., Weiss, D. S. (1993). In: *Organic Photoreceptors for Imaging Systems*. Marcel Dekker, New York.

Borsenberger, P. M., Weiss, D. S. (1998). In: *Organic Photoreceptors for Xerography*. Marcel Dekker, New York.

Borsenberger, P. M., Pautmeier, L., Bässler, H. (1991). *J. Chem. Phys. 94*:5447.

Borsenberger, P. M., Magin, E. H., Van der Auweraer, M., De Schryver, F. C. (1993). *Phys. Status Solidi 140*:9.

Borsenberger, P. M., Kan, H.-C., Vreeland, W. B. (1994). *Phys. Status Solidi (b) 142*:489.

Borsenberger, P. M., Kan, H.-C., Magin, E. H., Vreeland, W. B. (1995). *J. Imaging Sci. Technol. 39*:6.

Borsenberger, P. M., Magin, E. H., Detty, M. R. (1995a). *J. Imaging Sci. Technol. 39*:12.

Borsenberger, P. M., Gruenbaum, W. T., Wolf, U., Bässler, H. (1998). *Chem. Phys. 234*:277.

Brown, F. C. (1955). *Phys. Rev. 97*:355.

Burland, D. M., Schein, L. B. (1986). *Phys. Today 39*:46.

Champ, R. B. (1987). *SPIE 759*:40.

Champ, R. B., Shattuck, M. D. (1974). U.S. Patent 3,824,099.

Champ, R. B., Vollmer, R. L. (1987). U. S. Patent 4,677,045.

Chen, H.-Z., Wang, M., Feng, L.-X., Yang, S.-L. (1993). *J. Photochem. Photobiol. A: Chem. 70*: 179.

Chen, I. (1978). *J. Appl. Phys. 49*:1162.

Chen, I. (1990). *J. Imaging Sci. 34*:15.

Chen, I. (1993). *J. Imaging Sci. Technol. 37*:396.

Chen, W.-X, Xu, Z.-D, Li, W.-Z. (1995). *J. Photochem. Photobiol. A: Chem. 88*:179.

Chen, Y., Cai, R.-F., Huang, Z.-E., Bai, X., Yu, B.-C., Jin, W., Pan, D.-C., Wang, S.-T. (1996). *Polym. Bull. 36*:203.

Collins, L. F. (1974). *J. Appl. Phys. 45*:5356.

Cornelius, L. (1994). *R & R News*, July, p. 34.

Cort, L. F., Crawford, J. W., Vollmer, R. L. (1983). *IBM Tech. Discl. Bull. 26*:2855.

Coulson, J. M., Richardson, J. F. (1983). In: *Chemical Engineering*, Vol. 2. Pergamon Press, New York, Ch. 2.

Daimon, K., Nukada, K., Sakaguchi, Y., Igarashi, R. (1996). *J. Imaging Sci. Technol. 40*:249.

Decker, J., Fukae, K., Johnson, S., Kaieda, S., Yoshida, I. (1991). In: *Proceedings of the Seventh International Congress on Advances in Non-Impact Printing Technologies* (K. Pietrowski, ed.). IS&T, Springfield, VA, p. 328.

Delacote, G. M., Fillard, J. P., Marco, F. J. (1964). *Solid State Commun. 2*:373.

De Schamphelaere, L. A., and the Xeikon Team. (1994). In: *Proceedings of the Tenth International Congress on Advances in Non-Impact Printing Technologies* (A. Melnyk, ed.). IS&T, Springfield, VA, p. 1.

Detty, M. R., Eachus, R. S., Sinicropi, J. A., Lenhard, J. R., McMillan, M., Lanzafame, A. M., Luss, H. R., Young, R., Eilers, J. (1995). *J. Org. Chem. 60*:1674.

Dieckmann, A., Bässler, H., Borsenberger, P. M. (1993). *J. Chem. Phys. 99*:8136.

DiPaola-Baranyi, G., Hsiao, C. K., Hor, A. M. (1990). *J. Imaging Sci. 34*:224.

DiPaola-Baranyi, G., Hsiao, C. K., Kazmaier, P. M., Burt, R., Loutfy, R. O., Martin, T. I. (1988). *J. Imaging Sci. 32*:60.

Domes, H., Fischer, R., Haarer, D., Strohriegl, P. (1989). *Macromol. Chem. 190*:165.

Duff, J, Hor, A. M., Melnyk, A. R., Teney, D. (1990). *Proc. SPIE 1253*:183.

Duff, J. M., Hor, A. M., Allen, C. G., Melnyk, A., Teney, D. (1991). In: *Proceedings of the Seventh International Congress on Advances in Non-Impact Printing Technologies* (K. Pietrowski, ed.). IS&T, Springfield, VA, p. 284.

Dunitz, J. D., Bernstein, J. (1995). *Acc. Chem. Res. 28*:193.

Dunlap, D. H. (1995). *Phys. Rev. B 52*:939.

Dunlap, D. H. (1995a). *Proc. SPIE 2526*:2.

Dunlap, D. H. (1996). *Proc. SPIE 2850*:110.

Dunlap, D. H. (1997). *Proc. SPIE 3144*:80.

Dunlap, D. H., Kenkre, V. M. (1993). *Chem. Phys. 178*:67.

Dunlap, D. H., Parris, P. E., Kenkre, V. M. (1996). *Phys. Rev. Lett. 77*:542.

Ellis, W. H. (1986). In: *Federation Series on Coatings Technology*. Federation of Societies for Coatings Technology, Philadelphia, PA.

Endo, K., Miyaoka, S., Katsuya, Y. (1994). In: *Proceedings of the Tenth International Congress on Advances in Non-Impact Printing Technologies* (A. Melnyk, ed.). IS&T, Springfield, VA, p. 279.

Enokida, T., Ehashi, S. (1988). *Chem. Lett.:179*.

Enokida, T., Hirohashi, R. (1991). *Mol. Cryst. Liq. Cryst. 195*:265.

Enokida, T., Hirohashi, R. (1992). *J. Imaging Sci. Technol. 35*:135.

Enokida, T., Hirohashi, R., Nakamura, T. (1990). *J. Imaging Sci. 34*:234.

Enokida, T., Hirohashi, R., Kurata, R. (1991). *J. Appl. Phys. 70*:3242.

Enokida, T., Hirohashi, R., Mizukami, S. (1991a). *J. Imaging Sci. 35*:235.

Enokida, T., Hirohashi, R., Morohashi, N. (1991b). *Bull. Chem. Soc. Jpn. 64*:279.

Enokida, T., Yamamoto, S., Horohashi, R. (1993). In: *Proceedings of the Ninth International Congress on Advances in Non-Impact Printing Technologies* (M. Yokoyama, ed.). IS&T, Springfield, VA, p. 615.

Fox, S. J. (1974). In: *Proceedings of the Fourth International Congress on Electrophotography* (D. R. White, ed.). SPSE, New York, p. 170.

Frechet, J. M. J., Gautheir, S., Limburg, W. W., Loutfy, R. O., Murti, D. K., Spiewak, J. W. (1989). U.S. Patent 4,801,517.

Fujimaki, Y. (1991). In: *Proceedings of the Seventh International Congress on Advances in Non-Impact Printing Technologies* (K. Pietrowski, ed.). IS&T, Springfield, VA, p. 269.

Fujimaki, Y., Tadokoro, H., Oda, Y., Yoshioka, H., Homma, T., Moriguchi, H., Watanabe, K., Konishita, A., Hirose, N., Itami, A., Ikeuchi, S. (1991). *J. Imaging Technol. 17*:202.

Fujimori, M., Nagasawa, K., and Yorinobu, A. (1992). In: *Proceedings of the Eighth International Congress on Advances in Non-Impact Printing Technologies* (E. Hanson, ed.). IS&T, Springfield, VA, p. 240.

Gartstein, Yu. N., Conwell, E. M. (1994). *Chem. Phys. Lett. 217*:41.

Gartstein, Yu. N., Conwell, E. M. (1994a). *J. Chem. Phys. 100*:9175.

Gartstein, Yu. N., Conwell, E. M. (1995). *Phys. Rev. B 51*:6947.

Gartstein, Yu. N., Jeyadev, S., Conwell, E. M. (1994). *Phys. Rev. B 51*:4622.

Gavezzotti, A., Filippini, G. (1995). *J. Am. Chem. Soc. 117*:12299.

Goldman, A., Sigmond, R. S. (1985). *J. Electrochem. Soc.: Electrochem. Sci. Technol. 132*:2842.

Gould, R. D. (1985). *Thin Solid Films 125*:63.

Grammatica, S., Mort, J. (1981). *Appl. Phys. Lett. 38*:445.

Gregory, P. (1988). In: *Proceedings of the Fourth International Congress on Advances in Non-Impact Printing Technologies* (A. Jaffe, ed.). SPSE, Springfield, VA, p. 70.

Gregory, P. (1991). *High Technology Applications of Organic Colorants*. Plenum Press, New York.

Gruenbaum, W. T., Magin, E. H., Borsenberger, P. M. (1996). *J. Imaging Sci. Technol. 40*:310.

Gutoff, E. B., Cohen, E. D. (1995). *Coating and Drying Defects: Troubleshooting Operating Problems*. John Wiley, New York.

Hackett, C. F. (1971). *J. Chem. Phys. 55*:3178.

Hamann, C. (1968). *Phys. Status Solidi 26*:311.

Haridoss, S., Perlman, M. M., Carlone, C. (1982). *J. Appl. Phys. 53*:6106.

Hartenstein, B., Bässler, H. (1995). *J. Non-Cryst. Solids 190*:112.

Hartenstein, B., Bässler, H., Jakobs, A., Kehr, K. W. (1996). *Phys. Rev. B 54*:8574.

Hashimoto, M. (1986). *Denshi Shashin (Electrophotography) 25*:10.

Hayashi, N., Hayashida, S., Morishita, Y. (1992). *J. Imaging Sci. Technol. 36*:574.

Hayashida, S., Akimoto, T., Morishita, Y., Itagaki, M., Matsui, M. (1994). In: *Proceedings of the*

Tenth International Congress on Advances in Non-Impact Printing Technologies (A. Melnyk, ed.). IS&T, Springfield, VA, p. 249.

Haynes, J. R., Shockley, W. (1951). *Phys. Rev. 81*:835.

Hirano, A., Tsuruoka, E., Takeda, Y. (1995). In: *Proceedings of the Eleventh International Congress on Advances in Non-Impact Printing Technologies* (J. Anderson, ed.), IS&T, Springfield, VA, p. 60.

Hirao, A., Nishizawa, H. (1997). *Phys. Rev. B. 56*:R2904.

Hirao, A., Nishizawa, H., Sugiuchi, M. (1993). *J. Appl. Phys. 74*:1083.

Hirao, A., Nishizawa, H., Hosoya, M. (1994). *Jpn. J. Appl. Phys. 33*:1944.

Hirose, N., Hayata, H., Fujimaki, Y. (1992). In: *Proceedings of the Eighth International Congress on Advances in Non-Impact Printing Technologies* (E. Hanson, ed.). IS&T, Springfield, VA, p. 256.

Hor, A.-M., Loutfy, R. O. (1983). *Thin Solid Films 106*:291.

Hor, A.-M., Loutfy, R. O. (1986). U.S. Patent 4,587,189.

Hoshino, K., Ishibashi, O., Hiruta, S., Kokado, H., Yokoyama, S. (1990). In: *Proceedings of the Sixth International Congress on Advances in Non-Impact Printing Technologies* (R. Nash, ed.). IS&T, Springfield, VA, p. 429.

Hosoya, M., Miyamoto, H., Nishizawa, H., Hirao, A. (1995). In: *Proceedings of the Eleventh International Congress on Advances in Non-Impact Printing Technologies* (J. Anderson, ed.). IS&T, Springfield, VA, p. 51.

Hsiao, C. K., Murti, D. K., Hor, A. M. DiPaola-Baranyi, G., Liebermann, G. (1994). In: *Proceedings of the Tenth International Congress on Advances in Non-Impact Printing Technologies* (A. Melnyk, ed.). IS&T, Springfield, VA., p. 220.

Hsieh, B. R., Melnyk A. R. (1996). In: *Proceedings of the Twelfth International Congress on Digital Printing Technologies* (M. Hopper, ed.). IS&T, Springfield, VA., p. 461.

Hsieh, B. R., Melnyk, A. R. (1998). *Chem. Mater. 10*:2313.

Hu, C.-J., Oshima, R., Sato, S., Seno, M. (1988). *J. Polym. Sci., Polym. Lett. 26*:441.

Huang, T.-H., Sharp, J. H. (1982). *Chem. Phys. 65*:205.

Hung, Y., Klose, T. R., Regan, M. T., Rossi, L. J. (1987). U.S. Patent 4,701,396.

Hung, Y., Regan, M. T., Staudenmayer, W. J. (1987a). U.S. Patent 4,666,802.

Hunger, K., Merkle, K. (1983). *Org. Coatings Sci. Technol. 5*:91.

Ioannidis, A., Dodelet, J.-P. (1997). *J. Phys. Chem. 101*:891.

Ioannidis, A., Dodelet, J.-P. (1997a). *J. Phys. Chem. 101*:901.

Ioannidis, A., Dodelet, J.-P. (1997b). *J. Phys. Chem. 101*:5100.

Ioannidis, A., Lawrence, M. R., Kassi, H., Cote, R., Dodelet, J.-P., Leblanc, R. M. (1993). *Chem. Phys. Lett. 205*:46.

Ioannidis, A., Lawrence, M. R., Cote, R., Kassi, H., Dodelet, J.-P. (1994). *Mol. Cryst. Liq. Cryst. 252*:195.

Itami, A., Watanabe, K., Kinoshita, A., Suzuki, T., Takahashi, J. (1991). In: *Proceedings of the Seventh International Congress on Advances in Non-Impact Printing Technologies* (K. Pietrowski, ed.). IS&T, Springfield, VA., p. 302.

Ito, S., Yamashita, K., Yamamoto, M., Nishijima, Y. (1985). *Chem. Phys. Lett. 117*:171.

Jeyadev, S., Pai, D. M. (1995). In: *Proceedings of the Eleventh International Congress on Advances in Non-Impact Printing Technologies* (J. Anderson, ed.). IS&T, Springfield, VA., p. 141.

Jeyadev, S., Pai, D. M. (1996). *J. Imaging Sci. Technol. 40*:327.

Kaeding, J. E., Murray, B. J., Gruenbaum, W. T., Borsenberger, P. M. (1996). *J. Imaging Sci. Technol. 40*:245.

Kakuta, A., Mori, Y., Morishita, H. (1981). *IEEE Trans. Ind. Appl. IA-17*:382.

Kakuta, A., Mori, Y., Takano, S., Sawada, M., Shibuya, I. (1985). *J. Imaging Technol. 11*:7.

Kanemitsu, Y. (1992). *J. Appl. Phys. 71*:3033.

Kanemitsu, Y., Einami, J. (1990). *Appl. Phys. Lett. 57*:673.

Kanemitsu, Y., Funada, H. (1991). *J. Phys. D: Appl. Phys. 24*:1409.

Kanemitsu, Y., Imamura, S. (1987). *Jpn. J. Appl. Phys. 27*:235.

Kanemitsu, Y., Imamura, S. (1988). *Solid State Commun. 68*:701.

Kanemitsu, Y., Imamura, S. (1989). *Appl. Phys. Lett. 54*:872.

Kanemitsu, Y., Imamura, S. (1989a). *Appl. Surf. Sci. 41–42*:544.

Kanemitsu, Y., Imamura, S. (1990). *J. Appl. Phys. 67*:3728.

Kanemitsu, Y., Imanishi, D., Imamura, S. (1989). *J. Appl. Phys. 66*:4526.

Kanemitsu, Y., Funada, H., Imamura, S. (1990). *J. Appl. Phys. 67*:4152.

Kasap, S. O. (1989). *J. Electrostat. 22*:69.

Kasap, S. O., Baxendale, M., Juhasz, C. (1987). *J. Appl. Phys. 62*:171.

Kasap, S. O., Bhattacharyya, A., Liang, Z. (1992). *Jpn. J. Appl. Phys. 31*:72.

Kato, K., Shimokihara, S., Itakura, R., Mikawa, H., Yokoyama, M. (1983). *Denshi Shashin (Electrophotography) 21*:169.

Kato, M., Nishioka, Y., Kaifu, K., Kawamura, K., Ohno, S. (1985). *Appl. Phys. Lett. 46*:196.

Kato, M., Nishioka, Y., Kaifu, K. (1986). U.S. Patent 4,587,188.

Kato, M., Nishioka, Y., Kaifu, K. (1988). U.S. Patent 4,731,312.

Katsen, J. B., Himmelwright, R. S., Taylor, D. H. (1994). In: *Proceedings of the Tenth International Congress on Advances in Non-Impact Printing Technologies* (A. Melnyk, ed.). IS&T, Springfield, VA., p. 506.

Kazmaier, P. M., Baranyi, G., Loutfy, R. O. (1988). U.S. Patent 4,746,756.

Kazmaier, P. M., Burt, R. A., Baranyi, G. (1988a). U.S. Patent 4,751,327.

Kazmaier, P. M., Burt, R., DiPaola-Baranyi, G., Hsiao, C.-K. Loutfy, R. O., Martin, T. I., Hamer, G. K., Bluhm, T. L., Taylor, M. G. (1988b). *J. Imaging Sci. 32*:1.

Kazmaier, P. M., Burt, R. A., Hor, A.-M., and Hsiao, C.-K. (1988c). U.S. Patent 4,792,508.

Kenkre, V. M., Dunlap, D. H. (1992). *Philos. Mag. B 65*:831.

Kepler, R. G. (1960). *Phys. Rev. 119*:1226.

Kerr, J. W., Rokos, G. H. S. (1977). *J. Phys. D: Appl. Phys. 10*:1151.

Khe, N. C., Yokota, S., Takahashi, F. (1984). *Photogr. Sci. Eng. 28*:191.

Khe, N. C., Takenouchi, O., Kawara, T., Tanaka, H., Yokota, S. (1984a). *Photogr. Sci. Eng. 28*: 195.

Kin, S., Tanaka, H., Saeki, S., Torikoshi, K., and Pu, L. S. (1986). U.S. Patent 4,626,485.

Kishi, J., Inaba, Y., Takahashi, Y., Sakata, T. (1984). *Denshi Shashin (Electrophotography) 23*:203.

Kitamura, T., Yokoyama, M. (1991). *J. Appl. Phys. 69*:821.

Kitamura, T., Yoshimura, H. (1992). In: *Proceedings of the Eighth International Congress on Advances in Non-Impact Printing Technologies* (E. Hanson, ed.). IS&T, Springfield, VA., p. 237.

Kitamura, T., Imamura, S., Kawamata, M. (1988). *J. Imaging Sci. 14*:136.

Kobayashi, T., Wakita, K., Kubo, K., Nagae, S., Fujimoto, T., Koezuka, H. (1996). In: *Proceedings of the Twelfth International Congress on Digital Printing Technologies* (M. Hopper, ed.). IS&T, Springfield, VA., p. 480.

Kochelev, K. K., Zhylina, V. I., Khots, G. E., Kocheleva, O. K., Sleptsov, V. V. (1996). In: *Proceedings of the Twelfth International Congress on Digital Printing Technologies* (M. Hopper, ed.). IS&T, Springfield, VA., p. 483.

Kontani, T., Wada, T., Masui, M., Takeuchi, M. (1995). In: *Proceedings of the Eleventh International Congress on Advances in Non-Impact Printing Technologies* (J. Anderson, ed.). IS&T, Springfield, VA., p. 42.

Kubiak, R., Janczak, J., Ejsmont, K. (1995). *Chem. Phys. Lett. 245*:249.

Kubo, I., Hanna, J., Yamamoto, S., Kokado, H. (1988). *Jpn. J. Appl. Phys. 27*:1054.

Kuder, J. E., Pochan, J. M., Turner, S. R., Hinman, D. F. (1978). *J. Electrochem. Soc., 125*:1750.

Langevin, P. (1903). *Ann. Chem. Phys. 28*:289, 443.

Law, K. Y. (1987). *J. Imaging Sci. 31*:83.

Law, K. Y. (1988). *J. Phys. Chem. 92*:4226.

Law, K. Y. (1990). *J. Imaging Sci. 34*:38.

Law, K. Y. (1992). *J. Imaging Sci. Technol. 36*:567.

Law, K. Y. (1992a). *Chem. Mater. 4*:605.

Law, K.-Y. (1993). *Chem. Rev. 93*:449.

Law, K.-Y., Bailey, F. C. (1986). *Can. J. Chem. 64*:2267.

Law, K.-Y., Bailey, F. C. (1987). *J. Imaging Sci. 31*:172.

Law, K.-Y., Bailey, F. C. (1988). *Dyes Pigm. 9*:85.

Law, K.-Y., Bailey, F. C. (1991). *J. Chem. Soc., Chem. Commun.* 1156.

Law, K.-Y., Bailey, F. C. (1992). *Dyes Pigm. 20*:25.

Law, K.-Y., Bailey, F. C. (1993). *Dyes Pigm. 21*:1.

Law, K.-Y., Tarnawskyj, I. W. (1993). *J. Imaging Sci. Technol. 37*:22.

Law, K.-Y., Tarnawskyj, I. W. (1995). *J. Imaging Sci. Technol. 39*:126.

Law, K.-Y., Tarnawskyj, I. W. (1995a). *J. Imaging Sci. Technol. 39*:1.

Law, K.-Y., Facci, J. S., Bailey, F. C., Yanus, J. F. (1990). *J. Imaging Sci. 34*:31.

Law, K.-Y., Kaplan, S., Crandall, R., Tarnawskyj, I. W. (1993). *Chem. Mater. 5*:557.

Law, K.-Y., Tarnawskyj, I. W., Popovic, Z. D. (1994). *J. Imaging Sci. Technol. 38*:118.

Lawrance, R., Gibson, A. F. (1952). *Proc. Phys. Soc. B 65*:994.

Le Blanc, O. H., Jr. (1960). *J. Chem. Phys. 33*:626.

Lewis, P. A. (1988). In: *Federation Series on Coatings Technology.* Federation of Societies for Coatings Technology, Philadelphia, PA.

Liebermann, G., Hor, A.-M., Toth, A. E. J. (1988). U.S. Patent 4,771,133.

Lin, C.-W., Nozaki, T. (1995). In: *Proceedings of the Eleventh International Congress on Advances in Non-Impact Printing Technologies* (J. Anderson, ed.). IS&T, Springfield, VA., p. 138.

Lin, C.-W., Kutsuwada, N., Nakamura, Y. (1993). *J. Imaging Sci. Technol. 37*:476.

Lin, J. W., Dudek, L. P. (1986). U.S. Patent 4,628,018.

Lin, L.-B., Jenekhe, S. A., Borsenberger, P. M. (1996). *Appl. Phys. Lett. 69*:3495.

Limburg, W. W., Pai, D. M. (1982). U.S. Patent 4,330,608.

Limburg, W. W., Renfer, D. S. (1986). U.S. Patent 4,599,286.

Limburg, W. W., Renfer, D. S., Pai, D. M. (1983). U.S. Patent 4,397,931.

Loutfy, R. O. (1981). *Can. J. Chem. 59*:549.

Loutfy, R. O., Hsiao, C.-K. (1979). *Can. J. Chem. 57*:2546.

Loutfy, R. O., Ong, B. S. (1984). *Can. J. Chem. 62*:2546.

Loutfy, R. O., Hsiao, C. K., Ong, B. S., Keoshkerian, B. (1984). *Can. J. Chem. 62*:1877.

Loutfy, R. O., Hor, A. M., DiPaoloa-Baranyi, G., Hsiao, C. K. (1985). *J. Imaging Sci. 29*:116.

Loutfy, R. O. Hsiao, C. K., Hor, A. M., DiPaola-Baranyi, G. (1985a). *J. Imaging Sci. 29*:148.

Loutfy, R. O., Ong, B. S., Tadros, J. (1985b). *J. Imaging Sci. 29*:69.

Loutfy, R. O., Hor, A. M., Rucklidge, A. (1987). *J. Imaging Sci. 31*:31.

Loutfy, R. O., Hor, A.-M. Hsiao, C. K., Melnyk, A. (1988). In: *Proceedings of the Fourth International Congress on Advances in Non-Impact Printing Technologies* (A. Jaffe, ed.). SPSE, Springfield, VA., p. 52.

Loutfy, R. O., Hor, A.-M., Hsaio, C. K., Baranyi, G., Kazmaier, P. (1988a). *Pure Appl. Chem. 60*: 1047.

Loutfy, R. O., Hor, A. M., Kazmaier, P., Tam, M. (1989). *J. Imaging Sci. 33*:151.

Lutz, M. (1995). In: *Proceedings of the Eleventh International Congress on Advances in Non-Impact Printing Technologies* (J. Anderson, ed.). IS&T, Springfield, VA., p. 23.

Magin, E. H., Borsenberger, P. M. (1993). *J. Appl. Phys. 73*:787.

Makino, N., Horie, S., Watarai, S., Sato, H. (1988). U.S. Patent 4,724,192.

Martin, T. I., Mayo, J. D., Jennings, C. A., Gardner, S., Hsiao, C. K. (1995). In: *Proceedings of the Eleventh International Congress on Advances in Non-Impact Printing Technologies* (J. Anderson, ed.). IS&T, Springfield, VA., p. 30.

Matsui, M., Fukuyasu, K., Shibata, K., Muramatsu, H. (1993). *J. Chem. Soc. Perkin Trans. 2*:1107.

Matsumoto, S., Nakazawa, T. (1988). *OEP, October*:58.

Mayo, J. D. (1993). In: *Proceedings of the Ninth International Congress on Advances in Non-Impact Printing Technologies* (M. Yokoyama, ed.). IS&T, Springfield, VA., p. 652.

Mayo, J. D., Keoshkerian, B., Hsiao, C.-K., Gaynor, R. E., Gardner, S. J. (1994). In *Proceedings of the Tenth International Congress on Advances in Non-Impact Printing Technologies* (A. Melnyk, ed.). IS&T, Springfield, VA., p. 223.

McKay, R. B. (1989). *J. Oil Colour Chem. Assoc.* 72:89.

McMurty, D., Tinghitella, M., Svendsen, R. (1984). *IBM J. Res. Dev.* 28:257.

Melnyk, A. R., Pai, D. M. (1990). *SPIE 1253*:141.

Melnyk, A. R., Pai, D. M. (1993). In: *Physical Methods of Chemistry*, Vol. 3 (B. W. Rossiter, R. C. Baetzold, eds.). John Wiley, New York, p. 321.

Melz, P. J. (1972). *J. Chem. Phys.* 57:1694.

Melz, P. J., Champ, R. B., Chang, L. S., Chiou, C., Keller, G. S., Liclican, L. C., Nelman, R. R., Shattuck, M. D., Weiche, W. J. (1977). *Photogr. Sci. Eng.* 21:73.

Mey, W., Walker, E. I. P., Hosterey, D. C. (1979). *J. Appl. Phys.* 50:8090.

Miller, R. D., Michl, J. (1989). *Chem. Rev.* 89:1359.

Mishra, S., Pai, D. M. (1996). In: *Proceedings of the Twelfth International Congress on Digital Printing Technologies* (M. Hopper, ed.). IS&T, Springfield, VA., p. 464.

Miyazaki, H., Iuchi, K., Yamazaki, I., Takai, H., Matsumoto, M. (1990). In: *Proceedings of the Sixth International Congress on Advances in Non-Impact Printing Technologies* (R. Nash, ed.). IS&T, Springfield, VA., p. 327.

Mizuguchi, J. (1987). *Denshi Shashin (Electrophotography)* 26:216.

Mizuguchi, J., Rochat, A. C. (1988). *J. Imaging Sci.* 32:135.

Mizuta, Y., Kawahara, A., Miyamoto, E., Mutoh, N., Nakazawa, T. (1993). In: *Proceedings of the Ninth International Congress on Advances in Non-Impact Printing Technologies* (M. Yokoyama, ed.). IS&T, Springfield, VA., p. 663.

Mizuta, Y., Watanabe, Y., Sugai, F., Matsumoto, S., Kawaguchi, H., Akiba, N., Saitou, S., Okada, H., Goto, M., Nakazawa, T. (1996). In: *Proceedings of the Twelfth International Congress on Digital Printing Technologies* (M. Hopper, ed.). IS&T, Springfield, VA., p. 429.

Molaire, M. F., Magin, E. H., Borsenberger, P. M. (1997). *Proc. SPIE 3144*:26.

Mort, J. (1995). In: *Proceedings of the Eighth International School on Condensed Matter Physics, Electronic, Optoelectronic, and Magnetic Thin Films*. John Wiley, New York, p. 150.

Mort, J., Machonkin, M., Ziolo, R., Chen, I. (1992). *Appl. Phys. Lett.* 61:1829.

Moser, F. H., Thomas, A. L. (1963). *Phthalocyanine Compounds*. Reinhold, New York.

Moser, F. H., Thomas, A. L. (1983). *The Phthalocyanines: Vol. I, Properties; Vol. II, Manufacturing and Applications*. CRC Press; Boca Raton, FL.

Movaghar, B. (1987). *J. Molecular Electronics* 3:183.

Movaghar, B. (1991). In: *Condensed Systems of Low Dimensionality* (J. L. Beeby, ed.). Plenum Press, New York, p. 795.

Murayama, T., Aramaki, S., Matsuzaki, T., Itoh, T. (1988). In: *Proceedings of the Fourth International Congress on Advances in Non-Impact Printing Technologies* (A. Jaffe, ed.). SPSE, Springfield, VA., p. 15.

Murti, D. K., Kazmaier, P. M., DiPaola-Baranyi, G., Hsiao, C.-K., Ong, B. S. (1987). *J. Phys. D: Appl. Phys.* 20:1606.

Murti, D. K., McAneney, T. B., Popovic, Z. D., Ong, B. S., Loutfy, R. O. (1991). *J. Phys. D: Appl. Phys.* 29:953.

Murti, D. K., Popovic, Z. D., DiPaola-Baranyi, G., Loutfy, R. O., Gauthier, S. (1991a). In: *Proceedings of the Sixth International Congress on Advances in Non-Impact Printing Technologies* (R. Nash, ed.). IS&T, Springfield, VA., p. 306.

Mycielski, W., Ziolkowska, B., Lipinski, A. (1982). *Thin Solid Films 91*:335.

Nabeta, O., Kuroda, M., Furusho, N. (1990). *SPIE 1253*:155.

Nakanishi, K. (1987). *Nikkei New Mater.* March 9: 41.

Nakatani, K., Hana, J., Kodado, H. (1985). *Denshi Shashin (Electrophotography)* 24:2.

Nakazawa, T., Kawahara, A., Watanabe, Y., Mizuta, Y. (1994). *J. Imaging Sci. Technol.* 38:421.

Narushima, K., Takeuchi, M., Masui, M., Mase, H. (1996). In: *Proceedings of the Twelfth Interna-*

tional Congress on Digital Printing Technologies (M. Hopper, ed.). IS&T, Springfield, VA., p. 436.

Nashimoto, K. (1988). *J. Imaging Sci. 32*:205.

Nespurek, S., Lyons, L. E. (1981). *Mater. Sci. 7*:275.

Nguyen, K. C., Weiss, D. S. (1988). *Denshi Shashin (Electrophotography) 27*:2.

Nguyen, K. C., Weiss, D. S. (1989). *J. Imaging. Technol. 15*:158.

Niimi, T., Umeda, M. (1993). *J. Appl. Phys. 74*:465.

Niimi, T. Umeda, M. (1994). *J. Appl. Phys. 76*:1269.

Nikles, D. E., Kuder, J. E., Jasuta, J. A. (1992). *J. Imaging Sci. Technol. 36*:131.

Nishijima, H. (1985). In: *Proceedings of the First Photoreceptor Industry Conference.* Diamond Research Corporation, paper 4.

Nishino, T. K. Nogami, M., Hiramoto, Yokoyama, M. (1995). In: *Proceedings of the Eleventh International Congress on Advances in Non-Impact Printing Technologies* (J. Anderson, ed.). IS&T, Springfield, VA., p. 152.

Nogami, S., Mori, T., Iwabuchi, T. (1988). U.S. Patent 4,732,832.

Novikov, S. V., Vannikov, A. V. (1993). *Chem. Phys. 169*:21.

Novikov, S. V., Vannikov, A. V. (1994). *Chem. Phys. 187*:289.

Novikov, S. V., Vannikov, A. V. (1994a). *Chem. Phys. Lett. 224*:501.

Novikov, S. V., Dunlap, D. H., Kenkre, V. M. (1998). *Proc. SPIE 3471*:181.

Ogawa, I., Fujii, A., Maeda, S., Murayama, T. (1993). In: *Proceedings of the Ninth International Congress on Advances in Non-Impact Printing Technologies* (M. Yokoyama, ed.). SIST, Springfield, VA., p. 667.

Ohaku, K., Nakano, H., Aizawa, M. (1988). U.S. Patent 4,728,592.

Ohta, K. (1986). *Denshi Shashin (Electrophotography) 25*:83.

Oka, K., Okada, O. (1993). *J. Imaging Sci. Technol. 37*:607.

Okada, S. (1992). In: *Proceedings of the Eighth International Congress on Advances in Non-Impact Printing Technologies* (E. Hanson, ed.). IS&T, Springfield, VA., p. 261.

Okuda, M., Motomura, K., Naito,H., Matsushita, T., Nakau, T. (1982). *Jpn. J. Appl. Phys. 21*:1127.

Omote, A., Itoh, Y., Tsuchiya, S. (1995). *J. Imaging. Sci. Technol. 39*:271.

Ong, B. S., Keoshkerian, B., Martin, T. I., Hamer, G. K. (1985). *Can. J. Chem. 63*:147.

Ono, H., Takagishi, I., Matsuda, E., Murayama, T. (1991). In: *Proceedings of the Sixth International Congress on Advances in Non-Impact Printing Technologies* (R. Nash, ed.). IS&T, Springfield, VA., p. 312.

Onsager, L. (1934). *J. Chem. Phys. 2*:599.

Onsager, L. (1938). *Phys. Rev. B 54*:554.

O'Regan, M. B., Borsenberger, P. M., Magin, E. H., Zubil, T. (1995). *Proc. SPIE 2526*:54.

O'Regan. M. B., Borsenberger, P. M., Magin, E. H., Zubil, T. (1996). *J. Imaging Sci. Technol. 40*:1.

Oshima, R., Biswas, M., Wada, T., Uryu, T. (1985). *J. Polym. Sci.: Polym. Lett. Ed. 23*:151.

Oshima, R., Uryu, T., Seno, M. (1985a). *Macromolecules 18*:1043.

Otsuka, S., Murayama, T., Nagasaka, H. (1986). In: *Proceedings of the Third International Congress on Advances in Non-Impact Printing Technologies* (J. Gaynor, ed.). SPSE, Springfield, VA., p. 16.

Pacansky, J., Waltman, R. J. (1992). *J. Am. Chem. Soc. 114*:5813.

Pacansky, J., Coufal, H. C., Brown, D. W. (1987). *J. Photochem. 37*:293.

Pacansky, J., Waltman, R. J., Coufal, H., Cox, R. (1987a). *J. Radiat. Curing 14*:6.

Pacansky, J., Waltman, R. J., Grygier, R., Cox, R. (1991). *Chem. Mater. 3*:454.

Pai, D. M. (1985). In: *Physics of Disordered Materials* (D. Adler, H. Fritzsche, S. R. Ovshinsky, eds.). Plenum Press, New York, p. 579.

Pai, D. M. (1988). In: *Proceedings of the Fourth International Congress on Advances in Non-Impact Printing Technologies* (A. Jaffe, ed.). SPSE, Springfield, VA., p. 41.

Pai, D. M. (1991). In: *Frontiers of Polymer Research* (P. N. Prasad, J. K. Nigam, eds.). Plenum Press, New York, p. 315.

Pai, D. M. (1995). In: *Proceedings of the Eleventh International Congress on Advances in Non-Impact Printing Technologies* (J. Anderson, ed.). IS&T, Springfield, VA., p. 46.

Pai, D. M. Melnyk, A. R. (1986). *SPIE 617*:82.

Pai, D. M., Springett, B. E. (1993). *Rev. Mod. Phys. 65*:163.

Pai, D. M., Yanus, J. (1983). *Photogr. Sci. Eng. 27*:14.

Parris, P. E. (1995). *Proc. SPIE, 2526*:13.

Parris, P. E. (1996). *Proc. SPIE, 2850*:139.

Parris, P. E. (1997). *Proc. SPIE 3144*:92.

Parris, P. E. (1998). *Proc. SPIE 3471*:202.

Parris, P. E., Dunlap, D. H., Kenkre, V. M. (1997). *J. Polym. Sci.: Polym. Phys. 35*:2803.

Paxton, K. B. (1978). *Photogr. Sci. Eng. 22*:159.

Perlstein, J. H., and Borsenberger, P. M. (1982). In: *Extended Linear Chain Compounds* (J. S. Miller, ed.). Plenum Press, New York, p. 339.

Pierce, P. E., Schoff, C. K. (1988). In: *Federation Series on Coatings Technology*. Federation of Societies for Coatings Technology, Philadelphia, PA.

Pope, M. (1989). *Mol. Cryst. Liq. Cryst. 171*:89.

Pope, M., Swenberg, C. E. (1982). *Electronic Processes in Organic Crystals*. Oxford University Press, New York.

Pope, M., Swenberg, C. E. (1984). *Ann. Rev. Phys. Chem. 35*:613.

Popovic, Z. D., Iglesias, P., Parco, D., Robinette, S. (1991). *J. Imaging Technol. 17*:71.

Popovic, Z. D., Cowderg, R., Khan, I. M., Hor, A.-M., Goodman, J. (1999). *J. Imaging Technol. 43*:266.

Pritchard, D. L., Uehara, T. (1993). In: *Proceedings of the Ninth International Congress on Advances in Non-Impact Printing Technologies* (M. Yokoyama, ed.). IS&T, Springfield, VA., p. 685.

Rackovsky, S. (1991). *Chem. Phys. Lett. 178*:19.

Rackovsky, S., Scher, H. (1984). *Phys. Rev. Lett. 52*:453.

Rackovsky, S., Scher, H. (1988). *J. Chem. Phys. 89*:7242.

Rice, S. L., Balanson, R. D., Wingard, R. (1985). *J. Imaging Sci. 29*:7.

Ries, B., Bässler, H. (1987). *J. Mol. Electron. 3*:15.

Sahyun, M. R. V. (1984). *Photogr. Sci. Eng. 28*:185.

Sappok, R. (1978). *J. Oil Colour. Chem. Assoc. 61*:299.

Sasakawa, T., Ikeda, T., Tazuke, S. (1989). *J. Appl. Phys. 65*:2750.

Sasaki, M. (1988). U.S. Patent 4,777,296.

Sato, T., Hisada, H. (1996). *Denshi Shashin (Electrophotography) 35*:12.

Sato, T., Hisada, H., Shida, S. (1992). *Denshi Shashin (Electrophotography) 31*:40.

Schaffert, R. M. (1971). *IBM J. Res. Dev. 15*:75.

Schaffert, R. M. (1975). *Electrophotography*. Focal Press, London.

Schein, L. B. (1992). *Philos. Mag. B 65*:795.

Schein, L. B. (1995). *Electrophotography and Development Physics*. Laplacian Press, Morgan Hills, CA.

Scher, H. (1993). *Mol. Cryst. Liq. Cryst. 288*:41.

Schlosser, E.-G. (1978). *J. Appl. Photogr. Eng. 4*:118.

Schwartz, B. C. (1991). In: *Proceedings of the Sixth International Congress on Advances in Non-Impact Printing Technologies* (R. Nash, ed.). IS&T, Springfield, VA., p. 446.

Scott, J. C., Lo, G. S. (1990). In: *Proceedings of the Sixth International Congress on Advances in Non-Impact Printing Technologies* (R. Nash, ed.). IS&T, Springfield, VA., p. 403.

Scott, J. C., Skumanich, A., Shattuck, M. D., Nguyen, H. (1990). *SPIE 1253*:194.

Scozzafava, M., Chen, C. H., Reynolds, G. A., Perlstein, J. H. (1985). U.S. Patent 4,514,481.

Seki, K., Suzuki, Y., Yamanami, H. (1989). In: *Proceedings of the Fifth International Congress on Advances in Non-Impact Printing Technologies* (J. Moore, ed.). SPSE, Springfield, VA., p. 40.

Sharp, J. H., Abkowitz, M. (1973). *J. Phys. Chem. 77*:477.

Sharp, J. H., Lardon, M. (1968). *J. Phys. Chem. 72*:3230.

Shattuck, M. D., Vahtra, U. (1969). U.S. Patent 3,484,237.

Shimada, A., Anzai, M., Kakuta, A., Kawanishi, T. (1987). *Trans. IEEE Ind. Appl. IA-23*:804.

Silinsh, E. A. (1989). *Mol. Cryst. Liq. Cryst. 171*:135.

Silinsh, E. A., Capek, V. (1994). *Organic Molecular Crystals: Interaction, Localization and Transport Phenomena*. American Institute of Physics Press, New York.

Silinsh, E. A., Jurgis, A. J. (1985). *Chem. Phys. 94*:77.

Sim, J.-H., Ogino, K., Sato, H., Pei, Y. (1996). *J. Imaging Sci. Technol. 40*:164.

Slowik, J. H., Chen, I. (1983). *J. Appl. Phys. 54*:4467.

Smoluchowski, M. V. (1916). *Phys. Zeits. 15*:585.

Spear, W. E. (1957). *Proc. Phys. Soc. (London) B 70*:669.

Spear, W. E. (1960). *Proc. Phys. Soc. (London) B 76*:826.

Stasiak, J. W., Storch, T. J. (1996). *J. Imaging Sci. Technol. 40*:299.

Stasiak, J., Storch, T. J. (1996a). *Proc. SPIE 2850*:172.

Stasiak, J. W., Storch, T. J. (1996b). In: *Proceedings of the Twelfth International Congress on Digital Printing Technologies* (M. Hopper, ed.). SIST, Springfield, VA., p. 474.

Stasiak, J., Storch, T. J., Maeo, E. (1995). *Proc. SPIE 2526*:23.

Staudenmayer, W. J., Regan, M. T. (1988). U.S. Patent 4,719,163.

Stolka, M. (1995). In: *Special Polymers for Electronics and Optoelectronics* (J. A. Chilton and M. T. Goosey, eds.). Chapman and Hall, London, p. 284.

Stolka, M., Mort, J. (1994). In *Kirk-Othmer Encyclopedia of Chemical Technology*, 4th ed. John Wiley, New York, p. 245.

Sugiuchi, M., Nishizawa, H. (1993). *J. Imaging Sci. Technol. 37*:245.

Sugiuchi, M., Nishizawa, H., Uehara, T. (1990). In: *Proceedings of the Sixth International Congress on Advances in Non-Impact Printing Technologies* (R. Nash, ed.). IS&T, Springfield, VA., p. 298.

Sussman, A. (1967). *J. Appl. Phys. 38*:2738.

Suzuki, T., Murayama, T., Ono, H., Otsuka, S., Nozomi, M. (1988). U.S. Patent 4,725,519.

Takahashi, Y., Yamamoto, N. (1995). In: *Proceedings of the Eleventh International Congress on Advances in Non-Impact Printing Technologies* (J. Anderson, ed.). IS&T, Springfield, VA., p. 156.

Takano, S., Enokida, T., Kakuta, A., Mori, Y. (1984). *Chem. Lett.*:2037.

Takano, S., Mimura, Y., Matsui, N., Utsugi, K., Gotoh, T., Tani, C., Tateishi, K., Ohde, N. (1991). *J. Imaging Sci. Technol. 17*:46.

Takenouchi, S., Hirano, A., Yoshioka, H., Fujimaki, Y., Moriguchi, H. (1988). In: *Proceedings of the Fourth International Congress on Advances in Non-Impact Printing Technologies* (A. Jaffe, ed.). SPSE, Springfield, VA., p. 22.

Tanaka, H., Kin, S., Pu, L. S. (1987). U.S. Patents 4,700,001 and 4,707,427.

Tam, A. C. (1980). *Appl. Phys. Lett. 37*:978.

Thomas, A. L. (1990). *Phthalocyanine Research and Applications*. CRC Press, Boca Raton, FL.

Tokuhisa, H., Era, M., Tsutsui, T., Saito, S. (1995). *Appl. Phys. Lett. 66*:3433.

Treibs, A., Jacob, K. (1965). *Angew. Chem. Int. Ed. Engl. 4*:894.

Tse, M.-K. (1994). In: *Proceedings of the Tenth International Congress on Advances in Non-Impact Printing Technologies* (A. Melnyk, ed.). IS&T, Springfield, VA., p. 295.

Tse, M.-K. (1996). In: *Proceedings of the Twelfth International Congress on Digital Printing Technologies* (M. Hopper, ed.). IS&T, Springfield, VA., p. 343.

Tsuchiya, S., Omote, A., Murakami, M., Yoshimura, S. (1995). *J. Imaging Sci. Technol. 39*:294.

Ueda, H. (1988). U.S. Patent 4,769,302.

Umeda, M. (1994). In: *Proceedings of the Tenth International Congress on Advances in Non-Impact Printing Technologies* (A. Melnyk, ed.). IS&T, Springfield, VA., p. 239.

Umeda, M. (1998). *Proc. SPIE 3471*:212.

Umeda, M. (1999). *J. Imaging Sci. Technol. 43*:254.

Umeda, M., Hashimoto, M. (1992). *J. Appl. Phys. 72*:117.

Umeda, M., Niimi, T. (1994). *J. Imaging Sci. Technol. 38*:281.
Umeda, M., Niimi, T. (1994a). *Jpn. J. Appl. Phys. 33*:L1789.
Umeda, M., Yokoyama, M. (1997). *J. Appl. Phys. 81*:6179.
Umeda, M., Shimada, T., Aruga, T., Niimi, T., Sasaki, M. (1993). *J. Phys. Chem. 97*:8531.
Usov, N. N., Benderskii, V. A. (1970). *Phys. Status Solidi 37*:535.
Vahtra, U., Wolter, R. F. (1978). *IBM J. Res. Dev. 22*:34.
Valerian, H., Nespurek, S. (1993). *J. Appl. Phys. 73*:4370.
Visser, S. A., Gruenbaum, W. T., Magin, E. H., Borsenberger, P. M. (1999). *Chem. Phys. 240*:197.
Wagner, H. J., Loutfy, R. O., Hsiao, C.-K. (1982). *J. Mater. Sci. 17*:2781.
Wang, M., Chen, H. Z., Yang, S. L. (1991). *J. Photogr. Sci. 39*:25.
Wang, M., Pan, P. L., Shen, J. L., Zhou, M. J., Yang, S. L. (1993). *J. Photogr. Sci. 41*:126.
Wang, M., Chen, H. Z., Shen, J., L., Yang, S. L. (1996). *J. Photochem. Photobiol. A: Chem. 94*:2.
Wang, Y. (1992). *Nature 356*:585.
Watanabe, K., Itami, A., Akira, K., Fujimaki, Y. (1993). In: *Proceedings of the Ninth International Congress on Advances in Non-Impact Printing Technologies* (M. Yokoyama, ed.). IS&T, Springfield, VA., p. 659.
Weiss, D. S. (1990). *J. Imaging Sci. 34*:132.
Weiss, D. S. Chen, D. A. (1995). *J. Imaging Sci. Technol. 39*:425.
Weiss, D. S., Cowdery, J. R., Ferrar, W. T., Young, R. H. (1996). *J. Imaging Sci. Technol. 40*:322.
Weiss, D. S., Ferrar, W. T., Corvan, J. R., Parton, L. G., Miller, G. (1999). *J. Imaging Sci. Technol. 43*:280.
West, R. (1986). *J. Organomet. Chem. 300*:327.
Whitlock, J. B., Bird, G. R., Cox, M. D., Panayotatos, P. (1992). *Thin Solid Films 215*:84.
Wicks, Z. W. Jr. (1986). In: *Federation Series on Coatings Technology, Film Formation*. Federation of Societies for Coatings Technology, Philadelphia, PA.
Wicks, Z. W., Jr., Jones, F. N., Pappas, S. P. (1992). *Organic Coatings: Science and Technology. Vol. 1: Film Formation, Components, and Appearance*. John Wiley, New York.
Wicks, Z. W., Jr., Jones, F. N., Pappas, S. P. (1994). *Organic Coatings: Science and Technology. Vol. 2: Applications, Properties, and Performance*. John Wiley, New York.
Wiedemann, W., Spietschka, E., Troester, H. (1987). U.S. Patent 4,714,666.
Williams, E. M. (1984). *The Physics and Technology of Xerographic Process*. John Wiley, New York.
Wingard, R. E. (1982). *IEEE Ind. Appl.*:1251.
Wolf, U., Bässler, H., Borsenberger, P. M., Gruenbaum, W. T. (1997). *Chem. Phys. 222*:259.
Yamaguchi, Y., Yokoyama, M. (1991). *J. Appl. Phys. 70*:3726.
Yamaguchi, Y., Fujiyama, T., Tanaka, H., Yokoyama, M. (1990). *Chem. Mater. 2*:341.
Yamaguchi, Y., Fujiyama, T., Yokoyama, M. (1991). *J. Appl. Phys. 70*:855.
Yamamoto, K., Okugawa, Y., Ohmi, K., Yamada, Y. Sadamatsu, S. (1982). *Photogr. Sci. Eng. 26*: 179.
Yamamoto, K., Igarashi, R., Takegawa, I., Ojima, F. (1986). In: *Proceedings of the Third International Congress on Advances in Non-Impact Printing Technologies* (J. Gaynor, ed.). SPSE, Springfield, VA., p. 115.
Yanus, J. F., Limburg, W. W. (1986). U.S. Patent 4,606,986.
Yao, J., Yonehara, H., Pac, C. (1995). *Bull. Chem. Soc. Jpn. 68*:1001.
Yarmchuk, E. J., Keefe, G. E. (1989). *J. Appl. Phys. 66*:5435.
Yokoyama, K., Yokoyama, M. (1989). *Chem. Lett.*:1005.
Young, R. H. (1995). *Philos. Mag. B 72*:435.
Young, R. H., Fitzgerald, J. J. (1995). *J. Phys. Chem. 99*:4230.
Young, R. H., Sinicropi, J. A., Fitzgerald, J. J. (1995). *J. Phys. Chem. 99*:9497.
Yuh, H.-J., Pai, D. M. (1992). *J. Imaging Sci. Technol. 36*:477.
Ziólkowska-Pawlak, B., Mycielski, W., Lipinski, A. (1983). *J. Non-Cryst. Solids 55*:215.

11

Photoreceptors: Recent Imaging Applications for Amorphous Silicon

ROBERT JOSLYN

Kyocera Industrial Ceramics Corp., Vancouver, Washington

11.1 INTRODUCTION

A decade has passed since Joe Mort wrote what is now Chapter 16, but our understanding of the physics of amorphous silicon (a-Si) is little changed from what he presented there. Silicon, in its various forms, lies at the hearts of TFT liquid crystal displays, integrated circuits, solar power cells, sensor elements, and wide-area photoreceptors for images. These application areas are evolving separately in their different directions at their various rates, and it would be unusual and difficult to address more than one area in a single writing.

This update chapter is confined to wide-area amorphous silicon photoreceptor drums capable of receiving an electrostatic latent image for the purpose of development and transfer to a wide-area receiver—typically paper. For photoreceptor applications, a-Si is used only on rigid drums. There would be great demand for photoreceptor belts, but a-Si, at the thickness required, does not flex enough to be usable on a belt. Amorphous silicon in photoreceptor applications must be hydrogenated as explained in Chapter 16. This update chapter will therefore cover only advances in a-Si:H photoreceptor drums. For easier reading, "hydrogenated" and ":H" will be omitted and this material will be abbreviated as "a-Si" in the text.

Most of the recent advances in a-Si technology are related to its commercialization. Inorganic photoreceptor drums were codeveloped with photocopiers beginning in the 1950s. Low-cost organic photoconductors (OPC) enabled the development of less expensive laser printers and photocopiers beginning in the 1970s. These technologies defined the market for photoreceptors. When a-Si photoreceptor drums began to be manufactured in the 1980s, their commercialization depended upon them being perceived as useful components in existing or newly developed photocopiers and laser printers. The overwhelming benefit of a-Si drums is their long life, closely followed by their high process speed. The major barrier to their immediate adoption was their cost of manufacture, although a number of other characteristics of a-Si prevented its easy use within existing machines. The recent advances in a-Si photoreceptor technology have been mostly focused on reducing the cost of a-Si drums and on modifying other characteristics so that they operate more like selenium-based or OPC drums.

When a-Si drums became feasible, they generated a flurry of technical papers from universities and manufacturers. This interest peaked in 1989, but then many companies withdrew from merchant and internal manufacturing. Since 1993 most papers on a-Si photoreceptor drums have come from either Kyocera Corporation in Japan or from the partnership of AEG Elektrofotografie GmbH and Forschungs-und Applikationslabor Plasmatechnik GmbH in Germany. Ikeda et al. (1996) mentions that Kyocera began shipping a-Si drums in 1984 with cumulative output reaching 1 million drums in 1993.

11.2 THE BASIC a-Si PHOTORECEPTOR DRUM

a-Si photoreceptor drums are composed of at least three layers on a metal substrate, as shown in Fig. 11.1. The middle photosensitive layer performs the basic function of the drum. Under it is the carrier blocking layer, which prevents the injection of charge carriers from the metal substrate that would ruin the characteristics of the photosensitive layer. Outermost is the surface protection layer, which protects the photosensitive layer against abrasion and retains the surface charge. The following descriptions of advances will build upon this minimal structure.

In practice the metal substrate is an extruded aluminum tube machined and polished to an extremely smooth finish. Hu (1993) showed that the acceptance of surface potential of an a-Si:H photoreceptor is very sensitive to the microroughness of the substrate surface. However, when an infrared laser will expose the photoreceptor, then it is helpful if the substrate surface is carefully roughened to prevent internal reflections. Of course the substrates must be free of all contaminants before thin film deposition. Although exacting, these requirements are well within the capabilities of many vendors and will not be further discussed.

The main cost arises from the very slow deposition time of the relatively thick photosensitive layer. The plasma CVD process is used essentially as described in Chap-

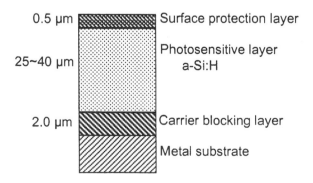

Figure 11.1   Basic layer structure of the a-Si drum.

ter 16. The blocking layer and surface protection layer are deposited inside the same reactor within the same pumpdown as the photosensitive layer.

11.3 IMPROVED CHARGE ACCEPTANCE

During its operation, the surface area of a photoreceptor drum is evenly charged in the dark, and then selected areas are discharged by laser or LED exposure light or by focused visible light from an original image. The latent image then consists of regions of voltage differences on the surface of the drum. The larger this voltage contrast, the easier to develop the image and ultimately, the better the image.

The classic capacitor model shows the factors involved in charging a drum:

$$V = \frac{d}{\varepsilon \, \varepsilon_0 S} Q$$

where V, d, ε, ε_0, S, and Q stand for the surface voltage, the dielectric layer thickness, the relative dielectric constant, the permittivity of free space, the drum surface area, and the charge amount, respectively.

One of the first applications for a-Si drums was to bring their extraordinary long life and nontoxicity to the high-speed copier industry, which was already using selenium based drums. The dielectric constant of a-Si is 12 to 13 compared to 6 to 9 for a-Se. In a replacement situation, Q and S would be fixed for an existing print engine. Clearly, the layer thickness d must be made thicker than for a-Se. However the *dark decay* rate of a-Si limits the utility of simply increasing d. The background thermal vibrations in the a-Si material are sufficient to excite enough carriers into the conduction band to cause a noticeable drop in surface potential. Of course this is a function of temperature, and the attainable surface voltage will drop by 5 to 10 V per °C of temperature increase. The point is that the number of thermally excited carriers increases with volume, and hence it increases with the increasing thickness of the a-Si dielectric layer.

Ikeda et al. (1992) described how to construct an a-Si drum so as to obtain surface potentials equivalent to amorphous selenium, more than 800 V at 42°C at a charge amount of 0.2 μC/cm². Their solution was to add a highly efficient carrier generation layer on top of the photosensitive layer, made of the same material, only deposited more slowly and called a *low-rate* layer. To visualize this layer structure, look ahead to Fig. 11.6.

The low-rate layer has a lower defect density and a smaller percentage of hydrogen. Figure 16.1(c) in Chapter 16 shows only single hydrogen atoms on the dangling bonds (Si-H bonds). More recent IR absorption measurements show that Si-H$_2$ bonds are also present and that the proportion of Si-H$_2$ bonds is less in the low-rate material than in the normally deposited s-Si. Other measurements show that the low-rate material has a lower concentration of hydrogen. All this suggests a lower defect density in the low-rate material. The thermal generation of carriers is presumed to happen at defects, so the thermal generation of carriers in the low-rate layer is less. The functional result is that the thin, low-rate, carrier generation layer inhibits dark decay conduction, trapping carriers at the interface, so that the main photosensitive layer can be made thicker to sustain a higher surface charge. Figure 11.2 shows the improvement in charge acceptance.

A thicker photosensitive layer also improves the sensitivity of the drum. From the capacitor model, when voltage is increased by increasing the layer thickness, the charge density Q/S does not increase, meaning less charge per volt. Fewer photogenerated charge

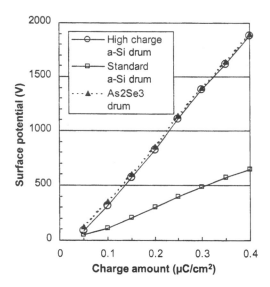

Figure 11.2 Charge acceptance of a-Si drums configured for visible light exposure.

carriers are required, so less exposure energy is required to discharge one volt, hence the higher sensitivity. Visible light is absorbed within the top 1 to 2 μm of the photosensitive layer, so most photogenerated charge carriers originate there.

11.4 BOTH CHARGE POLARITIES

Selenium-based photoreceptor drums were always positively charged because of the physics of selenium. This was an environmental advantage in that a positive charging corona generates much less ozone than a negative charging corona. As a replacement for a-Se, a-Si is also positively charged. This requires that the blocking layer be strongly doped with boron to prevent injection of the substrate metal's electrons into the photosensitive layer. The photosensitive layer is slightly doped with boron in order to achieve maximum photoconductivity, as explained in Section 2 of Chapter 16.

The physics of OPC, as first developed and commercialized, required negative charging. For compatibility with these print engines, a-Si for negative charging was developed. It was initially assumed that doping the blocking layer with phosphorous or another group III element would be necessary. However Ikeda et al. (1998) showed that this was unnecessary. a-Si is intrinsically an n-type semiconductor, and this turned out to be sufficient to prevent injection of carriers from the aluminum substrate. This finding was beneficial to manufacturing because it eliminated the use of dangerous phosphine (PH_3) gas.

The charge acceptance and the photoresponse properties of negatively charged a-Si are similar, in fact almost identical, to those of positively charged a-Si.

11.5 PHOTORESPONSE

Photoresponse is how quickly a charged photoreceptor loses its charge after it is exposed to light and is a most significant limitation on the speed of laser and LED printers. A pixel is exposed on a rotating drum. By the time that pixel reaches the development station,

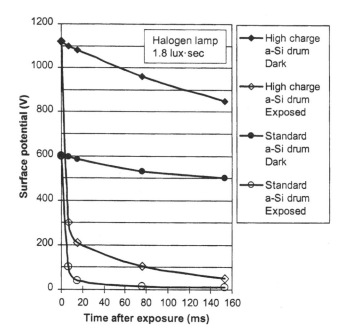

Figure 11.3 Photoresponse of a-Si drums configured for visible light exposure.

its voltage must have dropped enough to be developed. a-Si does this faster than the other photoreceptor technologies. For background, Fig. 11.3 shows the photoresponse to visible light of the high-charge and standard-charge drums previously discussed. Process speeds of less than 10 ms are possible. Dark decay curves are also shown, and within this high-speed time frame, the dark decay is not significant. Note that, for the standard drum, the surface potential quickly discharges to almost zero. From a 600 V charge, the development system sees a 550 V contrast voltage. By comparison, OPC discharges to a significantly higher residual voltage, which reduces the contrast voltage.

Refer to Borsenberger and Weiss (1998) for definitions of *emission-limited* discharge and *space-charge-limited* discharge. The sloping lines on the left of Fig. 11.4 show the emission-limited discharge of a-Si for different layer thicknesses. This sensitivity to exposure intensity exceeds that of selenium-based alloys and almost equals that of OPC. The emission-limited discharge of a-Si is linear, so it can be used for continuous-tone printing. The steeper slope reflects the higher sensitivity of thicker layers at higher surface potentials.

It is the space-charge-limited discharge of a-Si that makes it faster than OPC. Once the charge carriers are generated, they move quickly through a-Si. Adam et al. (1997) show that the charge carrier mobility of a-Si is three orders of magnitude higher than that of OPC. This is easily understandable because silicon is a true semiconductor.

11.6 FATIGUE AND TEMPERATURE EFFECTS

a-Si does not exhibit electrical fatigue. The graphs of the significant electrophotographic properties in Figs. 11.2 through 11.5 would be almost identical if measured after a million prints. This is in great contrast to the limited fatigue life of OPC.

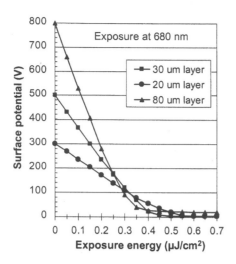

Figure 11.4   Photosensitivity as a function of layer thickness, for a-Si drums configured for visible light exposure.

It is possible to induce light fatigue in a-Si, but the effect is small. For example, Ikeda (1998b) shows a gain in surface potential of only 50 V when a drum is exposed to a fluorescent a few centimeters away, for 24 hours. This phenomenon is called the *Staebler-Wronski effect*. The mechanism is the creation of charge trapping sites. Routine incidental exposure to room light is simply not a problem.

The Staebler-Wronski effect can be reversed by annealing the drum at 200°C for 2 hours. In fact, a-Si drums can even be operated at temperatures in this range. There is no phase separation as with OPC. Temperatures up to 250°C are still below the deposition

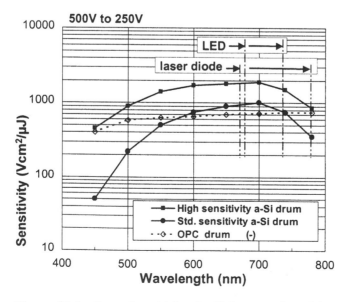

Figure 11.5   Spectral sensitivity of a-Si drums configured for near-infrared exposure.

temperature. This high-temperature capability can be used to dry out liquid toner or to accomplish other printer design objectives. Of course dark decay increases with temperature, but in a predictable manner. Attainable surface voltage declines by 5 to 10 volts/°C for standard thickness photosensitive layers. This sensitivity can be reduced to 1 volt/°C if inexpensive thin layers are used. Operation at 40°C is typical to prevent accumulation of surface moisture, as discussed in Section 8 Surface Protection Layer.

Another effect of temperature is that the aluminum substrate must be thick enough to not deform at the high deposition temperatures. This adds to the cost burden when compared to small, thin-walled OPC drums. But when large OPC drums are made with thick substrate walls for stability, then the cost difference is reduced.

11.7 SPECTRAL RESPONSE

a-Si has high sensitivity across the visible spectrum with a broad peak around 700 nm. At longer wavelengths, it is more sensitive than selenium-based photoreceptors, leading to more accurate, high-speed analog photocopiers. However, laser- and LED-based print engines, including digital copiers, only require sensitivity to their particular exposure device. a-Si has always been particularly well suited for LED print engines, because several low-cost diodes emit at wavelengths to which a-Si is very sensitive. However, the huge desktop laser printer market developed using low-cost 780 nm semiconductor lasers and OPC photoreceptors. The sensitivity of a-Si falls off past 740 nm, which initially discouraged its use. The development of thick-layer, high-sensitivity a-Si then raised its sensitivity at 780 nm to be equivalent to OPC, as shown in Fig. 11.5. The more recent emergence of low-cost 655 nm semiconductor lasers for the DVD market is expected to speed the adoption of standard thickness a-Si drums for future laser print engine designs.

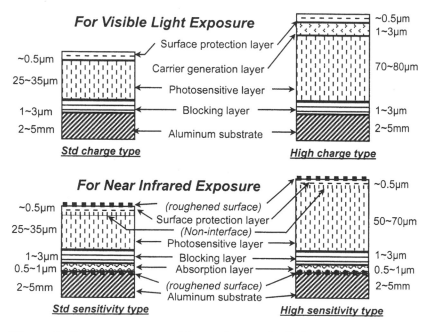

Figure 11.6 Layer structures of four variations of a-Si drums.

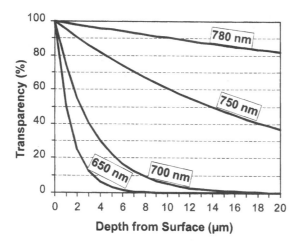

Figure 11.7   Penetration of exposure light into a-Si, as function of wavelength.

If a near-infrared laser is used on a standard a-Si drum, the result will be interference patterns from internal reflections. Interference patterns can be eliminated by incorporating some or all of the following modifications to the drum. First, begin with a roughened substrate surface, which results in a roughened drum surface. Second, use an absorption layer immediately next to the substrate surface, which prevents light from reaching the substrate surface. Third, reduce the discontinuity of the interface between the photosensitive layer and the surface protection layer, so that reflections from it are reduced. The other interface, between the blocking layer and the photosensitive layer, is not distinct enough to cause reflections. Figure 11.6 shows these modifications and includes the previously discussed visible light drums for comparison.

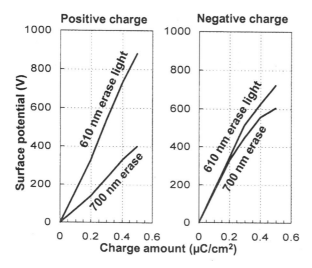

Figure 11.8   Effect of erase light wavelength on charge acceptance, for positive and negative charging.

Earlier it was stated that visible light was absorbed within 1 or 2 μm. Yet near-infrared light reaches the drum surface to cause reflections. Clearly, the depth to which light penetrates a-Si is a function of wavelength. Figure 11.7 illustrates this relationship. The big difference in transparency between 700 and 780 nm corresponds to the falloff of sensitivity in Fig. 11.5.

The choice of the wavelength of the erase light also affects the performance of a-Si drums. The purpose of the erase light is to uniformly discharge the entire surface of the photoreceptor so that no image remains. The charge carriers generated by the erase light can become trapped in the photosensitive layer and then emerge during charging to reduce the surface potential. The amount of this trapping varies with the penetration of the erase light, so it is more troublesome at longer wavelengths, seriously limiting the acceptance potential of drums for near-infrared applications. This is called the erase light memory effect. Figure 11.8 shows it and also shows that it is not as noticeable on negatively charged drums.

11.8   SURFACE PROTECTION LAYER

All current manufacturers of a-Si drums utilize an outer layer a-SiC:H. SiN is no longer used. As mentioned in Chapter 16, a layer is required to prevent the injection of charge carriers from the surface. a-SiC performs this function, and by depositing it very carefully, the field effect mechanism of lateral image spread is minimized.

a-SiC makes an extremely hard, abrasion-resistant surface. A 100 mm diameter drum will last over 4 million pages, and a 200 mm drum will double that to 8 million pages. SiC is impervious to the organic solvents in liquid toners and to the NO_x byproducts of corona charging.

Another type of lateral spreading of the latent electrostatic image can occur if a conducting film accumulates on the surface of the drum. Ozone generated by the corona oxidizes the silicon on the a-SiC surface into a hydrophilic oxide, which then absorbs water to form a conducting layer. Stahr et al. (1997) describe the process by

$$\text{Chemisorption:} \quad \text{Si}-\text{O}-\text{Si} + H_2O \leftrightarrow 2\,\text{SiOH}$$
$$\text{Physisorption:} \quad \text{SiOH} + H_2O \leftrightarrow \text{SiOH:OH}_2$$

First generation a-Si drums optimized the carbon content of the a-SiC:H layer to minimize dark conductivity and maximize a-SiC hardness. Water absorption was prevented by keeping the drum surface heated to approximately 40°C.

Ikeda et al. (1996) describe the optimization of the carbon content for sharp printing at 25°C. The idea is to eliminate Si atoms from the SiC on the surface of the drum. Carbon concentrations above 96% were found to eliminate image blurring, compared to the 85% of the first-generation a-SiC:H coatings. This *C-rich* layer is applied on top of the usual SiC layer. The deposition rate drops as the carbon content of the surface layer increases, from approximately 0.3 μm/h at 85% to approximately 0.1 μm/h at 97%. This production inefficiency was resolved by switching the 13.56 MHz RF power on and off, at 1 kHz and at a duty ratio of 1:1. This takes advantage of the fact that the energy efficiency of the C-rich layer deposition is higher while the RF power is being switched on, compared to continuous deposition with steady RF power.

Some polishing of any photoreceptor drum surface naturally occurs from the paper and cleaning blade. To build a heaterless print engine, it is not sufficient simply to use a

C-rich layer. The polishing effect must also be increased by modifying the toner and cleaning subsystem to be more abrasive.

11.9 OPTIMIZATION FOR PRINT ENGINES

Achieving a fine resolution requires a thin photosensitive layer. When a single dot is exposed, the generated charge carriers repel each other as they move to the surface. This causes the image of the dot to grow. The thicker the layer, the more time for lateral movement. However a thick layer is required for a high surface charge. Optimization of the layer thickness requires consideration of both these conflicting requirements.

Sasahara et al. (1998) describe a series of experiments to optimize the interface between the photosensitive layer and the surface protective layer. The CH_4/SiH_4 gas ratio, deposition time, and B_2H_6 dopant gas ratio were varied, and a photosensitive layer increase from 20 μm to 40 μm was achieved while maintaining a sharp dot size of 70 μm using dry toner. The results of this effort are shown in Fig. 11.9, where boundary curves show the simultaneous maximums of resolution and layer thickness. The tradeoff against surface potential is shown in Fig. 11.10.

The resolution, or minimum dot size, depends on the design of the print engine. Print resolution can be dramatically increased by using liquid toner, as shown in Fig. 11.9. Much smaller toner particles can be used, and the pigment content of liquid toner particles can be higher. It is feasible to use liquid toner with a-Si photoreceptors because the a-Si:H and a-SiC:H materials do not degrade in the toner solvent, whereas OPC does. With the more saturated color, reduced toner pileup, and increased resolution of liquid toner, it is possible to design color print-on-demand systems that have the image quality of offset presses.

11.10 MANUFACTURING a-Si PHOTORECEPTOR DRUMS

a-Si drums are manufactured by the plasma CVD process, essentially as shown in Figs. 4 and 5 of Chapter 16. Silane gas (SiH_4) is always the source of the silicon with methane

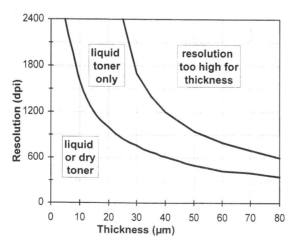

Figure 11.9 Achievable resolution and layer thickness combinations, by type of toner.

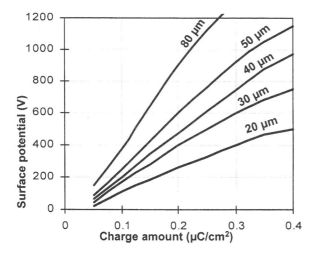

Figure 11.10   Charge acceptance of a-Si drums as a function of layer thickness.

(CH$_4$) added at the end as a source of carbon for the a-SiC:H surface layer. The drum substrate is heated to the range of 250 to 300°C.

The drum substrate typically is grounded and capacitively coupled to the counter-electrode, with plasma generated between the drum and the counterelectrode at radio frequencies. 13.56 MHz is mentioned by Ikeda et al. (1992), Lutz et al. (1996), Stahr et al. (1997), and Sasahara et al. (1998). 27.12 MHz is also mentioned by Lutz et al. (1996) and Stahr et al. (1997).

Increasing the speed of the a-Si deposition rate was the subject of research by several teams. Röhlccke et al. (1997) reported deposition rates of up to 18 µm/h by using helium to dilute the SiH$_4$. Takahashi et al. (1989) reported deposition rates of approximately 17 µm/h by using argon to dilute the SiH$_4$. Lutz et al. (1996) reported "With the optimizing of helium dilution, TEB doping and increasing of RF frequency the properties of photoreceptor are quite respectable with deposition rates in the range from 10 up to 15 µm/h." On the other hand, decreasing the deposition rate to 2 µm/h with helium dilution (Ikeda et al., 1992) or to 2.5 µm/h with H$_2$ (Sasahara et al., 1998) achieves a high-sensitivity charge generation layer.

When boron impurity doping is necessary, diborane (B$_2$H$_6$) is typically used, although Lutz et al. (1996) report that triethylboron (TEB) is effective and safer. Phosphine (PH$_3$) is typically used when phosphorous doping is required (Ikeda et al., 1998).

11.11   a-Si DRUM COSTS

a-Si is more costly than OPC because of the slow deposition rate. Manufacturers do not disclose production costs, but Ikeda (1998b) showed relative costs of a Kyocera a-Si drum from 1992 to 1998. Costs declined about 80% over these 6 years.

The longer life of an a-Si drum should mean a lower total cost of ownership, which, divided by page volume, should yield a lower cost per page. Lyra Research (1999) added up street prices of laser printers, consumables, and maintenance, comparing the printer using a-Si in each category against competitors using OPC. For the 16–18 page per minute category over 5 years, the cost per page for a-Si was 1.2 cents compared to 1.9 and 2.1

cents for OPC. For the 24–28 page per minute category over 5 years, the cost per page was 1.2 cents compared to 1.8 cents for the three OPC competitors, thus proving that use of a nonconsumable a-Si drum reduces the total costs by 33% at these print volumes.

11.12 SUMMARY

The benefits of a-Si photoreceptor drums are

> Long life
> Fast photoresponse
> No electrical fatigue
> Material safety

Available variations include

> Visible light or near-infrared exposure
> Positive or negative charging
> High or standard charge acceptance
> High or standard sensitivity

Their natural applications are high-speed or high-volume laser or LED print engines and office laser or LED printers where cost per page or reduction of consumables are important.

REFERENCES

Adam, D., Humpert, H., Dreihöfer, S., Pinsler, H., Lutz, M. (1997). *IS&T's NIP13: 1997 International Conference on Digital Printing Technologies*, pp. 245–247.

Borsenberger, P. M., Weiss, D. S. (1998). *Organic Photoreceptors for Xerography*, Marcel Dekker, Inc., New York. pp. 87–100.

Hu, J. (1993). *Proc. SPIE—Int. Soc. Opt. Eng. (USA)*, Vol. 1912 pp. 245–249.

Ikeda, A. (1998a). Proceedings of *The SEPJ 40th Anniversary Pan-Pacific Imaging Conference/ Japan Hardcopy '98*, p. 125.

Ikeda, A. (1998b). *Imaging News*, July/August 1998, pp. 67–76.

Ikeda, A., Kawakami, T., Ejima, K., Itoh, B., Sasaki, T., Shimono, Y., Wakita, K. (1992). *IS&T's Eighth International Congress on Advances in Non-Impact Printing Technologies*, pp. 227–232.

Ikeda, A., Fukunaga, H., Sasahara, M. (1996). *IS&T's NIP12: International Conference on Digital Printing Technologies*, pp. 444–451.

Ikeda, A., Fukunaga, H., Tsuda, M. (1998). *IS&T's NIP14: 1998 International Conference on Digital Printing Technologies*, pp. 532–534.

Lutz, M., Dreihöfer, S., Schade, K., Röhlecke, S., Kottwitz, A., Stahr, F. (1996). *IS&T's NIP12: International Conference on Digital Printing Technologies*, pp. 451–456.

Lyra Research, Inc. (1999). *Cost of Operation: A Study of Workgroup and Departmental Laser Printers*, November 1999.

Mort, J. (1991) *Handbook of Imaging Materials* (A. Diamond, ed.), Ch. 10, pp. 447–487.

Röhlecke, S., Steinke, O., Schade, K., Stahr, F., Albert, M., Deltschew, R., Kottwitz, A., Carius, R. (1997). *Amorphous and Microcrystalline Silicon Technology Materials Research Society Symposium Proceedings 467*:579–584.

Sasahara, M., Fukunaga, H., Ikeda, A. (1998). *IS&T's NIP14: 1998 International Conference on Digital Printing Technologies*, pp. 535–538.

Stahr, F., Kottwitz, A., Röhlecke, S., Schade, K., Lutz, M. (1997). *IS&T's NIP13: 1997 International Conference on Digital Printing Technologies*, pp. 233–237.

Stahr, F., Röhlecke, S., Schade, K., Lutz, M., Dreihöfer, S. (1999). *IS&T's NIP15: 1999 International Conference on Digital Printing Technologies*, pp. 728–731.

Takahashi, S., Nakanishi, T., Marukawa, Y., Yamazaki, T., and Moriguchi, H. (1989). *Electrophotography 28*(4):392–401.

12

Thermal Imaging Materials

KLAUS B. KASPER

Boulder Consultants, Boulder, Colorado

12.1 INTRODUCTION

Thermal printing is a direct printing process in which image information in electronic signals is converted to image modulated heat energy, which produces a printed image by either chemical or physical means, or by a combination of both.

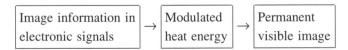

Unlike photography, which has been perfected over the years while maintaining the underlying silver halide chemistry, thermal imaging has evolved through a variety of chemical and physical processes and encompasses different printing as well as image forming mechanisms.

Thermal printing processes are characterized in terms of the chemical and physical mechanisms that form the image; direct printing and transfer printing. In direct printing, the image is formed directly in the print media. Transfer printing employs a donor ribbon that contains the colorant, which is transferred to the print medium during printing. The two thermal printing processes are generally referred to as ''direct thermal printing'' and ''thermal transfer printing,'' and this terminology will be used throughout this chapter.

The objective of this chapter is to familiarize the reader with the chemical and physical processes involved in forming the image and the media required for printing. A review of thermal media necessarily needs to consider the printer in which they are used. Media and printer constitute a complex imaging system that is carefully matched. The material presented here is a compilation of data and information published in the scientific, technical, and patent literature.

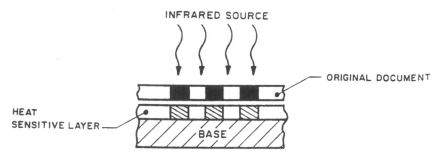

Figure 12.1   ThermoFax paper of the 1950s by 3M. (From Diamond, 1989.)

12.2 EARLY THERMAL PRINTING PROCESSES

Benjamin Franklin was probably the first person to use thermal energy to make an image. In 1759 he placed a number of square pieces of various color cloths on snow in bright sunshine. After a few hours he observed that the cloths had sunk into the snow to varying degrees, with darker cloths having sunk the deepest (Burlingame, 1967).

It took 200 years before the process of light-to-heat conversion, observed by Benjamin Franklin, was commercially utilized for printing. In the 1950s, 3M introduced ThermoFax, a document copying process the inventor had named Thermography (Harriman, 1978), the printing through a chemical reaction induced by heat. Hard copies are generated by an infrared source that causes differential heating in the original placed on top of a heat sensitive paper (Fig. 12.1). As a result, the thermal energy pattern is converted to a visible image in a heat sensitive layer, containing organometallic salts and reducing agents dispersed in a binder matrix. Heat causes reduction of the light colored salts to their dark, metallic state (Miller, 1956).

Earlier thermographic processes employed physical changes to create a visible image. Electrocardiograph chart printers, introduced in the 1930s (Fig. 12.2), used a heated metal stylus to create a line trace when moved over a coating of a white, waxy layer applied over a dark paper or interlayer. The white layer becomes transparent when heated, exposing the dark color underneath. The patent literature is replete with such blushed or obscuring layers that can be made transparent by heat fusing (Gold, 1964). These papers were pressure sensitive, gave poor definition, and were replaced by chemical thermographic papers. Upon heating chemical type papers above the conversion temperature, an irreversible chemical reaction takes place, and a colored component is produced (Jacobsen et al. 1976).

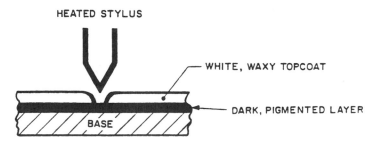

Figure 12.2   Electrocardiogram chart paper of the 1930s. (From Diamond, 1989.)

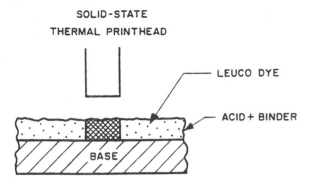

Figure 12.3 Thermal printer with electronically addressable printhead and direct thermal paper introduced by NCR in the 1960s. (From Diamond, 1989.)

In the 1960s, NCR introduced direct thermal print papers with a heat sensitive coating consisting of a ''colorless'' dye precursor and phenolic acid in a binder matrix (Talvalkar, 1969). When heated, the two colorless components react to form the colored leuco dye. The printer used a solid-state, electronically addressable printhead (Schroeder, 1964). Printer and media configuration are shown in Fig. 12.3. These initial coatings had low thermal sensitivity, and the rough paper surface adversely affected print quality. Over the years, the process has been steadily improved, and today leuco dye chemistry remains the technology of choice for direct thermal printing.

The printhead developed by Schroeder was also incorporated in a thermal transfer printer, called a thermal teleprinter, by Joyce of NCR and Homa of the U.S. Army (Joyce and Homa, 1967). The printer used a donor ribbon coated with a colored material that became tacky when heated, and the image could be transferred several times to plain paper under pressure (Chang, 1995). These NCR developments marked the beginning of thermal printing as we know it today.

12.3 THE THERMAL PRINTER

12.3.1 Thermal Printheads

The thermal printhead is the core of the thermal printing system. It consists of a fixed array of microscopic heating elements. When electrical energy is applied to the resistor elements, heat is generated in the area under the elements. The thermal printhead is in continuous contact with the thermosensitive recording material, moving against it under substantial pressure, while being subjected to the heating and cooling cycle thousands of times per minute. The most common thermal heads employed today are the thick film type and the thin film type. Thick film products were introduced by Hewlett Packard, NCR, and other major manufacturers in the early 1970s, followed by thin film heads later in the decade.

Figure 12.4 shows the structural differences between the two types of printheads. The fabrication process entails depositing successive layers under carefully controlled conditions. Thin film resistor layers are coated using vacuum deposition or sputtering for the resistor material, such as Ta_2N, resulting in extremely thin films measuring only 0.05–0.5 microns. Photolithography and microetching are used to form the individual resistive

THIN FILM TYPE

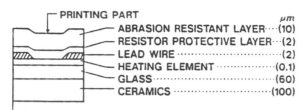

THICK FILM TYPE

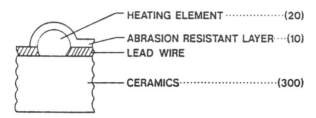

Figure 12.4 Structural differences between thick film and thin film printheads. (From Sodeyama, 1986.)

elements and to provide isolation between adjacent print elements. The thick film type resistors are screen coated with a paste of ruthenium oxide and sintered at high temperature, with film thicknesses ranging from 20 to 40 microns. Each resistor is a separate area. One side of the circuit is a common conductor touching all resistors, and the other side is made of individual conductors touching individual resistor elements.

The next issue is the location of the line of individual resistors, on the flat side of the head or close to or on the edge of the head. With resistors located on the flat side of the head, the print medium has to pass over a round platen. This led, in the early 1990s, to the introduction of edge or near edge type printheads that can use round or flat platens. Flat platens allow printing on rigid media and the design of smaller and more compact printers.

The printhead is made up of the individual resistor elements, numbering from a few to several thousands. The size of the heating elements varies from about 100 to 350 microns with spacing ranging from about 2.5 elements/mm for low-density up to 24 elements/mm (600 dpi) for high-density print heads. The element density is the most important factor determining the resolution of the printed image. Fig. 12.5 shows thermal head types in terms of the arrangement of the heating elements. The matrix type arrangement, popular initially, has largely been replaced with lower cost vertical serial printers, with the head moving across the print medium. Horizontal line heads are made in the width they will print, and only the paper moves, while the head remains in a fixed position. As a result, the printing unit is more reliable and easier to maintain.

Printing with resistive thermal heads is inherently a wasteful process. Much energy is lost in heating parts of the thermal head that are not used for printing. Designing heads with better thermal insulation has limitations, because individual heater elements have to cool down between each printing cycle to prevent printing without signal. The cooling time, much longer than the heating response, is the limiting factor on pulse repetition rate. Higher speeds can be achieved with "historical" correction. Because the initial temperature of each heating element varies with its heat history, electronic compensation methods are

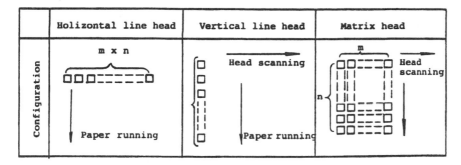

Figure 12.5 Dot element arrangement in thermal printheads: horizontal, vertical, and matrix, showing printhead scanning versus paper travel.

used to control the energy pulse duration based on the pulse history of that element. A similar technique may be needed to correct for the influence of adjacent data in the array.

Poorer thermal insulation between dots and greater thermal gradients within dot elements are found in thick film elements, resulting in greater dot bleed effects. The advantages of thick film heads are lower cost, higher heat transfer efficiency, and better abrasion resistance. This makes them suitable for operation in harsh industrial applications and cost sensitive designs such as fax machines and point of sales printers. The higher energy dissipation allows the use of low-sensitivity media.

In thin film heads, the low mass dot elements and good thermal insulation across the print head axis minimize dot bleed effects, yield inherently sharp and crisp individual dot images, and result in much shorter response time, the critical determinant of print speed. Thin film heads are capable of higher resolution [which has reached 600 dpi (Namiki, 1997)] and control of thermal uniformity and repeatability within one percent of temperature. This makes them better suited for grey scale and continuous tone printing.

12.3.2 The Thermal Printing Process

One of the outstanding features of the thermal printing process is its reliability. There are few moving parts that can break down. Thermal head life is usually measured in terms of thermal shock and kilometers of substrate travelled. In case of thermal shock, each element is subjected to a thermal rise from 25°C up to 425°C over approximately two ms (Srivastava, 1989). This is a very severe temperature change over a short time frame. Thick film heads are rated for 70–100 million and thin film heads for 30–50 million thermal pulses. After that, resistance of the element may change, which can result in deterioration of image quality. Printing a full page of text requires each element to be cycled approximately 66 times. In actual applications, thermal failure is extremely rare and is seldom catastrophic. From an abrasion standpoint, a thick film printhead is capable of imaging 75 to 100 km of paper under normal operating environments. This translates to about 225,000 to 300,000 prints or approximately 240 prints per day every working day for 5 years.

Thermal printing provides several unequaled combinations of print quality, speed, simplicity, reliability, and low cost. In bar codes and other labeling applications, no other technology can match the image sharpness of thermal printing anywhere near the price. The simplicity and reliability of the thermal printing process is a major reason it has found widespread applications where this is paramount, in fax machines that need to be operational on demand, unattended for 24 hours and 7 days a week, and in point of sales receipt and airline ticket printing and similar situations where failure would be highly disruptive.

12.4 DIRECT THERMAL PRINTING

Thermal printing on heat sensitive paper, or direct thermal printing, is the simplest printing technology today. The thermal printhead rests directly on the print paper. Electronic signals are converted to electric pulses and, through a driver circuit, selectively transferred to the heating elements of the printhead. The print paper carries a heat sensitive layer containing colorless components that react and form a colored dye at elevated temperatures. No waste is generated in the process. Because most of the technology is in the paper, rather than the printer, it is possible to build very low cost, compact, reliable printers that consist of little more than the thermal print head and rollers to feed the paper. Most of the world is now quite familiar with thermal printing because the technology is widely used in facsimile machines.

12.4.1 Direct Thermal Printing Systems

Leuco dye chemistry, developed and commercialized by NCR in the 1960s, has much improved over the years and is still the dominant direct thermal printing technology today. Other methods were largely replaced by leuco dye systems with superior stability, response, and contrast retention, or confined to specialized niche markets.

Diazo dye chemistry has been adapted to direct thermal printing. Diazo dyes are formed through the reaction of a diazonium salt compound and coupler in the presence of a sensitizer and base. Because of the high cost of the paper and large size of the printer, diazo based papers have not become widely accepted for facsimile printing, but they have found applications in medical diagnostic and scientific recording, and more recently for color printing.

12.4.2 Leuco Dye Direct Thermal Print Papers

Leuco dye based direct thermal papers are the current workhorse in commercial use. The heat sensitive coating consists of a number of active ingredients (Fig. 12.6). The color forming components are colorless leuco dye precursors of the triphenylmethane and fluorane type containing lactone rings and phenolic acid developers. Another critical ingredient in the color forming reaction is the sensitizer, a solid material that typically melts at a low temperature (<100°C). It acts as a solvent for the dye precursor and developer, forming a eutectic mixture during printing. Color develops by the phenolic acid contributing a proton that attacks the lactone ring of the dye precursor. This brings about a new compound in which a quinoidal resonance in the dye cation is responsible for the appearance of color. Depending on the nature of the leuco dye, various colors can be obtained, but in practice black is predominant. For bar code printing, leuco dyes that absorb in the near IR and can be read with LED and LD light sources have been developed (Higaki, 1990).

The color forming process is reversible, and antioxidants added to the coating, which act as stabilizers, move the equilibrium point of the color reaction toward the colored species (Goodwin, 1993). Other stabilizers include Novolak type epoxy compounds that react with the leuco dye (Watanabe et al., 1993), nonphenolic developer compounds containing sulfonylurea functional groups (Takahashi et al., 1994), and UV absorbing polyurethane/polyurea fillers (Mandoh et al., 1993). The stability of the printed image is affected by PVC plasticizers, marking pens, hand lotions, and a variety of chemicals. Microencapsulation of the dye precursor and salicylic acid developers increase resistance of the image to chemical attack (Miyamoto, 1991).

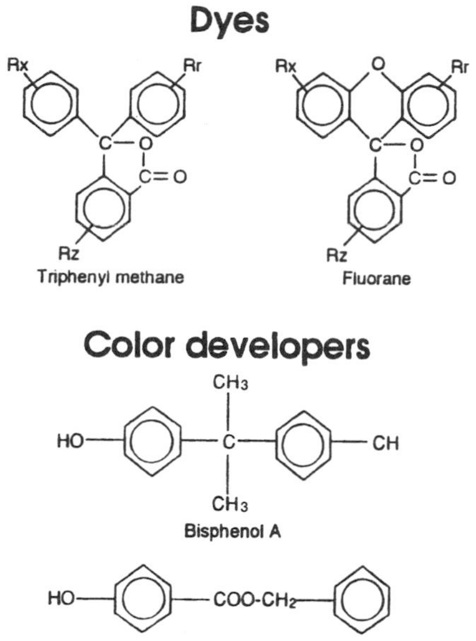

Dyes

Triphenyl methane

Fluorane

Color developers

Bisphenol A

Figure 12.6 Dye precursors and phenolic acid developers for leuco dye direct thermal printing. (Courtesy of Diamond Research Corporation, Ojai, California.)

While the chemicals participating in the color reaction are vital, other components are critical to the function and performance of thermal papers. Coating binders commonly used are high molecular weight water soluble or dispersible polymers, such as polyvinyl alcohol, polyvinyl acetate, and modified cellulose and starch derivatives. The binder should exhibit high pigment tolerance and adhere firmly to the paper. It must resist attack from oil, fat, and alcohol and should not resolubilize easily when wetted with water. Pigments and lubricants, such as zinc stearate, act as release and slip agents that prevent sticking of the paper to the printhead. Waxes added to the coating increase thermal sensitivity and cause the image to flow together so that more uniform print characters are formed. Antistatic agents help dissipate electrical charges that can be created as the paper feeds through the printer.

In addition to the active, color forming layer, other coating types are applied to direct thermal media that change its functionality. The base coat, applied directly on the base sheet, contains clay and binders and produces a level surface on which the active layer is coated. This facilitates good contact between paper and thermal head, assuring uniform image formation. It increases opacity and brightness of the sheet and acts as absorber for excess chemicals that may contaminate the printhead. Effective heat energy transfer to the color forming layer can be increased by introducing voids or air capsules into the base layer. This increases the thermal insulating properties, minimizes heat loss into the paper, and thereby increases thermal sensitivity (Lewis et al., 1992; Motosugi et al., 1992).

A third coating type used in direct thermal papers is the top coat. The top coating, primarily used for tags and labels, contains binders, clays, and lubricants, can impart pencil writability and erasability (Hara, 1991), and acts as a seal and protective coat for the heat sensitive layer. The final coating used in thermal papers is the back wet or the back coat. Its primary function is curl control and the imparting of antistatic properties. Only rarely are all of these coatings used on one sheet, but different coatings are combined to meet the requirements for specific applications. Thermal papers for pressure sensitive labels receive an adhesive back coating that is covered with a releasable liner (Arbree et al., 1986). More recently, special thermal printers were introduced that print on liner free thermal media. These self-wound, linerless media have the silicone release coating applied over the active thermal or top coating.

The performance of thermal papers has improved considerably over the years. Maximum density is affected not only by the composition and level but also by the particle size of the active ingredients (Usami, 1990). Increased sensitivity has resulted in higher printing speeds and reduced power consumption and load on the thermal head. Examples of papers exhibiting increasing thermal sensitivity are shown in Fig. 12.7.

Early calculator chart papers of the 1970s required heating levels that approached the combustion point of the paper. In today's facsimile sheets, heat from friction, generated by a fingernail scored across the surface, is sufficient to develop the color. Many parameters had to be optimized to bring about this improvement. High smoothness of the paper base, low eutectic temperature and viscosity combined with high dissolving power of the sensitizer, fine particle size of dye precursor and developer, as well as minimizing heat loss to the paper, all act to increase sensitivity. However, there are trade-offs. The increased sensitivity reduces shelf life and renders the product less environmentally stable. The high smoothness of the paper gives it an artificial appearance, quite unlike bond type office papers, a major reason direct thermal papers have not been accepted for correspondence quality printing.

The technology is capable of continuous tone reproduction with either density or area based gradation printing. Intimate contact between medium and thermal head is crucial. Smooth synthetic substrates such as polyester films and synthetic papers and paper laminates, which do not absorb moisture, rather than paper, are used as support. A dynamic tonal range of 64 grey levels with optical densities of up to 1.8 can be obtained.

12.4.3 Transparent Direct Thermal Print Media

Transparent thermal media are faced with the fact that leuco dye coatings, with dye precursor and developer dispersed in powder form in the binder matrix, represent a two-phase

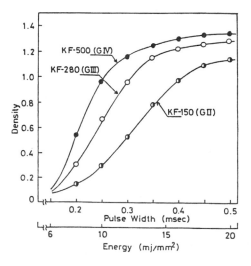

Figure 12.7 Dynamic sensitivity of typical direct thermal papers, sensitivity improved 15–30% each from GII to GIII and GIV. (From Higaki, 1990.)

system with different refractive indices. The coating layer also contains entrapped air, and the resulting light scattering renders the coating opaque.

This was overcome by first dissolving the developer in an organic solvent and emulsifying it in an aqueous solution of a water soluble polymer. The dye precursor, dissolved in oil and encapsulated in a heat sensitive microcapsule to prevent premature color development, is then dispersed in the pseudocontinuous developer phase. By making the refractive indices of the various components nearly equal, the coating can be made transparent (Usami et al., 1989). Maximum optical densities above 2.0 can be achieved, making the process suitable for medical recording (Ohga et al., 1991).

12.4.4 Diazo Dye Direct Thermal Print Media

The reaction mechanism for diazo dye thermal printing is similar to printing with leuco dye chemistry. Two colorless components, a diazonium salt compound and a coupler, react when heated in the presence of a sensitizer and an organic base, to form a colored diazo dye. To prevent color formation at room temperature, the diazonium salt is encapsulated in a heat sensitive microcapsule. The diazonium salt in the unprinted areas can be decomposed through ultraviolet radiation after printing. The dye formation and decomposition reactions are shown in Fig. 12.8.

Decomposition of the unreacted diazonium salt after printing results in a paper that is no longer heat sensitive and adds options for color printing to the process, not available with leuco dye chemistry.

12.4.5 Direct Thermal Color Print Media

Two different approaches have been developed for two-color printing with leuco dye chemistry; color addition and decolorization. Both use two separate color forming layers with different thermal sensitivities, an upper layer with color development taking place at lower temperature and a lower layer with color formed at higher temperature (Watanabe et al., 1993).

In both systems the upper color forming layer is first printed at a lower temperature. In the additive system a second printing at higher temperature produces color in both layers, with the two overlaid colors yielding a third color, i.e., red and green resulting in

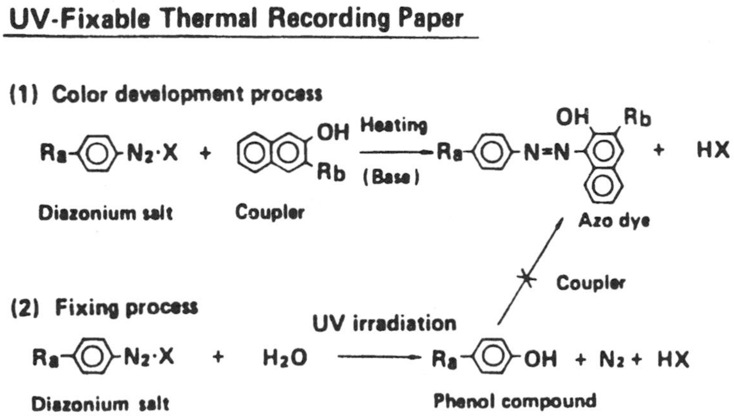

UV-Fixable Thermal Recording Paper

(1) Color development process

$$R_a\text{-}\bigcirc\text{-}N_2\text{·}X \ + \ \bigcirc\bigcirc \substack{\text{OH} \\ \text{Rb}} \xrightarrow[\text{(Base)}]{\text{Heating}} R_a\text{-}\bigcirc\text{-}N\text{=}N\text{-}\bigcirc\substack{\text{OH} \ \ \text{Rb}} \ + \ HX$$

Diazonium salt   Coupler   Azo dye

(2) Fixing process

$$R_a\text{-}\bigcirc\text{-}N_2\text{·}X \ + \ H_2O \xrightarrow{\text{UV irradiation}} R_a\text{-}\bigcirc\text{-}OH \ + \ N_2 + HX$$

Diazonium salt   Phenol compound

Figure 12.8 Components of UV-fixable, diazo dye direct thermal imaging systems. (From Sato et al., 1985. Reprinted with permission of IS&T: The Society for Imaging Science and Technology, sole copyright owners.)

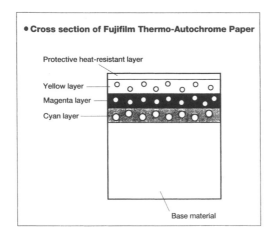

• **Cross section of Fujifilm Thermo-Autochrome Paper**

Protective heat-resistant layer

Yellow layer

Magenta layer

Cyan layer

Base material

1) The yellow layer reacts to low levels of thermal energy to generate the yellow portion of the image.

2) The entire print is irradiated by a UV lamp, which fixes the color by decomposing the undeveloped color elements in the yellow layer, destroying their color-synthesizing properties.

3) The magenta layer reacts to mid-range levels of thermal energy to generate the magenta portion of the image.

4) The entire print is irradiated by a UV lamp, which fixes the color by decomposing the undeveloped color elements in the magenta layer, destroying their color-synthesizing properties.

5) The cyan layer reacts to high levels of thermal energy to generate the cyan portion of the image.

Figure 12.9   Printing and fixing sequence of Fujifilm's Thermo-Autochrome Color Printing System. (Courtesy of Fuji Photo Film Co., Ltd.)

black. A different approach is used for decolorization media where an erasing layer is placed between the two color forming layers. The second printing at the higher temperature develops color in both layers, with the erasing layer at the same time destroying the dye formed in the upper layer, exposing the color in the layer underneath.

Diazo dye chemistry offers another option for color printing through the possibility of decomposing the diazonium compound with UV exposure. For two-color printing the thermosensitive layer contains a mixture of diazonium salts capable of forming green or red diazo dyes. A first thermal printing causes development of both colors, resulting in a black image. The green diazonium salt in the unprinted areas is then selectively decomposed through UV exposure of a certain wavelength. In a second printing step, the red colored image is recorded (Usami, 1990). The same result can be obtained with selected mixtures of cyan, magenta and yellow diazonium salts (Koyano et al., 1997).

Using a combination of diazo and leuco dye chemistry, Fuji Photo & Film Co. has developed a three-color, continuous tone direct thermal printing process. The system, called ThermoAutochrome, was introduced to the market in 1994. Figure 12.9 is a simplified cross section of the print paper structure. The yellow and magenta color forming layers contain diazonium salts and couplers and the bottom cyan layer a conventional leuco dye system. The diazonium salts and leuco dye precursor are encapsulated in microcapsules (Igarashi et al., 1994). Figure 12.10 shows the dynamic color forming response of each layer with reflective optical density plotted against recording energy. The three color layers are printed sequentially at increasing temperatures, and after each printing, the diazonium salt compounds in the yellow and magenta layers are selectively decomposed through UV exposure at wavelengths of 425 nm and 365 nm, respectively. Only the unprinted areas of the cyan layer are not "fixed" and remain heat sensitive after printing.

Encapsulation combined with UV fixing renders the image quite stable, but background density is likely to increase with exposure to light. While image stability of earlier products has been improved, additional work needs to be done (Sakai et al., 1999). A microscopic cross section of the ThermoAutochrome paper is shown in Fig. 12.11. The support is a microporous polyester coated with a TiO_2 filled polyethylene layer on the face side and a clear polyethylene layer on the back. The three color layers are protected by a heat resistant overcoat.

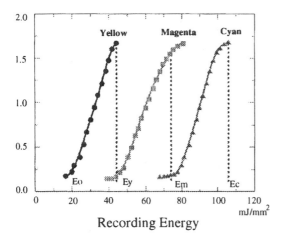

Recording Energy

Figure 12.10 Thermal recording characteristics of ThermoAutochrome Color Printing System. Optical density as a function of recording energy. (From Sato et al., 1994. Reprinted with permission of IS&T: The Society for Imaging Science and Technology, sole copyright owners.)

The system is aimed at the photographic, video printing, scientific, and medical recording markets. Printers are offered in resolutions of 150 dpi with 128 grey levels and 300 dpi and 254 grey levels. Quality is claimed to be comparable to thermal dye transfer printing, to which it offers an alternative that requires only a single consumable and does not generate waste.

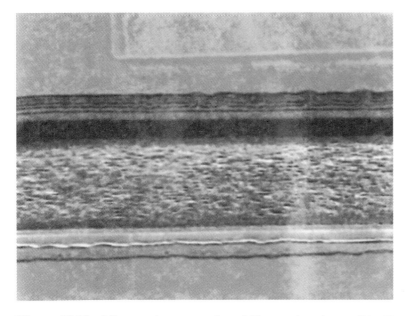

Figure 12.11 Microscopic cross section of Thermo-Autochrome Print Paper. From top: heat resistant layer; yellow, magenta and cyan color forming layers; pigmented polyethylene layer; microvoided polyester support; clear polyethylene layer; back coat. Total caliper 241 microns. (Courtesy of Felix Schoeller Technical Papers, Inc., Pulaski, New York.)

12.4.6   Future Outlook for Direct Thermal Printing

The ThermoAutochrome System is an impressive example of the technical potential of direct thermal printing and it shows that a technology considered mature can still produce new and unexpected results. For the largest application of direct thermal printing, facsimile, prospects do not look bright. The main reasons are (1) the overall decrease in fax traffic as users switch to e-mail or PC-based fax software and (2) the preference for plain paper fax machines which have become more competitively priced. For industrial, medical, retail, and point-of-sales printing, where media sensitivity is not acute and simplicity, reliable performance, and cost are primary concerns, direct thermal printing is well positioned. Another sector is the dominant output method on ships and aircraft for printing information from satellites, radar, and sonar.

12.5   THERMAL TRANSFER PRINTING

Thermal transfer printing uses modulated thermal energy to transfer a colorant from a donor ribbon to a receiver substrate. The process is the closest nonimpact analog to impact printing with dot matrix printers or typewriters. The ribbon, coated with the colorant material, releases and transfers the colorant to a receiver when heat is applied. Transfer can occur by two different physical processes, depending on the formulation of the ink layer. The image formation mechanisms for the two processes are sufficiently different, and they are treated as separate printing systems in the technical and commercial literature. There is no generally accepted terminology for these two printing processes, and in this chapter the terms thermal mass transfer and thermal dye transfer printing will be used, as they best describe the physical processes involved in forming the image.

  Thermal energy sources are most commonly thermal printheads, similar or identical to those used for direct thermal printing, but systems using focused laser beams have recently been introduced. The first thermal transfer printer, developed by NCR for the U.S. Army in the 1960s, was not suitable for commercial use. Thermal transfer products, developed by several Japanese companies, were introduced to the U.S. market in the early 1980s. By 1982, Fuji Kagakushi Kogyo marketed a wide range of color thermal transfer ribbons, and in 1985, 19 thermal transfer printer manufacturers offered a total of 37 printers, 26 of which were color printers. The first laser induced thermal dye transfer product from Kodak made its debut in 1991. An important thermal transfer product was IBM's Quietwriter, which used a resistive, self-heating ribbon instead of a thermal head. The product, introduced in 1984, was withdrawn from the market in 1991 because of unacceptably high consumables cost.

12.6   THERMAL MASS TRANSFER PRINTING

Figure 12.12 is a schematic of a thermal mass transfer printer. The printhead rests directly on the donor ribbon, which is held in close contact with the receiver substrate against a platen. Ribbon and receiver media are passed under the printhead simultaneously. In most printers, the paper and ribbon travel through the print mechanism at the same rate, so the cost per print is constant, independent of the area of ink coverage. This is particularly acute for color printing, which requires three or four separate color panels to produce one print. Some monochrome printers for label applications implement ribbon saving, stopping the motion of the ribbon when blank areas of the print are passing through, and taking the pressure off the head to avoid friction between donor ribbon and receiver.

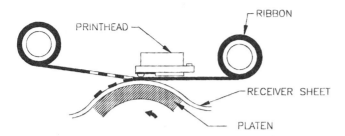

Figure 12.12   Schematic of thermal mass transfer printer. (Courtesy of Atlantek Inc., Wakefield, Rhode Island.)

For color printing, the different color layers on the donor ribbon are coated either in parallel stripes or in sequential patches. The parallel coating of ink is used for serial printers, and the colors are transferred in the printer in units of the various colors before the paper is moved to the next line. For horizontal line printers, the colors are coated in sequential patches on the ribbon. The first color is printed line by line on the whole page; the paper is then returned and the next color printed until all colors have been applied to the image.

The ink layer, coated on a support material, is formulated so that the colorant adheres firmly at room temperature but is transferred to the receiver when thermal energy is applied to the donor. The ink consists of a complex mixture of coloring pigment and binder, which melts at a certain temperature. When the ink layer reaches the melting point, the ink is transferred in a coherent mass to the receiver, hence the term mass transfer. Initially, transfer takes place through a combination of pressure and surface tack to the receiver. Once the entire thickness of the ink layer exceeds the melting point, transfer to a paper surface is aided by the capillary action of the paper fibers. Since the heating elements contact the side of the donor ribbon opposite the ink layer, heat transfer is less efficient compared to direct thermal printing, where the heat sensitive layer is in direct contact with the thermal head.

12.6.1 Thermal Mass Transfer Donor Ribbons

The two main components of the thermal transfer ribbon are the base film and the ink layer. In actual practice, thermal transfer print ribbons consist of several layers coated on a 2.5–6 micron thick polyester film, although condenser tissue paper and cellophane films have been considered (Anczurowski et al., 1987). The ink layer is usually applied over a release layer to assure complete transfer of the ink during printing. The side opposite the ink layer is coated with a heat resistant back layer, which provides release of the ribbon from the printhead and reduces head wear. The back layer may incorporate antistatic or conductive properties. Over/under ink coatings are used to impart special features such as antigloss, correctable print, or security.

The most common material used as binder is wax, and the process is frequently referred to as thermal wax transfer printing. Wax comes in a variety of mixtures of vegetable and mineral waxes from carnauba to paraffin. Resin and wax/resin mixtures are also popular. They are more expensive but provide a harder surface that is resistant to abrasion and scratching. In general, there is an inverse relationship between durability and cost and durability and energy required for printing. Wax is easy to scratch but melts readily for printing. Resin is tough but requires higher printing temperatures and is more expensive, with wax/resin ribbons falling somewhere in between. Wax is printed on a variety

Table 12.1 Thermal Mass Transfer Ink Compositions

	Standard (%)	Midrange (%)	Premium (%)
Wax	65–85	50–70	5–20
Pigments	5–20	5–20	5–20
Resin	5–15	10–25	50–80

Source: From Giga Information Group, Norwell, MA (1995).

of substrates, while resin ribbons are most suitable for printing on synthetic substrates. Reactive ribbons are a variant of resin ribbons, where the resin contains reactive groups attached to the polymer chain that cross-link during printing. The cross-linked polymer no longer melts and is insoluble in most solvents. Wax ribbons contain some resin to provide adhesion to the substrate and resin ribbons some wax to improve slip. Typical formulations are shown in Table 12.1.

Wax ribbons are also called standard ribbons and midrange ribbons wax/resin ribbons. Premium ribbons have a variety of names including 3rd generation, pure resin, and solvent resin.

For a complete transfer of the ink during printing, the ink should have a defined melting point with a sharp drop in viscosity. Melting points are not relevant for resins that go through a glass transition with a more or less broadly defined peak. Differential scanning calorimetry (DSC), in which the amount of power needed to cause a fixed rate of increase in temperature is plotted against temperature, is a convenient way of defining the thermal response of resinous materials. The DSC of a standard wax ribbon is shown in Fig. 12.13. It has a fairly well defined peak, a desirable feature for bimodal printing, as all the ink is transferred almost at once. For variable dot transfer a much broader peak is preferred, so that the amount of ink transferred can be controlled by varying the amount of energy supplied.

The other main component of the ink layer is the coloring component. Organic

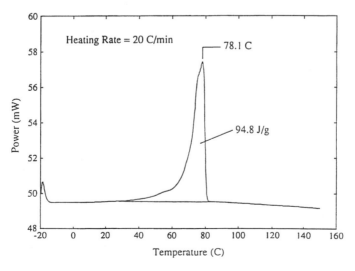

Figure 12.13 DSC of thermal wax transfer ink formulation. (From Giga Information Group, 1995. Contact Sara Martin, Giga Information Group, One Longwater Circle, Norwell, MA 02061, 1-800-874-9980 or 617-982-9500, fax 617-878-6650.)

C. I. Pigment Yellow 12

C. I. Pigment Red 57:1

C. I. Pigment Blue 15

Figure 12.14 Coloring pigments for thermal wax transfer inks. (From Seto et al., 1988. Reprinted with permission of IS&T: The Society for Imaging Science and Technology, sole copyright owners.)

pigment dyes are used for colored inks, and examples of pigments used are given in Fig. 12.14. Magnetic inks that contain magnetic components, such as iron oxide, are an important element in thermal transfer printing for MICR encoding (Talvalkar, 1992). Pigment dyes are considerably more environmentally stable than leuco dyes used for direct thermal printing, and black prints with carbon pigment are as stable as the paper support. One drawback is that smear, especially for wax based ribbons, remains a problem particularly at elevated temperatures.

To reduce ribbon costs, multipass thermal ribbons are offered that carry a number of ink layers separated by release layers. They are formulated so that one layer of ink will transfer to the receiver every time a dot is printed. The ribbons, mounted on reversible cartridges, are turned around for the next printing cycle and are conservatively rated at six printing cycles. Another approach uses a heavier ink layer and advances the ribbon at a slower speed than the print medium. The image printed is about four times longer than the ribbon and only a portion of the ink is transferred at each printing step (Giga, 1995).

A multiprinting thermal transfer ribbon, described by Maehashi et al. (1990), employs a single layer that consists of two nonmiscible components: a "supercooling" material and a thermoplastic ink. The two materials have different melting points and solidifying temperatures. At the multiprinting state, the supercooling material is still liquid, while the ink has solidified. Ink is transferred in incremental steps, with separation occurring in the liquid phase. This stage lasts long enough to allow sufficient peel time for most printers.

12.6.2 Thermal Mass Transfer Receiver Papers

In principle, printing should be possible on any medium to which the ink will adhere. In practice, the efficiency of the transfer depends on a combination of heat transfer conditions and the actual surface area contact in the printing nip. When printing is done on the same printer, the physical and mechanical properties of the medium predominantly influence the quality of the transferred image. Incomplete contact between donor and receiver in the printing nip results in print defects through dot fragmentation, density variations, and places with

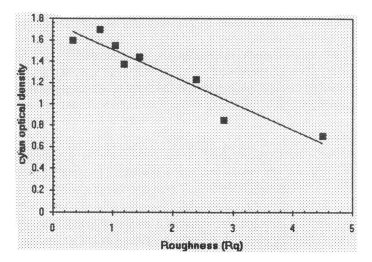

Figure 12.15 Cyan optical density as a function of surface roughness. (From Katsen et al., 1995. Reprinted with permission of IS&T: The Society for Imaging Science and Technology, sole copyright owners.)

no transfer at all. The surface topography of the paper is a major contributing factor (Katsen et al., 1995). In Fig. 12.15 surface roughness is plotted against optical print density, with smooth papers resulting in greatly increased optical density. Other factors are nip pressure, with higher pressure increasing contact area, and paper compressibility, which correlates with optical density (Fig. 12.16). High compressibility of the paper has the added benefit of compensating for nonuniformities in the pressure profile of the printing nip.

For effective transfer, it is important that the surface energy of the receiver be high enough to provide good wetting by the ink (>35 dyn/cm). Transfer of the ink begins at its

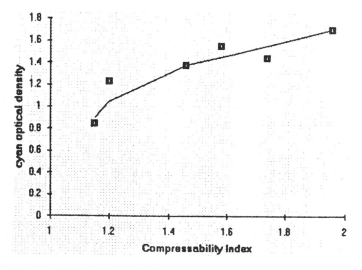

Figure 12.16 Cyan optical density as a function of media compressibility. (From Katsen, 1995. Reprinted with permission of IS&T: The Society for Imaging Science and Technology, sole copyright owners.)

glass transition temperature and peaks at its melting point. For optimum transfer, the cohesive force of the ink, the release of the ink from the carrier ribbon, and the surface tension between molten ink and medium surface need to be carefully balanced (Akutsu et al., 1992). This is no different from normal impact printing. In thermal transfer printing, the process is complicated by the rapid change in viscosity of the ink during cooling. Timing and geometry of separating the donor ribbon from the receiver is critical for obtaining good print quality. Sensitometric characterization of thermal mass transfer ribbons has been described by Swift et al. (1992) and methods for the evaluation of receivers by Spivak et al. (1992).

Many coated and machine finished papers are suitable for thermal mass transfer printing, but the requirement for a smooth paper is a drawback for office printing applications. The usual bond type office papers are not well suited, and users prefer the flexibility, familiarity, and "feel" of so-called plain paper. Improvements are achieved by optimizing printing pressure, timing, temperature, and print head design (OEP, 1992), but higher quality laser papers are still required. According to a study from Matsushita (Yoshikawa et al., 1993), high-quality printing is possible on any type of paper by first printing on an intermediate transfer roller and then by transfer to paper. Tektronix and its Japanese partners have taken a different approach and added an extra panel of a clear adhesivelike layer called ColorCoat to the three-color ribbon. Prior to printing the image, the ColorCoat is laid down on the paper in the areas where the image will be printed (Brandt et al., 1996). While this does not result in complete paper independence, printing is possible on a wider range of office papers, including copy papers.

12.6.3 Transparent Thermal Mass Transfer Media

Thermal mass transfer printing is well suited for printing of overhead transparencies. The base material is clear polyester film, which best meets the requirements for transparency, toughness, and heat resistance. Heat generated by the overhead projector can result in curl and cockle of the film. To prevent this, the film is subjected to a shrinking process, so that no subsequent shrinkage will occur.

Antistatic coatings applied to the back side provide low friction for processing through the printer. A surface coating is applied to assure adhesion of the ink (Kulkarni et al., 1995; Zawada, 1995). The coatings produce a slight haziness that does not visibly affect the projected image. Thermal mass transfer printing is one of the best choices for printing of overhead transparencies (Giga, 1993).

12.6.4 Variable Dot Thermal Mass Transfer Printing

Thermal mass transfer printing is a binary process, and resolution is limited by the spacing of the heating elements, at this time 300 dpi or 12 dot/mm for most thermal color devices. This is insufficient for high-quality pictorial printing. Variable dot size printing would considerably improve image quality. Changes in heater element and printer design to produce variable dot size have been suggested (Tsumura et al., 1988; Saito et al., 1993; Inui et al., 1993; Sonoda et al., 1993). Panasonic introduced a printer in 1990 that employed resistor elements with a "waist" in the middle, thereby increasing current density and producing a controllable dot size modulation capability (Degerstrom, 1990).

To produce controlled dot sizes, Fuji Photo Film Co. has combined the halftone modulating method named LOUVER with a thin layer thermal transfer ribbon in the design of their color proofer FIRST PROOF™. According to the company prints with high fidelity of conventional chemical proofs can be produced (Nakamura et al., 1997; Sawano et al., 1997).

One ink ribbon structure for variable dot size printing contains a porous filler in addition to colorant and binder. The porous material releases ink during printing in proportion to the energy applied (Kutami et al., 1990). Ink ribbons must transmit energy more efficiently, which is achieved in part by reducing the thickness of the ribbon. Ink coating weights are lower, but with higher concentration of pigment. The release layer properties are critical for uniform dot transfer. Recent product introductions by Fujicopian, Casio, and others employ a special printhead and ribbon. The Casio Printjoy, introduced to the U.S. in the fall of 1994, prints at 200×346 dpi and is capable of six bits of color. Geometry and spacing of heating elements are different from those of thermal dye transfer printheads. Resistors are square and exactly half the dot pitch in width, so that the gaps between resistors are as wide as the resistors (Giga, 1995).

Best results are obtained with a micropore receiver paper. A cross section of the paper is shown in Fig. 12.17. The support is a multilayer, biaxially oriented, voided polyolefine paper coated with a spongy mixture of two or more polymers not miscible with each other (Ichii et al., 1989). During printing, the molten ink is absorbed into the pores of the spongy polymer layer in amounts proportional to the energy supplied (Tanaka, 1993). Print quality does not quite reach the level of continuous tone thermal dye transfer prints but is much improved over binary thermal mass transfer. The lower printer and consumables costs make variable dot thermal printing a cost-effective alternative to thermal dye transfer printing in many applications. One drawback is that specialty receiver media are required. Alternative receivers include transparent and white films. They require more energy to print and give less uniform dots. Paper is entirely unsuitable, even when coated.

12.6.5 Resistive Ribbon Printing

An important thermal transfer product was IBM's Quietwriter, introduced as a monochrome serial printer in 1984. It was based on a unique self-heating, resistive ribbon technology. Instead of a print head, the Quietwriter ribbon has a built-in resistive layer that

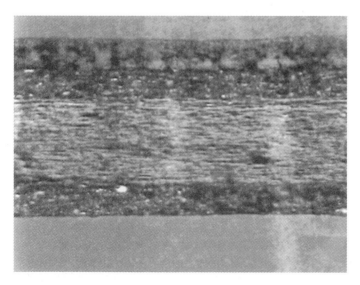

Figure 12.17   Microscopic cross section of thermal mass transfer receiver for variable dot size printing. From top: micropore receiver layer; three-layer, synthetic paper support (center layer is voided). Total caliper 163 microns. (Courtesy of Schoeller Technical Papers Inc., Pulaski, New York.)

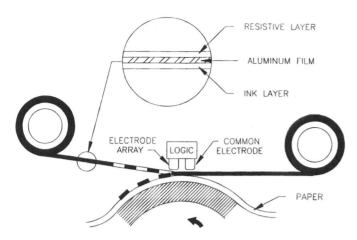

Figure 12.18 Schematic of resistive ribbon printer. (Courtesy of Atlantek Inc., Wakefield, Rhode Island.)

heats up when current is passed through it. Printer and ribbon configuration are illustrated in Fig. 12.18. The ribbon is a three-layer structure with a supporting substrate of an electrically conductive, carbon filled polycarbonate film that serves as the resistive element. The resistive support is coated with a thin layer of aluminum and an ink layer. A polyamide thermoplastic resin, loaded with carbon black, is a preferred ink composition (Crooks et al., 1986, 1989). During printing, electric current flows from the stylus array of the print electrodes through the resistive layer to the aluminum layer, causing localized heat to be generated in the resistor. When the ink is heated, it transfers from the ribbon to the paper. The mechanism is the same as for thermal mass transfer printing with a thermal head.

The advantages of a self-heating ribbon over a thermal printhead are many. There is no thermal cycle time, as with a thermal head, that limits print speed. Inks with a higher melting point, that do not smear, can be employed. The inks can be transferred at higher pressure, a feature that improves image resolution and makes the process virtually independent of the type of paper used (Diamond, 1989). It is possible to produce high resolution electrode arrays of 40 dot/mm or 1000 dpi (Lane et al., 1992).

The technology is applicable to thermal dye transfer printing, as studies by Miyawaki et al. (1990) and Taguchi et al. (1988, 1990) suggest. An eightfold increase in printing speed and 50% lower energy requirements were observed.

The Quietwriter had the best print quality of any thermal mass transfer product on the market, even on plain paper. Unfortunately, the machine also had an unsurpassed supplies cost of twenty cents per page. After launching two more generations of the Quietwriter, IBM withdrew the machine from the market in 1991.

12.6.6 Future Outlook for Thermal Mass Transfer Printing

Thermal mass transfer printing met with its greatest failure in the office color printing market. The high price of the thermal printers and consumables was a key factor while the market was being flooded with low cost color ink jet printers. Thermal mass transfer printing has been successful in other markets, becoming the dominant process for label printing, mainly of variable information and bar codes. The market has seen strong growth of print-

and-apply labeling on demand and new industrial marketing applications continue to evolve (Townsend, 1996). Major progress has been made in industrial grade printheads for bar code printing and miniature printheads for portable printers. For these applications, no other technology can match the performance, reliability, and cost advantages. For wide format signage, thermal transfer printing with resin-based ribbons can offer a number of advantages especially for outdoor exposure, such as durability, UV resistance, and weatherability.

Progress is being made with the introduction of variable dot thermal mass transfer printers (Terao et al., 2000) and microporous receiver papers (Maeda et al., 2000). Improved image quality is extending applications to ID card printing that enable photographs to be incorporated. Both thermal printing processes are expected to dominate this market (Giga 1998).

12.7 THERMAL DYE TRANSFER PRINTING

Thermal dye transfer printing, also known as dye diffusion thermal transfer, is usually called dye sublimation printing in the computer press. This confusion arose because early researchers believed that the process was related to sublimation dyeing of textiles, which was reasonable at the time (Honda, 1985). Since then, others have recognized that the process is one of diffusion, but the term dye sublimation printing persists (Hann et al., 1990; Hodge et al., 1991; Ozimek et al., 2000).

The process is capable of producing near photographic quality images, and the first thermal dye transfer printer, described in a presentation by Sony in 1982, was the Mavigraph color printer for use with the Mavica video camera. In 1985 Hitachi introduced a thermal dye transfer color printer with 64 grey levels per color. The Sony Mavigraph printer became commercially available in 1986, and in 1987 thermal dye transfer printers were introduced by Eastman Kodak, Hitachi, Sharp, ICI, and Dai Nippon Printing.

The imaging system configuration is similar to that of thermal mass transfer printers (Fig. 12.12). Heat is generated by energizing the thermal head in response to electronic image data while driving the donor ribbon and receiver paper underneath. Thermal dye transfer printers employ closely spaced heating elements on a thin film circuit. Eachelement is usually divided into two parallel sections in order to minimize visible structure in the final print. Thermal dye transfer printing requires more energy than thermal mass transfer or variable dot size mass transfer printing. A power supply that can deliver 2–4 J/cm^2 is required. Temperatures at the surface of each element can rise to 350°C on the surface of the thermal head and to 260°C at the interface between ribbon and receiver.

While the printer configuration for thermal mass and dye transfer printers is similar, and several manufacturers offer dual-purpose printers that can print on either medium the actual transfer mechanism is quite different. During thermal mass transfer printing, the entire ink layer, including binder, colorants, and additives, is transferred to the receiver during the printing process. As a result, thermal mass transfer printing is inherently a binary process.

During thermal dye transfer printing, only the coloring dye transfers, with the binder remaining attached to the donor ribbon. The transfer mechanism is based on the diffusion of the dye from the donor ribbon into a receiving layer on the receiver surface. The quantity of dye that is transferred, and thus the intensity of color at each image point, depends on the length and intensity of the heating pulse applied to the heating element, so that the process is intrinsically a form of continuous tone printing. The colors are coated on the ribbon as sequential panels. In the printing process, the ribbon moves forward continuously while the receiver is recycled underneath it. The entire image is written in yellow, magenta, and finally cyan, with black as an additional option. A truly continuous tone color image is built by overprinting of the three subtractive colors.

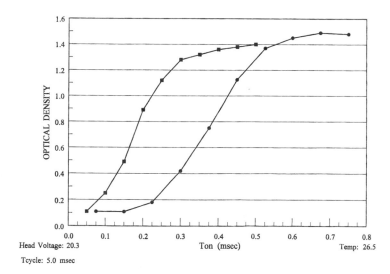

Figure 12.19 Thermal recording characteristics of bimodal thermal mass transfer (—■—) vs. continuous tone dye transfer printing (—●—). (Courtesy of Atlantek Inc., Wakefield, Rhode Island.)

Thermal dye transfer printing exhibits a wide dynamic range, similar to photography. Figure 12.19 illustrates the difference between the dynamic response of bimodal mass transfer and continuous tone dye transfer printing. Dot density variation is primarily a function of media chemistry and accuracy of the thermal control. The strategy is to operate in the "linear" region of the dynamic response. Print quality depends on media color gamut, media motion control, and accurate multiple pass dot registration (Hori et al., 1992). Dye thermal heads are designed to produce up to 256 different levels of color per dot, and by printing the three subtractive primary colors they can generate 16.7 million colors. The results are pictures that are all but indistinguishable from photographic prints to the untrained eye.

Figure 12.20 summarizes results of color research at RIT Research Corp., a spinoff of the Rochester Institute of Technology. The regions marked A and B are adequate in either spatial or color resolution to give the illusion of photographic realism. Area C is somewhat marginal but is adequate for spot color graphics and presentations. Area D is generally unacceptable. The wide boundary between C and D reflects differences in human perception. Moving upward and rightward in this chart represents an improvement in print quality. Much of the chart is empty of products, which have clustered around the top and bottom edges, that is, either at 8 bit/pixel (continuous tone) or 1 bit/pixel (monochrome or bimodal color). Thermal dye printing and ThermoAutochrome are at the top, and have increased to their present level of 8 bits of color and 300 dpi. Variable dot thermal mass transfer printing has been demonstrated at 6 or 7 bits, that is 64 or 128 tone levels, respectively, without giving up spatial resolution and bringing it closer to thermal dye transfer printing. Bimodal thermal mass transfer printers are at the bottom of the chart and, at 300 dpi, fall between areas D and C.

12.7.1 Thermal Dye Transfer Donor Ribbons

Donor ribbon and receiver for thermal dye transfer printing constitute a closely matched pair for optimum print quality. A ribbon/receiver configuration, depicted in Fig. 12.21, shows that the donor ribbon and receiver are multilayer coated structures. The donor rib-

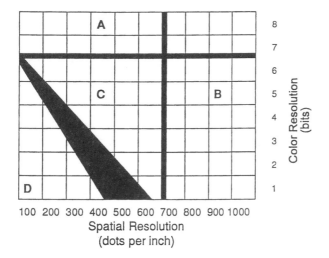

Figure 12.20 Spatial versus color resolution. The regions A and B are adequate in either spatial or color resolution to give the illusion of photographic realism. Area C is marginal but acceptable for presentations and spot color graphics. Area D is unacceptable. The wide boundary between areas C and D reflects differences in human perception. (From Giga Information Group, 1995. Contact Sara Martin, Giga Information Group, One Longwater Circle, Norwell, MA 02061, 1-800-874-9980 or 617-982-9500, fax 617-878-6650.)

bon is based on a biaxially oriented polyester film, typically 3.5–6 microns thick. On the side nearest the thermal head is a cross-linked slip coat, which assures that the ribbon does not stick and moves freely under the head during printing, even though the temperature at the surface of the head rises to 350°C or higher. As the base film is molten at this temperature, it cannot be relied upon to provide structural integrity, so that the slip coat must provide mechanical support during this part of the printing cycle (Sarkar et al., 1991; Hann et al., 1993; Hann, 1994; Tunney et al., 1997). On the other side of the base film is a priming coat, also called a subcoat or a barrier layer, to provide bonding between base film and dye layer deposited on top of it. Another requirement of the subcoat is low affinity/impermeability for the dye to provide a barrier for backward migration during printing (Henzel, 1991).

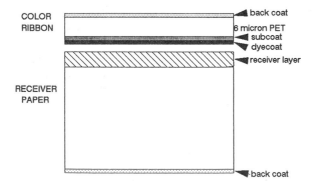

Figure 12.21 Thermal dye transfer donor/receiver schematic. (From Hann, 1994. Reprinted with permission of IS&T: The Society for Imaging Science and Technology, sole copyright owners.)

The active dye coat itself consists of a solid solution of dye in a binder polymer, such as cellulose derivatives (Vanier et al., 1987). The dye binder polymer acts as a solvent for the dye (no recrystallization) and must be heat stable, must not stick to the receiver, and must exhibit thermoplasticity. The dye and binder polymer are carefully designed mixtures in order to obtain the desired color, transfer, and stability properties (Katayama et al., 1991; Clifton et al., 1997).

The distinction between sublimation and diffusion is not just academic; it has an impact on the range of dyes that are suitable for thermal dye transfer printing. The early dyes were chosen from a small number of commercially available sublimable textile transfer dyes. The discovery that the process was primarily one of dye diffusion qualified a wider array of dyes as potential candidates for thermal dye transfer printing (Hann et al., 1991). However, in order to fulfill all the requirements, special dyes had to be designed (Bradbury, 1992; Etzbach et al., 1994; Vanmaele, 1994, 1995). Hundreds of patents have been granted describing thermal dye transfer dyes, but commercially used dyes have not been disclosed. A few suitable dyes are shown in Fig. 12.22. Necessary performance requirements are bright color (narrow absorption band), strong color (high extinction coefficient), and the capability of being easily transferred (Tomita et al., 1995), permanently fixed, light fast, and nontoxic. Some of these are at first sight mutually exclusive. It seems unreasonable to expect a dye to be easily transferable and permanently fixed, and in practice a compromise has to be made.

Multiuse ribbons have been disclosed by Ricoh (Mochizuki et al., 1989; Uemura et al., 1993) and Matsushita Electric (Taguchi et al., 1990a). The Ricoh ribbon goes by the acronym MUST, for multiuse sublimation transfer. The essential feature is that the dye is contained in a two-layer structure with different dye concentrations and glass transition temperatures. It is claimed that ten times the utilization can be reached without loss of print density. Matsushita's approach is to advance the donor at a slower speed than the receiver. To permit slippage of donor and receiver during printing it is essential that the modulus of elasticity of the donor ribbon does not change appreciably at the printing

Color	Class	Structure	Company
Yellow	Azo pyridone	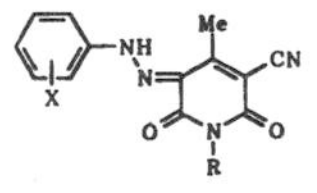	ICI/DNP/Kodak
Magenta	Azo	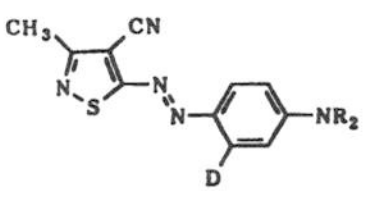	ICI/Kodak
Cyan	Azo	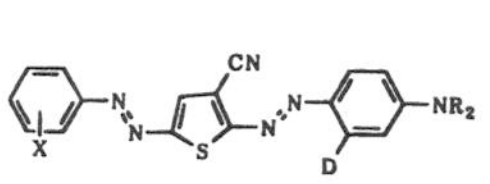	ICI
	Indoaniline	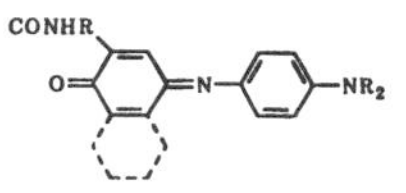	Kodak/Mitsubishi

Figure 12.22 Thermal dye transfer dyes. (From Calder, 1991. Reprinted with permission of IS&T: The Society for Imaging Science and Technology, sole copyright owners.)

temperatures. Aromatic polyamide films proved superior in this respect to PET films. No appreciable loss of print density was noticed at a $1:12$ speed ratio between donor and receiver, so that a utilization factor of ten seems attainable.

12.7.2 Thermal Dye Transfer Receivers

The receiver sheet consists of a supporting substrate with a functional layer coated on one side and a backing layer on the other. A number of substrates may be used. The preferred support material for reflection prints are rather complicated paper film laminates. Microscopic cross sections of two such laminates, disclosed in US patents from Dai Nippon Inatsu K. K. (Yoshikazu et al., 1988) and Kodak (Campbell, 1994, 1994a), are shown in Figs. 12.23 and 12.24, respectively. Both contain a paper core with a multilayer, voided, synthetic paper laminated to the face or printing side. The back side layer may be either a laminated synthetic paper (12.23) or an extruded polyethylene film (12.24). The lower thermal conductivity of these laminates, compared to other substrates, results in increased thermal sensitivity and optical print density as illustrated in Fig. 12.25. The good conformability and compliance of the voided component of the laminate also reduce printing artifacts caused by nonuniformities in the printing nip. An intermediate layer containing microcapsules, placed between substrate and receiving layer, produces similar effects (Ueno et al., 1993). Radiation curing of foamed coatings to produce microporous layers has been described by Mehnert et al. (1997) as an alternative to voided synthetic paper structures.

For transparencies, clear polyester film is a suitable support. As twice as much dye needs to be transferred to obtain a given density in transmission compared to reflection, there are special requirements for software, print drivers, and ribbon design, and double printing is an option. For printing of identification cards, driver's licenses, etc., that require a durable support, white opaque polyester is the material of choice. A recent introduction

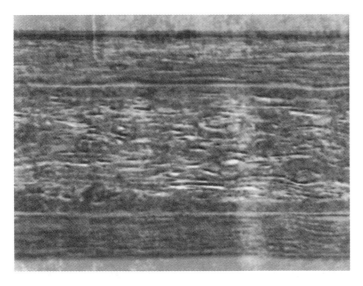

Figure 12.23 Microscopic cross section of thermal dye transfer receiver. From top: Receiver layer; synthetic paper; two-side coated paper; synthetic paper; back coat. Total caliper 192 microns. (Courtesy of Felix Schoeller Technical Papers Inc., Pulaski, New York.)

Figure 12.24 Microscopic cross section of thermal dye transfer receiver. From top: receiving layer; synthetic paper; pigmented polyethylene layer; paper; clear polyethylene layer; back coat. Total caliper 210 microns. (Courtesy of Felix Schoeller Technical Papers Inc., Pulaski, New York.)

from ICI Imagedata, for producing passport photos, has security features built into the receiver print medium.

The receiver layer must be receptive to the dye image (Shinozaki et al., 1993) and provide release from the donor after printing; it is typically made by coating a soluble polymer onto a suitable substrate. For thermal transfer dyes, derived from dyes used for dyeing polyester fibers, polyester resin serves as the receiver layer. Other suitable polymers include polycarbonates, polyurethanes, polyvinylchloride, and various copolymers. Thermal sensitivity and optical print density can be increased by modifying the viscoelasticity

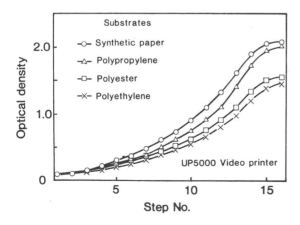

Figure 12.25 Printing characteristics of various thermal dye transfer receiver substrates. Optical density as a function of recording energy. (From Kato et al., 1990. Reprinted with permission of IS&T: The Society for Imaging Science and Technology, sole copyright owners.)

of the receiver resin through cross-linking, polymer blends, or polymer alloys and lightfastness can be affected by polymer-dye interactions (Kato et al., 1990a; Clifton et al., 2000). To prevent blocking or sticking between donor and receiver during printing, a silicone or similar release agent is either incorporated into the receiver layer or applied as an overcoat. The receiver layer can be coated directly on the substrate, or over a sublayer or undercoat. Incorporation of antistatic properties into the subcoat allows the dissipation of static charges, without impairing the release properties of the receiver (Hann et al., 1991a). The reverse side of the receiver has an antislip coating to aid feed through the printer.

The need for a highly specialized receiver is a drawback for wider applications of thermal dye transfer printing. An offset printing method, described by Matsushita, overcomes this problem by first printing the image onto an intermediate receiver layer, followed by heat lamination of the printed intermediate receiver to the final substrate (Fukui et al., 1993; Taguchi et al., 1992). Related processes, printing on an intermediate transfer recording medium followed by heat lamination to the final receiver, have been described (Oshima, 1998; Shiral, 1999), A heat seal and protective layer are part of the transfer assembly. These processes are particularly suited for transfer to polymer films and cards.

12.7.3   Thermal Dye Transfer Image Stability

Thermal dye transfer images, while almost indistinguishable from photographic prints, unfortunately fall short of the mark in the area of image stability. Compared to conventional photography, thermal dye transfer prints have inferior light stability and resistance to damage from fingerprints. They are aggressively attacked by plasticized polyvinylchloride sheets and easily retransfer to folders commonly found in office and home environments (Newmiller, 1991). With the dye transferred and absorbed into the receiver layer in a very short period of time during printing, it is not unexpected that retransfer can occur to another dye receptive material.

Reactive donor/ribbon combinations, in which components in the receiver layer react with the transferred dye to form a nondiffusing compound, have been described by Konica and Sony. Konica employs azo compounds capable of forming metallized dyes with a metal ion source in the receiver (Miura et al., 1993). Sony disclosed cationic dyes that form ionic bonds with a clay mineral incorporated in the receiver layer (Ito et al., 1993).

Kodak (Harrison et al., 1994) and Dai Nippon Printing (Oshima et al., 1995; Saito, 1999; Egashira, 2000) have applied a protective layer over the final print, which is integrated into the color ribbon as a fourth patch and laminated with the thermal head. ICI Imagedata has incorporated a protective layer and security features into a passport photo printer. Retransfer and resistance to fingerprints was greatly improved, but light stability improvement depends on the dye systems used (Bradbury, 2000). The addition of a protective layer moves the stability of thermal dye transfer prints much closer to conventional photographic prints.

12.7.4   Future Outlook for Thermal Dye Transfer Printing

The Mavica film camera and Mavigraph color printer were originally introduced as an alternative to conventional photography. Image quality of the early systems, however, with input from still video cameras tied to the resolution of commercial television signals, fell short of photographic prints. Since then, systems have been optimized; still video cameras were replaced by digital cameras; camera and printer prices have come down drastically; and printing speed has increased (Egashira et al. 1992). In the meantime, ink jet desktop printers can now produce photographic quality prints.

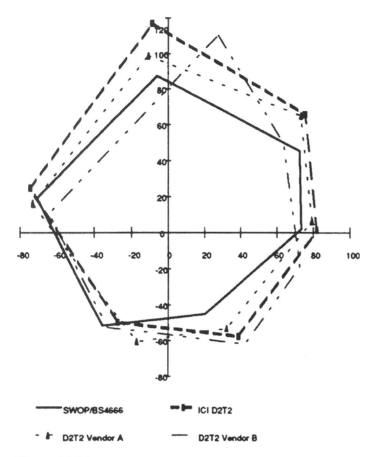

Figure 12.26 a*b* plane of maximum chroma at primary and secondary colors for thermal dye transfer proofs and BS4666/SWOP proofs and inks. (From Hann, 1994. Reprinted with permission of IS&T: The Society for Imaging Science and Technology, sole copyright owners.)

In the medical market, the introduction of new diagnostic tools and the expanded use of workstations have increased the need for a high quality, color output system. Environmental concerns and the need to simplify and expedite medical examinations have created opportunities for thermal dye transfer printing (Defieuw et al., 1995). Graphic arts are expanding their use of thermal dye transfer printers. The color gamut of thermal dye transfer proofers exceeds the SWOP standard, with only a part of the a*b* plane being inaccessible (Fig. 12.26).

Special printers, capable of printing text and images directly onto plastic cards, have established a use in security applications. The undesirable trait of the dye to retransfer is used in the production of novelty items through a second transfer of the printed image to another receptive coating applied to a ceramic tile, mug, or T-shirt.

12.8 LASER THERMAL PRINTING

Laser thermal printing is not new, dating back over three decades. The earliest report of a thermal imaging process is from an RCA worker in 1970. Then, a typewriter ribbon was used a color donor, with paper as the receiver. Focusing a laser beam through the base of an inked ribbon caused the illuminated ink to transfer (Levene, 1970). The fundamentals of the process have not changed much since. Meyers (1971) reported a laser writing process in which spots of metal one-third to two-thirds smaller than the nominal diameter of the laser beam were visibly altered by coalescence. In 1977 IBM workers reported laser thermal dye transfer images using off-the-shelf dyes such as crystal violet and methyl green coated in a nitrocellulose binder. They showed the future by using a computer to control the position and modulation of the laser beam (Bruce, 1977).

The first commercial product using a laser thermal process was the Crosfield Laser Mask, described at an SPIE meeting in 1980 (Gibbs, 1980; DeBoer, 1998). The product consisted of a polymer film base coated with finely dispersed graphite particles in an oxidizing binder such as nitrocellulose. It was exposed through the base with a YAG laser system that ablates the carbon layer transferring it imagewise to a paper receiver in contact. The result was a negative film and a positive proof on the paper. The negative could then be used to burn a lithographic printing plate. In the same year a UK patent application was filed by 3M (Baldock, 1980), describing a method of making direct laser printed images on a rotating drum with a YAG laser.

When high-powered diode lasers became available in the late 1980s, interest in this field was renewed. With these new lasers installed, maintenance-free laser thermal writing systems became a reality.

The field of photothermal and photoacoustic phenomena is an active one. The Gordon Research Conference is devoted to this topic. The field is mostly centered on relatively high-powered effects, where the physics of the material is nearly that of plasma. The regime occupied by graphic arts materials and processes is of considerably lower power, near the borderline between sublimation and ablation.

Compared to printing with a thermal head, the laser thermal process is a powerful approach to imaging. Both sublimation and ablation can be present in any given event. In some discussions it is useful to distinguish between the two processes, but for most events involving the transfer or movements of materials, DeBoer (1998) suggests that it is simpler to lump them together and call it "laser thermal ablation." This seems reasonable in light of the confusion that arose between sublimation and diffusion in the case of thermal dye transfer printing. As the actual physical and chemical mechanisms are better understood, processes can be more precisely defined (Kinoshita et al.,1998, 2000, 2000a; Koulikov et al., 2000, 2000a; Timpe, 1999). For our purpose, the general term laser thermal printing will be used.

Laser thermal systems are intrinsically capable of higher resolution than resistance head systems. Since the laser can be focused to form a very fine, diffraction-limited spot, this process can produce high quality images. Images with 8 bits of information at 1800 to 4000 dpi have been routinely generated (Patton, 1995) and higher resolutions have been demonstrated (Odai et al., 1996). In the early 1990s two laser thermal printing systems were introduced: Kodak's Approval™ Digital Color Proofing System and Polaroid's Helios Laser System. The printing mechanism of both systems are described in detail.

12.8.1 Laser Thermal Dye Transfer Printing

Kodak's Approval™ Digital Color Proofer is an example of laser thermal dye transfer printing. A schematic of the process is depicted in Fig. 12.27. The beam of a diode laser is focused onto a dye donor sheet, containing a color dye and an infrared dye in a binder of cellulose acetate, coated on a polyester support. The donor is overcoated with a layer

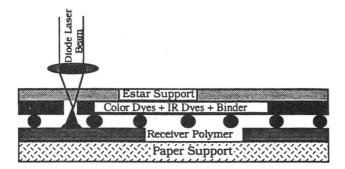

Figure 12.27 Laser transfer printing, dye donor/receiver schematic. (From Sarraf et al., 1993. Reprinted with permission of IS&T: The Society for Imaging Science and Technology, sole copyright owners.)

of polystyrene beads, a few microns in diameter, which are held in place by a small amount of binder (PVA emulsion). Both the size and the amount of the beads are important for optimum print sharpness and to prevent the donor from sticking to the receiver. The receiver is coated with a polymer that is a solvent for the dye, e.g., polycarbonate.

The transfer of the dye is a combination of ablation and sublimation mechanisms. Transfer efficiency was found to be a function of air pressure in the printing gap between donor and receiver. Lower than ambient pressure favors higher transfer efficiency, but below a certain level the gap collapses and loss of efficiency occurs.

Unlike thermal dye transfer printing, which is a continuous tone process, laser thermal dye transfer printing is capable of making continuous tone, halftone, and binary images. Images can be produced with a variety of equipment configurations. For high throughput of large format images, use of high-power lasers in a drum type system is favored. The Kodak Approval™ is based on this configuration. Throughput is met by using multiple lasers coupled to optical fibers. Fiber output is managed to provide up to 70 dot/mm (1800 dpi) halftone resolution using a high-power laser in each channel. The system is highly flexible. Screen rulings and screen angles can be specified, five different dot shapes chosen, and dot gain adjusted. The image is built up pixel by pixel through the successive transfer of the three primary subtractive colors and black to a receiver. The Approval™ system employs an intermediate receiver from which the four-color image is transferred to the actual printing stock (Pearce, 1994).

Comparing the energy efficiency of the laser dye transfer process with resistive head and resistive ribbon printing, one would expect the most efficient system to be the one where the heat is most localized in the imaging layer, i.e., the laser transfer system. This was confirmed by DeBoer (1991), who found that it takes about 300 mJ/cm^2 to print a density of 1.0 with laser dye transfer, which agrees well with earlier measurements by IBM workers. In comparison, energies necessary to print a 1.0 density with a 6 micron thick donor film were reported to be 4 J/cm^2, and for resistive ribbon printing 2 J/cm^2 were required.

12.8.2 Laser Phase Transition Printing

The image forming mechanism of the Polaroid Helios Laser System film is based on phase transition in a polymer matrix. The imaging layers, consisting of a laser sensitive layer adjoined by an imaging layer, are sandwiched between two transparent polyester substrates

(Chang, 1992). The imaging layer consists of carbon particles in a polymer matrix. The particles have a narrow size distribution that runs between 50 and 100 Å.

In the nonimaged state the sandwich can be separated between the laser sensitive and imaging layer. Imaging occurs when laser energy, produced by high-power gallium arsenide laser diodes, is focused onto the film. Absorbed energy causes a phase transition between the imaging layer and the laser sensitive layer, leading to a strong adhesion between the two layers, only at the site where the laser beam is absorbed. After the spots are formed, the film is "developed" by mechanically separating the sandwich, leaving the written image on one substrate and the negative on the other (Cargill et al., 1992; Habbal et al., 1994).

The Helios Laser System is a high-resolution imaging platform that lends itself to a number of applications (Miller et al., 1994). The film has a single threshold activation energy and writes spots or pixel elements with extremely high definition, as small as a few square microns. The system has been targeted at medical diagnostic applications. The full dynamic range of densities available are between a fixed maximum of 3.5 and a minimum optical density of 0.05. Since the number of available grey levels is over 12 bits, tone scales can be optimized to render the image content of any diagnostic modality.

12.8.3 Future Outlook for Laser Thermal Printing

The future of laser thermal printing seems assured. New fiber lasers offer promise of very high power at reasonable cost. The inherent advantages could make it the printing technology of choice for a variety of applications (Sarraf et al., 1993; Landsman, 1999). Future media for the graphic arts industry will likely be two areas: direct digital color proofing and computer-to-plate (CTP). Direct color proofing will be important for pre-press operations because CTP eliminates the film image, without which conventional proofs cannot be made.

Several direct digital color proofing methods have been demonstrated. One is Polaroid's SUNSPOT, a single-sheet process designed to produce a full-color, high-resolution, transparent image using a set of three diode lasers, each emitting at a different wavelength between 750 and 950 nm (Marshall, 1998). Polaroid has demonstrated a laser ablation color proofing system that can also be used to prepare printing plates from uncoated, grained, anodized aluminum. Konica has described both a laser thermal proofing system and a graphic arts film (DeBoer, 1997).

The Kodak DIRECT IMAGE recording film is a graphic arts film that can be handled under room light and does not require post-image processing (Neumann, 1997). Direct recording films occupy a place with the other direct imaging technologies such as direct digital color proofing, computer-to-plate, computer-to-press, and computer-to-print.

Accelerating patent activities have recently involved laser thermal imaging of lithographic printing plates. Huang et al. (1998) summarized these developments in a paper at IS&T's NIP 14. Kodak introduced the Professional Direct Image Thermal Printing Plate. In this product, the laser thermal heat generates an acid that changes the solubility of the ink receptive polymer layer in an alkaline developer. Recent advances in computer-to-plate technologies have triggered a series of research and development activities in the computer-to-waterless-plate area (CTWP) (Huang, 1999). Presstek is marketing a direct write laser thermal printing plate for the driographic press that uses no fountain solution. Many of the mechanistic details have been described by Hare et al. (1997, 1997a). With the current momentum of research, it is likely that laser thermal printing will become the primary technology for making direct lithographic plates and will give rise to other direct digital imaging systems.

ACKNOWLEDGMENT

I would like to thank Frederick A. Poole of Felix Schoeller Technical Papers Inc. for preparing the microscopic cross sections.

REFERENCES

Akutsu, E., et al. Analysis of polymer ink transfer phenomenon in thermal transfer printing technology. Proc. IS&T's Eighth International Congress on Advances in Non-Impact Printing Technologies, pp. 372–375 (1992).

Anczurowski, E., et al. Materials for thermal transfer printing. *J. Imaging Tech. 13*:97–102 (1987).

Arbree, R. R., et al. U.S. Patent 4,591,887 (1986).

Baldock, T. W., et al. GB 2,083,726A (1980).

Bradbury, R. Disazothiophene dyes in dye diffusion thermal transfer printing. Proc., IS&Ts Eighth International Congress on Advances in Non-Impact Printing Technologies, pp. 364–366 (1992).

Bradbury, R. Image stability of dye diffusion thermal transfer images, IS&T's NIP 16: Proc. Int'l. Conf. on Digital Printing Technologies, pp. 771–774 (2000).

Brandt, T. J., et al. U.S. Patents 5,512,930; 5,552,819 (1996).

Bruce, C. A., et al. *J. App. Photo. Eng. 3*:40–43 (1977).

Burlingame, R. Benjamin Franklin, envoy extraordinary. Coward-McCann, New York, p. 39 (1967).

Calder, A. Dyes in non-impact printing. Proc. Vol. I, IS&T's Seventh International Congress on Advances in Non-Impact Printing Technologies, pp. 3–24 (1991).

Campbell, B. C. Transfer efficiency in thermal dye transfer receivers. Proc., IS&T's 47th Annual Conference, pp. 814–818 (1994).

Campbell, B. C., et al. U.S. Patent 5,350,733 (1994a).

Cargill, E. B., et al. A report on the image quality characteristics of the polaroid helios laser system. Polaroid White Paper (1992).

Chang, K. C. U.S. Patent 5,155,003 (1992).

Chang, J. C. Evolution of thermal transfer imaging. Proc., IS&T's Eleventh International Congress on Advances in Non-Impact Printing Technologies, pp. 293–297 (1995).

Clifton, A. A., et al. Polymers for electronic imaging: The control of dye transport via dye-polymer interactions. Proc., IS&T's 50th Annual Conference, pp. 299–304 (1997).

Clifton, A. A., et al. Lightfastness performance of digital imaging: The role of dye-dye and dye-polymer interactions. IS&T's NIP 16: Proc. Int'l Conf. on Digital Printing Technologies, pp. 762–764, (2000).

Crooks, W., et al. Basic fundamentals of thermal transfer printing. Proc., International Congress of Photographic Science, Cologne, Germany, pp. 619–627 (1986).

Crooks, W., et al. An ink transfer mechanism and material considerations for a resistive ribbon high quality thermal-transfer printing process. *Proc., SPIE 1079*:317–328 (1989).

DeBoer, C. Digital imaging by laser induced transfer of volatile dyes. Proc., IS&T's Seventh International Congress on Advances in Non-Impact Printing Technologies, pp. 449–452 (1991).

DeBoer, C. High quality dry laser thermal printing technology. Proc., IS&T's 50th Annual Conference, p. 289 (1997).

DeBoer, C. Laser thermal media: The new graphic arts paradigm. *J. Imaging Science and Technology, 42*, pp. 63–69 (1998).

Defieuw, G., et al. DRYSTAR™—Advanced dye transfer printing for diagnostic medical imaging. Proc., IS&T's Eleventh International Congress on Advances in Non-Impact Printing Technologies, pp. 305–310 (1995).

Degerstrom, R. Breakthrough in color imaging: variable dot size thermal transfer. Fourth Annual Color Imaging Conference & Color Marathon, Diamond Research Corporation, pp. 1–15 (1990).

Diamond, A. S. Specialty papers for thermal imaging. Proceedings of White Papers & Office Automation Conference, Cambridge, MA (1989).

Egashira, N., et al. Heat transfer and printing characteristics in dye transfer printing. Proc., IS&T's Eighth International Congress on Advances in Non-Impact Printing Technologies, pp. 352–355 (1992).

Egashira, N. Durability of dye in thermal dye transfer printing. IS&T's NIP 16: Proc. Int'l Conf. on Digital Printing Technologies, pp. 759–761 (2000).

Etzbach, K. H., et al. Thermal dye transfer printing, the cyan challenge. Proc., IS&T's Tenth International Congress on Advances in Non-Impact Printing Technologies, pp. 334–336 (1994).

Fukui, Y., et al. Thermal offset printing employing dye-transfer type ink sheet (TOP-D). Proc., IST&T's Ninth International Congress on Advances in Non-Impact Printing Technologies, pp. 389–392 (1993).

Gibbs, J. H. Laser scanning and recording for graphic arts and publications. *Proc. SPIE 223*, (1980).

Giga Information Group. Report: Thermal Printing in the 1990s: Overview and Outlook, 1993 Edition.

Giga Information Group. Report: Thermal Printing 1995: New Products and Applications.

Giga Information Group. The 1998 European Electronic Printer Report Series. Vol. 2: Thermal Printing (1998).

Gold, R. Thermography—state-of-the-art review. SPSE Symposium on Unconventional Photographic Systems, pp. 1–52 (1964).

Goodwin, T. E. Direct thermal media. Advance Paper Summaries, IS&T's 46th Annual Conference, pp. 374–377 (1993).

Habbal, F., et al. Helios: a new hardcopy imaging platform. Proc. IS&T's Tenth International Congress on Advances in Non-Impact Printing Technologies, pp. 317–319 (1994).

Hann, R. A. Thermal dye transfer printing (D2T2)—the last seven years. Proc., IS&T's Tenth International Congress on Advances in Non-Impact Printing Technologies, pp. 343–345 (1994).

Hann, R. A., et al. Dye diffusion thermal transfer (D2T2) color printing. *J. Imaging Tech. 16*:238–241 (1990).

Hann, R. A., et al. Dye diffusion thermal transfer printing (D2T2)—dependence of print performance on dye structure. Proc., Vol. II, IS&T's Seventh International Congress on Advances in Non-Impact Printing Technologies, pp. 237–246 (1991).

Hann, R. A., et al. Use of a conductive sub-layer to improve antistatic properties of a D2T2 receiver sheet. Proc., Vol. II, IS&T's Seventh International Congress on Advances in Non-Impact Printing Technologies, pp. 386–389 (1991a).

Hann, R. A., et al. Control of D2T2 print quality by back coat friction properties. Proc., IS&T's Ninth International Congress on Advances in Non-Impact Printing Technologies, pp. 322–325 (1993).

Hann, R. A., et al. Design of the ribbon back coat for thermal dye transfer (D2T2) printing. Proc., IST's Tenth International Congress on Advances in Non-Impact Printing Technologies, pp. 368–370 (1994).

Hara, T. The layer structure and the surface roughness of thermal paper for pencil-writability and erasability. Proc., Vol. II., IS&T's Seventh International Congress on Advances in Non-Impact Printing Technologies, pp. 192–198 (1991).

Hare, D. E., et al. *J. Imaging Sci. Technol., 41*, p. 291.

Hare, D. E., et al. Pulse duration dependence for laser photothermal imaging media. Proc. IS&T's 50th Ann. Conf., pp. 290–295 (1997a).

Harriman, B. Thermography. 3M Company (1978).

Harrison, D. J., et al. Image stability advances in thermal dye transfer imaging. Proc., IS&T's Tenth International Congress on Advances in Non-Impact Printing Technologies, pp. 346–348 (1994).

Henzel, R. P. U.S. Patent 5,023,228 (1991).

Higaki, T. Trends of key materials for direct thermal papers. Hard Copy and Printing Materials, Media and Process. *Proc. SPIE 1253*:280–289 (1990).

Hodge, I. M., et al. Mass diffusion in resistive head thermal printing. Proc., Vol. II, IS&T's Seventh International Congress on Advances in Non-Impact Printing Technologies, pp. 226–231 (1991).

Honda, S. U.S. Patent 4,558,329 (1985).

Hori, Y., et al. Running characteristics of thermal transfer film and their effect on printing quality. Proc., IS&T's Eighth International Congress on Advances in Non-Impact Printing Technologies, pp. 356–360 (1992).

Huang, J. Technology overview on computer-to-waterless plates (CTWP). Proc. IS&T's NIP 15: Int'l Conf. on Digital Printing Technologies, pp. 213–216 (1999).

Huang, J., et al. Thermal imaging: Application in offset printing plate making. Proc. IS&T's NIP 14: Int'l. Conf. on Digital Printing Technologies, pp. 190–193 (1998).

Ichii, M., et al. U.S. Patent 4,849,457 (1998).

Igarashi, A., et al. The development of direct thermal full color recording material. Proc., IS&T's Tenth International Congress on Advances in Non-Impact Printing Technologies, pp. 323–326 (1994).

Inui, F., et al. Drive pulse control and electric circuit design for halftone color thermal printer "Louver." Proc., IS&T's Ninth International Congress on Advances in Non-Impact Printing Technologies, pp. 338–341 (1993).

Ito, K., et al. A novel method for fixing image of thermal dye transfer. Proc., IS&T's Ninth International Congress on Advances in Non-Impact Printing Technologies, pp. 306–309 (1993).

Jacobsen, K. I., et al. *Imaging Systems*. John Wiley, New York, pp. 136–138 (1976).

Joyce, R. D., Homa, S., Jr. High speed thermal transfer printer. AFIPS Conference Proc. of Fall Joint Computer Conference, pp. 261–267 (1967).

Katayama, S., et al. Sublimation transfer media. NITTO Technical Reports, pp. 112–122 (Jan. 1991).

Kato, M., et al. The role of receiving sheet substrates in thermal dye transfer printing. *J. Imaging Tech. 16*:242–244 (1990).

Kato, M., et al. Experimental investigation of dye transfer and diffusion thermal dye transfer receiver sheet. Proc., IS&T's Sixth International Congress on Advances in Non-Impact Printing Technologies, pp. 601–608 (1990a).

Katsen, B. J., et al. The fundamentals of thermal transfer process from the point of view of media and printer design as a system. Proc., IS&T's Eleventh International Congress on Advances in Non-Impact Printing Technologies, pp. 411–414 (1995).

Kinoshita, M., et al. Light-heat conversion material for dye thermal transfer by laser heating. IS&T's NIP 14: Proc. Int'l Conf. on Digital Printing Technologies, pp. 273–276 (1998).

Kinoshita, M., et al. Mechanism of dye thermal transfer from ink donor layer to receiving sheet by laser heating. *J. of Imaging Science and Tech.* 44, pp. 105–110 (2000).

Kinoshita, M., et al. Time resolved microscopic analysis of ink layer surface in laser dye thermal transfer printing. *J. of Imaging Science and Tech., 44*, 484–490 (2000a).

Koulikov, S. G., et al. Focus fluctuation in laser photothermal imaging. *J. of Imaging Science and Technology, 44*, pp. 1–12 (2000).

Koulikov, S. G., et al. Effects of energetic polymers on laser photothermal imaging materials. *J. of Imaging Science and Technology, 44*, pp. 111–119 (2000a).

Koyano, T., et al. Dual color direct thermal recording using diphenylether diazonium salt. Proc. IS&T's NIP 13: Int'l. Conf. on Digital Printing Technologies, pp. 750–755 (1997).

Kulkarni, S. K., et al. U.S. Patent 5,411,787 (1995).

Kutami, M., et al. A new thermal transfer ink sheet for continuous tone full color printer. *J. Imaging Tech. 16*:70–74 (1990).

Landsman, R. M. CTP productivity and recording design. Proc., IS&T's NIP 15, Int'l. Conf. on Digital Printing Technologies, pp. 204–208(1999).

Lane, R., et al. Forty dot/millimeter resistive ribbon thermal printhead. *J. Imaging Science Tech. 36*:93–98 (1992).

Levene, M. L., et al. *Appl. Opt., 9*, p. 2260 (1970).

Lewis, S. D., et al. EPA 0 512 696 (1992).

Maeda, S., et al. Development of paper having microporous layer for digital printing. *J. of Imaging Science and Tech., 44*, pp. 410–417 (2000).

Maehashi, T., et al. Thermal transfer printing process of multi-printing materials. Proc., Sixth International Congress on Advances in Non-Impact Printing Technologies, pp. 656–665 (1990).

Mandoh, R., et al. Improvement on stability to discoloration of direct thermal papers. Proc., IS&T's Ninth International Congress on Advances in Non-Impact Printing pp. 427–430 (1993).

Mehnert, R., et al. Radiation-cured layers for dye diffusion thermal transfer. Proc. IS&T's NIP 13: Int'l. Conf. on Digital Printing Technologies, pp. 418–419 (1997).

Meyers, W. C. SPSE Symposium III. Unconventional photographic systems (1971).

Miller, C. S. U.S. Patents 2,740,895; 2,740,896 (1956).

Miller, R. W., et al. Helios imaging applications. Proc., IS&T's Tenth International Congress on Advances in Non-Impact Printing Technologies, pp. 298–300 (1994).

Miura, N., et al. New materials for thermal transfer printing. Proc., IS&T's Ninth International Congress on Advances in Non-Impact Printing Technologies, pp. 314–317 (1993).

Miyamoto, A. Recent advances in direct thermal recording. Proc., Vol. II, IS&T's Seventh International Congress on Advances in Non-Impact Printing Technologies, pp. 185–191 (1991).

Miyawaki, K., et al. High speed and high quality printing by sublimation-type resistive sheet thermal transfer method. Proc., Sixth International Congress on Advances in Non-Impact Printing, pp. 634–643 (1990).

Mochizuki H., et al. Multi-use sublimation transfer (MUST) sheet. Proc., SPSE's Fifth International Congress on Advances in Non-Impact Printing Technologies, pp. 491–498 (1989).

Motosugi, T., et al. U.S. Patent 5,102,693 (1992).

Nakamura, H., et al. High quality halftone thermal imaging technology by thin-layer thermal transfer (3T) technology used for FIRST PROOF™. Proc. IS&T's NIP 13: Int'l. Conf. on Digital Printing Technologies, pp. 769–772 (1997).

Namiki, K., Thermal printhead technology and application. Proc., IS&T's NIP 13: Int'l. Conf. on Digital Printing Technologies, pp. 760–763 (1997).

Neumann, S. M. Laser dye removal technology for digital, dry imagesetting film. Proc., IS&T's 50th Annual Conference, pp. 296–297 (1997).

Newmiller, C. Thermal dye transfer color hard copy image stability. Printing Technologies for Images, Gray Scale and Color, SPIE 1458, p. 93 (1991).

OEP. Thermal print head prints well on rough paper, p. 36 (August, 1992).

Ohga, K., et al. The application of transparent thermal film for medical diagnosis image recording. Proc., Vol. II, IS&T's Seventh International Congress on Advances in Non-Impact Printing, pp. 255–259 (1991).

Oshima, K., et al. Additional function of dye image with protective layer. Proc., IS&T's Eleventh International Congress on Advances in Non-Impact Printing Technologies, pp. 408–410 (1995).Oshima, K. New thermal dye transfer recording method by using an intermediate transfer recording medium. Proc., IS&T's NIP 14: Int'l. Conf. on Digital Printing Technologies, pp. 187–189 (1998).

Oshima, K., et al. Additional function of dye image with protective layer. Proc. IS&T's Eleventh International Congress on Advances in Non-Impact Printing Technologies, pp. 408–410 (1995).

Patton, E. V. Advances in thermal printing. Proc., IS&T's Eleventh International Congress on Advances in Non-Impact Printing Technologies, pp. 311, 312 (1995).

Pearce, G. T. U.S. Patent 5,342,821 (1994).

Saito, H., et al. A half-tone color thermal printer "Louver" using wax transfer thermal printing method. Proc., IS&T's Ninth International Congress on Advances in Non-Impact Printing Technologies, pp. 334–337 (1993).

Saito, H. New thermal dye transfer media for digital photo usage. Proc., IS&T's NIP 15: Int'l. Conf. on Digital Printing Technologies, pp. 251–254 (1999).

Sakai, K., et al. Image stability of TA paper. Proc., IS&T's NIP 15: Int'l. Conf. on Digital Printing Technologies, pp. 235–238 (1999).

Sarkar, M., et al. U.S. Patent 5,001,012 (1991).

Sarraf, S., et al. Laser thermal printing. Proc., IS&T's Ninth International Congress on Advances in Non-Impact Printing Technologies, pp. 358–361 (1993).

Sato, M., et al. UV fixable thermal recording paper. *J. Imaging Tech.* *11*:137–142 (1985).

Sato, M., et al. "TA" (Thermo-Autochrome)—a continuous tone full color hardcopy system based on direct thermal. Proc., IS&T's Tenth International Congress on Advances in Non-Impact Printing Technologies, pp. 326–329 (1994).

Sawano, M., et al. Improved color consistency in halftone image by VR screen techonology used for First Proof™. Proc. IS&T's NIP 13: Int'l. Conf. on Digital Printing Technologies, pp. 779–783 (1997).

Schroeder, H. U. U.S. Patent 3,161,457 (1964).

Seto, T., et al. Effect of pigment dispersion on non-impact printing image. Proc., SPSE's Fourth International Congress on Advances in Non-Impact Printing Technologies, pp. 342–345 (1988).

Shinozaki, K., et al. Dye polymer affinity and its effect on thermal dye transfer printing. Proc., IS&T's Ninth International Congress on Advances in Non-Impact Printing Technologies, pp. 310–313 (1993).

Shirai, K. New thermal dye transfer printing applications by using an intermediate transfer printing method. Proc., IS&T's NIP 15: Int'l. Conf. on Digital Printing Technologies, pp. 255–257 (1999).

Sodeyama, H. Thermal paper and its applications. Proc. of FINA, The International Federation of Manufacturers and Converters of Adhesives and Heatseals on Paper and Other Base Materials (1986).

Sonoda, Y., et al. The effect of pixel layout on the quality of halftone images produced by "Louver". Proc., IS&T's Ninth International Congress on Advances in Non-Impact Printing Technologies, pp. 342–345 (1993).

Spivak, S. P., et al. Method for evaluation of thermal transfer receivers. Proc., IS&T's 45th Annual Conference, pp. 40–42 (1992).

Srivastava, O. P. Thermal printhead technology. Proc., IST&T's Fifth International Congress on Non-Impact Printing Technologies, pp. 723–726 (1989).

Swift, P. F., et al. Sensitometric characterization of thermal transfer ribbons. Proc., IS&T's 45th Annual Conference, pp. 35–37 (1992).

Taguchi, N., et al. Dye resistive sheet printing with high sensitivity printing speed and utilization of sheet. Proc., Fourth International Congress on Advances in Non-Impact Printing Technologies, pp. 338–341 (1988).

Taguchi, N., et al. High speed dye transfer printing with high dye utility sheet. SID '90 Digest, pp. 288–290 (1990).

Taguchi, N., et al, Image stability and multiusable printing characteristics of dye transfer sheet. *J. Imaging Tech.* *16*:33–37 (1990a).

Taguchi, N., et al. A high-speed dye-transfer printing process applicable to rough paper. *J. Imaging Science Tech.* *36*(2):171 (1992).

Takahashi, Y., et al. Novel developer for direct thermal paper. Proc., IS&T's Tenth International Congress on Advances in Non-Impact Printing Technologies, pp. 349–351 (1994).

Talvalkar, S. G. U.S. Patent 3,445,261 (1969).

Talvalkar, S. G. U.S. Patent 5,106,669 (1992).

Tanaka, H. High quality color printing by heat fusible penetrative method. Proc., IS&T's Ninth International Congress on Advances in Non-Impact Printing Technologies, pp. 326–329 (1993).

Terao, H., et al. Study of thermal print head for multi-level tone printing. Proc., IS&T's NIP 16: Int'l. Conf. on Digital Printing Technologies, pp. 775–778 (2000).

Timpe, H. Mechanical aspects of thermal plates. Proc., IS&T's NIP 15: Int'l. Conf. on Digital Printing Technologies, pp. 209–212 (1999).

Tomita, H., et al. The glass-transition temperature of dye molecules and its relation to printing density in D2T2. Proc., IS&T's Eleventh International Congress on Advances in Non-Impact Printing Technologies, pp. 323–326 (1995).

Townsend, A. Essential tools to aid in ribbon development for new thermal transfer printing applications. Proc., IS&T's 49th Annual Conference, pp. 371–377 (1996).

Tsumura, M., et al. Half tone wax transfer using a novel thermal head. Proc., SPSE's Fourth International Congress on Advances in Non-Impact Printing Technologies, pp. 273–276 (1988).

Tunney, S., et al. Polymide-polysiloxane copolymers as slipping layers for thermal dye donor. Proc., IS&T's NIP 13: Int'l. Conf. on Digital Printing Technologies, p. 764 (1997).

Uemura, H., et al. Image uniformity of multi-use sublimation transfer printing. Proceedings, IS&T's Ninth International Congress on Non-Impact Printing Technologies, pp. 385–388 (1993).

Ueno, T., et al. The effect of a cushion layer in dye-sublimation receiver materials. Proc., IS&T's Ninth International Congress on Advances in Non-Impact Printing Technologies, pp. 303–305 (1993).

Usami, T., et al. Transparent thermal film. Proc., IS&T's Fifth International Congress on Advances in Non-Impact Printing Technologies, pp. 485–490 (1989).

Usami, T. New frontiers in direct thermal media. Thermal Printing Conference Amsterdam, BIS CAP International (1990).

Vanier, N. R., et al. U.S. Patent 4,700,207 (1987).

Vanmaele, L. J. New dye diffusion thermal transfer dyes derived from malononitrile dimer. Proc., IS&T's Tenth International Congress on Advances in Non-Impact Printing Technologies, pp. 330–333 (1994).

Vanmaele, L. J. New heterocyclic triazene dyes for D2T2 printing. Proc., IS&T's Eleventh International Congress on Advances in Non-Impact Printing Technologies, pp. 298–301 (1995).

Watanabe, K., et al. Recent developments on thermal papers. Color Hardcopy and Graphic Arts II (Jan Bares, ed.). Proc., SPIE 1912, pp. 76–82 (1993).

Yoshikawa, M., et al. Thermal transfer printing with melt transfer type ink (TOP-M). Proc., IS&T's Ninth International Congress on Advances in Non-Impact Printing Technologies, pp. 330–333 (1993).

Yoshikazu, I., et al. U.S. Patent 4,778,782 (1988).

Zawada, R. C. U.S. Patent 5,427,847 (1995).

13

Photothermographic and Thermographic Imaging Materials

P. J. COWDERY-CORVAN and D. R. WHITCOMB

Eastman Kodak Company, Rochester, New York

13.1 INTRODUCTION

Great strides have been made in the development of the technology of thermally and photothermally generated dry imaging materials (TM and PTM) since the last published reviews. This article is intended to be a review update of the most recent articles (1–9), although it is also intended to be sufficiently complete to provide a basic understanding of the current state of knowledge to give a newcomer to the field a good starting point to begin exploitation of its capabilities.

It is a good time for a technology update, not only from the point of view of the significant advances in the published technology but also because of the substantial increase in industrial and commercial interest. The original promise of high resolution in combination with the demonstrated rapid and convenient format has now been realized in high-quality medical x-ray films, such as DryView™ Medical X-Ray film (Kodak), Dry CR™ DI-AL film (Fuji), DryPro™ (Konica), and Drystar™ (Agfa), and newly emerging products for image-setting films, such as DryView Recording Film™ (Kodak) and DX™ Facsimile Film (Fuji). Furthermore, advances made in the film attributes that previously limited acceptance by a broader market (such as film D_{min}, shelf life, light stability, and slow speed) have been overcome by innovative fixes. Aiding in propelling this technology into the next realm of applications have been the recent advances in (including the availability and manufacturability of) high-power diode lasers and high-resolution thermal heads. Customer appeal of dry film confirms the long-term presence of this technology.

In the most general sense, the basic component lists for thermographic and photothermographic formulations differ only by the presence of a light sensitive source (and sensitizing dyes, depending on the application). Therefore the discussions in the sections below tend to intermingle the specifics, keeping the great overlap between the two in mind. Thus even though a comment may be made for one, it is assumed to be evident that it often translates to the other. More specifically, the fundamental list comprises four main components: the light sensitive source (with associated sensitizing dyes), a silver ion source to generate the metallic silver particles of the image, a reducing agent for the silver ion source, and a binder to hold everything together on a substrate (opaque or transparent). In addition to this list, the finer properties of the resulting imaging materials for today's demanding applications are achieved by various other components in the formulation, which include ''toning agents'' (to enhance a black silver image over the natural brown one), stabilizers, antifoggants, supersensitizers, and chemical sensitizers. Finally, material property enhancers are part of the overall construction and include antihalation agents, surface matting agents, and protective topcoats. The advances made in the photographic properties of modern photothermographic imaging constructions by all of these components is summarized below.

The reaction that is fundamental to the image formation in TM, and on which the fundamental basis of all photothermographic imaging constructions are based, is simply the reduction of the silver source to metallic silver:

$$Ag^+ + e^- \text{ (via developer)} + \text{heat} \rightarrow Ag^\circ_{image} \text{ (\~500 \AA particles)}$$

The difference between the thermographic reaction above and photothermographic imaging is the presence of the latent image on silver halide, which enables the above reaction to proceed at lower temperature.

The fundamental practical difference between the photothermographic and thermographic portions of imaging materials, based on silver halide and an alternate silver source, is the temperature at which development occurs. A D vs. log E curve illustrating the point is shown for the photocatalytically induced image and the thermally induced image in the same film, in Fig. 13.1.

It can be seen that the thermally induced onset of image formation occurs at a higher temperature than the photocatalytically induced onset. The thermal separation between the curves provides the development latitude of the film. Ideally, the temperature at which

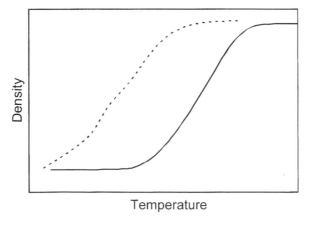

Figure 13.1 Idealized D log E curve for photothermographic, dashed line, and thermographic, solid line, imaging characteristics.

photocatalytically induced D_{max} occurs is well below the temperature of the thermally induced optical density. The AgX-free thermographic film clearly has just the one curve, which is typically adjusted to occur at the lowest temperature practically possible without adversely affecting the range of stability properties required.

A note is necessary regarding the focus on specific primary components in the photo-thermographic imaging construction discussed in this review. Silver bromide, silver carboxylates, hindered phenols, and polyvinylbutyral are, by far, the most important classes utilized for those individual fundamental properties listed above, respectively. As a result, while various silver halides and their mixtures are used as the light harvesting component (2), other silver sources are disclosed in the patent literature, silver benzotriazole (10) and acetylide being the second most cited sources (11–13). As a result, while other silver sources are disclosed in the patient literature, such as silver benzotriazole (10) and silver acetylide (11–13), far less information on these secondary choices is generally available outside of the references noted. The same can be said for polymers, which exhibit the properties needed as binders in these types of constructions (2,14), as well as for other classes of organic developers (2). Therefore this discussion will naturally gravitate to the most important components noted, although references and comments related to these others will be included whenever appropriate.

Finally, while the discussion below utilizes a compartmentalization approach for each component, it should be emphasized that, for the first time, chemical bonding interactions between certain components have been clearly demonstrated (15,16). These interactions mean that each individual component often cannot be regarded as an independent variable even though it may be discussed as such.

The cross section of a typical photothermographic film construction is shown in Fig. 13.2.

In general, the topcoat provides protection for the softer silver image layer beneath, especially during thermal processing and, depending on the T_g of the underlayers, only needs to be a few microns thick. The silver imaging layer contains all the components necessary for the formation of the silver image on thermal development and must be thick enough to provide sufficient D_{max}. This is the primary functional layer and therefore contains the silver source, a latent image capturing material, and developer. This layer also contains various components to maximize the image quality, including antifoggants, ton-

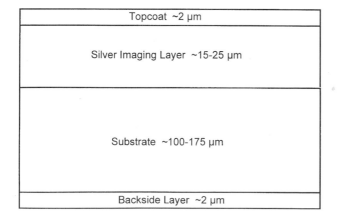

Figure 13.2 Generic cross section of thermally developed imaging materials.

ers, and stabilizers. The substrate layer is typically a clear film base, such as polyethylene terephthalate, whose thickness is dictated by the intended application; but it can be paper as well. The substrate may contain filter or antihalation dyes and may also be coated with a subbing layer to enhance coated layer adhesion. The backcoat is a thin polymer layer that incorporates components needed to improve physical handling such as antistats and matte particles.

A note on what is not covered in this review is also in order. Photothermographic and thermographic process technology are not included. The interested reader may be referred to some of the most significant patent literature on this topic (3,17). In addition, the color-based photothermographic imaging materials reported in the literature have been partially reviewed (5) and will be subjected to a limited update below. Also, thermally developed constructions based on silver halide and color couplers only, without a major portion of the construction providing silver ion from an alternative source, are not included.

With this overview, the following discussion demonstrates the advances made in the state-of-the-art of photothermographic and thermographic imaging materials and illustrates the potential for future applications.

13.2 PROPERTIES OF Ag$^+$ SOURCES USED IN TM AND PTM APPLICATIONS

Photothermographic and thermographic imaging materials utilize a non–silver halide source of silver ions to form the bulk of the black image (Ag°). Although many classes of silver compounds have been reported in the patent literature (18,19) there are only three classes that have found practical use, silver carboxylates (20), silver benzotriazoles (10), and silver acetylides (11–13). By far, silver carboxylates, nominally abbreviated as $Ag(O_2CR)$ but more accurately as $[Ag(O_2C_xH_{2x-1})]_2$ (see below), where $x = 12$–22, have been the source of choice, as can be seen by the number of companies utilizing it (20–25).

Silver carboxylates are often referred to as soaps, because of the long saturated hydrocarbon chain similarity with sodium and potassium soaps, typically C_{16}–C_{22}, although, as will be shown below, this is where the similarity ends. It should be further pointed out that the silver carboxylate used in the commercial formulations can be a mixture of primarily stearic ($HO_2C_{16}H_{31}$) and behenic ($HO_2C_{22}H_{43}$) acids with smaller fractions of palmitic ($HO_2C_{14}H_{27}$) and arachidic ($HO_2C_{20}H_{39}$) [for example, (20)] or pure silver behenate (21). In addition, there are always low levels of unsaturated carboxylic acids in the fatty acid source, which may contribute to printout instability in the nonexposed areas (26). Longer chain silver carboxylates prepared from these unsaturated carboxylic acids have been known for some time to exhibit poor stability (27,28). On the other hand, the structurally characterized silver tiglate (2-methyl-2-propenoate) complex has been found to be quite stable (29), as have the short chain di-carboxylates, silver fumarate, and silver maleate (30).

This discussion on the silver sources attempts to bring the physical properties of silver sources in line with their photothermographic properties, as illustrated for silver soaps in Table 13.1.

The fundamental properties of $[Ag(O_2C_xH_{2x-1})]_2$ that can be associated with the characteristics of photothermographic imaging chemistry (D_{max}, D_{min}, speed, stability, etc.) are summarized in Table 13.1 and are discussed in more detail under each category.

Table 13.1 Properties of $[Ag(O_2C_xH_{2x-1})]_2$ Related to Photothermographic and Thermographic Processes

Properties	Basis of property	Consequence of property
1. $[Ag(O_2C_xH_{2x-1})]_2$ dimer structure	Ag—O covalent bonds	Imaging reactivity characteristics
2. Three-coordinate silver	Two strong Ag—O bonds, one weaker Ag—O bond	Preferred crystal growth in a–b plane forms flat, tabular crystals; poor solubility
3. 2.9 Å Ag $\cdots$ Ag separation in $[Ag(O_2C_xH_{2x-1})]_2$: same as in metallic Ag°	Bridging carboxylates create the $[Ag(O_2C_xH_{2x-1})]_2$ dimer structure	Possible effect on growth of Ag° image or latent image
4. Reactivity	Coordinatively unsaturated Ag on crystal edge, or lattice defect sites	Reaction with compounds having an affinity for silver ion
5. Solubility limitations	Strong Ag $\cdots$ O interactions between $[Ag(O_2C_xH_{2x-1})]_2$ dimers	Poor solubility, small (500 Å) crystallite size, film stability
6. Nonlinear solubility temperature dependence	Possible micelle properties	Krafft temperature ~50–60°C
7. Thermal stability	Crystal defects, impurities, chain length	Possible effect on fog center formation, latent image centers
8. Multiple thermal transitions prior to and including the development temperature	Straight chain hydrocarbon and its chain length	Unreported
9. $[Ag(O_2C_xH_{2x-1})]_2$ color	λ_{max} = 250–260 nm	Colorless construction
10. Refractive index	Dominated by Ag and hydrocarbon content	Haze
11. Photostability	λ_{max} = 250–260 nm	Print stability
12. Conductivity	Log σ = −16	Insulator

13.2.1 The $[Ag(O_2C_xH_{2x-1})]_2$ Dimer Structure

In discussing silver carboxylates, it should be first emphasized that the "salt" nomenclature utilized in the imaging literature and patents implies an ionic $Ag^+(^-O_2CR)$ species and is misleading. The molecular structure of the silver stearate complex, $[Ag(O_2C_xH_{2x-1})]_2$, where $x = 18$, which was previously understood since 1949 only by its unit cell parameters (31), has recently been published (32). The structure consists of three-coordinate silver covalently bound in 8-membered dimer rings of $[Ag(O_2C_xH_{2x-1})]_2$ (see Fig. 13.3). The silver is three-coordinate because of two strong Ag—O bonds and a weaker Ag—O bond between dimers. The weaker, interdimer Ag—O bond may be more accurately labeled as a supramolecular synthon, as this interaction plays a major role in determining the properties of the material in the solid state (33,34). The chemical literature clearly shows that, within the range of typical linear carboxylic acids, the same fundamental 8-membered silver carboxylate dimer results in all cases (33 and references therein, 35) where there are no additional ligand groups. As a result, the "salt" nomenclature conflict is more than semantic, as the reaction chemistry related to these silver com-

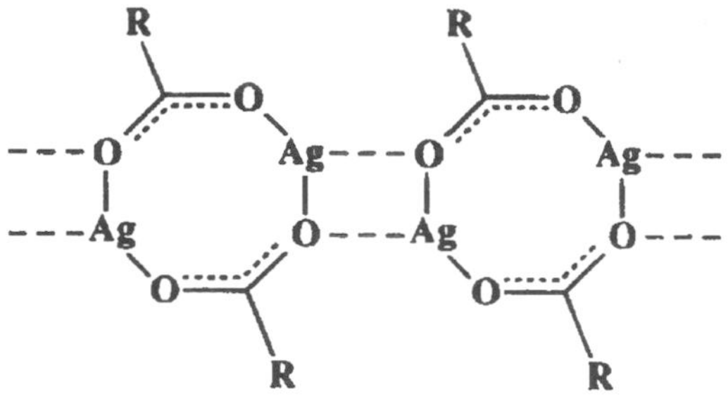

Figure 13.3   Ag—O dimer and neighbor bonding interactions in [Ag(O$_2$CR)]$_2$.

pounds becomes more clear and understandable when viewed as a covalently bonded [Ag(O$_2$C$_x$H$_{2x-1}$)]$_2$ dimer, as will be seen in the various individual property sections below.

Formally, there are six available coordination sites around the silver, although silver normally prefers 3–4 coordination sites occupied in molecular structures (35). The silver atoms in the dimer located on the crystal edge, therefore, are coordinatively unsaturated and easily accessible to compounds having the ability to coordinate to silver. From the photothermographic imaging chemistry point of view, components in the imaging formulation having the ability to coordinate to silver may react with the silver carboxylate at this location. The components having silver binding functionality include toners, development accelerators, and stabilizers. Specific examples of silver complexes of components contained within the coated film are discussed in the individual sections below.

By way of comparison, much less has been reported about the molecular structure of either of the other main commercially important silver sources, silver benzotriazoles and silver acetylides. The only silver benzotriazole structure reported in the literature contains the benzotriazole in its neutral, nondeprotonated form, {Ag(benzotriazole)$_2$ (NO$_3$)} (36); this clearly has only limited relevance to the silver benzotriazole used in photothermographic imaging materials. The poor solubility of silver benzotriazole, in which it has commonality to silver carboxylates, is considered to be a good indication of a polymeric structure. Auxiliary ligands, such as aromatic nitrogen heterocycles, thioureas, and phosphines, have been claimed as compounds suitable for improving the solubility of silver benzotriazoles in photothermographic imaging constructions (37). The molecularly simpler silver imidazole, which also is poorly soluble and is also used as a silver source (38), has recently had its molecular structure clarified by Rietveld analysis of powder diffraction data (39). In this case, the structure is well resolved and shows that the complex is, in fact, polymeric.

The molecular similarity to benzotriazole may be sufficient to conclude a similar connectivity and explain both solubility and the lack of thermal phase changes within the temperature range of the development processes.

Silver acetylides comprise a class of compounds for which analogs also must be used in order to explain properties relevant to photothermography. These complexes, similar to silver benzotriazoles and carboxylates, typically exhibit poor solubility and are thereby considered polymeric (40) or at least complex oligomers (41). Little more can be said regarding the structures of the silver acetylides such as those used in photothermography. An interesting variation on the silver acetylide theme is the incorporation of carboxylates where, in all probability, the coordination occurs through the carboxylate; but the triple bond may be important (40). Silver acetylides incorporating phosphine derivatives are more soluble, presumably by replacement of alkyne cross-linking groups with terminal phosphine ligands. From the molecular structures reported in the literature, the polymeric nature inferred for the silver acetylides used in photothermography can be assumed to be likely.

It is concluded that the thermo- and photothermographic consequences of the poor solubility of silver carboxylates may be considered relevant in the discussion of silver benzotriazoles and silver acetylides equally.

13.2.2 Three-Coordinate Silver in $[Ag(O_2C_xH_{2x-1})]_2$

In the solid state, three-coordinate silver is the rule, in which there are two strong Ag—O bonds and one weaker Ag—O bond [2.2 and 2.6 Å, respectively (32)]. The weaker Ag—O bond between $[Ag(O_2C_xH_{2x-1})]_2$ dimers (shown in Fig. 13.3) is common in silver carboxylates and is the source of many silver carboxylate solid state properties.

The silver soap structure can be considered to be a linear coordination polymer because of the extended Ag—O bond between $[Ag(O_2C_xH_{2x-1})]$ units. The consequence of the extended Ag—O bonding, in combination with the weak hydrocarbon interactions between terminal methyl groups in the structure, is highly preferred crystal growth in the a–b plane of the crystal lattice (that is, the plane resulting in sheets of Ag atoms). As a result, a large volume of hydrocarbon chains encompasses a thin sheet of silver atoms, but the attractive forces between the hydrocarbons is sufficiently weak that only extremely thin crystallites are formed in the preparation of silver carboxylates under aqueous conditions (42–45). The tabular plate structure of the silver carboxylate can be illustrated by recrystallized silver behenate (Fig. 13.4) and is typical, although not exclusive, for all long chain silver carboxylates.

Whereas there are strong interactions between $Ag \cdots O$ pairs in the a–b plane and weak interactions between the terminal methyl groups, the process of preparing the $[Ag(O_2C_xH_{2x-1})]_2$ soaps (from water at $>60°C$) leads to rapid precipitation of the insolu-

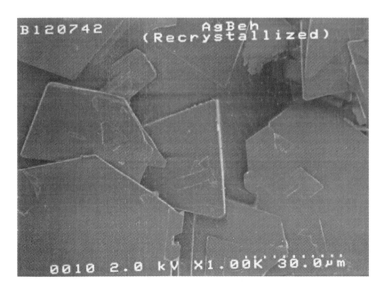

Figure 13.4   Recrystallized $[Ag(O_2C(CH_2)_{20}CH_3)]_2$ (99%)

ble solid. The ratio between the hydrocarbon surface, the (001) plane, and exposed silver atom surfaces, the (010) and (100) planes, therefore, becomes quite large, approximately 100:1, as can be quickly ascertained from scanning electron microscopy (SEM), such as in Fig. 13.4. This ratio suggests that reactivity of the silver soap is at least partially controlled by the low level of available silver atoms. Halidization of the silver soap, for example, would be expected to occur primarily on these silver exposed planes, and this is precisely what is seen (44,46).

Recently, imaging films having improved image tone have been claimed by increasing the percentage crystallinity of the silver carboxylate incorporated into the film (42,43). In this case, phosphonium compounds are incorporated into the silver carboxylate preparation procedure, which are claimed to promote the crystallinity of the resulting product. Crystallinity is defined as the ratio of the sum of the silver behenate diffraction lines to the sum of the diffraction lines of an Al_2O_3 reference. Crystallinity greater than 0.85 is preferred.

A slightly different approach to changing the structure of silver carboxylates is to focus on the odd-numbered chain lengths. A material based on a 9:1 mixture, by weight, of $C_{20}H_{41}COOAg$ and $C_{22}H_{45}COOAg$ gave slightly better D_{max} and D_{min} values than one based on silver behenate when fresh and much better values are found after incubation for 2 days (47).

13.2.3   Ag $\cdots$ Ag Separation in $[Ag(O_2C_xH_{2x-1})]_2$

An interesting consequence of the bridging nature of the carboxylate groups in $[Ag(O_2C_xH_{2x-1})]_2$ compounds is the location of the silver atoms at an unexpectedly close proximity to each other, 2.9 Å (34,48–50). While this distance is common for silver carboxylates, it happens to be the same as in metallic silver. The basis of this proximity is the fundamental M-M bridging properties of the carboxylate ligand in the $[Ag(O_2C_xH_{2x-1})]_2$ dimer structure, which is a well-established carboxylate bonding mode (51). However, this short distance and the similarity to the distance in metallic silver suggest possible bonding interactions between these metals in the dimer. The potential for bonding between the Ag^+ atoms is the subject of continuing debate (52–56). The consequence of such close Ag $\cdots$ Ag proximity is still uncertain in photothermographic imaging materials, although possible routes to the growth of Ag° have been postulated (1), because addition of one electron to the dimer could produce an incipient Ag_2^+ species, the well-known precursor to the latent image in conventional AgX systems (57).

13.2.4   Reactivity of $[Ag(O_2C_xH_{2x-1})]_2$

The open coordination site on the silver in the $[Ag(O_2C_xH_{2x-1})]_2$ dimer in the edge of the crystal lattice can be expected to be very reactive with compounds having the ability to bond silver. This type of reaction chemistry results in the facile formation of stable complexes, such as the formation of in situ AgX by the reaction of $[Ag(O_2C_xH_{2x-1})]_2$ with halides, and toners as described below. Furthermore, this reaction chemistry can also be illustrated by the extreme reactivity of neutral donors having strong affinity for silver, such as phosphine-based (58) and sulfur-based (59) ligands. Triphenylphosphine is a particularly good ligand to convert the insoluble $[Ag(O_2C_xH_{2x-1})]_2$ dimer to derivatives having differing reactivities and solubilities. Novel silver carboxylate silver sources have been disclosed as a result of the affinity of the silver ion for neutral donor ligands (60,61).

Derivatives of $[Ag(O_2C_xH_{2x-1})]_2$ and various ligands found in the photothermographic formulation have been reported and are discussed in the toner section below.

It would be useful to include a discussion of the relevant reactivity of silver benzotriazoles and silver acetylides with various photothermographic components, but there is no relevant reaction chemistry reported in the literature on these compounds.

13.2.5 Solubility

In general, the poor solubility of silver carboxylates, and silver soaps in particular, is directly attributable to the combination of the high molecular weight of the $[Ag(O_2C_xH_{2x-1})]_2$ dimer molecule (MW = 895.0 for pure silver behenate, 782.5 for pure silver stearate) and the strong $Ag \cdots O$ interactions between the $[Ag(O_2C_xH_{2x-1})]_2$ dimers. The solubility of silver carboxylates is typically very low, as illustrated in Table 13.2.

The only literature solubility data found for silver benzotriazole is 9.1×10^{-7} M (37), which confirms its poor observed solubility, but no literature references have been found for the solubility properties of thermographically important silver acetylide derivatives.

As noted above, larger crystals of silver carboxylates can be prepared by slow recrystallization from solvents containing coordinating ligands (44). Alternatively, properly designed types of carboxylate derivatives have been found to promote solubility of the resulting silver complex in organic solvents (65,66). In both cases, ether linkages in the hydrocarbon chain improve the solubility of the silver carboxylates, up to 10,000 times in one case (65). The rationale for the improved solubility, based on the molecular structure, is not discussed, but improved imaging properties are claimed as the result (65,66). Also, a "soluble" form of silver carboxylate, silver cyclohexanebutyrate, is commercially available for use as an ICPS standard. Solubility is claimed in xylene and 2-ethylhexanoic acid. The nominal structure, ignoring that it is most likely dimeric like all other known simple silver carboxylates, is

Higher solubility of silver carboxylates is claimed to be achieved in certain common organic solvents such as toluene (>0.1 M), although the "solubility" may be better described as forming gels (67,68). The poor solubility of silver benzotriazoles has been partly mitigated by certain auxiliary complexing agents as noted above (37).

Table 13.2 Solubility of $[Ag(O_2C_xH_{2x-1})]_2$ (10^{-5} moles/L)

X	H$_2$O	Acetone	MEK	MeOH	Toluene	Ether
2	6080 (62)	2.89 (62)	2.16 (62)	100 (62)		
12	1.13 (63)			195 (63) 23.9 (64)	1.19 (63)	22.6 (64)
14				19.9 (64)		18.7 (64)
16	58.6 (62)	2.85 (62)	2.31 (62)	46.4 (62) 16.0 (64)		17.3 (64)
18	31.4 (62) 0.7 (63)			17.9 (62)	2.12 (63)	12.4 (64)
				128 (63) 13.2 (64)		
Phenyl	1150 (62)	10.5 (62)	6.54 (62)			

A comment regarding the preparation of $[Ag(O_2C_xH_{2x-1})]_2$, which is commercially carried out in water, is warranted at this point. While the dry silver carboxylates are quite insoluble in water and clearly exhibit a waxy texture, the typical preparation of the silver soap involves precipitation from water at elevated temperatures. Under these conditions, the silver soap formed is initially quite hydrophilic and remains readily dispersed in the aqueous phase. Separation of the solid from the water yields a solid that still contains 70–80% water by weight (2). As long as the silver soap remains wet, it remains water dispersible; upon drying, this solid irreversibly becomes extremely hydrophobic. As noted below, some efforts have been successful in taking advantage of this hydrophilic property in the patent literature. Directly coating the freshly prepared hydrophilic silver carboxylate from aqueous media is the subject of multiple patent disclosures (69,70) or from a solvent dehydrated silver carboxylate (71,72), as discussed more fully in Section 13.7.

13.2.6 Nonlinear Solubility of $[Ag(O_2C_xH_{2x-1})]_2$

The solubility of the $[Ag(O_2C_xH_{2x-1})]_2$ soaps actually exhibits a sharp discontinuity in its otherwise linear temperature dependence (73): two distinct solubility regions exist. Silver stearate and silver laurate, which approximately represent the silver soaps used in photo-thermographic materials, for example, are poorly soluble in typical organic solvents and show a linear increase in solubility versus temperature until the discontinuity, labeled the Krafft temperature. At this point, the silver carboxylate solubility is seen to dramatically increase with temperature. Overall solubility remains quite low (10^{-4} mol/L), but on passing through the Krafft temperature a 10-fold increase in solubility occurs over a 60°C temperature range. For silver laurate in toluene, the Krafft temperature is 52.5°C.

In coordinating solvents, such as pyridine, however, there is no Krafft temperature. This is not surprising, as the solubilization process in pyridine most likely involves extraction of $[Ag(O_2C_xH_{2x-1})]_2$ dimers or pyridine complexes of fragments thereof. The solubilized form is probably better formulated as $\{[Ag(O_2C_xH_{2x-1})]_2 \cdot pyridine_n\}$, where the pyridine can be expected to be bound to the silver at the axial positions in which the pyridine takes the place of the $Ag \cdots O_{neighbor}$ interaction. Alternatively, analogous to the known triphenylphospine complex (58), intermediates based on $[Ag(O_2C_xH_{2x-1}) \cdot pyridine_2]$ may be expected to result. In either case, the soluble species is no longer the initial $[Ag(O_2C_xH_{2x-1})]_2$ complex. It is now apparent why extremely poor crystals (for x-ray characterization) of $[Ag(O_2C_xH_{2x-1})]_2$ result from recrystallization from EtOH/pyridine and why extensive measures were required to obtain the coordination sphere details in AgSt (32). Analogous to another silver carboxylate amine system (74) the pyridine is easily trapped in the lattice of the growing crystal and only slowly is lost.

13.2.7 Thermal Stability

The temperature at which thermal decomposition occurs in silver soap materials is nominally quite high, above 200°C (45,75,76), but crystal defects, impurities, and variable chain lengths (within the silver soap solid material) enable thermal decomposition to begin at substantially lower temperatures. Thermally induced decomposition of silver carboxylates has been reported (77–82). The first thermal transition observed in long chain silver carboxylates has been reported to be irreversible but unrelated to decomposition (79–81). Defect sites are considered to be the initiation centers for the decomposition of silver carboxylates (83) and could be the onset of fog center formation in an imaging material.

Conductivity measurements versus temperature have been reported and show that decomposition begins at relatively low temperatures (150°C), whereupon free fatty acid is released that subsequently affects conductivity (81). Repeated thermal cycling generates accumulations of fatty acid that increase conductivity. EXAFS study of silver stearate indicates major realignment of the silver atoms in relation to the carboxylates upon heating to relatively low temperatures (below the normal development temperature) (32). This realignment is argued to actually involve complete Ag—O bond cleavage in the $[Ag(O_2C_xH_{2x-1})]_2$ dimer and reformation of a separate Ag—O bond with the neighboring dimer, which is propagated through the crystal lattice. Consequently, the simple heating process may predispose the silver carboxylate to form undesirable defect sites, which may lead to fog centers. The observation of metallic silver and free fatty acid in samples heated extensively at 150°C confirms that decomposition occurs at these lower temperatures (80,81).

Some data regarding the thermal stability of the $[Ag(O_2C_xH_{2x-1})]_2$ dimer as a function of chain length have been published (45,84). In the first case, broad line NMR used to monitor decomposition at 120 and 130°C suggested that decomposition increases with decreasing chain length (for chains in the range of C_{22}–C_{12}). In the latter case, this trend is observed, Fig. 13.5, as measured by the rate of metallic silver deposition from alcohol reduction (over a broader chain length range), although the reactivity is clearly highest for C_4.

The thermal stability of silver benzotriazoles and silver acetylide derivatives is significant, although no specific literature reports are available that describe these properties in any detail.

13.2.8 Multiple Thermal Transitions

The thermal properties of $[Ag(O_2C_xH_{2x-1})]_2$ have been historically somewhat misconstrued, as standard data tables simply report the "melting" points to be above 200°C (75,76). Silver carboxylates having alkyl chains longer than 12–14 are far more compli-

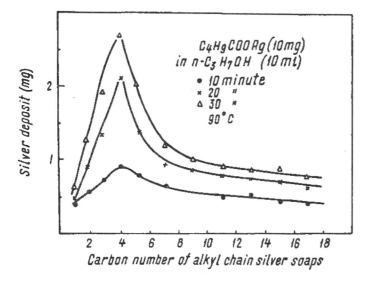

Figure 13.5 Silver deposit as a function of carbon chain length of silver carboxylates. (From Ref. 84.)

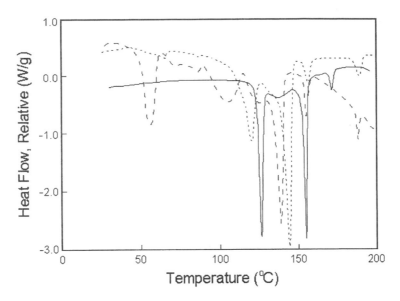

Figure 13.6   Thermal transitions in AgSt, commercial samples [· · · and - - -], and laboratory sample [—].

cated and begin to show multiple thermal transitions. The thermal transitions, as determined by DSC, consist of at least three transitions, as illustrated by silver stearate (80,81). These transitions can be deceiving, however, as shown in Fig. 13.6, for two different commercial sources of AgSt compared to a laboratory prepared sample. The first thermal transition, as measured by differential scanning calorimetry (DSC) (80,81), occurs around 100°C and steadily increases to around 130°C as the chain length is increased. It should be noted that transitions occurring below 100°C are usually due to residual free fatty acid, which is generally present after preparation but can be readily removed by a simple wash with solvent.

The first transition is thermally irreversible and occurs, coincidentally, below the temperature ranges typically utilized for thermal development of imaging materials. Kinetics of the development process, before and after this transition, show that it is unlikely to be involved in the development process (85). While this transition seems to be hydrocarbon based, some thermal decomposition does, however, occur at these lower temperatures (45,81). It is uncertain if the polymorphism exhibited in fatty acids (86), such as behenic acid, can translate to the silver complexed form, which would account for some of the observed phase changes.

13.2.9 $[Ag(O_2C_xH_{2x-1})]_2$ Color

It is important for the components in photothermographic imaging materials to be colorless, and most silver sources are conveniently quite white, including silver carboxylates, silver benzotriazoles, and some silver acetylide derivatives. The lack of visible color of silver carboxylate complexes is due to the absence of any absorption bands in the visible or the near UV, but there is an absorption occurring in the UV [λ_{max} = 250–260 nm (87,88)], which is a consequence of ligand-to-metal-charge-transfer transitions. Fortunately, most silver complexes containing toner ligands that have been isolated suggest that the intermediates in the imaging chemistry are also colorless (59,60). In addition, silver carboxylates will contribute very little to UV D_{min} in regions needed for subsequent UV exposure (>350 nm), an important feature for image-setting films (89).

13.2.10 Refractive Index

The refractive index of $[Ag(O_2C_xH_{2x-1})]_2$ is dominated by the content of silver in the complex and can be expected to increase as the chain length is decreased. For silver soaps used in photothermographic formulations (chain lengths of primarily C_{18} or C_{22}, or their mixtures), the refractive index has been reported to be greater than 1.6 (90). Considering that the refractive index of the binders is in the range of 1.48–1.50 (91,92), silver carboxylates similar to silver behenate will contribute some haze in the coating.

13.2.11 Photostability

Whereas the only inherent light absorption in $[Ag(O_2C_xH_{2x-1})]_2$ complexes is in the ultraviolet (λ_{max} = 250–260 nm), pure silver carboxylates can be expected to be generally quite photostable (87,88,93–97). The reaction by-products resulting from the thermal reaction of various components in the photothermographic formulation with $[Ag(O_2C_xH_{2x-1})]_2$, however, generate new complexes (see Section 13.6) that have not been thoroughly investigated. In addition, the radical by-product components of the Borodin–Hunsdieker reaction have been cited as potential fogging agents in postprocessed films (98).

13.2.12 Conductivity

The structure of $[Ag(O_2C_xH_{2x-1})]_2$ compounds, as noted above, is very conducive to forming layers of metal ions separated by the large hydrocarbon thickness. The layered structure is so regular that it has been proposed as an x-ray diffraction standard (99). The consequence of this arrangement is, in the ideal crystal, asymmetric conductivity. That is, metal ion mobility will be generally constricted to the metal ion layer as the hydrocarbon layer provides excellent insulation. Two-dimensional conductivity might be expected (100). Because large single crystals of long chain silver carboxylates have been difficult to obtain (44), there have been no reports to confirm this prediction.

Conductivity measurements have been measured on the bulk powders (81,101). Silver stearate, for example, shows extremely low conductivity at room temperature, $\sim 10^{-17}$ S/cm (81). The continuing conductivity increase observed on elevated temperature is directly attributed to the formation of free fatty acid from silver carboxylate decomposition. This overall low conductivity throughout the development process provides an indication of the need for alternate silver complexing agents that will enable the transport of the silver ion to the development centers for image formation.

In summary, it can be seen that many properties of the silver compound play significant roles in the choice of a proper source of silver for the metallic image in thermo- and photothermographic imaging materials. While significant portions of the properties of silver carboxylates can now be correlated to those properties important for imaging, the same cannot be said for other silver sources at this time. Considering the widespread use of silver carboxylates for thermographic and photothermographic applications, this discrepancy is likely to remain.

13.3 AgX PHOTOCATALYSTS

13.3.1 Silver Halide Grains

Very fine silver halide grains ($\sim$500 Å) are typically used as the light capturing materials that form the latent image in commercial PTM media (1,2). These materials are claimed to be used in "reactive association" (102,103), "intimate catalytic association" (104), and "synergistic association" (2). The non–silver halide thermographic versions consider the silver carboxylate and developer "in thermal working relationship" with each other (105). Although non–silver halide light capturing photocatalysts, such as ZnO (106), TiO$_2$ (107), and silver tetra-arylborate salts (108), have been reported in the literature, silver halides are overwhelmingly selected for this purpose because of their high efficiency, ease of dimension control, and property control over a wide variety of conditions/solvents. Probably most important is the vast base of experience with these materials in the photographic industry.

There are three general methods by which the silver halide particles are introduced into the PTM construction:

Ex situ: Addition of preprepared AgX particles, via aqueous or organic solvent preparation, to the silver soap dispersion.

In situ: Addition of halide ions to the silver soap dispersion. The silver halide particles are formed by the methathesis reaction between the soluble halide ions and the silver carboxylate:

$$[Ag(O_2C_xH_{2x-1})]_2 + 2X^- \rightarrow 2\,AgX + 2\,HO_2C_xH_{2x-1}$$

Preformed: Preparation of the silver carboxylate in the presence of pre-prepared aqueous precipitated silver halide particles.

In commercial PTM films, combinations of two of the above three methods are most often employed to achieve optimum sensitometric response.

Whereas most commercially available PTM films are presently coated using an organic solvent, and most of the large database of AgX technology is water/gel-based, there has been an understandable focus on making the AgX particles compatible with the nonaqueous binder. The ex situ preparation method mentioned above is often used for separately prepared AgX grains, which may be prepared in the same solvent that is used as the coating solvent, i.e., solvent made grain/solvent coated (109) or water made grain/water coated (69,70) or aqueous made grain/solvent coated (110).

Polyvinylbutyral has been commonly used as the peptizer for solvent precipitated AgX grains. Polyvinylbutyral is particularly suited for this latter application as it has both polar units (vinyl alcohol), which are considered to bond to the AgX particles (111), and solvent soluble units (vinylbutyral).

There are two subapproaches to the preparation of AgX via the ex situ method, conventional aqueous preparation and organic solvent preparation. In either case, the separately prepared AgX is simply added to the silver source formulation. For solvent grown AgX, solvent soluble sources of halide ion and silver ion are used, typically LiBr and silver trifluoroacetate, respectively, in acetone (102). In general, the AgX particle size is controlled by the same parameters used in conventional AgX/gel makes, that is, flow rates and reactor temperature. One of the disadvantages of solvent grown AgX grains usage is that only cubic grain morphologies have been reported by this precipitation method (102).

The in situ preparation method has advantages over the other methods in its simplicity. Typically, a soluble source of halide ions is added directly to a dispersion of the silver carboxylate or titrated into a dispersion with excess silver ion (72). The AgX particles form on the surface of the silver carboxylate particles, initially predominately on the lateral edges (44,46,112). Typical in situ grains are cubic and range in size from 500 to 1000 Å. The halide ion source material is usually a metal halide but may be an alkyl halide. During the in situ process, some of the AgX grains may remain epitaxially attached to the Ag-carboxylate particles. The interface between the AgX particles and silver stearate (AgSt) crystals have been investigated in detail by scanning and transmission electron microscopy (SEM and TEM, respectively) (46,112) (see Figs. 13.19 and 13.20 in Section 13.10). This interfacial layer is proposed to consist of $Ag_{1-x}M_xSt$, where M is an alkali metal cation. The morphology of the resulting developed image silver has been proposed to be influenced by the type of silver halide/silver carboxylate interface present in the film (112–114). In situ made AgX grains that contained the epitaxial layer were found to generate filamentary silver on development, whereas dendritic silver was produced on development of ex situ added AgX grains not having the epitaxial layer. It is suggested that this AgX/Ag-carboxylate epitaxy gives a boost to the photographic speed of coatings made with these materials.

Another type of in situ formed AgX is by exhaustive conversion of a portion of the silver carboxylate under preparation (115). In this method, a portion of the silver carboxylate is separated from the freshly prepared material, completely converted to the silver halide by a stoichiometric addition of a halide source, and then added back into the silver carboxylate preparation. After suitable mixing it is ready for the next stages of formulation.

In the preformed method, aqueous precipitated AgX grains have been incorporated into solvent coated PTM constructions using one of two methods. Preprepared AgX grains, typically peptized in gel, are added to a reactor as a dispersion; then the silver carboxylate, from silver nitrate and alkali metal carboxylate, is precipitated in the presence of these AgX grains. The resulting intimate mixture of AgX grains and Ag-carboxylate particles is washed, dried, and redispersed in the desired solvent (20). This method permits the use of any morphology of aqueous made grains without a step to remove the gel peptizer, which makes it attractive as a low-cost process. In another case, ex situ AgCl tabular grains, peptized in gelatin, are formulated into a PTM construction and coated from a solvent that yields a low-haze transparent film (116).

Core–shell AgX grains (preformed) have been disclosed in the preparation of a photothermographic film (117). As in the previous example, the primarily AgBr grains in this patent are prepared conventionally in a water/gel medium and have higher levels of iodide in the core (4–14 mol%) than in the shell (0–2 mol%).

Another process describes the use of aqueous made tabular AgX grains in a solvent coated film (118). In this process the aqueous made AgX grains with gel peptizer are redispersed in a cosolvent containing benzyl alcohol and polyvinylbutyral polymer using high-energy agitation. The resultant dispersion of AgX grains is combined with other typical development additives and coated from solvent.

Several patents have been issued over the past few years that claim that doping AgX grains with low levels of certain metal cation or complexes results in PTM films having improved contrast, lower D_{min}, etc. The metal ions or complexes are generally doped into preformed, aqueous made AgX grains. For example, films containing iridium doped silver halide grains are claimed to improve the speed, high-intensity reciprocity failure, and contrast on shelf aging versus a control without the iridium dopant (20). Hexacyanofer-rate(III) complexes have been doped into conventional AgBrI grains used in PTM imaging layers (119).

13.3.2 Chemical Sensitization

Just as in conventional silver halide photography, there is a considerable commercial interest in making photothermographic films faster in speed so that these films might find more applications that require the use of lower power lasers or other less energetic exposing devices. Photothermographic films are generally slower in speed than conventionally processed silver films having similarly sized AgX grains. Increasing the grain size to enhance the speed of these films has limited value because the grains remain behind after processing, which would be expected to contribute to film haze and cause poor print stability when the size exceeds about 0.2 μm. Additionally, AgX grain size has been reported to exert a relatively small impact on the speed of a film made from such grains (85).

Chemical sensitization of the AgX grains in a photothermographic film is a desirable route to higher speed films yet has remained an elusive goal. While chemical sensitization was first disclosed as early as 1984 (102), sulfur sensitization of PTM AgX grains has been observed to yield high fog levels (6). However, recently, it has been demonstrated that it is possible chemically to sensitize solvent made AgX grains using either gold or sulfur reagents (120,121). Sulfur sensitization of a mixture of silver carboxylate and AgX particles might be expected to be complicated by the competition of both AgX and silver carboxylate particles for the sulfur-containing compound. An ingenious solution to this problem has recently appeared in which a sulfur bearing sensitizing dye, seemingly having a preferential affinity for the AgX surface, is proposed to decompose on the grain surface to yield a high-speed coating (122). A novel high-speed photothermographic film construction that may find use in traditional x-ray applications has been recently disclosed (123). The tabular AgX grains in this construction are composed of greater than 70% chloride and have (100) major faces. The high chloride (100) tabular grains are claimed to provide lower minimum densities and higher levels of image discrimination after thermal processing than the high bromide (111) tabular grains typically used in radiographic elements that are intended for aqueous processing.

It is clear from the recent increase in patent activity and publications in extending the sensitivity of photothermographic films that there is significant interest in this area.

13.4 SPECTRAL SENSITIZATION

Spectral sensitization of PTM materials has some similarity to conventional silver halide media in that the same sensitizer families, such as cyanines and merocyanines, are typically used. As is the case with other elements of the PTM construction, little has been published outside of the patent literature on the spectral sensitization of these AgX grains.

There is, however, one recent paper on spectral sensitization (124). In this study, a series of carbocyanine sensitizing dyes having various ring substituents, ionic charges on the dye and methine bridge substitution were evaluated. All of the dyes in this study contained the trimethine group as the bridge between two heterocyclic rings derived from benzothiazole or benzoselenazole, as shown below (Fig. 13.7).

The results showed a strong preference for N-sulfopropyl groups on the benzothiazole ring over that of N-alkyl groups. Wedge spectrograms of films containing these dyes revealed that most of the dyes bearing the N-sulfopropyl groups exhibited intense, sharp spectral peaks at 650 nm, that were assigned as J-aggregate dye bands. Analogous dyes having N-alkyl groups exhibited no J-banding. The study indicated that anionic sensitizing dyes were more readily adsorbed by in situ prepared AgX grains than were cationic dyes in nonaqueous solvent, which is the reverse of the known tendency in aqueous media (57).

Recently, silver halide adsorption accelerators have been claimed in PTM (125). In this case, quarternary aromatic amines, with a general structure shown below, are combined with various phenolic developers for printing plate applications.

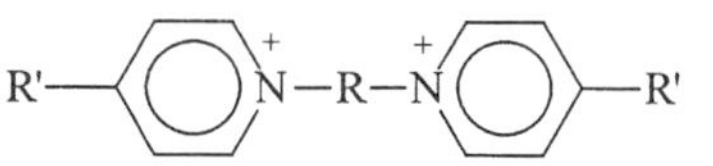

A series of cyanine sensitizing dyes having benzothiazole heterocyclic rings and N-sulfoalkyl groups bridged by a trimethine group, having similarities to the above report, has recently appeared (126). The novelty claimed here is that halogen substitution on the benzothiazole rings of the dye in combination with N-sulfoalkyl groups enhances both the speed and the shelf life characteristics of a PTM film over that of a film made using sensitizing dyes not having these features. In addition, this work was not done with conventionally grown AgX grains but rather with solvent grown AgX grains.

Similar results have been obtained using cyanine sensitizing dyes on preformed AgX grains that had been prepared using a conventional (water/gelatin) procedure (127). PTM elements may be sensitized to the near infrared by using heptacyanine dyes having a structure such as:

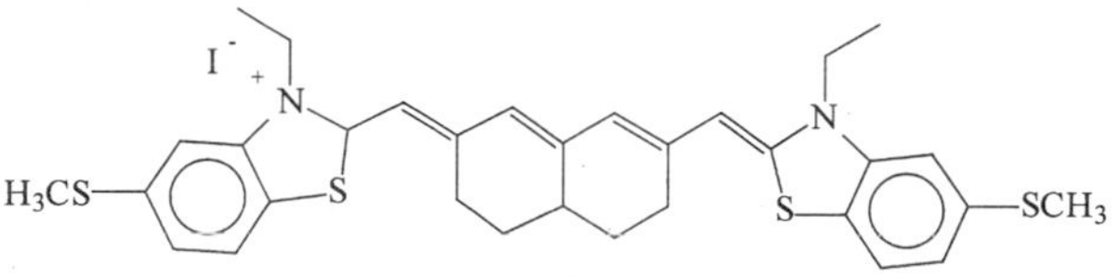

The novelty of this sensitizing dye structure is that the tetrahydronaphthyl group on the heptamethine chain imparts rigidity to the polymethine chain that, in turn, was found to enhance the spectral sensitivity and the shelf life in a coated film compared to a control film containing a nonrigidized dye. Also, in accord with similar studies (128,129), the addition of an alkyl carboxy group onto at least one of the nitrogen atoms of the benzothiazole rings produced further substantial increases in speed. This patent (128) also mentions that a significant speed boost may be attained in films made using these sensitizing dyes by the addition of a supersensitizer to the coating. Addition of supersensitizers such as benzimidazoles or benzothiazoles was found to provide speed increases of over 1.0 log *E* over similar coatings without a supersensitizer.

A further refinement of the above heptamethine sensitizing dye structure wherein thioalkyl groups are attached to the benzothiazole rings (R_1 in Fig. 13.7) has been de-

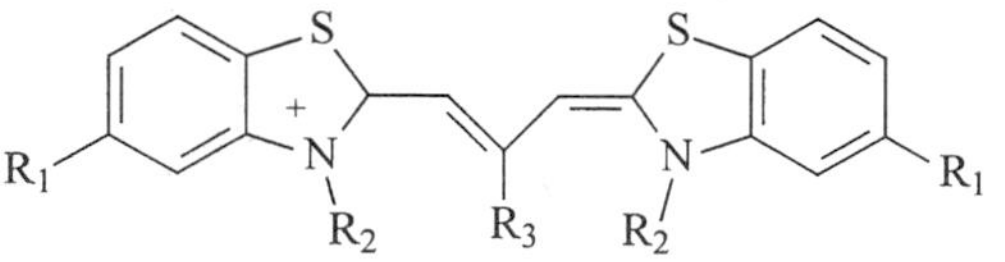

Figure 13.7 Benzothiazole carbocyanine dyes. R_1 = thioalkyl; R_2 = H or alkyl; R_3 = H, Me, Et; anion = $(CH_2)_3 SO_3^-$.

scribed (130). According to this patent, PTM films containing these rigidized dyes plus thioalkyl groups provide for enhanced speed, contrast, D_{min}, and extended solution pot life over that of a control not having the thioalkyl groups.

A PTM formulation containing novel infrared sensitizing dyes in combination with a series of substituted supersensitizers having a 4,4′-bis(triazinylamino)stilbene backbone has been reported (131). This combination of infrared sensitizing dye and supersensitizer is shown to enhance the photographic response over that of a similar film containing no supersensitizer for aqueous coated media.

In 1987, Fuji introduced the Pictrography 1000 digital color hardcopy printer line (132). The cyan, magenta, and yellow records for these media are exposed by three near-infrared laser diodes that range in wavelength from 680 to 810 nm. To achieve accurate color separations over this narrow wavelength range, three AgX sensitizing dye groups were devised that utilize novel PTM sensitizing dye technology. Each of the three AgX sensitizing dye groups form J-aggregate structures, which permit narrow spectral sensitivities with sharp drops in sensitivity at long wavelengths (133). The sensitizing dyes are based on pentamethine cyanines bearing substituted benzothiazole rings. After each color record is exposed, a small volume of water, 1 $\mu g/cm^2$, is coated on the donor sheet to activate a base generating chemistry. Then the donor is sandwiched with a receiver layer and thermally processed, generating the image dye forming dye releasing redox (DRR) compounds (see Section 13.5). The dyes diffuse through the donor and are mordanted to the receiver sheet.

It was also noted above, in the silver source properties sections, that the refractive index of silver carboxylates may play a role in the optical properties of a PTM, such as haze. While the refractive index of the $[Ag(O_2C_xH_{2x-1})]_2$ complex is not considered to be a significant contribution to film haze, it has recently been proposed to be a key factor in optimizing the photosensitivity of the photothermographic film (90,91). In the ideal case, the cubic AgBr is attached to the silver carboxylate crystal, although this has been observed only in model systems (46,112). The surfaces of both silver-containing compounds are, in this theory, completely covered with the sensitizing dye, as shown in Fig. 13.8.

In this case, the light not absorbed by the sensitizing dye is proposed to be internally reflected within the silver soap crystal and has a second opportunity to be absorbed, thereby increasing the optical sensitivity. Unfortunately, only under special circumstances were the reported well-formed $AgBr/[Ag(O_2C_xH_{2x-1})]_2$ crystals observed (46,112). Furthermore, despite a tantalizing early patent disclosure (134) and the claimed sensitization of silver carboxymethylcellulose (135), evidence for significant dye adsorption onto the silver carboxylate surface has not been independently reported.

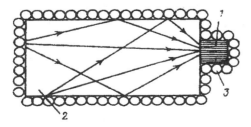

Figure 13.8 Light guide mechanism for the spectral sensitization of thermally developed photographic materials. 1 = AgBr, 2 = silver stearate, 3 = adsorbed dye molecules. (From Ref. 91.)

13.5 DEVELOPERS

13.5.1 Black and White

The role of the developer in PTG materials is to deliver electrons to the silver ion source for reduction to metallic silver at the latent image site. Similar to conventional silver halide developers, they are fundamentally simply reducing agents. The PTM developers tend to be somewhat less reactive than those used for aqueous development of AgX, which is partly related to the fact that they usually remain in the imaging construction after processing and must not contribute to subsequent increases in D_{min}. Because the main requirement is to provide electrons, large numbers of different classes of compounds can be used, and many have been disclosed in the patent literature. A few classes can be cited here as illustrations, such as phenolics, hydroxamic acids, ascorbic acid, and anilines. The phenolic class, however, is the class of choice for virtually all commercially available PTM. Indeed, since the last review of the literature (2), very little new has been revealed regarding developers both in terms of preferred developer properties as well as uniquely different classes or structures.

The following discussion logically begins with black-and-white developers. Because the developers used to produce colored images comprise a separate class (although often based on the same type of black-and-white developer structure), they are also highlighted below. The most obvious overlap between these classes is in the fundamental electron transfer process, from the developer, such as a phenol, to the intermediate silver complexes, ''AgL(O$_2$CR)'':

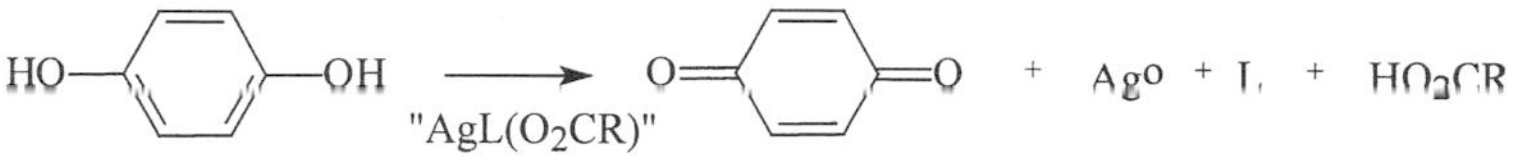

A wide variety of phenolic structures have been reported as developers, ranging from monophenols to polyphenols, with sterically hindered *bis*-phenols being preferred. For example, from the patent literature Fuji often discloses highly hindered *bis*-phenols in their formulations, such as that shown in the following (136), where L is a branched chain alkyl group and R is typically *ortho-t*-butyl groups.

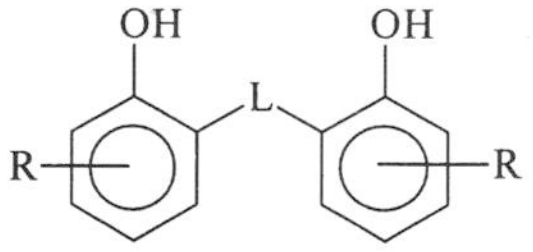

An Imation Corporation patent discloses the use of hindered *bis*-phenols (20), such as this sterically hindered structure:

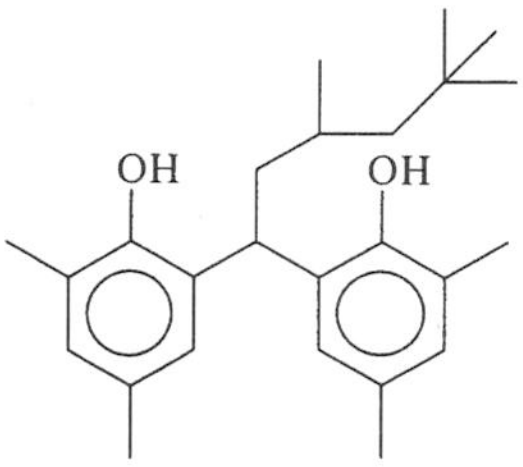

Agfa often incorporates very similar *bis*-phenols in imaging constructions (42). Kodak PTM products have also utilized the phenolic group, although monophenolic, most commonly as sulfonamidophenol derivatives (137), such as

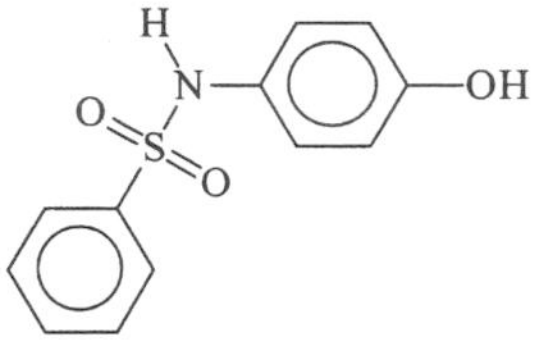

The sterically hindered *bis*-phenol class of developers is, by far, the best represented developer in the patent literature. A few additional examples may be illustrated below. Canon (138) claims *bis*-phenols having the structure

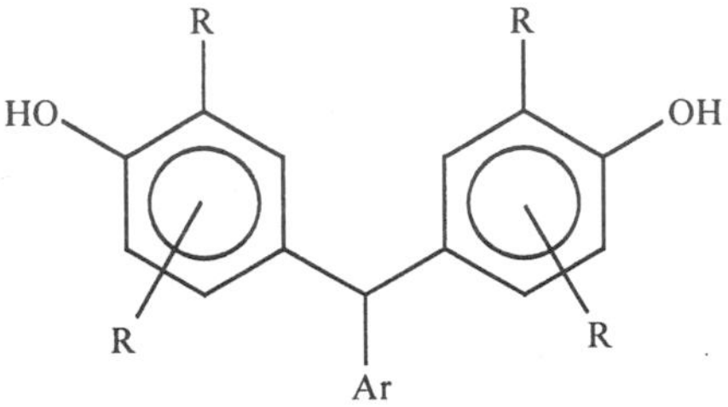

which is not too different from an ether bridged *bis*-phenol disclosed by Konishiroku (139):

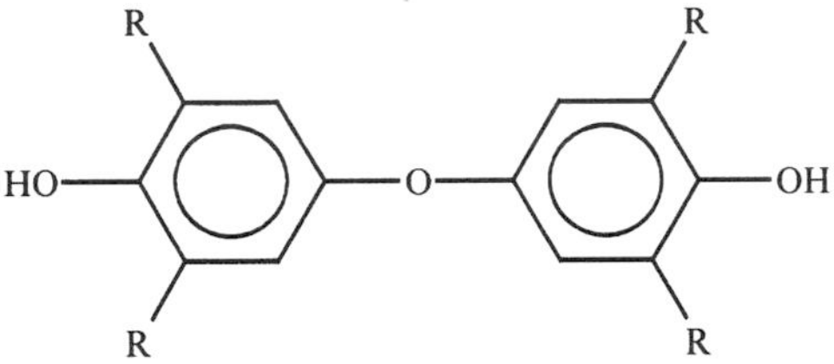

Kodak has further disclosed the use of auxiliary developers to be used in conjunction with weak reducing agents (140).

Heavily ballasted phenolic developers, such as long chain hydroquinones (141) are cited as particularly useful:

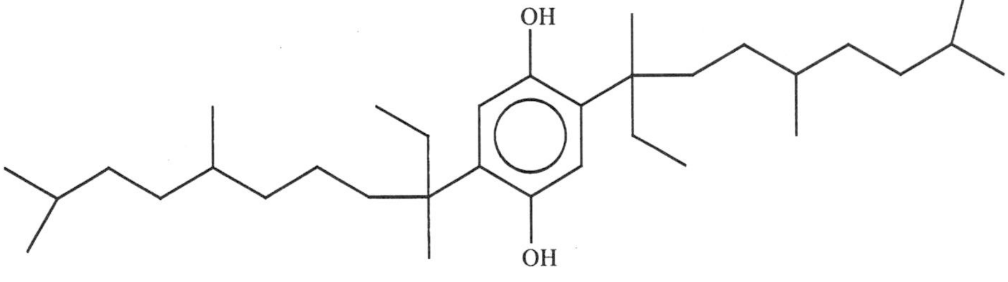

The more reactive class of *bis*-catechols has recently been rediscovered, such as (142), including tri-hydroxybenzene analogs. A new set of catechol derivatives has recently been claimed (143):

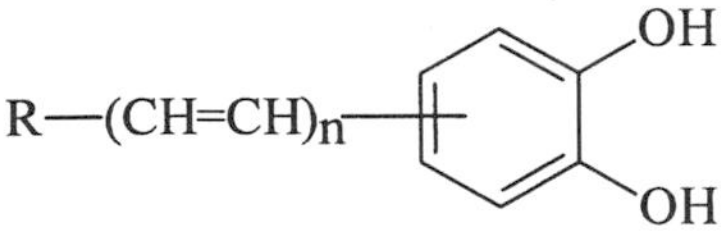

A spiro indane *bis*-catechol, which is a well-known reducing agent/metal complexing agent for iron based thermal imaging materials (144), has recently been applied to PTM (145):

The unique feature of this catechol is that the spiro linkage forces the catechol groups to remain perpendicular to each other, which is particularly beneficial to forming a highly hydrogen bonded, stable system (146).

Despite the importance of the role of the developer, very little has been published in the open literature regarding the mechanism of the silver reduction reaction. While it is generally considered that thermographic imaging based on silver carboxylate silver sources proceeds by physical development processes (1,2), the mechanistic details of electron transfer from the developer to the latent image/silver ion source are not well documented. Silver toner complexes, such as those described in Section 13.8.6, are presumed to be critical components in the delivery of the silver ion to the image site, but any analysis of the reaction of those complexes with the developers leading to the metallic silver image has not been reported. It may be of interest to note, however, that certain types of phenols have been reported to form stable complexes with silver ions. In these cases, the phenol is directly coordinated to the silver via the phenolic oxygen group (147), which might be the functional group interaction expected during the electron transfer step to form the metallic silver image. The phenols reported in these structures tend to have electron withdrawing groups attached, which probably facilitates the isolation of the complex by inhibiting any electron transfer reduction reaction.

The properties of developers that make them optimum for thermographic or photothermographic applications are also not well documented. Besides the need for available electrons, readily satisfied by the phenols, the requirements seem relatively flexible. Redox potentials, well documented for conventional silver halide systems (57) and presumably important in thermographic imaging systems, generally are not reported in conjunction with function (85). Other properties, such as solubility, and melting points, seem to cover fairly broad ranges. Steric constraints on the developer are generally considered as beneficial to control the preimage and postimage D_{min} stability and are illustrated in practice by the preponderant preference for sterically hindered *bis*-phenols in commercially available films.

One requirement claimed in the literature is that the developer must have good solubility within the silver carboxylate solid in order to reach the silver ions (5,98). An important consequence of this proposal is that the developer be extremely hydrophobic to accomplish this task. While most of the developers used in commercial films are often quite hydrophobic, the experimental confirmation of this proposal has not been reported.

In general, it may be concluded that wide varieties of electron sources are available for the preparation of black-and-white PTM. Depending on the silver source, and associated silver complexing agents, the developer properties can be arranged to match. While the direct formula for matching a specific silver source to the optimum specific developer class is not known, one can assume that there is a suitable phenolic developer readily available for most applications.

13.5.2 Color

The oxidized form of the reducing agents may also have a color associated with them. As noted in Section 13.6, for black-and-white imaging where the metallic silver composes the image, simple toning action by silver complexing agents enables the image color to be black. The color of the oxidized developer, in this case, typically contributes little to the overall image color. In the case where a color image is desirable, however, the color of the oxidized developer can be utilized to provide a full color imaging construction. Under these circumstances, however, the color provided by the metallic silver image may contribute to a muddying of the desired color, so transfer of the image from the image layer to a receptor is used to solve this problem.

A concise review of the technology related to the formation of color images from photothermographic processes has appeared recently (5). Consequently, only the highlights of the technology and recent advances need to be discussed below.

There are multiple technical routes to achieve a color image via PTM, and representative examples are noted in the following text. These routes, sketched below, include oxidation of leuco dyes (148–150) and the Fuji Pictrography™ system (151–156).

The leuco dye system is a good place to start, as it begins with the fundamental phenolic reducing agent. An initially colorless dye can be readily prepared from a phenolic group which, upon oxidation, generates good quality color images. A wide variety of yellow images can be obtained based on the structure.

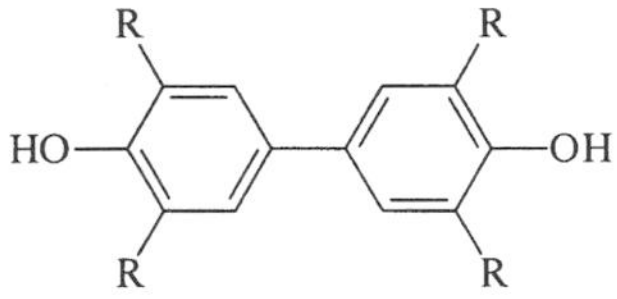

Other colors can be achieved by oxidative cleavage of protecting groups, such as the formation of cyan:

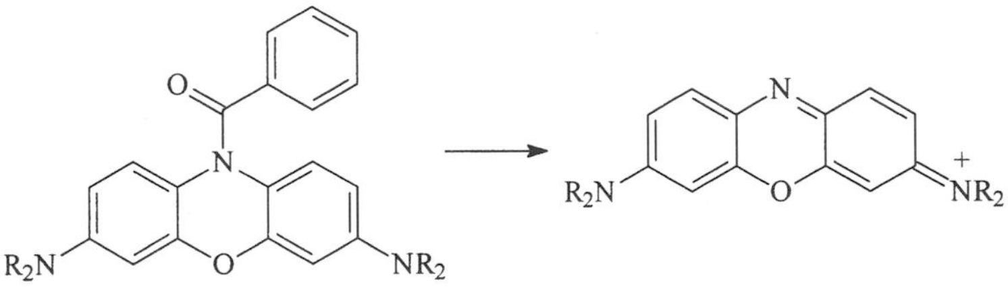

With magenta, obtained from a variation on the above structure, a full-color imaging construction is possible:

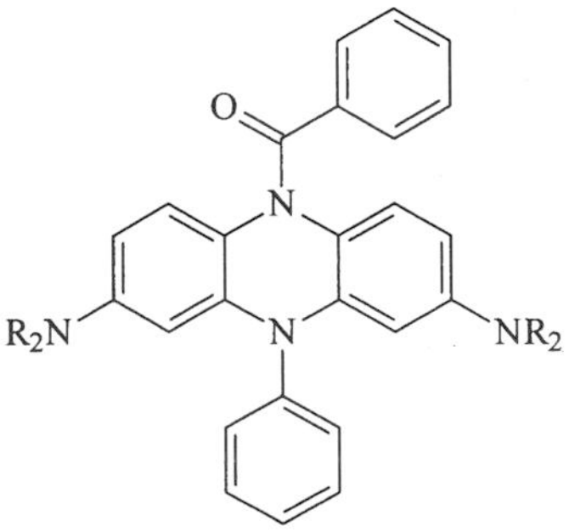

The major drawback to this system is the residual metallic silver formed in the image area, which prevents good color clarity. The solution has been to create a dye diffusion system with dyes such as those shown above. In this case, the dyes thermally transfer during the development step to a dye receptive layer, which then can be peeled away to give a clear full-color image. Not only is the metallic silver image left behind, the stability of the image is further improved as the developing chemistries are left behind in the donor sheet as well. Alternatively, the cationic dye generated by this chemistry can be prevented from diffusing away from the image using dye selective interlayers in the construction (150).

An alternate full-color system, based on photothermography, has been sold under the Pictrography name by Fuji (151–156). This system is composed of silver halide, a silver acetylide derivative, a hydrophilic basic metal compound, and a dye releasing redox (DRR) compound in a gelatin matrix that is laminated to a mordanting receiving layer containing a base generator in gelatin. The magenta DRR, for example, in the nondiffusible form is

The process involves exposure of the silver halide layer with specific wavelengths to correspond to the color of interest, followed by a uniform water wetting of the surface of the donor sheet just prior to lamination to the receptor sheet. The water, at $1 \mu g/cm^2$, transfers the chelating agent in the receptor layer to complex with the hydrophilic basic metal compound in the donor sheet in order to generate a strong base in the donor. The water also provides an environmentally friendly transfer medium that facilitates transfer of the dye from the donor to the receptor. During the thermal processing step, at 90°C for 20 seconds, the basic medium of the donor layer enables the silver ion redox chemistry to occur with the dye precursors in the image areas. The redox released dye then transfers to the receptor sheet. A cationic imidazole polymer is used further to mordant the dye in the receptor sheet. The silver acetylide compound is the silver ion source for the redox chemistry, much as silver behenate is for typical black-and-white photothermography. In this case, the silver acetylide is claimed to provide a broader processing latitude than silver carboxylates or other silver sources. In addition, the acetylide by-product is stated to be a stabilizer. As in all dye transfer systems, the receiving layer enables a clean full-color system to be achieved without the color being grayed by the metallic silver by-product.

A different photothermographic material, introduced as the Konica Dry Color System, utilizes a combination of steps ending in dye transfer to a receptor sheet (157). The thermal development system first generates a *p*-phenylenediamine color developer by decomposition of its precursor. The color developer is oxidized by the silver in silver benzotriazole, which reduces the color-forming couplers. Diffusion of the dyes formed from the system's photosensitive sheet to an image receptor sheet produces the final image. From analysis of the kinetics of the individual reactions, generation of the color developer was determined to be the rate-determining step. The high activation energy (>100 kJ/mol) is proposed to provide high stability of this material prior to development and good reactivity during thermal development.

Finally, a new noteworthy development in this area is the announcement of the first camera-speed color photothermographic imaging system (158). It is a photographically responsive, thermally developable layer that is panchromatically sensitized. It is claimed to be capable of concurrently forming silver and dye image densities. Unlike most photothermographic imaging constructions, the non–light sensitive silver source component in this case is silver 3-amino-5-benzylmercapto-1,2,4-triazole. Standard color coupler dye compounds are used as the color formers in this system.

In summary, developers built on phenols not only enable facile formation of a black metallic silver image for black-and-white imaging applications, they can be used as protecting groups in dye systems. In this function, they provide quality color images to form as well as liberate the dye from ballasting groups for enhanced mobility and transfer to a receptor sheet.

13.5.3 High Contrast

A fundamentally significant new development has been reported since the last reviews on PTM-based silver carboxylate systems, which involves infectious development, a key component for applications requiring high contrast, such as digital printing (89). An image-setting product (Kodak DryView recording film) has recently been introduced based on this technology (see, for example, Ref. 89), and efforts are underway in several laboratories to commercialize similar products (24,119,159–168). High-contrast films are preferred for halftone/text applications, such as image setting. In these materials, contrasts in the range of 20 to 30, and higher, have been obtained, compared to the low single digits for conventional PTM without the infectious developers. The typical D log E curve of a high-contrast film, compared to a conventional film, is shown in Fig. 13.9.

As with conventional silver halide systems used for image setting applications (57), it has been discovered that certain types of hydrazines and hydrazides can be used to generate high-contrast PTM imaging materials. These infectious developers or nucleators are represented by a formylarylhydrazine (A, B) (159,163) a trityl hydrazide (C) (163), and an acrylonitrile (D) (164).

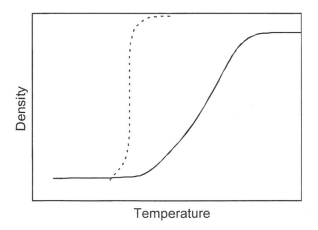

Figure 13.9 Idealized D log E curve of a high-contrast film compared to a conventional PTM film.

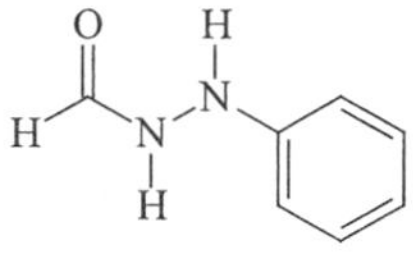

(A) 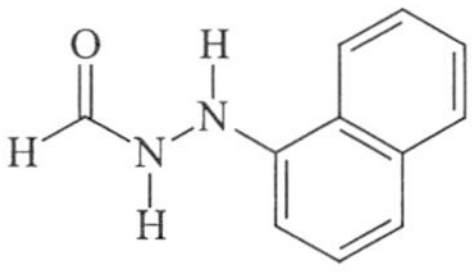   (B)

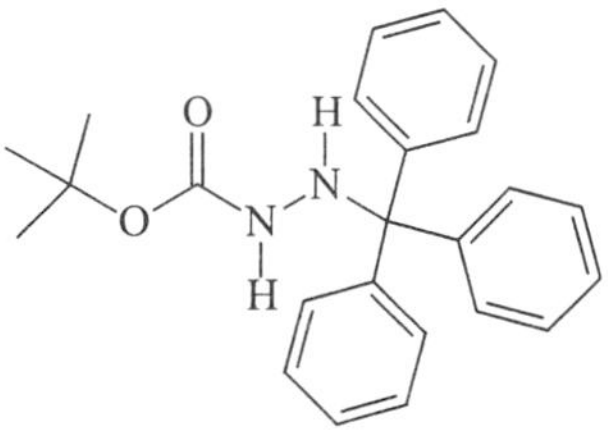

(C) 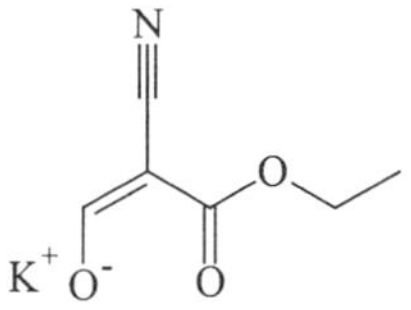   (D)

The mechanisms by which these high-contrast agents function are various and depend on the nucleator:

1. Hydrazide nucleation mechanism (170):
    a. $H-(C{=}O)-NH-NH-Ar + D_{ox} \rightarrow H-(C{=}O)-N{=}N-Ar +$ Developer
    b. $H-(C{=}O)-N{=}N-Ar + Nucleophile \rightarrow HN{=}N-Ar + H-(C{=}O)-Nuc$
        where Nucleophile = water or ROH.
    c. $HN{=}N-Ar + 2AgBe \rightarrow 2Ag° + [N{\equiv}N-Ar]^+$
2. Acrylonitrile nucleation mechanism (89):
    a. AgX hv $\rightarrow$ Ag° (latent image).
    b. Ag° (latent image) + AgBe + Developer Heat $\rightarrow$ Image Ag° + D_{ox} (oxidized developer).
3.

D_{ox} + 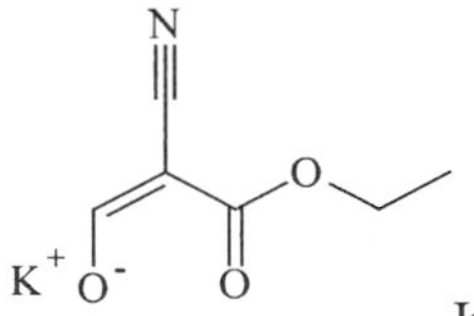   Image Ag°, Heat $\rightarrow HCO_2H +$
    Other by-products (cyanoacrylate nucleator) (trace water)
4. $HCO_2H + AgBe + Developer{-}Heat \rightarrow$ Nucleated image Ag° + HBe

Hydrazides are generally first activated to a diazene intermediate by oxidized developer formed around latent image from developed silver. Formic acid generated from the decomposition of the acrylonitrile nucleator has been proposed to be the active fogging agent (89). The addition of these infectious developers to photothermographic systems greatly changes the course of development. Films containing infectious developers typically require significantly less Ag to reach a given density than films without infectious developers. The morphology of the metallic silver that composes the nucleated image is unlike most of the conventional photothermographically formed silver (46,113,114,162). A recent report (168) has shown that in the case of the hydrazines, the infectious developers encourage the formation of extremely fine, approximately 0.028 μm size metal particles, well separated in the imaging layer, in contrast to the dendritic clusters or filaments typically seen (46). The result is substantial optical density and high contrast in the image area. This report discusses the development kinetics of high-contrast photothermographic films measured at various development temperatures using a spectrophotometric method. Particle size and morphology were found to correlate to light absorption and Ag covering power. Standard photothermographic development was shown to follow a sigmoidal, auto-catalytic rate equation. For such films, the thermodynamic parameters are similar for unex-posed and exposed samples. The calculated entropy of activation indicated a unimolecular transition state in the rate-determining step for films without infectious developers. Infec-tious development was shown to follow more complicated kinetics that did not generally fit simple mathematical models. In the presence of infectious developers, the "sphere of influence" (2,169) of each AgX grain depends more on the diffusion of fogging agent away from the grain than on the diffusion of Ag^+ to the grain. In the absence of infectious developers, nucleation sites are limited to the AgX bearing a latent image.

Photothermographic development without nucleators proceeds sequentially, whereas infectious development occurs in parallel. In parallel development, all Ag crystals form at the same time, while in sequential development the Ag particles form at different times (167). Because all the crystals develop at the same time in parallel (Lith) development, the slope of the overall development curve reflects the development of individual Ag particles and does not depend on exposure.

The basic formulation used for the high-contrast films is fundamentally the same as the conventional photothermographic film, including silver bromide in "catalytic associa-tion" with a silver source, developer, toners, and stabilizers. The addition of the infectious developer is primarily the only major change required. One difference in image-setting applications is that latent image stability is important, as the films may end up being pro-cessed hours after being exposed. Latent image stability is reported to be excellent for the Kodak DryView recording film as D_{max}, speed, and contrast are nearly invariant after a 24 h period between exposure and thermal processing (89). Also, one additional requirement for image setting is good UV transparency for the subsequent printing plate exposure. This is a potential problem, as there are several UV adsorbing materials present in the film, such as aromatic compounds and the process surviving AgX grains. However, UV transparency seems to be adequate, although it is higher than that of conventional image-setting films.

Finally, whereas all of the imaging components remain in the film, the question of image stability after processing also remains an issue. While the infectious developer may induce special stability needs, this does not appear to be a significant problem.

13.6 TONERS

A thermal imaging system, based on a simple mix of the silver source with the developer, does not produce the most esthetically appealing color upon development. Under these conditions, a yellow to dark brown metallic image is generally obtained. The color tone can be adjusted using certain types of organic additives, which have been called toners because of the effort of the formulators to create the proper image color, most desirably a deep black. As the chemical roles of the toners have become clearer, it is probably more appropriate to call these compounds development accelerators or promoters, which also happen to affect the image color. Nevertheless, the term toner will be retained here simply to remain consistent with the literature.

There is a wide choice of toners used in contemporary PTM; we will focus on those found to be most practical. Many toner classes have been reported over the years, although there is now little to report regarding new classes since the last reviews. Most commonly used are phthalazine (PHZ) (for example, 19,25,171–173), its carbonyl analog 1(2H)-phthalazinone (PAZ), and phthalimide (or succinimide), all shown here.

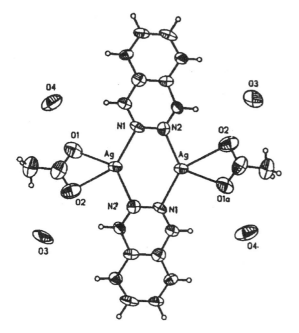

PHZ   PAZ   Phthalimide

A wide variety of new PAZ derivatives have been recently reported (174), which have provided enhanced properties, including stability improvements. Some preliminary observations regarding unusual photographic properties, such as long induction times followed by rapid development, were observed.

It is interesting to note that virtually all toners reported in the literature contain an aromatic nitrogen or a nitrogen capable of metal complexation. This feature in the toners has provided the basis for the proposal that the role of these compounds is to transport the silver ion to the reduction centers (2). The silver complexes that fulfill the role of intermediates by this mechanism, however, have never been reported. In the case of the most widely used toner, PHZ, for the first time since this type of toning compound was reported decades ago, several important compounds have been reported that provide the first evidence that toners can actually form stable complexes with silver that can be isolated and characterized (15,16,175,176).

The first of these complexes, the PHZ complex of silver acetate, showed the ability of the PHZ as a ligand to extract the silver from the dimeric silver carboxylate structure to form a stable, colorless complex, Fig. 13.10.

Figure 13.10 Molecular structure of $[Ag \cdot PHZ \cdot O_2CCH_3]_2$ (176). (Reprinted with permission from Kluwer Academic/Plenum Publisher.)

The acetate ligand, presumably a reasonable model for the longer chain carboxylates used in PTM, is now chelating the silver rather than bridging metal centers. It was not reported how this structure results in the formation of the black colored metallic silver upon addition of the reducing agent. It is clear from the preparation of this compound, however, how the solubility of the silver carboxylate was dramatically improved in the reaction solvent by the reaction with PHZ. Furthermore, a wide range of silver carboxylate complexes with PHZ, based on the chemistry reported for the acetate, have been claimed as suitable silver sources for photothermographic imaging materials (60,61). If complexation-induced solubility is important in PTM then these results are consistent.

In the case of photothermographic materials constructed from the phthalazine/ phthalic acid (PHZ/PA) toner system (20), a well-defined silver complex has also been reported (175). The structure of the silver phthalate complex with PHZ is shown in Fig. 13.11.

In this case, the same fundamental $[AgPHZ]_2^{2+}$ dimer forms completely analogous to the acetate version shown above. The main difference between these two complexes appears to be that the $[AgPHZ]_2^{2+}$ dimers stack in the form of a polymer on top of each other in the acetate complex while they are linearly polymeric in the phthalate complex. There is no report of the significance of this structural difference in these complexes regarding their reactivity, as they both precipitate black metallic silver in the presence of a developer. The solubility of this latter complex, while still poor compared to the starting silver phthalate, cannot be used as a good guide for silver ion transport, as it was prepared from water and contains water in the crystal lattice. In addition, the measured solubility and the anhydrous analogue of this complex were also not reported.

As noted above, mixed in with the toner class of compounds is that of development accelerators. It is worth noting that one member of that class, triphenylphosphine (37),

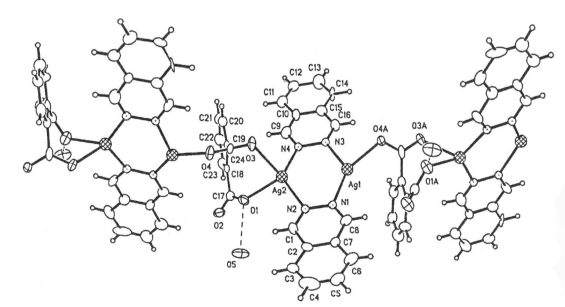

Figure 13.11   Molecular structure of $[(Ag \cdot PHZ)_2 \cdot (O_2C)_2C_6H_4]$ (175). (Reprinted with permission from Elsevier Science.)

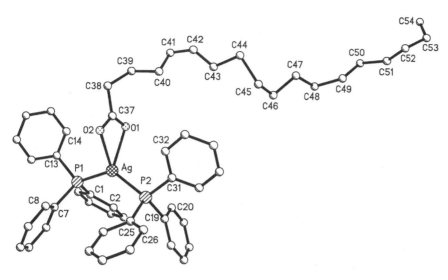

Figure 13.12 The molecular structure of [*bis*-(triphenylphosphine)-silver(I) stearate] (58). (Reprinted with permission from Kluwer Academic/Plenum Publisher.)

has been investigated as a ligand for silver carboxylate (58). The silver stearate complex with triphenylphosphine has been reported and is shown in Fig. 13.12.

The $2:1$ Ph$_3$P:Ag$^+$ complex is quite stable and soluble in the organic solvents investigated. The complex having this ratio is not very reactive with typical *bis*-phenolic developers, while the $1:1$ ratio complex, which was not isolated due to equilibrium preference for the $2:1$ deposition, apparently is. As in the cases where PHZ is used to complex silver, the Ph$_3$P ligand readily removes the silver from the otherwise stable $[Ag(O_2C_xH_{2x-1})]_2$ structure. The details for the solubilization of the silver stearate structure with this ligand are proposed in that report.

It is easily seen how the silver complexes resulting from reaction with these types of toning/development accelerator agents provide enhanced solubility of the silver carboxylate. As a result, one can see how the proposed mechanism for their function can be justified. So far, however, the story ends here, as the question of how these complexes deposit the metallic silver in a filamentary or dendritic form (Section 13.10), and the kinetics related thereto, has not been answered.

While the precise chemical role the toner/development accelerator plays in the development of photothermographic imaging materials has not been definitively elucidated, the recent progress made in understanding how those components can react with the silver source has provided new directions to improving the properties of these imaging materials (60,61). This initial work with silver carboxylates is encouraging in that a similar understanding of the reaction of toners/development accelerators with the other commercially important alternate silver sources should be possible. A rich reaction chemistry can be expected.

There is a further notable complication with at least the PHZ/PA type toner systems. Recently, stable $2:1$ and $2:3$ toner complexes of PHZ/PA have been shown to form easily (177,178), Figs. 13.13, 13.14. Simply modifying the crystallization conditions for the solutions of the mixtures precipitates compounds having these stoichiometries. Both complexes are simply the result of an acid/base interaction, although the hydrogen bonded

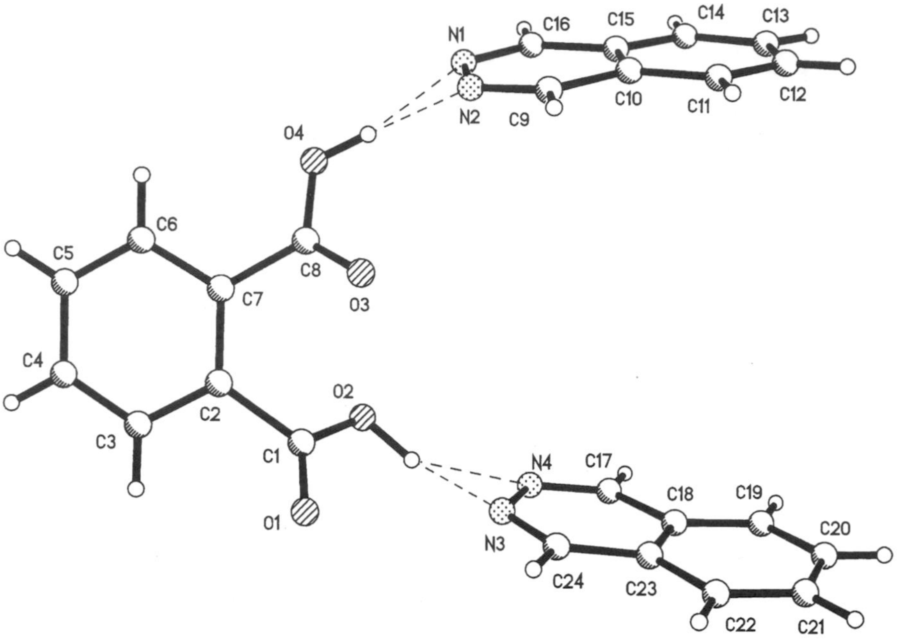

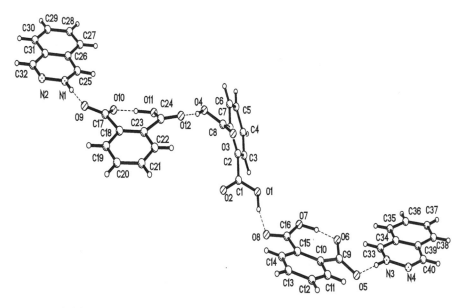

Figure 13.13   Structure of 2:1 PHZ:PA crystals (177). (Reprinted with permission from Elsevier Science.)

Figure 13.14   Structure of 2:3 PHZ:PA crystals (177). (Reprinted with permission from Elsevier Science.)

carboxylate hydrogen remains associated with the acid in the 2:1 complex, but is associated with the nitrogen in the 2:3 complex. The solid state melting point reflects this subtlety. Considering the ease by which these compounds were obtained, other ratios may be anticipated, although their significance in PTM is not known.

Parenthetically, it might be added that color tone has been reported to be controllable, or at least modified to a preferred bluer image tone, by preparing the silver carboxylate so as to change its crystallinity (42,43). Modification of the silver carboxylate preparation procedure, by the incorporation of certain types of phosphonium compounds and organic solvents, yields higher crystallinity solids, which provide the basis for affecting the eventual metallic silver tone.

13.7 POLYMER BINDERS

The polymeric binder of a PTM imaging layer typically performs several key roles in the overall performance of the film. Some of the properties the binder must possess to accomplish these key roles are

Transparency
Stabilization of submicron particle dispersions, AgX, and the silver source
T_g at or below thermal processing temperature
Soluble or dispersible in coating vehicle
Dissolves organic additives
Compatible with silver carboxylates and silver halides
Contains minimal levels of undesirable impurities
Cross-linkable

Most commercial PTM are coated from organic solvents (such as ketones) and require the use of binder polymers that are dissolved by such solvents. Polyvinyl acetal resins have been widely used as PTM binder polymers, as they meet these binder requirements (2). Polyvinyl acetal resins are terpolymers of vinyl acetate, vinyl alcohol, and a vinyl acetal, such as the generic structure shown below for polyvinyl butyral, and are available from several manufacturers.

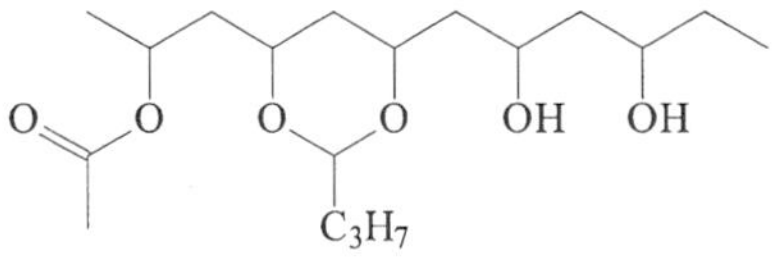

The polyvinylbutyral binders have a range of molecular weights, solubilities, and physical properties that are dictated, in part, by the ratios of the constituent monomers. Properties of representative polymers are summarized in Table 13.3.

Polyvinylbutyral resins are easily compatible with the long chain fatty acid Ag-carboxylates, are very soluble in ketone coating solvents, and generally do not adversely impact sensitometry. There is, in fact, some recently reported evidence that this type of binder actually enhances the photothermographic development reaction when used with silver carboxylates (180). Interestingly, polyvinylbutyral resins from various manufacturers have been found to exert a different impact on the developability of TM materials. These performance differences were correlated with the levels of impurities in the PVB resins. The impurities identified as being responsible for these variations of the TM films were traces of acid from the polymer synthesis and incorporation of stabilizers and antioxi-

Table 13.3   Selected Properties of Polyvinylbutyrals

Property/product	B-72	B-76	B-90
Molecular weight	170–250	90–120	70–100
Viscosity (10%)	1,600–2,500	200–450	200–400
Hydroxy content	17.5–20.0	11.0–13.0	18.0–20.00
Acetate content	0–2.5	0–1.5	0–1.5
Butyral content	80	88	80
T_g	72–78	62–72	72–78

Source: Ref.179.

dants for the binder. The overall significance of this work is that it echoes the message (26) that impurities, from polyvinylbutyral or any raw material used in the preparation of PTM or TM films, have a profound effect on the performance of the materials.

One of the disadvantages of the polyvinylbutyral binder system is that, due to its low glass transition temperature (T_g), relative to the processing temperature, it is subject to facile deformation at the processing temperature for PTM. This propensity for deformation (low modulus) of polyvinylbutyral coatings at processing temperatures can cause unwanted distortion or embossing of the surface of a PTM film. One common solution to the embossing problem is to overcoat the primary imaging layer with a protective layer, such as cellulose acetate butyrate (20).

During the past few years, patents have been issued that disclose methods for cross-linking polyvinyl acetal binder to enhance its modulus. The free hydroxyl groups in polyvinyl acetal resins present a point of chemical reactivity through which the resins may be cross-linked. There are several reagents known to react with the secondary alcohol functionality of the PVA unit to cross-link the polyvinyl acetal resin. For example, the polyvinyl acetal binders used in PTM have been reported to be hardened using boric acid or borate anion (181). Similarly, the use of boron alkoxides has been reported to cross-link polyvinyl acetal binders in PTM films (182). Di-isocyanates (20), dialdehydes, and epoxides have also been used to cross-link polyvinyl-acetal-containing materials.

For thirty years, PTM, such as Dry Silver, have been successfully coated from organic solvents. Recently, however, considerable patent activity in the area of aqueous coated PTM has appeared. Advantages from an environmental impact point of view, and the potential ease of coating, might be anticipated. The aqueous coated films employ either latex binder polymer dispersions or dissolved polymers such as gelatin or polyvinyl alcohol. One problem unique to aqueous coated PTM films is that additives such as developer, toners, and the like, that are freely soluble in organic solvents, are often insoluble in water and need to be milled into a solid particle dispersion prior to coating. For example, the use of a water soluble polyvinyl acetal as binder has been reported (180,183). Gelatin has been disclosed as a binder for aqueous coated PTM compositions (123,184). A water-dispersible polyvinylbutyral latex plus surfactant is reported that is used as the binder for a photothermographic emulsion layer (119). The use of certain Laponite clays, in combination with anionic polymers such as polystyrene sulphonic acid, are noted to be useful as gelatin thickeners to control the viscosity in coating fluids of aqueous PTM emulsion layers (186). The gelatin thickeners are reported to enable the coating of such fluids by curtain or extrusion coating techniques. Aqueous coated PTM imaging layers have recently been described that make use of styrene-butadiene copolymer latex binders, having a general composition: styrene:butadiene = 7:3 (187). Apparently, this type of latex binder copolymer is preferable to hydrophilic polymers, such as polyvinyl alcohol or gelatin, in that they adsorb less water and hence are less susceptible to fog in a high-humidity environment.

13.8 SUPPORTS, OVERCOATS AND BACKING LAYERS, ANTISTATS, AND ANTIHALATION MATERIALS

13.8.1 Supports

Polyethylene terephthalate (PET) has been the commercial film base material of choice for many years because of the large number of features that make it attractive for photographic properties, such as clarity, low UV adsorption, low humidity coefficient, good stiffness, high tear strength, and low cost. These outstanding film properties make this film base suitable for use in medical PTM and microfilms, for example. For some applications that require exposures in the red or near IR, it has been the practice to incorporate a process-surviving blue antihalation dye into the support. The antihalation dye is added to the molten film base polymer prior to extrusion. The blue antihalation dye in the support has a dual role in counteracting the generally warm image tone of the image silver. This approach is accompanied by an increase in the optical density of up to 0.10 in the red.

However, use of PTM on a PET support for image-setting applications is somewhat limited because of the potential for significant thermal distortion of PET base that can occur during thermal processing. This thermal distortion is brought about by the relatively high processing temperatures ($\sim$120°C) required for PTM, which are substantially above the T_g of this material (onset $\sim$100°C). The result of using standard PET as a support for PTM image-setting applications is additive, as dot registration is required for each color of a separation that is done. This thermal distortion problem can be minimized by careful control of PET support manufacturing conditions that anneal the support at high temperature under low windup tension (89). One solution claimed is by thermally annealing the film under a tension of below 4 kg/cm^2 so that the dimensional distortion in both the transverse and machine directions can be reduced (188).

The molecularly similar polyethylene naphthalenate (PEN) support has been reported to have advantages such as reduced thermal distortion and reduced curl over PET as a film base for PTM films (189). However, because of the additional aromatic ring, PEN has a significant absorption in the ultraviolet, which, in combination with the high cost of this film base, complicates the ability to utilize this material for PTM image-setting applications.

Alternatively, a tenfold reduction in thermal distortion in PTM image-setting films, compared to that exhibited by untreated PET, can be accomplished by using a polycarbonate support (119). The polycarbonate support has not achieved wide use at this time, as it must be coated using an aqueous emulsion priming layer, because solvents generally attack this film base.

13.8.2 Antistats

Static charge buildup on the surfaces of PTM film can lead to marking in the processed film, poor transport through manufacturing or processing equipment, and sticking together of processed sheets of film (''static cling''). To circumvent these problems, several different approaches have been utilized. The simplest solution is to avoid an active antistatic layer by balancing both sides of the film for triboelectric charge. This usually means having the outermost layers of the film composed of substantially the same materials although constructions based on this approach can have problems when the relative humidity is low. As in conventional films, electroconductive antistatic layers are typically coated on the side opposite the emulsion layer to actively dissipate static charge buildup. The electroconductive materials in the antistatic layer are either conductive metal oxides, such as antimony doped tin oxide or silver doped vanadium pentoxide dispersed in a polymeric binder, or are polymeric in nature. The antistatic layers are either ''buried,'' i.e., an outermost insulating layer covers the conductive layer (190), or else the electroconductive layer is part of the outermost layer (191). Static charge dissipation via a conductive outermost layer is the most effective way to provide protection against static related problems, as surface charge may be dissipated by conduction to ground.

PTM antistatic layer constructions have been disclosed that are composed of conductive polymers (192). Typically, these conductive polymer layers contain either an ionic copolymer in the form of particles dispersed in a binder layer or an electrically conductive polymer such as polythiophene (193), as shown.

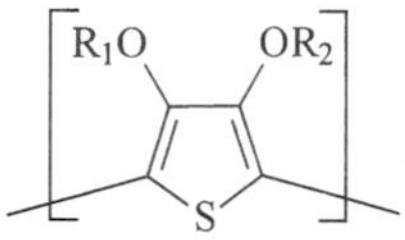

13.8.3 Overcoats and Backing Layers

The overcoat layer for PTM films plays an important role in the construction. It must, for example, protect the silver-containing layer from external influences, such as physical pressure during development or migration of undesirable reactants from the environment. Also, the overcoat must enhance the transport of the film, provide a barrier to transport of mobile chemicals from the silver-containing layer, and, in some cases, act as a reservoir for reactive materials used during thermal processing. Overcoat and backing layers are often composed of similar materials in order to balance physical property attributes such as humidity-induced curl and triboelectric differences. Several types of polymeric materials are used in overcoat and backing layers in thermographic and photothermographic films. Cellulose acetate polymers and their ester derivatives have been used for many years in PTM overcoats. This family of polymers has low thermal distortion, is reasonably tough, and may be solvent coated simultaneously with a nonaqueous silver-containing layer.

Overcoat compositions comprising polyvinyl alcohol reinforced by nanoparticulate silica sols have been disclosed (194). Similarly, overcoats composed of polyvinyl alcohol cross-linked by silica matte have also been claimed (195). In addition, the use of specific polycarbonate copolymers as binders in overcoat formulations for reducing surface deformations in TM materials, because of contact with the printhead, has been disclosed (196). The copolymers evaluated are composed of *bis*-phenol-A and the *bis*-phenol monomer shown below, which has a ring structure attached to the central carbon atom. By copolycondensation of *bis*-phenols of this type, it was possible to vary the T_g of the overcoat from 180 to 240°C. A high T_g for an overcoat binder is preferable for reducing thermal surface deformations.

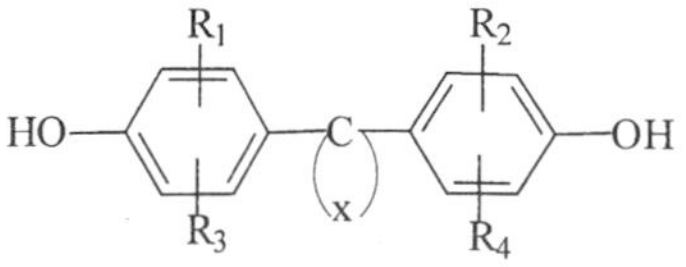

13.8.4 Antihalation Materials

Antihalation dyes are used in photothermographic films for the same reason as in conventional silver halide constructions (57): they improve the sharpness of the image. The antihalation dyes operate by absorbing stray light reflected at layer boundaries and hence reduce image artifacts. Photothermographic antihalation dyes fall into two general classes: those that survive thermal processing and those that are destroyed by the process or a postprocess light bleach step.

Two recent disclosures report the use of solubilized phthalocyanine compounds that can be used as process-surviving antihalation dyes in PTM. In one case, the dye is placed in a separate coated layer, while in the second, the dye is actually dissolved in the film support (197,198). The phthalocyanine dye class is attractive as a red and near-IR antihalation dye class because of its high extinction coefficient ($\sim$100 ppm dye dissolved in the support polymer is sufficient for good antihalation protection) and because the λ_{max} of the peak absorption is adjustable between the wavelengths of 630 and 770 nm by varying the center metal and/or the ring substituents. Also important is that metal phthalocyanines have a sharp absorption in the red to near IR (20–25 nm width at half height) with little other absorption in the visible spectrum (197,198). Because of the inherent absorption of the phthalocyanine antihalation dyes, they are limited to red light exposures.

The dihydroperimidine squarilium dye class has been reported to be effective when used as antihalation dyes in PTM films that are exposed with near IR ($\sim$ 750–850 nm) light (199).

A new type of thermally bleached antihalation system for photothermographic constructions has recently appeared (200). This antihalation system is coated as a separate layer and consists of a cyanine dye with special quaternizing groups and a base precursor. The dye is stable until the thermal processing step, shown in Fig. 13.15, during which the base generating chemistry is released, deprotonating the dye. The dye then cyclizes onto the polymethine chain, destroying the color. A wide range of cyanine dyes is claimed, along with any base precursor. The preferred base precursors are substituted *bis*-guanidines that are added as solid particle dispersions.

Another heat bleachable antihalation system has been patented that makes use of a free radical dye bleaching chemistry. In this system, the free radical generator is a substituted hexa-aryl-*bis*-imidazole that thermally decomposes, destroying a formazan dye (201).

A PTM acutance system based on microencapsulated dyes added to one of the layers of a photothermographic construction (usually the emulsion layer), to improve sharpness, image tone, and the light stability of the image, has also been described (202).

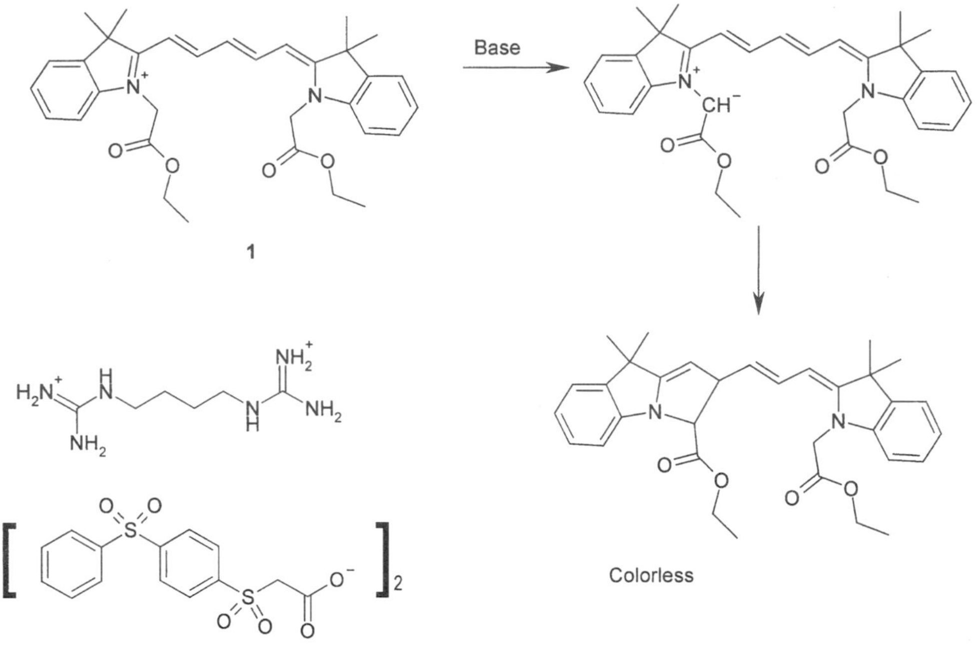

Figure 13.15   Thermal processing of polymethine dye.

The shelf life of thermally bleached antihalation dyes needs to be taken into consideration when formulating a PTM film construction, as there is a fine distinction between the thermal bleaching of the dye during the film coating operation, the slow decomposition of the antihalation dye on the shelf, and the intended bleaching of the dye during thermal processing of the film.

13.9 STABILIZERS

13.9.1 Stabilizing Compounds

Photothermographic imaging systems based on silver carboxylates require several types of stabilizers, such as antifoggants, shelf stabilizers, and print stabilizers. Early in the development of the photothermographic imaging technology, it was discovered that mercury can be added to these imaging materials, which provided the means for stable films (203). Presumably, the Hg(II) ion acted as a stabilizer by having the ability oxidatively to convert Ag° fog centers to relatively benign Ag^+, similar to conventional silver halide films (57). A second advantage was that the shelf life of the coated formulations was also improved significantly. As a result, most of the commercially available Dry Silver™ (3M Company) products incorporated mercury salts.

An alternative oxidant has been successfully incorporated into silver carboxylate–based constructions, pyridinium hydrobromide perbromide, pyridine · HBr_3 (204). The details of the mechanism are not reported, and in fact there is only sketchy information on the organic reactivity of this compound (205,206). Similar structures based on quarternary nitrogen, phosphorous, or sulfur polyhalides have since been claimed that are characterized by a specific use test (207).

In the drive to improve the overall print quality and environmental aspects of new films based on the silver carboxylate technology, mercury had to be replaced with alternative stabilizers. Unfortunately, the mechanism by which mercury actually produces its stability effects has been given relatively little attention (85), or so the lack of substantial data in the literature would suggest, and only a few reports exist on stability in general (208–211). The general statement regarding the Hg(II) driven oxidation of Ag° fog centers noted above is reasonable but unproven in PTM. The result of not understanding the details of stabilization complicates efforts to generate new materials having the effect of mercury without its disadvantages. Nevertheless, some impressive results have been achieved with the discovery of tribromomethyl-containing compounds (212), and as a result we have seen a wide range of highly effective functional derivatives (102,118,213–219). These include tribromomethyl-sulfonyl aromatic compounds, such as

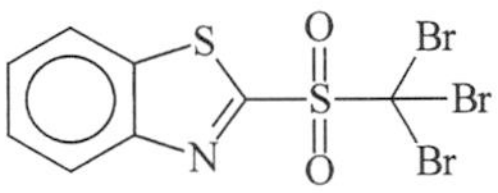

as well as nonaromatic versions such as

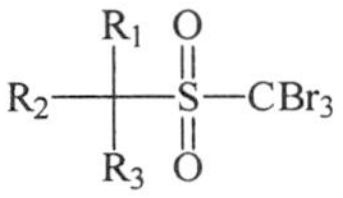

The antifoggant and stabilizing mechanism of tribromomethyl compounds in silver halide photographic materials is usually attributed to the equilibrium loss of bromine radicals, which then oxidize the metallic silver fog centers (2,220–222). Alternatively, photoinduced bromine elimination may generate the oxidizing materials (218). Such radical formation depends primarily on the overall stability or photoactivity of the tribromomethyl compound, which must be considered independent of its location in the film. Presumably, the process must be inherently inefficient, because a large portion of the radicals formed can be envisioned as scattered at locations unrelated to fog centers. In these cases, the proximity of silver carboxylate is such that the released bromide can readily exchange with the carboxylate to form additional silver bromide. Under these circumstances, the potential for the formation of the resulting additional photoactive sites might actually lead to less stability.

On the other hand, if the antifoggant could be attached directly to the silver halide crystal surface by coordination, the efficiency of the bromine elimination of fog centers should be noticeably improved. The recently reported chelating ability of α-bromo-alkylketone to silver, and its ability subsequently to generate AgBr (223), suggest that a similar Ag—Br bonded structure could from between antifoggants containing coordinating groups, such as 2-(tribromomethylsulfonyl)-benzothiazole, shown above, and a silver halide surface. The ability of this latter antifoggant, in particular, and similarly constructed analogs by implication, to coordinate to silver has been reported (210). The molecular structure of this complex is shown in Fig. 13.16.

While the bromine is not directly bound to the silver, as suggested by the reported chelation of α-bromo-alkylketone to the silver, the weak Ag ⋯ O bonding in this case can be expected to be in equilibrium with a transient Ag ⋯ Br form, as illustrated in Fig. 13.17.

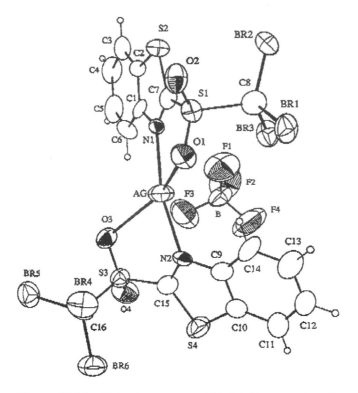

Figure 13.16 Molecular structure of *bis*-2-(tribromomethylsulfonyl)-benzothiazole-silver (210). (Reprinted with permission of the Society for Imaging Science and Technology, sole copyright owners of *The Journal of Imaging Science and Technology*.)

The aromatic nitrogen could serve to maintain the ligand position on the surface of the silver halide grain, enabling the equilibrium loss of Br• to occur somewhat near the fog center. Fog centers located on the silver halide crystal might then be eliminated through bromine transfer from the 2-(tribromomethylsulfonyl)-benzothiazole, with AgBr as the by-product. While attempts to prepare the direct intermediate for this hypothesis, a complex of 2-(tribromomethylsulfonyl)-benzothiazole with AgBr, were not successful, the ring nitrogen bound complex of AgBr-(2-amino-benzothiazole) has been (224), which provides some support for the hypothesis. It should be emphasized, however, that despite the large number of polyhalo compounds that contain a metal ligand group, the stabilizer's ability

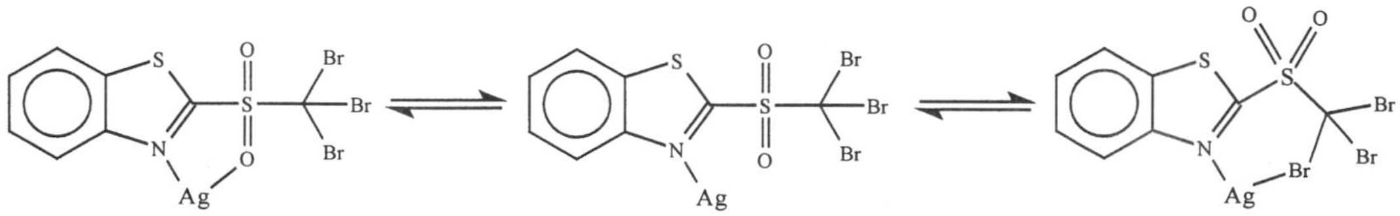

Figure 13.17 Possible equilibrium structures of coordinated aromatic amine on a silver halide surface (210). (Reprinted with permission of the Society for Imaging Science and Technology, sole copyright owners of *The Journal of Imaging Science and Technology*.)

to coordinate to the silver halide surface is not necessarily a prerequisite for effective stabilization by tribromomethyl compounds (225), and there is more to be learned in this area.

The inherent stability of tribromomethyl compounds varies greatly, and the more reactive versions may have undesirable health side effects. In order to minimize this potential, computational modeling has been reported to provide a good screening test prior to incorporation of these compounds into the imaging material (211). In this case, it was found that calculated octanol/water partition coefficients, $\log(P)$, were a good indication of undesirable compounds that might possess mutagenic properties. For example, compounds that had a calculated $\log(P)$ below 2.5 were likely to fail toxicology testing, while those having $\log(P)$ above 3.8 were found to be nonmutagenic. The middle ground between 2.5 and 3.8 required additional calculations. This latter region was found to be effectively predictable when correlated to ionization potentials. For example, calculated ionization potentials between 10.0 and 10.8 eV corresponded to nonmutagenic compounds. Calculations such as these have provided major advances for the predictions required to generate new, safe antifoggant compounds based on organic halides suitable for replacing mercury.

Other types of stabilizers having the ability simply to coordinate to silver ions, well known for conventional silver halide imaging materials (57), are also used in photothermographic materials. These include tetraazaindoles, such as 5-methyl-*s*-triazolo[1,5*a*]pyrimidin-7-ol or 4-hydroxy-6-methyl-1,3,3a,7-tetrazaindine (TAI), 1-Phenyl-1H-tetrazole-5-thiol (PMT), benzotriazoles (BZT), and 2-mercaptobenzimidazole derivatives (MBI), etc., such as illustrated here

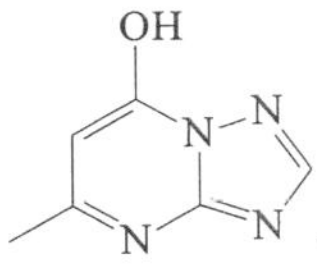

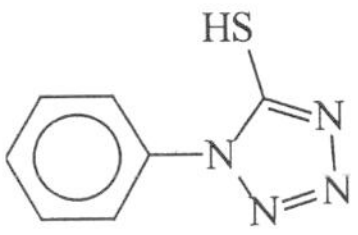

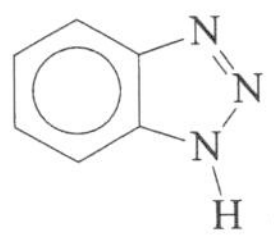

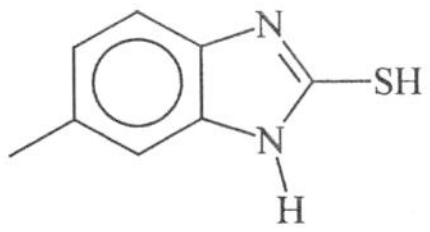

2-Mercapto-5-methylbenzimidazole

The mechanism of their activity in photothermographic materials is considered analogous to silver halide systems by removing residual Ag^+ ions from the reduction equation.

One of these, 2-mercapto-benzimidazole, has recently been revealed to form a highly complicated solid-state structure with AgBr under mild conditions in organic solvent (226). The repeating unit, based on connecting $[AgBr]_2$ dimers, is shown in Fig. 13.18.

In this case, the thione ligand readily extracts the silver bromide from the solid to form a light insensitive polymeric complex. While this ligand is reported to form AgMBI, via deprotonation (227,228), in aqueous media, this new complex retains the N-H bonds when prepared in organic solvent. The fact that such different species are formed from the same ligand and silver halide simply by changing the solvent suggests that there is more to be learned regarding the stabilizing mechanism by this and other metal complexing stabilizers. Because the photothermographic materials are often coated from organic solvents, this difference may be quite important.

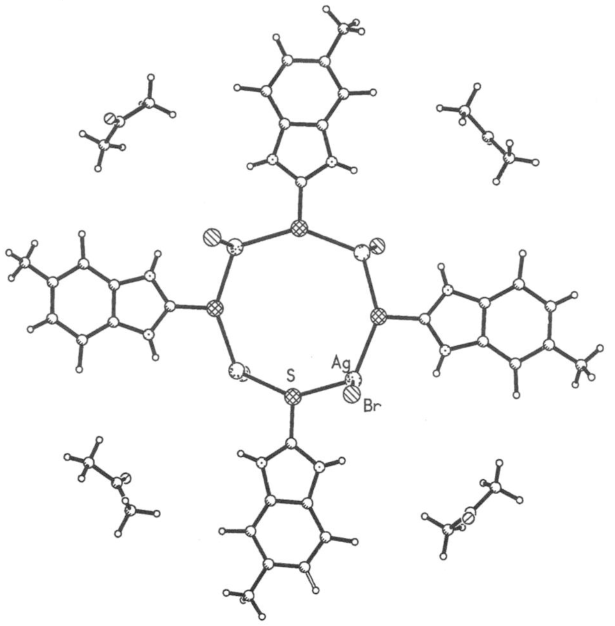

Figure 13.18   Molecular structure of 5-methyl-2-mercaptobenzimidazole silver bromide (226). (Reprinted with permission of the Society for Imaging Science and Technology, sole copyright owners of *The Journal of Imaging Science and Technology*.)

Halocyclopentadiene compounds have recently been claimed to exhibit good stabilization properties (229):

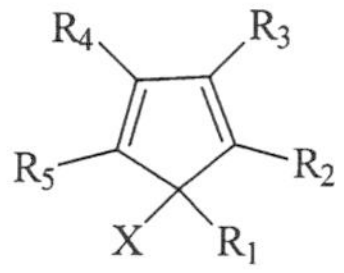

In this case, X is the halogen atom and the R groups are alkyl, such as a 1–20 carbons, and each may combine with each other to form a ring, such as a cyclopentene or cyclohexene, and a ring condensed with another ring. X is a halogen atom and includes F, Cl, Br, or I.

In the case of an aqueous coated TM, organophosphate esters are claimed to be particularly beneficial (230):

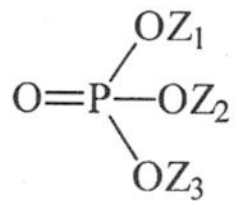

$$O{=}P\overset{\displaystyle OZ_1}{\underset{\displaystyle OZ_3}{-}OZ_2}$$

The phosphates must contain at least one $-$OH or $-$OM function for stabilizing activity. Furthermore, pH of the formulation is important; apparently the preferred pH is slightly acidic to neutral.

New sulphenimides and aryl-iodonium compounds are claimed as improved stabilizing components (231–233). These compounds are thought to act by complexation, followed by oxidation of Ag° fog centers.

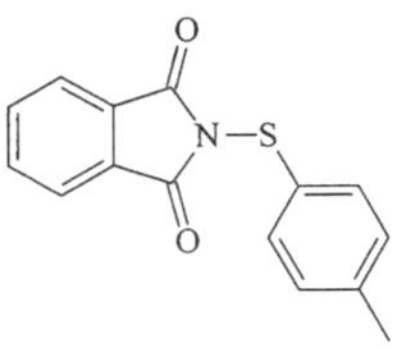

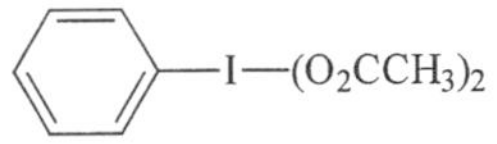

Organosulphur compounds containing the R_1-S-S-R_2 building block where R_1, R_2 = aryl, pyridyl or quinoline, etc., have been claimed as particularly useful postprocess stabilizers (234).

Even natural products (such as RNA) can be found that appear to provide stabilization for photothermographic materials (235).

Phthalimide or azlactone blocked postprocessing stabilizers (236,237) have recently been disclosed. In the case of the phthalimides, a light activated cleavage reaction occurs between the phthalimide group and the linkage to the stabilizer group **A**:

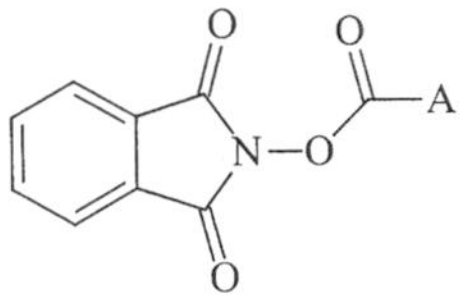

The by-products of the reaction are phthalimide, CO_2, and the stabilizer. The stabilizer is any 1M postprocess stabilizer containing an active hydrogen that enables it to be connected to the phthalimide protecting group such as benzotriazole, phenylmercaptotetrazole, and tetraazaindines.

The azlactone group readily undergoes a Michael addition, which can be effectively used to proton-labile compounds useful in stabilizing PTM to from a blocked stabilizer (237,238). Thermally, the azlactone is claimed to release reversibly, generating the stabilizer and preventing undesirable effects on preexposure sensitometry.

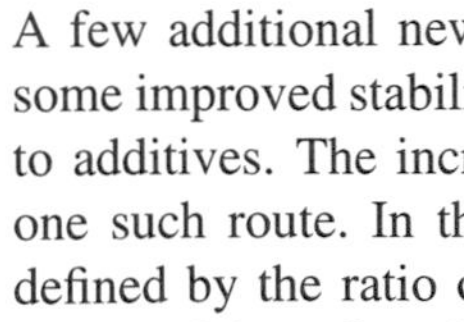

13.9.2 Stabilizing Processes

A few additional new developments in the stabilization of PTM are notable. There are some improved stability claims simply by control of certain physical attributes, as opposed to additives. The increase of crystallinity of silver carboxylate (42,43) is claimed to be one such route. In this case, it is proposed that the crystallinity of silver behenate, as defined by the ratio of the sum of the intensities of six silver behenate diffraction lines to that of three from NIST standard rhombohedral Al_2O_3, needs to be above 0.85. By this definition, conventionally prepared silver behenate exhibited crystallinity around 0.7, and silver behenate prepared under the conditions described (organic solvent included) showed crystallinity above 1.6. Under these circumstances, both image tone and print stability are claimed to be improved.

Also, improved stability is claimed by lowering the levels of formate in the preparation of the silver carboxylate (231). Current impurity levels of this material presumably contribute to Ag°-based fog formation.

In addition, since silver carboxylates based on unsaturated fatty acids are known to be unstable (27,28), apparently decomposing to Ag°, improved stability has been claimed by minimizing the concentration of such unsaturated fatty acids, which are commonly present in the natural product source (26).

Finally, Ag_2O has been proposed as an undesirable but easily incorporated by-product of the synthesis of silver carboxylates under aqueous alkaline conditions (239,240). Increasing the pH over a series of preparations, it was observed that D_{min} increases as well, even though standard chemical analysis detected no differences in the silver behenates prepared from these reactions. Alternatively, under high-speed mixing conditions, the sodium soap is added to the silver nitrate or silver acetate dispersion to avoid Ag_2O formation (240). In this latter case, the ratio of $NaO_2C_xH_{2x-1}$ to $HO_2C_xH_{2x-1}$ is specifically maintained at $1:1.005-1.04$ with 1.03 preferred. Yields of clean silver carboxylate greater than 95% for $x = 8-22$ are claimed.

13.9.3 Practical Stabilization Results

The confirmation of stabilizer utility in photothermographic imaging systems is the ability to prepare films that have clearly demonstrated stability under the conditions in which the materials were intended to be used. The dark and print stability of early Dry Silver papers were claimed to be exceptionally long (241) based on Arrhenius data under different conditions. For these films, archival stability was found to be in excess of 50 years (70°F and 30% RH), although print stability was far less due to a significant increase in D_{min}. Since this work, impressive gains have been made in improving the print stability of PTM. Recent studies carried out over a wide range of temperatures have indicated stability in excess of 300 years (242). The current data available, actual and not projected from Arrhenius plots, on the latest commercial films used in x-ray applications, have demonstrated exceptional stability (243,244). While not perfect (245), the improvements made in this technology have confirmed that it is at the stage required for the critical technical demands of such applications.

It may be concluded that there are a wide variety of approaches to provide stability to photothermographic films prepared from silver bromide and silver carboxylates. Unlike conventional silver halide film in which the development chemistry is separate from the film, photothermographic materials retain all the active chemistries (in general), so that stability approaching conventional silver halide is difficult to achieve. Because of the residual reaction chemistries after processing, photothermographic materials have a demanding, and generally ongoing, need to provide unique routes that enhance their stability under a wide range of conditions. considering the significant commercial breadth of the technology, one can expect new developments in this important area.

13.10 CHARACTERISTICS OF THE METALLIC SILVER OF THE IMAGE

Despite the significance of the nature of the metallic silver formed in the image area, particularly in its color tone and covering power, very little has been published on the characteristics of that silver. In the science of PTM this topic is just now being introduced. While this topic could be included in the discussion on toners, there are other factors influencing the metallic silver form, which are unrelated to toner properties that warrant consideration separately.

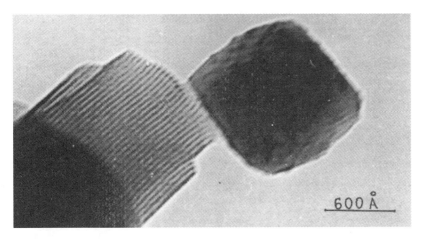

Figure 13.19 Epitaxial interface between AgBr and silver stearate (112). (Reprinted with permission of the Society for Imaging Science and Technology, sole copyright owners of *The Journal of Imaging Science and Technology*.)

In general, the metallic silver particles formed in the PTM image areas are dendritic or filamentary (46,112,113), similar to conventional silver halide materials (57). It has been observed recently, however, that the specific structure of the silver is often related to the nature of the silver halide used in the film (46,72,113). The special conditions of high-contrast films in image-setting film (168) yield extremely fine metallic silver particles, although they seem to be generally spherical.

For example, in photothermographic imaging materials prepared from silver bromide and silver carboxylate, using phthalazine and phthalic acid as toners, it was observed that

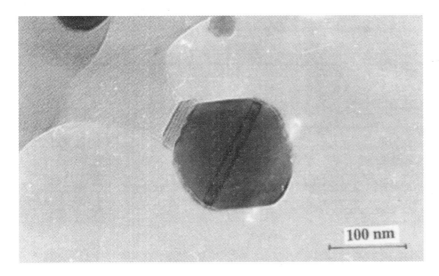

Figure 13.20 Epitaxial interface between AgBr and silver stearate (246). (Reprinted with permission of the Society for Imaging Science and Technology, sole copyright owners of *The Journal of Imaging Science and Technology*.)

the developed silver tends to be filamentary when the photothermographic material contains silver halide prepared by an in situ exchange reaction between silver carboxylate and a brominating agent. The brominating process for the formation of silver halide is discussed in Section 13.3 above. On the other hand, if previously synthesized silver halide crystals, such as via the preformed or ex situ AgBr route, are used to prepare the photothermographic material, the developed silver generally crystallizes in a dendritic from. The differences in the structure of the silver image are attributed to the interfaces formed between the silver halide and the silver fatty acid complex during the preparation processes. In the case of the in situ reaction, for example, a clearly defined epitaxial interface has been observed between the silver bromide and silver carboxylate by TEM in model and real systems (46,112,113). Excellent examples of that epitax are shown in Figs. 13.19 and 13.20.

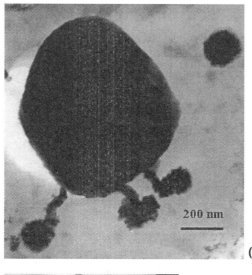

(a)

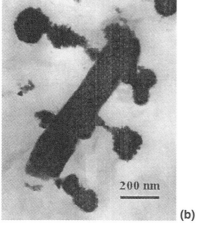

(b)

Figure 13.21 Dendritic and filamentary silver particles in thermally developed photothermographic material film, AgBr tabular grains edge on (a) and rotated 90° (b) (114). (Reprinted with permission of the Society for Imaging Science and Technology, sole copyright owners of *The Journal of Imaging Science and Technology*.)

In the case of photothermographic materials containing tabular grains, the formation of both dendritic and filamentary silver particles was observed (113,114). Regardless of the nature of the AgX surface, multiple particles, initiating as filaments and terminating as dendrites, are observed on the same silver bromide grain (Fig. 13.21).

By comparison, cubic preformed AgBr, in the maximum optical density area, was predominantly dendritic silver particles. Only a small number of filamentary crystals could be observed (Fig. 13.22). The size of dendritic silver clusters in the developed material is 100–150 nm.

The mechanism by which this metallic silver structural dichotomy occurs has not been reported, although it should be clearly stated that this model is the only formulation discussed in any detail in the literature. Other formulations, in which phthalimide or succinimide are used as toners, for example, generate metallic silver that does not completely correspond to the phthalazine/phthalic acid toner combination (1,72,114). In this latter case, the metallic silver appears to be exclusively spherical. More striking is the effect of infectious developers in this type of imaging system. Under these conditions, extremely fine silver spheres, highly dispersed in the image, are formed with no evidence of filaments (168). Therefore, any conclusion regarding whether these particle morphologies can be used to describe the mechanism of the respective AgBr source types as physical versus direct development processes, corresponding to the well-characterized mechanisms in conventional silver halide systems, remains somewhat premature. The preliminary evidence provided to date provides an intriguing insight into PTM mechanisms, particularly for the select set of conditions described so far. It would seem that the most general explanation, which would account for other major toner formulation results, is still open for discussion.

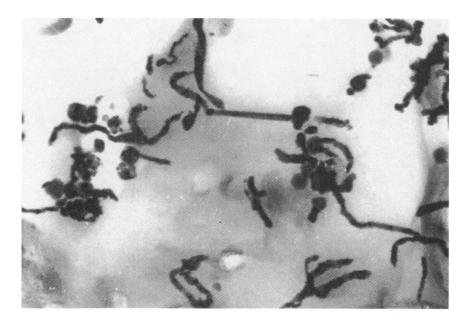

Figure 13.22 Filamentary metallic silver clusters in PTM (46). (Reprinted by permission of Wiley-Liss, Inc., a subsidiary of John Wiley & Sons, Inc.)

13.11   FILM PHOTOGRAPHIC PROPERTIES

It would appear that the logical place to incorporate a discussion of the photographic properties of PTM prepared from AgX and $[Ag(O_2C_xH_{2x-1})]_2$ is at the conclusion of the compendium of individual technologies that make up that film. The primary components composing those sections can now be put together into that working relationship for that discussion.

Inspection of a cross-sectional TEM of the light sensitive emulsion layer of an unprocessed photothermographic film gives insight into how such a film works. For the unexposed film, the two visible components in the image recording layer are the silver carboxylate particles, which are about 1 micron in size along the major axis, and the silver halide particles, which typically have cubic morphology and an edge length of 0.1 to 0.05 microns (see Fig. 13.23).

The development process for a photothermographic film begins with light exposure that generates latent image formation on the surface of the AgX grains. Following exposure, the film is heat processed, usually from 5 to 15 seconds at temperatures ranging from 110 to 125°C. The processing temperatures for thermographic films are generally much higher but have residence times that are on the order of milliseconds. Figure 13.24 is a TEM of an exposed and thermally processed photothermographic imaging layer in a D_{max} area.

It can be readily observed that the silver carboxylate particles have been consumed and that image silver has formed on the surface of the AgX particles. The latent image centers on the AgX grains act as catalytic sites for physical development using the silver carboxylate as the source of image silver density. Taking into account all of the individual sections above, the chemical equation for this photothermographic development process

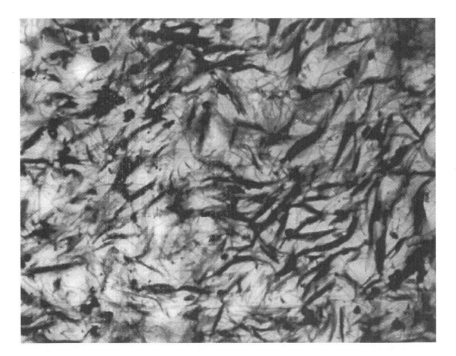

Figure 13.23   TEM cross section of PTM film, D_{max}.

can be described as follows. The silver source is thermally reduced with a developer upon delivery to the latent image center as a silver/toner complex. The metallic silver, filamentary, or dendritic or combination thereof, is formed in a matrix that now includes free fatty acid (melt) and oxidized developer.

$$\text{Ligand} \cdot [\text{Ag}(O_2C_xH_{2x-1})]_2 + R\!-\!OH \xrightarrow[\text{Heat}]{[\text{Ag}°]} 2\text{Ag}°$$

$$+ \ 2HO_2C_xH_{2x-1} + R'\!=\!O + \text{Ligand}$$

The developers used for PTM media are generally weaker reducing agents than those used for conventional photography and as such are not thought capable of substantially reducing the silver halide, only the silver carboxylate; but even this basic issue is not settled. The AgX particles have been reported to survive the development process unchanged (3), whereas AgX particles have also been claimed to be reduced in addition to the silver carboxylate particles (247). This discrepancy could be an instance where differences in PTM formulations may have a fundamental impact on experimental results or conclusions and serves as a warning that there may not be one general mechanism for latent image capture or thermal development that spans all formulations for photothermographic media.

The development accelerators function as silver carboxylate complexing agents that transport the silver carboxylate to the latent image site. Thermographic films do not contain silver halide particles with latent image to catalyze the development reaction and hence require higher image processing temperatures, depending on developer. Otherwise, they have the same basic components of silver carboxylate, developer, and development accelerator and the same overall chemical reaction process.

Photothermographic films have been constructed that have light sensitivities spanning the UV to the infrared, by the use of the appropriate spectral sensitizing dyes. In general, photothermographic films are considerably slower than conventional AgX films. This is because the AgX grains, which survive the development process and are not fixed out of the film, must be kept below ~0.1 micron edge length in order to inhibit subsequent D_{min} rise. Speeds for commercialized photothermographic films typically range from about 1000 to 20 erg/cm^2 (3,248). This trade-off of speed for print stability and D_{max} limits the usage of photothermographic films in camera-speed applications. Since photothermographic films are generally slower by necessity than conventional films, they must be exposed by more intense light sources such as IR or red laser diodes, HeNe lasers, and Xe flash lamps. Alternatively, large tabular grain silver halide crystals have recently been reported to be incorporated into a camera-speed photothermographic film (158).

One benefit of using very fine AgX grain sizes in PTM is that extremely high resolutions are possible. Under optimal exposure conditions, over 1000 line/mm resolving power have been observed (3,249).

One photographic requirement that has been especially challenging for PTM has been to achieve acceptable image stability, both shelf and print stability. The task of improving image stability is more difficult than in conventional photography, as both the developer and the silver halide grains are present in the pre- and postprocessed image. As noted above in Section 13.9, however, great strides have been made in this area, and excellent stability for the applications intended has been achieved.

The mechanistic details of the photophysics of PTM are important for significant improvements to be made in photographic speed, contrast, D_{max}, etc. Unlike the situation in conventional photography, however, relatively few papers have been published in the literature that explore the mechanisms of latent image capture and the development process for PTM. Until recently, the mechanism of latent image formation in PTM had been assumed to be analogous to that in conventional photography. However, in a landmark paper on latent image formation in PTM, it has been recognized that there are several elements of photothermographic technology that are inconsistent with conventional theory (1,85). In this paper it is shown that the usual AgX grain size to photographic speed relationships that hold for conventional photography do not apply to their photothermographic films, which contain AgBr(I) particles in intimate contact with Ag carboxylate particles. A probable explanation for the lack of dependence of AgX particle size on photospeed in this work is that the measured speed points for these films were made midway up the D log E curve, not at the toe points, and hence could be affected by contrast variations.

The model advanced in this paper (85) assumes that an epitaxial interface forms between the AgBr(I) and the silver carboxylate particles (TEMs of the epitaxial structures shown in Section 13.10) and that this interface functions like a heterojunction that creates a photographic diode. Figure 13.25 illustrates the proposed photocatalytic mechanism for formation of the latent image at the interface between AgBr(I) and silver carboxylate.

In this mechanism, the electron–hole pairs that are photogenerated in the AgBr crystal are separated at the interface. Consequently, holes are trapped at the I^- centers in the AgBr resulting in injected electrons into the interfacial zone [claimed to be the unique AgBr/[Ag($O_2C_xH_{2x-1}$)Na($O_2C_xH_{2x-1}$) epitax (114)] to reduce Ag^+ ions from the silver carboxylate. The specifics of how the Ag^+ ions are delivered are open, but a stepwise formation of silver clusters, Ag^0_n, results. Further details regarding this mechanism in

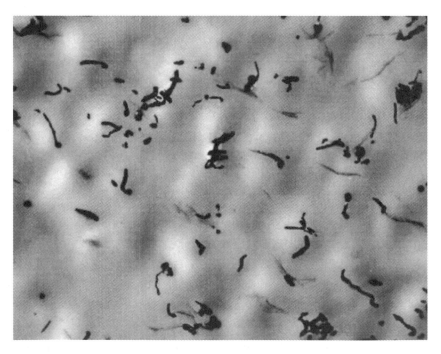

Figure 13.24 TEM cross section of PTM film, D_{min}.

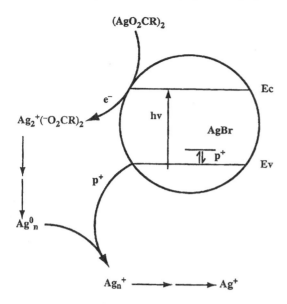

Figure 13.25 Photocatalytic mechanism (85). (Reprinted with permission of the Society for Imaging Science and Technology, sole copyright owners of *The Journal of Imaging Science and Technology*.)

systems in which the presence of such an epitaxial interface is questionable or uncertain, or not intentionally created, remain to be resolved.

More recently, further studies on the mechanism of photothermographic latent image formation show that the AgX/silver carboxylate interface effectively separates electron–hole pairs formed in the AgX phase, because no energetically favorable pathway exists for photoholes to transport into the silver carboxylate phase (1). Further, this paper proposes that the prior inability to achieve chemical sensitization of AgX grains in photothermographic films is that it is redundant when an AgX/silver carboxylate interface already exists. It is important to remember that the mechanisms for latent image formation proposed in this paper may not be universal for all photothermographic constructions. In fact, a contrary opinion has recently been published (252).

There have been some recent papers published on the important issue of the thermal development mechanism (1,72). Historically, the first model that attempted to explain the development characteristics of photothermographic films on a mathematical basis was developed in 1989 (2). It is commonly referred to as the Klosterboer–Rutledge model. The model has recently been extended and generalized (169).

The generalized Klosterboer–Rutledge relationship is given by

$$D = D_{max}(1 - \exp[-n\upsilon\overline{F}_{SH}(E)])$$

where D is the image density above D_{min}, D_{max}, is the image density above D_{min} achieved for a practical maximum exposure, n is the grain-number density, υ is the sphere of influence volume, and $\overline{F}_{SH}(E)$ is a function that contains all the exposure dependence up to the maximum exposure. The details of this model are discussed in these references and need not be covered here. However, it is instructive to review the highlights of the sphere of influence concept as it has recently been incorporated into patent disclosures (250).

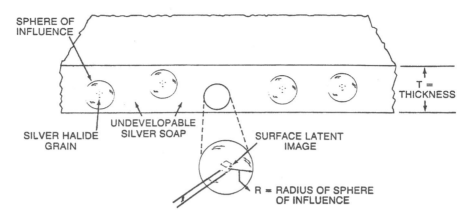

Figure 13.26 Sphere of influence of the AgX grain within a $[Ag(O_2C_xH_{2x-1})]_2$/binder matrix. (From Ref. 2.)

The Klosterboer–Rutledge model assumes a sphere of influence or a volume of silver carboxylate surrounding each AgX grain (Fig. 13.26). A TEM picture of the actual sphere of influence has since been published (169).

If the AgX grain is exposed, all the silver carboxylate in this sphere of influence is reduced to form image silver. The radius of the sphere of influence is determined by the diffusion range of silver ion from the silver carboxylate to the AgX grain to support the physical development reaction. It turns out that for optimal covering power, there should be one AgX grain per sphere unit volume in a coated film. If there are too few AgX grains per sphere of influence volume, then some of the silver carboxylate is undevelopable and wasted.

Using a series of predicted D log E curves, Klosterboer–Rutledge has shown (2) that as the (cubic) AgX particle size is increased, there is the expected increase in toe speed, using a constant coverage of AgX. However, once there are too few AgX grains per sphere of influence volume (grain number density gets too low), there is a rapid falloff in D_{max} and contrast. Therefore, attempts to produce a high-speed PTM film by merely increasing the AgX edge length is predicted to yield films having a low D_{max} and low contrast unless other formulation modifications are made. The Klosterboer–Rutledge relationship shows that increasing the sphere of influence does serve to shift the high-speed/density falloff point to higher speeds but does not specify what formulation changes, other than a reduction in the viscosity of the emulsion layer, would make this possible. Klosterboer–Rutledge does indicate that increasing the AgX coverage at constant particle size does increase the speed of a PTM film. Experimental data are disclosed that indicate an approximate upper limit in speed of ISO 5 for an AgX coating weight of 1.0 g/m^2 for a practical PTM film.

The Klosterboer–Rutledge model has significant limitations caused by the underlying assumptions, but these limitations have been substantially reduced by a refined version of the Klosterboer–Rutledge model (169). In this case, the AgX grain number density and overlapping spheres of influence restrictions have been mathematically corrected. As a result, the predictions of D_{max} versus AgX grain number density, for example, are excellent, as are the predictions for contrast.

In other attempts to improve imaging, pre-exposure heating is claimed to provide substantial advantages (251). By contrast, model systems do not appear to give improvements; in fact, pre-exposure heating has been reported to be detrimental to the development process (115).

13.12 CONCLUSIONS AND FUTURE TRENDS

In only a few years, major advances in the state of the art of TM and PTM, based on silver components, have been incorporated into new commercially manufactured TM and PTM materials. The consequences of these advances are evident not only in the dramatic improvements in the photographic properties of those films but also in the rate of acceptance in the marketplace. It is difficult to overstate the level of improvements in film qualities from the early paper and microfilm products to the current medical x-ray and image-setting films. It is clear that the "old" dry silver technology, based on silver halides and silver carboxylates, has matured into a true, fundamental imaging technology. While the basic foundation remains the same, the new developments have formed a basis for new applications. As demonstrated by the patent activity, significant efforts are clearly underway in multiple locations to drive the technology further. Additional insight into its reaction mechanisms, while still sketchy compared to that known for conventional AgX systems, can be expected to lay the groundwork for new advances.

We have tried to provide a highlight of the TM and PTM technology with an emphasis on the new developments in the field. There is more information that can be added, and it is certain that the trend will continue for the foreseeable future.

REFERENCES

1. M. R. V. Sahyun. *J. Imaging Sci. Tech.* *42*:23 (1998).
2. D. H. Klosterboer. Neblette's Eighth Edition: *Imaging Processes and Materials* (J. M. Sturge, V. Walworth, A. Shepp, eds.). Van Nostrand-Reinhold, New York, Ch. 9, pp. 279–291 (1989).
3. D. A. Morgan. *Handbook of Imaging Materials* (A. S. Diamond, ed.). Marcel Dekker, New York, Ch. 2, pp. 43–60 (1991).
4. V. M. Andreev, E. P. Fokin, Yu. I. Mikhailov, V. V. Boldyrev. *Zhurn. Nauch. i Prikl. Fotogr. i Kine.* *24*:311 (1979).
5. P. M. Zavlin, A. N. Dyakonov, S. S. Mnatsakanov, S. S. Tibilov, P. M. Velinzon, S. I. Gaft. *Tekh. Kino. i Televid. Nauka i Tekh.* *3*:12 (1991).
6. V. N. Bolshakov, M. A. Goryaev. *Optik. Tekh. Prom.* *11*:68 (1991).
7. T. E. Kekhva, V. N. Lebedev, I. V. Sokolova. *Zhur. Fiz. Khim.* *65*:2036 (1991).
8. M. A. Goryaev. *Zhur. Nauch. i Prikl. Fotogr. i Kine.* *36*:421 (1991).
9. P. M. Zavlin, A. N. Dyakonov, P. Z. Velinzon, S. I. Gaft, S. S. Tibilov. *Zhurn. Nauch. i Priklad. Fotogr.* *42*:21 (1997).
10. K. Sato, S. Ishikawa, K. Kawata. US Patent 4,835,272 (1989).
11. H. Hirai, H. Naito. US Patent 4,704,345 (1987).
12. S. Washisu, Y. Fukushige, K. Kawada, T. Usami. JP Patent 10166731 (1998).
13. H. Hirai, H. Hara, K. Kawata. US Patent 4,775,613 (1988).
14. M. Kobayashi, T. Fukui, O. Shimomaruko, M. Tamura, K. Kagami, M. Suzuki, K. Nishino. EP Patent 568,063 (1993).

15. D. R. Whitcomb, W. C. Frank, R. D. Rogers, B. P. Tolochko, S. V. Chernov, S. G. Nikitenko. IS&T 49th Ann. Conf., Minneapolis, MN, pp. 426–428 (1996).
16. D. R. Whitcomb, R. D. Rogers. *J. Imaging Sci. Tech.* *43*:517 (1999).
17. R. M. Biegler, R. E. Gronseth, R. J. Ryther, M. P. Juaire, J. Svendsen. US Patent 5,600,396 (1997) and references therein.
18. See, for many examples, K. Sashihara, S. Sakawaki, T. Kobashi, Y. Iwata, T. Asano, T. Masuda, S. Asaka. GB Patent 1,542,327 (1979).
19. S. Hirabayashi, T. Masukawa, W. Ishidawa, T. Harada. US Patent 4,451,561 (1984).
20. C. Zou, J. B. Phillip, S. M. Shor, M. C. Skinner, P. Zhou. US Patent 5,434,043 (1995).
21. L. M. Eshelman, M. E. Irving, D. H. Levy, K. R. C. Gisser. US Patent 5,843,632 (1998).
22. L. S. Harring, S. M. Simpson, F. H. Sansbury. US Patent 5,637,449 (1997).
23. K. Katoh, A. Hatakeyama. EP Patent 803,764 (1997).
24. H. Uytterhoeven, J. Loccufier. EP 851,284 (1997).
25. T. Shima, T. Nagasawa. EP 952,481 (1999).
26. J. P. Beese, B. J. Henne, J. L. Ekmanis, W. F. Smith. EP 857,999 (1998).
27. V. M. Andreev, L. P. Burleva, V. V. Boldyrev, Yu. I. Mikhailov. *Iz. Sib. Otdel. Akad. Nauk. SSSR, Ser. Khim. Nauk.* *4*:58–63 (1983).
28. L. P. Burleva, V. M. Andreev. *Izv. Sib. Otd., Akad. Sci. USSR, Ser. Khim. Nauk.* *17*:33–35 (1984).
29. V. Ogrodnik, D. A. Edwards, M. F. Mahon, K. C. Malloy. XXXIII Int. Conf. Coord. Chem., Florence, Italy, 30 August, (1998).
30. G. Smith, D. S. Sagatys, C. Dahlgren, D. E. Lynch, R. C. Bott, K. A. Byriel, C. H. L. Kennard. *Z. Kristal.* *210*:44 (1995).
31. V. Vand, A. Aitken, R. K. Campbell. *Acta Cryst.* *2*:398 (1949).
32. B. P. Tolochko, S. V. Chernov, S. G. Nikitenko, D. R. Whitcomb. *Nucl. Instr. Meth. Phys. Res. (A)* *405*:428–434 (1998).
33. D. R. Whitcomb, R. D. Rogers. 218th National Conference, Am. Chem. Soc., New Orleans. (1999).
34. F. Jaber, F. Charbonnier, M. Petit-Ramel, R. Faure. *Eur. J. Solid State Inorg. Chem.* *33*:429 (1996).
35. F. A. Cotton, G. Wilkenson. *Advanced Inorganic Chemistry*, 5th ed. John Wiley & Sons, New York (1988).
36. I. Søtofte, K. Nielsen. *Acta Chem. Scand.* *A37*:891 (1983).
37. J. R. Freedman, S. R. Sofen, K. M. Young. US Patent 5,328,799 (1994).
38. J. M. Winslow, D. H. Klosterboer. US Patent 4,260,677 (1981).
39. N. Masciocchi, M. Moret, P. Cairati, A. Sironi, G. A. Ardizzoia, G. La Monica. *J. Chem. Soc. Dalt. Trans.* 1671 (1995).
40. C. D. M. Beverwijk, G. J. M. van der Kerk, A. J. Leusink, J. G. Noltes. *Organometallic Chemistry Reviews* *A5*:215 (1970).
41. Y. H. Xu, L. Lin, J. G. Luo, B. K. Teo, Z. J. Xia, Y. H. Zou, H. Y. Chen. *Chin. Chem. Lett.* *10*:793 (1999).
42. S. Emmers, B. Horsten, I. Geuens, Y. Gilliams, A. Bellens, D. Bollen, I. Hoogmartens. EP Patent 848,286 (1997).
43. I. Geuens, I. Vanwelkenhuysen. *J. Imaging Sci. Technol.* *43*:521 (1999).
44. M. Ikeda, Y. Iwata. *Photogr. Sci. Eng.* *24*:273 (1980).
45. M. Ikeda. *Photogr. Sci. Eng.* *24*:277 (1980).
46. B. B. Bokhonov, L. P. Burleva, D. R. Whitcomb, M. R. V. Sahyun. *Microsc. Res. Tech.* *42*:152 (1998).
47. T. Masuda, H. Sato, H. Ono. JP Patent 52141222 (1977).
48. X.-M. Chen, T. C. W. Mak. *J. Chem. Soc., Dalton Trans.* 1219 (1991).
49. X.-M. Chen, T. C. W. Mak. *Polyhedron* *10*:1723 (1991).
50. W. E. Smith. *Coordination Chemistry Reviews* *35*:253 (1981).
51. R. C. Mehrotra, R. Bohra. *Metal Carboxylates*. Academic Press, New York (1983).

52. Y. Jiang, S. Alvarez, R. Hoffmann. *Inorg. Chem. 24*:749 (1985).
53. P. K. Mehrotra, R. Hoffman. *Inorg. Chem. 17*:2187 (1978).
54. K. M. Merz, Jr., R. Hoffmann. *Inorg. Chem. 27*:2120 (1988) and references therein.
55. A. Michaelides, S. Skoulika, V. Kiritsis, A. Aubre. *J. Chem. Soc., Chem. Commun.* 1415 (1995).
56. K. Singh, J. R. Long, P. Stavropoulos. *J. Am. Chem. Soc. 119*:2942 (1997).
57. T. H. James. *The Theory of the Photographic Process*, 4ᵗʰ ed. Macmillan, New York (1977).
58. D. R. Whitcomb, R. D. Rogers. *J. Chem. Cryst. 26*:99 (1996).
59. D. R. Whitcomb, R. D. Rogers. *J. Imaging Sci. Technol. 43*:504 (1999).
60. D. R. Whitcomb, W. C. Frank. US Patent 5,350,669 (1994).
61. D. R. Whitcomb, W. C. Frank. US Patent 5,466,804 (1995).
62. N. A. Izmailov, V. S. Chernyi. *Russian J. Phys. Chem. 34*:149 (1960).
63. W. U. Malik, A. K. Jain, O. P. Jhamb. *J. Chem. Soc. A.* 1514 (1971).
64. C. A. Jacobson, A. Holmes. *J. Biological Chem. 25*:29 (1916).
65. D. R. Whitcomb. US Patent 5,491,059 (1996).
66. M. Okazaki, M. Yabe, K. Kawatta. US Patent 4,943,515 (1990).
67. G. S. Whitby. *Science 53*:580 (1921).
68. P. N. Cheremisinoff. *J. Am. Oil Chem. Soc.* 278 (1951).
69. K. Katoh, A. Hatakeyama. EP Patent 803,764 (1997).
70. B. L. Marginean, S. R. Cuch, C. A. Whittaker, M. C. Patel. US Patent 5,424,182 (1995).
71. M. A. Goryaev, T. B. Kolesova, M. N. Timokhina, I. M. Gulkova. Russian Application 5051237/04 from 03.07.92.
72. Yu. E. Usanov, T. B. Kolesova. *J. Imaging Sci. Tech. 40*:104 (1996).
73. W. U. Malik, A. K. Jain, O. P. Jhamb. *J. Chem. Soc.* (A) 1514 (1971).
74. N. Kuzmina, S. Paramonov, R. Ivanov, V. Kezko, K. Polamo, S. Troyanov. *J. Phys. 4*:9. *Proceedings of the Twelfth European Conference on Chemical Vapour Deposition 2*:923 (1999).
75. R. C. Weast, ed. *Handbook of Chemistry and Physics*. CRC Press, Boca Raton, FL (1987).
76. K. S. Markley. *Fatty Acids*, Pt 2, p. 747, Interscience Publishers, New York (1960).
77. B. T. Usubaliev, Kh. N. Nadjafov, A. Adin, A. A. Musaev, S. Khudu, Kh. S. Mamedov. *Thermochim. Acta 93*:57 (1985).
78. V. M. Andreev, L. P. Burleva, V. V. Boldyrev. *J. Sib. Branch Acad. Sci. USSR 5*:3 (1984).
79. L. P. Burleva, V. M. Andreev, V. V. Boldyrev. *J. Therm. Anal. 33*:735 (1998).
80. M. Chadha, M. E. Dunnigan, M. R. V. Sahyun, T. Ishida. *J. Appl. Phys. 84*:887 (1998).
81. N. F. Uvarov, L. P. Burleva, M. B. Mizen, D. R. Whitcomb, C. Zou. *Solid State Ionics 107*: 31 (1998).
82. V. K. Zhuravlev, Yu. I. Mikhailov. *Kin. i Katal. 10*:89 (1969).
83. L. P. Burleva, V. M. Andreev, V. V. Boldyrev. *Izv. Sib. Otd. Akad. Nauk SSSR, Ser. Khim. Nauk 3*:31 (1988).
84. R. Gotoh, M. Arizawa, N. Kajikawa. *Tr. Mezhd. Kongr. Poverk.-Akt. Veshch.*, 7ᵗʰ, 2(II), 724 (1978).
85. C. Zou, M. R. V. Sahyun, B. Levy, N. Serpone. *J. Imaging Sci. Technol. 40*:94 (1996).
86. A. Leuthe, H. Riegler. *J. Phys. D: Appl. Phys. 25*:1786 (1992).
87. B. B. Bokhonov, B. I. Lomovskii, V. M. Andreev, V. V. Boldyrev. *Izv. Sib. Otd., Akad. Sci. USSR, Ser. Khim. Nauk. 5*:8 (1984).
88. V. M. Andreev, Yu. G. Golitzin, Yu. I. Mikhailov, V. V. Boldyrev. *Izv. Sib. Otd., Akad. Sci. USSR, Ser. Khim. Nauk. 2*:64 (1983).
89. S. M. Shor. 49ᵗʰ IS&T Conference, Springfield, VA, p. 442 (1996).
90. M. A. Goryaev. *Zhurn. Nauch. i Prikl. Fotogr. 43*:1 (1998).
91. M. A. Goryaev. *Pisma v Zhurn. Tekhn Fiziki 20*:40 (1994).
92. M. A. Goryaev. *Zhur. Nauch. Prikl. Fotogr. 41*:40 (1996).
93. B. B. Bokhonov, O. I. Lomovsky, V. M. Andreev, V. V. Boldyrev. *J. Solid State Chem. 58*: 170 (1985).

94. V. M. Andreev, L. P. Burleva, V. V. Boldyrev. *Izv. Sib. Otd., Akad. Sci. USSR, Ser. Khim. Nauk. 3*:37 (1988).
95. A. Iwasawa, H. Tabei, S. Hara, K. Matsuyama. *Photogr. Sci. Eng. 20*:246 (1976).
96. B. B. Bokanov, L. P. Burleva, D. R. Whitcomb, Y. E. Usanov. *J. Imaging Sci. Technol., 45*:259 (2001).
97. E. K. Fields, S. Meyerson. *J. Org. Chem. 41*:916 (1976).
98. P. M. Zavlin, A. N. Batrakov, P. Z. Velinzon, S. I. Gaft, L. L. Kuznetsov. *J. Imaging Sci. Technol. 43*:535 (1999).
99. T. N. Blanton, T. C. Huang, H. Toraya, C. R. Hubbard, S. B. Robie, D. Louër, H. E. Göbel, G. Will, R. Gilles, T. Raftery. *Powder Diff. 10*:91 (1995).
100. A. E. Gvozdev. *Ukrain. Fiz. Zhur. 24*:1856 (1979).
101. N. G. Khainovskii, E. F. Khairetdinov, V. M. Andreev. *Izv. Sib. Otd., Akad. Sci. USSR, Ser. Khim. Nauk. 5*:34 (1985).
102. J. W. Reeves. US Patent 4,435,499 (1984).
103. P. D. Knight, R. A. deMauriac, P. A. Graham. US Patent 4,128,557 (1978).
104. H. Vandenebeefe, H. Uytterhoeven. EP Patent 851,285 (1998).
105. B. Horsten, L. Leenders, A. Vankeerberghen. EP Patent 782043 (1997).
106. J. W. Shephard. *J. Appl. Photogr. Eng. 8*:210 (1982).
107. J. Robillard. FR Patent 2,254,047 (1975).
108. M. R. V. Sahyun, R. Patel, R. Muthyala. US Patent 5,260,180 (1993).
109. D. P. Sorensen, J. W. Shephard. US Patent 3,152,904 (1964).
110. M. J. Simons. US Patent 3,839,049 (1974).
111. G. Fleer, J. Lyklema. *J. Colloid Interface Sci. 46*:1 (1974).
112. B. B. Bokhonov, L. P. Burleva, W. C. Frank, J. R. Miller, M. B. Mizen, M. R. V. Sahyun, D. R. Whitcomb, J. M. Winslow, C. Zou. *J. Imaging Sci. Technol. 40*:85 (1996).
113. B. B. Bokhonov, L. P. Burleva, W. Frank, M. B. Mizen, M. R. V. Sahyun, D. R. Whitcomb, J. Winslow, C. Zou. *J. Imaging Sci. Technol. 40*:417 (1996).
114. B. B. Bokhonov, L. P. Burleva, D. R. Whitcomb. *J. Imaging Sci. Technol. 43*:505 (1999).
115. Yu. E. Usanov, T. B. Kolesova, I. M. Gulikova, L. P. Burleva, M. R. V. Sahyun, D. R. Whitcomb. *J. Imaging Sci. Technol. 43*:545 (1999).
116. P. Verrept, K. Elst. EP Patent 844,514 (1998).
117. S. M. Shor, C. Zou, P. Zhou, S. Aoki. US Patent 5,382,504 (1995).
118. J. W. Reeves. US Patent 4,264,725 (1981).
119. K. Katoh. EP Patent 803,765 (1997).
120. K. R. C. Gisser, L. M. Eshelman, M. E. Irving, D. H. Levy. US Patent 5,858,637 (1999).
121. K. R. C. Gisser, L. M. Eshelman, M. E. Irving, D. H. Levy. US Patent 5,843,632 (1998).
122. G. L. Featherstone, D. C. Lynch, J. R. Miller, M. C. Skinner, J. M. Winslow, S. M. Simpson. US Patent 5,891,615 (1999).
123. M. E. Irving, L. M. Eshelman, D. H. Levy, D. L. Hartsell. US Patent 5,876,905 (1999).
124. Y. Hayashi, S. Ogawa, M. Sanada, R. Hirohashi. *J. Imaging Sci. Technol. 33*:124 (1989).
125. K. Kato, JP Patent 9292674 (1997).
126. H. E. Dankosh. US Patent 5,510,236 (1996).
127. J. R. Miller, S. Kalousdian, B. C. Willett, J. M. Winslow, P. Zhou, C. Zou. US Patent 5,441,866 (1995).
128. R. W. Burrows, D. B. Oliff, J. B. Phillip. US Patent 5,393,654 (1994).
129. M. A. Goryaev, B. I. Shapiro. *Zhur. Nauch. i Priklad. Fotogr. 42*:65 (1997).
130. B. J. Kummeth, D. C. Lynch, J. R. Miller, B. C. Willett. US Patent 5,541,054 (1996).
131. H. Uytterhoeven, P. Callant, G. Deroover, J. Loccufier. EP Patent 821,271 (1998).
132. H. Hara, H. Naito, K. Sato. *J. Soc. Photogr. Sci. Technol. Jpn. 50*:402 (1987).
133. T. Hoshimiya, K. Yamada, T. Ezoe, K. Kawato, H. Sasaki, H. Suzuki. EP Patent 921,433 (1999).
134. A. von Konig, H. Kampfer, E. M. Brinckmann, F. C. Heugebaert. US Patent 3,933,507 (1976).

135. V. V. Komar, B. V. Konopleva. *Zhurn. Prikl. Khim.* *69*:1208 (1996).

136. H. Tsuzuki. US Patent 5,677,121 (1997).

137. A. J. Alton, J. P. Beese, J. L. Ekmanis, B. J. Henne, W. F. Smith. EP Patent 857999 (1998).

138. T. Fukui, M. Katayama, A. Mori, K. Isaka, S. Nakamura. JP Patent 3089245 (1991).

139. K. Komorida, K. Onodera. JP Patent 61267050 (1985).

140. D. F. Jennings, T. D. Weaver. EP Patent 849,625 (1998).

141. A. Matsushita, T. Koide. JP Patent 2178650 (1990).

142. H. Hiroyaki, H. Daimatsu. JP Patent 62085241 (1987).

143. D. Terrell, J. Loccufier, G. Defieuw, I. Hoogmartens. EP Patent 903,625 (1998).

144. D. R. Whitcomb, J. A. Bjork. US Patent 4,808,565 (1989).

145. C. Uyttendaele, H. Uytterhoeven, B. Horsten. US Patent 5,582,953 (1996).

146. J. A. Bjork, M. L. Brostrom, D. R. Whitcomb. *J. Chem. Cryst.* *27*:223 (1997).

147. G. Wulfsberg, D. Jackson, W. Ilsley, S. Dov, A. Weiss, J. Gagliardi, Jr. *Z. Naturforsch.* *47A*: 75 (1992).

148. R. A. Frenchik. US Patent 4,535,056 (1985).

149. T. Ishida. US Patent 4,594,307 (1986).

150. G. S. Prementine, T. Ishida. US Patent 5,238,792 (1993).

151. H. Hirai, H. Naito. US Patent 4,704,345 (1987).

152. H. Ozaki, H. Hirai, K. Kawata. US Patent 4,761,361 (1998).

153. H. Ozaki, K. Kawata, H. Ohmatsu. US Patent 5,089,378 (1992).

154. S. Sawada. *Soc. J. Photog. Sci. Eng.*, Annual Meeting, Tokyo, Japan (1988).

155. T. Yokokawa, T. Kamosaki, Y. Inagaki, Y. Aotsuka. Advances in Non-Impact Printing Technologies/Japan Hardcopy, 9th International Congress, IS&T/SEPJ (1993).

156. H. Ozaki, T. Yokokawa, Y. Inagaki. *J. Imaging Sci. Technol.*, 46th Annual Conference (1993).

157. Y. Suda, K. Ohbayashi, K. Onodera. *J. Imaging Sci. Technol.* *37*:598 (1993).

158. L. M. Eshelman, T. W. Stoebe. US Patent 6,040,131 (2000).

159. S. M. Simpson, L. S. Harring. US Patent 5,496,695 (1996).

160. S. Hirano, T. Kubo, K. Yamada. EP Patent 807,850 (1997).

161. L. Leenders, L. Van Rompuy. EP Patent 895,122 (1999).

162. J. R. Fyson, G. P. Levenson. *J. Photogr. Sci.* *25*:147 (1977).

163. S. M. Simpson, L. S. Harring. US Patent 5,558,983 (1996).

164. T. J. Murray, S. M. Simpson. US Patent 5,545,515 (1996).

165. D. Terrell, H. Strijckers, I. Hoogmartens. EP Patent 821,269 (1998).

166. I. Olivares, F. Bean, G. Haist. US Patent 3,782,949 (1974).

167. S. E. Hill, M. B. Mizen, M. R. V. Sahyun, Yu. E. Usanov. *J. Imaging Sci. Technol.* *40*:568 (1996).

168. M. B. Mizen. *J. Imaging Sci. Technol.* *43*:528 (1999).

169. S. H. Kong. *J. Imaging Sci. Tech.* *43*:509 (1999).

170. S.-X. Ji, F.-Y. Ma, P. Hu, X.-M. Ren. *Ganguang Kexue Yu Kuang Huaxue* (Photographic Science and Photochemistry), *9*:261 (1991).

171. A. Hatakeyama. JP Patent 9258366 (1997).

172. J. Loccufier, H. Uytterhoeven. EP Patent 851,284 (1998).

173. T. Ishizaka, H. Okamura, N. Asanuma. JP Patent 10339934 (1998).

174. W. Ball, H. Friedel. Int. Conf. Photogr. Sci., Antwerp, Belgium, 61 (1998).

175. D. R. Whitcomb, R. D. Rogers. *Inorg. Chim. Acta.* *256*:263 (1997).

176. D. R. Whitcomb, R. D. Rogers. *J. Chem. Cryst.* *25*:137 (1995).

177. R. D. Rogers, C. V. K. Sharma, D. R. Whitcomb. *Cryst. Eng.* *1*:255 (1998).

178. D. R. Whitcomb, R. D. Rogers. *J. Imaging Sci. Technol.* *43*:517–520 (1999).

179. Monsanto product bulletin.

180. F. Ruttens, *J. Imaging Sci. Technol.* *43*:535 (1999).

181. K. B. Sagawa. US Patent 4,558,003 (1985).

182. C. L. Bauer, R. B. Nielson, G. D. Young, R. Di Felice. US Patent 5,804,365 (1998).

183. T. Masaoka, K. Sakashita, Y. Ban. JP Patent 3197511 (1991).
184. I. Hoogmartens, F. Louwet, C. Uyttendaele, H. Van Aert. EP Patent 903,624 (1999).
185. I. Totani. JP Patent 8095191 (1996).
186. H. Vandenabeele. EP Patent 813,105 (1997).
187. K. Katoh, A. Hatakeyama. EP Patent 803764 (1997).
188. K. Katoh, M. Sakai, T. Arai, K. Hashimoto. US Patent 6 203 972 (2001).
189. Research Disclosure, 223–224, April (1994).
190. L. J. Markin, D. E. Kestner, W. M. Przezdziecki, P. J. Cowdery-Corvan. US Patent 5,310,640 (1994).
191. J. F. DeCory, C. C. Anderson, P. J. Cowdery-Corvan, S. M. Melpolder. US Patent 5,547,821 (1996).
192. D. M. Timmerman, H. Vandenabeele. US Patent 5,364,752 (1994).
193. J. Friedrich, K. Werner. EP Patent 440,957 (1991).
194. W. M. Przezdziecki. US Patent 4,741,992 (1988).
195. C. A. Uyttendaele, G. D. A. Jansen, B. C. Horsten, R. Schuerwegen. US Patent 5,536,696 (1996).
196. C. A. Uyttendaele, G. Defieuw. US Patent 5,759,752 (1998).
197. C. H. Weidner, D. T. Java. EP Patent 919,862 (1999).
198. C. H. Weidner, D. T. Java, E. K. Priebe. EP Patent 919,864 (1999).
199. R. H. Helland, W. D. Ramsden, C. W. Gomez, L. S. Harring, T. Van Thien. US Patent 5,380,635 (1995).
200. M. Noro, I. Fujiwara, M. Sakuranda, Y. Yabuki. EP Patent 911,693 (1999).
201. R. Goswami, R. J. Perry, P. A. Zielinski. US Patent 5,705,323 (1998).
202. I. Fujiwara, H. Okamura, H. Tsuzuki, Y. Yabuki. EP Patent 903629 (1999).
203. S. P. Birkeland. US Patent 3,589,903 (1971).
204. P. G. Skoug. US Patent 5,028,523 (1991).
205. K. Karunakaran, K. P. Elango. *J. Phys. Org. Chem.* *9*:105 (1996).
206. C. Giordano, L. Coppi. *J. Org. Chem.* *57*:2765 (1992).
207. J. Loccufier, I. Hoogmartens, H. Strijckers. EP Patent 821,268 (1998).
208. V. M. Andreev, V. V. Boldyrev, A. L. Kartuzhanskii, Yu. P. Larionov, S. I. Potapovich. *Zhur. Nauchn. Prikl. Fotogr. Kinemetog.* *26*:434 (1981).
209. V. M. Andreev, E. B. Knyazeva, S. I. Potapovich, A. F. Yurchenko. *Zhur. Nauchn. Prikl. Fotogr. Kinemetog.* *33*:365 (1988).
210. J. T. Blair, R. H. Patel, R. D. Rogers, D. R. Whitcomb. *J. Imaging Sci. Technol.* *40*:117 (1996).
211. J. Blair, G. LaBelle, F. Manganiello, K. Sakizadeh, D. Whitcomb. IS&T 50th Annual Conference, 581 (1996).
212. G. D. Tiers, J. A. Wiese. US Patent 3,707,377 (1972).
213. S. Swain. US Patent 4,756,999 (1988).
214. K. Sakizadeh. US Patent 5,464,737 (1995).
215. M. Hirotaka, I. Kimi. US Patent 5,958,668 (1999).
216. O. Hisashi, A. Naoki, T. Ichizo. US Patent 5,952,167 (1999).
217. T. Shima, T. Nagasawa. EP Patent 302,982 (1999).
218. L. F. Costa, J. A. Van Allan, F. Grum. US Patent 3,874,946 (1975).
219. J. W. Van den Houte. DE Patent 2306020 (1975).
220. Y. Hayashi, T. Akagi, S. Kinoshita, R. Hirohashi. *J. Soc. Photogr. Sci. Technol. Jpn.* *52*:21 (1989).
221. K. Itano, M. Nakano, M. Hashimoto. US Patent 3,667,954 (1972).
222. H. A. Hoyen, Jr., X. Wen. *Handbook of Photographic Science and Engineering*, 2nd ed. (C. N. Proudfoot, ed.). Soc. Sci. Eng. Technol. (1997).
223. See, for example, D. M. Van Seggen, O. P. Anderson, S. H. Strauss. *Inorg. Chem.* *31*:2987 (1992).

224. A. Giusti, G. Peyronel, E. Gilberti. *Spectrochim. Acta 38A*:1185 (1982).
225. K. Sakizadeh, J. T. Blair, D. T. Ask. US Patent 5,464,737 (1995).
226. D. R. Whitcomb, R. D. Rogers. *J. Imaging Sci. Technol. 43*:504 (1999).
227. T. Tani. *Photogr. Sci. Eng. 21*:317 (1977).
228. T. Araki, K. Seki, S. Narioka, H. Ishii, Y. Takata, T. Yokoyama, T. Ohta, T. Okajima, S. Watanabe, T. Tani. *Proc. 7*[th] Int. Conf. X-Ray Abs. Fine Struct., Kobe, Japan J. Appl. Phys. *32*(Suppl. 32-2):815 (1993).
229. H. Okada, I. Totani. JP Patent 928,1640 (1997).
230. K. R. Rush. US Patent 5,672,560 (1997).
231. A. J. Alton, J. P. Beese, P. J. Cowdery-Corvan, L. J. Magee, M. W. Martin. EP Patent 809141 (1997).
232. P. J. Cowdery-Corvan, R. L. Klaus, F. D. Saeva. EP Patent 880061 (1998).
233. P. J. Cowdery-Corvan, R. L. Klaus, F. D. Saeva. US Patent 5714311 (1998).
234. H. Okada, I. Totani, T. Kojima. JP Patent 9005926 (1997).
235. T. Ishida, F. J. Manganiello, K. Sakizadeh. US Patent 5,521,059 (1996).
236. R. Muthyala, R. J. Kenney, F. J. Manganiello, K. Sakizadeh, S. M. Simpson. US Patent 5,439,790 (1995).
237. L. R. Krepski, K. Sakizadeh, S. M. Simpson, D. R. Whitcomb. US Patent 5,175,081 (1992).
238. L. R. Krepski, K. Sakizadeh, S. M. Simpson, D. R. Whitcomb. US Patent 5,158,866 (1992).
239. Y. Hayashi, T. Akagi, S. Kinoshita, R. Hirohashi. *Nippon Shashin Gakkaishi* **52**:21 (1989).
240. L. A. Nesterova, Z. P. Bistrova, I. N. Orlova, V. G. Sevastyanov, G. R. Allakhverdov, B. I. Zhelnin, E. I. Volovich, L. N. Cinyaver, S. S. Tibilov, P. Z. Velinzon. SU Patent 1740369 (1992).
241. K. Kurttila. *J. Micrographics 10*:113 (1977).
242. American National Standard for Imaging Media, NAPM/ANSI IT9:19/049N (1994).
243. E. A. Krupinski. *Acad. Radiol. 3*:856 (1996).
244. Z. F. Lu, E. L. Nickoloff, T. Terilli. *Med. Phys. 26*:1817 (1999).
245. Report by P. Rauch, D. Peck, D. Hearshen, J. Windham, Dept. of Radiology, Henry Ford Hospital, Detroit, MI, ''Observed changes in images produced with dry silver film: influence of environmental conditions'' (1996).
246. B. B. Bokhonov, L. P. Burleva, D. R. Whitcomb, M. B. Mizen, M. R. V. Sahyun. 50th IS&T Conference, Springfield, VA, p. 38 (1997).
247. P. Zavlin, A. Dyakonov, S. Mnatsakanov, S. Tibilov, P. Velinzon, S. Gaft. *Tekh. Kino. Telev. Nauka i Tekhnike 9* (1990).
248. D. Morgan. SPSE International Tokyo Symposium, p. 52 (1973).
249. Eastman Kodak product publication # D-44 for DL microfilm 2474 (1992).
250. I. Totani, A. Hirano. JP Patent 10282601 (1998).
251. J. Bosschaerts, E. Daems, G. Halbedl, L. Leenders, J. Müller, L. Oelbrandt, R. Overmeer, H. Strijckers, F. Strumpf, T. Zehetmaier. EP Patent 836,116 (1998).
252. H. Strijckers, C. Van Roost. 2000 International Symposium on Silver Halide Technology, Quebec, p. 248 (2000).

14

Ink Jet Ink Technology

WALTER J. WNEK

DuPont, Inc., Wilmington, Delaware

**MICHAEL A. ANDREOTTOLA, PAUL F. DOLL,
and SEAN M. KELLY**

American Ink Jet Corporation, Billerica, Massachusetts

14.1 INTRODUCTION

Ink jet printing is a nonimpact printing process in which ink is ejected through very small orifices to form droplets that are directed to a medium to create an image. The mechanism by which a stream of liquid becomes unstable and breaks into droplets was first analyzed by Lord Rayleigh in 1878. The challenge for ink jet printing has been to control this process so that uniformly sized droplets can be produced reliably. Two main technologies have been developed to achieve this goal: continuous and drop-on-demand printing.

14.1.1 Continuous Ink Jet Systems

The continuous ink jet printing process is illustrated in Fig. 14.1. Ink is pumped continuously through an orifice plate to generate an array of jets. This plate is attached to a fluid cavity device to which piezoelectric crystals are bonded. These crystals are stimulated via a sinusoidal voltage to provide the ultrasonic energy for breaking the jets into streams of uniformly sized and evenly spaced droplets. The frequency of the vibration is that of the wave with the maximum growth rate, and its amplitude is sufficiently large to swamp out all other disturbances. For an inviscid liquid, this resonant wavelength is $\lambda_r = 4.51 D_j$, and its growth rate is proportional to $(\sigma/\rho D_j^3)^{1/2}$ where D_j is the jet diameter and ρ is the ink's density (Levich, 1962). σ is its dynamic surface tension in contrast to the equilibrium value. The volume of a drop is equal to that of a cylinder of diameter D_j and length λ_r

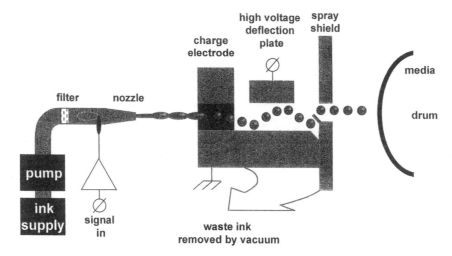

Figure 14.1 Continuous ink jet printing.

so that its diameter is $1.89D_j$. For a viscous jet, the maximum growth rate and resonant wavelength are modified to read (Levich, 1962):

$$q_{\mathrm{m}} = \left[\left(\frac{\sigma}{\rho D_j^3} \right)^{1/2} + 3\mu D_j/\sigma \right]^{-1} \tag{1}$$

$$\lambda_{\mathrm{r}} = 4.51 D_j \sqrt{1 + \frac{\mu}{2 \sqrt{\rho \sigma D_j}}} \tag{2}$$

where μ is the ink's viscosity. The effect of viscosity is to decrease q_{m}, make its peak less distinct, and have it occur at larger values of λ_{r}/D_j. The orifices must also be wetted uniformly around their perimeters so that there is no angular deflection of the jets caused by variations in the contact angle.

In the vicinity of the breakup point, whose length is proportional to the jet velocity divided by q_{m}, there is an array of electrodes, which induces a charge on the nonprint drops. The ink's electrical conductivity must be large enough so that the equilibrium charge is readily attained within the duration of the charging voltage pulse, which is approximately the reciprocal of the jet's vibration frequency. Thus it is important that the charging process and drop formation are highly synchronized. The resulting electrostatic force acting on these drops may also be enhanced by passing them between high-voltage deflection plates. The drops are thus directed into a catcher for disposal or are recycled back to the ink reservoir, which is held under vacuum.

This technology can generate drops in the range 15–400 microns at a rate of 50–1000 KHz. It has been commercialized in high-speed printers from Scitex and in high-quality fine art printers from Iris Graphics.

14.1.2 Drop-on-Demand Ink Jet Systems

With drop-on-demand (DOD) printing, a droplet is produced only when it is required to form a dot on the medium to create the image. Since there is no deflection of the drops,

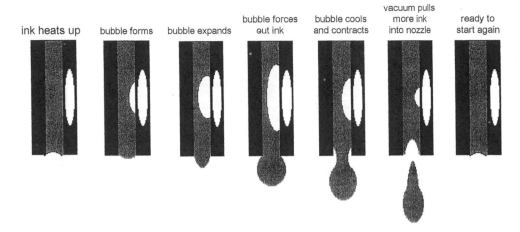

Figure 14.2 Thermal ink jet technology.

they are not charged, and they need not be conductive. The two main methods of ejecting the drops are piezo and thermal or bubble jet.

In the latter, a very small resistor or heating element is placed behind each nozzle as shown in Fig. 14.2. When an electrical pulse of a few microseconds is applied to it, the ink's temperature increases very quickly to around 300°C, approaching its critical temperature. This initiates thermal nucleation, and a bubble of vapor grows, which pushes a column of ink out from the nozzle. Once this heat has been dissipated, the bubble begins to collapse. The pressure reverses, and the column decelerates. It begins to neck down because its velocity lags behind that of the forming drop. Eventually, the column's velocity reverses, and the meniscus retracts. The column breaks when its length exceeds several times its diameter, and the drop detaches. This process of bubble formation and collapse lasts about 10 microseconds. Capillary action, driven by the ink's surface tension and contact angle, then draws ink from its reservoir to refill the orifice. The momentum of the refill ink can cause the meniscus to bulge out somewhat and then retract due to surface tension in a damped oscillatory motion. This is why DOD printers tend to need more viscous inks than continuous systems to assist in reaching equilibrium as soon as possible. The refill time is in the range of 80 to 200 microseconds. If a new cycle starts before the previous one has reached equilibrium, the resulting velocity variations can cause drop placement errors on the medium.

With piezo technology, the ink drop is ejected by a pressure wave created by the mechanical motion of a piezoelectric material in place of a bubble, as shown in Fig. 14.3. In a bend-mode design, these plates are bonded to a diaphragm and form an array of bilaminar electromechanical transducers. Another mode of operation is to contract the walls of the ink channels to squeeze out the ink.

Ink jet printers from Hewlett-Packard and Canon utilize thermal DOD technology, while those from Epson use the piezo method. Drop diameters are in the range of 18 to 50 microns at firing frequencies of 5 to 12 KHz.

14.1.3 Performance Requirements

The performance requirements that such a process places on an ink fall into several categories as listed in Table 14.1. Under print quality, aside from meeting the color specifications,

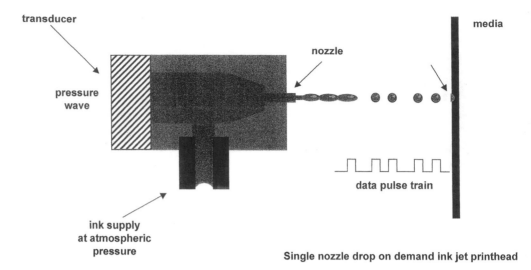

Figure 14.3 Piezo ink jet technology.

Table 14.1 Performance Requirements for Ink Jet Inks

I. Print quality	III. Machine runnability	IV. Machine compatibility
1. Match CIE Lab coordinates specifying color	1. Stable drop formation	1. No corrosion of print head
2. Minimal show-through	2. No crooked jets	2. No chemical attack of fluid system
3. Controlled dot spread with no intercolor bleed	3. No nozzle crusting or clogging	3. Easily washed off printer parts
4. No feathering	4. Readily starts up after a shutdown	
5. No splatter	5. If continuous system, also	**V. Stability**
6. No misting	Minimal foaming	1. Long shelf life
7. Gloss	Sufficiently conducive for deflection	2. Long machine life
8. Able to print on a range of media	Long filter life	3. No microbial growth
	No shorting of charge plate	4. No changes with heat/ freezing cycling
II. Archival	Wide stimulation voltage window	
1. Waterfast		**VI. Health and Safety**
2. Lightfast		1. Nontoxic
3. Dry and wet rub resistance		2. Nonflammable
4. Short drying and set time		3. Material Safety Data Sheet (MSDS)

the ink must not penetrate so deeply into the medium that it can be seen from the other side. Yet, it is just this property that is needed to reduce drying time, smearing and intercolor bleed. When the drop strikes the medium, it should not splatter to produce stray droplets around a character. A similar but more global print defect can occur if during drop formation a mist of smaller droplets is generated. Further, while some spreading of the drops is beneficial in connecting neighboring dots, going beyond that point will result in a loss of image resolution and intercolor bleed. Also, feathering, which results in a ragged dot profile, is undesirable, and of course the print should be both lightfast and waterfast to meet archival requirements.

To ensure machine runnability, the ink must be formulated to allow for stable drop formation/deflection and to maintain jet straightness over time under either continuous or intermittent operation. Viscosity and surface tension are the controlling physical properties for drop formation. With respect to the latter, its dynamic value is important because of the very short time of flight available for chemical species to migrate to the surface of the drop. Thus its value tends to be near that of water. The static value plays a role in preventing the ink from weeping out of the print head when not jetting and in wetting the medium for good penetration.

Viscosity affects jet dynamics via (1) viscous damping of the generated capillary wave, (2) the jet's velocity, which in turn changes its breakoff length, and more importantly (3) the voltage window over which stable stimulation occurs without satellite formation. Conductivity simply must be above a threshold value to allow complete charging of the drops for their uniform deflection. Dye solubility influences temporal behavior and is another area in which conflicting requirements arise, since the more soluble the dye the less tendency there is for it to deposit on an orifice; but then waterfastness is impaired.

Foam generation is a common problem experienced with continuous systems, and it is a result of using vacuum to recycle the unused ink, which causes the ink to mix with air. Even though there are filters of various ratings to remove particulates so as not to clog the orifices, the inks still must not contain an excessive amount that would cause short replacement schedules.

The ink, when it does dry on the orifice and charge plates, should be easily washed off, using either a manual or an automated procedure. It must also be compatible with the various components of the fluid system to ensure that there is no corrosion, swelling, or other adverse interactions such as residues building up on the electrodes of thermal ink jet printers, commonly known as *kogation*. Neither can the ink support microbial growth. Shipping and storage of the ink further require that it remain functional when exposed to freezing and elevated temperatures. Finally, the ink cannot pose any health or safety problems, such as toxicity or flammability.

The above performance requirements can be related to the set of physical properties and off-line tests listed in Table 14.2, which are used to screen an initial formulation prior to its optimization via printer evaluation. The resulting final formulation can consist of quite a few chemical components as shown in Table 14.3, and the rest of this review will be concerned with their selection and how they function.

Because colloid chemistry plays such an important role in the behavior of ink jet inks, we will begin with a somewhat detailed discussion of the colloidal properties of colorants. This will be followed by a discussion of their optical behavior and lightfastness. We will then cover each of the various components in an ink jet ink formulation and conclude with the manufacturing process.

Table 14.2 Important Physical Properties of Ink Jet Inks

1. Absorption spectra
2. Viscosity
3. Surface tension: static and dynamic
4. pH
5. Electrical conductivity
6. Density
7. Foaming test
8. Corrosion test
9. Filtration test
10. Humectancy/redissolvability
11. Penetration rate (Bristow)
12. Accelerated aging test at elevated temperatures
13. Drawdown evaluations of optical density, waterfastness, lightfastness, and show-through
14. Material Safety Data Sheet Information

14.2 COLLOIDAL PROPERTIES OF COLORANTS

Colloid chemistry plays a significant role in many of the properties of an ink jet ink. Among these are the solubility and aggregation of dyes and the dispersion stability of pigments, which in turn control the optical properties and nozzle clogging tendencies. This discipline also enters into how the ink interacts with the medium and why an ink can exhibit unwanted foaming. Thus an understanding of its basic principles will provide a foundation for the discussions of these phenomena to be presented in subsequent sections. Before we get into the details, an overview will be presented, which will serve as a roadmap to where we are heading.

Ionic groups that are attached to dyes or pigment surfaces undergo dissociation reactions in water and thus generate a charge. Acidic and anionic groups yield a negative charge, and conversely basic and cationic ones result in a positive value. The degree of their ionization depends on the pH or concentration of the potential-determining ions as

Table 14.3 Typical Components of an Ink Jet Ink Formulation

1. Colorant (dye or pigment)
2. Deionized water and cosolvents
3. Solubilizing agent/dispersant
4. Base
5. pH Buffer
6. Defoamer
7. Penetrant/wetting agent
8. Humectant
9. Chelating agent
10. Corrosion inhibitor
11. Biocide
12. Binder
13. Viscosity modifier

in any equilibrium reaction, but salts can also have an effect on the charge. The charge generates an electric field and a potential at the surface that attracts the oppositely charged ions or counterions, resulting in their surface concentrations being higher than the bulk values. This not only can shift the equilibrium reaction to suppress ionization but also sets up the conditions for ion binding to neutralize the charge. We will quantify these coupled phenomena and express the charge and potential as a function of the ionic site density, its dissociation constant, pH, and ionic strength.

We will then continue on to the interaction between charged entities as they approach each other in solution. These may be pigment particles or the dye's monomer and aggregates. Obviously, there is a repulsive force set up, but it is a result of the mutual electrostatic repulsion of the counterions as their concentration is increasing along with an increase in their entropy of mixing. Thus an osmotic pressure opposes their approach, not Coulombic repulsion between the particles. In solving the governing Poisson–Boltzmann equation, it has been generally assumed that either the surface potential or charge does not vary with separation. However, this implies that either the surface charge is decreasing or the surface potential is increasing, respectively. The actual situation lies between these two limits with both parameters changing, because the increasing surface concentration is driving the ions back onto the surface. This behavior is known as charge regulation, and the controlling variables are the dissociation constant, site density, and concentration of the absorbing ions. The constant potential case provides the lower bound, and the constant charge case gives the upper bound.

Another approach to creating a repulsive force between the particles is to attach nonionic surfactants and polymers to their surfaces. Such a dispersant consists of an anchor to hold it on the surface and a soluble chain that extends out from its surface. When the particles approach each other and the chains intermix, the resulting osmotic pressure drives the particles apart. Sterically stabilized dispersions offer certain advantages over charge stabilized ones, such as insensitivity to ionic strength or salts, reversible flocculation, and better freeze/thaw stability.

Another interaction acting between the particles is the attractive van der Waals force, which is also increasing with decreasing separation. In the constant potential case, the repulsive and attractive forces combine to give a maximum repulsion at separations on the order of several nanometers, depending on the ionic strength. At smaller separations, the van der Waals force dominates, and the interaction reverses to become attractive. Since the constant charge case exhibits higher repulsion, the maximum is greater and occurs at a smaller separation. In the steric case, a maximum is not exhibited.

Surface treatments such as the absorption of ionic surfactants can be quite effective in increasing the stability of pigments. Their ionic groups extend outward from the surface, and this shifting of the charge plane makes the repulsion larger and the attraction smaller, because the governing separation is now the distance between the charge planes instead of the surfaces. In fact, this effect can be so strong that net force is repulsive for all separations.

While this behavior imparts essentially total stability to the dispersion, it also promotes repeptization in which the pigment particles can lift off the medium upon wetting. Thus it appears that a constant charge situation is not optimal for waterfastness. What is needed is a charge regulated surface with a dissociation constant such that both a large maximum repulsion is generated at intermediate separations and a large attractive force at smaller values. It should be noted that even if the drying is not 100% complete, the resulting capillary forces will compress the particles into contact. Thus strongly dissociating groups such as sulfonic acid should be avoided, and a weaker acid with a pKa of 6.5 to 8 (e.g., carboxyl or a weakly basic amine) is preferred. The latter has the added advantage of imparting a positive charge to the pigment opposite to that of the paper and hence improving adhesion. But it should be noted that a positive charge by itself does not necessarily alleviate the need for strong cohesion within the mass of carbon black aggregates deposited on the paper.

An analogous approach has been applied to improve the water fastness of dyes that use sulphonic acid groups to impart water solubility (Kenyon, 1996). Some of these groups are replaced by carboxyl ones whose degree of ionization is more sensitive to pH. At the alkaline pH of the ink (7.5–10), these groups are sufficiently ionized to impart solubility. Upon contact with acidic paper (pH 4–6.5), their ionization is suppressed, resulting in lower solubility and dye aggregation.

In addition, the drying process itself can be used to enhance adhesion/cohesion. The loss of water causes the ionic strength to increase, resulting in a diminishing zeta potential and less stability. Redispersion would be hindered upon wetting as long as the dilution does not approach the original concentrations too closely. This phenomenon can be further enhanced if the dissociation constant is again of an intermediate value such that the decreasing pH suppresses ionization and drives the surface charge to zero. This only works if the initial pH is acidic. If it is basic, then just the opposite happens, but this effect can be reversed by the use of a volatile amine such as triethanolamine in combination with an acid, which is left behind after drying.

Another variable that one can consider manipulating is the valence of the counterions. The multivalent ones are more strongly attracted to a charged surface than monovalent ones such that they have even higher surface concentrations. Suppose that we have added a small amount of a multivalent salt such as $CaCl_2$ to the ink at a level that does not impact stability. As drying proceeds and the particles are drawn together, the concentration of Ca^{2+} at the surface progressively builds up until no more than a monolayer of Ca^{2+} is sandwiched between any two particles so that each cation is bound to a negative charge site on either surface. Thus Ca^{2+} at the appropriate concentration functions as a binder with its action being triggered by drying.

The complicating role that other ink components such as humectants, penetrants, and binders play in controlling repeptization will also be discussed in terms of the above spacing mechanism. Unfortunately, humectants in their role of preventing crusting of the ink jet nozzles also act as spacers in the dried state to compromise wet rub resistance.

Let us now proceed to develop the supporting details for reaching these conclusions.

14.2.1 Charge Acquisition by Chemical Groups

In order for chemical groups to ionize and become charged, they must be able to release their counterions into the solvent. This requires that its solvating power can overcome their binding energy to the surface. Water is particularly effective in solvating both cations and anions because oxygen's greater electronegativity draws electrons away from the two hydrogens resulting in charges of opposite sign on these atoms. For example, suppose that the surface groups are acidic at a site density of N_S and ionize to give a negative surface charge according to

$$AH \Leftrightarrow A^- + H^+$$

with a dissociation constant of K_d (Hunter, 1989). K_d is related to the adsorption free energy ΔG_{ads} per ion, which is the difference between the binding and solvation energies:

$$\Delta G_{ads} = -kT \ln K_d \tag{3}$$

where k is the Boltzmann constant and T absolute temperature. At equilibrium the surface concentrations are related through

$$K_d = \frac{[A^-][H^+]_S}{[AH]} \tag{4}$$

where $[H^+]_S$ is the molar concentration of the hydronium ion at the surface. $[A^-]$ may be readily solved for in combination with $N_S = [AH] + [A^-]$ to give the surface charge density $\sigma = -ze[A^-]$ (coulomb/m^2):

$$\sigma = \frac{-zeK_dN_S}{K_d + [H^+]_S} = \frac{\sigma_{max}}{1 + [H^+]_S/K_d} \tag{5}$$

where z is the valence and e the electronic charge (Healy and White, 1978). As a check, we see that for large K_d the charge is just determined by the number of surface sites and attains its maximum absolute value. When K_d equals $[H^+]_S$, half of the sites are ionized. It should be noted that Eq. (14.5) is not just valid for acidic sites but for any type of ionic groups with $[H^+]_S$ being replaced by the concentration of the dissociated ion.

Concentration of Counterions/Coions at Surface

It should be emphasized that $[H^+]_S$ is the concentration at the surface and is not necessarily equal to the total number of counterions nor the pH in the bulk solution. The reason for this is that Brownian motion allows the counterions to diffuse away from the surface, but this is hindered by electrostatic attraction back to it, resulting in a concentration profile from the surface to the bulk solution. The diffusional flux is D dc/dx where D is the diffusivity and x is the distance. The velocity (u) due to the electric field (E) is found from a force balance on an ion of valence z for which the electrostatic force zeE is equal to the frictional force $(kT/D)u$ as given by the Stokes–Einstein equation. At steady state, the diffusional and convection fluxes are equal:

$$D\frac{dc}{dx} = uc = \frac{zeE}{(kT/D)c} = -\left(\frac{zeD}{kT}\right)c\frac{d\Psi}{dx} \tag{6}$$

where c is the number density of ions and $\Psi(x)$ is the electric potential or work to bring a unit charge from infinity to position x. Rearranging this equation gives

$$d\ln c = -\left(\frac{ze}{kT}\right)d\Psi \tag{7}$$

which upon integration from the surface (s) to the bulk solution (b), where $\Psi = 0$, yields

$$c_s = c_b \exp\left(\frac{-ze\Psi_s}{kT}\right) \tag{8}$$

which is the Boltzmann distribution. Since Ψ_s is negative, the pH at the surface will be more acidic than its bulk value, and the effect is to suppress the ionization in comparison to the reaction occurring in the bulk. On the other hand, the concentration of the coions relative to that of the counterions at the surface is lower by the factor $e^{-ze\Psi_S/kT}$ or 7, 49, and 2397 for $\Psi_S = -25$, -50, and -100 mV at 25°C, respectively.

Surface Potential as a Function of Surface Charge

In order to complete the analysis, we need to relate Ψ_S to the surface charge. The electric field E generated by the space charge ρ per unit volume due to the counterions is given by Poisson's equation for any distance x from the surface (Hunter, 1987):

$$\varepsilon_0\varepsilon\frac{dE}{dx} = -\varepsilon_0\varepsilon\frac{d^2\Psi}{dx^2} = \rho \tag{9}$$

where ε_0 is the permittivity of free space and ε the solution's dielectric constant. The basis of this equation is the application of Coulomb's law to a point charge at position x and all its neighboring charges and then summing over these interactions to obtain the field at x (Jackson, 1975).

Equation (14.8) may now be used to calculate ρ by replacing the surface by any position x:

$$\rho(x) = \sum_i z_i ec_i(x) = \sum_i z_i ec_{bi} \exp\left[\frac{-z_i\, e\Psi(x)}{kT}\right] \tag{10}$$

where the sum includes both the counterions and the coions as well as any dissociated electrolytes such as salts, acids, or bases already in solution. It should be pointed out that there is a subtle distinction between the potentials as used in Eqs. (14.9) and (14.10). Because the ions are in constant random motion, the potential to be used in the Boltzmann equation is the potential of the mean force averaged over time, while in the Poisson equation it is the mean potential (Lyklema, 1991; Hunter, 1989). However, traditionally they have been assumed to be equal.

Integrating Eq. (14.9) from the surface to the bulk solution and using the electroneutrality condition that the sum of the surface charge density σ and the total space charge must equal zero, we obtain the deceptively simple relationship

$$\sigma = -\varepsilon_0\varepsilon\left(\frac{d\Psi}{dx}\right)_S \tag{11}$$

Next, let us take the space derivative of the concentration profile of each ion and sum them:

$$\sum_i\frac{dc_i}{dx} = -\sum_i\left(\frac{z_i ec_{bi}}{kT}\right)\exp\left[\frac{-z_i e\Psi(x)}{kT}\right]\frac{d\Psi}{dx} \tag{12}$$

Then, substitute Eq. (14.10) followed by Eq. (14.9):

$$\frac{d}{dx}\sum_i c_i = -\left(\frac{\rho}{kT}\right)\frac{d\Psi}{dx} = \left(\frac{\varepsilon_0\varepsilon}{kT}\right)\left(\frac{d\Psi}{dx}\right)\left(\frac{d^2\Psi}{dx^2}\right) = \left(\frac{\varepsilon_0\varepsilon}{2kT}\right)\frac{d}{dx}\left(\frac{d\Psi}{dx}\right)^2 \tag{13}$$

and integrate it from the surface to the bulk solution where $\Psi = d\Psi/dx = 0$, making use of Eq. (14.11):

$$\sum_i c_{Si} = \sum_i c_{bi} + \left(\frac{\varepsilon_0\varepsilon}{2kT}\right)\left(\frac{d\Psi}{dx}\right)_S^2 = \sum_i c_{bi} + \frac{\sigma^2}{2\varepsilon_0\varepsilon kT} \tag{14}$$

Finally, we solve for σ, replacing the c_{Si}'s with Eq. (14.8):

$$\sigma^2 = 2\varepsilon_0\varepsilon kT \sum_i c_{bi} \left[\exp\left(\frac{-z_i e\Psi_S}{kT}\right) - 1\right] \tag{15}$$

which relates the surface charge density, surface potential, and electrolyte concentrations in the bulk. It is interesting to note that we have obtained this relationship, known as the Grahame equation (Israelachvili, 1992), without having to integrate the complete Poisson–Boltzmann equation given by Eqs. (14.9) and (14.10).

Applying the Grahame Equation

As an example of the application of Eq. (14.15), let us consider the case of the mixed electrolyte HCl and NaCl in water: where H^+ is the potential-determining ion and the other ions are indifferent:

$$\sigma^2 = 2\varepsilon_0\varepsilon kT\{[H^+]_b e^{-e\Psi_S/kT} + [Na^+]_b e^{-e\Psi_S/kT} + [OH^-]_b e^{+e\Psi_S/kT}$$
$$+ [Cl^-]_b e^{+e\Psi_S/kT} - [H^+]_b - [Na^+]_b - [OH^-]_b - [Cl^-]_b\} \tag{16}$$

Noting that $[H^+]_b + [Na^+]_b = [OH^-]_b + [Cl^-]_b = c_b$ in order to satisfy charge neutrality in the bulk solution, the above equation becomes

$$\sigma = (8\varepsilon_0\varepsilon kT c_b)^{1/2} \sinh\left(\frac{e\Psi_S}{2kT}\right)$$

or

$$\sigma\left(\frac{C}{m^2}\right) = 0.117 c_b^{1/2}(M) \sinh\left[\frac{z\Psi_S(mV)}{51.4}\right] \qquad \text{at} \qquad 25°C \tag{17}$$

where

c_b = total ionic bulk concentration

$8\varepsilon_0\varepsilon kTN_0 = 8 \times 8.85 \times 10^{-12}$ C²/Jm $\times 78.5 \times 1.381 \times 10^{-23}$ J/K $\times 298K$

$\times 6.022 \times 10^{23}$/mol $\times 1000$ l/m³

$= 0.01377$ C²/m⁴M ($\sqrt{} = 0.117$ C/m²M$^{1/2}$),

N_0 = Avogadro's number.

Since the potential is accessible to measurement, this equation is plotted in Fig. 14.4 as Ψ_S against σ for a range of concentrations. It is seen that Ψ_S increases in a logarithmic fashion with σ but decreases with respect to $c_b^{1/2}$. If the electrolyte is multivalent as well as symmetric, then Ψ_S is multiplied by its value in Eq. (14.17). Thus for the same σ and c_b, Ψ_S would be reduced in half by increasing the valence from one to two. Further, the counterion concentration at the surface would be higher by the factor $e^{-e\Psi_S/kT}$ or 2.6, 7, and 49 for $\Psi_S = -25$, -50, and -100 mV, respectively, according to Eq. (14.8).

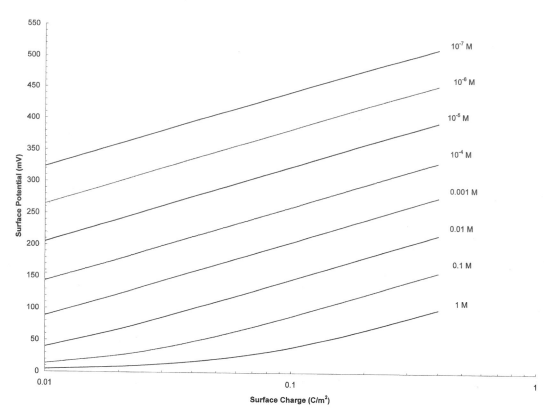

Figure 14.4 Surface potential (Ψ_s) vs. surface charge (σ) and ionic concentration.

The concentration dependence described by Eq. (14.17) is strictly valid for indifferent ions, which do not change the surface charge via adsorption or influence the extent of ionization of the surface groups. Since the pH falls in the latter category, we need to couple Eqs. (14.5) and (14.17) to account properly for its effect. Thus let us rewrite Eq. (14.5) in the form

$$\sigma = \frac{\sigma_{max}}{1 + 10^{-pH}e^{-\Psi_s(mV)/25.7}/K_d} \qquad \text{at} \qquad 25°C \tag{18}$$

We see that several phenomena are occurring when the pH is varied. First, the ionic strength of the solution changes, which in turn affects the surface potential according to Eq. (14.17). Second, the surface charge responds not only to the bulk pH but (third) also to the modified surface potential with both effects described by Eq. (14.18). Thus we set Eqs. (14.17) and (14.18) equal to each other, resulting in a nonlinear equation to be solved for Ψ_s:

$$2.303(pH - pK_d) = -\frac{\Psi_s(mV)}{25.7} - \ln\left[\frac{-1 + 8.547\sigma_{max}c_b^{-1/2}(M)}{\sinh\{\Psi_s(mV)/51.4\}}\right] \tag{19}$$

While this can be accomplished numerically, the following graphical method is useful in providing a pictorial representation of the process. We simply replot Eq. (14.17) as σ vs.

Ψ_S contours for a range of concentrations and superimpose Eq. (14.18) plotted in a similar manner over a range of pH's for fixed σ_{max} and K_d. The intersection of those two contours representing the total ionic concentration and pH yields the corresponding σ and Ψ_S.

This procedure is illustrated in Fig. 14.5 for $\sigma_{max} = -0.3$ C/m^2 and $K_d = 10^{-5}$ M. This value of σ_{max} corresponds to about two surface sites per nm^2, which is representative of a high charge density. Lower values would just shift the concentration contours downward since $\sigma_{max}/c_b^{1/2}$ is a natural grouping from Eq. (14.19). The value of K_d may be interpreted as the pH at the surface for which half the sites are dissociated. It was taken as 5, which is typical of a weak acidic group such as carboxyl —COOH and sulfite —OSO$_2$H (James and Parks, 1982). As can be seen from Eq. (14.19) (pH $-$ pK$_d$) is a natural scaling parameter so that the effect of other values is simply to relabel the pH contours. For example, increasing K_d to 10^{-4} M reduces the pH labels by one unit, and further increases bring the surface charge to its fully ionized state independent of pH and ionic strength. This is realized with strong acidic groups such as sulfonic —SO$_3$H and sulfate —OSO$_3$H (pK$_d < 1$). Similarly, a positive surface charge is imparted by basic groups such as weak amines —NH$_2$ and strong quaternary ammonium —N$^+$R$_3$.

A common characterization of colloids is to measure the potential as a function of the bulk pH at constant ionic strength by adjusting the salt concentration accordingly. If we follow any constant concentration contour, it is seen that both σ and Ψ_S become more negative with increasing pH, with the rate of change decreasing until they reach plateaus. On the other hand, if the ionic strength is increased at constant pH, σ becomes more negative while Ψ_S is driven toward zero. The reason for this is that the pH at the surface is increasing as a result of the decreasing $|\Psi_S|$, which allows for more ionization. Thus

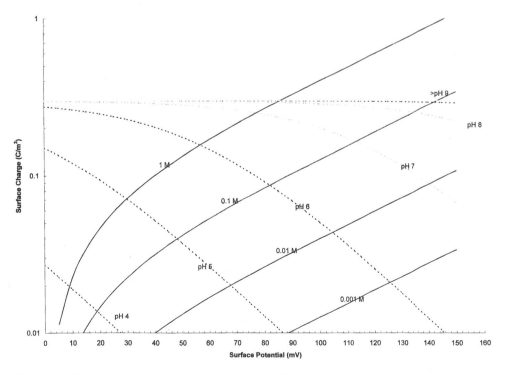

Figure 14.5 Surface charge and potential as a function of pH and ionic concentration.

widely varying combinations of σ and Ψ_S can be achieved via manipulation of pH and the salt concentration in addition to the degree of surface treatment (N_S) and type of chemical group (K_d).

14.2.2 Zeta Potential

We characterize the surface charge via electrophoresis experiments in which the colloidal particles migrate under the influence of an applied electrical field, and the resulting data are reported as an electrical potential. Unfortunately, it is not necessarily the value at the surface but that at some distance away from it because the particles carry a thin layer of stagnant fluid on the order of 10 angstroms along with themselves. This potential at the plane of shear is known as the zeta potential ς (Hunter, 1981), and its value also represents the potential at the effective location of the solid–liquid interface, which will be needed later in evaluating the electrostatic force when the particles approach each other.

The zeta potential has been related to the surface potential through various models, and their intent has been to relax certain limitations of the Poisson–Boltzmann equation. Some relevant concerns in the near vicinity of the surface are

1. The ions should not be treated as point charges because their finite size limits their maximum concentration at the surface and distance of closest approach to it. This also implies that there are no other forces involved in moving aside other ions and creating holes in the solvent beyond the electrostatic work of bringing an ion near the surface.

2. The high concentration of counterions at the surface could lead to a significant neutralization of the surface charge if they became bound to the surface, replacing some of the adsorbed water molecules.

3. The high electric field at the surface could cause some ordering of the water dipoles decreasing the dielectric constant.

4. While the surface charges are discrete, they have been implicitly assumed to be smeared out over the surface. Fortunately, this approximation has turned out to be a reasonable one (Israelachvili, 1992).

According to Eq. (14.14) a surface charge of -0.3 C/m^2 at room temperature (20°C) would yield an unreasonably high concentration of 26 M relative to the bulk for the counterions at the surface. The ionic radius of sodium is ≈ 1 angstrom, and multiplying these two numbers then leads to a counterion surface charge of 1.602×10^{-19} C $\times$ 1 $\times$ 10^{-10} m $\times$ 26 mol/L $\times$ 6.022 $\times$ 10^{23}/mol $\times$ 1000 L/m^3 = 0.25 C/m^2. On the other hand, a lower surface charge of -0.1 C/m^2 would only give a counterion surface charge of 0.03 C/m^2. Thus at high surface charges the counterions at the surface neutralize a significant amount of the surface charge as a consequence of their concentration being proportional to the square of the surface charge according to Eq. (14.14). In fact, the resulting net charge can even start to diminish after reaching a maximum with respect to the site density.

Counterion Binding

We can account for counterion binding onto the dissociated surface sites by characterizing the phenomenon analogous to the way that surface dissociation was treated earlier (James and Parks, 1982; James, 1987):

$$\mathrm{A^-} + \mathrm{Na^+} \overset{K_{\mathrm{Na}}}{\Leftrightarrow} \mathrm{ANa}$$

$$K_{\mathrm{Na}} = \frac{[\mathrm{ANa}]}{[\mathrm{A^-}][\mathrm{Na^+}]_S} \tag{20}$$

where the binding constant K_{Na} is on the order of the inverse of K_d. Another approach has been to use an adsorption isotherm such as Langmuir's for the counterions, but it seems more consistent to treat the hydrogen and sodium ions in the same way. Eqs. (14.4) and (14.20) are now solved together with $N_s = [AH] + [A^-] + [ANa]$, and Eq. (14.5) is modified to

$$\sigma = \frac{-zeN_s}{1 + K_{Na}[Na^+]_S + [H^+]_S/K_d} = \frac{\sigma_{max}}{1 + K_{Na}[Na^+]_S + [H^+]_S/K_d} \tag{21}$$

Thus we have the important result that the effect of counterion binding to neutralize the surface charge may be thought of as simply adding its surface concentration weighted by its equilibrium binding constant to the pH at the surface. This "effective" pH then becomes the label for the contour plots of Fig. 14.5. It should be noted that again the Boltzmann distribution is used to relate the surface concentrations to the bulk values. However, there is one subtle point of which one should be aware. It is possible that the plane of the adsorbed counterions does not exactly coincide with that of the hydrogen ions in the undissociated state. While some investigators have introduced two separate charge planes along with their different potentials, we shall forego this additional complexity at this time.

Orientation of Water Dipoles

The third concern about the possible orientation of the water dipoles means that close to a surface the dielectric constant of water is smaller than its bulk value. The traditional approach of handling this due to Stern (Hunter, 1987) has been to represent this region as an uncharged layer of thickness l of several angstroms with a reduced dielectric constant $\varepsilon_1 \approx 10-40$. Integrating Poisson's equation (14.9) with zero charge density gives the potential drop across it as $\sigma l/\varepsilon_0\varepsilon_1$. However, the trend today has been to question whether the electric field at the surface is really large enough to immobilize the water molecules adjacent to it, especially when counterion binding is occurring (Israelachvili, 1992).

Zeta Potential Model

It is generally accepted today that the Poisson–Boltzmann equation is valid for surface potentials less than 200 mV and ionic concentrations no greater than 1 M (Russel et al., 1989). Since our situation falls within these ranges, we shall assume that this equation can be used up to the surface and that the zeta potential may be calculated as the potential at a distance $\delta \approx 10$ Å away from the surface. This requires knowledge of how the potential varies with position, and it is obtained by integrating the Poisson–Boltzmann equation a second time. We can accomplish this by returning to Eqs. (14.14)–(14.17), which represent the first integration. Evaluating them at position x rather than at the surface allows us to write

$$\frac{d\Psi}{dx} = -\left(\frac{8kTc_b}{\varepsilon_0\varepsilon}\right)^{1/2} \sinh\left(\frac{e\Psi_x}{2kT}\right) \tag{22}$$

which is readily integrated to yield

$$\tanh\left(\frac{ze\Psi_\delta}{4kT}\right) = \tanh\left(\frac{ze\Psi_S}{4kT}\right) \exp(-\kappa\delta) \tag{23}$$

where

$$\kappa^2 = \sum_i \frac{c_{bi} z_i^2 e^2}{\varepsilon_0 \varepsilon k T} = 10.81 I \text{ (nm)}^{-2} \qquad \text{for water at} \qquad 25°C \qquad (24)$$

and I is the ionic strength $\frac{1}{2} \sum_i c_{bi} z_i^2$ (mol/L).

Inspection of the exponential term in the Eq. (14.23) shows that $1/\kappa$ is a characteristic distance over which the potential decays and the surface charge exerts an influence. It is known as the thickness of the diffuse double layer because it is also the distance that two capacitor plates subjected to a potential difference Ψ_s would have to be separated in order to induce surface charge densities $\pm\sigma$ on them. For NaCl solutions, it equals κ^{-1} (nm) = $0.304/[\text{NaCl}]^{1/2}$ or 96.1, 30.4, 9.6, 3.0, 0.96, and 0.30 nm for 10^{-5}, 10^{-4}, 10^{-3}, 0.01, 0.1, and 1 M, respectively. For 2:2 electrolytes, the values are decreased by a factor of 2. This exponential dependence on the position of the shear plane also explains why the zeta potential measurement can yield somewhat variable results.

In Fig. 14.6, the zeta potential is plotted against the surface potential according to Eq. (14.23) for these ionic strengths. It is seen that the zeta potential becomes increasingly smaller than the surface potential with increasing ionic strength because the double layer thickness is being compressed toward the shear plane and eventually is underneath it.

Some Solution Paths

This behavior suggests a potential solution path for maintaining dispersion stability of a pigment in water without sacrificing its cohesion and adhesion to the medium in the dried

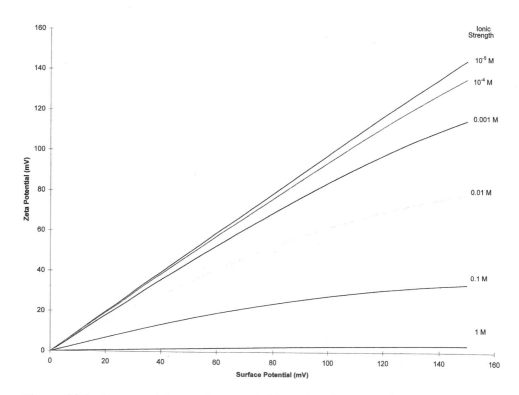

Figure 14.6 Zeta potential vs. surface potential for various ionic strengths.

state. During drying, the loss of water will cause the ionic strength to increase, resulting in a diminishing zeta potential and less stability. If this concentration effect is sufficiently great, then instability will be induced, and redispersion will be hindered upon rewetting, provided the dilution does not approach the original concentration too closely. This phenomenon can be further enhanced if the dissociation constant of the surface groups is of an intermediate value such that the increasing $[H^+]$ or decreasing pH suppresses ionization and drives the surface charge to zero. Of course, this assumes that the initial pH is acidic. If it is basic, then just the opposite happens, which can be reversed by the use of a volatile amine in combination with an acid that is left behind. Typically, triethanolamine has been used in dye-based inks to precipitate it onto the paper. The addition of a small amount of multivalent counterions could be of further benefit, and a divalent one such as calcium may even serve as a binder between two surface sites on neighboring particles (Visser, 1988, 1995). Its concentration would have to be carefully selected so that there would not be more than a monolayer in the dried state.

14.2.3 Electrical Double Layer Repulsion

Let us consider the forces that are acting on an increment of volume anywhere between the two surfaces. The electrostatic force on the space charge is $\rho E\,dx$ or $\rho\,d\psi$, and the osmotic pressure resulting from Brownian motion is $kT\,d\Sigma_i c_i$. Their sum is equal to the repulsive pressure dP of the counterions:

$$dP = \rho\,d\psi + kT\,d\Sigma_i c_i = -\varepsilon_0\varepsilon\left(\frac{d^2\Psi}{dx^2}\right)d\psi + kT\,d\Sigma_i c_i$$

$$= -\frac{1}{2}\varepsilon_0\varepsilon\,d\left(\frac{d\Psi}{dx}\right)^2 + kT\,d\Sigma_i c_i \tag{25}$$

We then integrate over the separation distance from infinity to D, keeping fixed the position x in the solution:

$$P_x(D) = -\frac{1}{2}\varepsilon_0\varepsilon\left(\frac{d\Psi}{dx}\right)^2_{x(D)} + kT\sum_i c_{xi}(D) + \frac{1}{2}\varepsilon_0\varepsilon\left(\frac{d\Psi}{dx}\right)^2_{x(\infty)} - kT\sum_i c_{xi}(\infty) \tag{26}$$

noting that $P_x(\infty)$ is zero. It is seen that the pressure consists of two components: (1) an electrostatic energy contribution, which is always attractive, and (2) the repulsive entropic or osmotic force, which drives the counterions apart. It is important to realize that contrary to intuition the nature of the repulsion is osmotic and not electrostatic, because the overall system is neutral (Israelachvili, 1992). Equation (14.26) can be considerably simplified for the case of identically charged surfaces by integrating Eq. (14.15) from their midplane (*m*) where $d\Psi/dx$ is zero to any point x:

$$\Sigma_i c_{xi} = \Sigma_i c_{mi} + \left(\frac{\varepsilon_0\varepsilon}{2kT}\right)\left(\frac{d\Psi}{dx}\right)^2_x \tag{27}$$

and then substituting it into Eq. (14.26) to cancel the derivative terms:

$$P(D) = kT[\Sigma_i c_{mi}(D) - \Sigma_i c_{bi}] \tag{28}$$

which shows that P is independent of x and also equals the pressure acting on the two surfaces. Thus P is just the excess osmotic pressure of the ions in the midplane over the bulk value and reduces the problem to determining the concentration of ions at midplane or the potential there in combination with the Boltzmann equation $c_{mi} = c_{bi} \exp(-z_i e \Psi_m / kT)$. This is accomplished by integrating Eq. (14.27) from midplane to the surface:

$$D = \int_{\psi_m}^{\psi_s} \frac{2d\psi}{[2\pi kT/\varepsilon_0 \varepsilon \Sigma_i c_{bi}\{\exp(-z_i e \Psi/kT) - \exp(-z_i e \Psi_m/kT)\}]^{1/2}} \tag{29}$$

for a range of ψ_m's to obtain D as function of ψ_m for a given ψ_s. Equation (14.28) is then used to yield the pressure vs. separation curve for two plates.

The more commonly used parameter is the interaction energy per unit area V_{R-pl} which is the work required to bring the plates from infinite separation to D, and it is readily calculated by integrating P with respect to D over these limits.

Two Spheres

The interaction of two spheres can be obtained from V_{R-pl} by dividing their surfaces into differential rings around a line connecting their centers and summing over these pairs of "plates" as originally proposed by Derjaguin (Hunter, 1987). The result is that the repulsive force F_{R-sp} and energy V_{R-sp} between two spheres of radius R are given by

$$F_{R-sp}(H_0) = -\pi R V_{R-pl}(H_0) \tag{30}$$

$$V_{R-sp}(H_0) = \pi R \int_{H_0}^{\infty} V_{R-pl}(D) dD \tag{31}$$

where H_0 is the shortest distance between them. These approximations are valid as long as the separation is much smaller than the radius.

Analytical Approximation

While all of these integrations can be performed numerically by the computer program STABIL developed by R. V. Linhart and J. H. Adair (1993), it is instructive to review first some of the available analytical solutions for trends (Verwey and Overbeek, 1948; Kruyt, 1952). Let us return to Eq. (14.28) for a 1:1 electrolyte such as NaCl and expand it in Taylor series around $\Psi_m = 0$, retaining terms up to the second power. It should be noted that the assumption made is that Ψ_m is small, not Ψ_s. But the limiting assumption is that Ψ_m is taken as the sum of the potentials from each surface for an isolated surface with Eq. (14.23) evaluated at $x = D/2$.

The end result for the interaction energy between two spheres is

$$V_{R-sp} \approx \left(\frac{64\pi kT R c_b v^2}{\kappa^2}\right) \exp(-\kappa H_0)$$

$$= 4.61 \times 10^{-11} R v^2 \exp(-\kappa H_0) \qquad \text{joules (for } z = 1\text{)} \tag{32}$$

where $\gamma = \tanh(ze\Psi_s/4kT)$.

This expression is accurate for separations greater than the double layer thickness. It is seen that the energy (1) decays exponentially with distance normalized by the double layer thickness, (2) increases nonlinearly to a plateau with Ψ_S, and (3) increases linearly with particle size.

Surface Conditions During Interaction

While STABIL provides numerical solutions that are valid for all separations, it is based on the boundary condition that the surface potential is held constant as the two surfaces come together. Since the potential at midplane increases toward Ψ_S as this occurs, the slope of the Ψ vs. x curve becomes less steep at the surface, and thus the surface charge, being proportional to $(d\Psi/dx)_S$, is driven toward zero [see Eq. (14.11)]. This behavior is most pronounced at separations smaller than the double layer thickness.

However, in our situation with strongly dissociated surface groups, a more appropriate boundary condition may be a constant surface charge, which would result in an increasing Ψ_S and thus higher repulsion and stability. In the limit as the separation becomes very small, the concentration of counterions becomes uniform within the gap and equal to their total number $2\sigma/ze$ divided by D or $2\sigma/zeD$. Then Eq. (14.28) gives the limiting pressure as $2\sigma kT/zeD$ for $D \to 0$, and it increases without bound. In the constant surface potential case, it reaches a plateau. On the other hand, for distances greater than the double layer thickness, there is little difference between the two cases because the interaction is weak (Israelachvili, 1992).

Yet the case of constant surface charge may not be completely descriptive of our situation. The increasing surface potential upon approach of the surfaces decreases the pH at the surface to reverse the ionization and increases the surface concentration of the counterions to promote their binding. Both effects act to neutralize the surface charge and minimize the change in the surface potential. This phenomenon has been termed surface charge regulation (Hunter, 1987), and its interaction energy lies between the constant surface charge case as an upper limit and the constant surface potential case as a lower limit. The differences increasingly grow for separations smaller than the double layer thickness, but all of these cases converge above it. This behavior presents an opportunity to manipulate the interaction through the dissociation constant via the type of surface group.

14.2.4 Attractive van der Waals Force

As a consequence of electronic fluctuations in atoms, instantaneous dipoles are present in solid materials, and their electric fields in turn induce dipoles in neighboring bodies. The interaction of these dipoles between two particles gives rise to the van der Waals attractive force $F_{\text{A-sp}}$ and energy $V_{\text{A-sp}}$ (Israelachvili, 1992):

$$F_{\text{A-sp}} = \frac{A_{131}R}{12H_0^2} \tag{33a}$$

$$V_{\text{A-sp}} = \frac{-A_{131}R}{12H_0} \tag{33b}$$

where A_{131} is the Hamaker constant for material 1 separated by liquid 3. It may be calculated from the individual values A_{ii} in vacuum as $A_{131} = (\sqrt{A_{11}} - \sqrt{A_{33}})^2$, which shows that an intervening liquid acts to reduce the force considerably. A_{131} is the range $0.01–2 \times 10^{-19}$ J where water is the medium.

While Hamaker constants are tabulated for many materials (Lyklema, 1991), they may also be calculated from their dielectric permittivity ε_i vs. frequency ω curves according to the Lifshitz theory (Israelachvili, 1992):

$$A_{131} = \frac{3}{2} kT \sum_{n=0}^{\infty}{}' \left[\frac{\varepsilon_1(i\omega) - \varepsilon_3(i\omega)}{\varepsilon_1(i\omega) + \varepsilon_3(i\omega)} \right]^2 \tag{34}$$

where $\omega_n = (2\pi kT/\hbar)n = n(2.45 \times 10^{14}$ radian/second) at 25°C and $\hbar$ is Planck's constant divided by 2π. The prime next to the summation sign denotes that the $n = 0$ term is multiplied by $1/2$. The frequencies of interest are zero (static case), none in the microwave ($\omega \approx 10^{11}$ rad/s), a few in the near infrared ($\omega \approx 10^{14}$ rad/s), somewhat more in the visible ($\omega \approx 3 \times 10^{15}$ rad/s), and many in the ultraviolet ($\omega > 10^{16}$ rad/s). Even though $\varepsilon_i(i\omega)$ is a function of imaginary frequencies, it is a real function that decreases monotonically and is measurable.

It may be observed that at zero spacing the attraction apparently goes to infinity, but actually the distance should be measured from the centers of the surface atoms of the two particles. Thus at contact the spacing Δ_A is equal to their diameter, which is around 1.5–4 Å.

It should be mentioned in passing that there is a related phenomenon known as the ion correlation effect. It is the result of van der Waals attraction between the highly polarized layers of counterions at each surface, and its force can become significant at small separations (< 4 nm), especially at high surface charges and with multivalent ions (Israelachvili, 1992).

Retardation

There is another phenomenon that should be taken into account. As the particles become further apart, the time of travel of an electromagnetic wave from one atom to another becomes comparable to the period of the oscillating dipoles. Its value may be estimated by the distance traveled by light during one rotation of a Bohr atom electron and is $\lambda \approx$ 100 nm. When the wave returns, it finds that the orientation of the dipole is no longer aligned, and this reduces the attraction. To include this retardation effect, Eq. (14.33) is modified to read (Schenkel and Kitchener, 1960; Wiese and Healy, 1970):

$$V_{\text{A-sp}} = -A_{131}R/12H_0 \times [1/(1 + 1.77p)] \quad \text{for} \quad p = 2\pi H_0/\lambda < 1 \tag{35a}$$

$$\begin{aligned} V_{\text{A-sp}} = -A_{131}R/H_0 \times [2.45/60p - 2.17/180p^2 \\ + 0.59/420p^3] \quad \text{for} \quad p = 2\pi H_0/\lambda > 1 \end{aligned} \tag{35b}$$

We are interested in separations on the order of nanometers, and the correction factors are 0.64, 0.47, 0.31, and 0.17 for 5, 10, 20, and 40 nm, respectively.

14.2.5 Charge Stabilization

As the last step in our analysis, we simply add the electrical double layer repulsion and van der Waals attraction together and plot the total interaction against the separation distance. This forms the basis of the classical DLVO theory for colloidal stability named after Derjaguin, Landau, Verwey, and Overbeek. It is generally believed today that when discrepancies are observed between experimental data and the theory, they are not due to a breakdown in the basic model itself but are rather the result of neglecting other forces such as ion correlation, solvation, or hydrophobic forces (Israelachvili, 1992). Short-range oscillatory solvation forces are a consequence of the liquid molecules being restructured into quasi-discrete layers within the constraining space between the surfaces when they are sufficiently smooth with a roughness less than a few angstroms. Surface interactions can also induce molecular order in the adjacent liquid, which results in a monotonically decreasing force with surface separation. The hydration force is repulsive when the water molecules are strongly bound to surfaces with hydrophilic groups, while it is attractive between hydrophobic surfaces. Disruption of the hydrogen bonding network is an additional possible factor.

The traditional interaction energy curve is shown in Fig. 14.7 along with the corresponding force curve in Fig. 14.8 as computed by STABIL. The first observation is that at small enough separations the van der Waals attraction will eventually exceed the double layer repulsion, which dominates at the larger separations. But at intermediate values, in the range 5–15 nm, there can be a peak in the curve that serves as an energy barrier to hinder any closer approach of the two particles. This maximum energy or zero force typically occurs in the vicinity of a separation of $1/\kappa$. Sometimes a weak secondary minimum in the energy can occur at larger separations leading to agglomeration.

It should be noted that the energy has been scaled with respect to the thermal energy kT because it is the driving force acting to push the particles over the barrier. Obviously, stability is favored by a maximum normalized energy much greater than 1, such as 25. Figure 14.9 shows how it increases with the zeta potential, while Fig. 14.10 shows that it decreases with ionic strength. It can also be seen from Eqs. (14.32) and (14.34) that it is proportional to the radius and decreases linearly with the Hamaker constant.

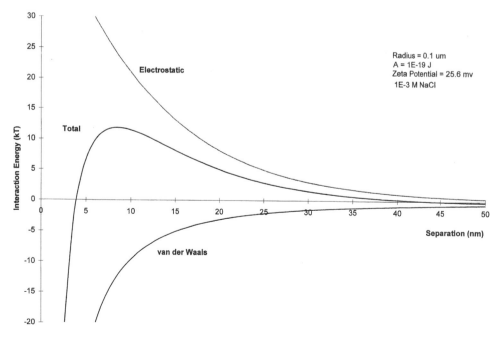

Figure 14.7   Interaction energy vs. separation.

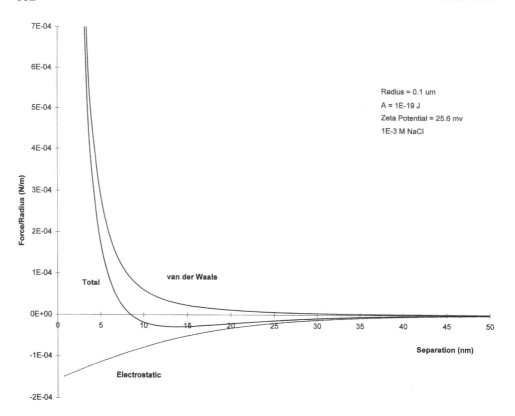

Figure 14.8 Interaction force vs. separation.

Agglomeration Rate

Let us go one step further and relate the energy maximum to the rate of agglomeration of the aggregates (Russel et al., 1989). The flux of particles toward any one is due to Brownian diffusion slowed down by the double layer repulsion [see Eq. (14.6)]:

$$j = -4\pi r^2 \left[D \frac{dN}{dr} + \left(\frac{N}{f} \right) \frac{dV_T}{dr} \right] \tag{36}$$

where N = number of aggregates or agglomerates per unit volume, V_T = sum of the double layer and van der Waals energies, r = radial distance from the particle surface, and f = friction factor = kT/D.

Integrating this equation from the particle surface ($N = 0$) to the bulk solution ($N = N_0$), one obtains the initial agglomeration rate:

$$J = \left(\frac{k_r}{W} \right) N_0^2 \tag{37}$$

where the rate constant for rapid agglomeration ($V_T = 0$) is

$$k_r = 8\pi R D = \frac{4kT}{3\mu} = 5.4 \times 10^{-12} \text{ cm}^3/\text{s} \quad \text{at} \quad 20°C \text{ for water} \tag{38}$$

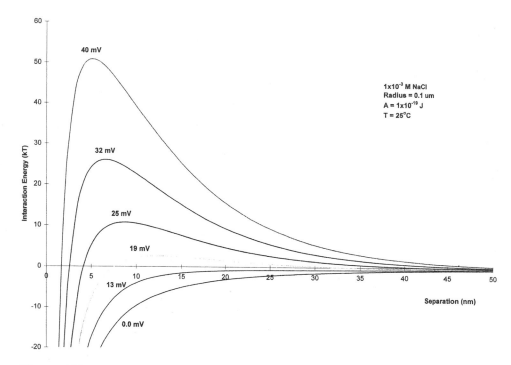

Figure 14.9   Total interaction curve as a function of the surface potential.

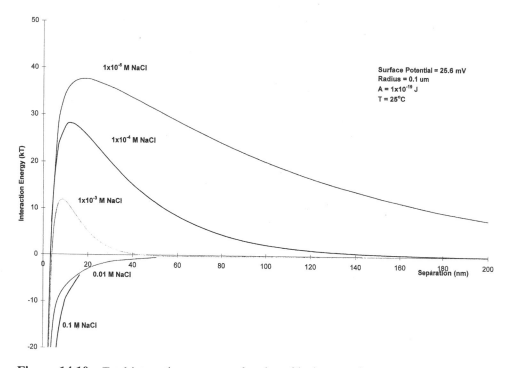

Figure 14.10   Total interaction curve as a function of ionic strength.

and the stability ratio that defines the reduction due to the energy barrier is

$$W = 2R \int_{2R}^{\infty} \exp\left(\frac{V_T}{kT}\right)\frac{dr}{r^2} \approx \exp\frac{(V_{max}/kT)}{2\kappa R} \tag{39}$$

The corresponding time constant for the initial particle concentration to be reduced in half is given by

$$\tau_W = W(k_r N_0)^{-1} = 1.85 \times 10^{11} W/N_0 = \tau_1 W \tag{40}$$

in seconds with N_0 in particle/cm^3.

The approximation in Eq. (14.40) is due to Reerink and Overbeek (1954), and it is arrived at by assuming that (1) $V_T \cong V_{max}$ over the double layer thickness κ^{-1}, otherwise $V_T = 0$, and (2) $2R \gg \kappa^{-1}$, so that $r \cong 2R$, which allows them to be taken outside the integral sign.

As an example, consider a dispersion of 5% by volume with particles of diameter 0.1 μm. Its concentration is $0.05/4/3\pi(1 \times 10^{-5})^3 = 1.19 \times 10^{13}$ particle/cm^3, which gives $\tau_1 = 15.5$ milliseconds. If the desired time constant is one year, then the stability ratio would have to be 2.03×10^9, and the required maximum energy is $V_{max} = kT \ln(2\kappa RW) \cong 21.4kT$, since $\ln W \gg \ln(2\kappa R)$. Then, with reference to Fig. 14.9, we can see that an appropriate potential at the charge plane is around -32 mV for an ionic strength of 10^{-3} M, which translates into a zeta potential of -29 mV according to Fig. 14.6.

14.2.6   Steric Stabilization

Nonionic surfactants and polymers may be attached to colloidal particles and provide stabilization via a steric mechanism. Such a disperkeep consists of two components: (1) an anchor to hold it on the particle's surface and (2) a soluble chain that extends out from its surface. The anchor may be absorbed as a result of (1) its insolubility in the solvent, (2) hydrogen bonding, (3) acid–base interaction, (4) electrostatic attraction, or (5) grafting. The chains must be sufficiently soluble so that they do not associate with each other when the particles approach and the chains intermix. They must also be long enough so that the resulting osmotic pressure can overcome the van der Waals force and keep the particles apart. If the chains contain ionic groups (as with polyelectrolytes), then charge stabilization will also play a role.

While the theory of steric stabilization is not developed to the same degree as for charge stabilization, the method offers several advantages for ink jet inks. Its stability is relatively insensitive to ionic strength or salts, while the concentration limit for charge stabilization is around 0.1 M. Sterically stabilized dispersions commonly exhibit reversible flocculation and better freeze/thaw stability. They also tend to maintain their stability at high pigment concentrations.

Stabilizing Chain

When two particles approach each other closely, the stabilizing chains attached to them will undergo intermixing. We apply the Flory–Huggins equation, which describes the free energy of mixing ΔG^M of polymer (2) with solvent (1), to a differential element of volume within this region (Napper, 1983):

$$\delta(\Delta G^M) = kT\{\delta n_1 \ln v_1 + \delta n_2 \ln v_2 + \chi_1 v_2 \delta n_1\} \tag{41}$$

where n = number of molecules per unit volume, v = volume fraction, k = Boltzmann constant, and T = temperature.

The first two terms represent the combinatorial entropy of mixing, and the third is the enthalpy of mixing due to contact dissimilarity. The term kT_{χ_1} is the difference in energy of a solvent molecule immersed in pure polymer compared with that in pure solvent, and it can be expressed in terms of solubility parameters δ^i, which are defined as the square root of the cohesion energy per unit volume (Barton, 1985):

$$\chi_1 = \chi_S + \frac{v_1}{kT} \sum_i (\delta^i_1 - \delta^i_2)^2 \tag{42}$$

where v_1 is the solvent's molecular volume and the entropic contribution $\chi_S \approx 0.34$. The superscripts denote the various contributions to δ, such as the van der Waals forces (d), dipole interaction (p), hydrogen bonding (h), and acid/base or electron donor/acceptor interactions (ab). Their values have been tabulated for a wide range of materials (Barton, 1985). The assumption behind this equation is that a geometric mean may be used to estimate the interaction energies. With the last two interactions, it should be noted that the two materials need to be paired so that one donates a proton or electron and the other can accept it. Otherwise, there is little interaction.

The change in steric energy, as the two surfaces move from infinity to a distance d apart, is found by integrating Eq. (14.41) over the volume between them. After some mathematical manipulations, one obtains the following expression for two flat plates (Napper, 1983):

$$\Delta^{FP}G^M = 2kT\left(\frac{v_S^2}{v_1}\right)\left(\frac{1}{2} - \chi_1\right)\upsilon^2 i^2 P^M(d) \tag{43}$$

where υ = number of chains per unit area, i = number of solvent sized segments in each chain, and v_S = volume of a segment.

The term P^M is known as the segment density integral and is given by

$$P^M = \frac{1}{2}\left[\int_0^d (\rho_d + \rho'_d)^2 dx - \int_0^\infty (\rho_\infty + \rho'_{d\infty})^2 dx\right] \tag{44}$$

where $\rho(x)$ is the normalized segment density function at a distance x from the first plate such that its integral from $x = 0$ to d is unity. The prime refers to the second plate, so that $\rho'(x) = \rho(d - x)$, and the subscripts denote the plate separation. P^M determines how $\Delta^{FP}G^M$ varies with separation distance given the conformation of the chains.

While their distribution function is Gaussian-like, a reasonable approximation is to treat it as a constant equal to $1/L_S$, where L_S is the thickness of the steric layer. This holds for the interpenetration domain where $L_S < d < 2L_S$. For $d < L_S$ where there is also compression, then ρ_d is changed to equal $1/d$. Fortunately, P^M is relatively insensitive to the exact details of ρ. Further, we will apply the Derjaguin procedure used previously for the electrostatic case to convert the interaction energy for plates to spheres by summing up the contributions of differential rings. The results for two spheres of radius R at a separation of H_0 are (Napper, 1983):

Interpenetration

$$\frac{\Delta^S G^M}{kT} = 4\pi R\left(\frac{v_S^2}{v_1}\right)\left(\frac{1}{2} - \chi_1\right)\upsilon^2 i^2\left(\frac{1 - H_0}{2L_S}\right)^2 \tag{45}$$

Interpenetration plus compression

$$\frac{\Delta^S G^M}{kT} = 2\pi R\left(\frac{v_S^2}{v_1}\right)\left(\frac{1}{2} - \chi_1\right)\upsilon^2 i^2\left\{3\ln\left(\frac{L_S}{H_0}\right) + 2\left(\frac{H_0}{L_S}\right) - 1.5\right\} \tag{46}$$

This domain would also be supplemented by the elastic free energy.

Thus we see that steric repulsion is maximized by driving χ_1 toward zero, which translates into matching the solubility parameters of the chains with those of the solvent. Otherwise, agglomeration would be favored. Their heat of absorption for the particle should also be negligible. Once these criteria are satisfied, the repulsion will also increase with the square of the number of chains per unit area and even faster with the chain length because of the segment density integral term. However, it should be noted that L_S is not necessarily equal to the actual chain length because of its conformation, which can vary with factors such as its solubility and area coverage. For example, close spacing would tend to force the chains to extend out straighter, while at low coverage the extension would be much less and vary with the square root of its molecular weight (Napper, 1983).

One can now combine these steric repulsion equations with that given for the van der Waals attraction [Eq. (14.35)] and compute the total interaction energy vs. separation curve as was done for the charge stabilization case. It has been found that a maximum is not exhibited as in charge stabilization, but a pseudosecondary minimum will occur at $H_0 = 2L_S$ because the steric layer is of finite length. As long as its value is less than the thermal energy kT, there will be steric stabilization (Fowkes, 1971). Thus we can set the van der Waals energy equal to kT and calculate the corresponding separation distance or read it off its graph. L_S may then be taken as half this value.

For example, an aqueous dispersion of 0.1 μm diameter particles with a Hamaker constant of 2×10^{-13} ergs would require $L_S = 65$ Å for $\lambda = 1000$ Å and 20°C. The bond length ℓ_0 is around 1.5 Å for carbon-based materials, and this leads to a fully stretched chain of $N_0 = 44$ molecules. This is sufficient when the chains are closely spaced. However, if they are far apart, then we have a flexible Gaussian chain whose mean-square end-to-end length is $\langle r^2 \rangle = N\ell^2$, not $(N\ell)^2$ (Napper, 1983). This means that the number of molecules in the chain would now have to be $N = 44^2 = 1936$. Actually, such a large number is not really needed because bond angles restrict the flexibility of the joints. Thus $N\ell^2$ is multiplied by a factor characteristic of the polymer in the range 4–10, and this reduces N to 484–194. Irrespective, one should strive for high area coverage to minimize the required chain length.

The selected dispersant may be a nonionic surfactant, or it may be a polymer, which contains multiple stabilizing chains and anchors. For example, they can be (1) comb graft copolymers, where the anchor is the backbone with the soluble chains spaced along it, or (2) block copolymers, where the anchor and soluble component alternate along its length. In the latter case, the stabilizing chains would form loops into the liquid between two anchors. Multiple anchors are also more effective than a single one, because their overall absorption energy can be high even if the individual ones are weak. This is a result of the low probability of all the anchors simultaneously lifting off the surface. Designed copolymers are to be preferred over random ones and homopolymers. Some examples of water-soluble chains are poly(ethylene oxide), poly(vinyl alcohol) and poly(vinyl pyrrolidone).

Attachment

The requirement for the anchor's attachment may be quantified by using the Langmuir isotherm, which expresses the amount of absorption in terms of its mole fraction x in solution and heat of absorption ΔG_{Ads} (Hunter, 1987):

$$\frac{n_{Ads}}{N_S} = \frac{x}{x + \exp(\Delta G_{Ads}/kT)} \tag{47}$$

where N_S = number of available surface sites per unit area, and n_{Ads} = number of sites absorbed onto.

ΔG_{Ads} may be calculated as the work required to exchange a unit area of liquid molecules (L) at the particle surface (P) for that of anchor ones (A) from the bulk liquid:

$$\Delta G_{Ads} = -\frac{2v_A^{2/3}}{kT} \left[\gamma_L + \sum_i \sqrt{\gamma_A^i \gamma_P^i} - \sqrt{\gamma_L^i \gamma_P^i} - \sqrt{\gamma_A^i \gamma_L^i} \right] \tag{48}$$

where γ is the surface energy. Analogous to how the solubility parameter was treated, the sum is taken over the various interactions, and again the values of γ^i have been tabulated for many materials (Panzer, 1973) or can be calculated from solubility parameters (Barton, 1985). The molecular area of the anchor has been taken as its molecular volume v_A raised to the 2/3 power, and its reciprocal determines N, which may be somewhat decreased for longer and branched soluble chains due to steric effects. Thus adsorption is favored by (1) large works of cohesion for the solvent and of adhesion between anchor and particle, (2) small works of adhesion between solvent and particle and between anchor and solvent, and (3) large anchor molecules. This makes ΔG_{Ads} more negative.

Anchor Size

Let us next consider what size of anchor (A) is needed to provide the proper spacing of the stabilizing chains (S). Since the chains must be greater than a certain minimum length and not spaced further apart than a certain maximum distance, the molecular volume ratio $R_v = v_A/v_S$ must in turn be less than a critical value $R_{crit} = v_{A,max}/v_{S,min}$. However, we can also express this ratio in terms of the tabulated hydrophile/lipophile balance (HLB) of a surfactant, which is defined as the weight percent of hydrophile divided by 5:

$$\mathrm{HLB} = \frac{20v_H\rho_H}{v_H\rho_H + v_L\rho_L} \tag{49}$$

where ρ is the mass density.

Eliminating the molecular volumes results in the criterion

$$\mathrm{HLB} > \frac{20}{(\rho_L/\rho_H)R_{crit} + 1} \tag{50}$$

when the hydrophile is the soluble component and lipophile the anchor as with polar solvents and

$$\text{HLB} < \frac{20}{(\rho_L/\rho_H)\,(1/R_{crit}) + 1} \tag{51}$$

for the reverse case with nonpolar solvents.

There can also be an upper and lower HLB bound on these two equations, respectively. As the anchor decreases in size, it will be less strongly adsorbed and eventually to an insufficient degree; as the soluble component increases in size, complete coverage is hindered.

HLB Kit

In order to evaluate R_{crit}, one may use the ATLAB HLB Kit available through Chem Service, Inc. (West Chester, PA), which consists of blends of Span 80® [sorbitan monooleate, HLB = 4.3] with Span 85® [sorbitan trioleate, HLB = 1.8], and Tween 80® [polyoxyethylene (20) sorbitan monooleate, HLB = 15], which is combined with Atlas G-2159 [polyoxyethylene (>50) monostearate, HLB = 18.8] to give an HLB range of 2 to 18. Using an excess of surfactant such as 10–20% of the mass of the particle phase to insure a monolayer coverage, dispersions are conveniently prepared over this range in test tubes with an ultrasonic probe. One then observes at what HLB value a stable dispersion is first obtained and whether this is a maximum or a minimum value. When this information is combined with the above equations, one knows the orientation of the surfactants and can calculate the value of R_{crit}.

Since surfactants with R_v less than the critical value can impart stability at a lower concentration, there is an optimum HLB value that requires a minimum concentration relative to the others. It is found by decreasing the concentration until one HLB value gives a dispersion better than the others. Usually, visual observation is sufficiently discriminating, but typical quantitative methods consist of monitoring turbidity and measuring sedimentation volumes.

The purpose of the surfactants in the HLB kit is to provide comparisons on a relative scale, and they do not necessarily represent the most appropriate ones for a given situation. Thus one should investigate other chemical families at the same HLB requirement with the goal of better satisfying the adsorption and solubility criteria set forth earlier. This is readily accomplished by referring to the annual edition of *McCutcheon's Detergents and Emulsifiers*, which contains an extensive compilation of commercially available surfactants and their HLB values, among other useful information. Typical HLB requirements for aqueous dispersions are around 15 with a wide latitude and are fairly independent of the pigment because the lipophilic anchor is rejected by the water and forced onto the pigment surface, even if there is a weak interaction. The more polar organic solvents behave in a similar manner but to a lesser degree. Ionic pigments in nonpolar solvents require HLB values around 5 with the hydrophile serving as a good anchor.

Difficulties are encountered with nonionic pigments in organic solvents because the rejection phenomenon plays a much smaller role and the pigment interacts only through its van der Waals surface energy. In the case of low-energy surfaces, the solvent strongly competes for the anchor to hinder its adsorption. On the other hand, a high-energy surface cannot sufficiently differentiate between the anchor and soluble components and adsorbs them both. In these situations, one does not have much leeway in satisfying the adsorption and solubility criteria as with the previous ones, and even then stability may not be attained. When this occurs, one must either find an anchor that undergoes a chemical reaction with the pigment surface or introduce ionic groups onto it.

14.2.7 Repeptization and Waterfastness

The phenomenon of repeptization is that coagulated colloidal particles can be redispersed simply by washing out or diluting the coagulating electrolyte. Yet it is not sufficient just to reduce its concentration to its coagulation value; it must be decreased well below it. As discussed by Frens and Overbeek (1972), the addition of potential determining ions to the wash water can assist repeptization. The presence of bulky adsorbed ions or molecules such as EDTA, acetone, phosphate, and citrate also can serve as repeptization aids. On the other hand, polyvalent counterions and strong counterion binding hinder redispersion.

While redispersion of pigments from the agglomerated state is a positive attribute for printer runnability, it can also result in them being easily rubbed off when the dried ink layer on the medium is wetted. Thus there is a trade-off between dispersion stability and repeptization for ink jet inks.

In the analysis of repeptization, emphasis is placed on separation distances smaller than where the maximum interaction energy occurs. The issue of constant potential vs. constant charge as boundary conditions then becomes more relevant. Further, it seems more natural to use the interaction force rather than energy, since redispersion involves overcoming the adhesive force.

Returning to Eq. (14.33) for the van der Waals force, we explicitly include the interatomic spacing in the form

$$F_{\text{A-sp}} = \frac{A_{131} R}{12(H_0 + \Delta_{\text{A}})^2} \tag{52}$$

which has the effect of decreasing the attraction.

Similarly, the plane in which the surface charge lies need not coincide with $H_0 = 0$; it may be displaced some distance Δ_{R} out from the surface, as is the case with adsorbed ionic surfactants.

We then modify Eq. (14.32) for the double layer repulsive force to read

$$F_{\text{R-sp}} = -\frac{\{64\pi kTRc_b v^2(\Psi_{\Delta R})/\kappa\} \exp[-\kappa(H_0 - 2\Delta_R)]}{1 \pm \exp[-\kappa(H_0 - 2\Delta_R)]} \tag{53}$$

where $\Psi_{\Delta R}$ is the potential at the charge plane for an isolated particle. With respect to the term in the denominator, the plus sign applies to approach at constant surface potential and the negative sign to approach at constant surface charge (van de Ven, 1989; Russel et al., 1989). The minimum separation is $H_0 = 2\Delta_R$, and as it is approached the two cases start to deviate with the latter increasing faster.

For aqueous solutions at 25°C, Eq. (14.53) may be written for the monovalent and divalent cases as

$$\frac{F_{\text{R-sp}}}{R} = -\frac{0.0482\pi[\text{NaCl}]^{1/2} \tanh^2\{\Psi_{\Delta R}(mV)/102.8\} \exp[-\kappa(H_0 - 2\Delta_R)]}{1 \pm \exp[-\kappa(H_0 - 2\Delta_R)]} \tag{54}$$

$$\frac{F_{\text{R-sp}}}{R} = -\frac{0.0211\pi[\text{CaCl}_2]^{1/2} \tanh^2\{2\Psi_{\Delta R}(mV)/102.8\} \exp[-\kappa(H_0 - 2\Delta_R)]}{1 \pm \exp[-\kappa(H_0 - 2\Delta_R)]} \tag{55}$$

where the unit of the concentrations in brackets is mol/L. Comparisons of Eq. (14.53) with STABIL for the constant surface potential case showed reasonable agreement, even at small separations. However, we would expect for the constant charge case that it over-estimates for the smaller values.

A Solution Path

The two cases of holding the potential or charge constant at the surfaces during their approach are compared in Fig. 14.11. It is seen that the interaction force is repulsive and of the same magnitude for both cases at the larger separations. In contrast, at smaller separations the force for the constant charge case becomes increasingly negative, while it reverses sign in the constant potential case becoming attractive. If the charge plane had not been displaced out from the surface, then the situation would have appeared as shown in Fig. 14.12 with the difference that the constant charge case would also have reversed sign but somewhat after the constant potential case did.

Thus the constant charge case alone provides additional repulsion over the constant potential case. However, when it is combined with the outward shifting of the charge plane, the van der Waals force is so overwhelmed that the total force is never attractive. This gives rise to essentially infinite stability and a coagulated dispersion that is readily repeptized. In fact, even if it were dried and the resulting capillary forces (which are on

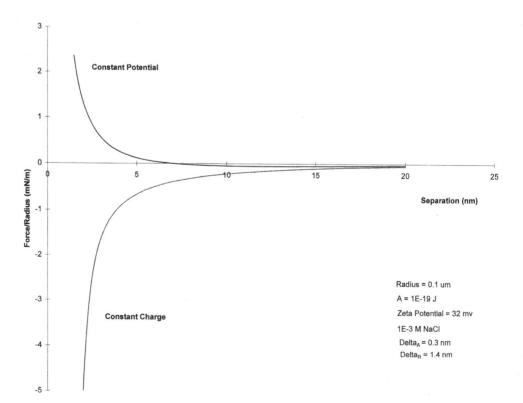

Figure 14.11 Constant surface potential vs. constant surface charge interaction for a shifted charge plane.

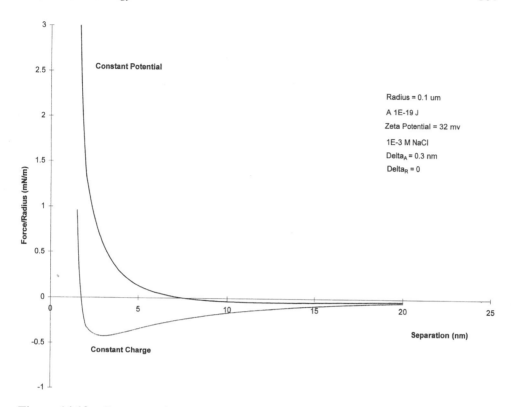

Figure 14.12 Constant surface potential vs. constant surface charge interaction for an unshifted charge plane.

the order of 72 mN/m per unit radius) compressed it into a cohesive mass, rewetting would simply generate back the double layer repulsion which would cause redispersion.

However, if the surface groups were not so strongly dissociated as to become discharged at the smaller separations (as happens in the constant potential case), then stability would still be maintained, but the ease of repeptization would be diminished. Thus good stability in the absence of repeptization is achievable. Poor wet abrasion resistance is the result of having so much stability to the point that adhesion is sacrificed.

Effect of Ink Formulation

It should also be recognized that an aqueous ink can contain other components such as humectants, penetrants, and polymeric binders and that they can play a role in repeptization. The purpose of the humectant is to prevent crusting at the nozzles of the print head, which results from evaporation of water when the printer is idle. Water-soluble glycols such as dipropylene and glycerine at a concentration of a few percent of the ink are commonly used in dye: based ink jet inks. They function by transforming into a viscous mass at low water contents during drying, which encapsulates the dye or pigment. This spacing action hinders close approach of the particles to reduce the van der Waals force, which will unfortunately have an adverse effect on wet rub resistance. If the resulting viscosity of these plugs is not too high, the ink jets can push them out of the nozzle tips upon startup.

The situation with penetrants is similar. They function by increasing the surface energy between ink and paper so that the higher capillary pressure drives the ink into the paper faster. Some examples are polypropylene glycol and glycol ethers such as butyl carbitol, and thus they can negatively impact wet rub resistance just like the humectants. They should be selected to assist the ink in penetrating deep enough into the medium to minimize rub-off without sacrificing a high optical density as a result of excessive light scattering.

On the other hand, a polymeric binder such as modified solution acrylics can provide both adhesion and cohesion if it adsorbs onto both the paper fibers and particles with the chains forming bridges between them. The resulting network acts to hold the particles together on the medium, thereby preventing repeptization. Of course, the surface sites must be available and not covered by other components such as the humectant. Again, triggers such as a volatile amine can be used to initiate the surface reactions or to make the polymer water insoluble.

14.3 OPTICAL BEHAVIOR OF COLORANTS

In this section we will discuss the relationships between the ink's color on the medium and the optical properties of its colorants. First, we will use the phenomenological Kubelka–Munk analysis to express the reflection spectrum in terms of the absorption coefficient, scattering coefficient, and thickness of the ink layer. Next, we will apply the more fundamental Mie theory to relate these coefficients to particle size, refractive index, and wavelength. Then we will connect these two theories to the CIE color system and calculate the tristimulus values to quantify the perceived color quality.

14.3.1 Kubelka–Munk Analysis

Let us consider a differential element of volume within the ink layer and follow what happens to the light flux as it passes through it. We will treat the case of two diffuse fluxes with one traveling downward and the other upward as shown in Fig. 14.13. While multiple flux models are available that allow for an angular dependence (Völz, 1995), this simpler model is still commonly used and will provide an easier route to understanding the physical phenomena governing absorption and scattering of the light (Judd and Wyszecki, 1975).

The downward flux i is decreased via absorption by the amount $Ki\,dx$, but scattering reverses the direction of a portion of it and adds the amount $Si\,dx$ to the upward flux j. K and S are the absorption and scattering coefficients, and they are wavelength dependent.

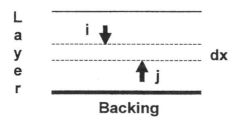

Figure 14.13 Kubelka–Munk analysis.

Similarly, the upward flux is reduced by the amounts $Kj\,dx$ and $Sj\,dx$ with the scattering term increasing i. The total changes in these fluxes are then

$$-di = -(K + S)i\,dx + Sj\,dx \tag{56a}$$

$$dj = -(K + S)j\,dx + Si\,dx \tag{56b}$$

and the reflectance R is given by the ratio j/i evaluated at the upper surface.

The solution to this set of linear differential equations was first worked out by Kubelka and Munk (1931) and may be expressed in the form

$$R = \frac{(R_g - R_\infty)/R_\infty - R_\infty(R_g - R_\infty^{-1})\exp[Sx(R_\infty^{-1} - R_\infty)]}{R_g - R_\infty - (R_g - R_\infty^{-1})\exp[Sx(R_\infty^{-1} - R_\infty)]} \tag{57}$$

where x is the layer's thickness and R_g the reflectance of the backing. R_∞ is the reflectance for a thickness so large that light does not reach the backing, and it is given by

$$R_\infty = 1 + \frac{K}{S} - \left(\frac{K^2}{S^2} + \frac{2K}{S}\right)^{1/2} \tag{58}$$

Thus we have the useful result that for a sufficiently thick ink layer the reflectance is just a function of the ratio of K and S at a given wavelength, and its value decreases with K/S. In the next section we shall see that the Mie theory allows us to relate these coefficients to the colorant's particle size.

For a mixture of colorants as well as the media, additivity of the individual contributions to absorption and scattering may be assumed as a first approximation:

$$\left(\frac{K}{S}\right)_{\text{Mixture}} = \frac{\sum_i c_i K_i}{\sum_i c_i S_i} \tag{59}$$

where c_i is the fractional concentration of the ith component. This equation helps explain the observation that the optical density can exhibit a plateau as the colorant's concentration is increased. At low concentrations, both colorant and medium contribute to scattering, but at higher concentrations the colorant can dominate both absorption and scattering so that its concentration in the numerator and denominator cancel.

One restriction of this analysis is that we have not taken into account partial reflections from the inside surface of the top boundary, which occur when there is a difference between the refractive indices of the layer and the surrounding medium. Given the reflectance coefficients on both sides of this surface, r_0 and r_2, we can calculate how much strength a light beam loses each time it is reflected from the surface back into the ink layer. With reference to Fig. 14.14, we first see that the actual light entering the layer is $i_0(1 - r_0)$. Applying the Kubelka–Munk (KM) equation, the intensity of the light reflected internally by the layer is $R_{\text{KM}}i_0(1 - r_0)$. Then some of this light is reflected back into the layer, and its intensity is $r_2 R_{\text{KM}}i_0(1 - r_0)$. This process repeats itself again, and the intensity of the light reflected a second time is $(r_2 R_{\text{KM}})^2 i_0(1 - r_0)$. This continues repeatedly, and the resulting geometric progression can be conveniently summed to yield the corrected reflectance in terms of R_{KM}, known as the Saunderson correction (Völz, 1995):

$$R_{\text{Corrected}} = r_0 + \frac{(1 - r_0)(1 - r_2)R_{\text{KM}}}{1 - r_2 R_{\text{KM}}} \tag{60}$$

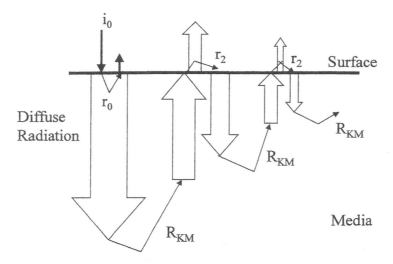

Figure 14.14   Saunderson correction.

r_0 and r_2 may be calculated from the Fresnel equations, which describe the fraction of light reflected at a surface in terms of the refractive index (m) and the angle of incidence (Judd and Wyszecki, 1975). For example, r_0 is equal to $(m - 1)^2/(m + 1)^2$ for perpendicular radiation, and $r_0 = 0.04$ for $m = 1.5$. For diffuse radiation, one takes an average over all angles to obtain $r_2 = 0.6$ for $m = 1.5$ (Judd, 1942). Thus this effect is not insignificant and assists in attaining higher optical densities.

14.3.2   Mie Theory

The Mie theory treats the problem of absorption and scattering from a more microscopic point of view and quantifies how light as an electromagnetic wave interacts with a single particle (Bohren and Huffman, 1983). One can think of the particle as consisting of a collection of harmonic oscillators or dipoles subject to the driving force of an oscillating electric field. The damping action dissipates the energy irreversibly and gives rise to the absorption. The oscillating dipoles act like antennae and reemit or scatter the incoming radiation in all directions. The mathematical approach is to solve the wave equation for a dielectric sphere with a complex refractive index ($m = m' + im''$). The real part characterizes the degree of scattering and the imaginary part the absorption.

While the mathematical details of the solution are somewhat involved, computer programs are available to perform the calculations (Barber and Hill, 1990). Their output gives the absorption and scattering efficiencies as a function of particle diameter (D), its complex refractive index, and the surrounding medium's real part. These efficiencies are defined as the amount of light absorbed or scattered divided by the amount intercepted by the particle's cross-sectional area.

A more useful parameter is obtained by multiplying these efficiencies by the cross-sectional area per unit mass ($3/2\rho D$) where ρ is the particle's density. When these parameters are then multiplied by the pigment mass per unit volume of the ink layer, these quantities describe the ink's total ability to absorb and scatter light. They are also proportional to the Kubelka–Munk coefficients. It should be noted that the mean path length for diffuse radiation is about twice as large as the layer thickness, which serves to increase these coefficients by the same amount. Further, as the concentration increases and the particles become closer together, light is back-scattered into them by their neighbors. Thus the scattering coefficient reaches a maximum and then decreases with concentration (Völz, 1995). These phenomena have the positive effect of increasing the optical density.

Figure 14.15 shows a typical plot of the absorption and scattering coefficients vs. particle size for several wavelengths, and other plots are available in Brockes (1964). Medalia and Richards (1972) have shown that these calculations are in agreement with data. The important observations to be made are

1.   As particle size increases, sensitivity to size and wavelength diminishes with values approaching the geometric limit.

2.   As particle size decreases, the absorption coefficient increases until it reaches a limiting value. The explanation is that light is initially absorbed in the surface layers. With decreasing size, the light penetrates further into the core of the particle. Since more of its mass is being utilized for absorption, the efficiency increases. Eventually, the light penetrates to the center, and then all the mass is absorbing light. This limit is inversely proportional to the wavelength.

3.   The scattering coefficient exhibits a maximum, and it decreases as the difference between the real parts of the refractive index for the particle and the surrounding medium becomes smaller. The explanation is that at sizes smaller than the wavelength the dipoles are close enough to oscillate in phase. As the size increases, there are more of them, and the coefficient increases directly with volume. It is also inversely proportional to the wavelength raised to the fourth power. Eventually, the dipoles become too far apart to be in phase, and they begin to cancel out each other.

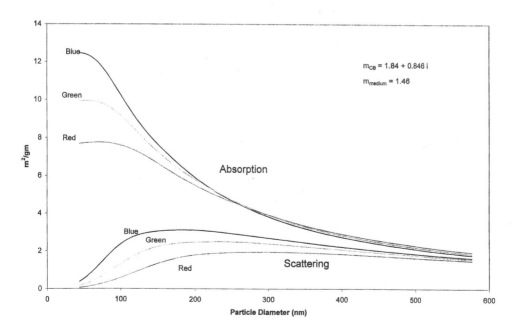

Figure 14.15   Optical cross-sectional area per unit mass vs. particle size.

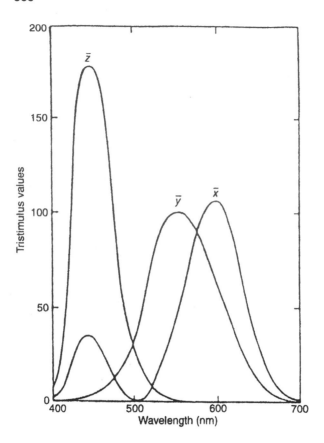

Figure 14.16 CIE color matching functions for the 10° observer.

14.3.3 CIE Color System

In 1931, the International Commission on Illumination (CIE) recommended a method to obtain parameters that quantify the color of a sample as seen under a standard source of illumination by a standard observer. The most commonly used illuminant is designated D65, and it simulates the spectral power distribution of average daylight having a color temperature of 6500°K. The 10° observer is also commonly used, and this means that the angle of vision is 10°, which is equivalent to viewing a sample 3″ in diameter from a distance of 18″. The spectral response of this observer's eye is shown in Fig. 14.16, and it is characterized by three color matching curves plotted against wavelength (λ). The $y(\lambda)$ curve was selected to match the eye's response curve to the total amount of power and provides information on the color's lightness. It peaks in midrange of the visible spectrum and decreases to zero at both of its ends similar to green. The $z(\lambda)$ curve is centered in the bluish range and $x(\lambda)$ in the reddish range.

The tristimulus values for a given wavelength are obtained by multiplying together the illuminant's power $P(\lambda)$, the sample's reflectance $R(\lambda)$, and each of the color matching parameters. Since the sample is not just reflecting light at one wavelength, we must add up all of its contributions or integrate over the entire visible wavelength:

$$X = k\!\int P(\lambda)R(\lambda)x(\lambda)\,d\lambda \tag{61a}$$

$$Y = k\!\int\! P(\lambda)R(\lambda)y(\lambda)\,d\lambda \tag{61b}$$

$$Z = k\!\int\! P(\lambda)R(\lambda)z(\lambda)\,d\lambda \tag{61c}$$

where k is a normalizing constant defined as

$$k = \frac{100}{\int P(\lambda)y(\lambda)\,d\lambda} \tag{61d}$$

to make Y equal to 100 for a perfect white object reflecting at 100% at all wavelengths.

While these tristimulus values contain all the necessary information to characterize the sample's color, there are two shortcomings that limit their use: their information is not readily visualized, and they are not linearly correlated to visual perception. Thus over the years investigators applied various transformations to remedy this situation. Today, the CIELAB system is the widely used one, and its set of equations is (Billmeyer and Saltzman, 1981):

$$a* = 500(X* - Y*) \tag{62a}$$

$$b* = 200(Y* - Z*) \tag{62b}$$

$$L* = 116Y* - 16 \tag{62c}$$

where $X* = (X/X_n)^{1/3}$, $Y* = (Y/Y_n)^{1/3}$, and $Z* = (Z/Z_n)^{1/3}$. The cube root function serves to give more physiologically uniform steps. If X/X_n, Y/Y_n, or Z/Z_n is less than 0.008856, then one uses $X* = 7.787\,(X/X_n) + 0.138$ with $Y*$ and $Z*$ similarly defined. The subscript n denotes the tristimulus values of the reference white for the selected illuminant, which are around 100.

The resulting CIELAB color space is illustrated in Fig. 14.17 and may be envisioned as a cylinder with the lightness $L*$ or gray value varying along its axis from zero to 100.

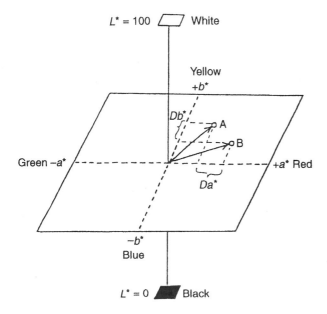

Figure 14.17   CIELAB color space.

At a given L^*, the a^*–b^* plane consists of circles whose radii vary with the saturation or chroma, and the hue varies along the perimeter of the circle. Since X is more weighted for the reddish wavelengths and Y for the green, positive values on the a^* axis indicate a red hue and negative values a green hue. Similarly, positive values on the b^* axis indicate a yellow hue and negative values a blue hue. While the negative values follow from Z^* being weighted in the blue range, a positive value for yellow is a little less obvious. Along the b^* axis, a^* is zero, which means $X^* = Y^*$ or a mixture of red and green to give yellow.

14.3.4 Color and Chemical Structure

Organic compounds become colored by absorbing electromagnetic radiation in the visible wavelength range (400–700 nm), and this excites an electron from its ground state into an orbital of higher energy. In order to move from the ultraviolet (<400 nm) to the visible range, dyes are conjugated so that they consist of alternating single and double bonds in chains and/or rings of carbon atoms. Other atoms such as nitrogen and oxygen may be substituted for some of the carbon ones. Conjugation allows the electrons to be delocalized over the chain/ring system, which has the desired effect of decreasing the excitation energy ΔE and increasing the absorption wavelength λ (McLaren, 1986):

$$\lambda = \frac{hc}{\Delta E} \tag{63}$$

where h is Planck's constant and c the velocity of light. The longer the conjugated chain is, the more bathochromic is the shift to lower energies. The transitions of interest for colorants are n $\rightarrow \pi^*$ and $\pi \rightarrow \pi^*$, where n denotes a lone pair or nonbonding electrons, as on an O or N atom, and π the pi bonding orbitals. The asterisk refers to the excited state.

Further shifts are produced by the presence of electron donor groups such as $-NH_2$, $-NMe_2$, $-OH$, and $-OR$, which release electrons into the conjugated system, and electron withdrawing groups such as $-NO_2$ and $-C=O$, which take electrons out of the system. There is a particularly high degree of electron delocalization when a donor and a withdrawing group are at opposite ends of a conjugated system. This has the effect of extending the conjugation (Nassau, 1998). In aqueous solutions, ionization of acids and anionic groups to impart a negative charge creates a strong electron donating group, while basic and cationic groups become positively charged and strongly electron withdrawing.

For example, consider a conjugated chain with an electron withdrawing group at one end and an amine group at the other. In acidic solution, the amine becomes protonated and changes from electron donating to also electron withdrawing. The result is that the two withdrawing groups fight each other, and the absorption shifts to a shorter wavelength. As another example, let the withdrawing group be a weak azo group ($N=N$) placed within the molecule. In slightly acidic solution, the azo group becomes protonated, transforming it into a stronger electron withdrawing group, and the absorption wavelength increases. However, when the solution is made more acidic, the amine group gets protonated and becomes electron withdrawing too. Now the absorption wavelength decreases. This is one reason there is not necessarily an exact correspondence between the ink's color in solution and as a dry solid on the medium.

As a means to quantify this inductive effect of substituents, one may make use of the Hammett constant, which is described in Section 14.4.4. It is an empirical measure of a substituent's electron donating or withdrawing strength and thus of the change in electron density for the atom in the chromogen, which in turn modifies its transition energy. Linear correlations have been obtained between the shift in wavelength and its Hammett constant, provided that the substituent is not directly involved in the excitation process, which would change its value (Griffiths, 1978). Irrespective, these plots exhibit a negative slope, even if not linear, and this indicates that electron withdrawing groups produce bathochromic shifts. Thus one can use this type of data to assist in designing colorants to obtain a desired hue.

A more theoretical approach is to use molecular modeling programs, as described in Section 14.4.4. They calculate the wavelength of maximum absorption as the excitation energy for a transition from the highest occupied molecular orbital (HOMO) to the lowest vacant molecular orbital (LVMO), and there is good agreement with measured values (Kogo, 1980; Naef, 1995). The oscillator strength is also provided so that information on the tinctorial strength is gained (Zollinger, 1991). The one characteristic that is difficult to calculate accurately is the shape of the absorption spectrum, which is integrated over the visible wavelength range to define the color as described in Section 14.3.3.

Colorants for Ink Jet Printing

Early ink jet printers were somewhat unreliable because the dyes employed then were textile grades such as direct dyes and acid dyes and thus not very pure (Kenyon, 1996). Their limited water solubility served prematurely to clog the nozzles with crystallized dye or insoluble impurities. The first generation ink jet dyes were still the same commercially available products, but they had been purified and made more water soluble. For example, Food Black 2, as shown in Fig. 14.18, solved the solubility problem with its four sulphonic

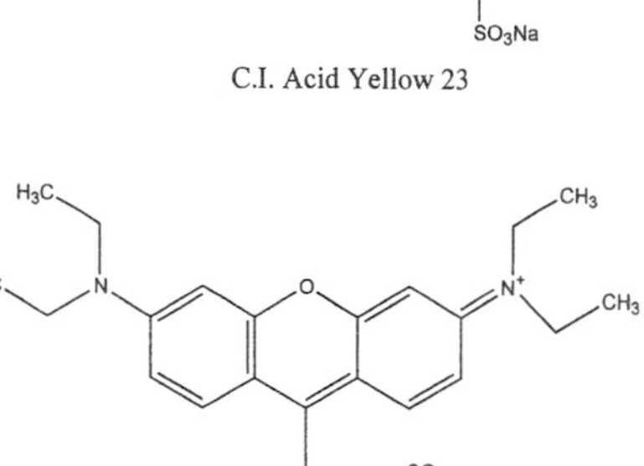

C.I. Acid Yellow 23

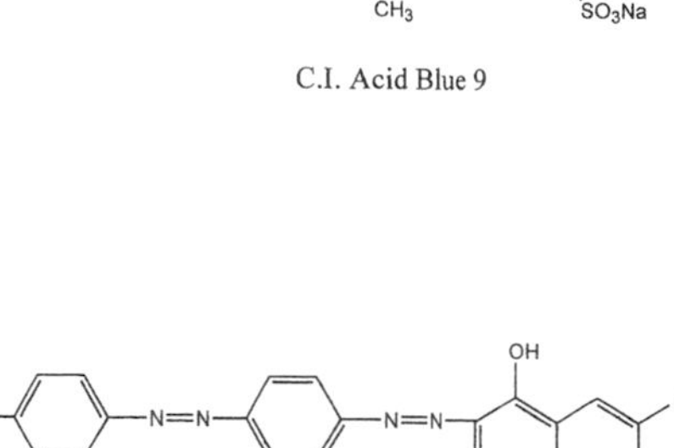

C.I. Acid Blue 9

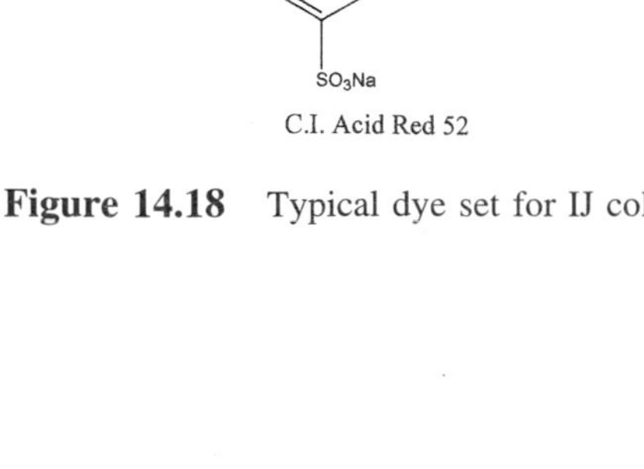

C.I. Acid Red 52

C.I. Food Black 2

Figure 14.18 Typical dye set for IJ color printing.

groups, but its waterfastness was quite poor, as discussed in Section 14.2. The second-generation black dye replaced some of these sulphonic groups with carboxylic ones to make it less water soluble so that it would aggregate upon contact with acidic paper. The next step was to metallize the dye, which improves lightfastness, as explained in Section 14.4.4. CI Reactive Black exemplifies this modification, as shown in Fig. 14.19 (Bauer et al., 1998). However, carbon black, as shown in Fig. 14.20, has now emerged as an effective replacement for black dyes, making pigmented inks a reality today. Reliability is achieved through a stable dispersion of submicron particles, as discussed in Section 14.2, and lightfastness is a given. However, waterfastness is not necessarily included if the pigment exhibits repeptization, as discussed in Section 14.2.7.

The first-generation cyan dyes were based on the triphenylmethane structure such as Acid Blue 9 in Fig. 14.18, but they have been superceded by the more lightfast copper phthalocyanines such as CI Direct Blue 199 in Fig. 14.19. It is interesting to note that copper phthalocyanine was first made in pigment form as shown in Fig. 14.20, and later it was transformed into a water soluble dye by incorporating sulphonic groups (Zollinger, 1991).

First-generation magenta dyes were xanthene dyes such as CI Acid Red 52 shown in Fig. 14.18 (Bauer et al., 1996). Because of their poor lightfastness, they have been superceded by azo H-acid dyes such as CI Reactive Red 180 shown in Fig. 14.19. As discussed in Section 14.4.2, the azo tautomer exhibits better lightfastness than the hydrazone one, and this leads to monoazo dyes such as CI Acid Red 37 which exist more in the azo form than CI Reactive Red 180. Again, lightfastness can be further improved by metallizing the dye as exemplified by CI Reactive Red 23, but unfortunately the shade becomes duller. Typical magenta pigments are quinacridone pigments such as CI Pigment Red 122 and Naphthol AS pigments such as CI Pigment Red 184 shown in Fig. 14.20. Descriptions of these pigments and others can be found in Herbst and Hunger (1997).

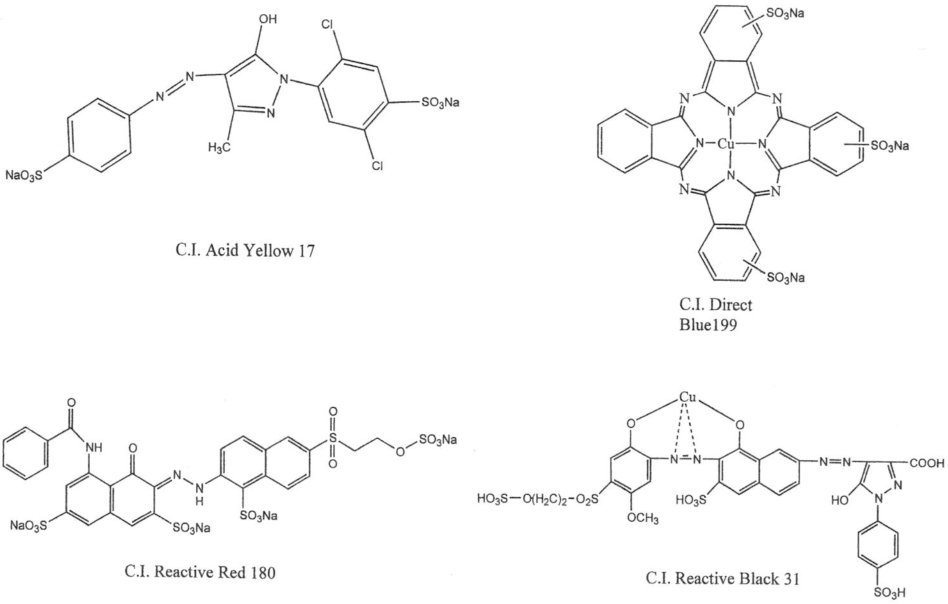

C.I. Acid Yellow 17

C.I. Direct Blue 199

C.I. Reactive Red 180

C.I. Reactive Black 31

Figure 14.19 Typical IJ dye set for lightfastness.

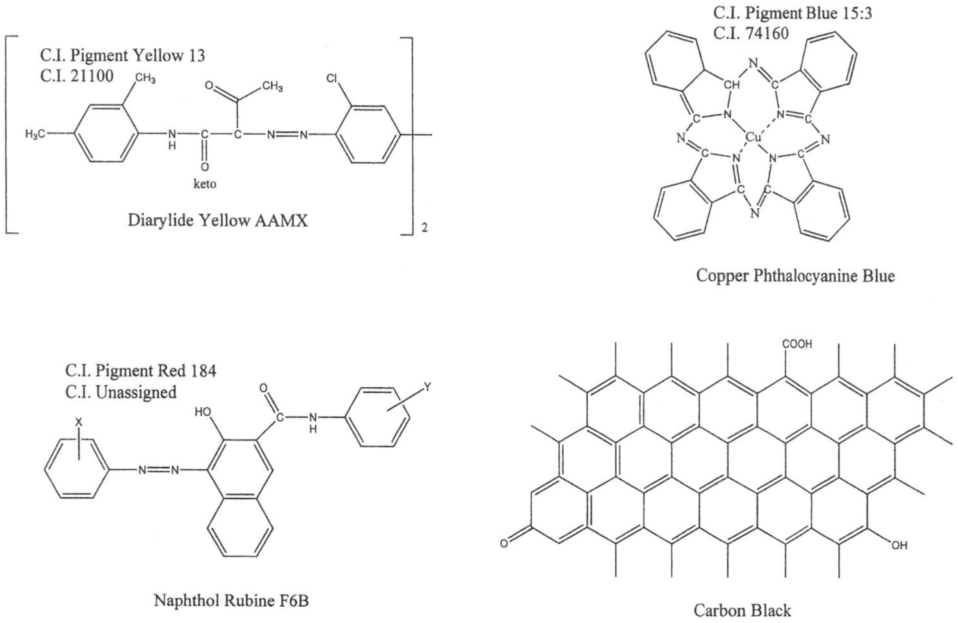

Figure 14.20 Typical pigment set for IJ color printing.

Typical yellow dyes are CI Direct Yellow 86, CI Direct Yellow 132, CI Acid Yellow 23, and CI Acid 17, with the latter two shown in Figs. 14.18 and 14.19, respectively. Representative yellow pigments are monazo yellows such as CI Pigment Yellow 74, benzimidazolone yellows such as CI Pigment Yellow 154, and diarylide ycllows such as CI Pigment Yellow 13 shown in Fig. 14.20.

14.4 LIGHTFASTNESS

Dyes generally resist fading in vacuo, but upon contact with the atmosphere, the medium, and other components in the ink they will fade to varying degrees. Unfortunately, our understanding of the controlling mechanisms is somewhat limited, and the approach to formulating ink jet inks with lightfastness characteristics sufficient for long-term display or archival purposes has been of an empirical nature. However, the problem of fading has been under investigation for some time for other systems such as dyeing of textiles and polymers, and the fading mechanisms identified in these studies are useful in providing guidance to improve the lightfastness of ink jet inks. Thus we shall review those mechanisms that appear to be pertinent for ink jet inks and discuss those parameters and guidelines that have been helpful in developing correlations with dye structure.

14.4.1 Reductive Fading Mechanism

Two pathways by which azo dyes can fade are photoreduction and photooxidation reactions, and these are dependent on the dye's chemical environment as determined by the medium, components in the ink, and surrounding atmosphere. Under anaerobic conditions, an azo dye can be reduced to its corresponding amine by abstracting a hydrogen atom from a hydrogen donor, as shown in Fig. 14.21 (Allen, 1987).

This reaction set is greatly accelerated when either the hydrogen donor or the dye is photoexcited, and its action spectrum is then determined by the absorption spectrum of the photoactive compound. Some examples of hydrogen donors are alcohols, amines, ketones, carboxylic acids, ethers, and esters, and these absorb in the ultraviolet range. That this is indeed the action spectrum has been confirmed experimentally by van Beek and Heertjes (1963). In fact, the wavelength of active light varied only with the hydrogen donor and did not change for different azo dyes using paper soaked with their aqueous solutions and air dried.

An example of how apparently slight structural variations in a hydrogen donor can have a surprisingly large impact on its photocatalytic activity is given by a series of alcohols with a quinone (Wells, 1961). The rate constant for hydrogen transfer increased in the order methanol, ethanol, *n*-propanol, *n*-butanol, and *sec*-butanol, but it was almost two orders of magnitude less for *tert*-butanol. It was reasoned that only the hydrogen atom on the alpha carbon was abstracted and that higher alkylation increased its reactivity via more electron donating. However, tert-butanol does not have a hydrogen atom on the

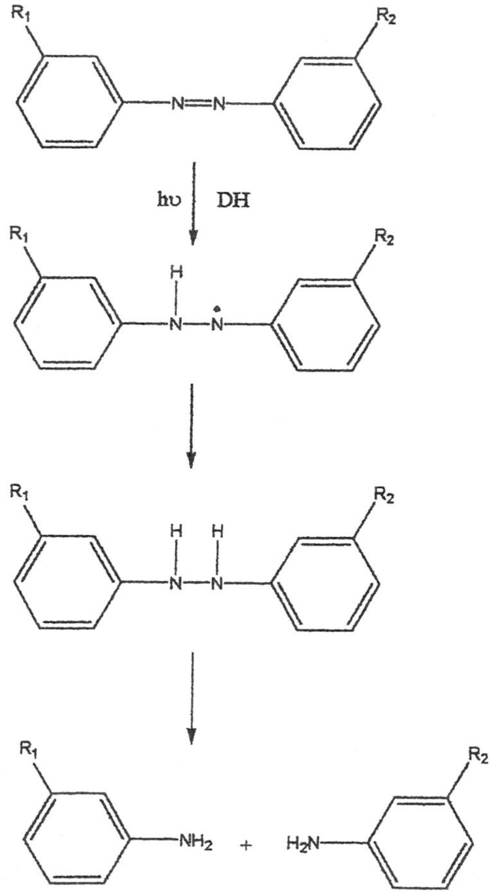

Figure 14.21 Reduction mechanism for azo dyes.

alpha carbon. Increasing the number of hydroxyl groups in an alcohol to form a glycol also decreased the reactivity, but not so drastically, because these groups are electron withdrawing. Increasing the alkylation again increased reactivity.

14.4.2 Azo vs. Hydrazone Tautomers

Mallet and Newbold (1974) have also found that there is a linear correlation of positive slope between the fading rate under nitrogen and the ratio of the hydrazone and azo tautomers in hydroxyazo dyes. The medium was polypropylene, which produces hydrogen radicals upon UV irradiation. This further supports the above reductive mechanism in that hydrazones are already in a partially reduced form.

The position and nature of substituents play a role in determining the state of the tautomeric equilibrium (Kishimoto et al., 1978). Since the azo group ($N=N$) is an electron acceptor, electron donating substituents, especially OCH_3, stabilize the azo tautomer. However, the imino group (NH) is an electron donor, and the hydrazone tautomer is favored by electron withdrawing substituents, especially NO_2.

Intramolecular hydrogen bonding can also provide additional stability to the hydrazone tautomer with certain structures (Ball and Nicholls, 1982). Typically, this occurs between an OH group at the ortho position and the further nitrogen of the azo group. Since oxygen is more electronegative than nitrogen, it forms a stronger hydrogen bond, and thus the hydrazone tautomer is stabilized more.

Intermolecular hydrogen bonding with the medium and components in the ink can have similar effects, and generally the more polar materials favor the hydrazone tautomer. Of course, the medium restricts free tautomeric interchange.

14.4.3 Oxidative Fading Mechanism

On the other hand, the oxidative fading of an azo dye has been attributed to the attack of singlet oxygen on its hydrazone tautomer as shown in Fig. 14.22 (Kuramoto, 1996). The initial ene reaction leads to the formation of an unstable peroxide which then undergoes decomposition. This reaction is promoted by singlet oxygen sensitizers such as anthraquinone dyes which upon excitation transfer their energy to the oxygen. Conversely, singlet oxygen quenchers such as 1,4-diazabicyclo[2,2,2]-octane (DABCO) and nickel-dibutyl-dithiocarbamate (NBC) suppress the fading. Such materials are not very water soluble and are more easily incorporated into the media than the ink.

14.4.4 Substituents for Controlling Fading

The photochemical activity of both the reductive and oxidative mechanisms is also a function of the substituents in both the dye and the hydrogen donor. Many studies have shown that a linear correlation exists between the logarithm of the fading rate and the Hammett constant of the substituent (Kuramoto, 1996). The Hammett equation is defined as (March, 1992)

$$\log\left(\frac{k}{k_0}\right) = \rho\sigma \tag{64}$$

where k_0 is the rate or equilibrium constant for the reference hydrogen substituent and k is the corresponding value for the R group in question

The Hammett constant σ characterizes the electron withdrawing or donating strength of group R, and its tabulated values are based on dissociation data for substituted benzoic

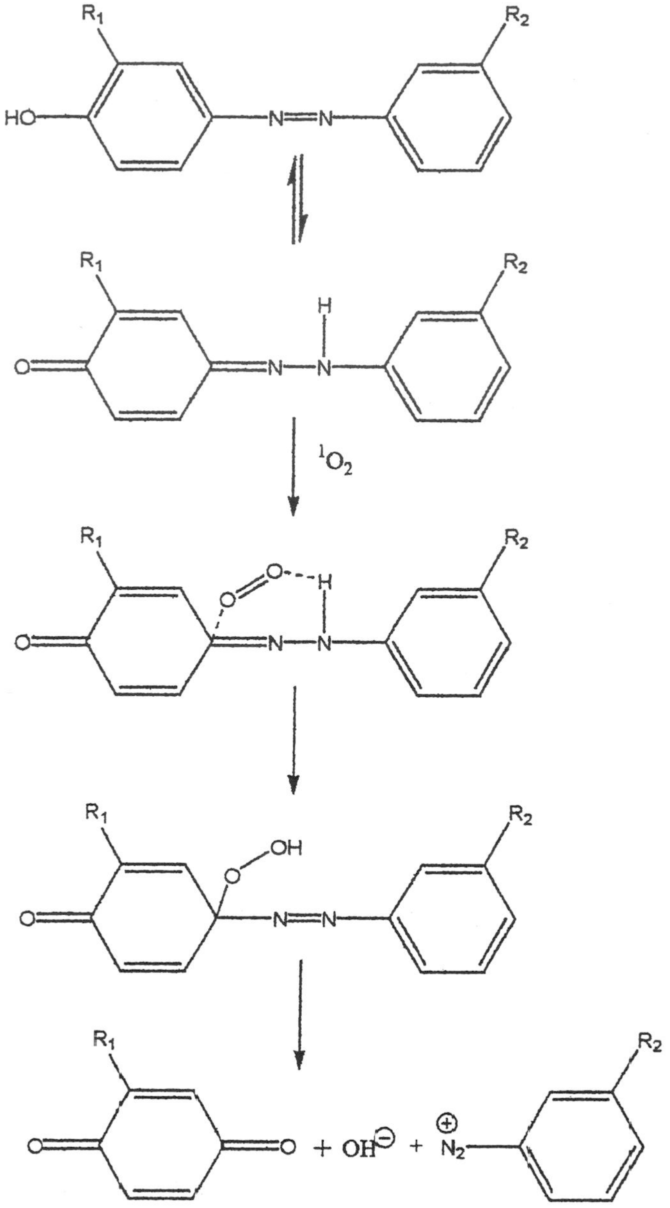

Figure 14.22 Oxidation mechanism for azo dyes.

acids. The constant ρ measures the sensitivity of the reaction in question to electrical effects and is influenced by conditions such as temperature and its surroundings.

Figure 14.23 illustrates a typical plot of the fading rate against the Hammett constant for various substituents in the para position. Since hydrogen is abstracted by electrophiles deficient in electrons, it is reasoned that the fading rate should increase with the more electron withdrawing substituents relative to hydrogen and thus with increasingly positive values of the Hammett constant, when reduction is the controlling mechanism.

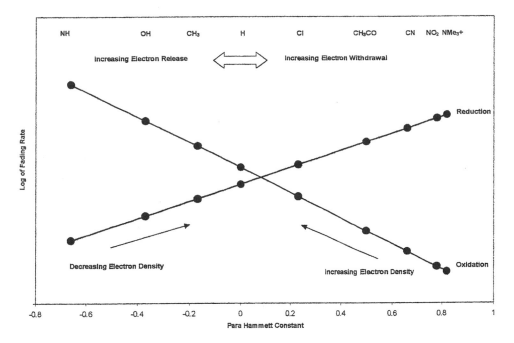

Figure 14.23 Schematic plot of the logarithm of fading rate vs. the Hammett constant of para-phenyl substituents.

Conversely, singlet oxygen is electrophilic and attacks positions of high electron density which is promoted by electron releasing substituents with negative Hammett constants in the case of oxidation. This is why metallized dyes with their electron withdrawing metal atom are particularly stable. Under moist conditions, singlet oxygen can also react with water to form hydrogen peroxide, and antioxidants can be used to decompose it.

Thus correlations with either positive or negative slope can be obtained, and if both mechanisms are operating, a transition from one to the other will occur. In fact, the two individual rates could compensate for each other, resulting in little variation in the overall fading rate. Performing experiments under nitrogen and oxygen is helpful in sorting out such effects.

The Hammett equation has also been applied to the example given earlier on the reactivity of alcohols (Wells, 1961). Taft's σ values for aliphatic molecules were used, and the total value $\Sigma\sigma$ for the alpha C—H bond was computed as a sum of the individual values of the constituent groups. Plotting the logarithm of the rate constant for hydrogen transfer against $\Sigma\sigma$ yielded quite a straight line over a hundredfold range.

While these correlations with the Hammett constant provide further support for the validity of these fading mechanisms, they do have their limitations. For example, only the substituent can be varied with the dye's structure held constant, and substituents in the ortho position generally do not correlate well. Since the partial charges on the atoms are what are really being varied, they can be calculated using molecular models. Fortunately, personal computers are now sufficiently powerful to handle such calculations for dye molecules. Using CambridgeSoft Chem3D, we carried out such calculations for substituted 1-phenylazo-2-naphthols as shown in Table 14.4. The structures were energy minimized by MOPAC using the AM1 potential function, and the Wang–Ford charge option was

Table 14.4 Partial Charges on Selected Atoms for Substituted 1-phenylazo-2-naphthols

Substituent	Hammett constant	Q on N7	Q on N8	Q on H9	Q on C5	Q on C10	Q on O20	Log of relative fading rate [8]
OCH$_3$	−0.28	0.06377	−0.16755	0.21201	−0.13609	−0.02855	−0.48313	0.21
CH$_3$	−0.14	0.02145	−0.17632	0.22917	−0.04291	−0.00172	−0.48444	0.12
H	0.00	−0.02654	−0.16387	0.24462	0.00623	0.00602	−0.48561	0.00
Cl	0.24	−0.05280	−0.16284	0.24835	0.07044	0.02304	−0.48030	−0.27
NO$_2$	0.81	−0.01142	−0.19208	0.23235	0.03505	0.05749	−0.46743	−0.82

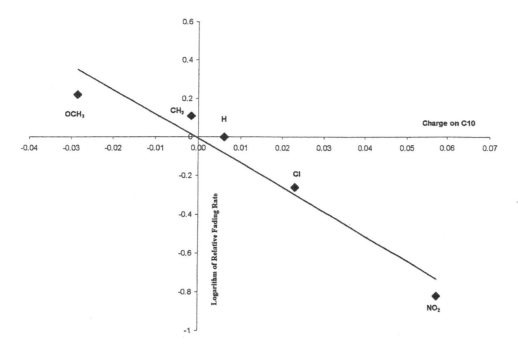

Figure 14.24 Fading rate vs. partial charge on carbon atom 10 for substituted 1-phenylazo-2-naphthols.

selected. Fading data under oxidative conditions is also included for comparison (Mustroph and Weiss, 1986).

It is seen that the partial charge on carbon atom 10 where the ene reaction takes place best follows the fading rate, and this plot is shown in Fig. 14.24. As its charge decreases and thus its electron density increases, this reaction is encouraged, resulting in faster fading.

14.4.5 Dyes as Photocatalysts

A troublesome problem encountered while developing our lightfast set of ink jet inks was that certain dye combinations exhibited less stability when together in a print area in comparison to being separately printed. As mentioned earlier, excited dyes can produce singlet oxygen resulting in oxidative fading, but they can also transfer their absorbed energy to another dye at a lower energy level to increase its radiative exposure and fading. This transfer tends to occur when the dyes are closer than 5 nm, and it has been found that using organic molecules (e.g., fatty acids) as spacers can hinder the transfer process (Kuhn, 1983).

Another way to combat this situation is to add a quencher that takes on this energy and dissipates it harmlessly. Ideally, it functions by forming a complex with the dye that undergoes charge transfer and emission of the excess energy. The molecular orbital scheme in Fig. 14.25 illustrates this process for the excited dye acting as an electron acceptor or donor. The resulting free energy change that we wish to be negative is (Weller, 1967):

$$\Delta F = - {}^1\Delta E_{-A} + IP_D - EA_A - C \tag{65}$$

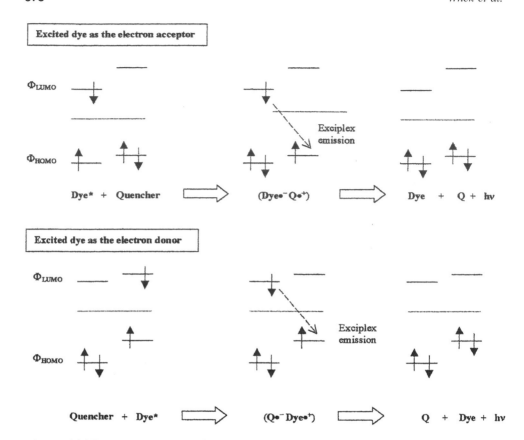

Figure 14.25   Exciplex formation via charge transfer.

where $^1\Delta E_A$ is the singlet excitation energy of the electron acceptor and C is the coulombic energy gained. In arriving at this equation, the energy of the highest occupied molecular orbital Φ_{HOMO} was set equal to the ionization potential of the electron donor IP_D and the energy of the lowest unoccupied orbital Φ_{LUMO} to the electron affinity of the electron acceptor EA_A. Since electron affinities are in general not accurately known, they may be approximated by polarographic reduction potentials and ionization potentials by oxidation potentials. Φ_{HOMO} and Φ_{LUMO} may also be calculated using Chem3D.

This equation has been evaluated in several studies on fluorescence quenching and was found to have validity. For example, amines as electron donors became more effective quenchers for rubicene as their ionization potential decreased (Nakajima and Akamatu, 1969). In contrast, a series of electron accepting quenchers became more effective for anthracene and pyrene as the electron affinity increased (Nakajima and Akamatu, 1968). For a given electron donating quencher, Young and Martin (1972) and Solomon et al. (1969) both showed that quenching correlated with the sum of the electron affinity and excitation energy for a set of aromatic materials. Thus each part of the equation has been checked.

14.4.6   Dye Aggregation and Pigments

As a general observation, dye aggregates have been found to be more resistant to fading than the monomolecular state. For example, Weissbein and Coven (1960) detected discrete particles of the more lightfast dyes in films of regenerated cellulose using an electron micro-scope. For decades, the time to fade to a certain percentage loss in optical density has been plotted against the initial dye concentration on log–log paper, which typically yields a linear relationship (Baxter et al., 1955). The assumption behind these plots, termed "characteristic fading-order" or CFO curves, is that the fading rate decreases with increasing dye aggrega-tion, which is promoted by higher concentrations. It is of further interest to note that a plot of surface tension vs. dye concentration is similar to that for a surfactant (Guo et al., 1994). It decreases until a plateau is reached where the surfactant associates into micelles.

The tendency of a dye to aggregate can also be characterized by measuring the change in the shape of its absorption spectrum in dilute solution. With increasing concen-tration, a second absorption band due to dimer formation appears at a shorter wavelength than that of the monomer whose intensity decreases (Coates, 1969). An additional band can appear at an even shorter wavelength, which is a result of higher aggregates. Steiger and Brugger (1998) have used this technique to study the lightfastness of ink jet inks.

The large effect that aggregation can have on lightfastness has been attributed to several factors. First, the surface area per unit mass of dye that is available for attack by reactive species is diminished. Since light is absorbed within the surface layers of the larger aggregates, there is a reservoir of unreacted dye in their interior. As the outer layer is degraded, reactants diffuse more slowly through it to reach the aggregate's interior. The fading rate varies with $1/r^2$ for large particles ($r =$ radius), with $1/r$ for intermediate ones, and is independent of size for small ones (Giles et al., 1977). It is also possible that the lifetime of the dye's excited state is shorter in the aggregated state, allowing it less time to react (Gordon and Gregory, 1987). All of this will serve to reduce the fading rate over time. It may be noted that these are the same arguments used to explain the better lightfastness of pigments over dyes.

Reducing the dye's solubility can also induce aggregation. This can be accomplished by the addition of a cosolvent of lesser solvating power; and solubility parameters can assist in making such a selection (Baughman et al., 1996). The amount added should be limited so that the resulting aggregate size is not so large as to clog the orifices of the printhead.

Since aqueous-based dyes typically contain acidic groups such as sulphonic and carboxylic to impart solubility, pH is another useful control variable. It functions by affect-ing the degree of ionization of the acidic groups, which results in a negative charge that creates electrostatic repulsion between the dye molecules. Making the pH less basic sup-presses their ionization and leads to aggregation. Since the aggregate's charge increases in absolute magnitude with its size, aggregation continues until the electrostatic repulsion is again greater than the attractive van der Waals force. This approach may also be used in conjunction with the media if it is acidic such that the dye precipitates upon contact.

The addition of salts is also well known to cause aggregation. They act as counter-ions to decrease the thickness of the electrical double layer and can be absorbed as well to neutralize the charge. Multivalent ions are particularly effective and can also act as a bridge between two charge groups to bind the dye molecules. Cationic surfactants can be used in a similar manner. In fact, both ionic and nonionic surfactants can be used to form mixed micelles with dyes (Garcia and Medel, 1986).

14.4.7 The Medium's Role

Ink jet media consist of an image-receiving layer coated onto a substrate, and this layer may be a polymeric matrix or a microporous structure of inorganic particle and a binder.

A backing layer serves to control properties such as friction and curl. A topcoat layer containing UV absorbers may be added and can also protect against exposure of the colorant to oxygen and moisture.

These UV absorbers function by strongly absorbing UV light and then releasing the energy nondestructively via a tautomeric process. A common structural feature is strong intramolecular hydrogen bonding (Chirinos-Padron and Allen, 1992). Typical chemical structures are benzophenones, benzotriazoles, and benzoates; and commercial products include Ciba-Geigy's TINUVIN 171 and 1130.

Commonly used materials for the polymeric receiving layer include water-soluble polymers such as methyl cellulose, carboxymethyl cellulose, polyvinyl alcohol (PVA), polyacrylic acid, polyacrylamide, polyvinyl pyrrolidone, and gelatin. A higher proportion of hydroxyl functionality tends to favor lightfastness, e.g., PVA and cellulose derivatives. It is believed that such materials can act as free radical scavengers (Yuan et al., 1997). PVA is also known to be a good barrier for oxygen.

The particles in a microporous receiving layer are inorganic oxides such as silica, alumina, and titanium dioxide. Their size can range from submicron to several microns, and the pore size is on the order of hundreds of angstroms. The same polymers listed above are used as binders. While the more porous structure enhances ink penetration, especially for pigmented inks, it is also more permeable to air and thus more conducive to colorant oxidation than a polymeric matrix.

UV absorption can also create excited electrons and positive holes, some of which migrate to the particle surface. Reactions with oxygen, water, and surface hydroxides can then lead to hydroperoxyl and hydroxyl radicals, which attack the colorants (Allen and McKellar, 1980).

Antioxidants may be incorporated into the medium and are of two types: free radical scavenger and peroxide decomposer. Primary antioxidants function by trapping free radicals or as labile hydrogen donors, which reduce alkylperoxyl radicals formed in the presence of oxygen to hydroperoxides. This stops possible chain reactions. Examples include hindered phenols with electron donating substituents and tetramethylpiperidine derivatives called HALS (hindered amine light stabilizer). Metal dithiocarbamates are not only antioxidants but can also act as UV absorbers and quenchers (Moura et al., 1997)

Secondary antioxidants function to reduce active hydroperoxides to inactive alcohols, and they can be synergistic when used in combination with primary antioxidants. Examples include phosphites and thiopropionates.

14.4.8 Lightfastness Testing

Xenon arc weatherometers are commonly used to provide accelerated fading data. They simulate the spectrum of sunlight fairly well, but at a higher intensity, and run at controlled temperature and humidity. A typical set of test conditions based on ANSI T9.9-1996 (Stability of Color Photographic Images—Methods for Measuring) are

> Simulated outdoor sunlight
> $765 \ W/m^2$ in 300–800 nm waveband;
> $75 \ W/m^2$ in 300–400 nm UV waveband;
> 100–150 Klux in visible range.
> Simulated indoor daylight through window glass
> UV portion reduced to $60 \ W/m^2$ via an optical filter
> 24°C and 50% RH

Light/dark cycle of 3.8 hours on and 1 hour off giving 5 cycles per day
 On: sample temperature of at least 35°C but not higher than 60°C; 40% RH $\pm$ 10
 Off: sample cooled to 25°C $\pm$ 5 within 15 minutes; 80% RH $\pm$ 5
Measure optical density and CIELAB coordinates of yellow, magenta, cyan, black,
 red, blue, and green patches periodically over time

Let us now consider a simple model for the kinetics of photodegradation and see how we should plot the above data to assist in extrapolating it. We assume that the amount of energy absorbed follows the Lambert–Beer equation:

$$I_A = I_0 \{1 - \exp(-\varepsilon\ell[c])\} \tag{66}$$

where I_A = intensity of radiation absorbed, I_0 = incident radiation intensity, $[c]$ = dye concentration, ℓ = film thickness, and ε = extinction coefficient.

The amount of dye degraded per energy quantum absorbed is found by multiplying by the quantum yield Φ of the reaction:

$$\frac{-d[c]}{dt} = \Phi I_0 \{1 - \exp(-\varepsilon\ell[c])\} \tag{67}$$

whose solution for small arguments is

$$[c] = [c]_0 \exp(-\Phi I_0 \varepsilon\ell t) \tag{68}$$

where t is time. But the optical density O.D. is defined as $\log_{10}(I_0/I_T)$, where I_T is the intensity of radiation transmitted, which equals $\varepsilon\ell[c]$. Substituting O.D. for $[c]$ then yields

$$\text{O.D.} = (\text{O.D.})_0 \exp(-\Phi I_0 \varepsilon\ell t) \tag{69}$$

which suggests that ln(O.D.) should be plotted against time of fading to obtain a straight line. We have found that the data does behave in this fashion and that the use of EXCEL's graphic options conveniently allows reasonable extrapolations to be made.

This equation has reciprocity built into it in that $I_0 t$ is the total energy absorbed, which is the basis for making projections for real-world situations where the radiation intensity is much less than in the weatherometer. However, one should keep in mind that this may not always be the case. Life is calculated to a first approximation by dividing the energy absorbed up to failure (e.g., a certain O.D. loss in the accelerated test) by the rate of incident energy in the place of interest. But what wavelength energy should be used: the total solar radiation or its UV portion, which varies from 3 to 6%? Usually, the latter is the better choice. Another caveat is that the higher intensities in the weatherometer can increase the fading rate and thus make this estimate a conservative one. Further, one should be watchful of temperature and humidity differences between the two. While reaction rates normally increase with temperature, this can result in a sample of lower water content, which acts to decrease the fading rate, leading to confusing and misleading trends.

14.4.9 Formulation Guidelines

There are a number of guidelines that can be drawn from the above discussion to assist in minimizing the fading of ink jet inks:

Carry out fading experiments under both a nitrogen and an oxygen atmosphere to
 help identify the contribution of the reduction and oxidation mechanisms. Doing

this at different wavelengths is a plus. Select the appropriate UV absorbers accordingly. Incorporate into the ink or medium, preferably as a topcoat.

Check for RH effects. Under moist conditions, singlet oxygen can react with water to form peroxides, and antioxidants can be used to decompose it. Incorporate into the ink or image receiving layer.

Avoid the presence of hydrogen donors and singlet oxygen sensitizers. Make use of substituents and the tautomers of azo dyes to minimize such interactions.

Use metallized dyes when feasible. Incorporate transition metal cations into medium. Color can be an issue.

Check the composite black and primary color images for dye catalyzed fading and the need of a quencher. It can be incorporated into the ink or medium.

Pigments and dye aggregation can greatly reduce fading. The latter can be controlled by higher concentrations, cosolvents, pH, surfactants, and salts. Incorporate aggregation agents into the medium.

Keep in mind the significant role that the medium plays in affecting lightfastness.

The image receiving layer may be a polymeric matrix or a microporous structure of inorganic particles and a binder.

A higher proportion of hydroxyl functionality in the polymer favors lightfastness.

Higher porous structures are more permeable to air and thus more conducive to colorant oxidation than a polymer matrix.

UV absorption can create excited electrons and positive holes in inorganic oxides, leading to the formation of hydroperoxyl and hydroxyl radicals, which attack the colorants.

14.5 INK FORMULATION

In designing an ink for a specific application, one begins by setting its performance requirements based on the desired print quality and the restrictions imposed by the printer's hardware and operating conditions (Table 14.1). These goals are then translated into a set of measurable physical properties of the ink to which operating windows are assigned through a combination of analysis and experimentation (Table 14.2). In the beginning, this set may be incomplete with wide specification ranges, but as the project matures more parameters enter into the picture and the ranges tighten.

These functional specifications become the target against which the chemical formulation (Table 14.3) is developed, and they later become part of the quality assurance plan for manufacturing the ink. At this stage, in order to be cost effective, the list may be abbreviated to include just those parameters that are most prone to vary. This is usually possible because now both the formulation and the process set points have been fixed.

Ideally, each functional requirement would be uniquely influenced by one component in the ink formulation, but any given component can play multiple roles, both positive and negative. A further complication is that the functional requirements themselves place conflicting demands on the ink, and hopefully the individual operating windows are overlapping and have sufficient latitude. Another part of the formulation work involves developing a qualification test for the raw materials and any intermediate products.

Finally, the laboratory procedure is scaled up to the pilot plant level where the appropriate process equipment is selected and the set points are optimized. The resulting process design is then passed on to manufacturing along with the functional and formulation specifications. After several successful production trials, the process is validated.

In the following sections, we will consider each of the ink's components and their interactions, drawing upon the information that has been given in the previous sections.

14.5.1 Colorant

The colorants first used for the ink jet printing of documents were water-soluble dyes as opposed to pigments because a solution better satisfied the runnability requirements than a dispersion. However, this is no longer the case, and pigmented inks are now available that do not clog up orifices and have a color gamut approaching that of dyes. Further, with today's emphasis on lightfastness and waterfastness, pigments are a natural choice. In fact, we have seen in Section 14.4.6 that aggregating the dye improves its lightfastness. Thus as dispersion technology is becoming more effective in producing stable dispersions of increasingly smaller particle sizes to minimize light scattering, dye technology is progressing in its ability to control the aggregation process. It may be that both technologies have the same target particle size and are approaching it from opposite ends. It should be noted that one method of making pigments is precipitating water-soluble dye ions via salt formation with an inorganic counterion.

Thus both dye aggregates and pigments will follow the behavior of a colloidal dispersion, as discussed in Section 14.2. So will the dye's monomer in a general way. Their ionizable groups give rise to a surface charge and a diffuse concentrated layer of counterions around it called the "electric double layer" such that on the whole charge neutrality is satisfied. When neighboring entities approach each other, the resulting electrostatic repulsion will repel them unless the van der Waals attraction is greater. The influence of the electrostatic repulsion is on the order of the double layer whose thickness decreases with salt concentration or more properly ionic strength, where the concentration is weighted by the square of the valence. This is because with the additional counterions a smaller volume of liquid is needed to satisfy the condition of neutrality. Thus one observes that both adding salts, especially multivalent ones, and decreasing the pH to suppress ionization of acidic groups result in instability. Further, using high-energy dispersion methods, such as sonification, increases absorbance and improves filterability at submicron ratings.

One problem in using commercial colorants, especially texile grades, lies in their inherent impurities and the potential lack of a consistent raw material. During their multistep synthesis, organic by-products are generated, surfactants are introduced, and inorganic salts are present in the process water and may also be used to precipitate the dye or pigment out of solution. The multivalent ions such as calcium are particularly effective at initiating precipitation at low concentrations of several ppm, since they can serve as the counterion at the anionic charge sites on more than just a single dye molecule and function like a bridge to join them. Chelating agents such as ethylenediaminetetraacetate (EDTA) may be added to the ink to tie up these ions as soluble complexes. The organic compounds even at low concentrations can function as foaming agents, which will be discussed in more detail later under "defoamer." Dialysis or ultrafiltration is effective in removing the inorganic and low molecular weight organic impurities by selecting a membrane impervious to the higher molecular weight dye. Solvent extraction is another useful method if the impurities or dye are preferentially soluble in some liquid. Foam fractionation is an efficient technique for removing surfactants, especially when present at low concentrations, and may be applied directly to the ink. Other treatments include passage through activated carbon beds and ion exchange columns. Thus it can be seen that a colorant can vary from batch to batch and among suppliers even though it is sold under the same name. For that reason, each lot received must undergo considerable qualification before it can be used.

14.5.2 Solubilizing Agent/Dispersant

Since lightfastness and waterfastness are favored by dyes of limited solubility, a solubilizing agent is usually the next component to be selected for dye-based inks. In the simplest case, it may be just a cosolvent such as an alcohol which compensates for the hydrophobic nature of the dye and allows the resulting mixture to match its solubility parameter. It should be noted that it is not sufficient just to match the overall value but also the individual components characterizing the hydrogen bonding, polar, and van der Waals interactions. Another simple approach is to control the acidity of the vehicle so as to have the dye in an ionized state by inducing an acid/base reaction. For example, anionic dyes would require a high pH, but not so high that the increased ionic strength collapses the electric double layer and thus shields the electrostatic repulsive force. Further, a buffer such as triethanolamine (TEA) should be present, since contact of the ink with air will cause carbon dioxide to be absorbed and a subsequent reduction in pH due to the formation of carboxylic acid.

For those dyes that do not contain a high proportion of ionic groups per molecule, they will associate until an aggregate size and charge are attained that provide sufficient electrostatic repulsion, since it increases faster with size than the van der Waals force. A common solution is to add an amine, which in addition to yielding a basic solution can form a more hydrophilic complex with the dye while in solution. However, upon drying, the amine evaporates off, leaving the hydrophobic dye on the paper and a waterfast print.

Another approach is to use the same dispersion methods that are applied to pigments as covered in Section 14.6.1. Here, the aggregates are broken down via high-energy or shear mixing such as a stator rotor or colloid mill. The resulting smaller aggregates are kept apart either by absorbing (1) an ionic surfactant or polyelectrolyte onto them for charge stabilization or (2) a nonionic surfactant or copolymer for steric stabilization. With dyes, the situation is probably more complex, since mixed aggregates or micelles can also be formed during the association process.

14.5.3 Defoamer

Foam is produced when air is entrained in the ink as a result of agitation. This can occur in manufacturing during mixing and filling operations, as well as in the ink jet printing process. It is particularly a problem with continuous ink jet systems that recycle the unprinted drops. Aside from runnability problems, print quality can be compromised if there are air bubbles in the deposited ink film giving rise to pin holes and craters.

Figure 14.26 illustrates the structure of foam as a honeycomb network of bubbles separated by thin liquid films that form three-way junctions. Their drainage is driven by the capillary pressure at these junctions, which is less than atmospheric, and surface viscosity provides the resistance to flow.

Resistance to rupture via stretching, which is initiated by sudden external disturbances, arises because the surface pressure in the stretched region has decreased due to the lower concentration of the surface active agent, as shown in Fig. 14.27 (Patton, 1979). The resulting pressure gradient causes surface transport of the surfactant, which drags along with itself a number of underlying water layers to restore the film to its original thickness. However, if the surfactant can be replenished by diffusing from the bulk of

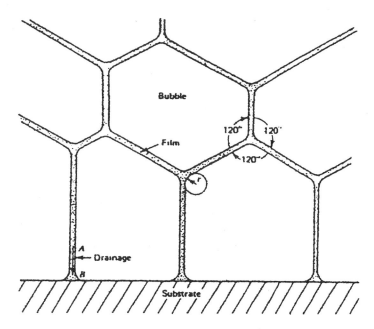

Figure 14.26 Structure of foam.

the film, then further thinning can occur leading to film rupture. Another film stabilizing mechanism results from the overlapping of the two electric double layers formed on either side of the film by absorbed ionic surfactants. The resulting electrostatic repulsion is analogous to that described previously for the dispersion of pigments and depends on ionic strength and pH in a similar manner. Unfortunately, the conditions that suppress foam also favor dye precipitation. In the case of nonionic surfactants, steric stabilization is the operating mechanism.

Thus the film can be thought of as possessing elasticity, and the lower its elasticity the faster the foam can be dissipated. This explains why pure liquids, even though they may have low surface tensions, do not foam: they are inelastic. It should be also noted that foaming agents in addition to their stabilizing effect lower surface tension and make foam generation even easier.

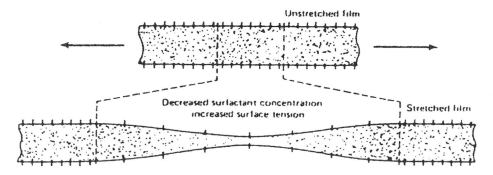

Figure 14.27 Stretched film of foam illustrating surface elasticity.

An inelastic surface can be achieved if the surface active agents minimize the van der Waals attraction between their hydrocarbon chains and lie flat at the air/water interface. This is favored by (1) chain branching, (2) having the polar group in the center of the molecule and scattering additional ones over its length, (3) bulky groups, and (4) minimizing the number of methylene groups required for insolubility (Osipow, 1972). Dyes tend to satisfy these criteria and inherently are not foaming agents. Also, the more soluble the dye, the less potential for foaming because of the diminished surface activity. It is their low molecular weight organic impurities, especially those straight carbon chains with polar groups at one end of the molecule, that are the foaming agents. Other sources include contaminated process equipment and ink containers, as well as surfactants in the formulation that serve some function such as wetting or penetration into the media. Further, foaming agents are effective at low concentrations, which makes them so troublesome to guard against.

The first approach in solving a foaming problem is to attack it at the source and minimize the introduction of organic impurities during the dye synthesis or remove them as a post operation such as dialysis or ultrafiltration. If this fails, one should select some defoamers using the above guidelines with the additional requirement that they be absorbed strongly enough at the air/water interface so that they can displace the foaming agents. Another criterion, especially for continuous ink jet systems, is that the defoamer not be absorbed by any components of the fluid handling system, such as filters, and lose their effectiveness. This behavior is observed because hydrophobicity is driving the defoamer out of the water phase to whatever interface is available. One way to try to compensate for this phenomenon is to use excess defoamer, especially above its critical micelle concentration, so that the micelles continue to replenish the defoamer as it is removed. Finally, defoamers may be classified into the following chemical types: (1) alcohols, (2) fatty acids and derived esters, (3) amides, (4) phosphate esters, (5) metallic soaps, (6) silicone oils, and (7) chemicals with multiple polar groups (Osipow, 1972).

14.5.4 Humectant

After a shutdown, the ink can dry to a hard, crystalline crust, and this can cause crooked jets or even block an orifice. Further, deposits can form on the charge plate and catcher of continuous printers, and they then interfere with their proper functioning. Thus the ink must exhibit sufficient humectancy that redissolution or redispersion can readily occur. Another desirable property would be a low evaporation rate, which also extends life by reducing the increase in colorant concentration for recycling systems. This problem can be controlled by including a cosolvent that is nonvolatile and not very viscous, such as a glycol. However, it should be kept in mind that the dry time of the image may be extended excessively, if too much is added.

14.5.5 Penetrant

In order to achieve fast drying and avoid smearing of the print, the ink must penetrate quickly into the media and not leave any of the colorant at its surface. This penetration driven by the capillary pressure can be described to first order by the Lucas–Washburn equation (Oliver, 1982):

$$L = \left(\frac{r_p \gamma \cos\theta_c}{2\mu} \right)^{1/2} t^{1/2} = K_A t^{1/2} \tag{70}$$

where L = penetration distance after time t, r_p = pore radius, γ = surface tension, θ_c = contact angle, μ = viscosity, and K_A = absorption coefficient.

K_A can be conveniently obtained by measuring the length of track produced when a given volume of ink is spread over a medium sample at different speeds using a Bristow instrument (Bristow, 1967). It can be seen that the solution is to drive the contact angle toward zero, but not necessarily all the way, since the dot size may become too large and its shape irregular, as well as optical density and show-through being adversely affected.

If the medium has a hydrophobic nature, an obvious approach is to add a nonaqueous cosolvent such as an alcohol. Otherwise, a surfactant must be added to the ink that is capable of absorbing onto the medium and making it hydrophilic upon contact with the ink. Nonionic surfactants are a good choice because their hydrophobic insoluble carbon chain serves as an anchor for absorption and the adjoining hydrophilic soluble chain such as polyethylene oxide confers hydrophilicity.

Another phenomenon to take into account when selecting a penetrant is the dynamic nature of wetting (Komor and Beiswanger, 1966). The wetting front is depleted of penetrant as a result of its absorption onto the medium and is replenished by diffusion. At concentrations above the critical micelle value, dissociation of the micelles into monomers and diffusion of both species are also involved. Such surfactants can be characterized by their HLB values as discussed in Section 14.2.6, and effective penetrants fall into the lower HLB range (e.g., 8 to 13). Thus their adsorption kinetics are of interest (Mourougou-Candoni et al., 1997), and measurements of dynamic surface tension vs. time are a useful screening tool (Ferri and Stebe, 1999).

As long as a drop is still at the medium's surface, its shape remains fairly uniform because there is a sufficient supply of ink for all the variously sized capillaries. However, once the drop has disappeared underneath the surface, the ink continues to spread by redistributing itself at the wetting front with ink flowing from the larger capillaries to the smaller ones because of the difference in capillary pressure between them. Ideally, one would like to stop the spreading at this point to prevent feathering, and pigmented inks offer this advantage by being filtered out by these pores. This is why media with uniform pore and chemical distribution are highly desirable. One approach is to cause the dye to precipitate out of solution, e.g., acidic medium or evaporation of an amine, which changes the pH from basic to acidic.

14.5.6 Binder

The binder is a polymer capable of film formation such as an acrylic resin, and it may be added to the formulation as a water soluble type or in an emulsion/dispersion form. In fact, if the polymer contains acidic/basic groups, varying the pH can transform it from one state to the other (Sauntson, 1975). For example, in the case of carboxyl groups low pH suppresses their ionization and the polymer is in the emulsion form because there is no electrostatic repulsion between the chains. As the pH increases, the acidic groups ionize, and the resulting repulsion forces the chains to separate and go into solution. Further increases in pH will eventually raise the ionic strength to the point that the electric double layer will start to collapse. This results in the chain curling and ultimately precipitation of the polymer. Another consequence of these configuration changes is that the viscosity rises up to a maximum, at which point the chains are fully extended and somewhat inflexible due to the electrostatic repulsion; and beyond this it will decline. Thus ionic polymers can also be used as viscosity modifiers.

The binder provides adhesion of the dye to the substrate by forming a film that encapsulates it upon drying. In the case of a solution polymer, this occurs by the polymer precipitating out to surround and immobilize the dye. On the other hand, an emulsion polymer accomplishes this by coalescence.

14.5.7 Corrosion Inhibitor

Corrosion is first avoided by selecting dyes that are soluble in a basic pH range, which is the region of minimum corrosivity for stainless steel and occurs by formation of an oxide film to passivate the surface. Other metals, such as nickel and copper, with which the ink may come in contact, are also evaluated, e.g., testing in an anodic polarization cell (Fontana and Greene, 1978). If additional protection is required, organic corrosion inhibitors may be added to the ink, which function by absorbing onto the surface to provide a protective film.

14.5.8 Biocide

An aqueous ink jet ink provides the required moisture and organic nutrients in the form of its components for bacterial growth to occur at room temperature. In fact, the components themselves can be sources of bacteria. Thus it is essential not only that a biocide be included in the formulation but that it be introduced into the deionized water as early as possible in the manufacturing process. The use of a clean room assists in minimizing airborne dust particles, which are carriers of bacteria. Further, mixing tanks, pumps, filters, bottles, and any other equipment that the ink comes in contact with must be kept clean to avoid contamination. Ideally, the biocide should have a high antimicrobial efficiency and a broad spectrum of antimicrobial activity. Commonly used biocides are 1,2 Benzi-sothiazolin-3-one available from Zeneca under the name PROXEL GXL and 2,6-Di-methyl-m-Dioxan-4-ol Acetate available from Angus Chemical under the name BIOBAN DXN. Working concentrations are 0.1–0.5% by weight of the ink. Compatibility of the biocide with ink components should always be checked for any adverse interactions.

A typical sterility test consists of the following steps:

1. Ink samples are inoculated with a mixed 24 hour culture of specified bacteria.
2. Samples are incubated at 30–32°C for 24 hours.
3. A nutrient agar streak plate is prepared from the samples, which will be read 48 hours later for a colony count.
4. Sample incubation continues for another 24 hours for a total of 48 hours incubation between inoculations.
5. Samples are reinoculated, incubated 24 hours, and restreaked.

This procedure is repeated for a number of cycles.

14.5.9 Optimization

The physical properties of an ink jet ink such as its absorbance, surface tension, and viscosity, vary with the concentrations of its components. If each property were only a function of one of the components, formulating an ink to meet a given specification would be straightforward. Unfortunately, just the opposite is usually the case, and the problem becomes a multivariable one in which each property depends on the concentration of more than one of the components. Thus any one property can be brought into specification by adjusting the concentration of one of the components, but then some of the others can possibly be driven out of specification with the process repeating itself. This can be particularly frustrating when multiple adjustments are being made on a manufacturing batch of ink and time is of the essence.

One approach to handling this type of situation is to treat it as a nonlinear programming problem. Each physical property is expressed as an equation in terms of the concentrations of the components that control its value. The data may be regressed using either theoretical relationships or simply empirical ones. The objective function to be minimized is the sum of the deviations that the physical properties are away from their specified centerline values. Alternately, it could be the cost of the ink. The constraints are defined by the upper and lower limits set on the physical properties, and the fractional concentrations must add up to one. Such a procedure is described in this section.

Optical Density

The absorption spectrum of the ink as a diluted solution is measured with a spectrophotometer, and the fraction of transmitted light (T) at one or more wavelengths (λ) commonly serves to specify its strength and color. The Lambert–Bear equation can be used to quantify this optical property for a mixture of dyes (Volz, 1995):

$$\log\left[\frac{1}{T(\lambda)}\right] = A = \sum \varepsilon_i(\lambda) l \ l \ C_i \tag{71}$$

where A = absorbance, ε_i = extinction coefficient of the ith dye, l = sample thickness, and C_i = concentration of the ith dye (mass/total liquids).

Viscosity

The viscosity of the ink's vehicle μ_v may be calculated from (Patton, 1979)

$$\log \mu v = \sum [w \log \mu]_i \tag{72}$$

where w_i and μ_i are the weight fraction and neat viscosity of the ith component. However, ink jet vehicles are aqueous mixtures of oxygenated organic solvents such as alcohols and glycols and undergo nonideal interactions such as hydrogen bonding. This behavior may be compensated for by using an effective viscosity for these components, and values for selected solvents have been tabulated. This approach is valid up to weight fractions of 0.3, after which certain nonlinear terms may be included. For example, w_i can be divided by $(1 - bw_i)$ to give a stronger upward trend to its viscosity curve, where b is a constant. This modification is particularly useful for polymers.

The effect of dye or pigment concentration $C_{colorant}$ is then expressed in the form

$$\mu_{ink} = \mu_v \ [l + f(C_{colorant})] \tag{73}$$

where the function f may be taken as a power series. For the low concentrations typically involved, just a linear term usually suffices. It should be noted that pH and ionic strength could also be factors, since they affect the degree of ionization of the dye's ionic groups.

Surface Tension

Surface tension is typically plotted against the logarithm of the concentration of surface active agent, yielding a curve that is initially flat at the lower concentrations and then decreases linearly until it levels out at the critical micelle concentration (Rosen, 1979). The Gibbs adsorption equation suggests that this linear portion may be represented by

$$\gamma - \gamma_{\text{Ref}} = \Sigma \, \beta_i \log \left(\frac{C_i}{C_i^{\text{Ref}}} \right) \tag{74}$$

where Ref denotes a reference point and the β_i's are constants. It should be noted that the slopes and CMC's could be different for mixtures in comparison to their values in the single state.

Solution Method

The problem may be stated in the form of a nonlinear programming problem:

Objective Function

Minimize the sum of the absolute deviations from the centerline specification values for the properties in question or the amount of components added.

Constraints

    The upper and lower bounds on each property.
    Sum of the weight fractions equals one.

Solver, which is an Excel Add-in, may be used to solve this set of equations. An example is given in Section 14.6.3 that shows how multiple adjustments of a manufacturing batch can be reduced to just a single one via this technique.

14.6 INK PROCESSING

Let us consider a typical process for manufacturing ink jet inks, which should be carried out under clean room conditions. The colorant is first purified and then dissolved or dispersed in water at the appropriate pH, which contains the solubilizing agent and the biocide. Heating may be required to assist the process. Ideally, the supplier has sufficiently purified the colorant and in the case of pigments is also providing a well dispersed product. In such a case, the ink manufacturer only needs to qualify the colorant at the laboratory level for acceptance and can rely on the raw material supplier for the first step. After the remaining components have been added, the ink is filtered and put into containers, which have been cleaned so as to be free of particulates, bacteria, and organic/inorganic contaminants. Similarly, precautions are taken to ensure that the process equipment is also free of any contamination before a run. Quality control checks are performed after each step, including inspection of the incoming raw materials, before proceeding to the next step. In this section, we will cover the key unit operations of mixing/dispersion and filtration, as well as a method for adjusting a manufacturing batch when it is initially out of specification.

14.6.1 Mixing and Dispersion

Mixing of the various components into the ink is usually a straightforward operation since they are either water soluble or come as an aqueous dispersion. A propeller mixer is typically used, and it creates an axial turbulent flow in the tank. Baffles may be added to guard against possible dead zones (Harnby et al., 1992).

On the other hand, if one is starting with the dry pigment, the situation is somewhat more challenging. The cohesive stress S holding an agglomerate together can be expressed as a product of three terms: (1) the adhesive force F per contact of any two particles of diameter D such as the van der Waals force, (2) the number of contacts J per agglomerate, and (3) the number of agglomerates N per unit area (Rumpf, 1962).

$$S = N(D, \varepsilon) \times J(\varepsilon) \times F \qquad (75)$$

where ε = the volume void fraction of the agglomerate and $N = (1 - \varepsilon)/\pi D^2$.

The coordination number is related to ε by $J = \pi/\varepsilon$, and ε can be measured by the pigment's oil absorption (Patton, 1979). Substituting the last two equations and Eq. (14.33a) into Eq. (14.75) yields the final expression

$$S \approx \frac{(\sqrt{A_1} - \sqrt{A_3})^2 \, (1/\varepsilon - 1)}{24 D H_0^2} \qquad (76)$$

Thus the agglomerate strength increases with decreasing particle size, but it decreases as its shape becomes more irregular, since the particles pack in a more open structure with higher porosity. For example, the various grades of carbon black are characterized by their primary particle size or surface area per unit mass and their structure by di(n-butyl) phthalate (DBP) absorption (Donnet et al., 1993). While the smaller sized grades are preferred for more blackness, this makes them more difficult to disperse, and the expected higher optical density may not be attained. However, one can compensate for this by selecting grades with higher DBP that are easier to disperse.

Good wetting, such that the vehicle penetrates into the interstices of the agglomerate and displaces the air, is effective in reducing the agglomerate strength, since the effective Hamaker constant is decreased. For example, A_1 is in the range 20–100 kT for solid materials, and A_3 equals 9.1 kT for water, resulting in A_{131} = 2.1–48.8 kT. The wetting process can be described by the Washburn equation given in Section 14.5.5 on penetrants.

The next step is to select the appropriate dispersion equipment that can generate sufficient shear stress to overcome the cohesive strength of the agglomerates. Since the shear stress is the product of viscosity and shear rate, both the disperser and the ink contribute to it. The device's flow pattern controls the shear rate through its velocity gradients, and the viscosity increases with the pigment concentration. As illustrated in Fig. 14.28,

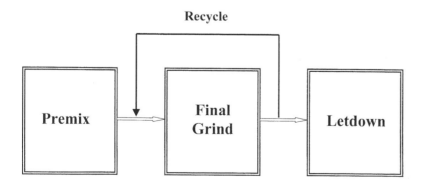

Figure 14.28 Staged milling for inks.

Table 14.5 High-speed Dispersers

Description

    Consists of a disk impeller with a saw-blade edge mounted on a shaft and centered in the tank. Different designs have these serrations bent at various angles.

    Disperses via shear with some attrition. Prefer laminar flow between disk and tank bottom. Tip speeds are in the range 4000–5200 fpm.

    Impeller size is 1/3 tank diameter and mounted 1/2 impeller diameter above tank bottom.

Advantages

    Simple to operate and easy to clean.

Limitations

    Inherent vortexing can cause foaming.

    Primarily used for premixing.

---

a dispersion line typically consists of three steps: (1) premixing, (2) milling with recycle, and (3) letdown (Schak, 1997).

    High-speed dispersers, as described in Table 14.5, are commonly used in the first step to break up the larger agglomerates into smaller ones that the subsequent mills can then handle. Impingement mills, as described in Table 14.6, are next in intensity and are also mounted on a shaft for ease of use. They can serve in both the first and second steps, either alone or in combination with a high-speed mixer. Media mills and microfluidizers, as described in Tables 14.7 and 14.8, are needed to achieve further size reduction and require good premixing so that the agglomerates will either fit into the interstices of the media particles or not plug the channels of the latter. For pigments that are particularly difficult to disperse, a roll mill or extruder can be used to provide a higher shear force, if the vehicle is sufficiently viscous, e.g., a polymeric dispersant (Patton, 1979).

    Because both the shear rate and the agglomerate strength have statistical distributions, there is a certain probability that the stronger agglomerates may not enter the highest shearing zones during one pass. Recycling is employed to account for such events. In order to maximize the shear force and throughput, the pigment is dispersed at loadings as high as feasible, and the resulting masterbatch is then diluted to the working concentration.

Table 14.6 Impingement Mill

Description

    Mill base is sucked into top and bottom of mill head and is thrown outward by rotating slots against a stationary slotted collar (e.g., 0.018 in.).

    Pigment agglomerates collide with stator wall and one another and are sheared by slots.

    Rotor tip speeds range from 5000 to 9000 ft/min.

Advantages

    Requires no premix.

    Capable of high throughput.

    Relatively easy to clean.

Limitations

    Can cause excessive turbulence at surface leading to foaming. A solution is flow assist via a propeller mixer.

Table 14.7 Media Mills

Description
A tank with an agitator, filled with the grinding medium such as steel, glass, or ceramic beads. Diameters range from 0.25 to 2 mm, and a narrow distribution is preferred. Shear stress varies inversely with diameter.
Disperses via media collisions and shearing action between the media. Rule of thumb: media size should be 10 times larger than initial agglomerates resulting in a possible fourfold size reduction.
Advantages
Disperses down into the submicron range.
Continuous or batch processing.
Jacketed with cooling water.
Limitations
Requires good premix.
Difficult to clean.

14.6.2 Filtration

Since the orifice diameter of a printhead is in the 10–40 μm range depending on the particular ink jet printer, the potential for clogging exists if particles larger than the submicron range are present in the ink. These particles can be dust, dye aggregates, undispersed pigment, or insolubles in the raw materials. Thus they must be removed via filtration, and this is a key process step in determining the functionality of an ink jet ink. There are two types of filters that are commonly used, surface and depth; and both can consist of layers of polymeric fibers such as polypropylene or microporous membranes such as polysulfone. However, their mechanisms of particle capture are quite different, and an understanding of them is of assistance in selecting a suitable filter system and trouble-shooting it in operation.

Surface Filtration

The main capture mechanism for surface filtration is simply sieving in which the pore size is smaller than the particles to be removed. Direct interception with the fibers and

Table 14.8 Microfluidizer

Description
Mill base is pressure fed through microchannels at high velocities up to hundreds of m/s.
It is then divided into two separate streams, which change direction and collide back into a single stream.
Shear, impact, and cavitation occur.
Advantages
Disperses down to the submicron range.
Jacketed for cooling.
Relatively easy cleanup.
Limitations
Requires a good premix with particle sizes of 40 to 50 μm. Rotor–stator mixer is recommended by company.

bridging of multiple particles across a pore also contribute, but to a much lesser degree. A narrow pore size distribution is needed in order to have an absolute removal rating for particles above a certain diameter (d_p). Treating the filter as a square array of fibers like a screen, the fraction of cross-sectional pore area available for particle collection is given by

$$\varepsilon_S = \frac{\ell_p^2}{(\ell_p + d_f)^2} = \frac{1}{(1 + d_f/\ell_p)^2} \tag{77}$$

where ℓ_p is the length of a pore and d_f the fiber's diameter. Multiple layers laid on each other have the effect of creating a smaller pore. For a fixed pore size, the filter's capacity to retain particles on its surface increases for smaller fiber diameters, and preferably it is smaller than the pore size for high efficiency. Unfortunately, for submicron particles it is difficult to produce such thin fibers, and even then such a filter will tend to plug up quickly and form a cake. Consequently, the pressure drop rises, and the flow rate decreases. The potential also exists for the colorant to be filtered out, especially for pigmented inks.

Depth Filtration

Depth filters are much more porous than surface filters; their pores can be orders of magnitude greater than the particles, and thus sieving is no longer the capture mechanism. If one were to follow a particle's path within the filter, it could be intercepted directly by a fiber or settle on it via gravity. Most likely, it will not be in line with the first fiber that it passes and will continue on until it approaches another fiber further downstream. Because the fluid's streamlines change direction while it flows around the fiber, smaller particles may not have enough inertia to continue in a straight line and will be dragged around the fiber. The criterion for this to happen is that the dimensionless Stokes number (Stk) is less than a threshold value such as 0.125 (Fuchs, 1964):

$$\text{Stk} = \frac{\text{stopping distance}}{d_f} = \frac{v\rho d_p^2}{9\mu d_f} \tag{78}$$

where v = interstitial velocity within the filter, ρ = density of the particle, and μ = fluid viscosity.

Above this threshold, the efficiency of inertial deposition increases with Stk.

Fortunately, submicron particles exhibit sufficient Brownian motion to diffuse across the streamlines and can still make contact with the fiber if the velocity is low enough. The efficiency E of this deposition rate J relative to straight line motion is given by (Fuchs, 1964)

$$E = \frac{J}{d_f v n_0} = \frac{4.64 D^{2/3}}{v_f^{2/3} d_q^{2/3} [4(2.002 - \ln \text{Re}_f)]^{1/3}} \tag{79}$$

where n_0 = particle concentration, D = Brownian diffusion coefficient, which is inversely proportional to the particle's diameter, and Re_f = fiber's Reynolds number.

This means that a plot of efficiency vs. particle size or velocity can exhibit a minimum in the micron range with diffusion controlling at the small particle end and inertia at the large end. Decreasing the fiber diameter shifts both curves upward. In order to compensate for this probabilistic mode of capture, depth filters have to be thicker than surface filters. Similar phenomena take place within the random network of channels in a membrane.

As deposition occurs within the filter, its pore size becomes smaller, which increases the pressure drop, leading to possible compression of the filter. Further, the higher interstitial velocity can create a shear force that is large enough to detach some of this deposit. Ideally, one would like the deposit to reach a certain maximum amount at the filter's entrance and then have its interior sections gradually take over until there is a uniform deposit across the length of the filter. This can be achieved by using a filter medium that has been surface treated to have ionic groups and thus a charge as discussed in Section 14.2. Preferably, these chemical groups are reacted into the polymer backbone in order to avoid possible contamination.

Since the majority of materials are negatively charged in water, the commercially available filters have been designed to be positively charged. As particles accumulate within the filter, its charge diminishes so that its ability to remove them decreases. Eventually, this charge will be reversed and become slightly negative resulting in electrostatic repulsion and no further deposition. Then the interior sections will remove more and more of the particles until they also become saturated. When the exit section of the filter finally becomes saturated, the filter can no longer remove any particles.

It should be noted that the electrostatic attraction does not extend out into the bulk liquid to draw the particles to the filter because the interaction is due to the overlapping of the electrical double layers. The charge serves to regulate the amount of deposition and provide an adhesive force on it. Analogous to dispersion stability discussed earlier, the filter's performance can also be dependent on pH and ionic strength, and the potential exists for anionic components in the ink such as surfactants to be filtered out. For a more detailed and quantitative discussion of these colloidal aspects of filtration, the reader is referred to the papers by Wnek et al.

Multiple and Pore Graded Filters

Since a filter collects particles at and above its rated size, a broad particle size distribution will prematurely clog the filter with the larger ones. The solution is to use a set of filters of varying ratings placed in series starting from the largest to the smallest rating. Typically, the last filter would be a surface filter, and the others would be depth filters. While the actual ratings should be optimized for the ink in question, one could start with 2 µm followed by 1 µm, 0.5 µm, and 0.2 µm depth filters, ending with a 0.2 µm surface filter as a polishing step for dye based inks. For a given velocity, the total flow rate may be increased by increasing the filter area via larger filters or adding filters in parallel.

On the other hand, filtering a pigmented ink efficiently presents a challenging design problem. Its mean particle size is around 0.1 µm, and the tail of the distribution should cut off no higher than 0.5 µm. The consequence of this is that depth filters may not exhibit sufficient size selectivity and may even be the reverse of what is needed. For example, at low velocities the 0.1 µm particles will preferentially deposit via diffusion with little inertial deposition of either size. As the velocity is increased, inertial deposition of the 0.5 µm particles will eventually start followed later by the 0.1 µm particles. Hopefully, before this happens, diffusion deposition of the 0.1 µm particles becomes negligible. Otherwise, a cascade of only surface filters may be the solution.

An alternate way to vary the pore size systematically is to use a filter that has a graded pore structure. These filter cartridges consist of an inner absolute rated section and multiple outer prefilter zones. The outer sections remove those particles larger than the rating, and the inner section removes the remaining particles at the stated rating. This is analogous to the above cascade of filters, and this type of filter is commercially available from companies such as Pall and US Filter.

14.6.3 Ink Adjustments in Manufacturing

As a result of the tight specifications required for the functionality of ink jet inks, small variations in their component materials can drive a manufacturing batch of ink out of specification, and this then involves adjusting the concentration of certain components. This may be done in an empirical fashion in which each property is brought into specification one at a time by varying the concentration of the component that controls it the most. The difficulty with this approach is that a previously adjusted property can be driven out of specification by a later adjustment. This can result in an iterative process and many adjustments. What is needed is a procedure that takes into account these interactions in which a property depends on more than just one component. Such a procedure was described in Section 14.5.9 (Optimization), and and an example of its application follows.

Table 14.9 gives the formulation and specifications for a certain ink jet ink along with the measured physical properties for a particular manufacturing batch. It is seen that the transmittance, viscosity, and surface tension are not in specification.

Table 14.9 Formulation, Specifications, and Measured Properties of an Ink Jet Ink

Component	Original Formulation Original wt. fraction	Original wt (kg)
Solvent—Water	0.7355	102.970
Cosolvent—Glycerine	0.1800	25.200
Surfactant	0.0550	7.700
Dye	0.0275	3.850
Biocide	0.0020	0.280
Total	1.000	140.000

Specifications			
Physical property	Minimum	Maximum	Target
Viscosity (cp)	2.10	2.30	2.20
Surface tension (dyne/cm)	34.50	35.50	35.00
%Transmittance (at 430 nm)	27.30	28.00	27.65
pH	7	9	8
Filtration rating		10	

Measured Physical Properties	
Physical property	Initial
Viscosity (cp)	2.71
Surface tension (dyne/cm)	36.50
%Transmittance (at 430 nm)	25.90
pH	7.58
Filtration rating	2.7

Table 14.10 Adjustments of Manufacturing Ink Batch

Adjustment step	Added water (kg)	Added dye (kg)	Added surfactant (kg)	%T	Viscosity (cp)	Surface tension (dyn/cm)
Initial				25.90	2.71	36.50
#1	5			26.83	2.55	
#2	2.5			28.39	2.34	
#3	1.5			28.82	2.32	
#4	2			29.17	2.30	
#5	2.5			29.69	2.30	
#6	3.5			30.40	2.17	
#7		0.27		28.03	2.18	
#8		0.04		27.84	2.24	36.50
#9			0.3	27.62	2.20	35.90
#10			0.3	28.12		35.80
#11			0.5	27.54	2.21	35.20

Table 14.10 shows the series of adjustments that were made to bring this batch of ink into specification. Since viscosity was the property most out of specification, it was adjusted first by adding water to decrease its value. Once it was in specification, it was found that the transmittance was too high, and dye was then added. Finally, the surfactant was added to bring the surface tension into specification.

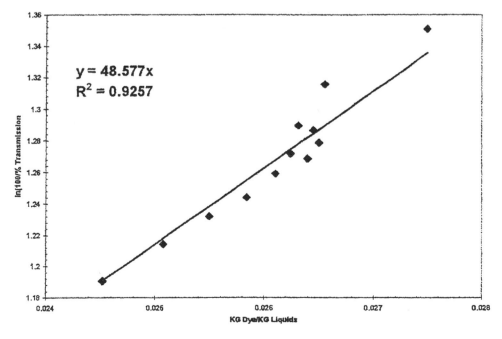

Figure 14.29 Logarithm of transmittance vs. dye concentration.

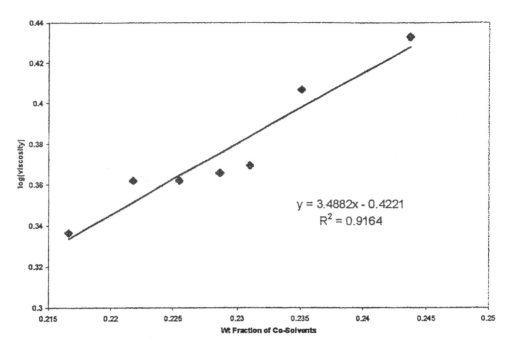

Figure 14.30   Logarithm of viscosity vs. weight fraction of cosolvent.

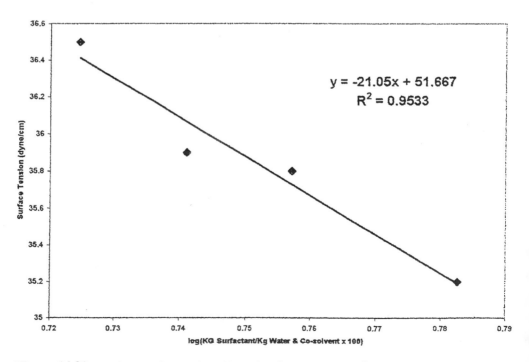

Figure 14.31   Surface tension vs. logarithm of surfactant concentration.

Table 14.11 Comparison of Calculated and Actual Ink Adjustments

Component	Predicted			Actual		
	Wt. fraction	kg to be added	New wt (kg)	Wt. fraction	kg added	New wt (kg)
Solvent—Water	0.7566	20.732	123.702	0.7573	17.00	119.97
Cosolvent—Glycerine	0.1541	0.000	25.200	0.1591	0.00	25.20
Surfactant	0.0616	2.368	10.068	0.0556	1.10	8.80
Dye	0.0260	0.393	4.243	0.0263	0.31	4.16
Biocide	0.0017	0.000	0.280	0.0018	0.00	0.28
Total	1.0000	23.493	163.493	1.0000	18.41	158.41

Physical property	Specification	Predicted	Actual
%Transmittance (at 430 nm)	27.30–28.00	28.00	27.54
Viscosity (cp)	2.10–2.30	2.30	2.21
Surface tension (dyn/cm)	34.5–35.5	35.50	35.20

Before we can use the proposed mathematical technique to solve simultaneously for all the concentration changes and component additions, we must verify that the model equations are representative of the data. The logarithm of transmittance is plotted against dye concentration in Fig. 14.29 and a good linear correlation is obtained in accordance with Eq. (14.71). Similarly, the logarithm of viscosity is plotted against the weight fraction of the cosolvent in Fig. 14.30, and again a good linear correlation is obtained in accordance with Eq. (14.72). Finally, surface tension is plotted against the logarithm of surfactant concentration in Fig. 14.31, and a good linear correlation is obtained in accordance with Eq. (14.74). Thus we are justified in using these model equations.

The objective function selected was to minimize the total amount of components that needed to be added. Table 14.11 shows the results of the calculation along with the actual manufacturing ones. The adjusted concentrations are in good agreement, including that no cosolvent should be added. It should be noted that while the added amounts are in agreement, they need not be because many such sets can result in the same adjusted concentrations.

Solver has been quite robust in its ability to converge using the initial concentrations as its starting point. When it reaches the boundary of the operating space, it may terminate its search. If one prefers a solution within the constraints, the search can be continued by minimizing the difference from a desired value.

REFERENCES

J. H. Adair, R. V. Linhart A generalized program to calculate interparticle interactions in a variety of suspensions conditions. In: *Handbook on Characterization Techniques for the Solid Solution Interface*. American Ceramic Society, Westerville, OH, 1993, pp. 69–84.

N. S. Allen. *Rev. Prog. Coloration* 17:61 (1987).

N. S. Allen, J. F. McKellar. *Photochemistry of Dyed and Pigmented Polymers*. Applied Science Publishers, London, 1980.

P. Ball, C. H. Nicholls. *Dyes and Pigments* 3:5 (1982).

P. W. Barber, S. C. Hill. *Light Scattering by Particles: Computational Methods*. World Science, Singapore, 1990.

A. F. M. Barton. *CRC Handbook of Solubility Parameters and Other Cohesion Parameters*. CRC Press, Boca Raton, FL, 1985.

W. Bauer, D. Baumgart, W. Zoller. IS&T's NIP 12: International Conference on Digital Printing Technologies, Toronto, October 28, 1996, p. 59.

W. Bauer, J. Geisenberger, H. Menzel. IS&T's NIP 14: International Conference on Digital Printing Technologies, Toronto, October 18, 1998, p. 99.

G. L. Baughman, S. Banerjee, T. A. Perenich. In: *Physico-Chemical Principles of Color Chemistry*, Advances in Color Chemistry Series, Vol. 4 (A. T. Peters, H. S. Freeman, eds.). Blackie, 1996, p. 145.

G. Baxter, C. H. Giles, M. N. McKee, N. Macaulay. *J.S.D.C. 71*:218 (1955).

F. W. Billmeyer, M. Saltzman. *Principles of Color Technology*. John Wiley, New York, 1981.

C. F. Bohren, D. R. Huffman. *Absorption and Scattering of Light by Small Particles*. John Wiley, New York, 1983.

J. A. Bristow. *Svensk Papperstiding 70*:623 (1967).

A. Brockes., *Optik 21*:550 (1964).

A. J. Chirinos-Padron, N. S. Allen. Aspects of polymer stabilization. In: *Handbook of Polymer Degradation* (S. H. Hamid, M. B. Amin, A. G. Maadhah, eds.). Marcel Dekker, New York, 1992.

E. Coates. *J.S.D.C. 85*:355 (1969).

J.-B. Donnnet, R. C. Bansal, M.-J. Wang. *Carbon Black*. Marcel Dekker, New York, 1993.

J. K. Ferri, K. J. Stebe. *J. Colloid Interface Sci. 209*:1 (1999); *Colloids Surfaces A:Physicochem. Eng. Aspects 156*:567 (1999).

M. G. Fontana, N. D. Greene. *Corrosion Engineering*. McGraw-Hill, New York, 1978.

F. M. Fowkes. In: *Chemistry and Physics of Interfaces-II* (D. E. Gushee, ed.). Am. Chem. Soc., Washington, D.C., 1971, p. 153.

G. Frens, J. Th. G. Overbeek. *J. Colloid Interface Sci. 38*:376 (1972).

N. A. Fuchs. *The Mechanics of Aerosols*. Pergamon, New York, 1964.

M. E. D. Garcia, A. S. Medel. *Talanta 33*:255 (1986).

C. H. Giles, D. J. Walsh, R. S. Sinclair. *J. Soc. Dyers and Colorists 93*:348 (1977).

P. F. Gordon, P. Gregory. *Organic Chemistry in Colour*. Springer-Verlag, Berlin, 1987.

J. Griffiths. *Colour and Constitution of Organic Molecules*. Academic Press, London, 1978.

L. N. Guo, M. Petit-Ramel, I. Arnaud, R. Gauthier, Y. Chevalier. *J.S.D.C. 110*:149 (1994).

N. Harnby, M. F. Edwards, A. W. Nienow. *Mixing in the Process Industries*. Butterworth-Heinemann, Oxford, 1992.

T. W. Healy, L. R. White. *Adv. Colloid Interface Sci. 9*:303 (1978).

W. Herbst, K. Hunger. *Industrial Organic Pigments*. VCH, Weinheim, 1997.

R. J. Hunter. *Zeta Potential in Colloid Science*. Academic Press, London, 1981.

R. J. Hunter. *Foundations of Colloid Science*. Vol. 1. Clarendon Press, Oxford, 1987.

R. J. Hunter. *Foundations of Colloid Science*. Vol. 2. Clarendon Press, Oxford, 1989.

J. Israelachvili. *Intermolecular and Surface Forces*. Academic Press, New York, 1992.

J. D. Jackson. *Classical Electrodynamics*. John Wiley, New York, 1975.

R. O. James. In: *Advances in Ceramics*. Vol. 21. *Ceramic Powder Science* (G. L. Messing et al., eds.). American Ceramic Society, Westerville, OH, 1987, pp. 349–410.

R. O. James, G. A. Parks. In: *Surface and Colloid Science* (E. Matijevic, ed). Vol. 12. Plenum Press, New York, 1982, pp. 119–216.

D. B. Judd. *J. Res. Natl. Bur. Std. 29*:329 (1942).

D. B. Judd, G. Wyszecki. *Color in Business, Science and Industry*. John Wiley, New York, 1975.

R. W. Kenyon. In: *Printing and Imaging Systems* (P. Gregory, ed.). Blackie, London, 1996.

S. Kishimoto, S. Kitahare, O. Manabe, H. Hiyama. *J. Org. Chem 43*:3882 (1978).

Y. Kogo, H. Kikuchi, M. Matsuoka, T. Kitao. *JSDC 96*:475 (1980).

J. A. Komor, J. P. G. Beiswanger. *J. Am. Oil Chem. Soc. 43*:435 (1966).

H. R. Kruyt. *Colloid Science*. Vol. 1. Elsevier, Amsterdam, 1952.

P. Kubelka, F. Munk. *Z. Tech. Physik. 12*:593 (1931).

H. Kubn.*Thin Solid Films 99*: 1 (1983).

N. Kuramoto. In: *Physico-Chemical Principles of Color Chemistry*. Advances in Color Chemistry Series, Vol. 4. A. T. Peters, H. S. Freeman, (eds.). Blackie, 1996, p. 296.

V. G. Levich. *Physicochemical Hydrodynamics*. Prentice-Hall, Englewood Cliffs, NJ, 1962.

J. Lyklema. *Fundamentals of Interface and Colloid Science*. Vol. 1. Academic Press, London, 1991.

V. N. Mallet, B. T. Newbold. *J. Soc. Dyers Colour 90*:4 (1974).

J. March. *Advanced Organic Chemistry*. John Wiley, New York, 1992.

K. McLaren. *The Colour Science of Dyes and Pigments*. Adam Hilger, Bristol, 1986.

A. I. Medialia, L. W. Richards. *J. Colloid Interface Sci. 40*:233 (1972).

J. C. V. P. Moura, A. M. F. Oliveira-Campos, J. Griffiths. *Dyes and Pigments 33*:173 (1997).

N. Mourougou-Candoni, B. Prunet-Foch, F. Legay, M. Vignes-Adler, K. Wong. *J. Colloid Interface Sci. 192*:129 (1997).

H. Mustroph, C. Weiss. *J. Prakt. Chem. 328*:937 (1986).

R. Naef. Computer-assisted dyestuff design and synthesis. In: *Modern Colorants: Synthesis and Structure* (A. T. Peters, H. S. Freeman, eds.). Blackie, London, 1995.

A. Nakajima, H. Akamatu. *Bull. Chem. Soc. Jp.* 41:1961 (1968); *42*:3030 (1969).

D. H. Napper. *Polymeric Stabilization of Colloidal Dispersions*. Academic Press, London, 1983.

K. Nassau. *Color for Science, Art and Technology*. Elsevier, Amsterdam, 1998.

J. F. Oliver. Wetting and penetration of paper surfaces. In: *Colloids and Surfaces in Reprographic Technology* (ACS Symposium Series 200). Washington, DC, American Chemical Society, 1982, p. 435.

L. I. Osipow. *Surface Chemistry*. Krieger, Huntington, NY, 1972.

J. Panzer. *J. Colloid Interface Sci 44*:142 (1973).

T. C. Patton. *Paint Flow and Pigment Dispersion*. Wiley-Interscience, New York, 1979.

H. Reerink, J. Th. G. Overbeek. *Discuss. Faraday Soc. 18*:74 (1954).

M. J. Rosen. *Surfactants and Interfacial Phenomena*. John Wiley, New York, 1989.

H. Rumpf. In: *Agglomeration* (W. A. Knepper, ed.). John Wiley, New York, 1962, p. 399.

W. B. Russel, D. A. Saville, W. R. Schowater. Colloidal Dispersions, Cambridge University Press, New York, 1989.

B. J. Sauntson. *Brit. Ink Maker 18*:26 (1975).

J. A. Schak. Dispersion of low viscosity water based inks. In: *Chemistry and Technology of Water Based Inks* (P. Laden, ed.). Blackie, London, 1997.

J. H. Schenkel, J. A. Kitchener. *Trans. Faraday Soc. 56*:161 (1960).

B. S. Solomon, C. Steel, Z. Weller. *Chem. Commun. 927* (1969).

R. Steiger, P. A. Brugger. IS&T's NIP 14: International Conference on Digital Printing Technologies, Toronto, October 18, 1998, p. 114.

H. C. A. van Beek, P. M. Heertjes. *J. Soc. Dyers Colour 79*:661 (1963).

T. G. M. Van de Ven. *Colloidal Hydrodynamics*. Academic Press, New York, 1989.

E. J. W. Verwey, J. Th. G. Overbeek. *Theory of Stability of Lyophobic Colloids*. Elsevier, Amsterdam, 1948.

J. Visser. In: *Fouling Science and Technology* (L. F. Melo, T. R. Bott, C. A. Bernardo, eds.). Kluwer Dordrecht, 1988, pp. 87–123.

J. Visser. *Particulate Sci. Technol. 13*:169 (1995).

H. G. Völz. *Industrial Color Testing: Fundamentals and Techniques*. VCH, Weinheim, 1995.

L. Weissbein, G. E. Coven. *Textile Research Journal 30*:58, 62 (1960).

A. Weller. *Fast Reactions and Primary Processes in Chemical Kinetics* (S. Claesson, ed.). Wiley-Interscience, New York, 1967, p. 413.

C. F. Wells. *Trans. Faraday Soc. 57*:1703, 1719 (1961).

G. R. Wiese. T. W. Healy. *Trans. Faraday Soc. 66*:490 (1970).

W. J. Wnek, D. Gidaspow, D. T. Wasan. *Chem. Eng. Sci. 30*:1035 (1975); W. J. Wnek. *Filtration Separation 11*:237 (1974); W. J. Wnek, R. Davies. *Colloid Interface Sci. 60*:361 (1977).

R. H. Young, R. L. Martin. *J. Amer. Chem. Soc. 94*:5183 (1972).

S. Yuan, S. Sargeant, J. Rundus, N. Jones, K. Nguyen. IS&T's NIP 13: International Conference on Digital Printing Technologies, Seattle, November 2, 1997, p. 413.

H. Zollinger. *Color Chemistry*. VCH, Weinheim, 1991.

15

Papers and Films for Ink Jet Printing

DOUGLAS E. BUGNER

Eastman Kodak Company, Rochester, New York

15.1 INTRODUCTION

In the previous chapter, an overview of ink jet printing technology was provided, including a description of both continuous ink jet (CIJ) and drop-on-demand (DOD) methods (1). In this chapter, we will discuss the requirements for the papers and films that receive the inks from such printers. We will be referring to these materials as ink jet *receivers*, as opposed to the more ambiguous term *media*. Laminating films, used to protect the surface of an ink jet print, are outside the scope of this chapter.

Before launching directly into this subject, it is necessary to review briefly the different ink technologies that are currently on the market, because receiver requirements are predicated on the type of ink being used. Figure 15.1 is one way to classify the different ink technologies in use today. By far, the most ubiquitous inks for desktop and larger format applications are water based. This is primarily for two reasons: aqueous inks are more environmentally friendly, and DOD thermal ink jet printheads, which account for the largest installed base of ink jet printers, require aqueous inks. Therefore the major focus of this chapter will be on receivers that are compatible with aqueous inks. We will also briefly discuss the unique requirements that are posed by the various nonaqueous ink technologies.

15.2 HISTORICAL PERSPECTIVE

Today we almost take for granted that ink jet printers should give acceptable, if not excellent, print quality on "plain" papers, but it was not always that way. The first commer-

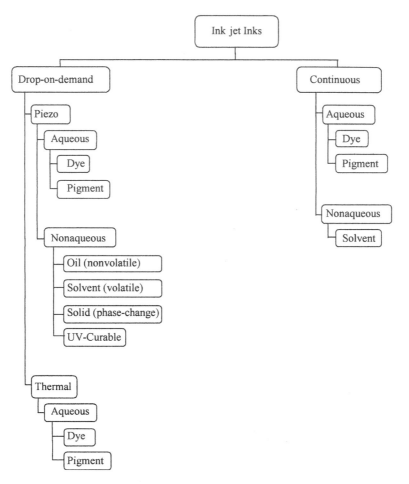

Figure 15.1   Ink taxonomy.

cially successful desktop ink jet printers required the use of a specially coated paper. This is because the inks used in the original HP ThinkJet printer (1985), and later the four-color PaintJet printer (1989), were relatively simple solutions of off-the-shelf, water-soluble dyes in a mixture of water and a less volatile cosolvent. These inks were rapidly absorbed into the uncoated paper fibers, leading to a phenomenon known as *wicking* or *feathering*. The net result was a fuzzy dot with low optical density.

In order to keep such inks from spreading and penetrating into the paper fibers, the paper was coated with a mixture of a silicate pigment in a binder. Such papers are generically referred to as *clay-coated* papers. The ionic nature of the pigment attracted the negatively charged dye molecules and kept them nearer to the surface, while allowing the ink vehicles to penetrate into the paper below. This resulted in darker, crisper text and graphics. This process of controlling the spread or migration of a dye molecule is often referred to as *mordanting*, and the additive or material used for such purposes is know as a *mordant*. This process will be discussed in detail below.

Among other things, the requirement for a special coated paper impeded the widespread adoption of ink jet printers as the preferred computer output device. This all changed with the introduction of the Hewlett-Packard DeskJet 500 series of plain paper ink jet printers (1992). Through a combination of dye modification and ink formulation, much better print quality could be obtained on most plain papers. However, for applications requiring even higher print quality, clay-coated papers were still preferred.

Although coated papers offered image quality and water resistance over plain papers, they were still limited in areas such as gloss and density. They also suffered from poor dimensional stability as a result of the interaction of the water from the ink and the underlying paper fibers. This resulted in a phenomenon known as *paper cockle*, in which the surface of the paper becomes wavy and uneven. In order to overcome these deficiencies, a new type of ''paper'' was developed, a white pigmented polyester film with a relatively thick coating of a water-absorbing polymer. At about the same time, similar coatings were applied to clear polyester films for use on overhead projectors.

The next development of historical significance was the Encad NovaJet series of wide format ink jet printers. Originally marketed as a 36-inch-wide, black-only machine for the CAD market, the NovaJet II series employed four of the HP DeskJet high-capacity black printheads in which the black ink had been replaced with cyan, magenta, and yellow inks in three of the four printheads. Although the drop volume ($\sim$100 picoliters) and addressability (300 dots per inch ''dpi'') of these printheads were inadequate for photographic quality images on $8\frac{1}{2} \times 11$ inch paper (A size), the results were quite stunning when printed at 34×44 inches (E size) on a NovaJet II printer.

In 1995, Epson laid first claim to photographic quality on the desktop with the introduction of the 720 dpi Epson Stylus Color printer. One of the interesting aspects of this printer was the offering of two different matte-coated papers: one for printing at 360 dpi, and one for printing at 720 dpi. The main difference between the two papers was the degree of drop spread that each allowed. Another first associated with the Epson Stylus Color printer was the introduction of a glossy white film that had a very thick, microporous coating of alumina instead of a nonporous, water-swellable polymer. The main advantage of the microporous coating was very fast dry times when compared to the nonporous receivers, such as Hewlett-Packard's glossy white film. This will be discussed in greater detail below.

Although the output from the Epson Stylus Color printer on either the 720-dpi paper or the white film could pass for photographic quality at first glance, neither offered the true look and feel of traditional silver halide photographs on resin-coated papers. It did not take long for Kodak and others to seize the opportunity to offer glossy resin-coated receivers for printing digital photographs on both desktop and wide-format printers.

To keep things interesting for the developers and manufacturers of ink jet receivers, the printer manufacturers have been continuously refining and modifying the inks. Of particular significance was the introduction of the DeskJet 600 and 800 series of printers (1995). These were the first DOD printers to feature pigment-based black ink, along with reformulated dye-based color inks. Especially challenging for the receiver formulator was the ability to design a coating that can handle both black pigment-based ink and colored dye-based inks on the same surface. Because of the particulate nature of the colorant in pigment-based inks, pigments tend to stay on or near the surface, even for porous receivers, while the dyes are absorbed well into the coatings. The reformulated dye-based inks were designed to minimize color-to-color bleed on plain papers. According to a Hewlett-Packard patent (2), as many as four different cosolvents were added in differing amounts in the yellow, magenta, and cyan inks. To complicate things further, magnesium nitrate was added to the magenta and cyan inks, while calcium nitrate was added to the yellow ink. Continuing a trend first noted in the DeskJet 500 series, the surface tensions of the black and color inks were also quite different. In addition, if that was not complex enough, both the cyan and magenta inks contained not one but two different dyes.

Another challenge to the receiver formulator has been the introduction of 6-ink printers, such as the original Hewlett-Packard PhotoSmart (1997), and the Epson Stylus Photo 700 printers (1998). These printers pioneered the use of light magenta and light cyan inks, which, when used in conjunction with dark cyan and dark magenta inks, produce a much smoother tone scale. The down side to this approach is that considerably higher volumes of ink per unit area are deposited upon the receiver, thus making fluid management and dry time a much harder task.

Other recent developments have actually made the design of ink jet receivers a bit easier. The trend to much smaller drop volumes, coupled with the ability to produce multiple drop sizes on the fly, have largely obviated the need to use light density inks to produce a smooth tone scale. The best example of such a printer is the Epson Stylus Color 900 (1999), which produces drop sizes of 3, 10, 11, 19, 23, and 29 picoliters. By using smaller drops and variable drop sizes, fluid management by the receiver becomes much easier. Another consequence is that near photographic image quality is achievable even on uncoated bond papers.

With drop volumes close to reaching the point of diminishing returns from an image quality perspective, printer manufactures have now turned their attention to print speed, coupled with inks that are capable of much higher levels of image stability. Thus there is a continuing need for further improvements in the receivers that need to be optimized to these printers and inks.

15.3 TAXONOMY AND ANATOMY OF INK JET RECEIVERS

Figure 15.2 provides a classification system for the different type of ink jet receivers available today. One way to envision an ink jet receiver is to look at it as comprising three main parts: a substrate, an ink-receiving layer, and a backside coating (Fig. 15.3). The simplest embodiment is obviously uncoated plain paper, which comprises just a substrate. While the ultimate goal of the ink formulator is to generate the best possible quality and durability on plain paper, there will always be a need for features and functionality that cannot be met with plain paper alone, for example, outdoor durability or transparency, to name just two. In this section, we will briefly describe the unique opportunities and features that can be realized by different combinations of substrates, ink-receptive layers, and backside coatings. We will also relate the key physical properties of the different receiver components and the receiver as a whole to the important customer observable features that they control.

15.3.1 Substrates

General Considerations

Usually the choice of a substrate is narrowed by the intended application. For example, if one were designing a new receiver for use in an overhead transparency, then an optically clear plastic film would be a requirement. Whereas the substrate forms the ''backbone'' of the ink jet receiver, the physical properties of the substrate are critical to the ultimate performance of the final design. In choosing a substrate for a specific application, the

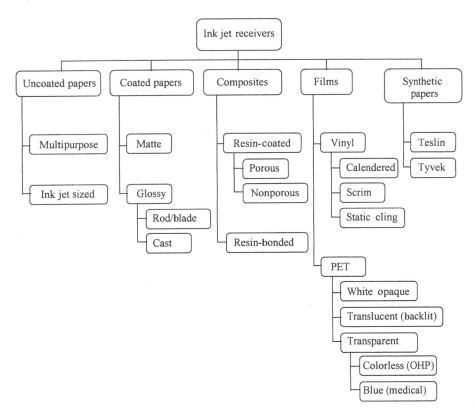

Figure 15.2 Receiver taxonomy.

following properties should be considered: opacity, dimensional stability, stiffness, caliper, basis weight, smoothness, colorimetry (including brightness and whiteness), and surface energy. Depending upon the application, additional considerations include recycled content, environmental stability, flame retardancy, cost, wide-roll dimensions, and availability.

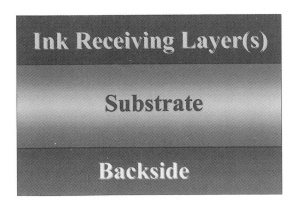

Figure 15.3 Anatomy of an ink jet receiver.

Papers

The most cost-effective and ubiquitous substrate is paper. The large multinational paper companies have clearly recognized the market opportunity that ink jet paper presents and have invested considerable effort in both redefining "plain paper" and improving the performance of paper as a substrate for specialized ink-receptive layers (3). Ink jet–optimized papers contain additives such as optical brighteners, minerals, and anticockle agents that yield better image quality and dimensional stability than ordinary copy paper. In any office superstore today one can find a wide variety of uncoated papers specifically designed for use in desktop ink jet printers, often selling up to three times the price of multipurpose grades.

Paper is most often used as a substrate for porous ink-receptive layers (see below), wherein the paper will be required to absorb a large portion of the ink vehicle. For these applications, the paper must not be sized or treated to the point that movement of the ink vehicle via capillary action from the coating is hindered. When the ink is water based, paper cockle is also a concern. This is particularly true for wide-format applications in which the waviness of the paper can interfere with the movement of the printhead as it passes back and forth across the receiver, resulting in a catastrophic defect known as *head strike*. In fact, paper cockle is still an issue that has not been dealt with cost effectively at present. Typical solutions include the incorporation of synthetic fibers, latex saturation, or the use of much thicker paper stock, all of which add to the cost of the substrate and/or reduce its ability to absorb the ink vehicles. For applications that require some degree of archivability, acid-free paper is a requirement unless the surface of the paper is sealed with an acid-impermeable layer (see the section on composites below).

Most of the large paper mills manufacture paper base for either direct ink jet printing or as a substrate for coating with ink-receptive layers, either internally or to third-party converters. A partial list includes Appleton, Boise Cascade, Consolidated Papers, Crown Vantage, Domtar, Felix Schoeller, Georgia Pacific, Hoffmann Englemann, International Paper, Kimberly Clark, Mead, Mitsubishi Paper, Neusiedler, Oji, P.H. Gladfelter, Renker, SAPPI, Stora Enso, Tomegawa, Westvaco, and Weyerhauser. Recent trends in the paper industry that bear watching include consolidation accompanied by cutbacks in and/or redirection of research and development (4). International Paper has recently acquired Union Camp and Champion. Likewise, SAPPI has acquired S. D. Warren.

Synthetic Papers

For special applications, such as outdoor display, synthetic papers such as Teslin (PPG), Tyvek (DuPont), Kimdura (Kimberly Clark), or Crisper (Toyobo) can be used as substrates. Synthetic papers, with respect to cost and performance, are more closely related to films (see below) than to traditional papers. Of the synthetic papers available today, Teslin is unique in that it has an open-cell microporous structure that will readily absorb ink solvents such as water, much like traditional paper. However, when ink is printed directly onto uncoated Teslin, the colorants tend to penetrate deep into the pores, resulting in very low optical densities. Therefore, in practice, Teslin requires a coating or surface treatment for optimum results. Tyvek is a synthetic fiber reinforced film that is valued for its high tear strength, high flexibility, and water resistance. It can also be sewn and grommeted, making it an ideal banner material. Kimdura is a closed-cell, voided poly(propylene) material, and Crisper is a similar closed-cell, voided polyester. Although they are both marketed as synthetic papers, they more closely resemble the opaque or translucent films described below in terms of their physical properties.

Films

The most common film materials used as ink jet substrates are poly(ethylene terephthalate) (PET) and poly(vinyl chloride), often referred to as just vinyl. PET is available as a transparent, translucent, or opaque substrate. The opacity is usually controlled by compounding varying levels of pigments such as titanium dioxide into the PET during the manufacturing process. Alternatively, the opacity can be controlled by microvoiding, as noted above. PET films are characterized as extremely smooth, dimensionally stable, tear resistant, and waterproof. PET films are capable of yielding a very high gloss surface when coated with the nonporous or microporous ink-receptive layers (more on this below). In order to coat ink-receptive layers on PET substrates, however, the PET needs to be treated first or primed to increase the adhesion of the coating.

PET-based ink jet receivers are used for both large-format and desktop applications. Clear and translucent PET films are commonly used for backlit signage in the large-format market. Although the white opaque PET is also used for outdoor signage, it is not as easily folded, stitched, or stretched as Tyvek or vinyl. Probably the largest application for PET as an ink jet substrate is for desktop overhead transparencies. White opaque PET-based ink jet receivers are also sold in the desktop market for use as glossy photographic-quality receivers. A special blue tinted version of clear PET is used for diagnostic medical imaging applications, such as digital radiography. Manufacturers of PET substrates include Diafoil, Du Pont-Teijin Films, Kimoto, SKC, Toray, and Toyobo. Eastman Kodak Company also manufactures PET substrates exclusively for internal use.

The two most common vinyl substrates are calendered vinyl and cast vinyl. Both are provided as white opaque films. Cast vinyls, as their name implies, are manufactured by casting the coating mixture onto a moving web. The dried and fused film is peeled from the web, yielding a very smooth, glossy surface. As with PET films, the tint and opacity are controlled by the addition of white pigments, such as titanium dioxide. In addition to white opaque, cast vinyls are also available in a wide variety of colors, as well as metallic, translucent, fluorescent, and transparent finishes. Calendered vinyls, also referred to as *scrim* vinyls, are prepared by pulling a heated mass of vinyl with plasticizers through polished heated rollers, under sufficient pressure to stretch and squeeze the vinyl into a 3–4 mil film. Compared to cast vinyl, scrim vinyl is a lower cost, lower quality alternative. Key feature trade-offs include dimensional stability, durability, UV resistance, opacity, and gloss. A third type of vinyl used as an ink jet substrate is ''static cling'' vinyl, used to temporarily adhere to a smooth surface such as glass. Suppliers of vinyl substrates include 3M, FLEXcon, Greystone, Tekra, and Spartan.

The primary application for vinyl-based inkjet receivers is for outdoor display. One of the key differentiating features between vinyl and alternative substrates such as Tyvek and PET is stretchability and the ability to conform to uneven surfaces. The best example of where these features are required is in vehicle signage. A consequence of these applications is that the ink-receptive layers for vinyl substrates must be weather-resistant and compatible with pigment-based inks. Another feature common to vinyl receivers commensurate with these applications is that they also typically comprise an adhesive backing. Ink-receptive layers and backside coatings will be discussed in more detail below.

Composites

Composites of paper and films are also widely used as substrates for ink jet receivers. *Resin-coated* paper (RC paper), long a standard in the photographic industry, is commonly used in both desktop and large-format printers for photographic quality output. RC papers comprise a core of paper that is coated on both sides with a polymeric resin. *Resin-bonded* paper (RB paper) is a recent development out of Mitsubishi Paper in Japan in which two webs of standard paper are laminated together with a layer of resin in between. The main advantage of RB paper is the virtual elimination of paper cockle.

RC papers fall into two categories, true photo-base and faux photo-base. The two are differentiated by cost, performance, and manufacturing process. True RC photo-base is manufactured by just a handful of companies, such as Kodak, Fuji Photo, Mitsubishi Paper, and Felix Schoeller. The core paper is made from very high quality nonphotoactive pulp and is characterized by a high degree of smoothness and extremely uniform density and tight formation. The front and back sides of the paper are coated with a polyolefinic resin such as poly(ethylene), poly(propylene), or blends thereof, usually by a melt-extrusion process. The still molten resin is cast against a chill roll that controls the surface smoothness and/or imparts a specific texture to the surface. The resin used on the front or image-bearing side includes additives such as pigments, optical brighteners, and stabilizers. The resin on the backside is typically unpigmented, allowing an optional back print on the paper to show through. Because the paper is essentially encapsulated with resin, paper cockle is not an issue. The encapsulation process also prevents any of the chemicals used in the paper making process from interacting with the image-bearing layers that are coated on top of the resin. The resin coating is also engineered to be nonyellowing and physically stable with respect to temperature, light, humidity, and air pollutants such as ozone. Thin subbing layers are often added during the resin-coating operation as adhesion promoters for the image-receptive and backside layers that are coated downstream and are discussed in greater detail below.

Faux photo-base typically starts with a less expensive, lower quality base paper. The paper is then coated by melt-extrusion, or it may also be solution coated with a water-impermeable barrier layer. Because the main driving force for faux photo-base is cost reduction, the quality and durability of paper and the barrier layer are usually compromised relative to true RC photo-base papers. Two companies that supply faux photo-base as a discrete substrate are Jen-Coat and Great Lakes. On the other hand, several integrated paper mills are capable of coating polymeric barrier layers either on- or off-machine that simulate resin-coated papers.

Specialty Substrates

It is also worth mentioning specialty substrates, such as natural and synthetic fabrics, that are also being employed as substrates for ink jet receivers. The custom printing of textiles is an area of very rapid growth. On the one hand, specialized systems and inks are being designed to print directly onto untreated fabrics (5). Alternatively, some fabrics are being coated or treated for use in existing DOD ink jet printers. One popular fabric substrate for printing with aqueous inks is canvas, especially in large-format fine art applications. In order to produce sharp, vibrant colors, the canvas is usually impregnated or coated with an ink-receptive formulation to minimize dot spread and to mordant the dyes. Other natural and synthetic fabrics, such as silk and polyester, have also been used as substrates. Fabric transfer receivers, in which the image is first formed on an intermediate receiver and then transferred via heat and pressure to fabric, will be discussed briefly in the context of specialty ink-receptive layers. Other specialty substrates include metal foils.

15.3.2 Ink Receiving Layers (IRLs)

Overview

There are two primary functions of an IRL: to absorb the ink vehicles, and to control the spread and penetration of the colorants. As mentioned above, the very first ink-receptive coatings for DOD ink jet printers comprised relatively large (1–10 μm) inorganic particles (silicate-based clays) in combination with a minor amount of polymeric binder coated directly onto paper base. The ink solvents, mostly water, penetrated the interstitial pores created by the particles and eventually into the underlying paper base. Because of the ionic nature of the inorganic particles, the penetration and spread of the dyes were inhibited. A limitation of these early coatings was that they tended to scatter light, and the surface was relatively rough, resulting in a nonglossy or *matte* appearance.

The first IRLs for films and RC papers, on the other hand, comprised nonporous coatings of polymers with a high capacity for swelling and absorbing the ink vehicles by molecular diffusion. Cationic substances were added to the coatings to serve as mordants for the anionic dyes. These coatings were optically transparent and very smooth, leading to high gloss "photo-grade" receivers, or, when coated onto clear PET films, transparencies suitable for overhead projection. Since then, porous particulate-based and nonporous polymer-based IRLs have both continued to evolve. In this section, we discuss the materials options for porous and nonporous IRL technologies, and the known performance trade-offs that currently exist: drying rate and mechanisms, gloss, image quality, and image stability.

Materials Options

Porous IRLs

The two primary ingredients in most porous IRLs are particles and polymeric binders. The key properties of the particles include chemical composition, size, shape, and intraparticle porosity. In addition to silicate-based clays, the most common inorganic particles in use today include silica (SiO_2), alumina (Al_2O_3), and calcium carbonate ($CaCO_3$). The manufacturing process determines the size, shape, and porosity of the particles.

Both natural (montmorillonite, hectorite, bentonite) and synthetic clays (Laponite) are available for use in IRL formulations (6). Synthetic clays are preferred because of their lack of color and low opacity. An interesting feature of clays is their ability to form swellable, binder-free coatings with a very regular "house of cards" porous structure. Suppliers of clays include Englehard Corporation, J. M. Huber, Kalamazoo Paper Chemicals, and Southern Clay Products.

Amorphous silica is probably the most ubiquitous inorganic particle in use today. It comes in a variety of types, including fumed, colloidal, precipitated, and gel (7). It is available in sizes ranging from less than 100 nm (colloidal, fumed) to greater than 1 μm (precipitated, gel) from vendors such as Cabot, Crosfield, Degussa, W. R. Grace, J. M. Huber, Nalco, Nissan Chemical Industries, PPG Industries, Rhodia, and Wacker-Chemie. In contrast to colloidal, precipitated, and fumed silica that comprise nonporous primary particles, particles of silica gel exist as rigid, porous, three-dimensional networks. Novel porous silica gels as small as 300 nm have been described in the literature (8).

Alumina has recently emerged as an alternative to silica, especially for high-performance glossy coatings. Like silica, alumina is also available in particle sizes ranging from less than 100 nm to greater than 1 μm. A special form of alumina, pseudoboehmite (AlO_2H), is available commercially as an aqueous dispersion of nanometer sized particles. It can also be prepared in situ by hydrolysis of aluminum *iso*-propoxide or *sec*-butoxide with acetic acid (9). The first examples of high gloss and transparent porous coatings on PET films were prepared from pseudoboehmite (10). Suppliers of alumina include Cabot, CONDEA Vista, Degussa, Lonza, Nissan Chemical Industries, PPG Industries, and Sumitomo.

Mixed oxides of silicon and aluminum are also available. Degussa offers particles prepared by cofuming $SiCl_4$ with a small amount of $AlCl_3$. Ludox CL is a colloidal silica with an alumina-modified surface available from Du Pont.

Although calcium carbonate has long been used as a filler in alkaline paper making, it has recently emerged as an alternative to inorganic oxides in porous IRLs. Calcium carbonates are available in both ground and precipitated forms. The latter tend to have smaller particle sizes but are also more expensive. Suppliers of calcium carbonate include Columbia River Carbonates, Dravo Lime, J. M. Huber, Omya, and Specialty Minerals.

Binders commonly used for porous particulate coatings include aqueous soluble hydrophilic polymers such as poly(vinyl alcohol) (PVA) and poly(vinyl pyrrolidone) (PVP), aqueous emulsions of styrene-acrylic and styrene-butadiene latexes, and solvent coatable hydrophilic polymers such as poly(vinyl acetate) (PVAc). Key suppliers of PVA and PVA derivatives are Air Products and Nippon Gohsei. International Specialty Products (ISP) specializes in PVP and PVP derivatives. Styrene-based latex emulsions are available from BASF, Dow Chemical, and Goodyear Tire and Rubber.

The optimum particle-to-binder ratio is a function of both the type of particle and the particle size. When the binder concentration is too high, it tends to fill in the interstitial pores and inhibits the absorption of the ink. When the binder concentration is too low, poor adhesion and/or cohesion results, and the particles tend to flake or dust off of the substrate (11). Key considerations in the design of porous particulate IRLs are (1) optimization of the surface energy, (2) optimization of the void volume, and (3) careful selection of both particle and pore size to balance rapid ink absorption and desired gloss levels. These properties will be discussed in more detail below.

When designing porous IRLs for use with inks containing anionic (negatively charged) dyes, a cationic (positively charged) mordant is usually included in the formulation. Commonly used commercially available mordants include poly(diallyldimethylammonium chloride) (polyDADMAC) and cationic-modified PVA. When used in combination with silica, polymeric mordants are usually added in relatively low concentrations due to their impact on rheology.

One of the more interesting recent developments is porous IRL technology embodied in the Konica QP ink jet receiver (12). This IRL is based on a hybrid organic–inorganic particle comprising a core of colloidal silica that has been encapsulated with a shell of proprietary cationic copolymer. In addition to functioning as a dye mordant, the charge density on the cationic shell is also tuned to balance van der Waals forces and optimize the pore structure and thus the void volume. This technology will be discussed further in the context of gloss and dry time.

The optimum thickness of a porous IRL will depend upon the pore density (void volume) and the type of substrate upon which it is coated. When coated on porous substrates such as paper, the coating can be relatively thin (10–20 μm). If the coating is too thin, however, the water from the ink will penetrate into the paper fibers, resulting in an objectionable level of paper cockle. When coated on a nonporous substrate such as PET film or RC paper, or when paper cockle must be eliminated, the entire volume of ink at maximum coverage must be absorbed within the pore structure of the coating. Typically, this requires the IRL thickness to be in the 30–50 μm range at a void fraction of about 0.5.

Glossy porous IRLs usually comprise multilayer structures. When coated on plain paper, a coarser fluid management layer is laid down underneath a finer gloss and color-control layer. In order to provide a smooth, glossy surface, special coating processes are often utilized, such as cast coating and film-transfer coating. Calendering with heat and pressure is also used in combination with conventional blade or rod coating on plain paper to produce gloss. The use of plastic pigments that melt and flow during calendering has been claimed to yield higher levels of opacity and gloss (13). For glossy porous IRLs on white PET film or RC-paper substrates, as many as three layers have been laid down (12).

In addition to porous IRLs based on inorganic particles, several novel approaches to porous ink jet receivers have been disclosed. One approach leverages technology originally developed to produce filtration membranes (14). Scientists at Celfa/Folex have commercialized porous ink jet receivers by a phase inversion coating process (15). In this process, a polymer is dissolved in a mixture of a less volatile good solvent and a more volatile poor solvent. As the more volatile solvent evaporates, the polymer begins to gel and contract, producing voids. Upon complete evaporation of the residual solvent, an open-cell porous layer is produced. Another novel approach to porous IRLs is based on the entrainment of air to produce a stable open-cell foam (16). The use of organic particles for general-purpose porous IRLs has also been described in the literature (17).

Nonporous IRLs

Nonporous IRLs require the use of polymers that can absorb the ink solvents. For water-based inks, naturally occurring polymers such as gelatins (18) and polysaccharides (cellulose derivatives) (19) have been found to swell rapidly and absorb water and the humectants in the ink. Synthetic hydrophilic polymers such as PVA (19a), PVP (20), poly(ethylene oxide) (PEO) (21), and poly(2-ethyl-2-oxazoline) (POx) (22) have also been found to be effective. In many cases, blends of the above and/or multilayer structures comprising several hydrophilic polymers have been used for nonporous IRLs. Yuan et al. provide an overview of hydrophilic polymers useful for nonporous IRLs along with a comparison of key performance factors across the different polymer types (23). Poerschke and Stumpf describe cross-linked blends of gelatin and PVA useful as IRLs (24). Gelatin suppliers include Croda Colloids, DGF Stoess, Kind and Knox Gelatine, Nitta, and SBI. Suppliers of PVA and PVP are listed above in the context of binders for porous IRLs. PEO is available from Hercules Incorporated or Meisei Chemical Industries. POx can be obtained from Polymer Chemistry Innovations.

It is relatively easy to obtain glossy ink jet receivers with nonporous IRLs, especially on smooth substrates such as RC paper or white PET film. As with glossy porous IRLs, it has become common practice to coat a relatively thick base coat to manage most of the ink solvents, and a thin top coat to optimize other receiver properties, such as dry time, gloss, and image quality. Nonporous IRLs with as many as five layers have been recently reported (25).

As with porous IRLs, mordants such as polymeric quaternary ammonium salts and cationic-modified PVAs are also commonly included in nonporous IRLs. Studies on multilayer IRLs have shown that both the type and the placement of the mordant can have a dramatic impact on the eventual location of the dye within the IRL structure, and that this will vary depending upon the specific dye set (26). Figure 15.4 shows how a blue patch comprising equal amounts of cyan and magenta inks interacts with a prototypical two-layer IRL comprising different mordants in the top and bottom layers (27b). A novel approach to mordanting dyes in nonporous IRLs involves the use of functionalized cyclo-dextrins as "molecular includants" (28).

Other common ingredients found in most nonporous IRLs are matting agents, such as silica or polymeric beads. These particles are typically in the 1–50 μm range and are present at relatively low concentrations relative to the balance of the IRL. The primary purpose of matte particles is to provide an optimum level of surface roughness without significantly impacting the perceived gloss of the receiver. The surface roughness helps to control the coefficient of friction for reliable feeding in the printer. The matte particles also provide for improved wet stacking as the prints exit the printer, as well as resistance to smudge and abrasion of the dry print. At higher concentrations, matte particles can be used to produce a semigloss or matte appearance in an inherently glossy nonporous IRL.

Drying Rates and Mechanisms

Perhaps one of the biggest performance differences between porous and nonporous IRLs is in the area of drying rates and mechanisms. In either case, it should be noted that aqueous inks dry by a two-step process. The initial step involves the relatively rapid absorption of the ink solvents into the IRL to produce a perceptually dry state, followed by a much

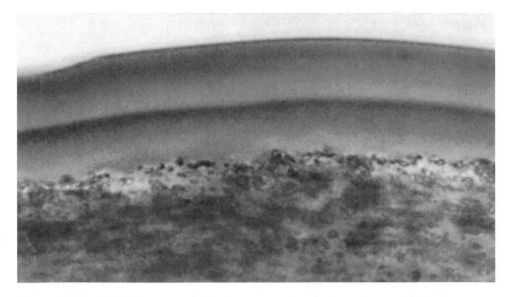

Figure 15.4   Optical photomicrograph of a cross section of a blue color patch printed on a two-layer nonporous IRL comprising different mordants in the top and bottom layers.

slower step in which the water evaporates and the nonvolatile cosolvents diffuse and equilibrate within the IRL structure. It is the initial absorption step that primarily differentiates porous and nonporous IRLs.

In the case of a porous IRL, the initial absorption step is dominated by capillary action. There have been a number of papers attempting to provide a mathematical model for how inks interact with porous surfaces such as plain papers (29). Usually, a modified form of the Lucas–Washburn equation is used to model the various parameters that affect the drying rate of a drop of ink on a porous layer. Equation (1) is one such representation that can be used to predict the apparent dry time t of a drop of ink of volume V, viscosity η, and surface tension γ on a surface of capillaries of radius r, assuming that there is perfect wetting of the capillaries by the ink.

$$ t = \left(\frac{2}{3}\right)^{4/3} \frac{2\eta}{\gamma r} \left(\frac{V}{\pi}\right)^{2/3} \tag{1} $$

Although in reality the capillary structure is not made up of perfectly uniform cylinders orthogonal to the surface of the ink jet receiver, this equation correctly predicts the relationship between the time it takes for a porous receiver to feel dry to the touch and the pertinent ink and receiver properties. In other words, bigger and more viscous drops with lower surface tensions take longer to dry on a given porous surface. Likewise, the smaller the diameter of the capillaries of a porous receiver, the longer the dry time for a given ink. Using values for V, η, γ, and r typical of ink jet inks and porous receivers, this equation predicts dry times of less than 1 second. This is consistent with the fact that most porous receivers appear dry to the touch immediately after exiting the printer. One consequence of this mechanism is that the void volume of the pore structure must accommodate the entire volume of ink in any given area of a print. This results in a dry thickness of porous IRLs in the 30–50 μm range when coated on nonporous substrates such as films and RC papers.

The modeling of the dry time of nonporous receivers is considerably more complex. Nonporous receivers absorb ink by diffusion of the ink solvents into the polymer matrix, resulting in a rapid swelling of the IRL in the areas where ink is deposited (Fick's law) (30). Compounding the situation is that the IRL will also absorb moisture from the atmosphere as a function of ambient humidity. Further complicating things is that the absorbed atmospheric moisture coupled with the ink solvents tends to plasticize the polymer matrix, lowering the glass transition temperature T_g to below room temperature. Although diffusion of the ink into the polymer matrix is accelerated in the glassy state, and the ink is typically absorbed within the first few seconds, the net result is a prolonged period of time during which the image feels tacky to the touch and is prone to smudging. Eventually the water evaporates, the ink cosolvents equilibrate within the IRL, and the T_g rises back above room temperature. At a relative humidity (RH) below about 50%, this all happens within a minute or less with many combinations of nonporous IRLs and aqueous inks. However, at higher RH and/or with inks that contain relatively high levels of cosolvents, it may take 5–10 min for the print to equilibrate and become smudge free. In any event, nonporous ink jet receivers are perceived to be much slower drying than porous receivers. On the other hand, because the polymers used in nonporous IRLs rapidly swell to many times their initial volume, even at high RH, the thickness of nonporous IRLs need only be in the range of 10–15 μm.

Gloss

Another feature that differentiates porous and nonporous IRLs is *gloss*. From an historical perspective, porous coatings have tended to be nonglossy (matte) in appearance, and nonporous coatings were developed specifically to produce high-gloss receivers, suitable for printing photographs, or in the case of overhead transparencies, to producing an optically clear haze-free receiver. Recently, relatively glossy porous ink jet receivers have been introduced, leveraging from coating processes originally developed for glossy offset printing papers.

Before discussing the specifics of porous vs. nonporous gloss, it should be pointed out that the measurement of gloss itself, as well as the understanding of the factors that influence the perception of gloss, is a complex subject that is outside the scope of this review (31). Gloss, also referred to as *specular gloss*, is the degree to which a surface appears mirrorlike. One commonly used instrument for measuring specular gloss is a glossmeter manufactured by BYK-Gardner. Measurements taken with such a device are referred to as Gardner gloss values. Suffice it to say that there is a strong angular dependence of both the perception and the measurement of gloss. Two common angles of measurement are 20° and 60° from a line perpendicular to the plane of the paper. In general, the higher the gloss, the lower the angle of measurement that is recommended. Although there is no universal preference for a specific range of gloss, glossy brochures, newsletters, magazines, and photographs are generally perceived to be of higher value than their nonglossy counterparts. One thing to keep in mind is that although it is relatively easy to "dial down" the gloss of an inherently glossy IRL, it is very difficult, if not impossible, to increase the gloss of a matte or semigloss IRL with out resorting to laminating or overcoating the image.

Numerous studies have attempted to understand the various aspects of pigment-based coated papers in the context of offset printing inks (32). With respect to porous ink jet receivers, Pesenti et al. have studied the effect of the shape of the pigment on surface microstructure and gloss (33). These and other studies have concluded that two of the major factors that impact the gloss of a porous surface are surface roughness and the size and relative refractive indices of the particles and pores that make up the porous matrix. In general, when the feature size, e.g., surface roughness or particle/pore dimension, falls below about half the wavelength of visible light (~200–400 nm), higher gloss values result (32a). The impact of pore and particle size on gloss can be further mitigated by closely matching their refractive indices. The glossiest porous receivers that have been observed to date comprise coatings of colloidal alumina and/or silica that (a) are coated on either a white PET film or an RC paper substrate, (b) are well under 100 nm in size, and (c) have very smooth surfaces.

Nonporous coatings, on the other hand, are inherently glossy when coated on smooth substrates such as white PET film or RC paper. The addition of matte particles, as noted above, will lower gloss slightly, but at the size and concentrations that are effective in adjusting the coefficient of friction, the effect of matte particles on observed gloss is minimal. At high enough levels of matte, however, a noticeable lowering of gloss will occur. Another way to lower the gloss of nonporous coatings is to use rougher substrates such as plain paper, or to use RC substrates in which a texture pattern has been intentionally applied to the resin coating itself.

Table 15.1 compares the Gardner gloss values at 20° and 60° for several representative porous and nonporous glossy IRLs on RC paper base. It can be seen that the nonporous

Table 15.1 20° and 60° Gardner Gloss Values for Representative Glossy IRLs on RC Paper base

Description	Type	20° gloss	60° gloss
Konica QP	Porous	16	36
Epson Premium RC	Porous	18	37
Hewlett-Packard Premium Plus Glossy	Nonporous	63	89
KODAK Premium Picture Paper	Nonporous	47	83

IRLs tend to yield higher levels of gloss than the porous IRLs. While this difference in measured gloss appears to be large, the differences in perceived gloss are not as apparent.

Print and Image Quality

Print quality and image quality vary considerably across the gamut of ink jet receivers ranging from uncoated plain papers to the photo-quality coated white PET films. For office documents, ink jet optimized plain papers yield levels of print quality on current ink jet printers that rival offset printing in many respects. When it comes to printing photographs, the perceived image quality is a function of both the look and the feel of the print. Although today's ink jet printers can do a respectable job of printing photographs on plain papers, the sharpness, saturation, stiffness, smoothness, and gloss suffer in comparison to the same images printed on glossy photo-quality ink jet receivers. For obvious reasons, glossy IRL coatings on RC paper substrates provide a very close match to the look and feel of traditional photographs. Of these, dye-based inks printed onto gelatin-based nonporous receivers provide the closest match to silver halide prints. On the other hand, if the goal is to produce a high-quality rendition of an oil painting, then a coated canvas receiver might do the trick. The beauty of ink jet is that there is an ink–receiver combination that will provide a high-quality result for just about any application.

Image Stability and Print Durability

Image stability refers to the permanence of the printed image, e.g., lightfastness, waterfastness, dark keeping. Print durability refers to the physical permanence of the receiver, e.g., tear and scratch resistance, wet strength, fingerprint resistance. A number of recent papers have touched on various aspects of image stability and print durability of ink jet prints (34–42). The primary factors that limit image stability are light, heat, humidity, and air quality, separately and in combination with each other. Additionally, if the print is stored in direct contact with glass, plastic, or the backside of another print, then the image stability will be further influenced by the nature of the material in contact with the print surface. Although the colorants and other additives used in the inks play a major role in the overall image stability of a print, the proper match of an ink jet receiver to a given ink set is critical to achieve the highest levels of image stability. Print durability, on the other hand, is primarily determined by the properties of the substrate and the IRL and to a lesser extent by the type of inks. The end-user expectations of image stability and print durability are largely determined by the intended use of an ink jet print. Clearly, a transient document printed on plain paper will have different stability and durability expectations than a large-format sign intended for outdoor display.

One of the unique image stability issues with ink jet prints is sensitivity to humidity. This phenomenon manifests itself as a gradual spreading or growth in the primary ink spots on the receiver. This in turn leads to an apparent increase in density and/or a hue shift, as well as a loss of sharpness or blurring of the image (43). As with ink drying mechanisms, the image stability of porous IRLs is much less sensitive to ambient relative humidity than that of nonporous IRLs. This observation can be attributed to the fact that porous IRLs absorb relatively little moisture from the air, and the moisture they do absorb does not effectively plasticize the mostly inorganic matrix. Nonporous IRLs, on the other hand, can be plasticized at high humidities to the extent that their T_g's fall below room temperature. This greatly enhances the ability of the dye molecules to diffuse laterally within the IRL, resulting in an observed increase in density and loss of sharpness. Because the different dyes in the ink set will diffuse at different rates, a noticeable hue shift is not uncommon. This effect can be minimized or even eliminated with the proper selection and placement of mordants within the IRL structure (26,27).

In addition to water vapor (humidity), ink jet prints are notorious for their lack of waterfastness. This is not all that surprising when you consider that the inks themselves are water based and that the IRLs are typically coated out of water. Nevertheless, some recent porous formulations have claimed to be very waterfast (12). Indeed, pure water is shown in the marketing literature to bead up on the surface of an inkjet print made on Konica QP. This paradox can be rationalized because the surface tension of pure water (70 dyn/cm) is significantly higher than that of typical ink jet inks (~30 dyn/cm). Thus the ability of some porous receivers to absorb ink readily but repel water is likely related to a differential wettability of the pores. In the case of nonporous IRLs, water resistance, like humidity resistance, is typically improved by employing water-insoluble latex binders and/or polymeric mordants (26,27).

The lightfastness of ink jet photographs has recently received a lot of marketing and trade attention. In particular, Epson has made claims of lightfastness of print made on the Stylus Photo 870/875/1270 series of printers as being comparable to "standard color photos" with print life estimates (based on accelerated testing) of 10 years on Epson's premium glossy RC paper. Subsequent studies (44) and trade reports (45) have questioned the validity of the accelerated test conditions given that prints displayed in typical ambient settings have exhibited a pronounced loss of cyan density (red shift). Epson has attributed the cyan fade of this particular ink–receiver combination to a pronounced sensitivity to airborne contaminants such as ozone, as opposed to a lightfastness issue (46). Heat and humidity appear to further accelerate this phenomenon. This is an excellent example of just how interrelated the various factors that impact image stability are. One simply cannot accurately estimate print life based on a single accelerated fade test.

Bugner and Suminsky have recently studied the effects of light intensity on the fading of a large number of photo-quality ink–receiver combinations (47). HP, Epson, Canon, and Lexmark inks were printed onto over 20 different porous and nonporous glossy ink jet receivers. They found that when the intensity of Plexiglas-filtered fluorescent lights was varied by a factor of about 12 (67 Klux vs. 5.4 Klux), a significant and highly variable *reciprocity* effect was observed. Specifically, the extent of fade at equivalent cumulative exposures was invariably higher at the 5.4 Klux exposure condition. Moreover, the porous receivers in general exhibited a consistently larger reciprocity effect across all of the ink sets than the nonporous receivers included in the study. This observation was attributed to the greater permeability of the porous receivers to oxygen and/or other airborne reactants.

Aspects of print durability such as fingerprint resistance and scratch resistance can be affected by the nature of the IRL. Glossy receivers are especially prone to fingerprinting, and particulate matte receivers are more prone to dusting and scratching than glossy receivers are. To date, however, there have been no detailed studies attempting to relate specific materials options to improvements in either of these attributes. Tear resistance, especially wet-strength, on the other hand, can be greatly enhanced by the choice of substrate. The most durable substrates in this regard are Tyvek and PET films (48).

The foregoing discussion on image stability and print durability of ink jet receivers assumes the absence of postimaging treatment options. For wide-format commercial applications, the use of conventional lamination technology is commonplace and provides very high levels of stability and durability (49). Alternatives to lamination involve the use of physically and/or chemically activated IRL technology. This will be discussed briefly in the following section.

Multifunctional Ink-Receptive Layers

In the context of IRL materials, it is worth a brief mention of materials used specifically in "multifunctional" IRLs, i.e., IRLs that do more than just receive the ink to make a document or print. One example of a multifunctional IRL is one that can be heat activated after printing either to fuse the image (15,50) or to adhere to another substrate (51). Along these same lines, IRLs that contain ink-reactive components that can be activated after printing by heat, light, and/or pressure have been disclosed (52). Another variation on this theme is an IRL composition that can be chemically treated after printing to enhance the durability of the print (53). Yet another example of a multifunctional IRL is one that luminesces or glows in the dark (54). This is an area of active research and development as companies search for higher value-added ink jet receivers.

15.3.3 Backside Coatings

Antistats

Essentially all ink jet receivers that comprise nonconductive substrates such as RC papers or PET films require an antistatic coating on the backside to eliminate the buildup of static charge. This is necessary to ensure consistent feeding of the receiver under low humidity conditions wherein static cling tends to result in multifeeds. There are also manufacturing issues associated with the buildup of static electricity. Although the use of antistatic coatings is not unique to ink jet receivers, there have been at least two patents that disclose materials useful in antistatic backside coatings for ink jet receivers. Useful antistatic materials include carboxylated or sulfonated polymers (55) and nonpolymeric materials such as sulfosuccinate esters, quaternary ammonium and phosphonium salts, and cationic sulfur-containing compounds (56).

Curl Control Layers

Most nonporous IRLs require a curl-control layer to counter the tendency of the IRL to expand or contract as a function of absorbed moisture. In the absence of a curl-control layer, this results in either positive or negative curl of the receiver. In the worst case, curl can adversely impact feeding of the receiver in the printer. Excessive curl can also interfere with the optical projection of overhead transparencies.

Perhaps the simplest solution is to coat a duplicate IRL on the backside so that the front and backsides respond identically to changes in ambient temperature and humidity. However, this is not necessarily the most cost-effective approach, especially when an ink jet–printable backside is not desired. The basic requirement for a curl-control layer is to respond to changes in temperature and humidity at a similar rate and to a similar degree as the IRL. This typically requires only a single layer of a hydrophilic polymer, with the thickness adjusted by experimentation to balance the particular IRL on the front side. Examples of suitable materials for use as curl-control layers for ink jet receivers have been disclosed (57). In an interesting twist, curl control can be provided to plain papers by metering an anticurl fluid onto the backside during the printing process (58).

Matting agents are often added to the backside layer to adjust the coefficient of friction and to provide enough ''tooth'' to be able to write easily on the backside with pen or pencil. Further, the functions of curl-control and antistatic control can be combined in a single layer.

Other Backside Layers and Coatings

Perhaps the third most common backside coating for inkjet receivers is an adhesive along with the requisite release liner. A detailed description of the various types of adhesives useful for the backside of ink jet receivers is outside the scope of this chapter. Suffice it to say that the different types of adhesives vary from less aggressive ''restickable'' to more aggressive and permanent. Applications include mailing labels, package labels, arts and crafts, and outdoor signage. A related application comprises a thin magnetic composite laminated to the backside of ink jet receivers so that the print can be adhered to surfaces such as refrigerators or vehicles.

15.4   RECEIVERS FOR NONAQUEOUS INKS

15.4.1   Oil-Based Inks

Oil-based inks are used primarily in D.O.D. wide-format printers for commercial applications such as signage and fleet graphics. At least part of the driving force for oil-based inks is the availability of nonpassivated piezoelectric printheads based on Xaar technology that are incompatible with aqueous inks. In addition, nonaqueous inks offer some advantages over aqueous inks: (1) lack of cockle when printed on lower cost coated, or uncoated paper substrates, (2) insensitivity to high humidities with respect to both dry time and image stability, and (3) better weather and water resistance without lamination.

In line with the intended applications, most ink jet receivers optimized for use with oil-based inks are coated onto durable substrates, such as films and wet-strength papers. By definition, nonvolatile oil-based inks dry by absorption of the hydrocarbon solvent into the ink jet receiver, as opposed to solvent-based inks, which dry by evaporation. Consequently, as with aqueous inks, faster drying and higher image quality are achieved by using IRL coatings that are engineered to absorb the ink vehicle, control drop spread, and fix the colorant. Although there have been a few isolated patent disclosures that claim glossy coatings compatible with oil-based inks (59), virtually all currently available IRLs are nonglossy and porous. In some cases, the same porous receivers can be used for both aqueous and oil-based inks. In spite of the advantages offered by oil-based inks, the market share for printers that use oil-based inks is still relatively small. This could be at least partly due to the obvious environmental concerns associated with hydrocarbon solvents used in most oil-based inks, and thus these types of printers are targeted primarily at commercial printing markets. Needless to say, the market share of ink jet receivers for oil-based inks is commensurate with the relatively small share of the total market that such printing systems as a whole enjoy. This limited market opportunity is reflected in the narrower range of ink jet receivers available and the much smaller number of patent and scientific publications when compared to inkjet receivers optimized for aqueous inks.

15.4.2 Solvent and UV-Curable Inks

Solvent-based inks dry by evaporation, and UV-curable inks are solidified by brief exposure to an UV source. Solvent-based and UV-curable inks are used primarily for commercial applications such as direct product labeling (date codes on glass or metal cans) and outdoor signage (direct onto uncoated films). The driving force for using these types of inks is to be able to print onto substrates that are not amenable to drying by absorption into an IRL. By definition, then, there really is no requirement for a specific ink-receptive paper or film for solvent or UV-curable inks.

15.4.3 Phase-Change Inks

Perhaps the most successful printers based on phase-change inks are the Tektronix (now Xerox) Phaser series. These printers are positioned as workgroup color printers, as an alternative to color laser printers. Phase-change inks are unique to piezoelectric D.O.D. printheads that can be operated at elevated temperatures. The ink at room temperature is a solid, waxy composition, similar to a crayon. When heated above the melting point of the wax, the ink becomes a low-viscosity liquid and behaves very much like an oil-based ink. In this case, the inks do not dry by absorption or by evaporation, but by ''freezing'' as they impact the surface of the receiver. However, a fusing step is typically required in order to drive the ink sufficiently into the receiver surface. This is necessary to improve the physical durability of the print, to enhance the image quality, and to eliminate the tactile relief pattern that the relatively high mounds of molten ink produce in an unfused state. The net result on plain paper is very similar in appearance to the fused toner images that are produced on color laser printers. Indeed, high image quality that is relatively insensitive to paper type is one of the key value drivers of this technology.

Despite the fact that phase-change inks do not require a specific ink-receptive coating, there are a number of specialty receivers that have been developed by Tektronix and others that have been cooptimized with the inks and the printing system. A recent publication by Korol and Stinson outlines the materials requirements for ink jet receivers optimized for phase-change inks that impact both print quality and print durability (60).

15.5 PHYSICAL PROPERTIES

15.5.1 Common Dimensions

There are different common dimensions used to describe the length, width, thickness (caliper), and basis weights of ink jet receivers. In North America, common sheet dimensions (Table 15.2) are in multiples of $8\frac{1}{2} \times 11$ inches (A,B,C,D,E), thickness is typically expressed in terms of thousandths of an inch (mil), and basis weights are in pounds per various square footages depending upon the application. Outside of North America, the common sheet dimensions are in multiples of 210 mm (8.27″) × 297 mm (11.69″) (A4, A3, A2, A1, A0), thickness is usually expressed in micrometers (1 mil = 25.4 μm), and basis weights are in grams per square meter. Sheet sizes as small as 4″ × 6″ (4R) and 105 mm × 148.5 mm (A6) are offered for applications such as postcards and digital snapshots. Rolls of ink jet receivers share standard widths of 24, 36, 42, 50, 54, 60, and

Table 15.2 Common Sheet Sizes

North America			Rest of world		
Size	Length (in.)	Width (in.)	Size	Length (mm)	Width (mm)
4R	6	4	A6	148.5	105
5R	7	5	A5	210	148.5
A	11	8½	A4	297	210
B	17	11	A3	420	297
C	22	17	A2	594	420
D	34	22	A1	840	594
E	44	34	A0	1188	840

72 inches. Outside of North America, these same widths are expressed in their metric equivalents (60.9, 91.4, 106.7, 127.0, 137.2, 152.4, and 182.9 cm). The length of the roll will be limited on the high end by the thickness of the receiver when the maximum roll diameter is reached. A common length is 100 ft (30 m). Rolls as short as 10 or 25 feet are sometimes sold as samples.

The caliper or thickness of most ink jet receivers ranges between about 4 mil (101.6 μm) and 10 mil (254.0 μm). Novelties such as magnetic-backed ink jet receivers can be as thick as 11 mil (279.4 μm) and sell for over $3 per sheet. Scrim vinyls can be up to 400 μm thick and greater than 900 g/m^2.

There are several commonly used measures for basis weight (61):

1. Pounds per 1300 square feet (500 sheets of C-size paper)
2. Pounds per 1800 square feet (500 sheets of 20″ × 26″ cover)
3. Pounds per 3000 square feet (500 sheets of 24″ × 36″ newsprint)
4. Pounds per 3300 square feet (500 sheets of 25″ × 38″ text)
5. Grams per square meter (g/m^2)

The industry as a whole is moving to method 5, the current TAPPI standard (61), to avoid the ambiguity of methods 1–4. It should be noted that basis weight comparisons are most useful within a given type of receiver, e.g., bond papers, PET films, resin-coated papers, etc. Table 15.3 summarizes common ranges of basis weights and calipers for different categories of ink jet receivers.

Table 15.3 Common Basis Weights and Calipers for Different Types of Inkjet Receivers

Receiver type	Basis weight range (g/m^2)	Caliper range (μm (mil))
Uncoated papers	75–150	101.6–152.4 (4–6)
Matte/glossy-coated papers	120–200	127.0–203.2 (5–8)
Coated synthetic papers	90–200	101.6–228.6 (4–9)
Resin-coated papers	150–300	127.0–254.0 (5–10)
PET films	140–230	101.6–177.8 (4–7)
Canvas/fabrics	80–240	203.2–635.0 (8–25)

Table 15.4 Relationship Between Selected Ink Jet Receiver Physical Properties, Engineering Metrics, and Customer Observable Features

Physical property	Engineering metric	Customer feature
Surface energy	Dry time	Productivity
	Dot spread	Image quality
Stiffness	Feedability	Reliability
	Foldability	Ease of use
Thickness	Feedability	Reliability
	Caliper	Feel
Smoothness	Gloss	Image quality
	Feedability	Reliability
Curl	Feedability	Reliability
Density	Basis weight	Feel
Coefficient of friction	Feedability	Reliability
Colorimetry	Tint, whiteness, D_{min}	Image quality
Opacity	D_{min}	Image quality
Surface resistivity	Feedability	Reliability

15.5.2 Critical-to-Function Properties

There are a number of critical-to-function physical properties that impact the performance of ink jet receivers. Table 15.4 summarizes the various physical properties, the engineering metrics that depends on these properties, and the customer observable features that are controlled by these properties. For example, the surface energy impacts both the dry time and the amount of drop spread. Dry time limits how quickly prints can be made, impacting productivity. Dot spread will affect the density and sharpness of an image, which in turn relate to customer perceived image quality.

15.6 SUMMARY

Ink jet technology spans a wide variety of printhead architectures, ink compositions, and receiver types, which together address a plethora of applications ranging from birthday cakes to giant billboards. In terms of common consumer applications, ink jet receivers range from multipurpose plain papers to semigloss and matte coated papers to high-gloss resin-coated papers and PET films. Typical applications include text-only and compound (picture plus text) documents, greeting cards, photographs, overhead transparency presentations, signage, and printing on fabrics. A key consideration of each specific application is the choice of substrate and ink-receptive coating. Substrates include plain bond papers, synthetic papers, PET and vinyl films, film–paper composites, fabrics, and foils.

For improved image quality, image stability, and physical durability, ink-receptive layers (IRLs) are coated on the various substrates. The two main types of IRLs are porous and nonporous. Porous IRLs are more versatile with respect to ink type and generally offer ''instant'' dry times and good water and humidity resistance, but they are limited in gloss, light stability, and scratch resistance. Nonporous IRLs provide high gloss, very high image quality, and light stability, but they tend to suffer from slow dry times and lower water and humidity resistance. Backside layers are commonly added to RC paper and film substrates in order to control static electricity and curl. Other backside layers for special applications include adhesives and magnetic laminates.

Special papers and films have also been developed for oil-based and phase-change inks. However, the diversity of receiver offerings for these types of inks pales in comparison to those that are available for aqueous inks. Because solvent-based and UV-curable inks are intentionally selected to print onto uncoated surfaces, there are even fewer receivers specifically designed for these types of inks.

ACKNOWLEDGMENTS

The author wishes to acknowledge the assistance of the following people in the preparation of this manuscript: Tom Nicholas, Lori Shaw-Klein, Brian Price, Terry Blake, Ian Newington, John Higgins, and Chuck Romano.

REFERENCES

1. For a more in-depth review, see H. P. Le. Progress and trends in ink-jet printing technology. *J. Imaging Sci. Tech. 42*(1):49–63 (1998).
2. L. E. Johnson, H. P. Lauw, N. E. Pawlowski, J. P. Shields, J. M. Skene. Thermal ink-jet inks having reduced black to color and color to color bleed. U.S. Patent 5,536,306 (1996).
3. D. I. Lunde. Tapping high-growth digital imaging market can boost mill profits. *Pulp and Paper 50–55* (February, 1999).
4. R. B. Phillips. Research and development in the pulp and paper industry: year 2000 and beyond. *TAPPI Journal 83*(1):42–46 (2000).
5. (a) B. Hunting, R. Puffer, S. Derby. Issues impacting the design and development of an ink jet printer for textiles. Proceedings of IS&T's Eleventh International Congress on Advances in Non-Impact Printing Technologies, 1995, pp. 374–377; (b) W. C. Tincher, Q. Hu, X. Li, Y. Tian, J. Zeng. Coloration systems for ink jet printing of textiles. Proceedings of IS&T's NIP14: 1998 International Conference on Digital Printing Technologies, pp. 243–246; (c) B. Hunting, S. Derby, R. Puffer, L. Loomie. Thermal ink jet printing of textiles. Recent Progress in Ink Jet Technologies II (E. Hanson, ed.). The Society for Imaging Science and Technology, Springfield, VA, 1999, pp. 568–573.
6. P. K. Jenness. Synthetic clay rheology modifiers for water based coatings. *Waterbone Coatings and Additives* (D. R. Karsa, W. D. Davies, eds.). Royal Society of Chemistry, Special Publication 165, pp. 217–231 (1995).
7. M. C. Whithiam. Silica pigment porosity effects on color ink jet printability. *Recent Progress in Ink Jet Technologies II* (E. Hanson, ed.). Society for Imaging Science and Technology, Springfield, VA, 1999, pp. 493–501.
8. D. M. Chapman, D. Michos. Novel sub-micron silica gels for glossy, ink-receptive coatings. Proceedings of IS&T's NIP15: 1999 International Conference on Digital Printing Technologies, pp. 164–168.
9. B. E. Yoldas. *J. Mater. Sci. 10*:1856–1860 (1975).
10. K. Misuda, H. Kijimuta, T. Hasegawa. U.S. Patent 5,104,730 (1992).
11. P. C. Adair. Ink jet coatings for pigmented inks. *Recent Progress in Ink Jet Technologies II* (E. Hanson, ed.). Society for Imaging Science and Technology, Springfield, VA, 1999, pp. 349–352.
12. K. Kasahara. A new quick-drying, high-water-resistant glossy ink jet paper. Recent Progress in Ink Jet Technologies II (E. Hanson, ed.). Society for Imaging Science and Technology, Springfield, VA, 1999, pp. 353–355.
13. H.-T. Chao. Inkjet recording sheet. U.S. Patent 5,919,558 (1999).

14. M. Dabral, L. F. Francis, L. E. Scriven. Structure evolution in asymmetric polymer coatings. Proceedings of the 9th International Coating Science and Technology Symposium, Newark, Delaware, May 17, 1998.

15. P. C. Walchli. A novel approach to IJ film (or paper) media granting highest resolution, fast drying and high durability. Recent Progress in Ink Jet Technologies I (R. Eschbach, I. Rezanka, eds.). Society for Imaging Science and Technology, Springfield, VA, 1999, pp. 253–256.

16. S. Maeda, T. Nakai, A. Nakamura, M. Hakomori, M. Kato. Development of a paper having micro-porous layer for digital printing. Proceedings of IS&T's NIP15: 1999 International Conference on Digital Printing Technologies, pp. 172–175.

17. H. Nogucji, Y. Satoh, S. Fujii, R. Hama, M. Satoh, M. Yamagishi. Organic, cationic, submicron particles for ink jet paper coatings. Recent Progress in Ink Jet Technologies II (E. Hanson, ed.). Society for Imaging Science and Technology, Springfield, VA, 1999, pp. 370–374.

18. R. Poerschke, J. Dolphin, Gelatine—a material for ink jet coatings. Recent Progress in Ink Jet Technologies II (E. Hanson, ed.). Society for Imaging Science and Technology, Springfield, VA, 1999, pp. 491–492.

19. (a) M. S. Viola. Ink jet transparency. U.S. Patent 4,575,465 (1986). (b) S. L. Malhotra. Ink jet transparencies with coating compositions thereover. U.S. Patent 4,592,954 (1986). (c) O. Farooq, D. W. Tweeten, M. Iqbal, S. K. Kulkarni. Crosslinked cellulose polymer/colloidal sol matrix and its use with ink jet recording sheets. U.S. Patent 5,686,602 (1997). (d) G. E. Missell, C. E. Romano. High uniform gloss ink-jet receivers. U.S. Patent 6,040,060 (2000).

20. J. R. Pinto, J. C. Hornby. Elucidating PVP dye-binding mechanisms in ink jet media. Recent Progress in Ink Jet Technologies II (E. Hanson, ed.). Society for Imaging Science and Technology, Springfield, VA, 1999, pp. 340–344.

21. (a) S. L. Malhotra. Ink jet transparencies with coating compositions thereover. U.S. Patent 4,592,954 (1986). (b) Y. Kojima, T. Omori, K. Nagai. Ink jet recording sheet. U.S. Patent 4,650,714 (1987).

22. S. Sargeant, N. Jones. Effect of hydrogel formation in IPN type inkjet receiver coatings on print quality and storage properties. Proceedings of IS&T's NIP15: 1999 International Conference on Digital Printing Technologies, pp. 169–171.

23. S. Yuan, S. Sergeant, J. Rundus, N. Jones, K. Nguyen. The development of receiving coatings for inkjet imaging applications. Proceedings of IS&T's NIP13: 1997 International Conference on Digital Printing Technologies, pp. 413–417.

24. R. Poerschke, A. Stumpf. Cross-linked ink jet layers of gelatin/polyvinyl alcohol mixtures. Proceedings of IS&T's NIP13: 1997 International Conference on Digital Printing Technologies, pp. 375–377.

25. L. J. Shaw-Klein. Ink jet recording element. U.S. Patent 6,110,585 (2000).

26. L. Shaw-Klein. Effects of mordant type and placement on inkjet receiver performance. Recent Progress in Ink Jet Technologies II (E. Hanson, ed.). Society for Imaging Science and Technology, Springfield, VA, 1999, pp. 335–339.

27. (a) L. Shaw-Klein, W. A. Light. Inkjet image recording elements with cationically modified cellulose ether layers. U.S. Patent 5,789,070 (1998). (b) G. E. Missell, D. E. Decker. Ink jet recording element. U.S. Patent 6,045,917 (2000).

28. R. Sinclair, J. G. MacDonald. Ink for ink jet printers. U.S. Patent 5,681,380 (1997).

29. See, for example, (a) L. Carreira, L. Agbezuge, A. Gooray. Correlation between drying time and ink jet print quality parameters. Recent Progress in Ink Jet Technologies I (R. Eschbach, I. Rezanka, eds.). Society for Imaging Science and Technology, Springfield, VA, 1999, pp. 1–4. (b) M. S. Selim, V. F. Yesavage, R. Chebbi, S. H. Sung, J. Borch, J. M. Olson. Drying of water-based inks on plain paper. *J. Imaging Sci. Tech.* 41(2):152–158 (1997).

30. See, for example, N. L. Thomas, A. H. Windle. A theory of case II diffusion. *Polymer 23*: 529–542 (1982).

31. See, for example, R. S. Hunter, R. W. Harold. *Measurement of Appearance.* John Wiley, 1987.

32. See, for example, (a) D. W. Donigian, J. N. Ishley, K. J. Wise. Coating pore structure and offset printed gloss. *TAPPI Journal 81*(5):163–172 (1997). (b) Y. Arai, K. Nojima. Coating structure for obtaining high print gloss. *TAPPI Journal 81*(5):213–221 (1998).

33. F. Pesenti, J. C. Hassler, P. Lepoutre. Influence of pigment morphology on microstructure and gloss of model coatings. Proceedings of IS&T's NIP12: International Conference on Digital Printing Technologies, pp. 405–408 (1996).

34. Steiger, R., Brugger, P.-A. Photochemical studies on the lightfastness of ink-jet systems. Recent Progress in Ink Jet Technologies II (E. Hanson, ed.). Society for Imaging Science and Technology, Springfield, VA, 1999, pp. 321–324.

35. A. Lavery, J. Provost, A. Sherwin, J. Watkinson. The influence of media on the light fastness of ink jet prints. ibid., pp. 329–334.

36. S. Sargeant, T. Chen, B. Parikh. Photoquality PQ and durability constraints for inkjet media. ibid., pp. 345–348.

37. A. Niemoeller, A. Becker. Interactions of ink jet inks with ink jet coatings. ibid., pp. 393–399.

38. A. Lavery, J. Provost. Color-media interactions in ink jet printing. ibid., pp. 400–405.

39. M. Fryberg, R. Hofmann, P.-A. Brugger. Permanence of ink-jet prints: a multi-aspect affair, ibid., pp. 419–423.

40. C. Lee, J. Urlaub, A. Bagwell, J. MacDonald, R. Nohr. Properties of inks containing novel lightfastness additives, ibid., pp. 444–446.

41. V. Kwan. Effect of resin/binders on lightfastness of colorants in inkjet inks. Proceedings of IS&T's NIP15: 1999 International Conference on Digital Printing Technologies, pp. 92–94.

42. J. Wang, T. Chen, O. Glass, S. Sargeant. Light fastness of large format ink jet media, ibid., pp. 183–186.

43. P. Hill, K. Suitor, P. Artz. Measurement of humidity effects on the dark keeping properties of inkjet photographic prints. Proceedings of IS&T's NIP16: International Conference on Digital Printing Technologies, pp. 70–73 (2000).

44. M. J. Carmody, S. Evans, S. Robinson. The image quality and lightfastness of photos from digital cameral appliance printing systems. Proceedings of IS&T's NIP16: International Conference on Digital Printing Technologies, pp. 124–127 (2000).

45. (a) *The Hard Copy Supplies Journal*, p. 35, July, 2000. (b) *The Hard Copy Observer*, p. 11, August, 2000.

46. http://www.epson.com/whatsnew/ygtsi/lightfast.html.

47. D. E. Bugner, C. Suminski. Filtration and reciprocity effects on the fade rate of inkjet photographic prints. Proceedings of IS&T's NIP16: International Conference on Digital Printing Technologies, pp. 90–94 (2000).

48. T. Cleary. Defining inkjet media attributes for imaging applications. Presentation given at CAP Ventures Communications Supplies and Industrial Marking Conference, April 2–4, 1997.

49. An in-depth discussion of laminate materials for ink jet prints is outside the scope of this chapter. For a discussion and further references, see T. Graczyk, B. Xie. Lamination study of ink jet media. *TAPPI Journal 83*(6): 63 (2000).

50. K. Hasegawa, M. Higuma, M. Sakaki, R. Arai, T. Akiya, N, Morohoshi. Image forming method. U.S. Patent 4,832,984 (1989).

51. (a) D. H. Brault, D. A. Cahill, R. S. Himmelwright, D. H. Taylor. Ink jet imaging process and recording element for use therein. U.S. Patent 5,795,425 (1998). (b) L. J. Shaw-Klein, A. A. Malcolm, D. E. Bugner. Low heat transfer material. U.S. Patent 6,036,808 (2000).

52. (a) R. P. Held, D. P. Fickes, J. E. Reardon, R. A. Work, III. Reactive media-ink system for ink jet printing. U.S. Patent 5,537,137 (1996). (b) S. G. Wu, J. G. Moehlmann, R. P. Held. Process for providing durable images on a printed medium. Eur. Pat. Appl., 0 775 596 A1 (1997).

53. See, for example, (a) D. Erdtmann, C. E. Romano, T. W. Martin. Pigmented ink jet prints on

gelatin overcoated with hardener. U.S. Patent 6,045,219 (2000). (b) X. Wen, D. Erdtmann, C. E. Romano. Printing apparatus with processing tank. U.S. Patent 6,082,853 (2000).

54. (a) D. L. Patton, A. M. Schwark, D. L. Cole. Glow-in-the-dark medium and method of making. U.S. Patent 5,965,242 (1999). (b) D. L. Patton, A. M. Schwark, D. L. Cole. Glow-in-the-dark medium and method of making. U.S. Patent 6,071,855 (2000).

55. S. J. Morganti, J. H. Thirtle. Element as a receptor for nonimpact printing. U.S. Patent 5,023,129 (1991).

56. S. L. Malhotra, K. N. Naik. Coated photographic papers. U.S. Patent 5,897,961 (1999).

57. S. L. Malhotra. Recording sheets. U.S. Patent 5,277,965 (1994).

58. L. M. Carriera, A. M. Gooray, K. C. Peter. Curl control of printed sheets. U.S. Patent 5,579,963 (1996).

59. Japanese Kokai, 98-562208 (1998).

60. S. Korol, R. Stinson. Media optimization for solid ink printing systems. Recent Progress in Ink Jet Technologies II (E. Hanson, ed.). Society for Imaging Science and Technology, Springfield, VA, 1999, pp. 406–412.

61. Grammage of paper and paperboard. TAPPI Test Method T 410 om-98, revised 1998.

16

Applications of Amorphous Silicon and Related Materials in Electronic Imaging

J. MORT

Xerox Corporation, Webster, New York

16.1 HISTORICAL REVIEW

One of the unifying characteristics of photoreceptors is their large area. It is for this reason that thin films are employed, whether for photographic emulsions or xerographic photoreceptors. The application as xerographic photoreceptors requires that such films be fabricated not only in large areas at an affordable cost but also with specific electrical and photoelectronic properties; see Williams (1984) and Borsenberger and Weiss (1998). These properties, as discussed in preceding chapters, include both high dark resistivity and photosensitivity to visible light and an ability to permit the passage of photogenerated charge carriers over distances significantly larger than the film thickness. One of the major developments in materials science in the past 25 years has been the recognition that certain amorphous materials can simultaneously possess these demanding photoelectronic and manufacturing requirements, together with the additional, associated chemical and mechanical stability required by diverse, often hostile, imaging environments (Williams, 1984; Borsenberger and Weiss, 1998). Moreover, the fundamental understanding of the electronic mechanisms critical for photoreceptor usage gleaned from research, as discussed in earlier chapters, also led to materials engineering of highly sophisticated photoreceptor composite systems with improved features compared with homogeneous photoreceptor materials.

Generally speaking the research and development of photoreceptor materials can, in retrospect, be seen to have occurred in a chronological sequence. For if selenium and its alloys, discussed in Chapter 9, emerged in the 1950s and 1960s (Felty, 1967), then the organic, polymeric-based photoreceptors (Mehl and Wolff, 1964; Gill, 1973; Mort and

Pfister, 1979), discussed in Chapter 10, were created during the 1960s and 1970s. More recently, plasma-deposited amorphous silicon (a-Si:H) and related materials have been developed (Mort et al., 1980; Shimizu et al., 1980; Yamamoto et al., 1981).

Now although, as we shall see, amorphous silicon (a-Si) can function as a photoreceptor like the other amorphous materials, it is important to recognize that it represents a fundamentally different type of this class of solids, being the only known amorphous extrinsic semiconductor (Spear, 1977). Accordingly, its applicability and value as an imaging material is much greater than that of its chalcogenide or organic relatives. This chapter discusses the nature and properties of amorphous silicon and related materials, their means of fabrication, and the broad range of potential and actual applications in electronic imaging, with particular emphasis on the functioning of a-Si as a photoreceptor. Chapter 11 discusses more recent development in a-Si photoreceptor technology.

We have seen that amorphous materials lack the long-range periodicity of the crystalline state (Mott and Davis, 1979). Thus, constraints on their growth are so relaxed that one common means of fabrication includes thermal evaporation of a material followed by condensation onto a cooled substrate. It has been known for many years that amorphous silicon can be produced in this fashion although, because of the relatively high boiling point of silicon, electron beam evaporation must be used. Unfortunately, the amorphous form of silicon so-formed (a-Si) has an electrical resistivity of $\sim 10^3$ Ω·cm, which makes it of little value for any applications.

In the mid-1960s, it was discovered that a form of amorphous silicon (henceforth termed plasma-deposited amorphous silicon and denoted by a-Si:H) could be produced by the glow-discharge dissociation of silane (SiH_4) gas and, unlike the thermally evaporated form (a-Si), had a room temperature resistivity of 10^9 Ω·cm or greater (Sterling and Swan, 1965; Chittick et al., 1969). This difference in the dark resistivity of these two types of amorphous silicon was related using the phenomenon of the field effect, discussed in more detail later, to variations in the density of localized states in the band gap (Spear, 1977). Based on this insight, it was next shown that a-Si:H could be extrinsically doped, by analogy with crystalline silicon, using either boron or phosphorus as the dopant to produce a p-type or n-type semiconductor, respectively (Spear and Le Comber, 1975). This opened up the possibility of producing, for the first time, large-area, monolithic electronic devices such as photovoltaic cells and, since much of this progress occurred in the early 1970s, it coincided with a growing awareness of an energy crisis. For this reason, the initial technological focus for a-Si:H was its potential use in photovoltaic solar cells (Carlson and Wronski, 1976; Kuwano, 1986). However, it was also widely appreciated that the requirements for such an application—namely, good electronic transport properties, high photosensitivity to visible light, and the low production costs needed for large areas—were precisely those required for xerographic photoreceptors and other imaging needs, such as large, linear arrays of p-n photodiodes, thin film transistors (TFTs) and charge-coupled devices (CCDs) for print bars and sensor applications in the growing field of electronic imaging (Böhm, 1988).

The next section discusses the properties of a-Si:H in more depth to provide a sufficient background for later sections. Section 16.3 deals in a comparative framework with the plasma deposition process and fabrication methods for various devices including, in addition to photoreceptors, TFTs. Section 16.4 treats the range of current applications including a-Si:H xerographic photoreceptors and other devices, while Section 16.5 considers in more detail the case of a-Si:H xerographic photoreceptors. A discussion of possible future developments constitutes the final section.

16.2 PROPERTIES OF PLASMA-DEPOSITED AMORPHOUS SILICON

Although a-Si:H is often referred to as though it were a single material, it should be emphasized at the outset that its properties can vary widely depending on the growth conditions. Indeed, as with most materials, the early studies were plagued with nonreproducibility because of insufficient understanding of important process/material property relationships (Adler, 1971). With the much more detailed, although by no means complete, understanding of today, material reproducibility is remarkably good (Spear, 1974).

As already mentioned, the properties of a-Si and a-Si:H are critically dependent on the density of localized states in the band gap. Such localized states, as in all amorphous materials, arise from both their inherent disorder and the presence of electrically active impurities (Mott and Davis, 1979). Not surprisingly, the dominant state in a-Si is the one associated with a broken silicon bond, commonly referred to as a dangling bond. It is now known that such dangling bonds and their associated localized state can be removed from the band gap by the attachment of atomic hydrogen. As shown in Figure 16.1, there is a much higher density of states at midgap for a-Si, about 10^{18} cm^{-3}, whereas in a-Si:H, depending on the substrate temperature employed, this density can be 10^{16} cm^{-3} or lower. Thus the much reduced density of gap states in a-Si:H, compared with a-Si, occurs because of the elimination of silicon dangling bonds by the ubiquitous hydrogen created

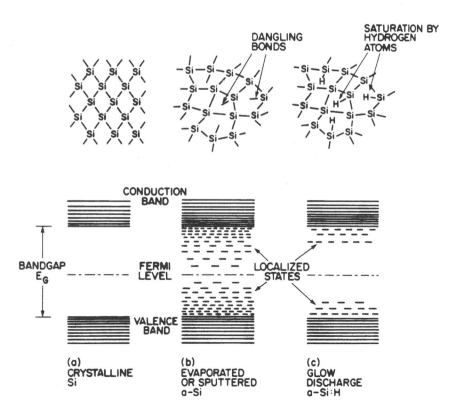

Figure 16.1 Structure and associated energy level diagrams. (a) Crystalline silicon. (b) Unhydrogenated amorphous silicon, a-Si, produced, for example, by thermal evaporation or sputtering. (c) Hydrogenated amorphous silicon, a-Si:H, produced by plasma deposition from the glow discharge of silane. (From Kuwano, 1986.)

from the dissociation of the silane gas (Mott and Davis, 1979; Fuhs, 1986). The degree of such hydrogenation is intimately associated with a number of process varibles, of which the substrate temperature is particularly critical (Spear, 1974). Typical films comprise about 7–10% by weight of hydrogen, and therefore a-Si:H should be considered to be an amorphous silicon–hydrogen alloy (hence the appellation, a-Si:H). Deferring a more detailed description of the fabrication process to the next section, let us now discuss some of the more important physical properties of a-Si:H as they relate to device performance.

The low density of midgap states achievable in a-Si:H deposited at 230°C results in the production of the best electronic grade a-Si:H, since under these conditions the highest ratio of photocurrents to dark currents is obtained (Spear, 1974). It also allows the production of p- and n-type material of varying resistivity by doping to the appropriate concentrations with either diborane or phosphine (Spear and Le Comber, 1975). Figure 16.2 shows the dependence of the room temperature resistivity of a-Si:H on doping. It is worth emphasizing that this should be viewed as a general guide only, since other reactors or fabrication conditions might yield curves that differ somewhat in detail. The doping changes the resistivity because the larger number of ionized acceptor or donor states increases the density of free carriers within the conduction and/or valence bands; in terms of a Fermi level description (the Fermi level is a mathematical parameter whose energetic position relative to the appropriate conduction band is used to compute the number of free carriers in that band), the Fermi level moves closer to the appropriate conduction

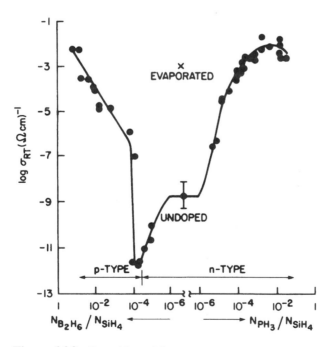

Figure 16.2 Logarithm of the room temperature electrical conductivity σ_{RT} of a-Si:H as a function of doping with diborane (B_2H_6) and phosphine (PH_3). The doping levels are defined by the ratios of gas flows referenced to that of the silane. In the center of the figure, data points are given for a nominally undoped a-Si:H sample and a thermally evaporated sample. (Adapted from Spear and Le Comber, 1975.)

band (Dekker, 1958). Nominally undoped a-Si:H has a relatively low, n-type resistivity at room temperature of about 10^9-10^{10} Ω·cm and is of limited value for practical applications. However, as can be seen in Figure 16.2, light doping (<10 ppm boron acceptors) results in sufficient compensation of the existing donor states to move the Fermi level back close to midgap (Spear, 1974). This corresponds to a situation in which the total number of free carriers is reduced so that a much more useful material with a higher room temperature resistivity of 10^{13} Ω·cm is obtained.

a-Si:H also has light absorption properties significantly different from those of crystalline silicon (Taylor, 1987). In the latter case, the band gap, hence the absorption edge, lie in the near-infrared region at a wavelength of about 1 μm; by contrast, the band gap of a-Si:H, although dependent on the degree of hydrogenation, is about 1.8 eV for electronic grade material, so the absorption edge lies at about 7000 Å or 0.7 μm. The quantum efficiency for the photogeneration of carriers η is essentially unity over the entire visible spectrum from 4000 to 7500 Å (Mort et al., 1980). Equally important, the extinction coefficient or strength of the light absorption is much greater in a-Si:H compared with crystalline silicon. As a result, a 1 μm film of a-Si:H is as effective in absorbing visible light as a 300 μm thick silicon crystal; hence the value of and the ability to utilize very thin films (Kuwano, 1986). Although some fine-tuning of the band gap, hence spectral response, is possible by controlling the degree of hydrogenation, for any significant shift into the infrared region, one must resort to silicon alloys such as those with germanium (Nishikawa et al., 1983).

By analogy with a-Si:H, it is possible to produce plasma-deposited films of a-Ge:H by the glow discharge dissociation of germane (GeH$_4$) (Paul, 1981). Studies of a-Ge:H show that it has a substantially lower room temperature dark resistivity. Although a lower dark resistivity is expected because the band gap in a-Ge:H is only half that of a-Si:H, it is also decreased because of a much higher density of band gap states, which leads to markedly inferior photoelectronic properties. These localized states are believed, by analogy with a-Si:H, to be associated with germanium dangling bonds. Unfortunately, in contrast to the case of a-Si:H, the Ge — H bond is much weaker, and the removal of dangling germanium bonds by hydrogen is much less effective. Some of these problems also carry over into the Si/Ge alloys formed by the glow-discharge of silane–germane gas mixtures. Invariably it is found that even slight additions of germane drastically reduce the photosensitivity of the resultant films as compared with that achievable with a-Si:H. This problem is exacerbated by the relatively weak dependence of the band gap of Si-Ge alloys on composition, which means that a significant shift of the band gap into the infrared requires alloys containing 20 atomic % or more of germanium.

The facility of charge motion is measured by the carrier mobility μ or velocity per unit field; at room temperature, the electron drift mobility in nominally undoped or lightly doped a-Si:H is about 0.5 cm^2/V·s, while for holes the value is about 10^{-3} cm^2/V·s (Spear, 1983). Carrier mobility is a critical parameter for transient applications or where absolute photocurrents are important, as in scanners and sensors. In xerographic photoreceptors, however, the speed with which carriers move is not usually a limitation, whereas how far carriers can move before becoming immobilized in deep traps often is. The parameter that determines this is known as the carrier range and equals μτ, where τ is the lifetime of the carrier before becoming trapped and immobilized in localized states lying near midgap (Dolezaek, 1976). The deep trapping lifetime for electrons in a-Si:H is typically about a microsecond or less, while for holes, values can be as high as milliseconds (Mort et al., 1980; Spear, 1983). As a result, the μτ values, in suitably doped material, for both electrons and holes of about 10^{-6} cm^2/V, are more than adequate for xerography (Mort et al., 1980). The qualification is important, since the electron lifetime and therefore range are drastically reduced in increasingly p-type material and similarly for holes in n-type a-Si:H (Street and Zesch, 1983). The $\mu\tau$ values in a-Ge:H are typically much lower, and the addition of germanium to form alloys with a-Si:H results in a severe degradation of this important parameter (Spear, 1977; Fuhs, 1986).

Just as the Fermi level can be moved by doping, it can be shifted by the application of an electric field, a phenomenon known as the field effect (Madan et al., (1976). Figure 16.3a

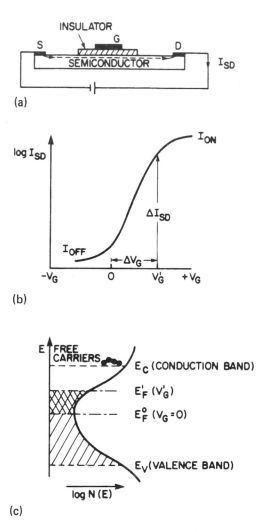

(a)

(b)

(c)

Figure 16.3   Elements of a field-effect device and the basis of the phenomenon. (a) Simple schematic: S and D are the source and drain electrodes, respectively; the current I_{SD} is controlled by the application of a voltage V_G applied to the gate electrode G. (b) Plot of the logarithm of I_{SD} versus V_G; at zero or negative V_G, in this example, the off-current I_{OFF} is extremely low, whereas as V_G becomes positive, I_{SD} rises by orders of magnitude to I_{ON}. (c) Conductivity is enhanced when V_G induces additional free carriers in the conduction band. This is described in terms of the energy shift in the Fermi level E_F; the amount of shift per unit applied field is small for a high density of localized band gap states.

shows a typical structure used for measuring this effect. The semiconductor to be studied is coated with a thin insulator, which in the case of a-Si:H is usually a plasma-deposited silicon nitride (a-SiN$_x$:H) layer produced by the glow discharge of silane–ammonia gas mixtures. Two ohmic contacts, known as the source and the drain, are made to the semiconductor, while a third electrode, the gate, is a metal layer deposited on the outer surface of the insulator. Typically a low constant voltage between source and drain establishes a so-called off-current I_{OFF}. As a voltage V_G, known as the gate voltage, is increased by only a few volts, the source–drain current I_{SD} can change by orders of magnitude (see Fig. 16.3b) to a maximum value known as the on-current, I_{ON}. The degree to which the applied electric field can move the Fermi level from its position at zero gate voltage, E_F^0 and modulate the electrical conductivity is determined by the background density of states in the band gap, which, as seen in Figure 16.3c, is some function of the electron energy $N(E)$. A high density of states means that it takes a lot of charge to fill the empty states lying about the Fermi level, so that for a given applied field the change in material resistivity and related Fermi level shift is small. The important initial measurement and monitoring of the density of states in a-Si:H was accomplished using the field effect (Madan et al., 1976). While a detailed analysis of the I_{SD} versus V_G can be used to map out the density of states in the band gap, it can also be the basis of an electronic three-terminal device known as a thin film transistor, TFT (also sometimes referred to as a field-effect transistor, FET) (Spear, 1981; Matsumura, 1986).

Like its parent crystalline silicon, a-Si:H has outstanding properties in terms of its thermal stability and mechanical hardness. However, there are some distinctions between a-Si:H and crystalline silicon due to the incorporated hydrogen. Thus, although a-Si:H itself is stable against crystallization up to temperatures of about 600°C, hydrogen evolution can occur at temperatures as low as 300°C with a concomitant degradation of photoelectronic properties (Fritsche, 1980). Even with this constraint, however, the thermal stability of a-Si:H is vastly superior to and gives a margin of stability compared with the more thermodynamically unstable chalcogenide and organic materials, which generally have the tendency, albeit under extreme conditions, toward either crystallization or phase separation (Felty, 1967; Borsenberger et al., (1978). However, the greatest advantage of a-Si:H and related materials lie in their hardness, which, although again degraded from the crystalline case by the incorporated hydrogen, manifests itself in vastly enhanced scratch and wear resistance as compared with the much softer chalcogenide and organic photoreceptor materials.

Other advantages claimed include nontoxicity and disposability of the devices (Nishikawa et al., 1983), and chemical stability, although in the latter instance, as discussed in Section 16.5, this initially proved to be a major area of concern for the application of a-Si:H as a photoreceptor.

16.3 METHODS OF FABRICATING AMORPHOUS SILICON DEVICES

The discussion that follows is a simplified treatment of a highly complex and rather poorly understood process. For the present purpose and, as it turns out, practical use too, this is more than adequate. For provided control and reproducibility of a deposition process can be established, one can proceed to use it without necessarily understanding it in detail. It has been empirically found that provided the basic process parameters, such as gas mixtures, flow rates, discharge power density, and substrate temperature, are controlled, the properties of a-Si:H are remarkably reproducible.

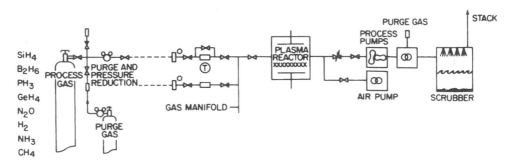

Figure 16.4 Basic features of a plasma deposition system. The pyrophoric nature of silane and the toxicity of diborane and phosphine require considerable care in the design and incorporation of safety features, not only for the operators, but also for the environment.

The basic elements of the plasma deposition process are illustrated in Figure 16.4. An electric discharge, either dc, rf, or microwave, between two electrodes in a reactor chamber is used to excite a precursor gas or mixture of gases. The precursor gas (e.g., silane, alone or in mixtures with hydrogen, diborane, ammonia, etc.) dissociates into radical fragments or species that are far and away predominantly neutral (Schmitt, 1983; Hirose, 1986). It is essential to understand that under normal circumstances the film growth, caused by the deposition of the radical fragments directly on the electrodes or a suitably mounted substrate, is a purely diffusive process driven by the radical density gradient. Therefore, an important contrast to thermal evaporation (of, e.g., the chalcogenides) is the fact that the deposition is not a line-of-sight process. This has the advantage of allowing deposition on a substrate immersed anywhere in the plasma, while it has the disadvantage that uncontrollable deposition on unwanted surfaces leads to an overall reduction in the achievable efficiency with which the precursor gases are utilized. However, other important virtues of plasma deposition are the remarkable range and types of material that can be produced by the simple turning on and off of valves. Thus within a single reactor, under the total control of a computer, semiconductors such as doped or undoped a-Si:H, its alloys with germanium, amorphous silicon oxides, nitrides, carbides, or amorphous carbon, in single layers or precisely controlled multilayers, can be produced at will by the selection of appropriate gases and their mixtures.

16.3.1 Xerographic Photoreceptors

While the basic principles outlined above are generally applicable, significant amendments must be made when specific applications are considered. Xerographic photoreceptor usage, a major emphasis of this chapter, constitutes an excellent illustration of the kinds of new issues that arise with respect to reactor design and scale-up, process flexibility, and the all-important question of economics (Jansen, 1986).

A major difference between this technological application and solar cell or thin film electronic imaging applications is the necessity to use relatively very thick films. This stems from the requirement to produce certain minimum contrast voltages during the imaging step to assure sufficiently large electrostatic attraction of the toner in the developer to produce the required image densities. This immediately raises questions as to whether the excellent photoelectronic properties of a-Si:H, originally found in thin (~1 μm) films, can be maintained in films 20–30 times thicker; the answer is in the affirmative. Moreover, because of such thickness requirements, the deposition rate and the reactor efficiency emerge as major concerns and must be taken into account in determining the economics of manufacturability.

In addition, in large-area electronic devices the individual elements often function merely as threshold switches, whereas solar cells are integrating collectors of light. Consequently, although film quality and uniformity are certainly important in solar cells, they are not vital for these applications. By contrast, xerographic photoreceptors are imaging elements and as such are exceptionally unforgiving, thus the highest film quality and uniformity are essential.

The characteristic drum configuration of photoreceptors, determined by the need for cyclical operation, requires the use of cylindrical plasma reactors that are both materials efficient and capable of producing the necessary film quality despite extended deposition times conditioned by the thickness requirements. Figure 16.5 shows a reactor design that has been developed to fabricate single a-Si:H photoreceptor drums (Jansen, 1986). Details of gas handling and its important safety aspects, power requirements, and discharge characteristics and diagnostics are fully discussed in the literature (Hirose, 1986; Jansen, 1986). Before briefly describing the features and functioning of this reactor, we must comment on the underlying concept and its important relationship to potential scale-up. The approach (Jansen, 1986) is to use a modular deposition cell in which the discharge plasma of the precursor gases in confined within the annulus formed by the drum and the encircling cylindrical counter electrode (Fig. 16.5a). The premise is that, if the discharge process conditions within such a cell can be optimized for material quality and xerographic performance, scale-up can be achieved by an ensemble of such cells connected, in terms of gas flow, in series and parallel. Such an approach eliminates the complications arising from nonuniformity or reproducibility effects associated with potential cross-talk between individual drums without such electrical isolation.

Methods for plasma deposition, described in the scientific and patent literature, can be characterized with respect to the method of electrical excitation and gas flow. Longitudinal gas flow patterns refer to gas flow in the direction of the cylinder axis. For the case in which the reaction gas is introduced perpendicular to the cylinder but exhausted at its end, the gas flow pattern is designated as mixed. In the case of transverse, or cross-flow and mixed flow, the substrate is generally rotated during deposition to ensure radially uniform properties of the deposit. In the reactor design shown in Figure 16.5b, this is accomplished using a combination of a rotary feed-through and slip-ring assembly to maintain the necessary electrical connections. The axial film uniformity depends on the way the gas is introduced, and the slits in the counterelectrode, labeled in Figure 16.5b, yield excellent thickness uniformity in the axial direction. In the case of longitudinal or mixed flow, the uniformity of film properties in the axial direction is a function of the discharge operating parameters, since reactor operation can cause the gas composition at the entrance of the module to be significantly different from that at the exhaust. Uniformity in this geometry therefore requires negligible consumption of silane gas over dimensions comparable to the drum length, and it is only at relatively low deposition rates that such depletion effects are likely to be avoided. For this reason, longitudinal gas flow patterns are considered the least desirable for xerographic drum manufacturing.

The deposition rate increases with flow rate to a maximum value determined by the electrical power density maintaining the plasma, so that the density of condensable radicals and the deposition rate can be increased by raising both the flow rate and the power density

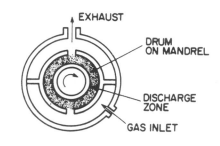

(a)

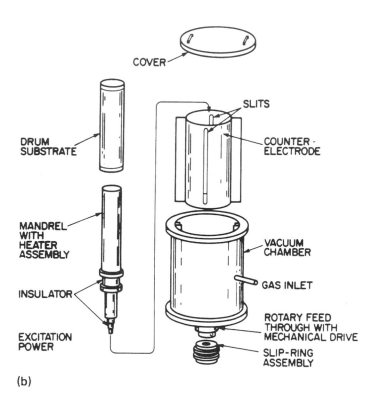

(b)

Figure 16.5   Exploded views of a single-drum reactor of the cross-flow type. (a) Vertical view into the reactor showing the plasma confinement within the annulus between the counterelectrode and the rotating drum substrate mounted on a heated mandrel. The slits in the counterelectrode, which allow the silane flow into the annulus are indicated and are also seen in (b), which is a more detailed view of the various components.

(Chen and Jansen, 1983). However, such an increase in deposition rate has to be considered in the context of the ultimate use of the produced film. For, eventually, the density of condensable species becomes so high that gas phase interactions commence, with associated degradation in the usefulness of the electrical and photoelectronic properties. Hence, there is a limit to the maximum deposition rate that is compatible with the required quality of the films. Maximum values reported for the deposition rate of amorphous silicon of acceptable xerographic quality in a dc discharge of silane are about 30 Å/s (10 μm/hr) at a reactor efficiency of about 20–30% (Jansen, 1986).

The relatively high temperature of deposition, 230°C, as well as the light emission from the plasma, help to minimize the resistivity of the amorphous silicon film. As a result, it is found that buildup of charge on the surface of the film is sufficiently small that a dc discharge can be stably operated under constant current and voltage conditions (Jansen, 1986). However, deposition conditions can affect the properties of plasma-deposited film, and as a result of ion bombardment of the film during growth, dc cathodic films (in which the drum substrate is the cathode) have a higher defect density than, for example, anodically deposited rf films (Morgan et al., 1984). Nonetheless, despite the relatively higher defect density, the resistivity of the dc material can still be maximized by doping and the required carrier range thus achieved.

The width of the band gap in the amorphous silicon, which increases with the hydrogen concentration, is of concern for xerographic applications, since it determines both the spectral response and dark decay of the photoreceptor. The amount of incorporated hydrogen, as previously mentioned, is a function of both the substrate temperature and the electrode on which the film is deposited. Cathodic films, for both rf- and dc-excited plasmas, contain less hydrogen than their anodic equivalents and so have a higher dark discharge rate at the same electric field than anodic films of the same thickness. Consequently, depending on the particular application, the dark conductivity of cathodic films can be undesirably high. In such a case, the band gap of the cathodic amorphous silicon film may be increased by the dilution of the silane gas in hydrogen. However, although the additionally incorporated hydrogen increases the resistivity of the film, the dilution results in a lower deposition rate (Jansen, 1986).

As a result of these various considerations, both high frequency (13.56 MHz) rf and dc excitation power can be used, although, particularly for manufacturing operations, the latter offers the major advantage of inherent simplicity. The applicability of dc glow discharge is, however, limited to the deposition of relatively narrow band gap and/or photoconductive materials such as a-Si:H or a-Si/Ge:H alloys, where significant charging of the film surface does not ultimately lead to the shutdown of the discharge. It cannot normally be used for insulating films such as a-SiO$_x$ or a-SiC$_x$ except that experimentally it has been found that the very thin passivation layers (of, e.g., SiN$_x$) can be produced by the dc method. In any case, even though there is additional complexity, rf discharge has been successfully scaled to manufacturing plant operations (Kuwano, 1986).

Since photoreceptors are light–electrical image transducers, the uniformity requirements of *bulk* film properties over large areas within and between deposition runs are extremely stringent. Mechanical integrity, defect concentration, and quality of their *surface*, particularly in view of the prolonged deposition conditions, are of equal concern. Nodular defects, frequently observed growing in a variety of photoreceptor materials, generally originate on surface irregularities such as dust particles and substrate asperities (Knights, 1980; Jansen, 1981). These surface defects can ultimately cause localized regions of high conductivity, which result in visible white spots (powder-deficient spots: PDSs) in the black background of xerographic images and negatively impact print quality. Conventional photoreceptor materials such as Se and As-Se based alloys are glasses deposited at substrate temperatures higher than their glass transition temperature (Felty, 1967). The viscous flow of such glasses tends to minimize the structural effect of surface irregularities, and relatively few of such defects are formed. Tetrahedral materials, on the other hand, are not glasses, and defect growth potentially poses more problems. Stringent clean-room conditions for substrate preparation and photoreceptor fabrication are essential for minimizing the defect density of the film.

16.3.2 Large-Area Electronic Devices

It is not the intent to give an in-depth exposition of large-area electronic device fabrication because the details are diverse, depending on the particular application. Rather, only the general principles are enumerated, with emphasis on the differences from those required for drum fabrication, for there are a number of distinctions. First, most of the electronic devices, whether they be photovoltaic cells, photosensors, or thin film transistors, are thin film devices with total device thicknesses of ~1 μm or less. Second, for imaging applications they are comprised of large-area, one- or two-dimensional, high density arrays of discrete devices, and therefore patterning processes are an inherent part of their fabrication. For purposes of illustration, the focus will be on TFTs.

Cross-sectional and vertical views of a TFT are shown in Figure 16.6. This is an example of the so-called inverted type in which the gate electrode and insulator a-SiN$_x$

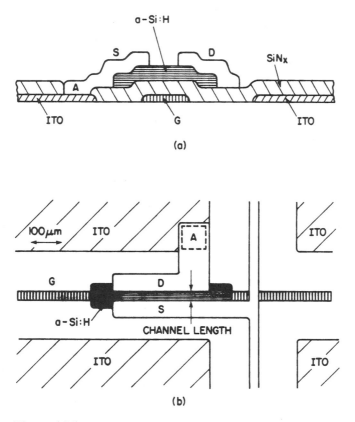

Figure 16.6 (a) Cross-sectional and (b) vertical views of an inverted-type a-Si:H thin film transistor (TFT). The name derives from the fact that the SiN$_x$ insulator, produced by the glow discharge of silane and ammonia mixtures, is deposited before the a-Si:H layer. In (b) the distance between the source and drain electrodes is termed the channel length. (Adapted from Snell et al., 1981.)

are deposited first, followed by the semiconductor a-Si:H, and source and drain electrodes last. For this configuration, it has been empirically determined that the field effect, and specifically the difference between I_{OFF} and I_{ON} for a given gate voltage, is larger than when the semiconductor is deposited before the insulator. Although this must be due to a difference in the densities of interface states between the two geometries, no detailed and fully satisfactory explanation has yet been proposed. It is also known that plasma-deposited silicon nitride (a-SiN$_x$) is superior to plasma-deposited silicon oxide (a-SiO$_x$) as the gate insulator, and that both the a-SiN$_x$ and Si:H films must be deposited within one vacuum pumpdown.

It has also been found that an n$^+$-doped (i.e., a heavily phosphorus doped) a-Si:H layer can be used for ohmic contacts (Madan et al., 1976; Matsumura, 1986). After the glass substrate has been cleaned, the gate metal is evaporated and patterned using photolithographic techniques. Then, the SiN$_x$, a-Si:H, and n$^+$-doped a-Si:H layers are successively deposited within one pumpdown, although to avoid contamination and cross-talk between the different gas mixtures, this involves using different chambers, accessed by vacuum load-locks. Typical thicknesses are 0.1–0.5, 0.2–0.6, and 0.1 μm, respectively. After the n$^+$ layer, just above the gate pattern, has been etched away, the a-Si:H and n$^+$ contact layers outside the transistor portion are etched off. Metal films (typically Al or NiCr) are then evaporated and patterned to form the source and drain electrodes and the interconnection line. Sometimes the n$^+$-doped a-Si:H deposition is omitted, and Al or NiCr evaporated directly on the a-Si:H layer; this simplifies the fabrication process but can result in a lower maximum I_{ON}. Since all the fabrication steps are carried out at a temperature of about 230°C, inexpensive glass substrates can be used. A typical channel length, the distance between S and D in Figure 16.6b, is several micrometers, although, by using even finer photoetching techniques, like those developed for very large scale integration (VLSI), shorter channels are possible.

16.4 APPLICATIONS IN ELECTRONIC IMAGING

In addition to its use as a photoreceptor in xerographic copiers, printers, and digital laser copies, a-Si:H is beginning to find applications in all aspects of electronic imaging. These range from the production of temporal images on liquid crystal video displays to image scanners and print bars, all of which together provide a bridge between the two worlds of electronic and paper documents. Now, and for the foreseeable future, both types of document will coexist, therefore it is important to ensure that interconversion can take place. Thus optical-to-electronic transducers (i.e., scanners) can enable document processing in which existing hard documents can be converted into digital electronic signals for modification using a workstation, followed by xerographic or electrographic printing of the revised document (Mort, 1989).

16.4.1 Xerographic Photoreceptors

A detailed discussion of the performance of a-Si:H as a photoreceptor will be found in Section 16.5, but for completeness some discussion is given here with respect to its value in the area of electronic imaging. Once a material has met the minimum specifications as a photoreceptor in terms of its photoelectronic properties, the image is largely independent of the particular photoreceptor. As a result, a-Si:H photoreceptors possess no intrinsic property that leads to single images of any greater quality than those produced by either chalcogenides or organic photoreceptors. However, differences can arise between photoreceptors in repetitive imaging, where image degradation can occur for any number of reasons. Thus with repeated cycling, the photoelectronic properties of a photoconductor can degrade or the dark decay increase. Surface damage or filming of a photoreceptor can occur because of possible severe mechanical interactions with toner particles or carrier beads. It is in this context that a-Si:H is felt to have an edge. As one example, because of its superior mechanical properties, it is possible to employ more stringent cleaning subsystems such as steel blade cleaning, and overall stability is such that a-Si:H photoreceptors are reported to consistently have useful lives of 750,000 or more images (Fritzsche, 1984). On the other hand, these advantages must be weighed against the costs of manufacturing as compared with the chalcogenide and organic options. Some perspectives and discussion of these issues are deferred to the final section.

16.4.2   Image Sensors

The p-i-n type of diode has been investigated extensively in commercial solar cells (Kuwano, 1986). In such an application a number of parameters (e.g., switching speed) are not particularly relevant. This is not the case when a-Si:H electronic devices are to be used in imaging. Here switching speed is critically important, as are the maximum achievable forward current density and rectification ratios. Forward current densities in suitably designed a-Si:H devices can exceed 10 A/cm^2 for forward voltages of only 2 V, whereas the reverse current can be minimal because of rectification ratios of more than 10^{10} (Nara et al., 1983). Schottky barrier diodes can also maintain these rectification characteristics at frequencies of 10 MHz. The combination of these capabilities with the simplicity in structure and fabrication of a-Si:H diodes and TFTs makes them attractive circuit elements for application in photodiode arrays for image sensors and other imaging technologies (Matsumura, 1986).

   An equivalent circuit of a TFT-addressed image sensor is shown in Figure 16.7. One picture element (pixel) is composed of an a-Si:H TFT, a storage capacitor, and a photosensitive element, such as an a-Si:H photodiode or an a-Si:H photoconductor. When the TFT is in the on-state (forward bias), the storage capacitor is discharged, and the photosensitive element is biased to V_{DD}, whereas with the TFT in the off state, the photocurrent flowing through the photosensitive element, on which light is falling, gradually charges up the capacitor. If the photocurrent in the photosensitive element is linear in light intensity over the range of exposures of interest, the incremental charge stored by the capacitor is proportional to the total amount of light that illuminates that particular pixel. When the TFT turns on again, the current pulse flowing through the TFT to restore the capacitor to its initial state can be detected by external circuits. By applying a sequence of pulsed voltages to the TFT array, the optical image is transformed into a serial output of current pulses, which by careful circuit design can be readout in times as short as 10 μs.

   An a-Si:H photodiode array having more than 1728 bits with 200 bits/in. has been evaluated. Such an array can operate at more than 1 MHz and can reproduce images of high quality (Suda et al., 1987). However, since each of the diode terminals is driven independently by specially designed integrated circuitry, a serious problem can arise because of the larger number of interconnects. Back-to-back connected diodes have been developed to alleviate this problem so that the number of external terminals can be reduced to less than 100 for a 720 bit linear array.

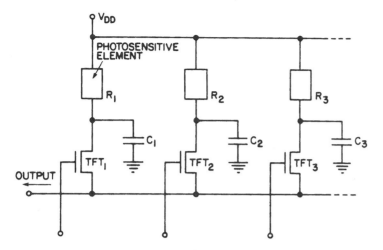

Figure 16.7 Equivalent circuit of an array of a-Si:H TFT-addressed photosensors. The photosensitive elements, which are either photoconductors with associated storage capacitors or photodiodes, are also made with a-Si:H. The magnitude of the output signals from each element is related to the light intensity that fell on the pixel and is captured as a sequence of serial output current pulses. (From Matsumura, 1986.)

16.4.3 Linear Print Bars

The a-Si:H TFT switching technology can be used in at least two ways to produce a page-wide, linear, high density array of pixels that acts as a print bar. The first involves the switching on and off of liquid crystal shutters, positioned in front of a light strip, so as to write with photons in an otherwise conventional way on a photoreceptor. The second, by contrast, involves the technology of ionography, where the electrostatic latent image is produced by the controlled, localized projection of an ion stream onto an insulating, nonphotosensitive dielectric receptor. In ionography, which actually predated xerography, the trajectory of corona-produced ions is controlled by voltages applied to what is in effect the grid of a triode tube (Fig. 16.8a). In the new technology, the voltages on a page-wide, bit/in. array of electrodes (Fig. 16.8b) are themselves controlled by an a-Si:H TFT matrix fabricated as an integral part of the write bar (Chuang et al., 1987). If a-Si:H image sensors are cofabricated on the same write bar, a multifunction scanner/printer capability can also be realized. It is even possible to use plasma-deposited materials such as a-SiC$_x$ (produced by the glow discharge of silane and a suitable hydrocarbon gas) as the dielectric receiver, since, for a high stoichiometry (large x), a highly insulating, photoinsensitive material with a low dielectric constant and outstanding hardness and wear properties can be produced (Kuhman et al., 1990).

16.4.4 Display Applications

Much research is directed toward the application of a-Si:H TFTs to panel displays, and pocket television receivers are already a reality (Snell et al., 1981; Matsumura, 1986). A schematic view of an a-Si:H TFT-addressed liquid crystal display scheme is shown in Figure 16.9 (Matsumura, 1986). The liquid crystal is in the gap between glass substrates,

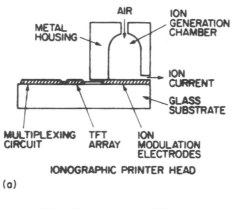

IONOGRAPHIC PRINTER HEAD

(a)

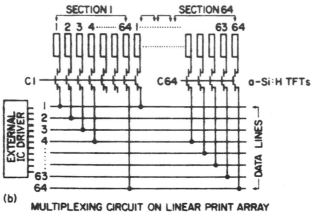

(b)
MULTIPLEXING CIRCUIT ON LINEAR PRINT ARRAY

Figure 16.8   Elements of a page-width linear image or write bar using the principle of ionography. (a) The flow of ions from a pixel is controlled by the application of voltage, using an a-Si:H TFT as the switch, to the modulation electrode. The emerging ions charge the surface of a highly insulating dielectric receiver or electroreceptor patternwise; this electrostatic image is then developed as in conventional xerography. (b) An array of several thousand such pixels can be addressed using a multiplexing scheme of this type. (From Chuang et al., 1987.)

one of which has a transparent electrode. On the other glass substrate, a-Si:H TFTs are fabricated in a two-dimensional array. The source electrodes of the TFTs are connected to each transparent electrode, while the gate electrodes, aligned in the same column, are connected together by a word line, and drain electrodes, aligned in the same row, are connected together by a bit line. The word and bit lines are driven by external logic circuits. It is only when one of the word lines is in a high state that parallel video signals are applied to bit lines simultaneously and flow through the TFTs to the transparent electrodes. The resultant charge is stored at capacitors, and when the word line falls to a low state, the stored charge remains at the pixel and continues to influence the optoelectrical properties of liquid crystal in the pixel. When the word line is in a low state, another word line takes a high state successfully, and all pixels are refreshed. A light shield above the gate electrode suppresses any light-enhanced leakage currents.

For satisfactory operation, the TFTs must have a sufficiently low off current for charges to remain at the pixel for a long time while the word line is in a low state. In

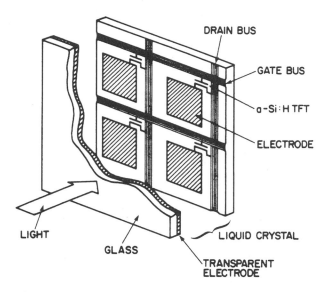

DRAIN BUS

GATE BUS

a-Si:H TFT

ELECTRODE

LIGHT

GLASS

LIQUID CRYSTAL

TRANSPARENT
ELECTRODE

Figure 16.9   Exploded schematic showing four pixels of an a-Si:H TFT-addressed liquid crystal display. (From Matsumura, 1986.)

addition, the TFT must have a sufficiently high on-current for charges to charge up the pixel in a short time while the word line is in a high state. The capacitance of a 1 mm $\times$ 1 mm pixel is about 10 pF, the duration of the high gate voltage state is about 100 μs, and the frame frequency is about 60 Hz. The maximum and minimum resistivities of the a-Si:H used for the TFTs should be more than 3×10^9 Ω·cm and less than 9×10^6 Ω·cm, respectively (Le Comber et al., 1979); these conditions can be easily satisfied by appropriately produced a-Si:H. Figure 16.10 shows the characteristics of state-of-the-art TFTs that easily meet these criteria.

Image quality obtained by using a 240 $\times$ 200 pixel flat-panel display with twisted nematic liquid crystal is limited by defects caused by dust during fabrication (Powell et al., 1988). Since such a panel has approximately the same number of transistors as a 64-kilobyte metal-oxide semiconductor dynamic random-access memory (MOS DRAM), clearly device fabrication in appropriate clean-room environments similar to those used for VLSI fabrication will be necessary.

16.5 XEROGRAPHIC PERFORMANCE

Before discussing specific results for a-Si:H photoreceptors, let us briefly review the basic photoelectronic parameters used to characterize photoreceptors. Since these parameters characterize only the electrical behavior of the photoreceptor material, they are necessary but insufficient. However, they can be determined on samples of rather small surface area without requiring fabrication of the large-area devices ultimately necessary for detailed print evaluation.

16.5.1 Photoelectrical Parameters

Figure 16.11 shows schematically how the photoelectronic parameters important for a xerographic photoreceptor are determined. The substrate, typically 2 in.^{2}, on which the a-Si:H

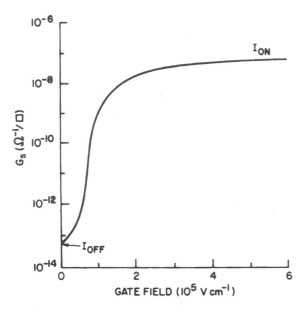

Figure 16.10   Representative data for the channel conductivity as a function of the electric field at the SiN_x–a-Si:H interface. A change in this field by only a factor of 2, corresponding to only a few volts, causes about 6 orders of magnitude change in the channel current. (Adapted from Snell et al., 1981.)

of thickness L has been deposited is mechanically moved under a corotron to initially charge the device and then to an exposure station. In this position, the sample becomes stationary, and its surface potential is measured using a calibrated transparent capacitive probe. By choice of an appropriate interval between stopping and the onset of illumination, the dark decay of the sample can be measured (Fig. 16.11b). When a shutter is opened, the device is exposed to light of variable wavelength and calibrated intensity, which causes the surface potential to decay as indicated. If the surface potential does not completely decay (dashed line), this means that $\mu\tau E$, the carrier *Schubweg* (average distance moved), is comparable to or less than the sample thickness. The initial discharge rate $(dV/dt)_{t=0}$, assuming no range limitation, equals $eF\eta(E)/C$, where F is the number of photons absorbed per unit area, C is the capacitance per unit area, and η is the efficiency of carrier generation and supply and is, in general, field dependent. This photodischarge curve is called the PIDC (photoinduced discharge curve), and it is possible to derive from it the change in surface potential as a function of exposure (Mort and Chen, 1976). This is known as the contrast potential curve.

   If the photoreceptor receives too little or too much exposure, the contrast potential is low. As a result, a maximum contrast potential V (Fig. 16.11c), can be achieved at an optimum exposure X_0 ($X_0 = Io_0t$). Operationally, an additional important parameter is the surface potential after this optimum exposure, since this defines the so-called background potential V_{BG}. To prevent the background potential from resulting in development of the exposed areas, which correspond to the white background of the document, the development system must be biased close to this background potential. Parenthetically, it should be pointed out that this relatively simple measurement is the most unambiguous technique for studying photogeneration and/or recombination processes in a depthwise, nonuni-

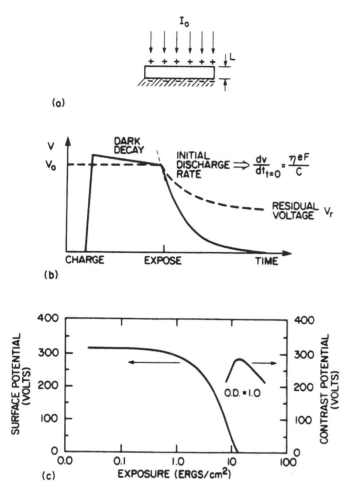

Figure 16.11 General principles for determining the critical electrical parameters of a xerographic photoreceptor. (a) The photoreceptor sample of thickness L is capacitively corona charged to an initial surface potential V_0 and after an interval exposed to light of intensity I_0. (b) The interval between charging and exposure allows the dark decay rate to be measured; the initial discharge rate $(dV/dt)_{t=0}$ is related to the light intensity I_0 and the efficiency with which photons are converted into charge carriers; the photogeneration efficiency is η. If the carrier range is too short, incomplete discharge occurs, leaving a residual voltage V_r. (c) Photoinduced discharge curve (PIDC) for an a-Si:H sample together with the contrast potential for an optical input density of unity; actual voltages achievable with a particular device depend on the charging efficiency of the corotron device and photoreceptor thickness.

formly illuminated photoconductor, since complications from secondary photocurrents associated with electrodes are essentially eliminated. Under certain conditions, it is possible to identify the discharge rate as being totally controlled by the photogeneration process itself.

For the purposes of establishing the basic xerographic parameters of a-Si:H, planar films are deposited on 2 in.² aluminum substrates. Typically, the films are 20 μm thick and are slightly boron doped to maximize the bulk resistivity. The boron doping level is about 10 ppm by weight. Figure 16.11 shows a typical PIDC for a planar a-Si:H photoreceptor (Mort et al., 1980). For films thinner than 20 μm, the charge acceptance is limited by either substrate or surface injection. The former can be significantly reduced by the interposition of a blocking layer between the substrate and the bulk film. This can be either an insulating film such as plasma-deposited SiN_x or SiO_x or a heavily boron-doped p-type layer (200 ppm B, 0.3 μm thick), if the device is to be charged with positive corona. For thicknesses in excess of 20 μm, the bulk conductivity of the a-Si:H becomes increasingly important and ultimately limiting. Charge acceptances of 20–30 V/μm represent maximum values that can be achieved, although it has been reported that in oxygen-doped material this can increase to about 50–70 V/μm (Yamamoto et al., 1981).

For the type of sample shown in Figure 16.11 the dark decay is acceptably low but does increase superlinearly with field. The data show essentially total light discharge with no residual, which in this case, since the corona polarity is positive, indicates that the *Schubweg* of holes exceeds 30 μm. In fact, values of the range, $\mu\tau \sim 10^{-6}$ cm²/V have been reported. It was such xerographic studies that first indicated that the hole range in a-Si:H was much higher than first adduced from studies on thin (i.e., <2 μm) films. Indeed, it is now known that the value for holes can be as large as that for electrons.

The optimum exposure is found to be about 7 ergs/cm² and the ratio of the maximum contrast potential to the initial surface potential is about 0.85. This ratio (assuming negligible dark decay) is related to the field dependence of the carrier generation/supply process, and such a high ratio is indicative of a relatively weak field dependence of these processes in a-Si:H (Mort and Chen, 1976). By making these PIDC measurements as a function of wavelength, one can map out the spectral sensitivity of a-Si:H. This is shown in Figure 16.12 as the reciprocal of the optimum exposure $(X_0)^{-1}$ plotted versus wavelength. The material, as was first established by photoconductivity studies, is sensitive from 4000–7000 Å (i.e., the entire visible spectrum). At fields in excess of 2×10^5 V/cm, the generation/supply of photocarriers occurs with essentially unit quantum efficiency.

As compared with other photoreceptor materials, a-Si:H has two minor disadvantages, namely its relatively high surface reflectivity and dielectric constant. Because of these properties, therefore, somewhat large optimum light exposures are required for the same field. By and large, however, the basic xerographic parameters of a-Si:H as they relate to charge acceptance, dark decay, and sensitometry are excellent and more than sufficient to qualify amorphous silicon as a significant new development in photoreceptor materials. Very early in the exploratory investigations, however, severe and somewhat unanticipated difficulties were encountered. These relate to the sensitivity of a-Si:H to the combined effects of corotron species and the humidity of the ambient (Mort and Jansen, 1986). The manifestation of these difficulties is illustrated in Figure 16.13a, which shows the results of monitoring the surface potential as a function of repeated charging and photodischarging. Under typical ambient humidity (40%), the width of the trace gradually increases, reflecting the growth of the electrical "noise" associated with highly localized surface breakdown. The observation that this "noise" could be essentially eliminated within a few minutes by replacing ambient with dry air suggested that the effects were caused by physically absorbed species rather than by chemical changes.

To eliminate these effects, it is necessary to encapsulate the bulk a-Si:H film by using a relatively thin (<0.5 μm) film of an insulator such as $a-Si:N_x$, $a-SiC_x$, $a-SiO_x$, or $a-C_x$. The choice of material, apart from its sensitivity to corona ambients, is determined by the requirement to maintain advantageous material properties such as hardness and disposability. Furthermore, the layer should be producible by the glow discharge process

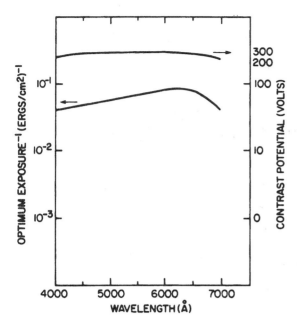

Figure 16.12 Spectral sensitivity of an a-Si:H photoreceptor. The spectral sensitivity is expressed as the reciprocal of the optimum exposure required to give the maximum contrast potential for an input density of unity. The constancy of the contrast potential with wavelength reflects the wavelength independence of the PIDC, an important fact for color copying.

so that the entire device can be fabricated in one process and a single pumpdown (Mort and Jansen, 1986). Figure 16.13b shows the result of overlaying the bulk material with a-SiN$_x$ film produced by the glow discharge of a silane–ammonia mixture. The conductivity of the SiN$_x$ layer is dependent on its stoichiometry, which is determined by the specific composition of the gas mixture employed. The value for x should be small enough to eliminate residual voltage buildup across the SiN$_x$ but high enough to avoid lateral conduction, which reduces surface potential differences induced by the nonuniform light exposure. With this combination of materials and structure, a device is producible that combines the excellent photoelectronic properties of amorphous silicon with the mechanical properties of an insulator such as SiN$_x$.

16.5.2 Print Testing

Thus an a-Si:H photoconductor coated with a passivating SiN$_x$ layer possesses excellent charge acceptance, photosensitivity, and charge transport properties, as well as excellent cyclic performance, which are the basic minimum parameters for it to function as a xerographic photoreceptor. However, the ultimate test of a photoreceptor is its ability to produce high quality images on paper, and, meeting the basic xerographic parameters just described does not in and of itself guarantee that a material will make a viable photoreceptor. The ultimate figure of merit of a photoreceptor is that it be capable of producing marks on paper to form an image of acceptable quality, and do so in a repetitive manner over many cycles.

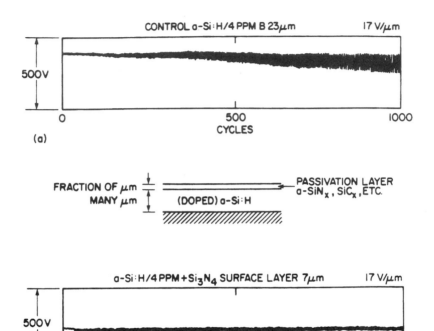

Figure 16.13 By itself, a-Si:H is sensitive to corona/humidity as evidenced by the growth of noise, due to incipient surface breakdown, with repeated cycling (a). The addition of a thin surface passivation layer of plasma-deposited SiN$_x$, (b) a-SiC$_x$ can also be used, which essentially eliminates such noise but can cause problems in print or copy quality.

It is, in principle, possible for a photoreceptor to exhibit excellent basic electrical xerographic behavior in terms of the parameters described in Section 16.5.1 and yet not be capable of producing a legible image. A trivial example would be a photoreceptor whose surface was coated with a metal film sufficiently thin to allow light transmission. Neglecting any contribution of dark injection of carriers from the metal into the photoreceptor, this would exhibit xerographic parameters identical to a noncoated device. Despite this, it would be impossible to develop any contrast potentials because of the equipotential surface produced by the metallic film. A further critical step therefore requires the production, evaluation, and, if necessary, improvement of prints. In addition, the production of images also gives the experimenter the use of a very high resolution electrostatic probe, namely, a toner particle.

While it is possible to produce such images in a single-shot mode (i.e., by using a single xerographic cycle using a flat-plate photoreceptor), evaluation in a cyclic mode gives much additional valuable information pertinent to the full commercialization of the material. Included in this information are data on unanticipated interactions with corotron species and developer materials, light fatigue, and mechanical abrasion or impact associated with critical subsystems in a copier machine. In the case of a-Si:H, the lesson that electrical performance is necessary but not sufficient is well exemplified. In particular, it is found that image degradation with a rather unusual origin is encountered, although it is undetectable in basic electrical evaluation. As noted earlier, a-Si:H is unusual in the sense that the Fermi level position in the gap can be changed by doping or by the application of an electric field. As just discussed, interactions of a-Si:H with corotron/humidity species requires the use of a passivation layer such as SiN_x or SiC_x, but although the electrical performance of such a device is more than adequate, other problems manifest themselves only in print testing.

Print testing is most conveniently and meaningfully done in a xerographic engine in a cyclic fashion. This requires the production of drum photoreceptors in which devices of the type evaluated in 2 in. squares are, instead, deposited on drum substrates. Figure 16.14 (top) shows xerographic prints made on a modified Xerox 3100 copier using positive corona charge (Mort et al., 1984). The aluminum drum substrate had a photoreceptor deposited by the glow discharge process over half of its dimensions, with the structure as indicated. The highly insulating a-SiN_x layer was 0.3 μm thick and the a-Si:H (4 ppm boron) bulk layer was 10 μm thick (Fig. 14a, bottom). Over this part of the drum it is

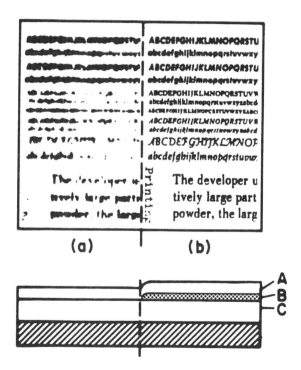

Figure 16.14 Prints made with a single drum coated with two different photoreceptor structures. (a) The structure (bottom) that produced this badly blurred image consisted of 10 μm of 4 ppm boron-doped a-Si:H, layer C, coated with 0.03 μm of insulating a-SiN_x, layer A. (b) The photoreceptor structure (bottom) that yielded the sharp image was identical to that in (a) except for the interposition of a 0.03 μm electron trapping layer of 100 ppm boron-doped a-Si:H, layer B.

seen that the image of the document is severely blurred even on the first copy cycle. Succeeding cycles result in increasing blurring, loss of image density, and ultimately, total deletion of the image. Figure 16.14b (top) shows the unblurred image that was obtained with a different photoreceptor structure, the development of which is now discussed.

Figure 16.15 suggests a mechanism by which image blurring could arise. In the light-exposed region, excess photogenerated electrons will remain in the silicon–silicon nitride interface region after the photogenerated holes have drifted across the bulk a-Si:H layer in the field associated with the positive corona. This exposure pattern leads to electrostatic fringe fields at the light/dark edges with a significant horizontal component. Under the influence of this horizontal field (Fig. 16.15), the excess electrons can spread into the unexposed regions and produce a redistribution of net surface charge as indicated. This produces in the light-exposed regions finite fields which, if of sufficient strength, develop out with toner. This partial development in the light-exposed region results in a visual spreading of what otherwise would have been a sharply defined demarcation. To explain the visually observed blurring shown in Figure 16.14a, however, the electron *Schubweg* would have to be more than 0.1 mm. This is inconsistent with the showing by independent studies that the electron range $\mu\tau E$ in a-Si:H doped with 4 ppm of boron is extremely short.

This inconsistency can be resolved by considering the a-Si:H photoreceptor coated with the insulating SiN_x layer to be essentially a giant TFT. As a result, the corona-produced electric field impressed on the silicon nitride coated a-Si:H produces a field effect in which the Fermi level is moved through the distribution of gap states toward the conduction band. This changes the occupancy of localized states, removes previously active electron traps, and leads to an increased lifetime, hence the required increase in $\mu\tau$. However, it is known that boron doping can dramatically reduce the electron lifetime. This suggests that a boron-doped trapping layer interposed between the top layer of the bulk a-Si:H and the $a\text{-}Si\text{:}N_x$ overcoat could remove the field-effect-induced blurring by inhibiting the lateral electron motion. The bottom part of Figure 16.14b shows such a structure where the trapping layer is a-Si:H doped with 100 ppm boron and is 0.03 μm thick. The bulk a-Si:H film in this device is part of the same layer used for the test in Figure 16.14a. As can be seen, in the presence of the additional electron trapping layer, the blurring is dramatically eliminated, giving support to the interpretation previously discussed (Mort et al., 1984).

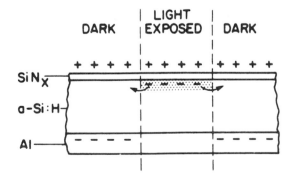

Figure 16.15 Schematic representation of the redistribution of surface and photogenerated charge at light/dark edges within the photoreceptor produced by lateral motion of electrons after the imaging step. This is caused by an enhanced electron lifetime near the surface due to filling of electron traps induced by the field effect.

These results show that the specific features of an extrinsic semiconductor relating to motion of the Fermi level with doping and field are observable in the xerographic process. The field effect leads to deleterious effects, whereas doping can be used to rectify the situation. The thickness of the trapping layer need be comparable only to that of the layer in which the conductivity is enhanced by the field effect, the so-called accumulation layer, which is only about 300 Å. Hence, the relatively high dark conductivity of the a-Si:H doped with 100 ppm boron has a negligible effect on the dark discharge characteristics of the photoreceptor. These fortuitous circumstances stem from the fact that xerographic photoreceptors, like field-effect transistors, are majority carrier devices, since the xerographic discharge for strongly absorbed light is controlled by only one carrier; in contrast, photovoltaic cells are minority carrier devices.

Alternative solutions to these problems associated with field-induced band bending have been suggested (Mort et al., 1984). Compensated a-Si:H up to doping levels of several hundred parts per million by weight (with a boron-to-phosphorus ratio 1.0) has been found to function xerographically, since even in this material adequate charge acceptance and hole range are maintained. Such compensated material has the additional virtue that, together with these high levels of donors/acceptors, deep-lying gap states are introduced. These pin the Fermi level, quench the field effect, and thereby eliminate the blurring/deletion forms of image degradation. Another approach is to modify the conductivity of the encapsulation layer by controlling the stoichiometry or by doping. The conductivity chosen must be such that dielectric relaxation allows the field across the overcoating to relax in times short compared to any process speed and yet to be insufficient to permit lateral conduction to cause image deletion.

In summary, it seems that the usefulness of a-Si:H by itself as a xerographic photoreceptor is limited by the reactivity of the material to corona species, particularly under high humidity conditions. This can be minimized by overcoating with passivating insulating layers such as a-SiN$_x$, a-SiC$_x$, or a-C:H$_x$. Such overcoating, in turn, can lead to field-effect phenomena and image degradation, which can, however, be eliminated by proper materials engineering. Fortunately the overcoatings employed are as hard as or harder than a-Si:H itself, thus preserving or even enhancing the prime advantage of this type of photoreceptor.

16.5.3 Manufacturing Issues

An area of concern often expressed with respect to a-Si:H photoreceptors is that of deposition rate and its relation to the question of the unit manufacturing cost (UMC) of the device. The accuracy with which one can arrive at a UMC for an embryonic technology, particularly in the absence of the specific production volumes to be ultimately used, is highly questionable. As a consequence, the discussion that follows is intended only to explore the factors that must be considered to arrive at such an estimate, and the specific numbers employed should be viewed in this context.

Figure 16.16 indicates the various factors that must be considered in the manufacturing process of a-Si:H photoreceptors, namely, materials, process, facilities, and labor. Beginning with a silane gas cost of 25¢ per gram and assuming that 30% of the initial precursor gas ends up on the drum substrate, the cost of, say, 30 μm of a-Si:H on the drum is roughly $10, which together with a cost of $15 for an aluminum drum substrate leads to a total material cost of around $25. Next the deposition process must be considered. Here we assume that the number of drums produced per reactor run, or reactor size,

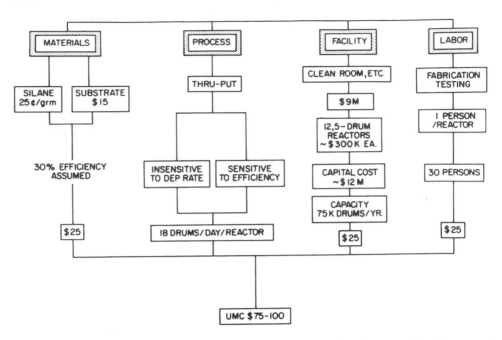

Figure 16.16   Some of the factors involved in estimating manufacturing costs. The figures used are only order-of-magnitude estimates, since specificity requires detailed reactor, facility design, and product volume scenarios.

is 5, and we estimate the throughput or the number of drums that can be manufactured per reactor per day. This requires an estimate of the manufacturing time per unit and involves the question of deposition rate.

A full fabrication cycle of a drum involves the loading of a drum, the evacuation of air from the reactor chamber, the heating of the drum substrate to the required 230°C, the deposition of a-Si:H, and the cooling of the substrate before removal. A reasonable estimate of this time is perhaps 2 hours plus the deposition time. If a deposition rate of 10 μm/h is used and a photoreceptor thickness of 30 μm is required, the total fabrication cycle would be 5 hours. Therefore with round-the-clock operation, about 20 drums per reactor per day could be produced. The interesting question is, How sensitive is this throughput to deposition rate? Using the approach just described, this is easy to estimate, and Figure 16.17 illustrates the intuitive result that because of the loading/heating/cooling/unloading interval requirements, the throughput becomes essentially insensitive for deposition rates in excess of about 10 μm/h.

The questions of facilities and labor costs are so specific that it is of little value to attempt any detailed estimate except to say that the manufacturing process is inherently not labor intensive, since it is particularly suited to automated computer control. However, making some reasonable estimates for the capital investment and ongoing labor costs and assuming a production volume of 100,000 drums per year for 10 years, one arrives at a total UMC per drum in the range of $75–$100. Although actual UMC costs of commercial photoreceptors are proprietary information, it is quite clear that the UMC of an a-Si:H

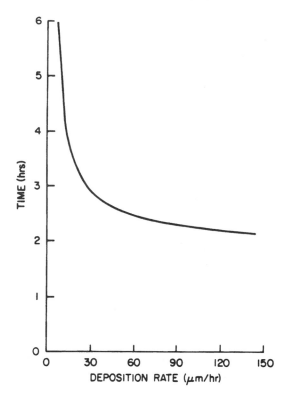

Figure 16.17   Plot of time for one fabrication cycle versus deposition rate for a-Si:H. The rapid decrease in fabrication time ceases beyond deposition rates of about 10 μm/h.

photoreceptor is significantly higher than is the case with the chalcogenides, and higher still when compared with organics.

16.6   FUTURE DEVELOPMENTS

Plasma-deposited amorphous silicon, and the plasma-deposition process itself, have come a long way in the past decade. Certainly they have already crossed the often insurmountable barrier between beguiling technology and the harsh realities of the commercial marketplace (Mort, 1989). This is not to say that problems do not remain to be solved, as undoubtedly they do, but new opportunities and ideas may also await discovery. This final section discusses some aspects of this continued exploration for improvement in the understanding of the materials, processes, and products.

Although extended life is the prime advantage anticipated for a-Si:H xerographic photoreceptors, photoreceptor life is an ill-defined term unless described within the context of a specific machine design with its related subsystems and intended application. As an example, consider the case of high speed copiers where, in one engineering approach, a flexible belt photoreceptor employs full document exposure using a light pulse. In this case, no matter what the advantages of amorphous silicon based materials, the high internal stress characteristics of the materials are incompatible with their production in the form of a flexible device. For this particular application, therefore, the advantages of a-Si:H, however real, are unrealizable.

Therefore, a-Si:H photoreceptors are constrained to situations in which drum configurations are architecturally acceptable and perhaps desirable in that mechanical integrity and stability are inherently maximized. Such requirements are particularly critical in high speed, high volume printers, including digital laser copiers.

As indicated in the preceding section, although such costs are difficult to quantify with a high degree of accuracy, there is no question that the UMCs of a-Si:H photoreceptors are significantly higher than those of organic photoreceptors. This leads to the conclusion that such a premium precludes the use of a-Si:H photoreceptors in lower volume office products, where current photoreceptors are universally viewed as replaceable subunits. In the marketplace, suppliers clearly have to optimize customer satisfaction and profits. This requires photoreceptors of a sufficiently long life to satisfy both customers and suppliers with respect to reliability; quite apart from the suppliers' interest in customer satisfaction, unreliability also results in expensive service costs, which adversely impact profits.

The UMC argument, however, takes on a different complexion if the photoreceptor is viewed as a life-of-the-machine part. This would have potential advantages for both customer and supplier. From the customer's perspective, the critical parameter, apart from an increase in reliability, would be the replacement cost of a photoreceptor measured in cost per thousand copies; this decreases linearly with photoreceptor life. On the other hand, the supplier can envisage substantial reduction in service calls and significant changes in product architecture, which could increase further overall product reliability. Photoreceptors based on a-Si:H and related materials offer some prospect of this development.

For printer applications, it is often desirable from both cost and design standpoints to employ solid state lasers. Since these commonly have output in the infrared region, there has been continued motivation to produce photoreceptors with extended spectral response into the near-infrared. In the case of a-Si:H, this is most likely achieved by utilizing an infrared sensitizing layer contiguous with a bulk transport layer, in analogy with most current organic-based photoreceptors (see Chapter 10). Several examples of this type of dual-layer photoreceptor have been reported (Nakayama et al., 1983; Nishikawa et al., 1983; Carasco, 1985).

One rather obvious, although far from trivial, approach to achieving an infrared sensitizing layer is the use of a-Si:H/Ge:H alloys produced by the glow discharge of silane–germane mixtures (Nakayama et al., 1982). A recent publication discusses this type of multilayered device, which shows high photosensitivity to GaAs laser wavelengths and has been utilized to function in a prototype high speed electronic printer capable of producing 200 pages per minute with a resolution of 300 spots per inch (Nishikawa, 1983). However, advances in semiconductor lasers make it more and more likely that visible solid state lasers will become the light sources of choice in future printers and laser copiers, thus obviating the need for infrared sensitizing layers. This would certainly lead to a enormous simplification in the manufacturing processes.

Even in these circumstances, there is still a need and value in multilayered plasma-deposited photoreceptors, since a-Si:H has another drawback. Although the achievable dark decay in a single-layer a-Si:H photoreceptor is sufficient for medium or high speed applications, it is too large for low speed applications. Intrinsically, this is associated with the relatively small band gap of a-Si:H and a residual density of localized band gap states of $\sim 10^{15}$ cm^{-3}, which has proven remarkably resistant to further reduction despite a variety of process variations. Given this state of affairs, the only remaining approach to achieving lower dark decay devices is to reduce the volume (and therefore total number of dark carriers) of the a-Si:H used. Thus, the ideal design would be a thin (<0.5 μm) sensitizing layer of a-Si:H from which photogenerated carriers could be injected with unit efficiency into a highly insulating transport layer. The resistivity, and dark decay, of the device would then be determined by the thicker transport layer, but the photoelectronic properties would be controlled by the sensitizing layer. Several reports of approaches to this type of multilayer device have been published (Carasco, 1985; Paasche and Bauer, 1985; Shimizu, 1985).

Possible candidates for the highly insulating transport layers producible by plasma deposition, which must be rf, include a-SiO$_x$:H, a-SiN$_x$:H, and a-C$_x$:H layers. Based on a conventional view of the relative positions of energy bands in the component layers (see Fig. 16.18a), the possibility of achieving unit injection efficiency, or close to it, would be zero because of the high energy barriers. In addition, on the basis of estimated μτ products in wide band gap materials such as high quality SiO$_x$, the achievable range μτE is likely to be inadequate because of the very small lifetime, τ, before being captured in deep-lying states. These dilemmas can, in principle, be resolved, as shown in Figure 16.18b (Carasco et al., 1985). In this simplified picture it is assumed that a discrete level of deep states lies isoenergetically with, for example, the conduction band of a-Si:H. High

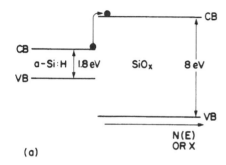

(a)

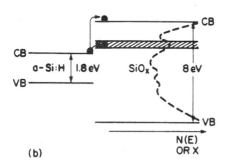

(b)

Figure 16.18 (a) Idealized band picture (CB = conduction band; VB = valence band) of a heterogeneous device structure of a-Si:H/a-SiO$_x$, where x is close to 2; the severe mismatch in the conduction states in the two materials leads to essentially no injection of photogenerated carriers from a-Si:H into the a-SiO$_x$ transport layer. (b) In reality, the band gap of a-SiO$_x$, or other wide band gap amorphous insulators, contains significant densities of localized states. With sufficient densities, determined by the value of x, wavefunction overlap can allow the transport of carriers by hopping between the localized states within the band gap; if such densities are also isoenergetic with the transport states in a-Si:H, efficient injection becomes possible.

injection efficiency and carrier range could then occur if the density of these states were, or could be made, high enough to produce sufficient wavefunction overlap for defect-hopping transport of the injected carriers to occur. This concept, where atomic defects play the role of dopant molecules, is an inorganic analogue with the molecularly doped polymer transport matrices discussed in Chapter 10.

Figure 16.19 shows xerographic discharge results on a device comprised of 0.5 μm of a-Si:H overlaying a 5 μm layer of $SiO_x:N_x:H$ produced by the glow discharge of a 12:1 gas mixture of N_2O and SiH_4. It is seen that very high injection efficiencies, approaching unity, are achieved at high field. The value of $\mu\tau E$ is essentially a maximum at this gas ratio. This is consistent with there being a correlation between the gas ratio and the defect density. Currently this is one of the most active areas of research worldwide, and other possible transport layers such as a-SiC$_x$ and a-C are being studied.

As we have seen, a number of attractive applications in electronic imaging exist for a-Si:H electronic devices produced in matrix arrays, and almost all have been fabricated or developed and in some instances commercialized. As a result, a-Si:H thin film transistors are now under advanced development for use as the pixel matrices for flat-panel video terminal displays, and a-Si:H photodiode arrays for contact-type image sensors and low cost facsimile products are already in production (Hamakawa, 1987; Matsumura, 1986). Based on such actual experience, the so-called learning curve, device performances have

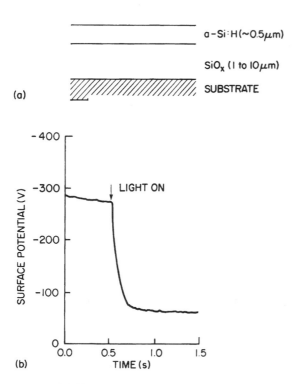

Figure 16.19 Prototype multilayered photoreceptor based on the concept illustrated in Figure 18. (a) The 0.5 μm a-Si:H acts as a sensitizing layer from which photogenerated carriers are injected into the underlying, and much thicker, a-SiO$_x$ transport layer. While the photodischarge (b) is controlled by the a-Si:H, the dark decay is now limited by the thicker, more insulating, wide band gap a-SiO$_x$.

significantly and steadily improved. Nonetheless the electrical performance of such a-Si:H electronic devices, unlike a-Si:H solar cells, are governed by a-Si:H–insulator interface properties, and the necessary large-scale and continued systematic studies in these areas are being carried out to obtain further significant improvements. As one example, when a dc voltage is applied to the gate of a TFT, the drain current can decrease, and the threshold voltage increase gradually with time (Powell and Nicholls, 1983). Thermal annealing can reverse these trends, but no detailed explanation for their cause yet exists, except that they are believed to be associated with the a-SiN$_x$–a-Si:H interface.

The dynamic performance of the present a-Si:H TFTs is determined by the large stray capacitance and long carrier transit times (Matsumura, 1986). To improve the dynamic performance, various technologies and structures have been proposed to date. These include self-aligned structures to reduce the stray capacitances and a reduced channel length L. The latter, in particular, is an effective way to improve the ultimate dynamic device performance of a TFT because the circuit response time is proportional to the transit time t_τ of electrons, which equals $L/\mu E$, and E is the applied voltage divided by L. Usually, L is determined by the photoetching process for delineating the gate pattern, and it is difficult to reduce it to less than 5 μm. For still smaller channels, new device structures are essential, and these are being developed where photoetching is not a limiting process. Using rather sophisticated ideas, such as vertical channels (Fig. 16.20), it may become possible to reduce transit times by about a factor of 100 compared with conventional, horizontal TFTs (Uchida and Matsumura, 1984).

It is hoped that the reader will have been struck by the remarkable degree to which the plasma deposition process and a-Si:H and related materials can contribute to almost every dimension of electronic imaging. However, lest euphoria carry the day, it is equally clear that in some critical aspects, notably in frequency or speed of scanning or writing, a-Si:H-based devices have significant deficiencies compared with their crystalline silicon analogues. Nonetheless, given the broad range of applications in which their speed is

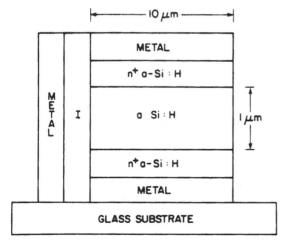

Figure 16.20 An advanced TFT concept in which a vertically built device allows the use of much thinner source–drain channel gaps. Since the speed of response is limited by the mobility of carriers in a-Si:H, and more specifically by their travel time across the channel, such vertical-type devices offer the prospect of response times that are faster by orders of magnitude. (From Matsumura, 1986.)

sufficient, it is perhaps not unreasonable to expect that this process and the materials and devices it enables will play an important and expanding role in electronic imaging in the 1990s and beyond. For, quite apart from inherent advantages of any particular device, considerable commercialization and manufacturing advantages accrue in having a compatible technological arsenal.

REFERENCES

Adler, D. (1971). *Amorphous Semiconductors*, Chemical Rubber Company, Cleveland, OH.

Böhm, M. (1988). *Solid State Technol.*, September, p. 125.

Borsenberger, P. M. Mey, W., and Chowdry, A. (1978). *J. Appl. Physics, 49*:273.

Borsenberger, P. M., Weiss, D. S. (1998). In: *Organic Photoreceptor for Xerography.* Marcel Dekker, Inc., New York.

Carasco, F., Mort, J., Jansen, F., and Grammatica, S. (1985). *J. Appl. Phys., 57*:5306.

Carlson, D. E., and Wronski, C. R. (1976). *Appl. Phys. Lett., 28*:671.

Chen, I., and Jansen, F. (1983). *J. Non-Cryst. Solids, 59–60*:695.

Chittick, R. C., Alexander, J. H., and Sterling, H. F. (1969). *J. Electrochem. Soc., 116*:77.

Chuang, T. C., Fennell, L. E., Jackson, W. B., Levine, J., Thompson, M. J., Tuan, H. C., Weisfield, R., Hamano, T., Itoh, H., Ozawa, T., and Tomiyama, S. (1987). *J. Non-Cryst. Solids, 97–98*: 301.

Dekker, A. J. (1958). *Solid State Physics*, Macmillan, London.

Dolezaek, F. K. (1976). In *Photoconductivity and Related Phenomena* (J. Mort and D. Pai, eds.), Elsevier, Amsterdam, p. 27.

Felty, E. J. (1967). "New photoconductors for xerography," in *Proceedings of the Second International Congress on Reprography*, Cologne (O. Helwich, ed.).

Fritzsche, H. (1980). *Solar Energy Mater., 53*: 447.

Fritzsche, H. (1984). *Phys. Today, 37*:34.

Fuhs, W. (1986). In *Plasma Deposited Thin Films* (J. Mort and F. Jansen, eds.), CRC Press, Boca Raton, FL, p. 45.

Gill, W. D. (1973). *J. Appl. Phys., 43*:5033.

Hamakawa, Y. (1987). In *Noncrystalline Solids*, Vol. 1 (M. Pollak, ed.), CRC Press, Boca Raton, FL, p. 229.

Hirose, M. (1986). In *Plasma Deposited Thin Films* (J. Mort and F. Jansen, eds.), CRC Press, Boca Raton, FL, p. 21.

Jansen, F. (1981). *Thin Solid Films, 78*: 15.

Jansen, F. (1986). In *Plasma Deposited Thin Films* (J. Mort and F. Jansen, eds.), CRC Press, Boca Raton, FL, p. 1.

Knights, J. C. (1980). *J. Non-Cryst. Solids, 35–36*:159.

Kuhman, D., Grammatica, S., and Jansen, F. (1989). *Thin Solid Films, 177*: 253.

Kuwano, Y. (1986). In *Plasma Deposited Thin Films* (J. Mort and F. Jansen, eds.), CRC Press, Boca Raton, FL, p. 161.

LeComber, P. G., Spear, W. E., and Ghaith, A. (1979). *Electron. Lett., 15*:179.

Madan, A., Le Comber, P. G., and Spear, W. E. (1976). *J. Non-Cryst. Solids, 20*:239.

Matsumura, M. (1986). In *Plasma Deposited Thin Films* (J. Mort and F. Jansen, eds.), CRC Press, Boca Raton, FL, p. 205.

Mehl, W., and Wolff, N. E. (1965). *J. Phys. Chem. Solids, 25*:1221.

Morgan, M., Jansen, F., Grammatica, S., Kuhman, D., and Mort, J. (1984). *J. Non-Cryst. Solids, 66*:77.

Mort, J. (1989). *The Anatomy of Xerography*, McFarland Press, Jefferson, NC.

Mort, J., and Chen, I. (1976). In *Applied Solid State Science*, Vol. 5 (R. Wolfe, ed.), Academic Press, New York, p. 69.

Mort, J., and Jansen, F. (1986). In *Plasma Deposited Thin Films* (J. Mort and F. Jansen, eds.), CRC Press, Boca Raton, FL, p. 193.

Mort, J., Grammatica, S., Knights, J. C., and Lujan, R. (1980). *Photogr. Sci. Eng., 24*:241.

Mort, J., and Pfiste, G. (1979). *Polym.–Plas., Tech. and Eng., 12*:89.

Mort, J., Jansen, F., Grammatica, S., Morgan, M., and Chen, I. (1984). *J. Appl. Phys., 55*:3197.

Mott, N. F., and Davis, E. A. (1979). *Electronic Processes in Non-Crystalline Materials*, Oxford University Press, Oxford.

Nakayama, Y., Natsuhara, T., Nakano, M., Yamamoto, N., and Kawamura, T. (1982). *Proceedings of the Fourteenth Conference on Solid State Devices*, Tokyo, p. 453.

Nakayama, Y., Wakita, K., Nakano, M., and Kawamura, T. (1983). *J. Non-Cryst. Solids, 59–60*: 1231.

Nara, Y., Kudou, Y., and Matsumura, M. (1983). *J. Non-Cryst. Solids, 59–60*: 1175.

Nishikawa, S., Kakinuma, H., Watanabe, T., and Kaminishi, K. (1983). *J. Non-Cryst. Solids, 59–60*:1235.

Paasche, S. M., and Bauer, G. H. (1985). *J. Non-Cryst. Solids, 77–78*:1433.

Paul, W. (1981). In *Solid State Sciences*, Vol. 5 (Y. Yonezawa, ed.), Springer-Verlag, Berlin, p. 72.

Powell, M. J., and Nicholls, D. H. (1983). *Proc. IEEE, 130*:2.

Powell, M. J., Chapman, J. A., Knapp, A. G., French, I. D., Hughes, J. R., Pearson, A. D., Edwards, M. J., Ford, R. A., Hemings, M. C., Hill, O. F., Nicholls, D. H., and Wright, N. K. (1988). *Proceedings of the Society for Information Display, 29*: 227.

Schmitt, J. P. M. (1983). *J. Non-Cryst. Solids, 59–60*: 649.

Shimizu, I. (1985). *J. Non-Cryst. Solids, 77–78*:1363.

Shimizu, I., Komatsu, T., Santo, K., and Inoue, E. (1980). *J. Non-Cryst. Solids, 35–36*:773.

Snell, A. J., Mackenzie, K. D., Spear, W. E., and Le Comber, P. G. (1981). *Appl. Phys., 24*:357.

Spear, W. E. (1974). *Proceedings Fifth International Conference on Amorphous and Liquid Semiconductors, Vol. 2*, Garmisch-Partenkirchen (J. Stuke and W. Brenig, eds.) Taylor and Francis, London, p. 1015.

Spear, W. E. (1977). *Adv. Phys., 26*:811.

Spear, W. E. (1981). In *Solid State Sciences*, Vol. 5 (Y. Yonezawa, ed.), Springer-Verlag, Berlin, p. 40.

Spear, W. E. (1983). *J. Non-Cryst. Solids, 59–60*:1.

Spear, W. E., and Le Comber, P. G. (1975). *Solid State Commun., 17*:1193.

Sterling, H. F., and Swan, R. C. G. (1965). *Solid State Electron., 8*:653.

Street, R. A., and Zesch, J. (1983). *J. Non-Cryst. Solids, 59–60*:449.

Suda, Y., Suzuki, K., Nakai, T., Takayama, S., Mori, K., and Saito, T. (1987). *Proceedings of the Society for Information Display, 28*:317.

Taylor, P. C. (1987). In *Noncrystalline Solids*, Vol. 1 (M. Pollak, ed.), CRC Press, Boca Raton, FL, p. 20.

Uchida, Y., and Matsumura, M. (1984). *IEEE Electron Dev. Lett., 5*:105.

Williams, E. M. (1984). *The Physics and Technology of Xerographic Processes*, Wiley-Interscience, New York.

Yamamoto, N., Wakita, N., Nakayama, K., and Kawamura, T. (1981). *J. Phys. (Paris), C4, 42*:495.

Index